SOFTWARE TO ENHANCE YOUR LINEAR ALGEBRA COURSE

- *Interactive Applications for Linear Algebra*

- *Linear Kit*

WILEY
Publishers Since 1807

(See other side for details)

SOFTWARE TO ENHANCE YOUR LINEAR ALGEBRA COURSE

Interactive Applications for Linear Algebra software by IntelliPro consists of 10 application modules that provide an interactive environment for students to explore, modify and interpret mathematical models. For PCs running Windows (3.1 or higher).

Linear Kit software performs most of the basic operations of linear algebra using either fractions or decimals, including Matrix Operations (up to size 6 x 6) and Vector Operations (up to 6 components). For PCs running DOS.

To purchase a copy of either *Interactive Applications for Linear Algebra* software *or Linear-Kit*, bring this card to your campus bookstore and ask if they carry or will order them. If they are not available through your bookstore, you may purchase them directly from John Wiley & Sons, Inc. To order:

WILEY

1. **Call 1-800-CALL-WILEY (1-800-225-5945)** — refer to author/title/ISBN # below. Or,

2. **Complete the postage-paid form below, tear at perforation and mail.**

PLEASE SEND ME:

❏ **Interactive Applications for Linear Algebra** ISBN 0-471-00827-3.................$31.95
❏ **Linear-Kit Software and User's Manual** ISBN 0-471-61998-1 ....$29.95

❏ Save Money! Check enclosed (Wiley pays postage and handling)
❏ Bill my credit card: plus sales tax, postage and handling:
     ❏ Visa   ❏ Mastercard   ❏ American Express

Card #_____ Expiration Date: _____
Signature _____
(offer invalid unless signed)

SEND TO:

Name_____
Address _____
City_____ State_____ Zip _____

Prices and terms subject to change without notice.

ELEMENTARY LINEAR ALGEBRA

SEVENTH EDITION

HOWARD ANTON
Drexel University

JOHN WILEY & SONS, INC.
New York • Chichester • Brisbane • Toronto • Singapore

ACQUISITIONS EDITOR Barbara Holland
DEVELOPMENTAL EDITOR Joan Carrafiello
MARKETING MANAGER Susan Elbe
PRODUCTION EDITOR Nancy Prinz
PRODUCTION SUPERVISOR Hudson River Studio
DESIGN SUPERVISOR Maddy Lesure
COVER AND TEXT DESIGN Hudson River Studio
MANUFACTURING MANAGER Andrea Price
COPY EDITOR Lilian Brady
ILLUSTRATION COORDINATOR Sigmund Malinowski
ELECTRONIC ART Techsetters, Inc.

This book was set in Times New Roman by CRWaldman Graphic Communications
and printed and bound by Von Hoffmann Press, Inc. The cover was printed by
Phoenix Color Corp.

Library of Congress Cataloging-in-Publication Data:

Anton, Howard.
    Elementary linear algebra / Howard Anton. — 7th ed.
        p. cm.
    Includes index.
    ISBN 0-471-58742-7
    1. Algebras, Linear.   I. Title.
    QA184.A57 1994
    512′.5—dc20
                                       93-28546
                                       CIP

Printed in the United States of America

10 9 8 7 6 5 4 3

To my wife Pat
and
my children
Brian, David, and Lauren

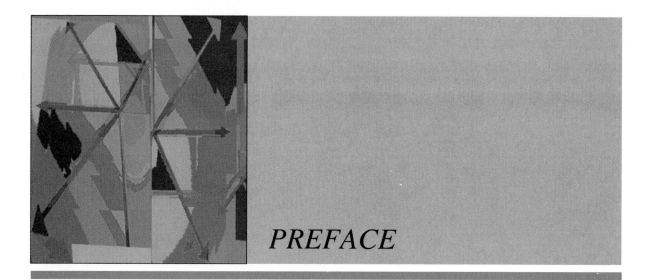

PREFACE

As with previous editions, this new edition gives an elementary treatment of linear algebra that is suitable for students in their freshman or sophomore year. My aim is to present the fundamentals of linear algebra in the clearest possible way—pedagogy is the main consideration. Calculus is not a prerequisite, but there are exercises and examples for students with calculus backgrounds; those exercises and examples are clearly marked as such and can be omitted with no loss of continuity.

SUMMARY OF CHANGES IN THIS EDITION

Although this edition has much in common with its predecessors, it is a substantial revision. I have tried to maintain the clarity and style of the earlier editions, yet reflect the changing needs of a new generation of students. To this end I have implemented a number of the recommendations of the *Linear Algebra Curriculum Study Group*. I have also made some organizational changes that should make it easier for instructors to cover the fundamentals of *all* major topics, even with severe time constraints. A chapter-by-chapter description of the changes is given later in this preface, but here is a summary of the more noteworthy changes:

- **Greater Emphasis on Relationships Between Concepts:** One of the important goals of a course in linear algebra is to establish the intricate thread of relationships between systems of equations, matrices, determinants, vectors, linear transformations, and eigenvalues. In this edition that thread of relationships is developed through the following crescendo of theorems that link each new idea with ideas that preceded it: 1.5.3, 1.6.4, 2.3.6, 4.3.4, 5.6.9, 6.2.7, 6.4.5, 7.1.5. These theorems not only bring a coherence to the linear algebra landscape but also serve as a constant source of review.

- **Smoother Transition to Abstraction:** The transition from R^n to general vector spaces is traumatic for most students, so I have tried to smooth out that transition by discussing R^n in detail with more emphasis on the underlying geometry before proceeding to general vector spaces.

- **Early Exposure to Linear Transformations and Eigenvalues:** To ensure that the material on linear transformations and eigenvalues does not get lost at the end of the course, some of the basic concepts relating to those topics are developed earlier in the text and then reviewed when the topic is developed in more depth in the later part of the text. For example, characteristic equations are discussed briefly in the section on determinants. Linear transformations from R^n to R^m are discussed immediately after R^n is introduced and reviewed later in the context of general linear transformations. These revisions will help ensure that students are exposed to the fundamentals of all major topics, even when time is tight.

- **Greater Emphasis on Visualization:** In keeping with the current interest in visualization and the growing applications of linear algebra to graphics, I have placed greater emphasis on the geometric aspects of rotations, projections, and reflections in R^2 and R^3.

- **New Material on Least Squares and *QR*-Decomposition:** New material on least squares and the *QR*-decomposition has been added in response to the growing interest in those topics.

- **More Proofs:** A number of proofs that were previously omitted have been added. All proofs in the text are written in a style tailored for beginners, and special care has been exercised to ensure that the accessibility and friendliness of the text has not been adversely affected by the additional proofs. Those who want a tighter course mathematically will find the new edition better suited for that purpose, and those who want a more conceptual course will have a greater choice in the proofs to include or exclude.

DETAILS OF THE CHANGES IN THE SEVENTH EDITION

The wide acceptance of the first six editions has been most gratifying, and I am appreciative of the many constructive suggestions received from users and reviewers. Portions of the text have been revised for greater clarity, and substantial changes in content and organization have been made in response to both reviewer and user suggestions as well as recommendations of the *Linear Algebra Curriculum Study Group*.

There are many ways in which one can order the material in a linear algebra course; the ordering of the chapters that I have selected reflects my adherence to the axiom that one should proceed from the familiar to the unfamiliar and from the concrete to the abstract.

Here is a chapter-by-chapter summary of the major changes in the seventh edition.

- **Chapter 1** There is a new section on matrices with special forms: diagonal, triangular, and symmetric. By reorganizing the material slightly, the number of sections in this chapter has not increased.

- **Chapter 2** New introductory material on eigenvalues, eigenvectors, and characteristic equations has been added to this determinant chapter. This material is reviewed and then discussed in more detail in Chapter 7. The proof of the formula $\det(AB) = \det(A)\det(B)$ has been added.

- **Chapter 3** There is new material on the vector equations of lines and planes and the geometric interpretation of 2×2 and 3×3 determinants.

- **Chapter 4** This is a new chapter devoted exclusively to R^n. Basic concepts are developed, and there is an introduction to linear transformations from R^n to R^m with emphasis on the geometry of projections, rotations, and reflections. Unlike the previous edition, this material now comes *before* the development of general vector spaces. The material in this chapter is reexamined later in the context of general vector spaces.

- **Chapter 5** This chapter corresponds to Chapter 4 in the previous edition. Many of the proofs that were omitted in the previous edition have been added. There is also new material on the Wronskian for students who have studied calculus, and there is new material on the four fundamental spaces of a matrix.

- **Chapter 6** This chapter corresponds to Chapter 5 in the previous edition. There is new material on orthogonal complements, QR-decomposition, and least squares.

- **Chapter 7** This chapter corresponds to Chapter 6 in the previous edition. Material developed earlier on eigenvalues and eigenvectors is reviewed. There is new material on geometric and algebraic multiplicity and an improved explanation of the requirements for diagonalizability.

- **Chapter 8** This chapter corresponds to Chapter 7 in the previous edition. The material has been rewritten substantially to reflect the fact that linear transformations from R^n to R^m were introduced in Chapter 4.

- **Chapter 9** This chapter corresponds to Chapter 8 and Sections 9.1 and 9.2 in the previous edition. The section on the geometry of linear operators on R^2 has been rewritten to build on concepts developed in Section 4.2.

- **Chapter 10** This chapter corresponds to Chapter 10 in the previous edition. The changes are minor.

ABOUT THE EXERCISES

Each exercise set begins with routine drill problems, progresses to problems with more substance, and concludes with theoretical problems. At the end of most chapters there is a set of Supplementary Exercises that are more challenging and force the student to draw ideas from an entire chapter rather than a specific section.

SUPPLEMENTARY MATERIALS

FOR THE STUDENT

- *Student Solutions Manual Charles A. Grobe, Jr. (Bowdoin College) and Elizabeth M. Grobe* provides detailed solutions to most theoretical exercises and many computational exercises in the text. Included is a solution to at least one nonroutine exercise of every type. (ISBN: 0-471-30622-3)

FOR THE INSTRUCTOR

- *Test Bank Randy Schwartz (Schoolcraft College)* includes approximately 50 free-form questions and 5 essay questions for each chapter and one sample cumulative final exam. Also given are worked out solutions to each question in the test bank. (ISBN: 0-471-30848-X)

- *Computerized Test Bank* provides instructors with the entire test bank on disk and the capability to create customized tests by scrambling questions or choosing specific questions and sequences. This software is available for IBM PC's and compatibles and Macintosh computers. (IBM ISBN: 0-471-00824-9; Macintosh ISBN: 0-471-00826-5)

FOR A MICROCOMPUTER-ASSISTED COURSE

- *Linear Algebra with Derive Benny Evans and Jerry Johnson (Oklahoma State University)* is a supplementary manual designed to help students use the Derive program as a tool to solve problems in the course, and to provide unusual problems that encourage exploration and discovery. (ISBN: 0-471-59194-7)

- *Linear Algebra Applications Software (IntelliPro)* consists of 10 application modules featuring graphic representations, animations, and simulation problems requiring a high degree of student interaction. Available for IBM PC's and compatibles running *Windows*. This software may be purchased independently of the text (ISBN: 0-471-00827-3) or shrinkwrapped with the text at a substantial discount price (ISBN: 0-471-00828-1).

- *LINEAR-KIT* is a software package that can perform most of the basic linear algebra computations using either fractions or decimals and will save the student many hours of homework time that can be used for other pedagogical purposes. It is available free to departments adopting this text or can be purchased by individuals. This software is available for IBM PC's and compatibles. (ISBN: 0-471-61998-1)

If any of these supplements is not in your bookstore, ask the bookstore manager to order a copy for you.

A GUIDE FOR THE INSTRUCTOR

POSSIBLE SCHEDULES FOR A STANDARD COURSE

I have reviewed a large number of course outlines for linear algebra courses. The variation between institutions is wide, but courses tend to fall into two categories—those with about 25–30 lectures (excluding tests and reviews) and those with about 35–40 lectures (excluding tests and reviews). Based on my examination of the course outlines, I have provided two templates for constructing your course outline. These will have to be adjusted to reflect your local interests and requirements, but they may be helpful as a starting point. In the long template, it is assumed that all sections in the chapter are covered, and in the short template it is assumed that the instructor selects material to fit the available time.

Two changes in the organization of the text make it easier to construct shorter courses: the brief introduction to eigenvalues and eigenvectors that occurs in Sections 2.3 and 4.3 and the earlier placement of linear transformations from R^n to R^m in Chapter 4. These changes ensure that the student will have some familiarity with these basic concepts, even if the time available for Chapters 7 and 8 is limited. Note also that Chapter 3 can be omitted without loss of continuity for students who are already familiar with the material.

	Long Template	Short Template
Chapter 1	7 lectures	6 lectures
Chapter 2	4 lectures	3 lectures
Chapter 4	3 lectures	3 lectures
Chapter 5	8 lectures	7 lectures
Chapter 6	6 lectures	3 lectures
Chapter 7	4 lectures	3 lectures
Chapter 8	6 lectures	2 lectures
Total	38 lectures	27 lectures

VARIATIONS OF THE STANDARD COURSE

Many variations of the standard course are possible. For example, one might create an alternative long template by following the time allocations in the short template and devoting the remaining 11 lectures to some of the topics in Chapters 9 and 10.

APPLICATIONS-ORIENTED COURSE

Chapter 9 contains selected applications of linear algebra that are mostly of a mathematical nature. Instructors who are interested in a wider variety of applications may want to consider the alternative version of this text, ***Elementary Linear Algebra, Applications Version,*** by Howard Anton and Chris Rorres. That text provides numerous applications to business, biology, engineering, economics, the social sciences, and the physical sciences.

Linear Algebra Applications Software by IntelliPro consists of 10 application modules featuring graphic representations, animations, and simulation problems requiring a high degree of student interaction. For IBM PC's and compatibles running *Windows*, this software may be purchased independently of the text (ISBN: 0-471-00827-3) or shrinkwrapped with the text at a substantial discount price (ISBN: 0-471-00828-1).

ACKNOWLEDGEMENTS

I express my appreciation for the helpful guidance provided by the following people:

REVIEWERS AND CONTRIBUTORS TO EARLIER EDITIONS

Steven C. Althoen, *University of Michigan–Flint*
C. S. Ballantine, *Oregon State University*
Erol Barbut, *University of Idaho*
William A. Brown, *University of Maine*
Joseph Buckley, *Western Michigan University*
Thomas Cairns, *University of Tulsa*
Douglas E. Cameron, *University of Akron*
Bomshik Chang, *University of British Columbia*
Peter Colwell, *Iowa State University*
Carolyn A. Dean, *University of Michigan*
Ken Dunn, *Dalhousie University*
Bruce Edwards, *University of Florida*
Murray Eisenberg, *University of Massachusetts*
Harold S. Engelsohn, *Kingsborough Comm. College*
Garret Etgen, *University of Houston*
Marjorie E. Fitting, *San Jose State University*
Dan Flath, *University of South Alabama*
David E. Flesner, *Gettysburg College*
Mathew Gould, *Vanderbilt University*
Ralph P. Grimaldi, *Rose–Hulman Institute*

William W. Hager, *University of Florida*
Collin J. Hightower, *University of Colorado*
Joseph F. Johnson, *Rutgers University*
Robert L. Kelley, *University of Miami*
Arlene Kleinstein
Myren Krom, *California State University*
Lawrence D. Kugler, *University of Michigan*
Charles Livingston, *Indiana University*
Nicholas Macri, *Temple University*
Roger H. Marty, *Cleveland State University*
Patricia T. McAuley, *SUNY–Binghamton*
Robert M. McConnel, *University of Tennessee*
Douglas McLeod, *Drexel University*
Michael R. Meck, *Southern Connecticut State Univ.*
Craig Miller, *University of Pennsylvania*
Donald P. Minassian, *Butler University*
Hal G. Moore, *Brigham Young University*
Thomas E. Moore, *Bridgewater State College*
Robert W. Negus, *Rio Hondo Junior College*
Bart S. Ng, *Purdue University*

James Osterburg, *University of Cincinnati*
Michael A. Penna, *Indiana–Purdue University*
Gerald J. Porter, *University of Pennsylvania*
F. P. J. Rimrott, *University of Toronto*
C. Ray Rosentrater, *Westmont College*
Kenneth Schilling, *University of Michigan–Flint*
William Scott, *University of Utah*
Donald R. Sherbert, *University of Illinois*
Bruce Solomon, *Indiana University*
Mary T. Treanor, *Valparaiso University*

William F. Trench, *Trinity University*
Joseph L. Ullman, *University of Michigan*
W. Vance Underhill, *East Texas State University*
James R. Wall, *Auburn University*
Arthur G. Wasserman, *University of Michigan*
Evelyn J. Weinstock, *Glassboro State College*
Rugang Ye, *Stanford University*
Frank Zorzitto, *University of Waterloo*
Daniel Zwick, *University of Vermont*

REVIEWERS AND CONTRIBUTORS TO THE SEVENTH EDITION

Mark B. Beintema, *Southern Illinois University*
Paul Wayne Britt, *Louisiana State University*
David C. Buchthal, *University of Akron*
Keith Chavey, *University of Wisconsin–River Falls*
Stephen L. Davis, *Davidson College*
Blaise DeSesa, *Drexel University*
Dan Flath, *University of South Alabama*
Peter Fowler, *California State University*
Marc Frantz, *Indiana–Purdue University*
Sue Friedman, *Bernard M. Baruch College, CUNY*
William Golightly, *College of Charleston*
Hugh Haynsworth, *College of Charleston*
Tom Hern, *Bowling Green State University*
J. Hershenov, *Queens College, CUNY*
Steve Humphries, *Brigham Young University*
Steven Kahan, *Queens College, CUNY*

Andrew S. Kim, *Westfield State College*
John C. Lawlor, *University of Vermont*
M. Malek, *California State University at Hayward*
J. J. Malone, *Worcester Polytechnic Institute*
William McWorter, *Ohio State University*
Valerie A. Miller, *Georgia State University*
Hal G. Moore, *Brigham Young University*
S. Obaid, *San Jose State University*
Ira J. Papick, *University of Missouri–Columbia*
Donald Passman, *University of Wisconsin*
Robby Robson, *Oregon State University*
David Ryeburn, *Simon Fraser University*
Ramesh Sharma, *University of New Haven*
David A. Sibley, *Pennsylvania State University*
Donald Story, *University of Akron*
Michael Tarabek, *Southern Illinois University*

PROBLEM SOLUTIONS, PROOFREADING, AND INDEX

Michael Dagg, *Numerical Solutions, Inc.*
Susan L. Friedman, *Bernard M. Baruch College, CUNY*
Maureen Kelley, *Northern Essex Community College*
Randy Schwartz, *Schoolcraft College*
Daniel Traster (*Student*), *Yale University*

SUPPLEMENTS

Benny Evans, *Oklahoma State University*
Charles A. Grobe, Jr., *Bowdoin College*
Elizabeth M. Grobe
IntelliPro, Inc.
Jerry Johnson, *Oklahoma State University*
Randy Schwartz, *Schoolcraft College*

OTHER CONTRIBUTIONS

Special thanks to the following professors who read the text material in depth and made significant contributions to the quality of the mathematics and the exposition:

Stephen Davis, *Davidson College*
Blaise DeSesa, *Drexel University*
Dan Flath, *University of South Alabama*
Marc Frantz, *Indiana–Purdue University*
William McWorter, *Ohio State University*
Donald Passman, *University of Wisconsin*
David Ryeburn, *Simon Fraser University*
Lois Craig Stagg, *University of Wisconsin–Milwaukee*

I would also like to thank:

Barbara Holland, my editor, who helped me mold the concept of this new edition and whose enthusiasm even made the hard work fun (once in awhile).

Ann Berlin, Lucille Buonocore, and Nancy Prinz of the Wiley Production Department for caring so much about the quality of this work and for providing me with an extraordinary level of support.

Lilian Brady, whose eye for detail and unerring sense of aesthetics greatly enhanced the accuracy of the text and beauty of the typography.

Joan Carafiello and Sharon Prendergast for superb work in coordinating the myriad of details that magically produced the answers and supplements on time.

The group at Hudson River Studio for dealing so tactfully with a picky author.

Mildred Jaggard, my assistant, who coordinated every detail of the text from proofreading to index with consummate skill and who tolerated my idiosyncrasies with saintly patience.

HOWARD ANTON

CONTENTS

CHAPTER **1**

SYSTEMS OF LINEAR EQUATIONS AND MATRICES

1.1 INTRODUCTION TO SYSTEMS OF LINEAR EQUATIONS

The study of systems of linear equations and their solutions is one of the major topics in linear algebra. In this section we shall introduce some basic terminology and discuss a method for solving such systems.

LINEAR EQUATIONS

A line in the xy-plane can be represented algebraically by an equation of the form

$$a_1 x + a_2 y = b$$

An equation of this kind is called a linear equation in the variables x and y. More generally, we define a ***linear equation*** in the n variables $x_1, x_2, \ldots, x_n$ to be one that can be expressed in the form

$$a_1 x_1 + a_2 x_2 + \cdots + a_n x_n = b$$

where $a_1, a_2, \ldots, a_n$, and b are real constants. The variables in a linear equation are sometimes called the ***unknowns***.

Example 1   The following are linear equations:

$$x + 3y = 7 \qquad\qquad x_1 - 2x_2 - 3x_3 + x_4 = 7$$
$$y = \tfrac{1}{2}x + 3z + 1 \qquad\qquad x_1 + x_2 + \cdots + x_n = 1$$

Observe that a linear equation does not involve any products or roots of variables. All variables occur only to the first power and do not appear as arguments for trigonometric, logarithmic, or exponential functions. The following are *not* linear equations:

$$x + 3y^2 = 7 \qquad\qquad 3x + 2y - z + xz = 4$$
$$y - \sin x = 0 \qquad\qquad \sqrt{x_1} + 2x_2 + x_3 = 1$$

1

A **solution** of a linear equation $a_1x_1 + a_2x_2 + \cdots + a_nx_n = b$ is a sequence of n numbers $s_1, s_2, \ldots, s_n$ such that the equation is satisfied when we substitute $x_1 = s_1$, $x_2 = s_2, \ldots, x_n = s_n$. The set of all solutions of the equation is called its **solution set** or sometimes the **general solution** of the equation.

Example 2 Find the solution set of

(a) $4x - 2y = 1$ (b) $x_1 - 4x_2 + 7x_3 = 5$

Solution (*a*). To find solutions of (a), we can assign an arbitrary value to x and solve for y, or choose an arbitrary value for y and solve for x. If we follow the first approach and assign x an arbitrary value t, we obtain

$$x = t, \qquad y = 2t - \tfrac{1}{2}$$

These formulas describe the solution set in terms of the arbitrary parameter t. Particular numerical solutions can be obtained by substituting specific values for t. For example, $t = 3$ yields the solution $x = 3$, $y = \tfrac{11}{2}$; and $t = -\tfrac{1}{2}$ yields the solution $x = -\tfrac{1}{2}$, $y = -\tfrac{3}{2}$.
   If we follow the second approach and assign y the arbitrary value t, we obtain

$$x = \tfrac{1}{2}t + \tfrac{1}{4}, \qquad y = t$$

Although these formulas are different from those obtained above, they yield the same solution set as t varies over all possible real numbers. For example, the previous formulas gave the solution $x = 3$, $y = \tfrac{11}{2}$ when $t = 3$, while the formulas immediately above yield that solution when $t = \tfrac{11}{2}$.

Solution (*b*). To find the solution set of (b) we can assign arbitrary values to any two variables and solve for the third variable. In particular, if we assign arbitrary values s and t to x_2 and x_3, respectively, and solve for x_1, we obtain

$$x_1 = 5 + 4s - 7t, \qquad x_2 = s, \qquad x_3 = t \quad \blacktriangle$$

LINEAR SYSTEMS

A finite set of linear equations in the variables $x_1, x_2, \ldots, x_n$ is called a **system of linear equations** or a **linear system**. A sequence of numbers $s_1, s_2, \ldots, s_n$ is called a **solution** of the system if $x_1 = s_1$, $x_2 = s_2, \ldots, x_n = s_n$ is a solution of every equation in the system. For example, the system

$$4x_1 - x_2 + 3x_3 = -1$$
$$3x_1 + x_2 + 9x_3 = -4$$

has the solution $x_1 = 1$, $x_2 = 2$, $x_3 = -1$ since these values satisfy both equations. However, $x_1 = 1$, $x_2 = 8$, $x_3 = 1$ is not a solution since these values satisfy only the first of the two equations in the system.
   Not all systems of linear equations have solutions. For example, if we multiply the second equation of the system

$$x + y = 4$$
$$2x + 2y = 6$$

by $\frac{1}{2}$, it becomes evident that there are no solutions since the resulting equivalent system

$$x + y = 4$$
$$x + y = 3$$

has contradictory equations.

A system of equations that has no solutions is said to be ***inconsistent***; if there is at least one solution of the system, it is called ***consistent***. To illustrate the possibilities that can occur in solving systems of linear equations, consider a general system of two linear equations in the unknowns x and y:

$$a_1x + b_1y = c_1 \qquad (a_1, b_1 \text{ not both zero})$$
$$a_2x + b_2y = c_2 \qquad (a_2, b_2 \text{ not both zero})$$

The graphs of these equations are lines; call them l_1 and l_2. Since a point (x, y) lies on a line if and only if the numbers x and y satisfy the equation of the line, the solutions of the system of equations correspond to points of intersection of l_1 and l_2. There are three possibilities (Figure 1):

- The lines l_1 and l_2 may be parallel, in which case there is no intersection and consequently no solution to the system.
- The lines l_1 and l_2 may intersect at only one point, in which case the system has exactly one solution.
- The lines l_1 and l_2 may coincide, in which case there are infinitely many points of intersection and consequently infinitely many solutions to the system.

Although we have considered only two equations with two unknowns here, we will show later that the same three possibilities hold for arbitrary linear systems:

Every system of linear equations has either no solutions, exactly one solution, or infinitely many solutions.

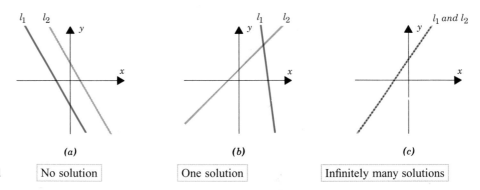

(a)	(b)	(c)
No solution	One solution	Infinitely many solutions

Figure 1

An arbitrary system of m linear equations in n unknowns can be written as

$$a_{11}x_1 + a_{12}x_2 + \cdots + a_{1n}x_n = b_1$$
$$a_{21}x_1 + a_{22}x_2 + \cdots + a_{2n}x_n = b_2$$
$$\vdots \qquad \vdots \qquad \qquad \vdots \qquad \vdots$$
$$a_{m1}x_1 + a_{m2}x_2 + \cdots + a_{mn}x_n = b_m$$

where $x_1, x_2, \ldots, x_n$ are the unknowns and the subscripted a's and b's denote constants. For example, a general system of three linear equations in four unknowns can be written as

$$a_{11}x_1 + a_{12}x_2 + a_{13}x_3 + a_{14}x_4 = b_1$$
$$a_{21}x_1 + a_{22}x_2 + a_{23}x_3 + a_{24}x_4 = b_2$$
$$a_{31}x_1 + a_{32}x_2 + a_{33}x_3 + a_{34}x_4 = b_3$$

The double subscripting on the coefficients of the unknowns is a useful device that is used to specify the location of the coefficient in the system. The first subscript on the coefficient a_{ij} indicates the equation in which the coefficient occurs, and the second subscript indicates which unknown it multiplies. Thus, a_{12} is in the first equation and multiplies unknown x_2.

AUGMENTED MATRICES

If we mentally keep track of the location of the $+$'s, the x's, and the $=$'s, a system of m linear equations in n unknowns can be abbreviated by writing only the rectangular array of numbers:

$$\begin{bmatrix} a_{11} & a_{12} & \cdots & a_{1n} & b_1 \\ a_{21} & a_{22} & \cdots & a_{2n} & b_2 \\ \vdots & \vdots & & \vdots & \vdots \\ a_{m1} & a_{m2} & \cdots & a_{mn} & b_m \end{bmatrix}$$

This is called the **augmented matrix** for the system. (The term *matrix* is used in mathematics to denote a rectangular array of numbers. Matrices arise in many contexts, which we will consider in more detail in later sections.) For example, the augmented matrix for the system of equations

$$x_1 + x_2 + 2x_3 = 9$$
$$2x_1 + 4x_2 - 3x_3 = 1$$
$$3x_1 + 6x_2 - 5x_3 = 0$$

is

$$\begin{bmatrix} 1 & 1 & 2 & 9 \\ 2 & 4 & -3 & 1 \\ 3 & 6 & -5 & 0 \end{bmatrix}$$

REMARK. When constructing an augmented matrix, the unknowns must be written in the same order in each equation and the constants must be on the right.

The basic method for solving a system of linear equations is to replace the given system by a new system that has the same solution set but which is easier to solve. This new system is generally obtained in a series of steps by applying the following three types of operations to eliminate unknowns systematically.

1. Multiply an equation through by a nonzero constant.
2. Interchange two equations.
3. Add a multiple of one equation to another.

Since the rows (horizontal lines) of an augmented matrix correspond to the equations in the associated system, these three operations correspond to the following operations on the rows of the augmented matrix.

1. Multiply a row through by a nonzero constant.
2. Interchange two rows.
3. Add a multiple of one row to another row.

ELEMENTARY ROW OPERATIONS

These are called *elementary row operations*. The following example illustrates how these operations can be used to solve systems of linear equations. Since a systematic procedure for finding solutions will be derived in the next section, it is not necessary to worry about how the steps in this example were selected. The main effort at this time should be devoted to understanding the computations and the discussion.

Example 3 In the left column below we solve a system of linear equations by operating on the equations in the system, and in the right column we solve the same system by operating on the rows of the augmented matrix.

$$
\begin{aligned}
x + y + 2z &= 9 \\
2x + 4y - 3z &= 1 \\
3x + 6y - 5z &= 0
\end{aligned}
\qquad
\begin{bmatrix}
1 & 1 & 2 & 9 \\
2 & 4 & -3 & 1 \\
3 & 6 & -5 & 0
\end{bmatrix}
$$

Add -2 times the first equation to the second to obtain

Add -2 times the first row to the second to obtain

$$
\begin{aligned}
x + y + 2z &= 9 \\
2y - 7z &= -17 \\
3x + 6y - 5z &= 0
\end{aligned}
\qquad
\begin{bmatrix}
1 & 1 & 2 & 9 \\
0 & 2 & -7 & -17 \\
3 & 6 & -5 & 0
\end{bmatrix}
$$

Add -3 times the first equation to the third to obtain

Add -3 times the first row to the third to obtain

$$
\begin{aligned}
x + y + 2z &= 9 \\
2y - 7z &= -17 \\
3y - 11z &= -27
\end{aligned}
\qquad
\begin{bmatrix}
1 & 1 & 2 & 9 \\
0 & 2 & -7 & -17 \\
0 & 3 & -11 & -27
\end{bmatrix}
$$

Multiply the second equation by $\frac{1}{2}$ to obtain

$$\begin{aligned} x + y + 2z &= 9 \\ y - \tfrac{7}{2}z &= -\tfrac{17}{2} \\ 3y - 11z &= -27 \end{aligned}$$

Multiply the second row by $\frac{1}{2}$ to obtain

$$\begin{bmatrix} 1 & 1 & 2 & 9 \\ 0 & 1 & -\tfrac{7}{2} & -\tfrac{17}{2} \\ 0 & 3 & -11 & -27 \end{bmatrix}$$

Add -3 times the second equation to the third to obtain

$$\begin{aligned} x + y + 2z &= 9 \\ y - \tfrac{7}{2}z &= -\tfrac{17}{2} \\ -\tfrac{1}{2}z &= -\tfrac{3}{2} \end{aligned}$$

Add -3 times the second row to the third to obtain

$$\begin{bmatrix} 1 & 1 & 2 & 9 \\ 0 & 1 & -\tfrac{7}{2} & -\tfrac{17}{2} \\ 0 & 0 & -\tfrac{1}{2} & -\tfrac{3}{2} \end{bmatrix}$$

Multiply the third equation by -2 to obtain

$$\begin{aligned} x + y + 2z &= 9 \\ y - \tfrac{7}{2}z &= -\tfrac{17}{2} \\ z &= 3 \end{aligned}$$

Multiply the third row by -2 to obtain

$$\begin{bmatrix} 1 & 1 & 2 & 9 \\ 0 & 1 & -\tfrac{7}{2} & -\tfrac{17}{2} \\ 0 & 0 & 1 & 3 \end{bmatrix}$$

Add -1 times the second equation to the first to obtain

$$\begin{aligned} x + \tfrac{11}{2}z &= \tfrac{35}{2} \\ y - \tfrac{7}{2}z &= -\tfrac{17}{2} \\ z &= 3 \end{aligned}$$

Add -1 times the second row to the first to obtain

$$\begin{bmatrix} 1 & 0 & \tfrac{11}{2} & \tfrac{35}{2} \\ 0 & 1 & -\tfrac{7}{2} & -\tfrac{17}{2} \\ 0 & 0 & 1 & 3 \end{bmatrix}$$

Add $-\frac{11}{2}$ times the third equation to the first and $\frac{7}{2}$ times the third equation to the second to obtain

$$\begin{aligned} x &= 1 \\ y &= 2 \\ z &= 3 \end{aligned}$$

Add $-\frac{11}{2}$ times the third row to the first and $\frac{7}{2}$ times the third row to the second to obtain

$$\begin{bmatrix} 1 & 0 & 0 & 1 \\ 0 & 1 & 0 & 2 \\ 0 & 0 & 1 & 3 \end{bmatrix}$$

The solution

$$x = 1, \qquad y = 2, \qquad z = 3$$

is now evident. ▲

EXERCISE SET 1.1

1. Which of the following are linear equations in x_1, x_2, and x_3?

(a) $x_1 + 5x_2 - \sqrt{2}x_3 = 1$ (b) $x_1 + 3x_2 + x_1x_3 = 2$ (c) $x_1 = -7x_2 + 3x_3$

(d) $x_1^{-2} + x_2 + 8x_3 = 5$ (e) $x_1^{3/5} - 2x_2 + x_3 = 4$ (f) $\pi x_1 - \sqrt{2}x_2 + \tfrac{1}{3}x_3 = 7^{1/3}$

2. Given that k is a constant, which of the following are linear equations?

(a) $x_1 - x_2 + x_3 = \sin k$ (b) $kx_1 - \dfrac{1}{k}x_2 = 9$ (c) $2^k x_1 + 7x_2 - x_3 = 0$

3. Find the solution set of each of the following linear equations.
   (a) $7x - 5y = 3$ (b) $3x_1 - 5x_2 + 4x_3 = 7$
   (c) $-8x_1 + 2x_2 - 5x_3 + 6x_4 = 1$ (d) $3v - 8w + 2x - y + 4z = 0$

4. Find the augmented matrix for each of the following systems of linear equations.
   (a) $3x_1 - 2x_2 = -1$ (b) $2x_1 \qquad + 2x_3 = 1$ (c) $x_1 + 2x_2 \qquad - x_4 + x_5 = 1$ (d) $x_1 \qquad\qquad = 1$
   $\quad 4x_1 + 5x_2 = \quad 3$ $\quad 3x_1 - x_2 + 4x_3 = 7$ $\quad 3x_2 + x_3 \qquad - x_5 = 2$ $\qquad x_2 \qquad = 2$
   $\quad 7x_1 + 3x_2 = \quad 2$ $\quad 6x_1 + x_2 - \ x_3 = 0$ $\qquad\quad x_3 + 7x_4 \qquad\ = 1$ $\qquad\qquad x_3 = 3$

5. Find a system of linear equations corresponding to the augmented matrix.

   (a) $\begin{bmatrix} 2 & 0 & 0 \\ 3 & -4 & 0 \\ 0 & 1 & 1 \end{bmatrix}$ (b) $\begin{bmatrix} 3 & 0 & -2 & 5 \\ 7 & 1 & 4 & -3 \\ 0 & -2 & 1 & 7 \end{bmatrix}$

   (c) $\begin{bmatrix} 7 & 2 & 1 & -3 & 5 \\ 1 & 2 & 4 & 0 & 1 \end{bmatrix}$ (d) $\begin{bmatrix} 1 & 0 & 0 & 0 & 7 \\ 0 & 1 & 0 & 0 & -2 \\ 0 & 0 & 1 & 0 & 3 \\ 0 & 0 & 0 & 1 & 4 \end{bmatrix}$

6. (a) Find a linear equation in the variables x and y that has the general solution $x = 5 + 2t$, $y = t$.
   (b) Show that $x = t$, $y = \frac{1}{2}t - \frac{5}{2}$ is also the general solution of the equation in part (a).

7. The curve $y = ax^2 + bx + c$ shown in Figure 2 passes through the points (x_1, y_1), (x_2, y_2), and (x_3, y_3). Show that the coefficients a, b, and c are a solution of the system of linear equations whose augmented matrix is

   $\begin{bmatrix} x_1^2 & x_1 & 1 & y_1 \\ x_2^2 & x_2 & 1 & y_2 \\ x_3^2 & x_3 & 1 & y_3 \end{bmatrix}$

   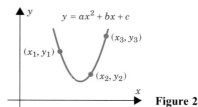

   Figure 2

8. For which value(s) of the constant k does the following system of linear equations have no solutions? Exactly one solution? Infinitely many solutions?

   $$x - \ y = 3$$
   $$2x - 2y = k$$

9. Consider the system of equations

   $$ax + by = k$$
   $$cx + dy = l$$
   $$ex + fy = m$$

   Discuss the relative positions of the lines $ax + by = k$, $cx + dy = l$, and $ex + fy = m$ when
   (a) the system has no solutions
   (b) the system has exactly one solution
   (c) the system has infinitely many solutions

10. Show that if the system of equations in Exercise 9 is consistent, then at least one equation can be discarded from the system without altering the solution set.

11. Let $k = l = m = 0$ in Exercise 9; show that the system must be consistent. What can be said about the point of intersection of the three lines if the system has exactly one solution?

12. Consider the system of equations

$$
\begin{aligned}
x + y + 2z &= a \\
x \quad\ + z &= b \\
2x + y + 3z &= c
\end{aligned}
$$

Show that in order for this system to be consistent, a, b, and c must satisfy $c = a + b$.

13. Prove: If the linear equations $x_1 + kx_2 = c$ and $x_1 + lx_2 = d$ have the same solution set, then the equations are identical.

1.2 GAUSSIAN ELIMINATION

In this section we shall give a systematic procedure for solving systems of linear equations; it is based on the idea of reducing the augmented matrix to a form that is simple enough that the system of equations can be solved by inspection.

REDUCED ROW-ECHELON FORM

In Example 3 of the preceding section, we solved the given linear system by reducing the augmented matrix to

$$
\begin{bmatrix}
1 & 0 & 0 & 1 \\
0 & 1 & 0 & 2 \\
0 & 0 & 1 & 3
\end{bmatrix}
$$

from which the solution of the system was evident. This is an example of a matrix that is in *reduced row-echelon form*. To be of this form, a matrix must have the following properties.

> **1.** *If a row does not consist entirely of zeros, then the first nonzero number in the row is a 1. (We call this a **leading 1**.)*
> **2.** *If there are any rows that consist entirely of zeros, then they are grouped together at the bottom of the matrix.*
> **3.** *In any two successive rows that do not consist entirely of zeros, the leading 1 in the lower row occurs farther to the right than the leading 1 in the higher row.*
> **4.** *Each column that contains a leading 1 has zeros everywhere else.*

A matrix having properties **1, 2,** and **3** (but not necessarily 4) is said to be in *row-echelon form*.

Example 1   The following matrices are in reduced row-echelon form.

$$\begin{bmatrix} 1 & 0 & 0 & 4 \\ 0 & 1 & 0 & 7 \\ 0 & 0 & 1 & -1 \end{bmatrix}, \begin{bmatrix} 1 & 0 & 0 \\ 0 & 1 & 0 \\ 0 & 0 & 1 \end{bmatrix}, \begin{bmatrix} 0 & 1 & -2 & 0 & 1 \\ 0 & 0 & 0 & 1 & 3 \\ 0 & 0 & 0 & 0 & 0 \\ 0 & 0 & 0 & 0 & 0 \end{bmatrix}, \begin{bmatrix} 0 & 0 \\ 0 & 0 \end{bmatrix}$$

The following matrices are in row-echelon form but not in reduced row-echelon form.

$$\begin{bmatrix} 1 & 4 & 3 & 7 \\ 0 & 1 & 6 & 2 \\ 0 & 0 & 1 & 5 \end{bmatrix}, \begin{bmatrix} 1 & 1 & 0 \\ 0 & 1 & 0 \\ 0 & 0 & 0 \end{bmatrix}, \begin{bmatrix} 0 & 1 & 2 & 6 & 0 \\ 0 & 0 & 1 & -1 & 0 \\ 0 & 0 & 0 & 0 & 1 \end{bmatrix}$$

The reader should check to see that each of the matrices above satisfies all the necessary requirements. ▲

REMARK. As the preceding example illustrates, a matrix in row-echelon form has zeros below each leading 1, whereas a matrix in reduced row-echelon form has zeros both above and below each leading 1.

If, by a sequence of elementary row operations, the augmented matrix for a system of linear equations is put in reduced row-echelon form, then the solution set of the system will be evident by inspection or after a few simple steps. The next example illustrates this.

Example 2   Suppose that the augmented matrix for a system of linear equations has been reduced by row operations to the given reduced row-echelon form. Solve the system.

(a) $$\begin{bmatrix} 1 & 0 & 0 & 5 \\ 0 & 1 & 0 & -2 \\ 0 & 0 & 1 & 4 \end{bmatrix}$$
(b) $$\begin{bmatrix} 1 & 0 & 0 & 4 & -1 \\ 0 & 1 & 0 & 2 & 6 \\ 0 & 0 & 1 & 3 & 2 \end{bmatrix}$$

(c) $$\begin{bmatrix} 1 & 6 & 0 & 0 & 4 & -2 \\ 0 & 0 & 1 & 0 & 3 & 1 \\ 0 & 0 & 0 & 1 & 5 & 2 \\ 0 & 0 & 0 & 0 & 0 & 0 \end{bmatrix}$$
(d) $$\begin{bmatrix} 1 & 0 & 0 & 0 \\ 0 & 1 & 2 & 0 \\ 0 & 0 & 0 & 1 \end{bmatrix}$$

Solution (*a*). The corresponding system of equations is

$$\begin{aligned} x_1 \qquad\qquad &= 5 \\ x_2 \qquad &= -2 \\ x_3 &= 4 \end{aligned}$$

By inspection, $x_1 = 5$, $x_2 = -2$, $x_3 = 4$.

Solution (b). The corresponding system of equations is

$$x_1 \qquad\qquad\quad + 4x_4 = -1$$
$$x_2 \qquad + 2x_4 = \quad 6$$
$$x_3 + 3x_4 = \quad 2$$

Since x_1, x_2, and x_3 correspond to leading 1's in the augmented matrix, we call them **leading variables**. The nonleading variables (in this case x_4) are called **free variables**. Solving for the leading variables in terms of the free variable gives

$$x_1 = -1 - 4x_4$$
$$x_2 = \quad 6 - 2x_4$$
$$x_3 = \quad 2 - 3x_4$$

From this form of the equations we see that the free variable x_4 can be assigned an arbitrary value, say t, which then determines the values of the leading variables x_1, x_2, and x_3. Thus there are infinitely many solutions, and the general solution is given by the formulas

$$x_1 = -1 - 4t, \qquad x_2 = 6 - 2t, \qquad x_3 = 2 - 3t, \qquad x_4 = t$$

Solution (c). The corresponding system of equations is

$$x_1 + 6x_2 \qquad\qquad + 4x_5 = -2$$
$$x_3 \qquad + 3x_5 = \quad 1$$
$$x_4 + 5x_5 = \quad 2$$

Here the leading variables are x_1, x_3, and x_4, and the free variables are x_2 and x_5. Solving for the leading variables in terms of the free variables gives

$$x_1 = -2 - 6x_2 - 4x_5$$
$$x_3 = \quad 1 - 3x_5$$
$$x_4 = \quad 2 - 5x_5$$

Since x_5 can be assigned an arbitrary value, t, and x_2 can be assigned an arbitrary value, s, there are infinitely many solutions. The general solution is given by the formulas

$$x_1 = -2 - 6s - 4t, \qquad x_2 = s, \qquad x_3 = 1 - 3t, \qquad x_4 = 2 - 5t, \qquad x_5 = t$$

Solution (d). The last equation in the corresponding system of equations is

$$0x_1 + 0x_2 + 0x_3 = 1$$

Since this equation cannot be satisfied, there is no solution to the system. ▲

GAUSSIAN ELIMINATION

We have just seen how easy it is to solve a system of linear equations once its augmented matrix is in reduced row-echelon form. Now we shall give a step-by-step procedure that can be used to reduce any matrix to reduced row-echelon form. As we state each step in the procedure, we shall illustrate the idea by reducing the following matrix to reduced row-echelon form.

$$\begin{bmatrix} 0 & 0 & -2 & 0 & 7 & 12 \\ 2 & 4 & -10 & 6 & 12 & 28 \\ 2 & 4 & -5 & 6 & -5 & -1 \end{bmatrix}$$

Step 1. Locate the leftmost column that does not consist entirely of zeros.

$$\begin{bmatrix} 0 & 0 & -2 & 0 & 7 & 12 \\ 2 & 4 & -10 & 6 & 12 & 28 \\ 2 & 4 & -5 & 6 & -5 & -1 \end{bmatrix}$$

↑
—Leftmost nonzero column

Step 2. Interchange the top row with another row, if necessary, to bring a nonzero entry to the top of the column found in Step 1.

$$\begin{bmatrix} 2 & 4 & -10 & 6 & 12 & 28 \\ 0 & 0 & -2 & 0 & 7 & 12 \\ 2 & 4 & -5 & 6 & -5 & -1 \end{bmatrix}$$

The first and second rows in the preceding matrix were interchanged.

Step 3. If the entry that is now at the top of the column found in Step 1 is a, multiply the first row by $1/a$ in order to introduce a leading 1.

$$\begin{bmatrix} 1 & 2 & -5 & 3 & 6 & 14 \\ 0 & 0 & -2 & 0 & 7 & 12 \\ 2 & 4 & -5 & 6 & -5 & -1 \end{bmatrix}$$

The first row of the preceding matrix was multiplied by $\frac{1}{2}$.

Step 4. Add suitable multiples of the top row to the rows below so that all entries below the leading 1 become zeros.

$$\begin{bmatrix} 1 & 2 & -5 & 3 & 6 & 14 \\ 0 & 0 & -2 & 0 & 7 & 12 \\ 0 & 0 & 5 & 0 & -17 & -29 \end{bmatrix}$$

-2 times the first row of the preceding matrix was added to the third row.

Step 5. Now cover the top row in the matrix and begin again with Step 1 applied to the submatrix that remains. Continue in this way until the *entire* matrix is in row-echelon form.

$$\begin{bmatrix} 1 & 2 & -5 & 3 & 6 & 14 \\ 0 & 0 & -2 & 0 & 7 & 12 \\ 0 & 0 & 5 & 0 & -17 & -29 \end{bmatrix}$$

↑
**—Leftmost nonzero column in
the submatrix**

$$\begin{bmatrix} 1 & 2 & -5 & 3 & 6 & 14 \\ 0 & 0 & 1 & 0 & -\frac{7}{2} & -6 \\ 0 & 0 & 5 & 0 & -17 & -29 \end{bmatrix}$$

The first row in the submatrix was multiplied by $-\frac{1}{2}$ to introduce a leading 1.

$$\begin{bmatrix} 1 & 2 & -5 & 3 & 6 & 14 \\ 0 & 0 & 1 & 0 & -\frac{7}{2} & -6 \\ 0 & 0 & 0 & 0 & \frac{1}{2} & 1 \end{bmatrix}$$

-5 times the first row of the submatrix was added to the second row of the submatrix to introduce a zero below the leading 1.

$$\begin{bmatrix} 1 & 2 & -5 & 3 & 6 & 14 \\ 0 & 0 & 1 & 0 & -\frac{7}{2} & -6 \\ 0 & 0 & 0 & 0 & \frac{1}{2} & 1 \end{bmatrix}$$

The top row in the submatrix was covered, and we returned again to Step 1.

**Leftmost nonzero column
in the new submatrix**

$$\begin{bmatrix} 1 & 2 & -5 & 3 & 6 & 14 \\ 0 & 0 & 1 & 0 & -\frac{7}{2} & -6 \\ 0 & 0 & 0 & 0 & 1 & 2 \end{bmatrix}$$

The first (and only) row in the new submatrix was multiplied by 2 to introduce a leading 1.

The *entire* matrix is now in row-echelon form. To find the reduced row-echelon form we need the following additional step.

Step 6. Beginning with the last nonzero row and working upward, add suitable multiples of each row to the rows above to introduce zeros above the leading 1's.

$$\begin{bmatrix} 1 & 2 & -5 & 3 & 6 & 14 \\ 0 & 0 & 1 & 0 & 0 & 1 \\ 0 & 0 & 0 & 0 & 1 & 2 \end{bmatrix}$$

$\frac{7}{2}$ times the third row of the preceding matrix was added to the second row.

$$\begin{bmatrix} 1 & 2 & -5 & 3 & 0 & 2 \\ 0 & 0 & 1 & 0 & 0 & 1 \\ 0 & 0 & 0 & 0 & 1 & 2 \end{bmatrix}$$

-6 times the third row was added to the first row.

$$\begin{bmatrix} 1 & 2 & 0 & 3 & 0 & 7 \\ 0 & 0 & 1 & 0 & 0 & 1 \\ 0 & 0 & 0 & 0 & 1 & 2 \end{bmatrix}$$

5 times the second row was added to the first row.

The last matrix is in reduced row-echelon form.

The above procedure for reducing a matrix to reduced row-echelon form is called **Gauss–Jordan elimination*** (see page 13). If we use only the first five steps, the procedure produces a row-echelon form and is called **Gaussian elimination**.

REMARK. It can be shown that *every matrix has a unique reduced row-echelon form*; that is, one will arrive at the same reduced row-echelon form for a given matrix no matter how the row operations are varied. (A proof of this result can be found in the article ''The Reduced Row Echelon Form of a Matrix is Unique: A Simple Proof,'' by Thomas Yuster, *Mathematics Magazine*, Vol. 57, No. 2, 1984, pp. 93–94.) In contrast, *a row-echelon form of a given matrix is not unique*: different sequences of row operations can produce different row-echelon forms.

Example 3   Solve by Gauss–Jordan elimination.

$$
\begin{aligned}
x_1 + 3x_2 - 2x_3 \quad\quad\;\; + 2x_5 \quad\quad\quad &= 0 \\
2x_1 + 6x_2 - 5x_3 - \; 2x_4 + 4x_5 - \; 3x_6 &= -1 \\
5x_3 + 10x_4 \quad\quad + 15x_6 &= 5 \\
2x_1 + 6x_2 \quad\quad + \; 8x_4 + 4x_5 + 18x_6 &= 6
\end{aligned}
$$

The augmented matrix for the system is

$$
\begin{bmatrix}
1 & 3 & -2 & 0 & 2 & 0 & 0 \\
2 & 6 & -5 & -2 & 4 & -3 & -1 \\
0 & 0 & 5 & 10 & 0 & 15 & 5 \\
2 & 6 & 0 & 8 & 4 & 18 & 6
\end{bmatrix}
$$

---

* *Karl Friedrich Gauss* (1777–1855) was a German mathematician and scientist. Sometimes called the ''prince of mathematicians,'' Gauss ranks with Isaac Newton and Archimedes as one of the three greatest mathematicians who ever lived. In the entire history of mathematics there may never have been a child so precocious as Gauss—by his own account he worked out the rudiments of arithmetic before he could talk. One day, before he was even three years old, his genius became apparent to his parents in a very dramatic way. His father was preparing the weekly payroll for the laborers under his charge while the boy watched quietly from a corner. At the end of the long and tedious calculation, Gauss informed his father that there was an error in the result and stated the answer, which he had worked out in his head. To the astonishment of his parents, a check of the computations showed Gauss to be correct!

In his doctoral dissertation Gauss gave the first complete proof of the fundamental theorem of algebra, which states that every polynomial equation has as many solutions as its degree. At age 19 he solved a problem that baffled Euclid, inscribing a regular polygon of seventeen sides in a circle using straightedge and compass; and in 1801, at age 24, he published his first masterpiece, *Disquisitiones Arithmeticae*, considered by many to be one of the most brilliant achievements in mathematics. In that paper Gauss systematized the study of number theory (properties of the integers) and formulated the basic concepts that form the foundation of that subject.

Among his myriad achievements, Gauss discovered the Gaussian or ''bell-shaped'' curve that is fundamental in probability, gave the first geometric interpretation of complex numbers and established their fundamental role in mathematics, developed methods of characterizing surfaces intrinsically by means of the curves that they contain, developed the theory of conformal (angle-preserving) maps, and discovered non-Euclidean geometry 30 years before the ideas were published by others. In physics he made major contributions to the theory of lenses and capillary action, and with Wilhelm Weber he did fundamental work in electromagnetism. Gauss invented the heliotrope, bifilar magnetometer, and an electrotelegraph.

Gauss was deeply religious and aristocratic in demeanor. He mastered foreign languages with ease, read extensively, and enjoyed minerology and botany as hobbies. He disliked teaching and was usually cool and discouraging to other mathematicians, possibly because he had already anticipated their work. It has been said that if Gauss had published all of his discoveries, the current state of mathematics would be advanced by 50 years. He was without a doubt the greatest mathematician of the modern era.

Wilhelm Jordan (1842–1899) was a German engineer who specialized in geodesy. His contribution to solving linear systems appeared in his popular book, *Handbuch der Vermessungskunde* (*Handbook of Geodesy*), in 1888.

Adding -2 times the first row to the second and fourth rows gives

$$\begin{bmatrix} 1 & 3 & -2 & 0 & 2 & 0 & 0 \\ 0 & 0 & -1 & -2 & 0 & -3 & -1 \\ 0 & 0 & 5 & 10 & 0 & 15 & 5 \\ 0 & 0 & 4 & 8 & 0 & 18 & 6 \end{bmatrix}$$

Multiplying the second row by -1 and then adding -5 times the new secoɪ. ˙ row to the third row and -4 times the new second row to the fourth row gives

$$\begin{bmatrix} 1 & 3 & -2 & 0 & 2 & 0 & 0 \\ 0 & 0 & 1 & 2 & 0 & 3 & 1 \\ 0 & 0 & 0 & 0 & 0 & 0 & 0 \\ 0 & 0 & 0 & 0 & 0 & 6 & 2 \end{bmatrix}$$

Interchanging the third and fourth rows and then multiplying the third row of the resulting matrix by $\frac{1}{6}$ gives the row-echelon form

$$\begin{bmatrix} 1 & 3 & -2 & 0 & 2 & 0 & 0 \\ 0 & 0 & 1 & 2 & 0 & 3 & 1 \\ 0 & 0 & 0 & 0 & 0 & 1 & \frac{1}{3} \\ 0 & 0 & 0 & 0 & 0 & 0 & 0 \end{bmatrix}$$

Adding -3 times the third row to the second row and then adding 2 times the second row of the resulting matrix to the first row yields the reduced row-echelon form

$$\begin{bmatrix} 1 & 3 & 0 & 4 & 2 & 0 & 0 \\ 0 & 0 & 1 & 2 & 0 & 0 & 0 \\ 0 & 0 & 0 & 0 & 0 & 1 & \frac{1}{3} \\ 0 & 0 & 0 & 0 & 0 & 0 & 0 \end{bmatrix}$$

The corresponding system of equations is

$$\begin{aligned} x_1 + 3x_2 \quad\quad + 4x_4 + 2x_5 &= 0 \\ x_3 + 2x_4 \quad\quad\quad &= 0 \\ x_6 &= \tfrac{1}{3} \end{aligned}$$

(We have discarded the last equation, $0x_1 + 0x_2 + 0x_3 + 0x_4 + 0x_5 + 0x_6 = 0$, since it will be satisfied automatically by the solutions of the remaining equations.) Solving for the leading variables, we obtain

$$\begin{aligned} x_1 &= -3x_2 - 4x_4 - 2x_5 \\ x_3 &= -2x_4 \\ x_6 &= \tfrac{1}{3} \end{aligned}$$

If we assign the free variables x_2, x_4, and x_5 arbitrary values r, s, and t, respectively, the general solution is given by the formulas

$$x_1 = -3r - 4s - 2t, \quad x_2 = r, \quad x_3 = -2s, \quad x_4 = s, \quad x_5 = t, \quad x_6 = \tfrac{1}{3} \quad \blacktriangle$$

**BACK-
SUBSTITUTION**

Example 4   It is sometimes preferable to solve a system of linear equations by using Gaussian elimination to bring the augmented matrix into row-echelon form without continuing all the way to the reduced row-echelon form. When this is done, the corresponding system of equations can be solved by a technique called ***back-substitution***. We shall illustrate this method using the system of equations in Example 3.

From the computations in Example 3, a row-echelon form of the augmented matrix is

$$\begin{bmatrix} 1 & 3 & -2 & 0 & 2 & 0 & 0 \\ 0 & 0 & 1 & 2 & 0 & 3 & 1 \\ 0 & 0 & 0 & 0 & 0 & 1 & \frac{1}{3} \\ 0 & 0 & 0 & 0 & 0 & 0 & 0 \end{bmatrix}$$

To solve the corresponding system of equations

$$x_1 + 3x_2 - 2x_3 \quad\quad + 2x_5 \quad\quad\quad = 0$$
$$x_3 + 2x_4 \quad\quad + 3x_6 = 1$$
$$x_6 = \tfrac{1}{3}$$

we proceed as follows:

Step 1.   Solve the equations for the leading variables.

$$x_1 = -3x_2 + 2x_3 - 2x_5$$
$$x_3 = 1 - 2x_4 - 3x_6$$
$$x_6 = \tfrac{1}{3}$$

Step 2.   Beginning with the bottom equation and working upward, successively substitute each equation into all the equations above it.

Substituting $x_6 = \tfrac{1}{3}$ into the second equation yields

$$x_1 = -3x_2 + 2x_3 - 2x_5$$
$$x_3 = -2x_4$$
$$x_6 = \tfrac{1}{3}$$

Substituting $x_3 = -2x_4$ into the first equation yields

$$x_1 = -3x_2 - 4x_4 - 2x_5$$
$$x_3 = -2x_4$$
$$x_6 = \tfrac{1}{3}$$

Step 3.   Assign arbitrary values to the free variables, if any.

If we assign x_2, x_4, and x_5 the arbitrary values r, s, and t, respectively, the general solution is given by the formulas

$$x_1 = -3r - 4s - 2t, \qquad x_2 = r, \qquad x_3 = -2s, \qquad x_4 = s, \qquad x_5 = t, \qquad x_6 = \tfrac{1}{3}$$

This agrees with the solution obtained in Example 3. ▲

REMARK. The arbitrary values that are assigned to the free variables are often called **parameters**. Although we shall generally use the letters r, s, t, ... for the parameters, any letters that do not conflict with the variable names may be used.

Example 5 Solve

$$x + y + 2z = 9$$
$$2x + 4y - 3z = 1$$
$$3x + 6y - 5z = 0$$

by Gaussian elimination and back-substitution.

Solution. This is the system in Example 3 of Section 1.1. In that example we converted the augmented matrix

$$\begin{bmatrix} 1 & 1 & 2 & 9 \\ 2 & 4 & -3 & 1 \\ 3 & 6 & -5 & 0 \end{bmatrix}$$

to the row-echelon form

$$\begin{bmatrix} 1 & 1 & 2 & 9 \\ 0 & 1 & -\tfrac{7}{2} & -\tfrac{17}{2} \\ 0 & 0 & 1 & 3 \end{bmatrix}$$

The system corresponding to this matrix is

$$x + y + 2z = 9$$
$$y - \tfrac{7}{2}z = -\tfrac{17}{2}$$
$$z = 3$$

Solving for the leading variables yields

$$x = 9 - y - 2z$$
$$y = -\tfrac{17}{2} + \tfrac{7}{2}z$$
$$z = 3$$

Substituting the bottom equation into those above yields

$$x = 3 - y$$
$$y = 2$$
$$z = 3$$

and substituting the second equation into the top yields

$$x = 1$$
$$y = 2$$
$$z = 3$$

This agrees with the result found by Gauss–Jordan elimination in Example 3 of Section 1.1. ▲

HOMOGENEOUS LINEAR SYSTEMS

A system of linear equations is said to be ***homogeneous*** if the constant terms are all zero; that is, the system has the form

$$a_{11}x_1 + a_{12}x_2 + \cdots + a_{1n}x_n = 0$$
$$a_{21}x_1 + a_{22}x_2 + \cdots + a_{2n}x_n = 0$$
$$\vdots \qquad \vdots \qquad \qquad \vdots \qquad \vdots$$
$$a_{m1}x_1 + a_{m2}x_2 + \cdots + a_{mn}x_n = 0$$

Every homogeneous system of linear equations is consistent, since all such systems have $x_1 = 0, x_2 = 0, \ldots, x_n = 0$ as a solution. This solution is called the ***trivial solution***; if there are other solutions, they are called ***nontrivial solutions***.

Because a homogeneous linear system always has the trivial solution, there are only two possibilities for its solutions:

- The system has only the trivial solution.
- The system has infinitely many solutions in addition to the trivial solution.

In the special case of a homogeneous linear system of two equations in two unknowns, say

$$a_1 x + b_1 y = 0 \qquad (a_1, b_1 \text{ not both zero})$$
$$a_2 x + b_2 y = 0 \qquad (a_2, b_2 \text{ not both zero})$$

the graphs of the equations are lines through the origin, and the trivial solution corresponds to the point of intersection at the origin (Figure 1).

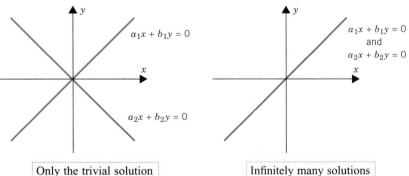

Figure 1

Only the trivial solution   Infinitely many solutions

There is one case in which a homogeneous system is assured of having nontrivial solutions, namely, whenever the system involves more unknowns than equations. To see why, consider the following example of four equations in five unknowns.

Example 6   Solve the following homogeneous system of linear equations by Gauss–Jordan elimination.

$$
\begin{aligned}
2x_1 + 2x_2 - x_3 + x_5 &= 0 \\
-x_1 - x_2 + 2x_3 - 3x_4 + x_5 &= 0 \\
x_1 + x_2 - 2x_3 - x_5 &= 0 \\
x_3 + x_4 + x_5 &= 0
\end{aligned} \tag{1}
$$

Solution. The augmented matrix for the system is

$$
\begin{bmatrix}
2 & 2 & -1 & 0 & 1 & 0 \\
-1 & -1 & 2 & -3 & 1 & 0 \\
1 & 1 & -2 & 0 & -1 & 0 \\
0 & 0 & 1 & 1 & 1 & 0
\end{bmatrix}
$$

Reducing this matrix to reduced row-echelon form, we obtain

$$
\begin{bmatrix}
1 & 1 & 0 & 0 & 1 & 0 \\
0 & 0 & 1 & 0 & 1 & 0 \\
0 & 0 & 0 & 1 & 0 & 0 \\
0 & 0 & 0 & 0 & 0 & 0
\end{bmatrix}
$$

The corresponding system of equations is

$$
\begin{aligned}
x_1 + x_2 + x_5 &= 0 \\
x_3 + x_5 &= 0 \\
x_4 &= 0
\end{aligned} \tag{2}
$$

Solving for the leading variables yields

$$
\begin{aligned}
x_1 &= -x_2 - x_5 \\
x_3 &= -x_5 \\
x_4 &= 0
\end{aligned}
$$

Thus, the general solution is

$$
x_1 = -s - t, \qquad x_2 = s, \qquad x_3 = -t, \qquad x_4 = 0, \qquad x_5 = t
$$

Note that the trivial solution is obtained when $s = t = 0$. ▲

Example 6 illustrates two important points about solving homogeneous systems of linear equations. First, none of the three elementary row operations alters the final column of zeros in the augmented matrix, so that the system of equations corresponding to the reduced row-echelon form of the augmented matrix must also be a homogeneous system [see system (2)]. Second, depending on whether the reduced row-echelon form of the augmented matrix has any zero rows, the number of equations in the reduced system is the same as or less than the number of equations in the original system [compare systems (1) and (2)]. Thus, if the given homogeneous system has m equations in n unknowns with $m < n$, and if there are r nonzero rows in the reduced row-echelon form of the augmented matrix, we will have $r < n$. It follows that the system of equations corresponding to the reduced row-echelon form of the augmented matrix will have the form

$$
\begin{aligned}
\cdots x_{k_1} \qquad\qquad\quad + \Sigma(\ ) &= 0 \\
\cdots x_{k_2} \qquad\quad + \Sigma(\ ) &= 0 \\
\cdots\ \ \ddots \qquad\ \ \vdots \\
x_{k_r} + \Sigma(\ ) &= 0
\end{aligned}
\tag{3}
$$

where $x_{k_1}, x_{k_2}, \ldots, x_{k_r}$ are the leading variables and $\Sigma(\ )$ denotes sums (possibly all different) that involve the $n - r$ free variables [compare system (3) with system (2) above]. Solving for the leading variables gives

$$
\begin{aligned}
x_{k_1} &= -\Sigma(\ ) \\
x_{k_2} &= -\Sigma(\ ) \\
&\ \ \vdots \\
x_{k_r} &= -\Sigma(\ )
\end{aligned}
$$

As in Example 6, we can assign arbitrary values to the free variables on the right-hand side and thus obtain infinitely many solutions to the system.

In summary, we have the following important theorem.

Theorem 1.2.1. *A homogeneous system of linear equations with more unknowns than equations has infinitely many solutions.*

REMARK. Note that Theorem 1.2.1 applies only to homogeneous systems. A nonhomogeneous system with more unknowns than equations need not be consistent (Exercise 34); however, if the system is consistent, it will have infinitely many solutions. This will be proved later.

COMPUTER SOLUTION OF LINEAR SYSTEMS

In applications it is not uncommon to encounter large linear systems that must be solved by computer. Most computer algorithms for solving such systems are based on Gaussian elimination or Gauss–Jordan elimination, but the basic procedures are often modified to deal with such issues as

- Reducing roundoff errors
- Minimizing the use of computer memory space
- Solving the system with maximum speed

Some of these matters will be considered in Chapter 9. For hand computations fractions are an annoyance that often cannot be avoided. However, in some cases it is possible to avoid them by varying the elementary row operations in the right way. Thus, once the methods of Gaussian elimination and Gauss–Jordan elimination have been mastered, the reader may wish to vary the steps in specific problems to avoid fractions (see Exercise 18).

EXERCISE SET 1.2

1. Which of the following 3×3 matrices are in reduced row-echelon form?

(a) $\begin{bmatrix} 1 & 0 & 0 \\ 0 & 1 & 0 \\ 0 & 0 & 1 \end{bmatrix}$ (b) $\begin{bmatrix} 1 & 0 & 0 \\ 0 & 1 & 0 \\ 0 & 0 & 0 \end{bmatrix}$ (c) $\begin{bmatrix} 0 & 1 & 0 \\ 0 & 0 & 1 \\ 0 & 0 & 0 \end{bmatrix}$ (d) $\begin{bmatrix} 1 & 0 & 0 \\ 0 & 0 & 1 \\ 0 & 0 & 0 \end{bmatrix}$ (e) $\begin{bmatrix} 1 & 0 & 0 \\ 0 & 0 & 0 \\ 0 & 0 & 1 \end{bmatrix}$

(f) $\begin{bmatrix} 0 & 1 & 0 \\ 1 & 0 & 0 \\ 0 & 0 & 0 \end{bmatrix}$ (g) $\begin{bmatrix} 1 & 1 & 0 \\ 0 & 1 & 0 \\ 0 & 0 & 0 \end{bmatrix}$ (h) $\begin{bmatrix} 1 & 0 & 2 \\ 0 & 1 & 3 \\ 0 & 0 & 0 \end{bmatrix}$ (i) $\begin{bmatrix} 0 & 0 & 1 \\ 0 & 0 & 0 \\ 0 & 0 & 0 \end{bmatrix}$ (j) $\begin{bmatrix} 0 & 0 & 0 \\ 0 & 0 & 0 \\ 0 & 0 & 0 \end{bmatrix}$

2. Which of the following 3×3 matrices are in row-echelon form?

(a) $\begin{bmatrix} 1 & 0 & 0 \\ 0 & 1 & 0 \\ 0 & 0 & 1 \end{bmatrix}$ (b) $\begin{bmatrix} 1 & 2 & 0 \\ 0 & 1 & 0 \\ 0 & 0 & 0 \end{bmatrix}$ (c) $\begin{bmatrix} 1 & 0 & 0 \\ 0 & 1 & 0 \\ 0 & 2 & 0 \end{bmatrix}$ (d) $\begin{bmatrix} 1 & 3 & 4 \\ 0 & 0 & 1 \\ 0 & 0 & 0 \end{bmatrix}$

(e) $\begin{bmatrix} 1 & 5 & -3 \\ 0 & 1 & 1 \\ 0 & 0 & 0 \end{bmatrix}$ (f) $\begin{bmatrix} 1 & 2 & 3 \\ 0 & 0 & 0 \\ 0 & 0 & 1 \end{bmatrix}$

3. In each part determine whether the matrix is in row-echelon form, reduced row-echelon form, both, or neither.

(a) $\begin{bmatrix} 1 & 2 & 0 & 3 & 0 \\ 0 & 0 & 1 & 1 & 0 \\ 0 & 0 & 0 & 0 & 1 \\ 0 & 0 & 0 & 0 & 0 \end{bmatrix}$ (b) $\begin{bmatrix} 1 & 0 & 0 & 5 \\ 0 & 0 & 1 & 3 \\ 0 & 1 & 0 & 4 \end{bmatrix}$ (c) $\begin{bmatrix} 1 & 0 & 3 & 1 \\ 0 & 1 & 2 & 4 \end{bmatrix}$

(d) $\begin{bmatrix} 1 & -7 & 5 & 5 \\ 0 & 1 & 3 & 2 \end{bmatrix}$ (e) $\begin{bmatrix} 1 & 3 & 0 & 2 & 0 \\ 1 & 0 & 2 & 2 & 0 \\ 0 & 0 & 0 & 0 & 1 \\ 0 & 0 & 0 & 0 & 0 \end{bmatrix}$ (f) $\begin{bmatrix} 0 & 0 \\ 0 & 0 \\ 0 & 0 \end{bmatrix}$

4. In each part suppose that the augmented matrix for a system of linear equations has been reduced by row operations to the given reduced row-echelon form. Solve the system.

(a) $\begin{bmatrix} 1 & 0 & 0 & -3 \\ 0 & 1 & 0 & 0 \\ 0 & 0 & 1 & 7 \end{bmatrix}$ (b) $\begin{bmatrix} 1 & 0 & 0 & -7 & 8 \\ 0 & 1 & 0 & 3 & 2 \\ 0 & 0 & 1 & 1 & -5 \end{bmatrix}$

(c) $\begin{bmatrix} 1 & -6 & 0 & 0 & 3 & -2 \\ 0 & 0 & 1 & 0 & 4 & 7 \\ 0 & 0 & 0 & 1 & 5 & 8 \\ 0 & 0 & 0 & 0 & 0 & 0 \end{bmatrix}$ (d) $\begin{bmatrix} 1 & -3 & 0 & 0 \\ 0 & 0 & 1 & 0 \\ 0 & 0 & 0 & 1 \end{bmatrix}$

5. In each part suppose that the augmented matrix for a system of linear equations has been reduced by row operations to the given row-echelon form. Solve the system.

(a) $\begin{bmatrix} 1 & -3 & 4 & 7 \\ 0 & 1 & 2 & 2 \\ 0 & 0 & 1 & 5 \end{bmatrix}$ (b) $\begin{bmatrix} 1 & 0 & 8 & -5 & 6 \\ 0 & 1 & 4 & -9 & 3 \\ 0 & 0 & 1 & 1 & 2 \end{bmatrix}$

(c) $\begin{bmatrix} 1 & 7 & -2 & 0 & -8 & -3 \\ 0 & 0 & 1 & 1 & 6 & 5 \\ 0 & 0 & 0 & 1 & 3 & 9 \\ 0 & 0 & 0 & 0 & 0 & 0 \end{bmatrix}$ (d) $\begin{bmatrix} 1 & -3 & 7 & 1 \\ 0 & 1 & 4 & 0 \\ 0 & 0 & 0 & 1 \end{bmatrix}$

6. Solve each of the following systems by Gauss–Jordan elimination.

(a) $\begin{aligned} x_1 + x_2 + 2x_3 &= 8 \\ -x_1 - 2x_2 + 3x_3 &= 1 \\ 3x_1 - 7x_2 + 4x_3 &= 10 \end{aligned}$ (b) $\begin{aligned} 2x_1 + 2x_2 + 2x_3 &= 0 \\ -2x_1 + 5x_2 + 2x_3 &= 1 \\ 8x_1 + x_2 + 4x_3 &= -1 \end{aligned}$

(c) $\begin{aligned} x - y + 2z - w &= -1 \\ 2x + y - 2z - 2w &= -2 \\ -x + 2y - 4z + w &= 1 \\ 3x \qquad\qquad - 3w &= -3 \end{aligned}$ (d) $\begin{aligned} -2b + 3c &= 1 \\ 3a + 6b - 3c &= -2 \\ 6a + 6b + 3c &= 5 \end{aligned}$

7. Solve each of the systems in Exercise 6 by Gaussian elimination.

8. Solve each of the following systems by Gauss–Jordan elimination.

(a) $\begin{aligned} 2x_1 - 3x_2 &= -2 \\ 2x_1 + x_2 &= 1 \\ 3x_1 + 2x_2 &= 1 \end{aligned}$ (b) $\begin{aligned} 3x_1 + 2x_2 - x_3 &= -15 \\ 5x_1 + 3x_2 + 2x_3 &= 0 \\ 3x_1 + x_2 + 3x_3 &= 11 \\ -6x_1 - 4x_2 + 2x_3 &= 30 \end{aligned}$

(c) $\begin{aligned} 4x_1 - 8x_2 &= 12 \\ 3x_1 - 6x_2 &= 9 \\ -2x_1 + 4x_2 &= -6 \end{aligned}$ (d) $\begin{aligned} 10y - 4z + w &= 1 \\ x + 4y - z + w &= 2 \\ 3x + 2y + z + 2w &= 5 \\ -2x - 8y + 2z - 2w &= -4 \\ x - 6y + 3z &= 1 \end{aligned}$

9. Solve each of the systems in Exercise 8 by Gaussian elimination.

10. Solve each of the following systems by Gauss–Jordan elimination.

(a) $\begin{aligned} 5x_1 - 2x_2 + 6x_3 &= 0 \\ -2x_1 + x_2 + 3x_3 &= 1 \end{aligned}$ (b) $\begin{aligned} x_1 - 2x_2 + x_3 - 4x_4 &= 1 \\ x_1 + 3x_2 + 7x_3 + 2x_4 &= 2 \\ x_1 - 12x_2 - 11x_3 - 16x_4 &= 5 \end{aligned}$ (c) $\begin{aligned} w + 2x - y &= 4 \\ x - y &= 3 \\ w + 3x - 2y &= 7 \\ 2u + 4v + w + 7x &= 7 \end{aligned}$

11. Solve each of the systems in Exercise 10 by Gaussian elimination.

12. Without using pencil and paper, determine which of the following homogeneous systems have nontrivial solutions.

(a) $\begin{aligned} 2x_1 - 3x_2 + 4x_3 - x_4 &= 0 \\ 7x_1 + x_2 - 8x_3 + 9x_4 &= 0 \\ 2x_1 + 8x_2 + x_3 - x_4 &= 0 \end{aligned}$ (b) $\begin{aligned} x_1 + 3x_2 - x_3 &= 0 \\ x_2 - 8x_3 &= 0 \\ 4x_3 &= 0 \end{aligned}$

(c) $\begin{aligned} a_{11}x_1 + a_{12}x_2 + a_{13}x_3 &= 0 \\ a_{21}x_1 + a_{22}x_2 + a_{23}x_3 &= 0 \end{aligned}$ (d) $\begin{aligned} 3x_1 - 2x_2 &= 0 \\ 6x_1 - 4x_2 &= 0 \end{aligned}$

13. Solve the following homogeneous systems of linear equations by any method.

    (a) $\begin{aligned} 2x_1 + x_2 + 3x_3 &= 0 \\ x_1 + 2x_2 \qquad &= 0 \\ x_2 + x_3 &= 0 \end{aligned}$ (b) $\begin{aligned} 3x_1 + x_2 + x_3 + x_4 &= 0 \\ 5x_1 - x_2 + x_3 - x_4 &= 0 \end{aligned}$ (c) $\begin{aligned} 2x + 2y + 4z &= 0 \\ w \qquad - y - 3z &= 0 \\ 2w + 3x + y + z &= 0 \\ -2w + x + 3y - 2z &= 0 \end{aligned}$

14. Solve the following homogeneous systems of linear equations by any method.

    (a) $\begin{aligned} 2x - y - 3z &= 0 \\ -x + 2y - 3z &= 0 \\ x + y + 4z &= 0 \end{aligned}$ (b) $\begin{aligned} v + 3w - 2x &= 0 \\ 2u + v - 4w + 3x &= 0 \\ 2u + 3v + 2w - x &= 0 \\ -4u - 3v + 5w - 4x &= 0 \end{aligned}$ (c) $\begin{aligned} x_1 + 3x_2 \qquad + x_4 &= 0 \\ x_1 + 4x_2 + 2x_3 \qquad &= 0 \\ -2x_2 - 2x_3 - x_4 &= 0 \\ 2x_1 - 4x_2 + x_3 + x_4 &= 0 \\ x_1 - 2x_2 - x_3 + x_4 &= 0 \end{aligned}$

15. Solve the following systems by any method.

    (a) $\begin{aligned} 2I_1 - I_2 + 3I_3 + 4I_4 &= 9 \\ I_1 \qquad - 2I_3 + 7I_4 &= 11 \\ 3I_1 - 3I_2 + I_3 + 5I_4 &= 8 \\ 2I_1 + I_2 + 4I_3 + 4I_4 &= 10 \end{aligned}$ (b) $\begin{aligned} Z_3 + Z_4 + Z_5 &= 0 \\ -Z_1 - Z_2 + 2Z_3 - 3Z_4 + Z_5 &= 0 \\ Z_1 + Z_2 - 2Z_3 \qquad - Z_5 &= 0 \\ 2Z_1 + 2Z_2 - Z_3 \qquad + Z_5 &= 0 \end{aligned}$

16. Solve the following systems, where a, b, and c are constants.

    (a) $\begin{aligned} 2x + y &= a \\ 3x + 6y &= b \end{aligned}$ (b) $\begin{aligned} x_1 + x_2 + x_3 &= a \\ 2x_1 \qquad + 2x_3 &= b \\ 3x_2 + 3x_3 &= c \end{aligned}$

17. For which values of a will the following system have no solutions? Exactly one solution? Infinitely many solutions?

    $$\begin{aligned} x + 2y - \qquad 3z &= 4 \\ 3x - y + \qquad 5z &= 2 \\ 4x + y + (a^2 - 14)z &= a + 2 \end{aligned}$$

18. Reduce

    $$\begin{bmatrix} 2 & 1 & 3 \\ 0 & -2 & 7 \\ 3 & 4 & 5 \end{bmatrix}$$

    to reduced row-echelon form without introducing any fractions.

19. Find two different row-echelon forms of

    $$\begin{bmatrix} 1 & 3 \\ 2 & 7 \end{bmatrix}$$

20. Solve the following system of nonlinear equations for the unknown angles α, β, and γ, where $0 \le \alpha \le 2\pi$, $0 \le \beta \le 2\pi$, and $0 \le \gamma < \pi$.

    $$\begin{aligned} 2\sin\alpha - \cos\beta + 3\tan\gamma &= 3 \\ 4\sin\alpha + 2\cos\beta - 2\tan\gamma &= 2 \\ 6\sin\alpha - 3\cos\beta + \tan\gamma &= 9 \end{aligned}$$

21. Solve the following system of nonlinear equations for x, y, and z.

$$x^2 + y^2 + z^2 = 6$$
$$x^2 - y^2 + 2z^2 = 2$$
$$2x^2 + y^2 - z^2 = 3$$

22. Show that the following nonlinear system has eighteen solutions if $0 \le \alpha \le 2\pi$, $0 \le \beta \le 2\pi$, and $0 \le \gamma \le 2\pi$.

$$\sin \alpha + 2 \cos \beta + 3 \tan \gamma = 0$$
$$2 \sin \alpha + 5 \cos \beta + 3 \tan \gamma = 0$$
$$- \sin \alpha - 5 \cos \beta + 5 \tan \gamma = 0$$

23. For which value(s) of γ does the following system of equations have nontrivial solutions?

$$(\lambda - 3)x + \quad y = 0$$
$$x + (\lambda - 3)y = 0$$

24. Consider the system of equations

$$ax + by = 0$$
$$cx + dy = 0$$
$$ex + fy = 0$$

Discuss the relative positions of the lines $ax + by = 0$, $cx + dy = 0$, and $ex + fy = 0$ when
(a) the system has only the trivial solution (b) the system has nontrivial solutions

25. Figure 2 shows the graph of a cubic equation $y = ax^3 + bx^2 + cx + d$. Find the coefficients a, b, c, and d.

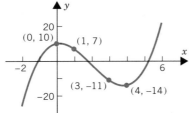

Figure 2

26. Recall from plane geometry that three points, not all lying on a straight line, determine a unique circle. It is shown in analytic geometry that a circle in the xy-plane has an equation of the form

$$ax^2 + ay^2 + bx + cy + d = 0$$

Find an equation of the circle shown in Figure 3.

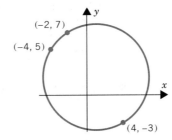

Figure 3

27. Describe the possible reduced row-echelon forms of

$$\begin{bmatrix} a & b & c \\ d & e & f \\ g & h & i \end{bmatrix}$$

28. Show that if $ad - bc \neq 0$, then the reduced row-echelon form of

$$\begin{bmatrix} a & b \\ c & d \end{bmatrix} \quad \text{is} \quad \begin{bmatrix} 1 & 0 \\ 0 & 1 \end{bmatrix}$$

29. Use Exercise 28 to show that if $ad - bc \neq 0$, then the system

$$ax + by = k$$
$$cx + dy = l$$

has exactly one solution.

30. Solve the system

$$2x_1 - x_2 \qquad = \lambda x_1$$
$$2x_1 + x_2 + x_3 = \lambda x_2$$
$$-2x_1 + 2x_2 + x_3 = \lambda x_3$$

for x_1, x_2, and x_3 if
(a) $\lambda = 1$ (b) $\lambda = 2$

31. Consider the system of equations

$$ax + by = 0$$
$$cx + dy = 0$$

(a) Show that if $x = x_0$, $y = y_0$ is any solution of the system and k is any constant, then $x = kx_0$, $y = ky_0$ is also a solution.
(b) Show that if $x = x_0$, $y = y_0$ and $x = x_1$, $y = y_1$ are any two solutions, then $x = x_0 + x_1$, $y = y_0 + y_1$ is also a solution.

32. Consider the systems of equations

(I) $ax + by = k$ (II) $ax + by = 0$
$\quad\ cx + dy = l$ $\quad\ cx + dy = 0$

(a) Show that if $x = x_1$, $y = y_1$ and $x = x_2$, $y = y_2$ are both solutions of I, then $x = x_1 - x_2$, $y = y_1 - y_2$ is a solution of II.
(b) Show that if $x = x_1$, $y = y_1$ is a solution of I and $x = x_0$, $y = y_0$ is a solution of II, then $x = x_1 + x_0$, $y = y_1 + y_0$ is a solution of I.

33. (a) In the system of equations numbered (3), explain why it would be incorrect to denote the leading variables by $x_1, x_2, \ldots, x_r$ rather than $x_{k_1}, x_{k_2}, \ldots, x_{k_r}$ as we have done.
(b) The system of equations numbered (2) is a specific case of (3). What value does r have in this case? What are $x_{k_1}, x_{k_2}, \ldots, x_{k_r}$ in this case? Write out the sums denoted by $\Sigma(\ )$ in (3).

34. Find an inconsistent linear system that has more unknowns than equations.

1.3 MATRICES AND MATRIX OPERATIONS

Rectangular arrays of real numbers arise in many contexts other than as augmented matrices for systems of linear equations. In this section we shall consider such arrays as objects in their own right and develop some of their properties for use in our later work.

MATRIX NOTATION AND TERMINOLOGY

> **Definition.** A **matrix** is a rectangular array of numbers. The numbers in the array are called the **entries** in the matrix.

Example 1 Some examples of matrices are

$$\begin{bmatrix} 1 & 2 \\ 3 & 0 \\ -1 & 4 \end{bmatrix}, \quad [2 \quad 1 \quad 0 \quad -3], \quad \begin{bmatrix} -\sqrt{2} & \pi & e \\ 3 & \frac{1}{2} & 0 \\ 0 & 0 & 0 \end{bmatrix}, \quad \begin{bmatrix} 1 \\ 3 \end{bmatrix}, \quad [4]$$

The **size** of a matrix is described in terms of the number of rows (horizontal lines) and columns (vertical lines) it contains. For example, the first matrix in Example 1 has three rows and two columns, so its size is 3 by 2 (written 3×2). In a size description, the first number always denotes the number of rows and the second denotes the number of columns. The remaining matrices in Example 1 have sizes 1×4, 3×3, 2×1, and 1×1, respectively. A matrix with only one column is called a **column matrix** (or a **column vector**), and a matrix with only one row is called a **row matrix** (or a **row vector**). Thus, in Example 1 the 2×1 matrix is a column matrix, the 1×4 matrix is a row matrix, and the 1×1 matrix is both a row matrix and a column matrix. (The term *vector* has another meaning that we will discuss in subsequent chapters.)

REMARK. It is common practice to omit the brackets on a 1×1 matrix. Thus, we might write 4 rather than [4]. Although this makes it impossible to tell whether 4 denotes the number ''four'' or the 1×1 matrix whose entry is ''four,'' this rarely causes problems, since it is usually possible to tell which is meant from the context in which the symbol appears.

We shall use capital letters to denote matrices and lowercase letters to denote numerical quantities; thus, we might write

$$A = \begin{bmatrix} 2 & 1 & 7 \\ 3 & 4 & 2 \end{bmatrix} \quad \text{or} \quad C = \begin{bmatrix} a & b & c \\ d & e & f \end{bmatrix}$$

When discussing matrices, it is common to refer to numerical quantities as **scalars**. Unless stated otherwise, *scalars will be real numbers*; complex scalars will be considered in Chapter 10.

The entry that occurs in row i and column j of a matrix A will be denoted by a_{ij}.

Thus, a general 3×4 matrix might be written as

$$A = \begin{bmatrix} a_{11} & a_{12} & a_{13} & a_{14} \\ a_{21} & a_{22} & a_{23} & a_{24} \\ a_{31} & a_{32} & a_{33} & a_{34} \end{bmatrix}$$

and a general $m \times n$ matrix as

$$A = \begin{bmatrix} a_{11} & a_{12} & \cdots & a_{1n} \\ a_{21} & a_{22} & \cdots & a_{2n} \\ \vdots & \vdots & & \vdots \\ a_{m1} & a_{m2} & \cdots & a_{mn} \end{bmatrix} \tag{1}$$

When compactness of notation is desired, the preceding matrix can be written as

$$[a_{ij}]_{m \times n} \qquad \text{or} \qquad [a_{ij}]$$

the first notation being used when it is important in the discussion to know the size and the second when the size need not be emphasized. Usually, we shall match the letter denoting a matrix with the letter denoting its entries; thus, for a matrix B we would generally use b_{ij} for the entry in row i and column j and for a matrix C we would use c_{ij}.

The entry in row i and column j of a matrix A is also commonly denoted by the symbol $(A)_{ij}$. Thus, for matrix (1) above, we have

$$(A)_{ij} = a_{ij}$$

and for the matrix

$$A = \begin{bmatrix} 2 & -3 \\ 7 & 0 \end{bmatrix}$$

we have $(A)_{11} = 2$, $(A)_{12} = -3$, $(A)_{21} = 7$, and $(A)_{22} = 0$.

Row and column matrices are of special importance, and it is common practice to denote them by boldface lowercase letters rather than capital letters. For such matrices double subscripting of the entries is unnecessary. Thus, a general $1 \times n$ row matrix $\mathbf{a}$ and a general $m \times 1$ column matrix $\mathbf{b}$ would be written as

$$\mathbf{a} = [a_1 \quad a_2 \quad \cdots \quad a_n] \qquad \text{and} \qquad \mathbf{b} = \begin{bmatrix} b_1 \\ b_2 \\ \vdots \\ b_m \end{bmatrix}$$

A matrix A with n rows and n columns is called a ***square matrix of order n***, and the entries $a_{11}, a_{22}, \ldots, a_{nn}$ are said to be on the ***main diagonal*** of A (see the shaded entries in Figure 1).

Figure 1
$$\begin{bmatrix} a_{11} & a_{12} & \cdots & a_{1n} \\ a_{21} & a_{22} & \cdots & a_{2n} \\ \vdots & \vdots & & \vdots \\ a_{n1} & a_{n2} & \cdots & a_{nn} \end{bmatrix}$$

OPERATIONS ON MATRICES

So far, we have used matrices to abbreviate the work in solving systems of linear equations. For other applications, however, it is desirable to develop an "arithmetic of matrices" in which matrices can be added, subtracted, and multiplied in a useful way. The remainder of this section will be devoted to developing this arithmetic.

Definition. Two matrices are defined to be *equal* if they have the same size and their corresponding entries are equal.

In matrix notation, if $A = [a_{ij}]$ and $B = [b_{ij}]$ have the same size, then $A = B$ if and only if $(A)_{ij} = (B)_{ij}$, or equivalently, $a_{ij} = b_{ij}$ for all i and j.

Example 2 Consider the matrices

$$A = \begin{bmatrix} 2 & 1 \\ 3 & x \end{bmatrix} \qquad B = \begin{bmatrix} 2 & 1 \\ 3 & 5 \end{bmatrix} \qquad C = \begin{bmatrix} 2 & 1 & 0 \\ 3 & 4 & 0 \end{bmatrix}$$

If $x = 5$, then $A = B$, but for all other values of x the matrices A and B are not equal, since not all of their corresponding entries are equal. There is no value of x for which $A = C$ since A and C have different sizes. ▲

Definition. If A and B are matrices of the same size, then the *sum* $A + B$ is the matrix obtained by adding the entries of B to the corresponding entries of A, and the *difference* $A - B$ is the matrix obtained by subtracting the entries of B from the corresponding entries of A. Matrices of different sizes cannot be added or subtracted.

In matrix notation, if $A = [a_{ij}]$ and $B = [b_{ij}]$ have the same size, then

$$(A + B)_{ij} = (A)_{ij} + (B)_{ij} = a_{ij} + b_{ij} \quad \text{and} \quad (A - B)_{ij} = (A)_{ij} - (B)_{ij} = a_{ij} - b_{ij}$$

Example 3 Consider the matrices

$$A = \begin{bmatrix} 2 & 1 & 0 & 3 \\ -1 & 0 & 2 & 4 \\ 4 & -2 & 7 & 0 \end{bmatrix} \qquad B = \begin{bmatrix} -4 & 3 & 5 & 1 \\ 2 & 2 & 0 & -1 \\ 3 & 2 & -4 & 5 \end{bmatrix} \qquad C = \begin{bmatrix} 1 & 1 \\ 2 & 2 \end{bmatrix}$$

Then

$$A + B = \begin{bmatrix} -2 & 4 & 5 & 4 \\ 1 & 2 & 2 & 3 \\ 7 & 0 & 3 & 5 \end{bmatrix} \quad \text{and} \quad A - B = \begin{bmatrix} 6 & -2 & -5 & 2 \\ -3 & -2 & 2 & 5 \\ 1 & -4 & 11 & -5 \end{bmatrix}$$

The expressions $A + C$, $B + C$, $A - C$, and $B - C$ are undefined. ▲

Definition. If A is any matrix and c is any scalar, then the *product* cA is the matrix obtained by multiplying each entry of A by c.

In matrix notation, if $A = [a_{ij}]$, then

$$(cA)_{ij} = c(A)_{ij} = ca_{ij}$$

Example 4 For the matrices

$$A = \begin{bmatrix} 2 & 3 & 4 \\ 1 & 3 & 1 \end{bmatrix} \qquad B = \begin{bmatrix} 0 & 2 & 7 \\ -1 & 3 & -5 \end{bmatrix} \qquad C = \begin{bmatrix} 9 & -6 & 3 \\ 3 & 0 & 12 \end{bmatrix}$$

we have

$$2A = \begin{bmatrix} 4 & 6 & 8 \\ 2 & 6 & 2 \end{bmatrix} \qquad (-1)B = \begin{bmatrix} 0 & -2 & -7 \\ 1 & -3 & 5 \end{bmatrix} \qquad \tfrac{1}{3}C = \begin{bmatrix} 3 & -2 & 1 \\ 1 & 0 & 4 \end{bmatrix}$$

It is common practice to denote $(-1)B$ by $-B$. ▲

If $A_1, A_2, \ldots, A_n$ are matrices of the same size and $c_1, c_2, \ldots, c_n$ are scalars, then an expression of the form

$$c_1 A_1 + c_2 A_2 + \cdots + c_n A_n$$

is called a ***linear combination*** of $A_1, A_2, \ldots, A_n$ with ***coefficients*** $c_1, c_2, \ldots, c_n$. For example, if A, B, and C are the matrices in Example 4, then

$$2A - B + \tfrac{1}{3}C = 2A + (-1)B + \tfrac{1}{3}C$$

$$= \begin{bmatrix} 4 & 6 & 8 \\ 2 & 6 & 2 \end{bmatrix} + \begin{bmatrix} 0 & -2 & -7 \\ 1 & -3 & 5 \end{bmatrix} + \begin{bmatrix} 3 & -2 & 1 \\ 1 & 0 & 4 \end{bmatrix}$$

$$= \begin{bmatrix} 7 & 2 & 2 \\ 4 & 3 & 11 \end{bmatrix}$$

is the linear combination of A, B, and C with scalar coefficients 2, -1, and $\tfrac{1}{3}$.

Thus far we have defined multiplication of a matrix by a scalar but not the multiplication of two matrices. Since matrices are added by adding corresponding entries and subtracted by subtracting corresponding entries, it would seem natural to define multiplication of matrices by multiplying corresponding entries. However, it turns out that such a definition would not be very useful for most problems. Experience has led mathematicians to the following less natural but more useful definition of matrix multiplication.

Definition. If A is an $m \times r$ matrix and B is an $r \times n$ matrix, then the ***product*** AB is the $m \times n$ matrix whose entries are determined as follows. To find the entry in row i and column j of AB, single out row i from the matrix A and column j from the matrix B. Multiply the corresponding entries from the row and column together and then add up the resulting products.

Example 5 Consider the matrices

$$A = \begin{bmatrix} 1 & 2 & 4 \\ 2 & 6 & 0 \end{bmatrix} \qquad B = \begin{bmatrix} 4 & 1 & 4 & 3 \\ 0 & -1 & 3 & 1 \\ 2 & 7 & 5 & 2 \end{bmatrix}$$

Since A is a 2×3 matrix and B is a 3×4 matrix, the product AB is a 2×4 matrix. To determine, for example, the entry in row 2 and column 3 of AB, we single out row 2 from A and column 3 from B. Then, as illustrated below, we multiply corresponding entries together and add up these products.

$$\begin{bmatrix} 1 & 2 & 4 \\ 2 & 6 & 0 \end{bmatrix} \begin{bmatrix} 4 & 1 & 4 & 3 \\ 0 & -1 & 3 & 1 \\ 2 & 7 & 5 & 2 \end{bmatrix} = \begin{bmatrix} \square & \square & \square & \square \\ \square & \square & \boxed{26} & \square \end{bmatrix}$$

$$(2 \cdot 4) + (6 \cdot 3) + (0 \cdot 5) = 26$$

The entry in row 1 and column 4 of AB is computed as follows.

$$\begin{bmatrix} 1 & 2 & 4 \\ 2 & 6 & 0 \end{bmatrix} \begin{bmatrix} 4 & 1 & 4 & 3 \\ 0 & -1 & 3 & 1 \\ 2 & 7 & 5 & 2 \end{bmatrix} = \begin{bmatrix} \square & \square & \square & \boxed{13} \\ \square & \square & \square & \square \end{bmatrix}$$

$$(1 \cdot 3) + (2 \cdot 1) + (4 \cdot 2) = 13$$

The computations for the remaining products are

$$\begin{aligned} (1 \cdot 4) + (2 \cdot 0) + (4 \cdot 2) &= 12 \\ (1 \cdot 1) - (2 \cdot 1) + (4 \cdot 7) &= 27 \\ (1 \cdot 4) + (2 \cdot 3) + (4 \cdot 5) &= 30 \\ (2 \cdot 4) + (6 \cdot 0) + (0 \cdot 2) &= 8 \\ (2 \cdot 1) - (6 \cdot 1) + (0 \cdot 7) &= -4 \\ (2 \cdot 3) + (6 \cdot 1) + (0 \cdot 2) &= 12 \end{aligned} \qquad AB = \begin{bmatrix} 12 & 27 & 30 & 13 \\ 8 & -4 & 26 & 12 \end{bmatrix}$$

The definition of matrix multiplication requires that the number of columns of the first factor A be the same as the number of rows of the second factor B in order to form the product AB. If this condition is not satisfied, the product is undefined. A convenient way to determine whether a product of two matrices is defined is to write down the size of the first factor and, to the right of it, write down the size of the second factor. If, as in Figure 2, the inside numbers are the same, then the product is defined. The outside numbers then give the size of the product.

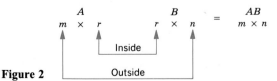

Figure 2

Example 6 Suppose that A, B, and C are matrices with the following sizes:

$$\begin{array}{ccc} A & B & C \\ 3 \times 4 & 4 \times 7 & 7 \times 3 \end{array}$$

Then AB is defined and is a 3×7 matrix; CA is defined and is a 7×4 matrix; and BC is defined and is a 4×3 matrix. The products AC, CB, and BA are all undefined.

If $A = [a_{ij}]$ is a general $m \times r$ matrix and $B = [b_{ij}]$ is a general $r \times n$ matrix, then as illustrated by the shading in Figure 3, the entry $(AB)_{ij}$ in row i and column j of AB is given by

$$(AB)_{ij} = a_{i1}b_{1j} + a_{i2}b_{2j} + a_{i3}b_{3j} + \cdots + a_{ir}b_{rj} \tag{2}$$

$$AB = \begin{bmatrix} a_{11} & a_{12} & \cdots & a_{1r} \\ a_{21} & a_{22} & \cdots & a_{2r} \\ \vdots & \vdots & & \vdots \\ a_{i1} & a_{i2} & \cdots & a_{ir} \\ \vdots & \vdots & & \vdots \\ a_{m1} & a_{m2} & \cdots & a_{mr} \end{bmatrix} \begin{bmatrix} b_{11} & b_{12} & \cdots & b_{1j} & \cdots & b_{1n} \\ b_{21} & b_{22} & \cdots & b_{2j} & \cdots & b_{2n} \\ \vdots & \vdots & & \vdots & & \vdots \\ b_{r1} & b_{r2} & \cdots & b_{rj} & \cdots & b_{rn} \end{bmatrix}$$

Figure 3

PARTITIONED MATRICES

A matrix can be subdivided or *partitioned* into smaller matrices by inserting horizontal and vertical rules between selected rows and columns. For example, below are three possible partitions of a general 3×4 matrix A—the first is a partition of A into four *submatrices* A_{11}, A_{12}, A_{21}, and A_{22}; the second is a partition of A into its row matrices $\mathbf{r}_1$, $\mathbf{r}_2$, and $\mathbf{r}_3$; and the third is a partition of A into its column matrices $\mathbf{c}_1$, $\mathbf{c}_2$, $\mathbf{c}_3$, and $\mathbf{c}_4$:

$$A = \left[\begin{array}{ccc|c} a_{11} & a_{12} & a_{13} & a_{14} \\ a_{21} & a_{22} & a_{23} & a_{24} \\ \hline a_{31} & a_{32} & a_{33} & a_{34} \end{array} \right] = \begin{bmatrix} A_{11} & A_{12} \\ A_{21} & A_{22} \end{bmatrix}$$

$$A = \left[\begin{array}{cccc} a_{11} & a_{12} & a_{13} & a_{14} \\ \hline a_{21} & a_{22} & a_{23} & a_{24} \\ \hline a_{31} & a_{32} & a_{33} & a_{34} \end{array} \right] = \begin{bmatrix} \mathbf{r}_1 \\ \mathbf{r}_2 \\ \mathbf{r}_3 \end{bmatrix}$$

$$A = \left[\begin{array}{c|c|c|c} a_{11} & a_{12} & a_{13} & a_{14} \\ a_{21} & a_{22} & a_{23} & a_{24} \\ a_{31} & a_{32} & a_{33} & a_{34} \end{array} \right] = [\mathbf{c}_1 \quad \mathbf{c}_2 \quad \mathbf{c}_3 \quad \mathbf{c}_4]$$

MATRIX MULTIPLICATION BY COLUMNS AND BY ROWS

Sometimes it may be desirable to find a particular row or column of a matrix product AB without computing the entire product. The following results, whose proofs are left as exercises, are useful for that purpose:

$$j\text{th column matrix of } AB = A[\,j\text{th column matrix of } B\,] \tag{3}$$

$$i\text{th row matrix of } AB = [\,i\text{th row matrix of } A\,]B \tag{4}$$

Example 7 If A and B are the matrices in Example 5, then from (3) the second column matrix of AB can be obtained by the computation

$$\begin{bmatrix} 1 & 2 & 4 \\ 2 & 6 & 0 \end{bmatrix} \begin{bmatrix} 1 \\ -1 \\ 7 \end{bmatrix} = \begin{bmatrix} 27 \\ -4 \end{bmatrix}$$

Second column of B **Second column of AB**

and from (4) the first row matrix of AB can be obtained by the computation

$$\begin{bmatrix} 1 & 2 & 4 \end{bmatrix} \begin{bmatrix} 4 & 1 & 4 & 3 \\ 0 & -1 & 3 & 1 \\ 2 & 7 & 5 & 2 \end{bmatrix} = \begin{bmatrix} 12 & 27 & 30 & 13 \end{bmatrix}$$

First row of A **First row of AB**

If $\mathbf{a}_1, \mathbf{a}_2, \ldots, \mathbf{a}_m$ denote the row matrices of A and $\mathbf{b}_1, \mathbf{b}_2, \ldots, \mathbf{b}_n$ denote the column matrices of B, then it follows from Formulas (3) and (4) that

$$AB = A[\mathbf{b}_1 \mid \mathbf{b}_2 \mid \cdots \mid \mathbf{b}_n] = [A\mathbf{b}_1 \mid A\mathbf{b}_2 \mid \cdots \mid A\mathbf{b}_n] \tag{5}$$

(***AB* computed column by column**)

$$AB = \begin{bmatrix} \mathbf{a}_1 \\ \mathbf{a}_2 \\ \vdots \\ \mathbf{a}_m \end{bmatrix} B = \begin{bmatrix} \mathbf{a}_1 B \\ \mathbf{a}_2 B \\ \vdots \\ \mathbf{a}_m B \end{bmatrix} \tag{6}$$

(***AB* computed row by row**)

REMARK. Formulas (5) and (6) are special cases of a more general procedure for multiplying partitioned matrices (see Exercises 15–17).

MATRIX PRODUCTS AS LINEAR COMBINATIONS

Row and column matrices provide an alternative way of thinking about matrix multiplication. For example, suppose that

$$A = \begin{bmatrix} a_{11} & a_{12} & \cdots & a_{1n} \\ a_{21} & a_{22} & \cdots & a_{2n} \\ \vdots & \vdots & & \vdots \\ a_{m1} & a_{m2} & \cdots & a_{mn} \end{bmatrix} \quad \text{and} \quad \mathbf{x} = \begin{bmatrix} x_1 \\ x_2 \\ \vdots \\ x_n \end{bmatrix}$$

Then

$$A\mathbf{x} = \begin{bmatrix} a_{11}x_1 + a_{12}x_2 + \cdots + a_{1n}x_n \\ a_{21}x_1 + a_{22}x_2 + \cdots + a_{2n}x_n \\ \vdots & \vdots & & \vdots \\ a_{m1}x_1 + a_{m2}x_2 + \cdots + a_{mn}x_n \end{bmatrix} = x_1 \begin{bmatrix} a_{11} \\ a_{21} \\ \vdots \\ a_{m1} \end{bmatrix} + x_2 \begin{bmatrix} a_{12} \\ a_{22} \\ \vdots \\ a_{m2} \end{bmatrix} + \cdots + x_n \begin{bmatrix} a_{1n} \\ a_{2n} \\ \vdots \\ a_{mn} \end{bmatrix}$$

$$\tag{7}$$

In words, (7) tells us that *the product A**x** of a matrix A with a column matrix **x** is a linear combination of the column matrices of A with the coefficients coming from the matrix **x**.* In the exercises we ask the reader to show that *the product **y**A of a 1 × m matrix **y** with an m × n matrix A is a linear combination of the row matrices of A with scalar coefficients coming from **y**.*

Example 8   The matrix product

$$\begin{bmatrix} -1 & 3 & 2 \\ 1 & 2 & -3 \\ 2 & 1 & -2 \end{bmatrix} \begin{bmatrix} 2 \\ -1 \\ 3 \end{bmatrix} = \begin{bmatrix} 1 \\ -9 \\ -3 \end{bmatrix}$$

can be written as the linear combination

$$2\begin{bmatrix} -1 \\ 1 \\ 2 \end{bmatrix} - 1\begin{bmatrix} 3 \\ 2 \\ 1 \end{bmatrix} + 3\begin{bmatrix} 2 \\ -3 \\ -2 \end{bmatrix} = \begin{bmatrix} 1 \\ -9 \\ -3 \end{bmatrix}$$

and the matrix product

$$\begin{bmatrix} 1 & -9 & -3 \end{bmatrix} \begin{bmatrix} -1 & 3 & 2 \\ 1 & 2 & -3 \\ 2 & 1 & -2 \end{bmatrix} = \begin{bmatrix} -16 & -18 & 35 \end{bmatrix}$$

can be written as the linear combination

$$1\begin{bmatrix} -1 & 3 & 2 \end{bmatrix} - 9\begin{bmatrix} 1 & 2 & -3 \end{bmatrix} - 3\begin{bmatrix} 2 & 1 & -2 \end{bmatrix} = \begin{bmatrix} -16 & -18 & 35 \end{bmatrix}$$ ▲

It follows from (5) and (7) that *the jth column matrix of a product AB is a linear combination of the column matrices of A with the coefficients coming from the jth column of B.*

Example 9   We showed in Example 5 that

$$AB = \begin{bmatrix} 1 & 2 & 4 \\ 2 & 6 & 0 \end{bmatrix} \begin{bmatrix} 4 & 1 & 4 & 3 \\ 0 & -1 & 3 & 1 \\ 2 & 7 & 5 & 2 \end{bmatrix} = \begin{bmatrix} 12 & 27 & 30 & 13 \\ 8 & -4 & 26 & 12 \end{bmatrix}$$

The column matrices of *AB* can be expressed as linear combinations of the column matrices of *A* as follows:

$$\begin{bmatrix} 12 \\ 8 \end{bmatrix} = 4\begin{bmatrix} 1 \\ 2 \end{bmatrix} + 0\begin{bmatrix} 2 \\ 6 \end{bmatrix} + 2\begin{bmatrix} 4 \\ 0 \end{bmatrix}$$

$$\begin{bmatrix} 27 \\ -4 \end{bmatrix} = \begin{bmatrix} 1 \\ 2 \end{bmatrix} - \begin{bmatrix} 2 \\ 6 \end{bmatrix} + 7\begin{bmatrix} 4 \\ 0 \end{bmatrix}$$

$$\begin{bmatrix} 30 \\ 26 \end{bmatrix} = 4\begin{bmatrix} 1 \\ 2 \end{bmatrix} + 3\begin{bmatrix} 2 \\ 6 \end{bmatrix} + 5\begin{bmatrix} 4 \\ 0 \end{bmatrix}$$

$$\begin{bmatrix} 13 \\ 12 \end{bmatrix} = 3 \begin{bmatrix} 1 \\ 2 \end{bmatrix} + \begin{bmatrix} 2 \\ 6 \end{bmatrix} + 2 \begin{bmatrix} 4 \\ 0 \end{bmatrix}$$ ▲

MATRIX FORM OF A LINEAR SYSTEM

Matrix multiplication has an important application to systems of linear equations. Consider any system of m linear equations in n unknowns.

$$\begin{aligned} a_{11}x_1 + a_{12}x_2 + \cdots + a_{1n}x_n &= b_1 \\ a_{21}x_1 + a_{22}x_2 + \cdots + a_{2n}x_n &= b_2 \\ &\vdots \\ a_{m1}x_1 + a_{m2}x_2 + \cdots + a_{mn}x_n &= b_m \end{aligned}$$

Since two matrices are equal if and only if their corresponding entries are equal, we can replace the m equations in this system by the single matrix equation

$$\begin{bmatrix} a_{11}x_1 + a_{12}x_2 + \cdots + a_{1n}x_n \\ a_{21}x_1 + a_{22}x_2 + \cdots + a_{2n}x_n \\ \vdots \qquad \vdots \qquad \qquad \vdots \\ a_{m1}x_1 + a_{m2}x_2 + \cdots + a_{mn}x_n \end{bmatrix} = \begin{bmatrix} b_1 \\ b_2 \\ \vdots \\ b_m \end{bmatrix}$$

The $m \times 1$ matrix on the left side of this equation can be written as a product to give

$$\begin{bmatrix} a_{11} & a_{12} & \cdots & a_{1n} \\ a_{21} & a_{22} & \cdots & a_{2n} \\ \vdots & \vdots & & \vdots \\ a_{m1} & a_{m2} & \cdots & a_{mn} \end{bmatrix} \begin{bmatrix} x_1 \\ x_2 \\ \vdots \\ x_n \end{bmatrix} = \begin{bmatrix} b_1 \\ b_2 \\ \vdots \\ b_m \end{bmatrix}$$

If we designate these matrices by A, $\mathbf{x}$, and $\mathbf{b}$, respectively, the original system of m equations in n unknowns has been replaced by the single matrix equation

$$A\mathbf{x} = \mathbf{b}$$

The matrix A in this equation is called the ***coefficient matrix*** of the system. The augmented matrix for the system is obtained by adjoining $\mathbf{b}$ to A as the last column; thus the augmented matrix is

$$[A \vdots \mathbf{b}] = \begin{bmatrix} a_{11} & a_{12} & \cdots & a_{1n} & \vdots & b_1 \\ a_{21} & a_{22} & \cdots & a_{2n} & \vdots & b_2 \\ \vdots & \vdots & & \vdots & \vdots & \vdots \\ a_{m1} & a_{m2} & \cdots & a_{mn} & \vdots & b_m \end{bmatrix}$$

TRANSPOSE OF A MATRIX

We conclude this section by defining two matrix operations that have no analogs in the real numbers.

> **Definition.** If A is any $m \times n$ matrix, then the ***transpose of A***, denoted by A^T, is defined to be the $n \times m$ matrix that results from interchanging the rows and columns of A; that is, the first column of A^T is the first row of A, the second column of A^T is the second row of A, and so forth.

Example 10 The following are some examples of matrices and their transposes.

$$A = \begin{bmatrix} a_{11} & a_{12} & a_{13} & a_{14} \\ a_{21} & a_{22} & a_{23} & a_{24} \\ a_{31} & a_{32} & a_{33} & a_{34} \end{bmatrix} \quad B = \begin{bmatrix} 2 & 3 \\ 1 & 4 \\ 5 & 6 \end{bmatrix} \quad C = \begin{bmatrix} 1 & 3 & 5 \end{bmatrix} \quad D = \begin{bmatrix} 4 \end{bmatrix}$$

$$A^T = \begin{bmatrix} a_{11} & a_{21} & a_{31} \\ a_{12} & a_{22} & a_{32} \\ a_{13} & a_{23} & a_{33} \\ a_{14} & a_{24} & a_{34} \end{bmatrix} \quad B^T = \begin{bmatrix} 2 & 1 & 5 \\ 3 & 4 & 6 \end{bmatrix} \quad C^T = \begin{bmatrix} 1 \\ 3 \\ 5 \end{bmatrix} \quad D^T = \begin{bmatrix} 4 \end{bmatrix} \quad ▲$$

Observe that not only are the columns of A^T the rows of A, but the rows of A^T are columns of A. Thus, the entry in row i and column j of A^T is the entry in row j and column i of A; that is,

$$(A^T)_{ij} = (A)_{ji} \tag{8}$$

Note the reversal of the subscripts.

In the special case where A is a square matrix, the transpose of A can be obtained by interchanging entries that are symmetrically positioned about the main diagonal (Figure 4). Stated another way, A^T can be obtained by "reflecting" A about its main diagonal.

$$A = \begin{bmatrix} 1 & -2 & 4 \\ 3 & 7 & 0 \\ -5 & 8 & 6 \end{bmatrix} \rightarrow \begin{bmatrix} 1 & -2 & 4 \\ 3 & 7 & 0 \\ -5 & 8 & 6 \end{bmatrix} \rightarrow A^T = \begin{bmatrix} 1 & 3 & -5 \\ -2 & 7 & 8 \\ 4 & 0 & 6 \end{bmatrix}$$

Interchange entries that are symmetrically positioned about the main diagonal.

Figure 4

TRACE OF A SQUARE MATRIX

Definition. If A is a square matrix, then the *trace of A*, denoted by tr(A), is defined to be the sum of the entries on the main diagonal of A. The trace of A is undefined if A is not a square matrix.

Example 11 The following are examples of matrices and their traces.

$$A = \begin{bmatrix} a_{11} & a_{12} & a_{13} \\ a_{21} & a_{22} & a_{23} \\ a_{31} & a_{32} & a_{33} \end{bmatrix} \quad B = \begin{bmatrix} -1 & 2 & 7 & 0 \\ 3 & 5 & -8 & 4 \\ 1 & 2 & 7 & -3 \\ 4 & -2 & 1 & 0 \end{bmatrix}$$

$$\text{tr}(A) = a_{11} + a_{22} + a_{33} \qquad \text{tr}(B) = -1 + 5 + 7 + 0 = 11 \quad ▲$$

EXERCISE SET 1.3

1. Suppose that A, B, C, D, and E are matrices with the following sizes:

A	B	C	D	E
(4×5)	(4×5)	(5×2)	(4×2)	(5×4)

Determine which of the following matrix expressions are defined. For those which are defined, give the size of the resulting matrix.

(a) BA (b) $AC + D$ (c) $AE + B$ (d) $AB + B$

(e) $E(A + B)$ (f) $E(AC)$ (g) $E^T A$ (h) $(A^T + E)D$

2. Solve the following matrix equation for a, b, c, and d.

$$\begin{bmatrix} a - b & b + c \\ 3d + c & 2a - 4d \end{bmatrix} = \begin{bmatrix} 8 & 1 \\ 7 & 6 \end{bmatrix}$$

3. Consider the matrices

$$A = \begin{bmatrix} 3 & 0 \\ -1 & 2 \\ 1 & 1 \end{bmatrix}, \quad B = \begin{bmatrix} 4 & -1 \\ 0 & 2 \end{bmatrix}, \quad C = \begin{bmatrix} 1 & 4 & 2 \\ 3 & 1 & 5 \end{bmatrix}, \quad D = \begin{bmatrix} 1 & 5 & 2 \\ -1 & 0 & 1 \\ 3 & 2 & 4 \end{bmatrix}, \quad E = \begin{bmatrix} 6 & 1 & 3 \\ -1 & 1 & 2 \\ 4 & 1 & 3 \end{bmatrix}$$

Compute the following (where possible).

(a) $D + E$ (b) $D - E$ (c) $5A$ (d) $-7C$

(e) $2B - C$ (f) $4E - 2D$ (g) $-3(D + 2E)$ (h) $A - A$

(i) $\operatorname{tr}(D)$ (j) $\operatorname{tr}(D - 3E)$ (k) $4 \operatorname{tr}(7B)$ (l) $\operatorname{tr}(A)$

4. Using the matrices in Exercise 3, compute the following (where possible).

(a) $2A^T + C$ (b) $D^T - E^T$ (c) $(D - E)^T$ (d) $B^T + 5C^T$

(e) $\frac{1}{2}C^T - \frac{1}{4}A$ (f) $B - B^T$ (g) $2E^T - 3D^T$ (h) $(2E^T - 3D^T)^T$

5. Using the matrices in Exercise 3, compute the following (where possible).

(a) AB (b) BA (c) $(3E)D$ (d) $(AB)C$

(e) $A(BC)$ (f) CC^T (g) $(DA)^T$ (h) $(C^T B)A^T$

(i) $\operatorname{tr}(DD^T)$ (j) $\operatorname{tr}(4E^T - D)$ (k) $\operatorname{tr}(C^T A^T + 2E^T)$

6. Using the matrices in Exercise 3, compute the following (where possible).

(a) $(2D^T - E)A$ (b) $(4B)C + 2B$ (c) $(-AC)^T + 5D^T$

(d) $(BA^T - 2C)^T$ (e) $B^T(CC^T - A^T A)$ (f) $D^T E^T - (ED)^T$

7. Let

$$A = \begin{bmatrix} 3 & -2 & 7 \\ 6 & 5 & 4 \\ 0 & 4 & 9 \end{bmatrix} \quad \text{and} \quad B = \begin{bmatrix} 6 & -2 & 4 \\ 0 & 1 & 3 \\ 7 & 7 & 5 \end{bmatrix}$$

Use the method of Example 7 to find

(a) the first row of AB (b) the third row of AB (c) the second column of AB

(d) the first column of BA (e) the third row of AA (f) the third column of AA

8. Let A and B be the matrices in Exercise 7.

(a) Express each column matrix of AB as a linear combination of the column matrices of A.

(b) Express each column matrix of BA as a linear combination of the column matrices of B.

9. Let

$$\mathbf{y} = [y_1 \ y_2 \ \cdots \ y_m] \quad \text{and} \quad A = \begin{bmatrix} a_{11} & a_{12} & \cdots & a_{1n} \\ a_{21} & a_{22} & \cdots & a_{2n} \\ \vdots & \vdots & & \vdots \\ a_{m1} & a_{m2} & \cdots & a_{mn} \end{bmatrix}$$

Show that the product $\mathbf{y}A$ can be expressed as a combination of the row matrices of A with the scalar coefficients from $\mathbf{y}$.

10. Let A and B be the matrices in Exercise 7.
   (a) Use the result in Exercise 9 to express each row matrix of AB as a linear combination of the row matrices of B.
   (b) Use the result in Exercise 9 to express each row matrix of BA as a linear combination of the row matrices of A.

11. Let C, D, and E be the matrices in Exercise 3. Using as few computations as possible, determine the entry in row 2 and column 3 of $C(DE)$.

12. (a) Show that if AB and BA are both defined, then AB and BA are square matrices.
   (b) Show that if A is an $m \times n$ matrix and $A(BA)$ is defined, then B is an $n \times m$ matrix.

13. In each part find matrices A, $\mathbf{x}$, and $\mathbf{b}$ that express the given system of linear equations as a single matrix equation $A\mathbf{x} = \mathbf{b}$.

   (a) $\begin{aligned} 2x_1 - 3x_2 + 5x_3 &= 7 \\ 9x_1 - x_2 + x_3 &= -1 \\ x_1 + 5x_2 + 4x_3 &= 0 \end{aligned}$
   (b) $\begin{aligned} 4x_1 \quad\quad - 3x_3 + x_4 &= 1 \\ 5x_1 + x_2 \quad\quad - 8x_4 &= 3 \\ 2x_1 - 5x_2 + 9x_3 - x_4 &= 0 \\ 3x_2 - x_3 + 7x_4 &= 2 \end{aligned}$

14. In each part, express the matrix equation as a system of linear equations.

   (a) $\begin{bmatrix} 3 & -1 & 2 \\ 4 & 3 & 7 \\ -2 & 1 & 5 \end{bmatrix} \begin{bmatrix} x_1 \\ x_2 \\ x_3 \end{bmatrix} = \begin{bmatrix} 2 \\ -1 \\ 4 \end{bmatrix}$

   (b) $\begin{bmatrix} 3 & -2 & 0 & 1 \\ 5 & 0 & 2 & -2 \\ 3 & 1 & 4 & 7 \\ -2 & 5 & 1 & 6 \end{bmatrix} \begin{bmatrix} w \\ x \\ y \\ z \end{bmatrix} = \begin{bmatrix} 0 \\ 0 \\ 0 \\ 0 \end{bmatrix}$

15. If A and B are partitioned into submatrices, for example,

$$A = \left[\begin{array}{c|c} A_{11} & A_{12} \\ \hline A_{21} & A_{22} \end{array} \right] \quad \text{and} \quad B = \left[\begin{array}{c|c} B_{11} & B_{12} \\ \hline B_{21} & B_{22} \end{array} \right]$$

then AB can be expressed as

$$AB = \left[\begin{array}{c|c} A_{11}B_{11} + A_{12}B_{21} & A_{11}B_{12} + A_{12}B_{22} \\ \hline A_{21}B_{11} + A_{22}B_{21} & A_{21}B_{12} + A_{22}B_{22} \end{array} \right]$$

provided the sizes of the submatrices of A and B are such that the indicated operations can be performed. This method of multiplying partitioned matrices is called **block multiplication**. In each part compute the product by block multiplication. Check your results by multiplying directly.

   (a) $A = \left[\begin{array}{cc|cc} -1 & 2 & 1 & 5 \\ 0 & -3 & 4 & 2 \\ \hline 1 & 5 & 6 & 1 \end{array} \right]$, $B = \left[\begin{array}{cc|c} 2 & 1 & 4 \\ -3 & 5 & 2 \\ \hline 7 & -1 & 5 \\ 0 & 3 & -3 \end{array} \right]$

   (b) $A = \left[\begin{array}{cc|cc} -1 & 2 & 1 & 5 \\ \hline 0 & -3 & 4 & 2 \\ 1 & 5 & 6 & 1 \end{array} \right]$, $B = \left[\begin{array}{cc|c} 2 & 1 & 4 \\ -3 & 5 & 2 \\ \hline 7 & -1 & 5 \\ 0 & 3 & -3 \end{array} \right]$

16. Adapt the method of Exercise 15 to compute the following products by block multiplication.

(a) $\left[\begin{array}{cc|c} 3 & -1 & 0 \\ 2 & 1 & 4 \end{array}\,\begin{array}{c} -3 \\ 5 \end{array}\right]\left[\begin{array}{ccc} 2 & -4 & 1 \\ 3 & 0 & 2 \\ 1 & -3 & 5 \\ \hline 2 & 1 & 4 \end{array}\right]$

(b) $\left[\begin{array}{cc} 2 & -5 \\ 1 & 3 \\ 0 & 5 \\ \hline 1 & 4 \end{array}\right]\left[\begin{array}{cccc} 2 & -1 & 3 & -4 \\ 0 & 1 & 5 & 7 \end{array}\right]$

(c) $\left[\begin{array}{ccc|cc} 1 & 0 & 0 & 0 & 0 \\ 0 & 1 & 0 & 0 & 0 \\ 0 & 0 & 1 & 0 & 0 \\ \hline 0 & 0 & 0 & 2 & 0 \\ 0 & 0 & 0 & -1 & 2 \end{array}\right]\left[\begin{array}{cc} 3 & 3 \\ -1 & 4 \\ 1 & 5 \\ \hline 2 & -2 \\ 1 & 6 \end{array}\right]$

17. In each part determine whether block multiplication can be used to compute AB from the given partitions If so, compute the product by block multiplication.

(a) $A = \left[\begin{array}{ccc|c} -1 & 2 & 1 & 5 \\ 0 & -3 & 4 & 2 \\ \hline 1 & 5 & 6 & 1 \end{array}\right]$, $B = \left[\begin{array}{cc|c} 2 & 1 & 4 \\ -3 & 5 & 2 \\ \hline 7 & -1 & 5 \\ 0 & 3 & -3 \end{array}\right]$

(b) $A = \left[\begin{array}{cccc} -1 & 2 & 1 & 5 \\ 0 & -3 & 4 & 2 \\ \hline 1 & 5 & 6 & 1 \end{array}\right]$, $B = \left[\begin{array}{c|c|c} 2 & 1 & 4 \\ -3 & 5 & 2 \\ 7 & -1 & 5 \\ 0 & 3 & -3 \end{array}\right]$

18. (a) Show that if A has a row of zeros and B is any matrix for which AB is defined, then AB also has a row of zeros.

(b) Find a similar result involving a column of zeros.

19. Let A be any $m \times n$ matrix and let 0 be the $m \times n$ matrix each of whose entries is zero. Show that if $kA = 0$, then $k = 0$ or $A = 0$.

20. Let I be the $n \times n$ matrix whose entry in row i and column j is

$$\begin{cases} 1 & \text{if} \quad i = j \\ 0 & \text{if} \quad i \neq j \end{cases}$$

Show that $AI = IA = A$ for every $n \times n$ matrix A.

21. In each part find a 6×6 matrix $[a_{ij}]$ that satisfies the stated condition. Make your answers as general as possible by using letters rather than specific numbers for the nonzero entries.

(a) $a_{ij} = 0$ if $i \neq j$ (b) $a_{ij} = 0$ if $i > j$ (c) $a_{ij} = 0$ if $i < j$ (d) $a_{ij} = 0$ if $|i - j| > 1$

22. Find the 4×4 matrix $A = [a_{ij}]$ whose entries satisfy the stated condition.

(a) $a_{ij} = i + j$ (b) $a_{ij} = i^{j-1}$ (c) $a_{ij} = \begin{cases} 1 & \text{if} \quad |i-j| > 1 \\ -1 & \text{if} \quad |i-j| \leq 1 \end{cases}$

23. Prove: If A is an $m \times n$ matrix, then

$$\text{tr}(AA^T) = \text{tr}(A^TA) = s$$

where s is the sum of the squares of the entries of A.

24. Use the result in Exercise 23 to prove the following.

(a) If A is an $m \times n$ matrix such that $AA^T = 0$ or $A^TA = 0$, then $A = 0$.

(b) If A is an $n \times n$ matrix such that $A = A^T$ and $A^2 = 0$, then $A = 0$.

1.4 INVERSES; RULES OF MATRIX ARITHMETIC

In this section we shall discuss some properties of the arithmetic operations on matrices. We shall see that many of the basic rules of arithmetic for real numbers also hold for matrices but a few do not.

PROPERTIES OF MATRIX OPERATIONS

For real numbers a and b, we always have $ab = ba$, which is called the *commutative law for multiplication*. For matrices, however, AB and BA need not be equal. Equality can fail to hold for three reasons. It can happen, for example, that AB is defined but BA is undefined. This is the case if A is a 2×3 matrix and B is a 3×4 matrix. Also, it can happen that AB and BA are both defined but have different sizes. This is the situation if A is a 2×3 matrix and B is a 3×2 matrix. Finally, as Example 1 shows, it is possible to have $AB \neq BA$ even if both AB and BA are defined and have the same size.

Example 1   Consider the matrices

$$A = \begin{bmatrix} -1 & 0 \\ 2 & 3 \end{bmatrix} \qquad B = \begin{bmatrix} 1 & 2 \\ 3 & 0 \end{bmatrix}$$

Multiplying gives

$$AB = \begin{bmatrix} -1 & -2 \\ 11 & 4 \end{bmatrix} \qquad BA = \begin{bmatrix} 3 & 6 \\ -3 & 0 \end{bmatrix}$$

Thus, $AB \neq BA$. ▲

Although the commutative law for multiplication is not valid in matrix arithmetic, many familiar laws of arithmetic are valid for matrices. Some of the most important ones and their names are summarized in the following theorem.

Theorem 1.4.1. *Assuming that the sizes of the matrices are such that the indicated operations can be performed, the following rules of matrix arithmetic are valid.*

(a) $A + B = B + A$ (***Commutative law for addition***)

(b) $A + (B + C) = (A + B) + C$ (***Associative law for addition***)

(c) $A(BC) = (AB)C$ (***Associative law for multiplication***)

(d) $A(B + C) = AB + AC$ (***Left distributive law***)

(e) $(B + C)A = BA + CA$ (***Right distributive law***)

(f) $A(B - C) = AB - AC$

(g) $(B - C)A = BA - CA$

(h) $a(B + C) = aB + aC$

(i) $a(B - C) = aB - aC$

(j) $(a + b)C = aC + bC$

(k) $(a - b)C = aC - bC$

(l) $a(bC) = (ab)C$

(m) $a(BC) = (aB)C = B(aC)$

To prove the equalities in this theorem we must show that the matrix on the left side has the same size as the matrix on the right side and that corresponding entries on the two sides are equal. With the exception of the associative law in part (*c*), the proofs all follow the same general pattern. We shall prove part (*d*) as an illustration. The proof of the associative law, which is more complicated, is outlined in the exercises.

Proof (*d*). We must show that $A(B + C)$ and $AB + AC$ have the same size and that corresponding entries are equal. To form $A(B + C)$, the matrices B and C must have the same size, say $m \times n$, and the matrix A must then have m columns, so its size must be of the form $r \times m$. This makes $A(B + C)$ an $r \times n$ matrix. It follows that $AB + AC$ is also an $r \times n$ matrix and, consequently, $A(B + C)$ and $AB + AC$ have the same size.

Suppose that $A = [a_{ij}]$, $B = [b_{ij}]$, and $C = [c_{ij}]$. We want to show that corresponding entries of $A(B + C)$ and $AB + AC$ are equal; that is,

$$[A(B + C)]_{ij} = [AB + AC]_{ij}$$

for all values of i and j. But from the definitions of matrix addition and matrix multiplication we have

$$\begin{aligned}
[A(B + C)]_{ij} &= a_{i1}(b_{1j} + c_{1j}) + a_{i2}(b_{2j} + c_{2j}) + \cdots + a_{im}(b_{mj} + c_{mj}) \\
&= (a_{i1}b_{1j} + a_{i2}b_{2j} + \cdots + a_{im}b_{mj}) + (a_{i1}c_{1j} + a_{i2}c_{2j} + \cdots + a_{im}c_{mj}) \\
&= [AB]_{ij} + [AC_{ij}] = [AB + AC]_{ij} \quad \blacksquare
\end{aligned}$$

REMARK. Although the operations of matrix addition and matrix multiplication were defined for pairs of matrices, associative laws (*b*) and (*c*) enable us to denote sums and products of three matrices as $A + B + C$ and ABC without inserting any parentheses. This is justified by the fact that no matter how parentheses are inserted, the associative laws guarantee that the same end result will be obtained. In general, *given any sum or any product of matrices, pairs of parentheses can be inserted or deleted anywhere within the expression without affecting the end result.*

Example 2 As an illustration of the associative law for matrix multiplication, consider

$$A = \begin{bmatrix} 1 & 2 \\ 3 & 4 \\ 0 & 1 \end{bmatrix} \qquad B = \begin{bmatrix} 4 & 3 \\ 2 & 1 \end{bmatrix} \qquad C = \begin{bmatrix} 1 & 0 \\ 2 & 3 \end{bmatrix}$$

Then

$$AB = \begin{bmatrix} 1 & 2 \\ 3 & 4 \\ 0 & 1 \end{bmatrix}\begin{bmatrix} 4 & 3 \\ 2 & 1 \end{bmatrix} = \begin{bmatrix} 8 & 5 \\ 20 & 13 \\ 2 & 1 \end{bmatrix} \qquad \text{and} \qquad BC = \begin{bmatrix} 4 & 3 \\ 2 & 1 \end{bmatrix}\begin{bmatrix} 1 & 0 \\ 2 & 3 \end{bmatrix} = \begin{bmatrix} 10 & 9 \\ 4 & 3 \end{bmatrix}$$

Thus,

$$(AB)C = \begin{bmatrix} 8 & 5 \\ 20 & 13 \\ 2 & 1 \end{bmatrix}\begin{bmatrix} 1 & 0 \\ 2 & 3 \end{bmatrix} = \begin{bmatrix} 18 & 15 \\ 46 & 39 \\ 4 & 3 \end{bmatrix}$$

and

$$A(BC) = \begin{bmatrix} 1 & 2 \\ 3 & 4 \\ 0 & 1 \end{bmatrix} \begin{bmatrix} 10 & 9 \\ 4 & 3 \end{bmatrix} = \begin{bmatrix} 18 & 15 \\ 46 & 39 \\ 4 & 3 \end{bmatrix}$$

so $(AB)C = A(BC)$, as guaranteed by Theorem 1.4.1c. ▲

ZERO MATRICES   A matrix, all of whose entries are zero, such as

$$\begin{bmatrix} 0 & 0 \\ 0 & 0 \end{bmatrix} \qquad \begin{bmatrix} 0 & 0 & 0 \\ 0 & 0 & 0 \\ 0 & 0 & 0 \end{bmatrix} \qquad \begin{bmatrix} 0 & 0 & 0 & 0 \\ 0 & 0 & 0 & 0 \end{bmatrix} \qquad \begin{bmatrix} 0 \\ 0 \\ 0 \\ 0 \end{bmatrix} \qquad [0]$$

is called a ***zero matrix***. A zero matrix will be denoted by *0*; if it is important to emphasize the size, we shall write $0_{m \times n}$ for the $m \times n$ zero matrix.

If A is any matrix and *0* is the zero matrix with the same size, it is obvious that $A + 0 = 0 + A = A$. The matrix *0* plays much the same role in these matrix equations as the number 0 plays in the numerical equations $a + 0 = 0 + a = a$.

Since we already know that some of the rules of arithmetic for real numbers do not carry over to matrix arithmetic, it would be foolhardy to assume that all the properties of the real number zero carry over to zero matrices. For example, consider the following two standard results in the arithmetic of real numbers.

- If $ab = ac$ and $a \neq 0$, then $b = c$. (This is called the *cancellation law*.)
- If $ad = 0$, then at least one of the factors on the left is 0.

As the next example shows, the corresponding results are not generally true in matrix arithmetic.

Example 3   Consider the matrices

$$A = \begin{bmatrix} 0 & 1 \\ 0 & 2 \end{bmatrix} \qquad B = \begin{bmatrix} 1 & 1 \\ 3 & 4 \end{bmatrix} \qquad C = \begin{bmatrix} 2 & 5 \\ 3 & 4 \end{bmatrix} \qquad D = \begin{bmatrix} 3 & 7 \\ 0 & 0 \end{bmatrix}$$

Here

$$AB = AC = \begin{bmatrix} 3 & 4 \\ 6 & 8 \end{bmatrix}$$

Although $A \neq 0$, it is *incorrect* to cancel the A from both sides of the equation $AB = AC$ and write $B = C$. Thus, the cancellation law fails to hold for matrices. Also, $AD = 0$, yet $A \neq 0$ and $D \neq 0$. ▲

In spite of the above example, there are a number of familiar properties of the real number 0 that *do* carry over to zero matrices. Some of the more important ones are summarized in the next theorem. The proofs are left as exercises.

Theorem 1.4.2. *Assuming that the sizes of the matrices are such that the indicated operations can be performed, the following rules of matrix arithmetic are valid.*

(a) $A + 0 = 0 + A = A$

(b) $A - A = 0$

(c) $0 - A = -A$

(d) $A0 = 0; \quad 0A = 0$

IDENTITY MATRICES

Of special interest are square matrices with 1's on the main diagonal and 0's off the main diagonal, such as

$$\begin{bmatrix} 1 & 0 \\ 0 & 1 \end{bmatrix}, \quad \begin{bmatrix} 1 & 0 & 0 \\ 0 & 1 & 0 \\ 0 & 0 & 1 \end{bmatrix}, \quad \begin{bmatrix} 1 & 0 & 0 & 0 \\ 0 & 1 & 0 & 0 \\ 0 & 0 & 1 & 0 \\ 0 & 0 & 0 & 1 \end{bmatrix}, \quad \text{and so on.}$$

A matrix of this form is called an ***identity matrix*** and is denoted by I. If it is important to emphasize the size, we shall write I_n for the $n \times n$ identity matrix.

If A is an $m \times n$ matrix, then, as illustrated in the next example,

$$AI_n = A \quad \text{and} \quad I_m A = A$$

Thus, an identity matrix plays much the same role in matrix arithmetic as the number 1 plays in the numerical relationships $a \cdot 1 = 1 \cdot a = a$.

Example 4 Consider the matrix

$$A = \begin{bmatrix} a_{11} & a_{12} & a_{13} \\ a_{21} & a_{22} & a_{23} \end{bmatrix}$$

Then

$$I_2 A = \begin{bmatrix} 1 & 0 \\ 0 & 1 \end{bmatrix} \begin{bmatrix} a_{11} & a_{12} & a_{13} \\ a_{21} & a_{22} & a_{23} \end{bmatrix} = \begin{bmatrix} a_{11} & a_{12} & a_{13} \\ a_{21} & a_{22} & a_{23} \end{bmatrix} = A$$

and

$$AI_3 = \begin{bmatrix} a_{11} & a_{12} & a_{13} \\ a_{21} & a_{22} & a_{23} \end{bmatrix} \begin{bmatrix} 1 & 0 & 0 \\ 0 & 1 & 0 \\ 0 & 0 & 1 \end{bmatrix} = \begin{bmatrix} a_{11} & a_{12} & a_{13} \\ a_{21} & a_{22} & a_{23} \end{bmatrix} = A$$

As the next theorem shows, identity matrices arise naturally in studying reduced row-echelon forms of *square* matrices.

Theorem 1.4.3. *If R is the reduced row-echelon form of an $n \times n$ matrix A, then either R has a row of zeros or R is the identity matrix I_n.*

Proof. Suppose that the reduced row-echelon form of A is

$$R = \begin{bmatrix} r_{11} & r_{12} & \cdots & r_{1n} \\ r_{21} & r_{22} & \cdots & r_{2n} \\ \vdots & \vdots & & \vdots \\ r_{n1} & r_{n2} & \cdots & r_{nn} \end{bmatrix}$$

Either the last row in this matrix consists entirely of zeros or it does not. If not, the matrix contains no zero rows, and consequently each of the n rows has a leading entry of 1. Since these leading 1's occur progressively further to the right as we move down the matrix, each of these 1's must occur on the main diagonal. Since the other entries in the same column as one of these 1's are zero, R must be I_n. Thus, either R has a row of zeros or $R = I_n$. ▮

INVERSE OF A MATRIX

Definition. If A is a square matrix, and if a matrix B of the same size can be found such that $AB = BA = I$, then A is said to be **invertible** and B is called an **inverse** of A.

Example 5 The matrix

$$B = \begin{bmatrix} 3 & 5 \\ 1 & 2 \end{bmatrix} \quad \text{is an inverse of} \quad A = \begin{bmatrix} 2 & -5 \\ -1 & 3 \end{bmatrix}$$

since

$$AB = \begin{bmatrix} 2 & -5 \\ -1 & 3 \end{bmatrix} \begin{bmatrix} 3 & 5 \\ 1 & 2 \end{bmatrix} = \begin{bmatrix} 1 & 0 \\ 0 & 1 \end{bmatrix} = I$$

and

$$BA = \begin{bmatrix} 3 & 5 \\ 1 & 2 \end{bmatrix} \begin{bmatrix} 2 & -5 \\ -1 & 3 \end{bmatrix} = \begin{bmatrix} 1 & 0 \\ 0 & 1 \end{bmatrix} = I \quad \blacktriangle$$

Example 6 The matrix

$$A = \begin{bmatrix} 1 & 4 & 0 \\ 2 & 5 & 0 \\ 3 & 6 & 0 \end{bmatrix}$$

is not invertible. To see why, let

$$B = \begin{bmatrix} b_{11} & b_{12} & b_{13} \\ b_{21} & b_{22} & b_{23} \\ b_{31} & b_{32} & b_{33} \end{bmatrix}$$

be any 3×3 matrix. The third column of BA is

$$\begin{bmatrix} b_{11} & b_{12} & b_{13} \\ b_{21} & b_{22} & b_{23} \\ b_{31} & b_{32} & b_{33} \end{bmatrix} \begin{bmatrix} 0 \\ 0 \\ 0 \end{bmatrix} = \begin{bmatrix} 0 \\ 0 \\ 0 \end{bmatrix}$$

Thus,

$$BA \neq I = \begin{bmatrix} 1 & 0 & 0 \\ 0 & 1 & 0 \\ 0 & 0 & 1 \end{bmatrix} \quad \blacktriangle$$

**PROPERTIES OF
INVERSES**

It is reasonable to ask whether an invertible matrix can have more than one inverse. The next theorem shows that the answer is no—*an invertible matrix has exactly one inverse.*

Theorem 1.4.4. *If B and C are both inverses of the matrix A, then B = C.*

Proof. Since B is an inverse of A, we have $BA = I$. Multiplying both sides on the right by C gives $(BA)C = IC = C$. But $(BA)C = B(AC) = BI = B$, so that $C = B$. ▌

As a consequence of this important result, we can now speak of "the" inverse of an invertible matrix. If A is invertible, then its inverse will be denoted by the symbol A^{-1}. Thus,

$$AA^{-1} = I \quad \text{and} \quad A^{-1}A = I$$

The inverse of A plays much the same role in matrix arithmetic that reciprocal a^{-1} plays in the numerical relationships $aa^{-1} = 1$ and $a^{-1}a = 1$.

In the next section we shall develop a method for finding inverses of invertible matrices of any size; however, the following theorem gives conditions under which a 2×2 matrix is invertible and provides a simple formula for the inverse.

Theorem 1.4.5. *The matrix*

$$A = \begin{bmatrix} a & b \\ c & d \end{bmatrix}$$

is invertible if $ad - bc \neq 0$, in which case the inverse is given by the formula

$$A^{-1} = \frac{1}{ad - bc} \begin{bmatrix} d & -b \\ -c & a \end{bmatrix} = \begin{bmatrix} \dfrac{d}{ad - bc} & -\dfrac{b}{ad - bc} \\ -\dfrac{c}{ad - bc} & \dfrac{a}{ad - bc} \end{bmatrix}$$

Proof. We leave it for the reader to verify that $AA^{-1} = I_2$ and $A^{-1}A = I_2$. ▌

Theorem 1.4.6. *If A and B are invertible matrices of the same size, then:*
(a) *AB is invertible.*
(b) *$(AB)^{-1} = B^{-1}A^{-1}$.*

Proof. If we can show that $(AB)(B^{-1}A^{-1}) = (B^{-1}A^{-1})(AB) = I$, then we will have simultaneously shown that the matrix AB is invertible and that $(AB)^{-1} = B^{-1}A^{-1}$. But $(AB)(B^{-1}A^{-1}) = A(BB^{-1})A^{-1} = AIA^{-1} = AA^{-1} = I$. A similar argument shows that $(B^{-1}A^{-1})(AB) = I$. ∎

Although we will not prove it, this result can be extended to include three or more factors; that is,

A product of any number of invertible matrices is invertible, and the inverse of the product is the product of the inverses in the reverse order.

Example 7   Consider the matrices

$$A = \begin{bmatrix} 1 & 2 \\ 1 & 3 \end{bmatrix} \qquad B = \begin{bmatrix} 3 & 2 \\ 2 & 2 \end{bmatrix} \qquad AB = \begin{bmatrix} 7 & 6 \\ 9 & 8 \end{bmatrix}$$

Applying the formula in Theorem 1.4.5, we obtain

$$A^{-1} = \begin{bmatrix} 3 & -2 \\ -1 & 1 \end{bmatrix} \qquad B^{-1} = \begin{bmatrix} 1 & -1 \\ -1 & \frac{3}{2} \end{bmatrix} \qquad (AB)^{-1} = \begin{bmatrix} 4 & -3 \\ -\frac{9}{2} & \frac{7}{2} \end{bmatrix}$$

Also,

$$B^{-1}A^{-1} = \begin{bmatrix} 1 & -1 \\ -1 & \frac{3}{2} \end{bmatrix} \begin{bmatrix} 3 & -2 \\ -1 & 1 \end{bmatrix} = \begin{bmatrix} 4 & -3 \\ -\frac{9}{2} & \frac{7}{2} \end{bmatrix}$$

Therefore, $(AB)^{-1} = B^{-1}A^{-1}$ as guaranteed by Theorem 1.4.6. ▲

POWERS OF A MATRIX

Next, we shall define powers of a square matrix and discuss their properties.

Definition. If A is a square matrix, then we define the nonnegative integer powers of A to be

$$A^0 = I \qquad A^n = \underbrace{AA \cdots A}_{n \text{ factors}} \qquad (n > 0)$$

Moreover, if A is invertible, then we define the negative integer powers to be

$$A^{-n} = (A^{-1})^n = \underbrace{A^{-1}A^{-1} \cdots A^{-1}}_{n \text{ factors}}$$

Because this definition parallels that for real numbers, the usual laws of exponents hold. (We omit the details.)

> **Theorem 1.4.7.** *If A is a square matrix and r and s are integers, then*
>
> $$A^r A^s = A^{r+s} \qquad (A^r)^s = A^{rs}$$

The next theorem provides some useful properties of negative exponents.

> **Theorem 1.4.8.** *If A is an invertible matrix, then:*
> (a) A^{-1} *is invertible and* $(A^{-1})^{-1} = A$.
> (b) A^n *is invertible and* $(A^n)^{-1} = (A^{-1})^n$ *for* $n = 0, 1, 2, \ldots$
> (c) *For any nonzero scalar* k, *the matrix* kA *is invertible and* $(kA)^{-1} = \dfrac{1}{k} A^{-1}$.

Proof

(a) Since $AA^{-1} = A^{-1}A = I$, the matrix A^{-1} is invertible and $(A^{-1})^{-1} = A$.
(b) This part is left as an exercise.
(c) If k is any nonzero scalar, results (*l*) and (*m*) of Theorem 1.4.1 enable us to write

$$(kA)\left(\frac{1}{k}A^{-1}\right) = \frac{1}{k}(kA)A^{-1} = \left(\frac{1}{k}k\right)AA^{-1} = (1)I = I$$

Similarly, $\left(\dfrac{1}{k}A^{-1}\right)(kA) = I$ so that kA is invertible and $(kA)^{-1} = \dfrac{1}{k}A^{-1}$.

Example 8 Let A and A^{-1} be as in Example 7, that is,

$$A = \begin{bmatrix} 1 & 2 \\ 1 & 3 \end{bmatrix} \qquad \text{and} \qquad A^{-1} = \begin{bmatrix} 3 & -2 \\ -1 & 1 \end{bmatrix}$$

Then

$$A^3 = \begin{bmatrix} 1 & 2 \\ 1 & 3 \end{bmatrix}\begin{bmatrix} 1 & 2 \\ 1 & 3 \end{bmatrix}\begin{bmatrix} 1 & 2 \\ 1 & 3 \end{bmatrix} = \begin{bmatrix} 11 & 30 \\ 15 & 41 \end{bmatrix}$$

$$A^{-3} = (A^{-1})^3 = \begin{bmatrix} 3 & -2 \\ -1 & 1 \end{bmatrix}\begin{bmatrix} 3 & -2 \\ -1 & 1 \end{bmatrix}\begin{bmatrix} 3 & -2 \\ -1 & 1 \end{bmatrix} = \begin{bmatrix} 41 & -30 \\ -15 & 11 \end{bmatrix} \quad \blacktriangle$$

POLYNOMIAL EXPRESSIONS INVOLVING MATRICES

If A is a square matrix, say $m \times m$, and if

$$p(x) = a_0 + a_1 x + \cdots + a_n x^n \tag{1}$$

is any polynomial, then we define

$$p(A) = a_0 I + a_1 A + \cdots + a_n A^n$$

where I is the $m \times m$ identity matrix. In words, $p(A)$ is the $m \times m$ matrix that results when A is substituted for x in (1) and a_0 is replaced by $a_0 I$.

Example 9 If

$$p(x) = 2x^2 - 3x + 4 \qquad \text{and} \qquad A = \begin{bmatrix} -1 & 2 \\ 0 & 3 \end{bmatrix}$$

then

$$p(A) = 2A^2 - 3A + 4I = 2\begin{bmatrix} -1 & 2 \\ 0 & 3 \end{bmatrix}^2 - 3\begin{bmatrix} -1 & 2 \\ 0 & 3 \end{bmatrix} + 4\begin{bmatrix} 1 & 0 \\ 0 & 1 \end{bmatrix}$$

$$= \begin{bmatrix} 2 & 8 \\ 0 & 18 \end{bmatrix} - \begin{bmatrix} -3 & 6 \\ 0 & 9 \end{bmatrix} + \begin{bmatrix} 4 & 0 \\ 0 & 4 \end{bmatrix} = \begin{bmatrix} 9 & 2 \\ 0 & 13 \end{bmatrix} \quad \blacktriangle$$

PROPERTIES OF THE TRANSPOSE

The next theorem lists the main properties of the transpose operation.

> **Theorem 1.4.9.** *If the sizes of the matrices are such that the stated operations can be performed, then*
>
> *(a)* $((A)^T)^T = A$
> *(b)* $(A + B)^T = A^T + B^T$ *and* $(A - B)^T = A^T - B^T$
> *(c)* $(kA)^T = kA^T$, *where k is any scalar*
> *(d)* $(AB)^T = B^T A^T$

Keeping in mind that transposing a matrix interchanges its rows and columns, parts (*a*), (*b*), and (*c*) should be self-evident. For example, part (*a*) states that interchanging rows and columns twice leaves a matrix unchanged; part (*b*) asserts that adding and then interchanging rows and columns yields the same result as first interchanging rows and columns, then adding; and part (*c*) asserts that multiplying by a scalar and then interchanging rows and columns yields the same result as first interchanging rows and columns, then multiplying by the scalar. Part (*d*) is not so obvious, so we give its proof.

Proof (d). Let

$$A = [a_{ij}]_{m \times r} \qquad \text{and} \qquad B = [b_{ij}]_{r \times n}$$

so that the products AB and $B^T A^T$ can both be formed. We leave it for the reader to check that $(AB)^T$ and $B^T A^T$ have the same size, namely $n \times m$. Thus, it only remains to show that corresponding entries of $(AB)^T$ and $B^T A^T$ are the same; that is,

$$\left((AB)^T\right)_{ij} = (B^T A^T)_{ij} \tag{2}$$

Applying Formula (8) of Section 1.3 to the left side of this equation and using the definition of matrix multiplication, we obtain

$$\left((AB)^T\right)_{ij} = (AB)_{ji} = a_{j1}b_{1i} + a_{j2}b_{2i} + \cdots + a_{jr}b_{ri} \tag{3}$$

To evaluate the right side of (2) it will be convenient to let a'_{ij} and b'_{ij} denote the *ij*th entries of A^T and B^T, respectively, so

$$a'_{ij} = a_{ji} \qquad \text{and} \qquad b'_{ij} = b_{ji}$$

From these relationships and the definition of matrix multiplication we obtain

$$
\begin{aligned}
(B^T A^T)_{ij} &= b'_{i1}a'_{1j} + b'_{i2}a'_{2j} + \cdots + b'_{ir}a'_{rj} \\
&= b_{1i}a_{j1} + b_{2i}a_{j2} + \cdots + b_{ri}a_{jr} \\
&= a_{j1}b_{1i} + a_{j2}b_{2i} + \cdots + a_{jr}b_{ri}
\end{aligned}
$$

This, together with (3), proves (2).

Although we shall not prove it, part (d) of this theorem can be extended to include three or more factors; that is,

> *The transpose of a product of any number of matrices is equal to the product of their transposes in the reverse order.*

REMARK. Note the similarity between this result and the result following Theorem 1.4.6 about the inverse of a product of matrices.

INVERTIBILITY OF A TRANSPOSE

The following theorem establishes a relationship between the inverse of an invertible matrix and the inverse of its transpose.

> **Theorem 1.4.10.** *If A is an invertible matrix, then A^T is also invertible and*
>
> $$(A^T)^{-1} = (A^{-1})^T \tag{4}$$

Proof. We can prove the invertibility of A^T and obtain (4) by showing that

$$A^T(A^{-1})^T = (A^{-1})^T A^T = I$$

But from part (d) of Theorem 1.4.9 and the fact that $I^T = I$ we have

$$
\begin{aligned}
A^T(A^{-1})^T &= (A^{-1}A)^T = I^T = I \\
(A^{-1})^T A^T &= (AA^{-1})^T = I^T = I
\end{aligned}
$$

which completes the proof.

Example 10 Consider the matrices

$$
A = \begin{bmatrix} -5 & -3 \\ 2 & 1 \end{bmatrix} \qquad A^T = \begin{bmatrix} -5 & 2 \\ -3 & 1 \end{bmatrix}
$$

Applying Theorem 1.4.5 yields

$$
A^{-1} = \begin{bmatrix} 1 & 3 \\ -2 & -5 \end{bmatrix} \qquad (A^T)^{-1} = \begin{bmatrix} 1 & -2 \\ 3 & -5 \end{bmatrix}
$$

As guaranteed by Theorem 1.4.10, these matrices satisfy (4). ▲

EXERCISE SET 1.4

1. Let

$$A = \begin{bmatrix} 2 & -1 & 3 \\ 0 & 4 & 5 \\ -2 & 1 & 4 \end{bmatrix}, \quad B = \begin{bmatrix} 8 & -3 & -5 \\ 0 & 1 & 2 \\ 4 & -7 & 6 \end{bmatrix}, \quad C = \begin{bmatrix} 0 & -2 & 3 \\ 1 & 7 & 4 \\ 3 & 5 & 9 \end{bmatrix}, \quad a = 4, \quad b = -7$$

Show that
(a) $A + (B + C) = (A + B) + C$ (b) $(AB)C = A(BC)$ (c) $(a + b)C = aC + bC$
(d) $a(B - C) = aB - aC$

2. Using the matrices and scalars in Exercise 1, show that
(a) $a(BC) = (aB)C = B(aC)$ (b) $A(B - C) = AB - AC$ (c) $(B + C)A = BA + CA$
(d) $a(bC) = (ab)C$

3. Using the matrices and scalars in Exercise 1, show that
(a) $(A^T)^T = A$ (b) $(A + B)^T = A^T + B^T$ (c) $(aC)^T = aC^T$ (d) $(AB)^T = B^T A^T$

4. Use Theorem 1.4.5 to compute the inverses of the following matrices.

$$A = \begin{bmatrix} 3 & 1 \\ 5 & 2 \end{bmatrix}, \quad B = \begin{bmatrix} 2 & -3 \\ 4 & 4 \end{bmatrix}, \quad C = \begin{bmatrix} 2 & 0 \\ 0 & 3 \end{bmatrix}$$

5. Verify that the three matrices A, B, and C in Exercise 4 satisfy the relationships $(AB)^{-1} = B^{-1}A^{-1}$ and $(ABC)^{-1} = C^{-1}B^{-1}A^{-1}$.

6. Let A and B be square matrices of the same size. Is $(AB)^2 = A^2B^2$ a valid matrix identity? Justify your answer.

7. In each part use the given information to find A.

(a) $A^{-1} = \begin{bmatrix} 2 & -1 \\ 3 & 5 \end{bmatrix}$ (b) $(7A)^{-1} = \begin{bmatrix} -3 & 7 \\ 1 & -2 \end{bmatrix}$

(c) $(5A^T)^{-1} = \begin{bmatrix} -3 & -1 \\ 5 & 2 \end{bmatrix}$ (d) $(I + 2A)^{-1} = \begin{bmatrix} -1 & 2 \\ 4 & 5 \end{bmatrix}$

8. Let A be the matrix

$$\begin{bmatrix} 2 & 0 \\ 4 & 1 \end{bmatrix}$$

Compute A^3, A^{-3}, and $A^2 - 2A + I$.

9. Let A be the matrix

$$\begin{bmatrix} 3 & 1 \\ 2 & 1 \end{bmatrix}$$

In each part find $p(A)$.
(a) $p(x) = x - 2$ (b) $p(x) = 2x^2 - x + 1$ (c) $p(x) = x^3 - 2x + 4$

10. Let $p_1(x) = x^2 - 9$, $p_2(x) = x + 3$, and $p_3 = x - 3$.
(a) Show that $p_1(A) = p_2(A)p_3(A)$ for the matrix A in Exercise 9.
(b) Show that $p_1(A) = p_2(A)p_3(A)$ for any square matrix A.

11. Find the inverse of

$$\begin{bmatrix} \cos\theta & \sin\theta \\ -\sin\theta & \cos\theta \end{bmatrix}$$

12. (a) Find 2×2 matrices A and B such that $(A + B)^2 \neq A^2 + 2AB + B^2$.
(b) Show that if A and B are square matrices such that $AB = BA$, then

$$(A + B)^2 = A^2 + 2AB + B^2$$

(c) Find an expansion of $(A + B)^2$ that is valid for all square matrices A and B having the same size.

13. Consider the matrix

$$A = \begin{bmatrix} a_{11} & 0 & 0 & \cdots & 0 \\ 0 & a_{22} & 0 & \cdots & 0 \\ \vdots & \vdots & \vdots & & \vdots \\ 0 & 0 & 0 & \cdots & a_{nn} \end{bmatrix}$$

where $a_{11}a_{22}\cdots a_{nn} \neq 0$. Show that A is invertible and find its inverse.

14. Show that if a square matrix A satisfies $A^2 - 3A + I = 0$, then $A^{-1} = 3I - A$.

15. (a) Show that a matrix with a row of zeros cannot have an inverse.
(b) Show that a matrix with a column of zeros cannot have an inverse.

16. Is the sum of two invertible matrices necessarily invertible?

17. Let A and B be square matrices such that $AB = 0$. Show that if A is invertible, then $B = 0$.

18. In Theorem 1.4.2 why didn't we write part (d) as $A0 = 0 = 0A$?

19. The real equation $a^2 = 1$ has exactly two solutions. Find at least eight different 3×3 matrices that satisfy the matrix equation $A^2 = I_3$. [**Hint.** Look for solutions in which all the entries off the main diagonal are zero.]

20. (a) Find a nonzero 3×3 matrix A such that $A^T = A$.
(b) Find a nonzero 3×3 matrix A such that $A^T = -A$.

21. A square matrix A is called **symmetric** if $A^T = A$ and **skew-symmetric** if $A^T = -A$. Show that if B is a square matrix, then
(a) BB^T and $B + B^T$ are symmetric   (b) $B - B^T$ is skew-symmetric

22. If A is a square matrix and n is a positive integer, is it true that $(A^n)^T = (A^T)^n$? Justify your answer.

23. Let A be the matrix

$$\begin{bmatrix} 1 & 0 & 1 \\ 1 & 1 & 0 \\ 0 & 1 & 1 \end{bmatrix}$$

Determine whether A is invertible, and if so, find its inverse. [**Hint.** Solve $AX = I$ by equating corresponding entries on the two sides.]

24. Prove:
(a) part (b) of Theorem 1.4.1   (b) part (i) of Theorem 1.4.1   (c) part (m) of Theorem 1.4.1

25. Apply parts (*d*) and (*m*) of Theorem 1.4.1 to the matrices A, B, and $(-1)C$ to derive the result in part (*f*).

26. Prove Theorem 1.4.2.

27. Consider the laws of exponents $A^r A^s = A^{r+s}$ and $(A^r)^s = A^{rs}$.
    (a) Show that if A is any square matrix, these laws are valid for all nonnegative integer values of r and s.
    (b) Show that if A is invertible, these laws hold for all negative integer values of r and s.

28. Show that if A is invertible and k is any nonzero scalar, then $(kA)^n = k^n A^n$ for all integer values of n.

29. (a) Show that if A is invertible and $AB = AC$, then $B = C$.
    (b) Explain why part (a) and Example 3 do not contradict one another.

30. Prove part (*c*) of Theorem 1.4.1. [**Hint.** Assume that A is $m \times n$, B is $n \times p$, and C is $p \times q$. The *ij*th entry on the left side is $l_{ij} = a_{i1}[BC]_{1j} + a_{i2}[BC]_{2j} + \cdots + a_{in}[BC]_{nj}$ and the *ij*th entry on the right side is $r_{ij} = [AB]_{i1}c_{1j} + [AB]_{i2}c_{2j} + \cdots + [AB]_{ip}c_{pj}$. Verify that $l_{ij} = r_{ij}$.]

1.5 ELEMENTARY MATRICES AND A METHOD FOR FINDING A^{-1}

In this section we shall develop an algorithm for finding the inverse of an invertible matrix, and we shall discuss some basic properties of invertible matrices.

ELEMENTARY MATRICES

Definition. An $n \times n$ matrix is called an *elementary matrix* if it can be obtained from the $n \times n$ identity matrix I_n by performing a single elementary row operation.

Example 1 Listed below are four elementary matrices and the operations that produce them.

$$\begin{bmatrix} 1 & 0 \\ 0 & -3 \end{bmatrix} \qquad \begin{bmatrix} 1 & 0 & 0 & 0 \\ 0 & 0 & 0 & 1 \\ 0 & 0 & 1 & 0 \\ 0 & 1 & 0 & 0 \end{bmatrix} \qquad \begin{bmatrix} 1 & 0 & 3 \\ 0 & 1 & 0 \\ 0 & 0 & 1 \end{bmatrix} \qquad \begin{bmatrix} 1 & 0 & 0 \\ 0 & 1 & 0 \\ 0 & 0 & 1 \end{bmatrix}$$

Multiply the second row of I_2 by -3.	Interchange the second and fourth rows of I_4.	Add 3 times the third row of I_3 to the first row.	Multiply the first row of I_3 by 1.

When a matrix A is multiplied on the *left* by an elementary matrix E, the effect is to perform an elementary row operation on A. This is the content of the following theorem, the proof of which is left for the exercises.

Theorem 1.5.1. *If the elementary matrix E results from performing a certain row operation on I_m and if A is an $m \times n$ matrix, then the product EA is the matrix that results when this same row operation is performed on A.*

Example 2 Consider the matrix

$$A = \begin{bmatrix} 1 & 0 & 2 & 3 \\ 2 & -1 & 3 & 6 \\ 1 & 4 & 4 & 0 \end{bmatrix}$$

and consider the elementary matrix

$$E = \begin{bmatrix} 1 & 0 & 0 \\ 0 & 1 & 0 \\ 3 & 0 & 1 \end{bmatrix}$$

which results from adding 3 times the first row of I_3 to the third row. The product EA is

$$EA = \begin{bmatrix} 1 & 0 & 2 & 3 \\ 2 & -1 & 3 & 6 \\ 4 & 4 & 10 & 9 \end{bmatrix}$$

which is precisely the same matrix that results when we add 3 times the first row of A to the third row. ▲

REMARK. Theorem 1.5.1 is primarily of theoretical interest and will be used for developing some results about matrices and systems of linear equations. Computationally, it is preferable to perform row operations directly rather than multiplying on the left by an elementary matrix.

If an elementary row operation is applied to an identity matrix I to produce an elementary matrix E, then there is a second row operation that, when applied to E, produces I back again. For example, if E is obtained by multiplying the ith row of I by a nonzero constant c, then I can be recovered if the ith row of E is multiplied by $1/c$. The various possibilities are listed in Table 1.

TABLE 1

Row Operation on I That Produces E	Row Operation on E That Reproduces I
Multiply row i by $c \neq 0$	Multiply row i by $1/c$
Interchange rows i and j	Interchange rows i and j
Add c times row i to row j	Add $-c$ times row i to row j

The operations on the right side of this table are called the ***inverse operations*** of the corresponding operations on the left.

Example 3   In each of the following, an elementary row operation is applied to the 2×2 identity matrix to obtain an elementary matrix E, then E is restored to the identity matrix by applying the inverse row operation.

$$\begin{bmatrix} 1 & 0 \\ 0 & 1 \end{bmatrix} \longrightarrow \begin{bmatrix} 1 & 0 \\ 0 & 7 \end{bmatrix} \longrightarrow \begin{bmatrix} 1 & 0 \\ 0 & 1 \end{bmatrix}$$

| Multiply the second row by 7. | | Multiply the second row by $\frac{1}{7}$. |

$$\begin{bmatrix} 1 & 0 \\ 0 & 1 \end{bmatrix} \longrightarrow \begin{bmatrix} 0 & 1 \\ 1 & 0 \end{bmatrix} \longrightarrow \begin{bmatrix} 1 & 0 \\ 0 & 1 \end{bmatrix}$$

| Interchange the first and second rows. | | Interchange the first and second rows. |

$$\begin{bmatrix} 1 & 0 \\ 0 & 1 \end{bmatrix} \longrightarrow \begin{bmatrix} 1 & 5 \\ 0 & 1 \end{bmatrix} \longrightarrow \begin{bmatrix} 1 & 0 \\ 0 & 1 \end{bmatrix}$$

| Add 5 times the second row to the first. | | Add -5 times the second row to the first. |

The next theorem gives an important property of elementary matrices.

Theorem 1.5.2.   *Every elementary matrix is invertible, and the inverse is also an elementary matrix.*

Proof.   If E is an elementary matrix, then E results from performing some row operation on I. Let E_0 be the matrix that results when the inverse of this operation is performed on I. Applying Theorem 1.5.1 and using the fact that inverse row operations cancel the effect of each other, it follows that

$$E_0 E = I \quad \text{and} \quad EE_0 = I$$

Thus, the elementary matrix E_0 is the inverse of E. ▮

The next theorem establishes some fundamental relationships between invertibility, homogeneous linear systems, reduced row-echelon forms, and elementary matrices. These results are extremely important and will be used many times in later sections.

> **Theorem 1.5.3.** *If A is an $n \times n$ matrix, then the following statements are equivalent, that is, all true or all false.*
> (*a*) *A is invertible.*
> (*b*) *$A\mathbf{x} = \mathbf{0}$ has only the trivial solution.*
> (*c*) *The reduced row-echelon form of A is I_n.*
> (*d*) *A is expressible as a product of elementary matrices.*

Proof. We shall prove the equivalence by establishing the chain of implications: $(a) \Rightarrow (b) \Rightarrow (c) \Rightarrow (d) \Rightarrow (a)$.

$(a) \Rightarrow (b)$: Assume A is invertible and let $\mathbf{x}_0$ be any solution of $A\mathbf{x} = \mathbf{0}$; thus, $A\mathbf{x}_0 = \mathbf{0}$. Multiplying both sides of this equation by the matrix A^{-1} gives $A^{-1}(A\mathbf{x}_0) = A^{-1}\mathbf{0}$, or $(A^{-1}A)\mathbf{x}_0 = \mathbf{0}$, or $I\mathbf{x}_0 = \mathbf{0}$, or $\mathbf{x}_0 = \mathbf{0}$. Thus, $A\mathbf{x} = \mathbf{0}$ has only the trivial solution.

$(b) \Rightarrow (c)$: Let $A\mathbf{x} = \mathbf{0}$ be the matrix form of the system

$$
\begin{aligned}
a_{11}x_1 + a_{12}x_2 + \cdots + a_{1n}x_n &= 0 \\
a_{21}x_1 + a_{22}x_2 + \cdots + a_{2n}x_n &= 0 \\
\vdots \qquad \vdots \qquad\qquad \vdots \quad\ \vdots & \\
a_{n1}x_1 + a_{n2}x_2 + \cdots + a_{nn}x_n &= 0
\end{aligned}
\tag{1}
$$

and assume that the system has only the trivial solution. If we solve by Gauss–Jordan elimination, then the system of equations corresponding to the reduced row-echelon form of the augmented matrix will be

$$
\begin{aligned}
x_1 \qquad\qquad\qquad\quad &= 0 \\
x_2 \qquad\qquad\quad &= 0 \\
\ddots \qquad\quad & \\
x_n &= 0
\end{aligned}
\tag{2}
$$

Thus, the augmented matrix

$$
\begin{bmatrix}
a_{11} & a_{12} & \cdots & a_{1n} & 0 \\
a_{21} & a_{22} & \cdots & a_{2n} & 0 \\
\vdots & \vdots & & \vdots & \vdots \\
a_{n1} & a_{n2} & \cdots & a_{nn} & 0
\end{bmatrix}
$$

for (1) can be reduced to the augmented matrix

$$
\begin{bmatrix}
1 & 0 & 0 & \cdots & 0 & 0 \\
0 & 1 & 0 & \cdots & 0 & 0 \\
0 & 0 & 1 & \cdots & 0 & 0 \\
\vdots & \vdots & \vdots & & \vdots & \vdots \\
0 & 0 & 0 & \cdots & 1 & 0
\end{bmatrix}
$$

for (2) by a sequence of elementary row operations. If we disregard the last column (of zeros) in each of these matrices, we can conclude that the reduced row-echelon form of A is I_n.

$(c) \Rightarrow (d)$: Assume that the reduced row-echelon form of A is I_n, so that A can be reduced to I_n by a finite sequence of elementary row operations. By Theorem 1.5.1 each of these operations can be accomplished by multiplying on the left by an appropriate elementary matrix. Thus, we can find elementary matrices $E_1, E_2, \ldots, E_k$ such that

$$E_k \cdots E_2 E_1 A = I_n \tag{3}$$

By Theorem 1.5.2, $E_1, E_2, \ldots, E_k$ are invertible. Multiplying both sides of Equation (3) on the left successively by $E_k^{-1}, \ldots, E_2^{-1}, E_1^{-1}$ we obtain

$$A = E_1^{-1} E_2^{-1} \cdots E_k^{-1} I_n = E_1^{-1} E_2^{-1} \cdots E_k^{-1} \tag{4}$$

By Theorem 1.5.2, this equation expresses A as a product of elementary matrices.

$(d) \Rightarrow (a)$: If A is a product of elementary matrices, then from Theorems 1.4.6 and 1.5.2 the matrix A is a product of invertible matrices, and hence is invertible. ▌

ROW EQUIVALENCE

If a matrix B can be obtained from a matrix A by performing a finite sequence of elementary row operations, then obviously we can get from B back to A by performing the inverses of these elementary row operations in reverse order. Matrices that can be obtained from one another by a finite sequence of elementary row operations are said to be ***row equivalent***. With this terminology it follows from parts (a) and (c) of Theorem 1.5.3 that an $n \times n$ matrix A is invertible if and only if it is row equivalent to the $n \times n$ identity matrix.

A METHOD FOR INVERTING MATRICES

As our first application of Theorem 1.5.3, we shall establish a method for determining the inverse of an invertible matrix. Inverting the left and the right sides of (4) yields $A^{-1} = E_k \cdots E_2 E_1$, or equivalently,

$$A^{-1} = E_k \cdots E_2 E_1 I_n \tag{5}$$

which tells us that A^{-1} can be obtained by multiplying I_n successively on the left by the elementary matrices $E_1, E_2, \ldots, E_k$. Since each multiplication on the left by one of these elementary matrices performs a row operation, it follows, by comparing Equations (3) and (5), that *the sequence of row operations that reduces A to I_n will reduce I_n to A^{-1}.* Thus, we have the following result:

To find the inverse of an invertible matrix A, we must find a sequence of elementary row operations that reduces A to the identity and then perform this same sequence of operations on I_n to obtain A^{-1}.

A simple method for carrying out this procedure is given in the following example.

Read
this

Example 4 Find the inverse of

$$A = \begin{bmatrix} 1 & 2 & 3 \\ 2 & 5 & 3 \\ 1 & 0 & 8 \end{bmatrix}$$

Solution. We want to reduce A to the identity matrix by row operations and simultaneously apply these operations to I to produce A^{-1}. To accomplish this we shall adjoin the identity matrix to the right side of A, thereby producing a matrix of the form

$$[A \mid I]$$

then we shall apply row operations to this matrix until the left side is reduced to I; these operations will convert the right side to A^{-1}, so that the final matrix will have the form

$$[I \mid A^{-1}]$$

The computations are as follows:

Can't do two operations at once to get elementary matrices

$$\begin{bmatrix} 1 & 2 & 3 & \vdots & 1 & 0 & 0 \\ 2 & 5 & 3 & \vdots & 0 & 1 & 0 \\ 1 & 0 & 8 & \vdots & 0 & 0 & 1 \end{bmatrix}$$

$$\begin{bmatrix} 1 & 2 & 3 & \vdots & 1 & 0 & 0 \\ 0 & 1 & -3 & \vdots & -2 & 1 & 0 \\ 0 & -2 & 5 & \vdots & -1 & 0 & 1 \end{bmatrix} E_1^{-1}$$

We added -2 times the first row to the second and -1 times the first row to the third.

$$\begin{bmatrix} 1 & 2 & 3 & \vdots & 1 & 0 & 0 \\ 0 & 1 & -3 & \vdots & -2 & 1 & 0 \\ 0 & 0 & -1 & \vdots & -5 & 2 & 1 \end{bmatrix} E_2^{-1}$$

We added 2 times the second row to the third.

$$\begin{bmatrix} 1 & 2 & 3 & \vdots & 1 & 0 & 0 \\ 0 & 1 & -3 & \vdots & -2 & 1 & 0 \\ 0 & 0 & 1 & \vdots & 5 & -2 & -1 \end{bmatrix}$$

We multiplied the third row by -1.

$$\begin{bmatrix} 1 & 2 & 0 & \vdots & -14 & 6 & 3 \\ 0 & 1 & 0 & \vdots & 13 & -5 & -3 \\ 0 & 0 & 1 & \vdots & 5 & -2 & -1 \end{bmatrix}$$

We added 3 times the third row to the second and -3 times the third row to the first.

$$\begin{bmatrix} 1 & 0 & 0 & \vdots & -40 & 16 & 9 \\ 0 & 1 & 0 & \vdots & 13 & -5 & -3 \\ 0 & 0 & 1 & \vdots & 5 & -2 & -1 \end{bmatrix}$$

We added -2 times the second row to the first.

Thus,

$$A^{-1} = \begin{bmatrix} -40 & 16 & 9 \\ 13 & -5 & -3 \\ 5 & -2 & -1 \end{bmatrix}$$

Often it will not be known in advance whether a given matrix is invertible. If an $n \times n$ matrix A is not invertible, then it cannot be reduced to I_n by elementary row operations [part (c) of Theorem 1.5.3]. Stated another way, the reduced row-echelon form of A has at least one row of zeros. Thus, if the procedure in the last example is attempted on a matrix that is not invertible, then at some point in the computations a row of zeros will occur on the *left side*. It can then be concluded that the given matrix is not invertible, and the computations can be stopped.

Example 5 Consider the matrix

$$A = \begin{bmatrix} 1 & 6 & 4 \\ 2 & 4 & -1 \\ -1 & 2 & 5 \end{bmatrix}$$

Applying the procedure of Example 4 yields

$$\left[\begin{array}{ccc:ccc} 1 & 6 & 4 & 1 & 0 & 0 \\ 2 & 4 & -1 & 0 & 1 & 0 \\ -1 & 2 & 5 & 0 & 0 & 1 \end{array}\right]$$

$$\left[\begin{array}{ccc:ccc} 1 & 6 & 4 & 1 & 0 & 0 \\ 0 & -8 & -9 & -2 & 1 & 0 \\ 0 & 8 & 9 & 1 & 0 & 1 \end{array}\right]$$

We added -2 times the first row to the second and added the first row to the third.

$$\left[\begin{array}{ccc:ccc} 1 & 6 & 4 & 1 & 0 & 0 \\ 0 & -8 & -9 & -2 & 1 & 0 \\ 0 & 0 & 0 & -1 & 1 & 1 \end{array}\right]$$

We added the second row to the third.

Since we have obtained a row of zeros on the left side, A is not invertible.

Example 6 In Example 4 we showed that

$$A = \begin{bmatrix} 1 & 2 & 3 \\ 2 & 5 & 3 \\ 1 & 0 & 8 \end{bmatrix}$$

is an invertible matrix. From Theorem 1.5.3 it follows that the system of equations

$$\begin{aligned} x_1 + 2x_2 + 3x_3 &= 0 \\ 2x_1 + 5x_2 + 3x_3 &= 0 \\ x_1 \quad\quad + 8x_3 &= 0 \end{aligned}$$

has only the trivial solution.

EXERCISE SET 1.5

1. Which of the following are elementary matrices?

(a) $\begin{bmatrix} 1 & 0 \\ -5 & 1 \end{bmatrix}$ (b) $\begin{bmatrix} -5 & 1 \\ 1 & 0 \end{bmatrix}$ (c) $\begin{bmatrix} 1 & 0 \\ 0 & \sqrt{3} \end{bmatrix}$ (d) $\begin{bmatrix} 0 & 0 & 1 \\ 0 & 1 & 0 \\ 1 & 0 & 0 \end{bmatrix}$

(e) $\begin{bmatrix} 1 & 1 & 0 \\ 0 & 0 & 1 \\ 0 & 0 & 0 \end{bmatrix}$ (f) $\begin{bmatrix} 1 & 0 & 0 \\ 0 & 1 & 9 \\ 0 & 0 & 1 \end{bmatrix}$ (g) $\begin{bmatrix} 2 & 0 & 0 & 2 \\ 0 & 1 & 0 & 0 \\ 0 & 0 & 1 & 0 \\ 0 & 0 & 0 & 1 \end{bmatrix}$

2. Find a row operation that will restore the given elementary matrix to an identity matrix.

(a) $\begin{bmatrix} 1 & 0 \\ -3 & 1 \end{bmatrix}$ (b) $\begin{bmatrix} 1 & 0 & 0 \\ 0 & 1 & 0 \\ 0 & 0 & 3 \end{bmatrix}$ (c) $\begin{bmatrix} 0 & 0 & 0 & 1 \\ 0 & 1 & 0 & 0 \\ 0 & 0 & 1 & 0 \\ 1 & 0 & 0 & 0 \end{bmatrix}$ (d) $\begin{bmatrix} 1 & 0 & -\frac{1}{7} & 0 \\ 0 & 1 & 0 & 0 \\ 0 & 0 & 1 & 0 \\ 0 & 0 & 0 & 1 \end{bmatrix}$

3. Consider the matrices

$$A = \begin{bmatrix} 3 & 4 & 1 \\ 2 & -7 & -1 \\ 8 & 1 & 5 \end{bmatrix}, \quad B = \begin{bmatrix} 8 & 1 & 5 \\ 2 & -7 & -1 \\ 3 & 4 & 1 \end{bmatrix}, \quad C = \begin{bmatrix} 3 & 4 & 1 \\ 2 & -7 & -1 \\ 2 & -7 & 3 \end{bmatrix}$$

Find elementary matrices E_1, E_2, E_3, and E_4 such that
(a) $E_1 A = B$ (b) $E_2 B = A$ (c) $E_3 A = C$ (d) $E_4 C = A$

4. In Exercise 3 is it possible to find an elementary matrix E such that $EB = C$? Justify your answer.

In Exercises 5–7 use the method shown in Examples 4 and 5 to find the inverse of the given matrix if the matrix is invertible and check your answer by multiplication.

5. (a) $\begin{bmatrix} 1 & 4 \\ 2 & 7 \end{bmatrix}$ (b) $\begin{bmatrix} -3 & 6 \\ 4 & 5 \end{bmatrix}$ (c) $\begin{bmatrix} 6 & -4 \\ -3 & 2 \end{bmatrix}$

6. (a) $\begin{bmatrix} 3 & 4 & -1 \\ 1 & 0 & 3 \\ 2 & 5 & -4 \end{bmatrix}$ (b) $\begin{bmatrix} -1 & 3 & -4 \\ 2 & 4 & 1 \\ -4 & 2 & -9 \end{bmatrix}$ (c) $\begin{bmatrix} 1 & 0 & 1 \\ 0 & 1 & 1 \\ 1 & 1 & 0 \end{bmatrix}$ (d) $\begin{bmatrix} 2 & 6 & 6 \\ 2 & 7 & 6 \\ 2 & 7 & 7 \end{bmatrix}$ (e) $\begin{bmatrix} 1 & 0 & 1 \\ -1 & 1 & 1 \\ 0 & 1 & 0 \end{bmatrix}$

7. (a) $\begin{bmatrix} \frac{1}{5} & \frac{1}{5} & -\frac{2}{5} \\ \frac{1}{5} & \frac{1}{5} & \frac{1}{10} \\ \frac{1}{5} & -\frac{4}{5} & \frac{1}{10} \end{bmatrix}$ (b) $\begin{bmatrix} \sqrt{2} & 3\sqrt{2} & 0 \\ -4\sqrt{2} & \sqrt{2} & 0 \\ 0 & 0 & 1 \end{bmatrix}$ (c) $\begin{bmatrix} 1 & 0 & 0 & 0 \\ 1 & 3 & 0 & 0 \\ 1 & 3 & 5 & 0 \\ 1 & 3 & 5 & 7 \end{bmatrix}$

(d) $\begin{bmatrix} -8 & 17 & 2 & \frac{1}{3} \\ 4 & 0 & \frac{2}{5} & -9 \\ 0 & 0 & 0 & 0 \\ -1 & 13 & 4 & 2 \end{bmatrix}$ (e) $\begin{bmatrix} 0 & 0 & 2 & 0 \\ 1 & 0 & 0 & 1 \\ 0 & -1 & 3 & 0 \\ 2 & 1 & 5 & -3 \end{bmatrix}$

8. Find the inverse of each of the following 4×4 matrices, where k_1, k_2, k_3, k_4, and k are all nonzero.

(a) $\begin{bmatrix} k_1 & 0 & 0 & 0 \\ 0 & k_2 & 0 & 0 \\ 0 & 0 & k_3 & 0 \\ 0 & 0 & 0 & k_4 \end{bmatrix}$ (b) $\begin{bmatrix} 0 & 0 & 0 & k_1 \\ 0 & 0 & k_2 & 0 \\ 0 & k_3 & 0 & 0 \\ k_4 & 0 & 0 & 0 \end{bmatrix}$ (c) $\begin{bmatrix} k & 0 & 0 & 0 \\ 1 & k & 0 & 0 \\ 0 & 1 & k & 0 \\ 0 & 0 & 1 & k \end{bmatrix}$

9. Consider the matrix

$$A = \begin{bmatrix} 1 & 0 \\ -5 & 2 \end{bmatrix}$$

(a) Find elementary matrices E_1 and E_2 such that $E_2E_1A = I$.

(b) Write A^{-1} as a product of two elementary matrices.

(c) Write A as a product of two elementary matrices.

10. In each part perform the stated row operation on

$$\begin{bmatrix} 2 & -1 & 0 \\ 4 & 5 & -3 \\ 1 & -4 & 7 \end{bmatrix}$$

by multiplying A on the left by a suitable elementary matrix. Check your answer in each case by performing the row operation directly on A.

(a) Interchange the first and third rows.

(b) Multiply the second row by $\frac{1}{3}$.

(c) Add twice the second row to the first row.

11. Express the matrix

$$A = \begin{bmatrix} 0 & 1 & 7 & 8 \\ 1 & 3 & 3 & 8 \\ -2 & -5 & 1 & -8 \end{bmatrix}$$

in the form $A = EFGR$, where E, F, and G are elementary matrices, and R is in row-echelon form.

12. Show that if

$$A = \begin{bmatrix} 1 & 0 & 0 \\ 0 & 1 & 0 \\ a & b & c \end{bmatrix}$$

is an elementary matrix, then at least one entry in the third row must be a zero.

13. Show that

$$A = \begin{bmatrix} 0 & a & 0 & 0 & 0 \\ b & 0 & c & 0 & 0 \\ 0 & d & 0 & e & 0 \\ 0 & 0 & f & 0 & g \\ 0 & 0 & 0 & h & 0 \end{bmatrix}$$

is not invertible for any values of the entries.

14. Prove that if A is an $m \times n$ matrix, there is an invertible matrix C such that CA is in reduced row-echelon form.

15. Prove that if A is an invertible matrix and B is row equivalent to A, then B is also invertible.

16. (a) Prove: If A and B are $m \times n$ matrices, then A and B are row equivalent if and only if A and B have the same reduced row-echelon form.

    (b) Show that A and B are row equivalent, and find a sequence of elementary row operations that produces B from A.

$$A = \begin{bmatrix} 1 & 2 & 3 \\ 1 & 4 & 1 \\ 2 & 1 & 9 \end{bmatrix} \qquad B = \begin{bmatrix} 1 & 0 & 5 \\ 0 & 2 & -2 \\ 1 & 1 & 4 \end{bmatrix}$$

17. Prove Theorem 1.5.1.

1.6 FURTHER RESULTS ON SYSTEMS OF EQUATIONS AND INVERTIBILITY

In this section we shall establish more results about systems of linear equations and invertibility of matrices. Our work will lead to a totally new method for solving n equations in n unknowns.

A BASIC THEOREM

We begin by proving a fundamental result about linear systems that was anticipated in the first section of this text.

> **Theorem 1.6.1.** *Every system of linear equations has either no solutions, exactly one solution, or infinitely many solutions.*

Proof. If $A\mathbf{x} = \mathbf{b}$ is a system of linear equations, exactly one of the following is true: (a) the system has no solutions, (b) the system has exactly one solution, or (c) the system has more than one solution. The proof will be complete if we can show that the system has infinitely many solutions in case (c).

Assume that $A\mathbf{x} = \mathbf{b}$ has more than one solution, and let $\mathbf{x}_0 = \mathbf{x}_1 - \mathbf{x}_2$, where $\mathbf{x}_1$ and $\mathbf{x}_2$ are any two distinct solutions. Because $\mathbf{x}_1$ and $\mathbf{x}_2$ are distinct, the matrix $\mathbf{x}_0$ is nonzero; moreover,

$$A\mathbf{x}_0 = A(\mathbf{x}_1 - \mathbf{x}_2) = A\mathbf{x}_1 - A\mathbf{x}_2 = \mathbf{b} - \mathbf{b} = \mathbf{0}$$

If we now let k be any scalar, then

$$A(\mathbf{x}_1 + k\mathbf{x}_0) = A\mathbf{x}_1 + A(k\mathbf{x}_0) = A\mathbf{x}_1 + k(A\mathbf{x}_0)$$
$$= \mathbf{b} + k\mathbf{0} = \mathbf{b} + \mathbf{0} = \mathbf{b}$$

But this says that $\mathbf{x}_1 + k\mathbf{x}_0$ is a solution of $A\mathbf{x} = \mathbf{b}$. Since $\mathbf{x}_0$ is nonzero and there are infinitely many choices for k, the system $A\mathbf{x} = \mathbf{b}$ has infinitely many solutions. ▉

SOLVING LINEAR SYSTEMS BY MATRIX INVERSION

Thus far, we have studied two methods for solving linear systems: Gaussian elimination and Gauss–Jordan elimination. The following theorem provides a new method for solving certain linear systems.

> **Theorem 1.6.2.** *If A is an invertible n × n matrix, then for each n × 1 matrix **b**, the system of equations A**x** = **b** has exactly one solution, namely, **x** = A^{-1}**b**.*

Proof. Since $A(A^{-1}\mathbf{b}) = \mathbf{b}$, it follows that $\mathbf{x} = A^{-1}\mathbf{b}$ is a solution of $A\mathbf{x} = \mathbf{b}$. To show that this is the only solution, we will assume that $\mathbf{x}_0$ is an arbitrary solution and then show that $\mathbf{x}_0$ must be the solution $A^{-1}\mathbf{b}$.

If $\mathbf{x}_0$ is any solution, then $A\mathbf{x}_0 = \mathbf{b}$. Multiplying both sides by A^{-1}, we obtain $\mathbf{x}_0 = A^{-1}\mathbf{b}$. ▊

Example 1   Consider the system of linear equations

$$
\begin{aligned}
x_1 + 2x_2 + 3x_3 &= 5 \\
2x_1 + 5x_2 + 3x_3 &= 3 \\
x_1 \qquad\quad + 8x_3 &= 17
\end{aligned}
$$

In matrix form this system can be written as $A\mathbf{x} = \mathbf{b}$, where

$$
A = \begin{bmatrix} 1 & 2 & 3 \\ 2 & 5 & 3 \\ 1 & 0 & 8 \end{bmatrix} \qquad \mathbf{x} = \begin{bmatrix} x_1 \\ x_2 \\ x_3 \end{bmatrix} \qquad \mathbf{b} = \begin{bmatrix} 5 \\ 3 \\ 17 \end{bmatrix}
$$

In Example 4 of the preceding section we showed that A is invertible and

$$
A^{-1} = \begin{bmatrix} -40 & 16 & 9 \\ 13 & -5 & -3 \\ 5 & -2 & -1 \end{bmatrix}
$$

By Theorem 1.6.2 the solution of the system is

$$
\mathbf{x} = A^{-1}\mathbf{b} = \begin{bmatrix} -40 & 16 & 9 \\ 13 & -5 & -3 \\ 5 & -2 & -1 \end{bmatrix} \begin{bmatrix} 5 \\ 3 \\ 17 \end{bmatrix} = \begin{bmatrix} 1 \\ -1 \\ 2 \end{bmatrix}
$$

or $x_1 = 1, x_2 = -1, x_3 = 2$. ▲

REMARK.   Note that the method of Example 1 applies only when the system has as many equations as unknowns and the coefficient matrix is invertible.

SOLVING MULTIPLE LINEAR SYSTEMS WITH A COMMON COEFFICIENT MATRIX

Frequently, one is concerned with solving a sequence of systems

$$
A\mathbf{x} = \mathbf{b}_1, \quad A\mathbf{x} = \mathbf{b}_2, \quad A\mathbf{x} = \mathbf{b}_3, \quad \ldots, \quad A\mathbf{x} = \mathbf{b}_k
$$

each of which has the same square coefficient matrix A. If A is invertible, then the solutions

$$
\mathbf{x}_1 = A^{-1}\mathbf{b}_1, \quad \mathbf{x}_2 = A^{-1}\mathbf{b}_2, \quad \mathbf{x}_3 = A^{-1}\mathbf{b}_3, \quad \ldots, \quad \mathbf{x}_k = A^{-1}\mathbf{b}_k
$$

can be obtained with one matrix inversion and k matrix multiplications. However, a more efficient method is to form the matrix

$$
[A \mid \mathbf{b}_1 \mid \mathbf{b}_2 \mid \cdots \mid \mathbf{b}_k] \tag{1}
$$

in which the coefficient matrix A is "augmented" by all k of the matrices $\mathbf{b}_1, \mathbf{b}_2, \ldots, \mathbf{b}_k$. By reducing (1) to reduced row-echelon form we can solve all k systems at once by Gauss–Jordan elimination. This method has the added advantage that it applies even when A is not invertible.

Example 2 Solve the systems

(a) $x_1 + 2x_2 + 3x_3 = 4$ (b) $x_1 + 2x_2 + 3x_3 = 1$

   $2x_1 + 5x_2 + 3x_3 = 5$ $2x_1 + 5x_2 + 3x_3 = 6$

   $x_1 \qquad + 8x_3 = 9$ $x_1 \qquad + 8x_3 = -6$

Solution. The two systems have the same coefficient matrix. If we augment this coefficient matrix with the columns of constants on the right sides of these systems, we obtain

$$\begin{bmatrix} 1 & 2 & 3 & \vdots & 4 & \vdots & 1 \\ 2 & 5 & 3 & \vdots & 5 & \vdots & 6 \\ 1 & 0 & 8 & \vdots & 9 & \vdots & -6 \end{bmatrix}$$

Reducing this matrix to reduced row-echelon form yields (verify)

$$\begin{bmatrix} 1 & 0 & 0 & \vdots & 1 & \vdots & 2 \\ 0 & 1 & 0 & \vdots & 0 & \vdots & 1 \\ 0 & 0 & 1 & \vdots & 1 & \vdots & -1 \end{bmatrix}$$

It follows from the last two columns that the solution of system (a) is $x_1 = 1$, $x_2 = 0$, $x_3 = 1$ and of system (b) is $x_1 = 2$, $x_2 = 1$, $x_3 = -1$. ▲

PROPERTIES OF INVERTIBLE MATRICES

Up to now, to show that an $n \times n$ matrix A is invertible, it has been necessary to find an $n \times n$ matrix B such that

$$AB = I \qquad \text{and} \qquad BA = I$$

The next theorem shows that if we produce an $n \times n$ matrix B satisfying *either* condition, then the other condition holds automatically.

> **Theorem 1.6.3.** *Let A be a square matrix.*
> (*a*) *If B is a square matrix satisfying BA = I, then $B = A^{-1}$.*
> (*b*) *If B is a square matrix satisfying AB = I, then $B = A^{-1}$.*

We shall prove part (*a*) and leave part (*b*) as an exercise.

Proof (*a*). Assume that $BA = I$. If we can show that A is invertible, the proof can be completed by multiplying $BA = I$ on both sides by A^{-1} to obtain

$$BAA^{-1} = IA^{-1} \qquad \text{or} \qquad BI = IA^{-1} \qquad \text{or} \qquad B = A^{-1}$$

To show that A is invertible, it suffices to show that the system $A\mathbf{x} = \mathbf{0}$ has only the trivial solution (see Theorem 1.5.3). Let $\mathbf{x}_0$ be any solution of this system. If we multiply both sides of $A\mathbf{x}_0 = \mathbf{0}$ on the left by B, we obtain $BA\mathbf{x}_0 = B\mathbf{0}$ or $I\mathbf{x}_0 = \mathbf{0}$ or $\mathbf{x}_0 = \mathbf{0}$. Thus, the system of equations $A\mathbf{x} = \mathbf{0}$ has only the trivial solution.

We are now in a position to add two more statements that are equivalent to the four given in Theorem 1.5.3.

> **Theorem 1.6.4.** *If A is an n × n matrix, then the following are equivalent.*
>
> (a) *A is invertible.*
> (b) *Ax = 0 has only the trivial solution.*
> (c) *The reduced row-echelon form of A is I_n.*
> (d) *A is expressible as a product of elementary matrices.*
> (e) *Ax = b is consistent for every n × 1 matrix b.*
> (f) *Ax = b has exactly one solution for every n × 1 matrix b.*

Proof. Since we proved in Theorem 1.5.3 that (a), (b), (c), and (d) are equivalent, it will be sufficient to prove that $(a) \Rightarrow (f) \Rightarrow (e) \Rightarrow (a)$.

$(a) \Rightarrow (f)$: This was already proved in Theorem 1.6.2.

$(f) \Rightarrow (e)$: This is self-evident: If $A\mathbf{x} = \mathbf{b}$ has exactly one solution for every $n \times 1$ matrix $\mathbf{b}$, then $A\mathbf{x} = \mathbf{b}$ is consistent for every $n \times 1$ matrix $\mathbf{b}$.

$(e) \Rightarrow (a)$: If the system $A\mathbf{x} = \mathbf{b}$ is consistent for every $n \times 1$ matrix $\mathbf{b}$, then in particular, the systems

$$A\mathbf{x} = \begin{bmatrix} 1 \\ 0 \\ 0 \\ \vdots \\ 0 \end{bmatrix}, \quad A\mathbf{x} = \begin{bmatrix} 0 \\ 1 \\ 0 \\ \vdots \\ 0 \end{bmatrix}, \quad \dots, \quad A\mathbf{x} = \begin{bmatrix} 0 \\ 0 \\ 0 \\ \vdots \\ 1 \end{bmatrix}$$

are consistent. Let $\mathbf{x}_1, \mathbf{x}_2, \dots, \mathbf{x}_n$ be solutions of the respective systems, and let us form an $n \times n$ matrix C having these solutions as columns. Thus, C has the form

$$C = [\mathbf{x}_1 \mid \mathbf{x}_2 \mid \cdots \mid \mathbf{x}_n]$$

As discussed in Section 1.3, the successive columns of the product AC will be

$$A\mathbf{x}_1, A\mathbf{x}_2, \dots, A\mathbf{x}_n$$

Thus,

$$AC = [A\mathbf{x}_1 \mid A\mathbf{x}_2 \mid \cdots \mid A\mathbf{x}_n] = \begin{bmatrix} 1 & 0 & \cdots & 0 \\ 0 & 1 & \cdots & 0 \\ 0 & 0 & \cdots & 0 \\ \vdots & \vdots & & \vdots \\ 0 & 0 & \cdots & 1 \end{bmatrix} = I$$

By part (b) of Theorem 1.6.3 it follows that $C = A^{-1}$. Thus, A is invertible. ∎

We know from earlier work that invertible matrix factors produce an invertible product. The following theorem looks at the converse: It shows that if the product of square matrices is invertible, then the factors themselves must be invertible.

Theorem 1.6.5. *Let A and B be square matrices of the same size. If AB is invertible, then A and B must also be invertible.*

In our later work the following fundamental problem will occur frequently in various contexts.

A Fundamental Problem. Let A be a fixed $m \times n$ matrix. Find all $m \times 1$ matrices **b** such that the system of equations $A\mathbf{x} = \mathbf{b}$ is consistent.

If A is an invertible matrix, Theorem 1.6.2 completely solves this problem by asserting that for *every* $m \times 1$ matrix **b**, the linear system $A\mathbf{x} = \mathbf{b}$ has the unique solution $\mathbf{x} = A^{-1}\mathbf{b}$. If A is not square, or if A is square but not invertible, then Theorem 1.6.2 does not apply. In these cases the matrix **b** must satisfy certain conditions in order for $A\mathbf{x} = \mathbf{b}$ to be consistent. The following example illustrates how Gaussian elimination can be used to determine such conditions.

Example 3 What conditions must b_1, b_2, and b_3 satisfy in order for the system of equations

$$x_1 + x_2 + 2x_3 = b_1$$
$$x_1 \quad + x_3 = b_2$$
$$2x_1 + x_2 + 3x_3 = b_3$$

to be consistent?

Solution. The augmented matrix is

$$\begin{bmatrix} 1 & 1 & 2 & b_1 \\ 1 & 0 & 1 & b_2 \\ 2 & 1 & 3 & b_3 \end{bmatrix}$$

which can be reduced to row-echelon form as follows.

$$\begin{bmatrix} 1 & 1 & 2 & b_1 \\ 0 & -1 & -1 & b_2 - b_1 \\ 0 & -1 & -1 & b_3 - 2b_1 \end{bmatrix}$$

−1 times the first row was added to the second and −2 times the first row was added to the third.

$$\begin{bmatrix} 1 & 1 & 2 & b_1 \\ 0 & 1 & 1 & b_1 - b_2 \\ 0 & -1 & -1 & b_3 - 2b_1 \end{bmatrix}$$

The second row was multiplied by −1.

$$\begin{bmatrix} 1 & 1 & 1 & b_2 \\ 0 & 1 & 1 & b_1 - b_2 \\ 0 & 0 & 0 & b_3 - b_2 - b_1 \end{bmatrix}$$

The second row was added to the third.

It is now evident from the third row in the matrix that the system has a solution if and only if b_1, b_2, and b_3 satisfy the condition

$$b_3 - b_2 - b_1 = 0 \qquad \text{or} \qquad b_3 = b_1 + b_2$$

To express this condition another way, $A\mathbf{x} = \mathbf{b}$ is consistent if and only if $\mathbf{b}$ is a matrix of the form

$$\mathbf{b} = \begin{bmatrix} b_1 \\ b_2 \\ b_1 + b_2 \end{bmatrix}$$

where b_1 and b_2 are arbitrary. ▲

Example 4 What conditions must b_1, b_2, and b_3 satisfy in order for the system of equations

$$\begin{aligned} x_1 + 2x_2 + 3x_3 &= b_1 \\ 2x_1 + 5x_2 + 3x_3 &= b_2 \\ x_1 \qquad\qquad + 8x_3 &= b_3 \end{aligned}$$

to be consistent?

Solution. The augmented matrix is

$$\begin{bmatrix} 1 & 2 & 3 & b_1 \\ 2 & 5 & 3 & b_2 \\ 1 & 0 & 8 & b_3 \end{bmatrix}$$

Reducing this to reduced row-echelon form yields (verify)

$$\begin{bmatrix} 1 & 0 & 0 & -40b_1 + 16b_2 + 9b_3 \\ 0 & 1 & 0 & 13b_1 - 5b_2 - 3b_3 \\ 0 & 0 & 1 & 5b_1 - 2b_2 - b_3 \end{bmatrix} \qquad (2)$$

In this case there are no restrictions on b_1, b_2, and b_3; that is, the given system $A\mathbf{x} = \mathbf{b}$ has the unique solution

$$x_1 = -40b_1 + 16b_2 + 9b_3, \quad x_2 = 13b_1 - 5b_2 - 3b_3, \quad x_3 = 5b_1 - 2b_2 - b_3 \quad (3)$$

for all $\mathbf{b}$. ▲

REMARK. Because the system $A\mathbf{x} = \mathbf{b}$ in the preceding example is consistent for all $\mathbf{b}$, it follows from Theorem 1.6.4 that A is invertible. We leave it for the reader to verify that the formulas in (3) can also be obtained by calculating $\mathbf{x} = A^{-1}\mathbf{b}$.

EXERCISE SET 1.6

In Exercises 1–8 solve the system by inverting the coefficient matrix and using Theorem 1.6.2.

1. $\begin{aligned} x_1 + x_2 &= 2 \\ 5x_1 + 6x_2 &= 9 \end{aligned}$

2. $\begin{aligned} 4x_1 - 3x_2 &= -3 \\ 2x_1 - 5x_2 &= 9 \end{aligned}$

3. $\begin{aligned} x_1 + 3x_2 + x_3 &= 4 \\ 2x_1 + 2x_2 + x_3 &= -1 \\ 2x_1 + 3x_2 + x_3 &= 3 \end{aligned}$

4. $5x_1 + 3x_2 + 2x_3 = 4$
$3x_1 + 3x_2 + 2x_3 = 2$
$x_2 + x_3 = 5$

5. $x + y + z = 5$
$x + y - 4z = 10$
$-4x + y + z = 0$

6. $-x - 2y - 3z = 0$
$w + x + 4y + 4z = 7$
$w + 3x + 7y + 9z = 4$
$-w - 2x - 4y - 6z = 6$

7. $3x_1 + 5x_2 = b_1$
$x_1 + 2x_2 = b_2$

8. $x_1 + 2x_2 + 3x_3 = b_1$
$2x_1 + 5x_2 + 5x_3 = b_2$
$3x_1 + 5x_2 + 8x_3 = b_3$

9. Solve the following general system by inverting the coefficient matrix and using Theorem 1.6.2.

$$x_1 + 2x_2 + x_3 = b_1$$
$$x_1 - x_2 + x_3 = b_2$$
$$x_1 + x_2 = b_3$$

Use the resulting formulas to find the solution if
(a) $b_1 = -1, \quad b_2 = 3, \quad b_3 = 4$ (b) $b_1 = 5, \quad b_2 = 0, \quad b_3 = 0$ (c) $b_1 = -1, \quad b_2 = -1, \quad b_3 = 3$

10. Solve the three systems in Exercise 9 using the method of Example 2.

In Exercises 11–14 use the method of Example 2 to solve the systems in all the parts simultaneously.

11. $x_1 - 5x_2 = b_1$
$3x_1 + 2x_2 = b_2$

(a) $b_1 = 1, \quad b_2 = 4$
(b) $b_1 = -2, \quad b_2 = 5$

12. $-x_1 + 4x_2 + x_3 = b_1$
$x_1 + 9x_2 - 2x_3 = b_2$
$6x_1 + 4x_2 - 8x_3 = b_3$

(a) $b_1 = 0, \quad b_2 = 1, \quad b_3 = 0$
(b) $b_1 = -3, \quad b_2 = 4, \quad b_3 = -5$

13. $4x_1 - 7x_2 = b_1$
$x_1 + 2x_2 = b_2$

(a) $b_1 = 0, \quad b_2 = 1$
(b) $b_1 = -4, \quad b_2 = 6$
(c) $b_1 = -1, \quad b_2 = 3$
(d) $b_1 = -5, \quad b_2 = 1$

14. $x_1 + 3x_2 + 5x_3 = b_1$
$-x_1 - 2x_2 = b_2$
$2x_1 + 5x_2 + 4x_3 = b_3$

(a) $b_1 = 1, \quad b_2 = 0, \quad b_3 = -1$
(b) $b_1 = 0, \quad b_2 = 1, \quad b_3 = 1$
(c) $b_1 = -1, \quad b_2 = -1, \quad b_3 = 0$

15. The method of Example 2 can be used for linear systems with infinitely many solutions. Use that method to solve the systems in both parts at the same time.
(a) $x_1 - 2x_2 + x_3 = -2$
$2x_1 - 5x_2 + x_3 = 1$
$3x_1 - 7x_2 + 2x_3 = -1$
(b) $x_1 - 2x_2 + x_3 = 1$
$2x_1 - 5x_2 + x_3 = -1$
$3x_1 - 7x_2 + 2x_3 = 0$

In Exercises 16–19 find conditions that b's must satisfy for the system to be consistent.

16. $6x_1 - 4x_2 = b_1$
$3x_1 - 2x_2 = b_2$

17. $x_1 - 2x_2 + 5x_3 = b_1$
$4x_1 - 5x_2 + 8x_3 = b_2$
$-3x_1 + 3x_2 - 3x_3 = b_3$

18. $x_1 - 2x_2 - x_3 = b_1$
$-4x_1 + 5x_2 + 2x_3 = b_2$
$-4x_1 + 7x_2 + 4x_3 = b_3$

19. $x_1 - x_2 + 3x_3 + 2x_1 = b_1$
$-2x_1 + x_2 + 5x_3 + x_1 = b_2$
$-3x_1 + 2x_2 + 2x_3 - x_1 = b_3$
$4x_1 - 3x_2 + x_3 + 3x_1 = b_4$

20. Consider the matrices

$$A = \begin{bmatrix} 2 & 1 & 2 \\ 2 & 2 & -2 \\ 3 & 1 & 1 \end{bmatrix} \quad \text{and} \quad \mathbf{x} = \begin{bmatrix} x_1 \\ x_2 \\ x_3 \end{bmatrix}$$

(a) Show that the equation $A\mathbf{x} = \mathbf{x}$ can be rewritten as $(A - I)\mathbf{x} = \mathbf{0}$ and use this result to solve $A\mathbf{x} = \mathbf{x}$ for $\mathbf{x}$.

(b) Solve $A\mathbf{x} = 4\mathbf{x}$.

21. Solve the following matrix equation for X.

$$\begin{bmatrix} 1 & -1 & 1 \\ 2 & 3 & 0 \\ 0 & 2 & -1 \end{bmatrix} X = \begin{bmatrix} 2 & -1 & 5 & 7 & 8 \\ 4 & 0 & -3 & 0 & 1 \\ 3 & 5 & -7 & 2 & 1 \end{bmatrix}$$

22. In each part determine whether the homogeneous system has a nontrivial solution (without using pencil and paper); then state whether the given matrix is invertible.

(a)
$$\begin{aligned} 2x_1 + x_2 - 3x_3 + x_4 &= 0 \\ 5x_2 + 4x_3 + 3x_4 &= 0 \\ x_3 + 2x_4 &= 0 \\ 3x_4 &= 0 \end{aligned} \qquad \begin{bmatrix} 2 & 1 & -3 & 1 \\ 0 & 5 & 4 & 3 \\ 0 & 0 & 1 & 2 \\ 0 & 0 & 0 & 3 \end{bmatrix}$$

(b)
$$\begin{aligned} 5x_1 + x_2 + 4x_3 + x_4 &= 0 \\ 2x_3 - x_4 &= 0 \\ x_3 + x_4 &= 0 \\ 7x_4 &= 0 \end{aligned} \qquad \begin{bmatrix} 5 & 1 & 4 & 1 \\ 0 & 0 & 2 & -1 \\ 0 & 0 & 1 & 1 \\ 0 & 0 & 0 & 7 \end{bmatrix}$$

23. Let $A\mathbf{x} = \mathbf{0}$ be a homogeneous system of n linear equations in n unknowns that has only the trivial solution. Show that if k is any positive integer, then the system $A^k\mathbf{x} = \mathbf{0}$ also has only the trivial solution.

24. Let $A\mathbf{x} = \mathbf{0}$ be a homogeneous system of n linear equations in n unknowns, and let Q be an invertible $n \times n$ matrix. Show that $A\mathbf{x} = \mathbf{0}$ has just the trivial solution if and only if $(QA)\mathbf{x} = \mathbf{0}$ has just the trivial solution.

25. Let $A\mathbf{x} = \mathbf{b}$ be any consistent system of linear equations, and let $\mathbf{x}_1$ be a fixed solution. Show that every solution to the system can be written in the form $\mathbf{x} = \mathbf{x}_1 + \mathbf{x}_0$, where $\mathbf{x}_0$ is a solution to $A\mathbf{x} = \mathbf{0}$. Show also that every matrix of this form is a solution.

26. Use part (*a*) of Theorem 1.6.3 to prove part (*b*).

1.7 DIAGONAL, TRIANGULAR, AND SYMMETRIC MATRICES

In this section we shall consider certain classes of matrices that have special forms. The matrices that we study in this section are among the most important in linear algebra, and they will arise in many different settings throughout this text.

DIAGONAL MATRICES

A square matrix in which all of the entries off the main diagonal are zero is called a *diagonal matrix*; some examples are

$$\begin{bmatrix} 2 & 0 \\ 0 & -5 \end{bmatrix} \qquad \begin{bmatrix} 1 & 0 & 0 \\ 0 & 1 & 0 \\ 0 & 0 & 1 \end{bmatrix} \qquad \begin{bmatrix} 6 & 0 & 0 & 0 \\ 0 & -4 & 0 & 0 \\ 0 & 0 & 0 & 0 \\ 0 & 0 & 0 & 8 \end{bmatrix}$$

A general $n \times n$ diagonal matrix D can be written as

$$D = \begin{bmatrix} d_1 & 0 & \cdots & 0 \\ 0 & d_2 & \cdots & 0 \\ \vdots & \vdots & & \vdots \\ 0 & 0 & \cdots & d_n \end{bmatrix} \tag{1}$$

A diagonal matrix is invertible if and only if all of its diagonal entries are nonzero; in this case the inverse of (1) is

$$D^{-1} = \begin{bmatrix} 1/d_1 & 0 & \cdots & 0 \\ 0 & 1/d_2 & \cdots & 0 \\ \vdots & \vdots & & \vdots \\ 0 & 0 & \cdots & 1/d_n \end{bmatrix}$$

The reader should verify that $DD^{-1} = D^{-1}D = I$.

Powers of diagonal matrices are easy to compute; we leave it for the reader to verify that if D is the diagonal matrix (1) and k is a positive integer, then

$$D^k = \begin{bmatrix} d_1^{\,k} & 0 & \cdots & 0 \\ 0 & d_2^{\,k} & \cdots & 0 \\ \vdots & \vdots & & \vdots \\ 0 & 0 & \cdots & d_n^{\,k} \end{bmatrix}$$

Example 1   If

$$A = \begin{bmatrix} 1 & 0 & 0 \\ 0 & -3 & 0 \\ 0 & 0 & 2 \end{bmatrix}$$

then

$$A^{-1} = \begin{bmatrix} 1 & 0 & 0 \\ 0 & -\tfrac{1}{3} & 0 \\ 0 & 0 & \tfrac{1}{2} \end{bmatrix} \quad A^5 = \begin{bmatrix} 1 & 0 & 0 \\ 0 & -243 & 0 \\ 0 & 0 & 32 \end{bmatrix} \quad A^{-5} = \begin{bmatrix} 1 & 0 & 0 \\ 0 & -\tfrac{1}{243} & 0 \\ 0 & 0 & \tfrac{1}{32} \end{bmatrix} \; \blacktriangle$$

Matrix products that involve diagonal factors are especially easy to compute. For example,

$$\begin{bmatrix} d_1 & 0 & 0 \\ 0 & d_2 & 0 \\ 0 & 0 & d_3 \end{bmatrix} \begin{bmatrix} a_{11} & a_{12} & a_{13} & a_{14} \\ a_{21} & a_{22} & a_{23} & a_{24} \\ a_{31} & a_{32} & a_{33} & a_{34} \end{bmatrix} = \begin{bmatrix} d_1 a_{11} & d_1 a_{12} & d_1 a_{13} & d_1 a_{14} \\ d_2 a_{21} & d_2 a_{22} & d_2 a_{23} & d_2 a_{24} \\ d_3 a_{31} & d_3 a_{32} & d_3 a_{33} & d_3 a_{34} \end{bmatrix}$$

$$\begin{bmatrix} a_{11} & a_{12} & a_{13} \\ a_{21} & a_{22} & a_{23} \\ a_{31} & a_{32} & a_{33} \\ a_{41} & a_{42} & a_{43} \end{bmatrix} \begin{bmatrix} d_1 & 0 & 0 \\ 0 & d_2 & 0 \\ 0 & 0 & d_3 \end{bmatrix} = \begin{bmatrix} d_1 a_{11} & d_2 a_{12} & d_3 a_{13} \\ d_1 a_{21} & d_2 a_{22} & d_3 a_{23} \\ d_1 a_{31} & d_2 a_{32} & d_3 a_{33} \\ d_1 a_{41} & d_2 a_{42} & d_3 a_{43} \end{bmatrix}$$

In words, *to multiply a matrix A on the left by a diagonal matrix D, one can multiply successive rows of A by the successive diagonal entries of D, and to multiply A on the right by D one can multiply successive columns of A by the successive diagonal entries of D.*

TRIANGULAR MATRICES

A square matrix in which all the entries above the main diagonal are zero is called *lower triangular*, and a square matrix in which all the entries below the main diagonal are zero is called *upper triangular*. A matrix that is either upper triangular or lower triangular is called *triangular*.

Example 2

$$\begin{bmatrix} a_{11} & a_{12} & a_{13} & a_{14} \\ 0 & a_{22} & a_{23} & a_{24} \\ 0 & 0 & a_{33} & a_{34} \\ 0 & 0 & 0 & a_{44} \end{bmatrix} \qquad \begin{bmatrix} a_{11} & 0 & 0 & 0 \\ a_{21} & a_{22} & 0 & 0 \\ a_{31} & a_{32} & a_{33} & 0 \\ a_{41} & a_{42} & a_{43} & a_{44} \end{bmatrix}$$

A general 4 × 4 upper triangular matrix

A general 4 × 4 lower triangular matrix

REMARK. Observe that diagonal matrices are both upper triangular and lower triangular since they have zeros below and above the main diagonal. Observe also that a *square* matrix in row-echelon form is upper triangular since it has zeros below the main diagonal.

The following are four useful characterizations of triangular matrices. The reader will find it instructive to verify that the matrices in Example 2 have the stated properties.

- A square matrix $A = [a_{ij}]$ is upper triangular if and only if the ith row starts with at least $i - 1$ zeros.
- A square matrix $A = [a_{ij}]$ is lower triangular if and only if the jth column starts with at least $j - 1$ zeros.
- A square matrix $A = [a_{ij}]$ is upper triangular if and only if $a_{ij} = 0$ for $i > j$.
- A square matrix $A = [a_{ij}]$ is lower triangular if and only if $a_{ij} = 0$ for $i < j$.

The following theorem lists some of the basic properties of triangular matrices.

Theorem 1.7.1

(a) *The transpose of a lower triangular matrix is upper triangular, and the transpose of an upper triangular matrix is lower triangular.*

(b) *The product of lower triangular matrices is lower triangular, and the product of upper triangular matrices is upper triangular.*

(c) *A triangular matrix is invertible if and only if its diagonal entries are all nonzero.*

(d) *The inverse of an invertible lower triangular matrix is lower triangular, and the inverse of an invertible upper triangular matrix is upper triangular.*

Part (a) is evident from the fact that transposing a square matrix can be accomplished by reflecting the entries about the main diagonal; we omit the formal proof. We will prove (b), but we will defer the proofs of (c) and (d) to the next chapter, where we will have the tools to prove those results more efficiently.

Proof (b). We will prove the result for lower triangular matrices; the proof for upper triangular matrices is similar. Let $A = [a_{ij}]$ and $B = [b_{ij}]$ be lower triangular $n \times n$ matrices, and let $C = [c_{ij}]$ be the product $C = AB$. From the remark preceding this theorem, we can prove that C is lower triangular by showing that $c_{ij} = 0$ for $i < j$. But from the definition of matrix multiplication,

$$c_{ij} = a_{i1}b_{1j} + a_{i2}b_{2j} + \cdots + a_{in}b_{nj}$$

If we assume that $i < j$, then the terms in this expression can be grouped as follows:

$$c_{ij} = \underbrace{a_{i1}b_{1j} + a_{i2}b_{2j} + \cdots + a_{ij-1}b_{j-1j}}_{\substack{\text{Terms in which the row} \\ \text{number of } b \text{ is less than the} \\ \text{column number of } b}} + \underbrace{a_{ij}b_{jj} + \cdots + a_{in}b_{nj}}_{\substack{\text{Terms in which the row} \\ \text{number of } a \text{ is less than} \\ \text{the column number of } a}}$$

In the first grouping all of the b factors are zero since B is lower triangular, and in the second grouping all of the a factors are zero since A is lower triangular. Thus, $c_{ij} = 0$, which is what we wanted to prove. ▌

Example 3 Consider the upper triangular matrices

$$A = \begin{bmatrix} 1 & 3 & -1 \\ 0 & 2 & 4 \\ 0 & 0 & 5 \end{bmatrix} \qquad B = \begin{bmatrix} 3 & -2 & 2 \\ 0 & 0 & -1 \\ 0 & 0 & 1 \end{bmatrix}$$

The matrix A is invertible, since its diagonal entries are nonzero, but the matrix B is not. We leave it for the reader to calculate the inverse of A by the method of Section 1.5 and show that

$$A^{-1} = \begin{bmatrix} 1 & -\frac{3}{2} & \frac{7}{5} \\ 0 & \frac{1}{2} & -\frac{2}{5} \\ 0 & 0 & \frac{1}{5} \end{bmatrix}$$

This inverse is upper triangular, as guaranteed by part (d) of Theorem 1.7.1. We also leave it for the reader to check that the product AB is

$$AB = \begin{bmatrix} 3 & -2 & -2 \\ 0 & 0 & 2 \\ 0 & 0 & 5 \end{bmatrix}$$

This product is upper triangular, as guaranteed by part (b) of Theorem 1.7.1. ▲

SYMMETRIC MATRICES

A square matrix A is called **symmetric** if $A = A^T$.

Example 4 The following matrices are symmetric, since each is equal to its own transpose (verify).

$$\begin{bmatrix} 7 & -3 \\ -3 & 5 \end{bmatrix} \quad \begin{bmatrix} 1 & 4 & 5 \\ 4 & -3 & 0 \\ 5 & 0 & 7 \end{bmatrix} \quad \begin{bmatrix} d_1 & 0 & 0 & 0 \\ 0 & d_2 & 0 & 0 \\ 0 & 0 & d_3 & 0 \\ 0 & 0 & 0 & d_4 \end{bmatrix} \quad ▲$$

It is easy to recognize symmetric matrices by inspection: The entries on the main diagonal may be arbitrary, but "mirror images" of entries across the main diagonal must be equal (Figure 1).

Figure 1

This follows from the fact that transposing a square matrix can be accomplished by interchanging entries that are symmetrically positioned about the main diagonal. Expressed in terms of the individual entries, a matrix $A = [a_{ij}]$ is symmetric if and only if $a_{ij} = a_{ji}$ for all values of i and j. As illustrated in Example 4, all diagonal matrices are symmetric.

The following theorem lists the main algebraic properties of symmetric matrices. The proofs are direct consequences of Theorem 1.4.9 and are left as exercises.

Theorem 1.7.2. *If A and B are symmetric matrices with the same size, and if k is any scalar, then:*

(a) A^T *is symmetric.*
(b) $A + B$ *and* $A - B$ *are symmetric.*
(c) kA *is symmetric.*

REMARK. It is not true, in general, that the product of symmetric matrices is symmetric. To see why this is so, let A and B be symmetric matrices with the same size. Then from part (d) of Theorem 1.4.9 and the symmetry we have

$$(AB)^T = B^T A^T = BA$$

Since AB and BA are not usually equal, it follows that AB will not usually be symmetric. However, in the special case where $AB = BA$, the product AB will be symmetric. If A

and B are matrices such that $AB = BA$, then we say that A and B **commute**. In summary: *The product of two symmetric matrices is symmetric if and only if the matrices commute.*

Example 5 The first of the following equations shows a product of symmetric matrices that *is not* symmetric, and the second shows a product of symmetric matrices that *is* symmetric. We conclude that the factors in the first equation do not commute, but those in the second equation do. We leave it for the reader to verify that this is so.

$$\begin{bmatrix} 1 & 2 \\ 2 & 3 \end{bmatrix} \begin{bmatrix} -4 & 1 \\ 1 & 0 \end{bmatrix} = \begin{bmatrix} -2 & 1 \\ -5 & 2 \end{bmatrix}$$

$$\begin{bmatrix} 1 & 2 \\ 2 & 3 \end{bmatrix} \begin{bmatrix} -4 & 3 \\ 3 & -1 \end{bmatrix} = \begin{bmatrix} 2 & 1 \\ 1 & 3 \end{bmatrix} \blacktriangle$$

In general, a symmetric matrix need not be invertible; for example, a square zero matrix is symmetric, but not invertible. However, if a symmetric matrix is invertible, then that inverse is also symmetric.

Theorem 1.7.3. *If A is an invertible symmetric matrix, then A^{-1} is symmetric.*

Proof. Assume that A is symmetric and invertible. From Theorem 1.4.10 and the fact that $A = A^T$ we have

$$(A^{-1})^T = (A^T)^{-1} = A^{-1}$$

which proves that A^{-1} is symmetric. ▮

MATRICES OF THE FORM AA^T AND A^TA

Matrix products of the form AA^T and A^TA arise in a variety of applications. If A is an $m \times n$ matrix, then A^T is an $n \times m$ matrix, so the products AA^T and A^TA are both square matrices; the matrix AA^T has size $m \times m$ and the matrix A^TA has size $n \times n$. Such products are always symmetric since

$$(AA^T)^T = (A^T)^T A^T = AA^T \qquad \text{and} \qquad (A^TA)^T = A^T(A^T)^T = A^TA$$

Example 6 Let A be the 2×3 matrix

$$A = \begin{bmatrix} 1 & -2 & 4 \\ 3 & 0 & -5 \end{bmatrix}$$

Then

$$A^TA = \begin{bmatrix} 1 & 3 \\ -2 & 0 \\ 4 & -5 \end{bmatrix} \begin{bmatrix} 1 & -2 & 4 \\ 3 & 0 & -5 \end{bmatrix} = \begin{bmatrix} 10 & -2 & -11 \\ -2 & 4 & -8 \\ -11 & -8 & 41 \end{bmatrix}$$

$$AA^T = \begin{bmatrix} 1 & -2 & 4 \\ 3 & 0 & -5 \end{bmatrix} \begin{bmatrix} 1 & 3 \\ -2 & 0 \\ 4 & -5 \end{bmatrix} = \begin{bmatrix} 21 & -17 \\ -17 & 34 \end{bmatrix}$$

Observe that A^TA and AA^T are symmetric as expected. ▲

Later in this text, we will obtain general conditions on A under which AA^T and A^TA are invertible. However, in the special case where A is *square* we have the following result.

Theorem 1.7.4. *If A is an invertible matrix, then AA^T and A^TA are also invertible.*

Proof. Since A is invertible, so is A^T by Theorem 1.4.10. Thus, AA^T and A^TA are invertible, since they are the products of invertible matrices. ▪

EXERCISE SET 1.7

1. Determine whether the matrix is invertible; if so, find the inverse by inspection.

(a) $\begin{bmatrix} 2 & 0 \\ 0 & -5 \end{bmatrix}$
(b) $\begin{bmatrix} 4 & 0 & 0 \\ 0 & 0 & 0 \\ 0 & 0 & 5 \end{bmatrix}$
(c) $\begin{bmatrix} -1 & 0 & 0 \\ 0 & 2 & 0 \\ 0 & 0 & \frac{1}{3} \end{bmatrix}$

2. Compute the product by inspection.

(a) $\begin{bmatrix} 3 & 0 & 0 \\ 0 & -1 & 0 \\ 0 & 0 & 2 \end{bmatrix}\begin{bmatrix} 2 & 1 \\ -4 & 1 \\ 2 & 5 \end{bmatrix}$
(b) $\begin{bmatrix} 2 & 0 & 0 \\ 0 & -1 & 0 \\ 0 & 0 & 4 \end{bmatrix}\begin{bmatrix} 4 & -1 & 3 \\ 1 & 2 & 0 \\ -5 & 1 & -2 \end{bmatrix}\begin{bmatrix} -3 & 0 & 0 \\ 0 & 5 & 0 \\ 0 & 0 & 2 \end{bmatrix}$

3. Find A^2, A^{-2}, and A^{-k} by inspection.

(a) $A = \begin{bmatrix} 1 & 0 \\ 0 & -2 \end{bmatrix}$
(b) $A = \begin{bmatrix} \frac{1}{2} & 0 & 0 \\ 0 & \frac{1}{3} & 0 \\ 0 & 0 & \frac{1}{4} \end{bmatrix}$

4. Which of the following matrices are symmetric?

(a) $\begin{bmatrix} 2 & -1 \\ 1 & 2 \end{bmatrix}$
(b) $\begin{bmatrix} 3 & 4 \\ 4 & 0 \end{bmatrix}$
(c) $\begin{bmatrix} 2 & -1 & 3 \\ -1 & 5 & 1 \\ 3 & 1 & 7 \end{bmatrix}$
(d) $\begin{bmatrix} 0 & 0 & 1 \\ 0 & 2 & 0 \\ 3 & 0 & 0 \end{bmatrix}$

5. By inspection, determine whether the given triangular matrix is invertible.

(a) $\begin{bmatrix} -1 & 2 & 4 \\ 0 & 3 & 0 \\ 0 & 0 & 5 \end{bmatrix}$
(b) $\begin{bmatrix} 0 & 1 & -2 & 5 \\ 0 & 1 & 5 & 6 \\ 0 & 0 & -3 & 1 \\ 0 & 0 & 0 & 5 \end{bmatrix}$

6. Find all values of a, b, and c for which A is symmetric.

$$A = \begin{bmatrix} 2 & a - 2b + 2c & 2a + b + c \\ 3 & 5 & a + c \\ 0 & -2 & 7 \end{bmatrix}$$

7. Find all values of a and b for which A and B are both not invertible.

$$A = \begin{bmatrix} a + b - 1 & 0 \\ 0 & 3 \end{bmatrix}, \quad B = \begin{bmatrix} 5 & 0 \\ 0 & 2a - 3b - 7 \end{bmatrix}$$

8. Use the given equation to determine by inspection whether the matrices on the left commute.

(a) $\begin{bmatrix} 1 & -3 \\ -3 & 2 \end{bmatrix}\begin{bmatrix} 4 & 1 \\ 1 & 2 \end{bmatrix} = \begin{bmatrix} 1 & -5 \\ -10 & 1 \end{bmatrix}$ (b) $\begin{bmatrix} 2 & -1 \\ -1 & 3 \end{bmatrix}\begin{bmatrix} 3 & 2 \\ 2 & 1 \end{bmatrix} = \begin{bmatrix} 4 & 3 \\ 3 & 1 \end{bmatrix}$

9. Show that A and B commute if $a - d = 7b$.

$$A = \begin{bmatrix} 2 & 1 \\ 1 & -5 \end{bmatrix}, \qquad B = \begin{bmatrix} a & b \\ b & d \end{bmatrix}$$

10. Find a diagonal matrix A that satisfies

(a) $A^5 = \begin{bmatrix} 1 & 0 & 0 \\ 0 & -1 & 0 \\ 0 & 0 & -1 \end{bmatrix}$ (b) $A^{-2} = \begin{bmatrix} 9 & 0 & 0 \\ 0 & 4 & 0 \\ 0 & 0 & 1 \end{bmatrix}$

11. (a) Factor A into the form $A = BD$, where D is a diagonal matrix.

$$A = \begin{bmatrix} 3a_{11} & 5a_{12} & 7a_{13} \\ 3a_{21} & 5a_{22} & 7a_{23} \\ 3a_{31} & 5a_{32} & 7a_{33} \end{bmatrix}$$

(b) Is your factorization the only one possible? Explain.

12. Verify Theorem 1.7.1*b* for the product AB, where

$$A = \begin{bmatrix} -1 & 2 & 5 \\ 0 & 1 & 3 \\ 0 & 0 & -4 \end{bmatrix}, \qquad B = \begin{bmatrix} 2 & -8 & 0 \\ 0 & 2 & 1 \\ 0 & 0 & 3 \end{bmatrix}$$

13. Verify Theorem 1.7.1*d* for the matrices A and B in Exercise 12.

14. Verify Theorem 1.7.3 for the given matrix A.

(a) $A = \begin{bmatrix} 2 & -1 \\ -1 & 3 \end{bmatrix}$ (b) $A = \begin{bmatrix} 1 & -2 & 3 \\ -2 & 1 & -7 \\ 3 & -7 & 4 \end{bmatrix}$

15. Let A be a symmetric matrix.
(a) Show that A^2 is symmetric.
(b) Show that $2A^2 - 3A + I$ is symmetric.

16. Let A be a symmetric matrix.
(a) Show that A^k is symmetric if k is any nonnegative integer.
(b) If $p(x)$ is a polynomial, is $p(A)$ necessarily symmetric? Explain.

17. Let A be an upper triangular matrix and let $p(x)$ be a polynomial. Is $p(A)$ necessarily upper triangular? Explain.

18. Prove: If $A^TA = A$, then A is symmetric and $A = A^2$.

19. What is the maximum number of distinct entries that an $n \times n$ symmetric matrix can have?

20. Let $A = [a_{ij}]$ be an $n \times n$ matrix. Determine whether A is symmetric.
(a) $a_{ij} = i^2 + j^2$ (b) $a_{ij} = i^2 - j^2$
(c) $a_{ij} = 2i + 2j$ (d) $a_{ij} = 2i^2 + 2j^3$

21. Based on your experience with Exercise 20, devise a general test that can be applied to a formula for a_{ij} to determine whether $A = [a_{ij}]$ is symmetric.

22. A square matrix A is called **skew-symmetric** if $A^T = -A$. Prove:
   (a) If A is an invertible skew-symmetric matrix, then A^{-1} is skew-symmetric.
   (b) If A and B are skew-symmetric, then so are A^T, $A + B$, $A - B$, and kA for any scalar k.
   (c) Every square matrix can be expressed as the sum of a symmetric matrix and a skew-symmetric matrix.

23. We showed in the text that the product of symmetric matrices is symmetric if and only if the matrices commute. Is the product of commuting skew-symmetric matrices skew-symmetric? Explain. [**Note.** See Exercise 22 for terminology.]

24. If the $n \times n$ matrix A can be expressed as $A = LU$, where L is a lower triangular matrix and U is an upper triangular matrix, then the linear system $A\mathbf{x} = \mathbf{b}$ can be expressed as $LU\mathbf{x} = \mathbf{b}$ and can be solved in two steps:

   Step 1. Let $U\mathbf{x} = \mathbf{y}$, so that $LU\mathbf{x} = \mathbf{b}$ can be expressed as $L\mathbf{y} = \mathbf{b}$. Solve this system.
   Step 2. Solve the system $U\mathbf{x} = \mathbf{y}$ for $\mathbf{x}$.

   In each part use this two-step method to solve the given system.

   (a) $\begin{bmatrix} 1 & 0 & 0 \\ -2 & 3 & 0 \\ 2 & 4 & 1 \end{bmatrix} \begin{bmatrix} 2 & -1 & 3 \\ 0 & 1 & 2 \\ 0 & 0 & 4 \end{bmatrix} \begin{bmatrix} x_1 \\ x_2 \\ x_3 \end{bmatrix} = \begin{bmatrix} 1 \\ -2 \\ 0 \end{bmatrix}$

   (b) $\begin{bmatrix} 2 & 0 & 0 \\ 4 & 1 & 0 \\ -3 & -2 & 3 \end{bmatrix} \begin{bmatrix} 3 & -5 & 2 \\ 0 & 4 & 1 \\ 0 & 0 & 2 \end{bmatrix} \begin{bmatrix} x_1 \\ x_2 \\ x_3 \end{bmatrix} = \begin{bmatrix} 4 \\ -5 \\ 2 \end{bmatrix}$

SUPPLEMENTARY EXERCISES

1. Use Gauss–Jordan elimination to solve for x' and y' in terms of x and y.

   $x = \frac{3}{5}x' - \frac{4}{5}y'$
   $y = \frac{4}{5}x' + \frac{3}{5}y'$

2. Use Gauss–Jordan elimination to solve for x' and y' in terms of x and y.

   $x = x' \cos \theta - y' \sin \theta$
   $y = x' \sin \theta + y' \cos \theta$

3. Find a homogeneous linear system with two equations that are not multiples of one another and such that

   $x_1 = 1, \quad x_2 = -1, \quad x_3 = 1, \quad x_4 = 2$

   and

   $x_1 = 2, \quad x_2 = 0, \quad x_3 = 3, \quad x_4 = -1$

   are solutions of the system.

4. A box containing pennies, nickels, and dimes has 13 coins with a total value of 83 cents. How many coins of each type are in the box?

5. Find positive integers that satisfy

$$x + y + z = 9$$
$$x + 5y + 10z = 44$$

6. For which value(s) of a does the following system have zero, one, infinitely many solutions?

$$x_1 + x_2 + x_3 = 4$$
$$x_3 = 2$$
$$(a^2 - 4)x_3 = a - 2$$

7. Let

$$\begin{bmatrix} a & 0 & b & 2 \\ a & a & 4 & 4 \\ 0 & a & 2 & b \end{bmatrix}$$

be the augmented matrix for a linear system. For what values of a and b does the system have

(a) a unique solution,   (b) a one-parameter solution,
(c) a two-parameter solution,   (d) no solution?

8. Solve for x, y, and z.

$$xy - 2\sqrt{y} + 3zy = 8$$
$$2xy - 3\sqrt{y} + 2zy = 7$$
$$-xy + \sqrt{y} + 2zy = 4$$

9. Find a matrix K such that $AKB = C$ given that

$$A = \begin{bmatrix} 1 & 4 \\ -2 & 3 \\ 1 & -2 \end{bmatrix}, \quad B = \begin{bmatrix} 2 & 0 & 0 \\ 0 & 1 & -1 \end{bmatrix}, \quad C = \begin{bmatrix} 8 & 6 & -6 \\ 6 & -1 & 1 \\ -4 & 0 & 0 \end{bmatrix}$$

10. How should the coefficients a, b, and c be chosen so that the system

$$ax + by - 3z = -3$$
$$-2x - by + cz = -1$$
$$ax + 3y - cz = -3$$

has the solution $x = 1$, $y = -1$, and $z = 2$?

11. In each part solve the matrix equation for X.

(a) $X \begin{bmatrix} -1 & 0 & 1 \\ 1 & 1 & 0 \\ 3 & 1 & -1 \end{bmatrix} = \begin{bmatrix} 1 & 2 & 0 \\ -3 & 1 & 5 \end{bmatrix}$ (b) $X \begin{bmatrix} 1 & -1 & 2 \\ 3 & 0 & 1 \end{bmatrix} = \begin{bmatrix} -5 & -1 & 0 \\ 6 & -3 & 7 \end{bmatrix}$

(c) $\begin{bmatrix} 3 & 1 \\ -1 & 2 \end{bmatrix} X - X \begin{bmatrix} 1 & 4 \\ 2 & 0 \end{bmatrix} = \begin{bmatrix} 2 & -2 \\ 5 & 4 \end{bmatrix}$

12. (a) Express the equations

$$\begin{aligned} y_1 &= x_1 - x_2 + x_3 \\ y_2 &= 3x_1 + x_2 - 4x_3 \\ y_3 &= -2x_1 - 2x_2 + 3x_3 \end{aligned} \quad \text{and} \quad \begin{aligned} z_1 &= 4y_1 - y_2 + y_3 \\ z_2 &= -3y_1 + 5y_2 - y_3 \end{aligned}$$

in the matrix forms $Y = AX$ and $Z = BY$. Then use these to obtain a direct relationship $Z = CX$ between Z and X.

(b) Use the equation $Z = CX$ obtained in (a) to express z_1 and z_2 in terms of x_1, x_2, and x_3.

(c) Check the result in (b) by directly substituting the equations for y_1, y_2, and y_3 into the equations for z_1 and z_2 and then simplifying.

13. If A is $m \times n$ and B is $n \times p$, how many multiplication operations and how many addition operations are needed to calculate the matrix product AB?

14. Let A be a square matrix.
(a) Show that $(I - A)^{-1} = I + A + A^2 + A^3$ if $A^4 = 0$.
(b) Show that $(I - A)^{-1} = I + A + A^2 + \cdots + A^n$ if $A^{n+1} = 0$.

15. Find values of a, b, and c so that the graph of the polynomial $p(x) = ax^2 + bx + c$ passes through the points $(1, 2)$, $(-1, 6)$, and $(2, 3)$.

16. (*For readers who have studied calculus.*) Find values of a, b, and c so that the graph of the polynomial $p(x) = ax^2 + bx + c$ passes through the point $(-1, 0)$ and has a horizontal tangent at $(2, -9)$.

17. Let J_n be the $n \times n$ matrix each of whose entries is 1. Show that if $n > 1$, then

$$(I - J_n)^{-1} = I - \frac{1}{n-1} J_n$$

18. Show that if a square matrix A satisfies $A^3 + 4A^2 - 2A + 7I = 0$, then so does A^T.

19. Prove: If B is invertible, then $AB^{-1} = B^{-1}A$ if and only if $AB = BA$.

20. Prove: If A is invertible, then $A + B$ and $I + BA^{-1}$ are both invertible or both not invertible.

21. Prove that if A and B are $n \times n$ matrices, then
(a) $\text{tr}(A + B) = \text{tr}(A) + \text{tr}(B)$ (b) $\text{tr}(kA) = k\,\text{tr}(A)$ (c) $\text{tr}(A^T) = \text{tr}(A)$ (d) $\text{tr}(AB) = \text{tr}(BA)$

22. Use Exercise 21 to show that there are no square matrices A and B such that

$$AB - BA = I$$

23. Prove: If A is an $m \times n$ matrix and B is the $n \times 1$ matrix each of whose entries is $1/n$, then

$$AB = \begin{bmatrix} \bar{r}_1 \\ \bar{r}_2 \\ \vdots \\ \bar{r}_m \end{bmatrix}$$

where $\bar{r}_i$ is the average of the entries in the ith row of A.

24. (*For readers who have studied calculus.*) If the entries of the matrix

$$C = \begin{bmatrix} c_{11}(x) & c_{12}(x) & \cdots & c_{1n}(x) \\ c_{21}(x) & c_{22}(x) & \cdots & c_{2n}(x) \\ \vdots & \vdots & & \vdots \\ c_{m1}(x) & c_{m2}(x) & \cdots & c_{mn}(x) \end{bmatrix}$$

are differentiable functions of x, then we define

$$\frac{dC}{dx} = \begin{bmatrix} c'_{11}(x) & c'_{12}(x) & \cdots & c'_{1n}(x) \\ c'_{21}(x) & c'_{22}(x) & \cdots & c'_{2n}(x) \\ \vdots & \vdots & & \vdots \\ c'_{m1}(x) & c'_{m2}(x) & \cdots & c'_{mn}(x) \end{bmatrix}$$

Show that if the entries in A and B are differentiable functions of x and the sizes of the matrices are such that the stated operations can be performed, then

(a) $\dfrac{d}{dx}(kA) = k\dfrac{dA}{dx}$ (b) $\dfrac{d}{dx}(A + B) = \dfrac{dA}{dx} + \dfrac{dB}{dx}$ (c) $\dfrac{d}{dx}(AB) = \dfrac{dA}{dx}B + A\dfrac{dB}{dx}$

25. (*For readers who have studied calculus.*) Use part (c) of Exercise 24 to show that

$$\frac{dA^{-1}}{dx} = -A^{-1}\frac{dA}{dx}A^{-1}$$

State all the assumptions you make in obtaining this formula.

26. Find the values of A, B, and C that will make the equation

$$\frac{x^2 + x - 2}{(3x - 1)(x^2 + 1)} = \frac{A}{3x - 1} + \frac{Bx + C}{x^2 + 1}$$

an identity. [**Hint.** Multiply through by $(3x - 1)(x^2 + 1)$ and equate the corresponding coefficients of the polynomials on each side of the resulting equation.]

27. If P is an $n \times 1$ matrix such that $P^T P = 1$, then $H = I - 2PP^T$ is called the corresponding **Householder matrix** (named after the American mathematician A. S. Householder).
    (a) Verify that $P^T P = 1$ if $P^T = [\frac{3}{4} \ \frac{1}{6} \ \frac{1}{4} \ \frac{5}{12} \ \frac{5}{12}]$ and compute the corresponding Householder matrix.
    (b) Prove that if H is any Householder matrix, then $H = H^T$ and $H^T H = I$.
    (c) Verify that the Householder matrix found in part (a) satisfies the conditions proved in part (b).

28. Assuming that the stated inverses exist, prove the following equalities.
    (a) $(C^{-1} + D^{-1})^{-1} = C(C + D)^{-1}D$ (b) $(I + CD)^{-1}C = C(I + DC)^{-1}$
    (c) $(C + DD^T)^{-1}D = C^{-1}D(I + D^T C^{-1}D)^{-1}$

29. (a) Show that if $a \neq b$, then

$$a^n + a^{n-1}b + a^{n-2}b^2 + \cdots + ab^{n-1} + b^n = \frac{a^{n+1} - b^{n+1}}{a - b}$$

(b) Use the result in part (a) to find A^n if

$$A = \begin{bmatrix} a & 0 & 0 \\ 0 & b & 0 \\ 1 & 0 & c \end{bmatrix}$$

[**Note.** This exercise is based on a problem by John M. Johnson, *The Mathematics Teacher*, Vol. 85, No. 9, 1992.]

CHAPTER **2**

DETERMINANTS

2.1 THE DETERMINANT FUNCTION

*We are all familiar with functions such as $f(x) = \sin x$ and $f(x) = x^2$, which associate a real number $f(x)$ with a real value of the variable x. Since both x and $f(x)$ assume only real values, such functions are described as "real-valued functions of a real variable." In this section we shall study the **determinant function**, which is a "real-valued function of a matrix variable" in the sense that it associates a real number $f(X)$ with a matrix X. Our work on determinant functions will have important applications to the theory of systems of linear equations and will also lead us to an explicit formula for the inverse of an invertible matrix.*

Recall from Theorem 1.4.5 that the matrix

$$A = \begin{bmatrix} a & b \\ c & d \end{bmatrix}$$

is invertible if $ad - bc \neq 0$. The expression $ad - bc$ occurs so frequently in mathematics that it has a name; it is called the *determinant* of the 2×2 matrix A, and it is denoted by the symbol $\det(A)$. With this notation the inverse of A can be expressed as

$$A^{-1} = \frac{1}{\det(A)} \begin{bmatrix} d & -b \\ -c & a \end{bmatrix}$$

One of the goals in this chapter is to obtain analogs of this formula for higher order matrices. This will require that we extend the concept of a determinant to higher order matrices. For this purpose we will need some preliminary results about permutations.

PERMUTATIONS

> **Definition.** A ***permutation*** of the set of integers $\{1, 2, \ldots, n\}$ is an arrangement of these integers in some order without omissions or repetitions.

Example 1 There are six different permutations of the set of integers $\{1, 2, 3\}$. These are

$$(1, 2, 3) \quad (2, 1, 3) \quad (3, 1, 2)$$
$$(1, 3, 2) \quad (2, 3, 1) \quad (3, 2, 1)$$

One convenient method of systematically listing permutations is to use a *permutation tree*. This method is illustrated in our next example.

Example 2 List all permutations of the set of integers $\{1, 2, 3, 4\}$.

Solution. Consider Figure 1. The four dots labeled 1, 2, 3, 4 at the top of the figure represent the possible choices for the first number in the permutation. The three branches emanating from these dots represent the possible choices for the second position in the permutation. Thus, if the permutation begins $(2, -, -, -)$, the three possibilities for the second position are 1, 3, and 4. The two branches emanating from each dot in the second position represent the possible choices for the third position. Thus, if the permutation begins $(2, 3, -, -)$, the two possible choices for the third position are 1 and 4. Finally, the single branch emanating from each dot in the third position represents the only possible choice for the fourth position. Thus, if the permutation begins with $(2, 3, 4, -)$, the only choice for the fourth position is 1. The different permutations can now be listed by tracing out all the possible paths through the "tree" from the first position to the last position. We obtain the following list by this process.

$$(1, 2, 3, 4) \quad (2, 1, 3, 4) \quad (3, 1, 2, 4) \quad (4, 1, 2, 3)$$
$$(1, 2, 4, 3) \quad (2, 1, 4, 3) \quad (3, 1, 4, 2) \quad (4, 1, 3, 2)$$
$$(1, 3, 2, 4) \quad (2, 3, 1, 4) \quad (3, 2, 1, 4) \quad (4, 2, 1, 3)$$
$$(1, 3, 4, 2) \quad (2, 3, 4, 1) \quad (3, 2, 4, 1) \quad (4, 2, 3, 1)$$
$$(1, 4, 2, 3) \quad (2, 4, 1, 3) \quad (3, 4, 1, 2) \quad (4, 3, 1, 2)$$
$$(1, 4, 3, 2) \quad (2, 4, 3, 1) \quad (3, 4, 2, 1) \quad (4, 3, 2, 1)$$

From this example we see that there are 24 permutations of $\{1, 2, 3, 4\}$. This result could have been anticipated without actually listing the permutations by arguing as follows. Since the first position can be filled in four ways and then the second position in three ways, there are $4 \cdot 3$ ways of filling the first two positions. Since the third posi-

Figure 1

tion can then be filled in two ways, there are $4 \cdot 3 \cdot 2$ ways of filling the first three positions. Finally, since the last position can then be filled in only one way, there are $4 \cdot 3 \cdot 2 \cdot 1 = 24$ ways of filling all four positions. In general, the set $\{1, 2, \ldots, n\}$ will have $n(n - 1)(n - 2) \cdots 2 \cdot 1 = n!$ different permutations.

To denote a general permutation of the set $\{1, 2, \ldots, n\}$, we shall write $(j_1, j_2, \ldots, j_n)$. Here, j_1 is the first integer in the permutation, j_2 is the second, and so on. An **inversion** is said to occur in a permutation $(j_1, j_2, \ldots, j_n)$ whenever a larger integer precedes a smaller one. The total number of inversions occurring in a permutation can be obtained as follows: (1) find the number of integers that are less than j_1 and that follow j_1 in the permutation; (2) find the number of integers that are less than j_2 and that follow j_2 in the permutation. Continue this counting process for $j_3, \ldots, j_{n-1}$. The sum of these numbers will be the total number of inversions in the permutation.

Example 3 Determine the number of inversions in the following permutations:

(a) $(6, 1, 3, 4, 5, 2)$ (b) $(2, 4, 1, 3)$ (c) $(1, 2, 3, 4)$

Solution.

(a) The number of inversions is $5 + 0 + 1 + 1 + 1 = 8$.
(b) The number of inversions is $1 + 2 + 0 = 3$.
(c) There are no inversions in this permutation. ▲

Definition. A permutation is called **even** if the total number of inversions is an even integer and is called **odd** if the total number of inversions is an odd integer.

Example 4 The following table classifies the various permutations of $\{1, 2, 3\}$ as even or odd.

Permutation	Number of Inversions	Classification
$(1, 2, 3)$	0	even
$(1, 3, 2)$	1	odd
$(2, 1, 3)$	1	odd
$(2, 3, 1)$	2	even
$(3, 1, 2)$	2	even
$(3, 2, 1)$	3	odd

▲

DEFINITION OF A DETERMINANT

By an **elementary product** from an $n \times n$ matrix A we shall mean any product of n entries from A, no two of which come from the same row or same column.

Example 5 List all elementary products from the matrices

(a) $\begin{bmatrix} a_{11} & a_{12} \\ a_{21} & a_{22} \end{bmatrix}$ (b) $\begin{bmatrix} a_{11} & a_{12} & a_{13} \\ a_{21} & a_{22} & a_{23} \\ a_{31} & a_{32} & a_{33} \end{bmatrix}$

Solution (a). Since each elementary product has two factors, and since each factor comes from a different row, an elementary product can be written in the form

$$a_{1\_}a_{2\_}$$

where the blanks designate column numbers. Since no two factors in the product come from the same column, the column numbers must be $\underline{1}\,\underline{2}$ or $\underline{2}\,\underline{1}$. Thus, the only elementary products are $a_{11}a_{22}$ and $a_{12}a_{21}$.

Solution (b). Since each elementary product has three factors, each of which comes from a different row, an elementary product can be written in the form

$$a_{1\_}a_{2\_}a_{3\_}$$

Since no two factors in the product come from the same column, the column numbers have no repetitions; consequently, they must form a permutation of the set $\{1, 2, 3\}$. These $3! = 6$ permutations yield the following list of elementary products.

$$a_{11}a_{22}a_{33} \qquad a_{12}a_{21}a_{33} \qquad a_{13}a_{21}a_{32}$$
$$a_{11}a_{23}a_{32} \qquad a_{12}a_{23}a_{31} \qquad a_{13}a_{22}a_{31}$$

As this example points out, an $n \times n$ matrix A has $n!$ elementary products. They are the products of the form $a_{1j_1}a_{2j_2}\cdots a_{nj_n}$, where $(j_1, j_2, \ldots, j_n)$ is a permutation of the set $\{1, 2, \ldots, n\}$. By a ***signed elementary product from A*** we shall mean an elementary product $a_{1j_1}a_{2j_2}\cdots a_{nj_n}$ multiplied by $+1$ or -1. We use the $+$ if $(j_1, j_2, \ldots, j_n)$ is an even permutation and the $-$ if $(j_1, j_2, \ldots, j_n)$ is an odd permutation.

Example 6   List all signed elementary products from the matrices.

$$\text{(a)}\ \begin{bmatrix} a_{11} & a_{12} \\ a_{21} & a_{22} \end{bmatrix} \qquad \text{(b)}\ \begin{bmatrix} a_{11} & a_{12} & a_{13} \\ a_{21} & a_{22} & a_{23} \\ a_{31} & a_{32} & a_{33} \end{bmatrix}$$

Solution.

(a)

Elementary Product	Associated Permutation	Even or Odd	Signed Elementary Product
$a_{11}a_{22}$	$(1, 2)$	even	$a_{11}a_{22}$
$a_{12}a_{21}$	$(2, 1)$	odd	$-a_{12}a_{21}$

(b)

Elementary Product	Associated Permutation	Even or Odd	Signed Elementary Product
$a_{11}a_{22}a_{33}$	$(1, 2, 3)$	even	$a_{11}a_{22}a_{33}$
$a_{11}a_{23}a_{32}$	$(1, 3, 2)$	odd	$-a_{11}a_{23}a_{32}$
$a_{12}a_{21}a_{33}$	$(2, 1, 3)$	odd	$-a_{12}a_{21}a_{33}$
$a_{12}a_{23}a_{31}$	$(2, 3, 1)$	even	$a_{12}a_{23}a_{31}$
$a_{13}a_{21}a_{32}$	$(3, 1, 2)$	even	$a_{13}a_{21}a_{32}$
$a_{13}a_{22}a_{31}$	$(3, 2, 1)$	odd	$-a_{13}a_{22}a_{31}$

We are now in a position to define the determinant function.

Definition. Let A be a square matrix. The ***determinant function*** is denoted by ***det***, and we define $\det(A)$ to be the sum of all signed elementary products from A. The number $\det(A)$ is called the ***determinant of*** A.

**EVALUATING
2 × 2 AND 3 × 3
DETERMINANTS**

Example 7 Referring to Example 6, we obtain

(a) $\det \begin{bmatrix} a_{11} & a_{12} \\ a_{21} & a_{22} \end{bmatrix} = a_{11}a_{22} - a_{12}a_{21}$

(b) $\det \begin{bmatrix} a_{11} & a_{12} & a_{13} \\ a_{21} & a_{22} & a_{23} \\ a_{31} & a_{32} & a_{33} \end{bmatrix} = a_{11}a_{22}a_{33} + a_{12}a_{23}a_{31} + a_{13}a_{21}a_{32}$
$$- a_{13}a_{22}a_{31} - a_{12}a_{21}a_{33} - a_{11}a_{23}a_{32}$$

To avoid memorizing these unwieldy expressions we suggest using the mnemonic devices described in Figure 2. The first formula in Example 7 is obtained from Figure 2*a* by multiplying the entries on the rightward arrow and subtracting the product of the entries on the leftward arrow. The second formula in Example 7 is obtained by re-copying the first and second columns as shown in Figure 2*b*. The determinant is then computed by summing the products on the rightward arrows and subtracting the products on the leftward arrows.

Figure 2   (*a*)   (*b*)

Example 8   Evaluate the determinants of

$$A = \begin{bmatrix} 3 & 1 \\ 4 & -2 \end{bmatrix} \quad \text{and} \quad B = \begin{bmatrix} 1 & 2 & 3 \\ -4 & 5 & 6 \\ 7 & -8 & 9 \end{bmatrix}$$

Solution.   Using the method of Figure 2a gives

$$\det(A) = (3)(-2) - (1)(4) = -10$$

Using the method of Figure 2b gives

$$\det(B) = (45) + (84) + (96) - (105) - (-48) - (-72) = 240$$

$$\begin{bmatrix} 1 & 2 & 3 \\ -4 & 5 & 6 \\ 7 & -8 & 9 \end{bmatrix} \begin{matrix} 1 & 2 \\ -4 & 5 \\ 7 & -8 \end{matrix}$$

Warning.   We emphasize that the methods shown in Figure 2 do not work for determinants of 4×4 matrices or higher.

Evaluating determinants directly from the definition leads to computational difficulties. Indeed, evaluating a 4×4 determinant directly would involve computing $4! = 24$ signed elementary products, and a 10×10 determinant would involve $10! = 3,628,800$ signed elementary products. Even the fastest of digital computers cannot handle the computation of a 25×25 determinant by this method in a practical amount of time. Much of the remainder of this chapter is devoted, therefore, to developing properties of determinants that will simplify their evaluation.

COMMENTS ON NOTATION AND TERMINOLOGY

We conclude this section with some comments about terminology and notation. First, we note that the symbol $|A|$ is an alternative notation for $\det(A)$. For example, the determinant of a 3×3 matrix can be written as

$$\det \begin{bmatrix} a_{11} & a_{12} & a_{13} \\ a_{21} & a_{22} & a_{23} \\ a_{31} & a_{32} & a_{33} \end{bmatrix} \quad \text{or} \quad \begin{vmatrix} a_{11} & a_{12} & a_{13} \\ a_{21} & a_{22} & a_{23} \\ a_{31} & a_{32} & a_{33} \end{vmatrix}$$

In the latter notation the determinant of the matrix A in Example 8 would be written as

$$\begin{vmatrix} 3 & 1 \\ 4 & -2 \end{vmatrix} = -10$$

REMARK.   Strictly speaking, the determinant of a matrix is a number. However, it is common practice to "abuse" the terminology slightly and use the term "determinant" to refer to the *matrix* whose determinant is being computed. Thus, we might refer to

$$\begin{vmatrix} 3 & 1 \\ 4 & -2 \end{vmatrix}$$

as a 2×2 determinant and call 3 the entry in the first row and first column of the determinant.

Finally, we note that the determinant of A is often written symbolically as

$$\det(A) = \sum \pm a_{1\,j_1} a_{2\,j_2} \cdots a_{n\,j_n} \tag{1}$$

where $\sum$ indicates that the terms are to be summed over all permutations $(j_1, j_2, \ldots, j_n)$ and the $+$ or $-$ is selected in each term according to whether the permutation is even or odd. This notation is useful when the definition of a determinant needs to be emphasized.

EXERCISE SET 2.1

1. Find the number of inversions in each of the following permutations of $\{1, 2, 3, 4, 5\}$.
   (a) $(4\ 1\ 3\ 5\ 2)$ (b) $(5\ 3\ 4\ 2\ 1)$ (c) $(3\ 2\ 5\ 4\ 1)$ (d) $(5\ 4\ 3\ 2\ 1)$ (e) $(1\ 2\ 3\ 4\ 5)$ (f) $(1\ 4\ 2\ 3\ 5)$

2. Classify each of the permutations in Exercise 1 as even or odd.

In Exercises 3–12 evaluate the determinant.

3. $\begin{vmatrix} 3 & 5 \\ -2 & 4 \end{vmatrix}$ **4.** $\begin{vmatrix} 4 & 1 \\ 8 & 2 \end{vmatrix}$ **5.** $\begin{vmatrix} -5 & 6 \\ -7 & -2 \end{vmatrix}$ **6.** $\begin{vmatrix} \sqrt{2} & \sqrt{6} \\ 4 & \sqrt{3} \end{vmatrix}$ **7.** $\begin{vmatrix} a-3 & 5 \\ -3 & a-2 \end{vmatrix}$ **8.** $\begin{vmatrix} -2 & 7 & 6 \\ 5 & 1 & -2 \\ 3 & 8 & 4 \end{vmatrix}$

9. $\begin{vmatrix} -2 & 1 & 4 \\ 3 & 5 & -7 \\ 1 & 6 & 2 \end{vmatrix}$ **10.** $\begin{vmatrix} -1 & 1 & 2 \\ 3 & 0 & -5 \\ 1 & 7 & 2 \end{vmatrix}$ **11.** $\begin{vmatrix} 3 & 0 & 0 \\ 2 & -1 & 5 \\ 1 & 9 & -4 \end{vmatrix}$ **12.** $\begin{vmatrix} c & -4 & 3 \\ 2 & 1 & c^2 \\ 4 & c-1 & 2 \end{vmatrix}$

13. Find all values of λ for which $\det(A) = 0$.

   (a) $\begin{bmatrix} \lambda-2 & 1 \\ -5 & \lambda+4 \end{bmatrix}$ (b) $\begin{bmatrix} \lambda-4 & 0 & 0 \\ 0 & \lambda & 2 \\ 0 & 3 & \lambda-1 \end{bmatrix}$

14. Classify each permutation of $\{1, 2, 3, 4\}$ as even or odd.

15. Use the results in Exercise 14 to construct a formula for the determinant of a 4×4 matrix.

16. Use the formula obtained in Exercise 15 to evaluate

$$\begin{vmatrix} 4 & -9 & 9 & 2 \\ -2 & 5 & 6 & 4 \\ 1 & 2 & -5 & -3 \\ 1 & -2 & 0 & -2 \end{vmatrix}$$

17. Use the determinant definition to evaluate

   (a) $\begin{vmatrix} 0 & 0 & 0 & 0 & -3 \\ 0 & 0 & 0 & -4 & 0 \\ 0 & 0 & -1 & 0 & 0 \\ 0 & 2 & 0 & 0 & 0 \\ 5 & 0 & 0 & 0 & 0 \end{vmatrix}$ (b) $\begin{vmatrix} 5 & 0 & 0 & 0 & 0 \\ 0 & 0 & 0 & 0 & -4 \\ 0 & 0 & 3 & 0 & 0 \\ 0 & 0 & 0 & 1 & 0 \\ 0 & -2 & 0 & 0 & 0 \end{vmatrix}$

18. Solve for x.

$$\begin{vmatrix} x & -1 \\ 3 & 1-x \end{vmatrix} = \begin{vmatrix} 1 & 0 & -3 \\ 2 & x & -6 \\ 1 & 3 & x-5 \end{vmatrix}$$

19. Show that the value of the determinant

$$\begin{vmatrix} \sin\theta & \cos\theta & 0 \\ -\cos\theta & \sin\theta & 0 \\ \sin\theta - \cos\theta & \sin\theta + \cos\theta & 1 \end{vmatrix}$$

does not depend on θ.

20. Prove that if a square matrix A has a row or column of zeros, then $\det(A) = 0$.

21. Prove that the matrices

$$A = \begin{bmatrix} a & b \\ 0 & c \end{bmatrix} \quad \text{and} \quad B = \begin{bmatrix} d & e \\ 0 & f \end{bmatrix}$$

commute if and only if

$$\begin{vmatrix} b & a-c \\ e & d-f \end{vmatrix} = 0$$

2.2 EVALUATING DETERMINANTS BY ROW REDUCTION

In this section we shall show that the determinant of a matrix can be evaluated by reducing the matrix to row-echelon form. This method is important since it avoids the lengthy computations involved in directly applying the determinant definition.

A BASIC THEOREM

We begin with a fundamental theorem about determinants.

Theorem 2.2.1. *Let A be a square matrix.*
(a) If A has a row of zeros or a column of zeros, then $\det(A) = 0$.
(b) $\det(A) = \det(A^T)$.

Proof (a). Since every signed elementary product from A has one factor from each row and one factor from each column, every signed elementary product would of necessity have a factor from a zero row or a factor from a zero column. In such cases, every signed elementary product is zero, and $\det(A)$, which is the sum of all the signed elementary products, is zero. ▮

We omit the proof of part (b), but recall that an elementary product has one factor from each row and each column, so it is evident that A and A^T have precisely the same set of elementary products. With the help of some theorems on permutations, which would take us too far afield to discuss, it can be shown that A and A^T actually have the same set of *signed* elementary products. This implies that $\det(A) = \det(A^T)$.

REMARK. Because of Theorem 2.2.1b, nearly every theorem about determinants that contains the word "row" in its statement is also true when the word "column" is substituted for "row." To prove a column statement one need only transpose the matrix in question to convert the column statement to a row statement, and then apply the corresponding known results for rows.

DETERMINANTS OF TRIANGULAR MATRICES

The following theorem makes it easy to evaluate the determinant of a triangular matrix, regardless of its size.

Theorem 2.2.2. *If A is an $n \times n$ triangular matrix (upper triangular, lower triangular, or diagonal), then $\det(A)$ is the product of the entries on the main diagonal; that is, $\det(A) = a_{11}a_{22}\cdots a_{nn}$.*

For simplicity of notation, we will prove the result for a 4×4 lower triangular matrix

$$A = \begin{bmatrix} a_{11} & 0 & 0 & 0 \\ a_{21} & a_{22} & 0 & 0 \\ a_{31} & a_{32} & a_{33} & 0 \\ a_{41} & a_{42} & a_{43} & a_{44} \end{bmatrix}$$

The argument in the $n \times n$ case is similar. A proof for upper triangular matrices can be obtained by applying Theorem 2.2.1b and observing that the transpose of an upper triangular matrix is a lower triangular matrix with the same diagonal entries.

Proof of Theorem 2.2.2 (4×4 *lower triangular case*). The only elementary product from A that can be nonzero is $a_{11}a_{22}a_{33}a_{44}$. To see that this is so, consider a typical elementary product $a_{1j_1}a_{2j_2}a_{3j_3}a_{4j_4}$. Since $a_{12} = a_{13} = a_{14} = 0$, we must have $j_1 = 1$ in order to have a nonzero elementary product. If $j_1 = 1$, we must have $j_2 \neq 1$, since no two factors come from the same column. Further, since $a_{23} = a_{24} = 0$, we must have $j_2 = 2$ in order to have a nonzero product. Continuing in this way, we obtain $j_3 = 3$ and $j_4 = 4$. Since $a_{11}a_{22}a_{33}a_{44}$ is multiplied by $+1$ in forming the signed elementary product, we obtain

$$\det(A) = a_{11}a_{22}a_{33}a_{44} \quad \blacksquare$$

Example 1

$$\begin{vmatrix} 2 & 7 & -3 & 8 & 3 \\ 0 & -3 & 7 & 5 & 1 \\ 0 & 0 & 6 & 7 & 6 \\ 0 & 0 & 0 & 9 & 8 \\ 0 & 0 & 0 & 0 & 4 \end{vmatrix} = (2)(-3)(6)(9)(4) = -1296 \quad \blacktriangle$$

EFFECT OF ELEMENTARY ROW OPERATIONS ON A DETERMINANT

The next theorem shows how an elementary row operation on a matrix affects the value of its determinant.

Theorem 2.2.3. *Let A be an $n \times n$ matrix.*

(a) *If B is the matrix that results when a single row or single column of A is multiplied by a scalar k, then $\det(B) = k \det(A)$.*

(b) *If B is the matrix that results when two rows or two columns of A are interchanged, then $\det(B) = -\det(A)$.*

(c) *If B is the matrix that results when a multiple of one row of A is added to another row or when a multiple of one column is added to another column, then $\det(B) = \det(A)$.*

A proof of this theorem can be obtained by using Formula (1) of Section 2.1 to compute the determinants involved, and then verifying the equalities. We omit the proof but give the following example that illustrates the theorem for 3×3 determinants.

Example 2

Relationship	Operation
$\begin{vmatrix} ka_{11} & ka_{12} & ka_{13} \\ a_{21} & a_{22} & a_{23} \\ a_{31} & a_{32} & a_{33} \end{vmatrix} = k \begin{vmatrix} a_{11} & a_{12} & a_{13} \\ a_{21} & a_{22} & a_{23} \\ a_{31} & a_{32} & a_{33} \end{vmatrix}$ $\det(B) = k \det(A)$	The first row of A is multiplied by k.
$\begin{vmatrix} a_{21} & a_{22} & a_{23} \\ a_{11} & a_{12} & a_{13} \\ a_{31} & a_{32} & a_{33} \end{vmatrix} = - \begin{vmatrix} a_{11} & a_{12} & a_{13} \\ a_{21} & a_{22} & a_{23} \\ a_{31} & a_{32} & a_{33} \end{vmatrix}$ $\det(B) = -\det(A)$	The first and second rows of A are interchanged.
$\begin{vmatrix} a_{11}+ka_{21} & a_{12}+ka_{22} & a_{13}+ka_{23} \\ a_{21} & a_{22} & a_{23} \\ a_{31} & a_{32} & a_{33} \end{vmatrix} = \begin{vmatrix} a_{11} & a_{12} & a_{13} \\ a_{21} & a_{22} & a_{23} \\ a_{31} & a_{32} & a_{33} \end{vmatrix}$ $\det(B) = \det(A)$	A multiple of the second row of A is added to the first row.

We will verify the equation in the last row of the table and leave the first two for the reader. With the help of Example 7 in Section 2.1 we obtain

$$\det(B) = (a_{11} + ka_{21})a_{22}a_{33} + (a_{12} + ka_{22})a_{23}a_{31} + (a_{13} + ka_{23})a_{21}a_{32}$$
$$- a_{31}a_{22}(a_{13} + ka_{23}) - a_{33}a_{21}(a_{12} + ka_{22}) - a_{32}a_{23}(a_{11} + ka_{21})$$
$$= \det(A) + k(a_{21}a_{22}a_{33} + a_{22}a_{23}a_{31} + a_{23}a_{21}a_{32}$$
$$- a_{31}a_{22}a_{23} - a_{33}a_{21}a_{22} - a_{32}a_{23}a_{21})$$
$$= \det(A) + 0 = \det(A) \ \blacktriangle$$

REMARK. As illustrated by the first equation in Example 2, part (*a*) of Theorem 2.2.3 allows us to bring a "common factor" from any row (or column) through the determinant sign.

DETERMINANTS OF ELEMENTARY MATRICES

Recall that an elementary matrix results from performing a single elementary row operation on an identity matrix; thus, if we let $A = I_n$ in Theorem 2.2.3 [so that we have $det(A) = det(I_n) = 1$], then the matrix B is an elementary matrix, and the theorem yields the following result about determinants of elementary matrices.

Theorem 2.2.4. *Let E be an $n \times n$ elementary matrix.*

(*a*) *If E results from multiplying a row of I_n by k, then $det(E) = k$.*

(*b*) *If E results from interchanging two rows of I_n, then $det(E) = -1$.*

(*c*) *If E results from adding a multiple of one row of I_n to another, then $det(E) = 1$.*

Example 3 The following determinants of elementary matrices, which are evaluated by inspection, illustrate Theorem 2.2.4.

$$\begin{vmatrix} 1 & 0 & 0 & 0 \\ 0 & 3 & 0 & 0 \\ 0 & 0 & 1 & 0 \\ 0 & 0 & 0 & 1 \end{vmatrix} = 3 \qquad \begin{vmatrix} 0 & 0 & 0 & 1 \\ 0 & 1 & 0 & 0 \\ 0 & 0 & 1 & 0 \\ 1 & 0 & 0 & 0 \end{vmatrix} = -1 \qquad \begin{vmatrix} 1 & 0 & 0 & 7 \\ 0 & 1 & 0 & 0 \\ 0 & 0 & 1 & 0 \\ 0 & 0 & 0 & 1 \end{vmatrix} = 1$$

The second row of I_4 was multiplied by 3.	**The first and last rows of I_4 were interchanged.**	**7 times the last row of I_4 was added to the first row.**

DETERMINANTS WITH PROPORTIONAL ROWS OR COLUMNS

If a square matrix A has two proportional rows, then a row of zeros can be introduced by adding a suitable multiple of one of the rows to the other. Similarly for columns. But adding a multiple of one row or column to another does not change the determinant, so from Theorem 2.2.1*a*, we must have $det(A) = 0$. This proves the following theorem.

Theorem 2.2.5. *If A is a square matrix with two proportional rows or two proportional columns, then $det(A) = 0$.*

Example 4 The following computation illustrates the introduction of a row of zeros when there are two proportional rows:

$$\begin{vmatrix} 1 & 3 & -2 & 4 \\ 2 & 6 & -4 & 8 \\ 3 & 9 & 1 & 5 \\ 1 & 1 & 4 & 8 \end{vmatrix} = \begin{vmatrix} 1 & 3 & -2 & 4 \\ 0 & 0 & 0 & 0 \\ 3 & 9 & 1 & 5 \\ 1 & 1 & 4 & 8 \end{vmatrix} = 0$$

> The second row is 2 times the first, so we added −2 times the first row to the second to introduce a row of zeros.

Each of the following matrices has two proportional rows or columns; thus, by inspection, each has a determinant of zero.

$$
\begin{bmatrix} -1 & 4 \\ -2 & 8 \end{bmatrix}
\qquad
\begin{bmatrix} 1 & -2 & 7 \\ -4 & 8 & 5 \\ 2 & -4 & 3 \end{bmatrix}
\qquad
\begin{bmatrix} 3 & -1 & 4 & -5 \\ 6 & -2 & 5 & 2 \\ 5 & 8 & 1 & 4 \\ -9 & 3 & -12 & 15 \end{bmatrix} \; \blacktriangle
$$

EVALUATING DETERMINANTS BY ROW REDUCTION

We shall now give a method for evaluating determinants that involves substantially less computation than applying the determinant definition directly. The idea of the method is to reduce the given matrix to upper triangular form by elementary row operations, then compute the determinant of the upper triangular matrix (an easy computation), then relate that determinant to that of the original matrix. Here is an example.

Example 5 Evaluate $\det(A)$ where

$$
A = \begin{bmatrix} 0 & 1 & 5 \\ 3 & -6 & 9 \\ 2 & 6 & 1 \end{bmatrix}
$$

Solution. We will reduce A to row-echelon form (which is upper triangular) and apply Theorem 2.2.3:

$$
\det(A) = \begin{vmatrix} 0 & 1 & 5 \\ 3 & -6 & 9 \\ 2 & 6 & 1 \end{vmatrix} = - \begin{vmatrix} 3 & -6 & 9 \\ 0 & 1 & 5 \\ 2 & 6 & 1 \end{vmatrix}
$$

| The first and second rows of A were interchanged. |

$$
= -3 \begin{vmatrix} 1 & -2 & 3 \\ 0 & 1 & 5 \\ 2 & 6 & 1 \end{vmatrix}
$$

| A common factor of 3 from the first row was taken through the det sign. |

$$
= -3 \begin{vmatrix} 1 & -2 & 3 \\ 0 & 1 & 5 \\ 0 & 10 & -5 \end{vmatrix}
$$

| -2 times the first row was added to the third row. |

$$
= -3 \begin{vmatrix} 1 & -2 & 3 \\ 0 & 1 & 5 \\ 0 & 0 & -55 \end{vmatrix}
$$

| -10 times the second row was added to the third row. |

$$
= (-3)(-55) \begin{vmatrix} 1 & -2 & 3 \\ 0 & 1 & 5 \\ 0 & 0 & 1 \end{vmatrix}
$$

| A common factor of -55 from the last row was taken through the det sign. |

$$
= (-3)(-55)(1) = 165 \; \blacktriangle
$$

REMARK. The method of row reduction is well suited for computer evaluation of determinants because it is systematic and easily programmed. However, in subsequent sections we will develop methods that are often easier for hand computation.

Example 6 Compute the determinant of

$$A = \begin{bmatrix} 1 & 0 & 0 & 3 \\ 2 & 7 & 0 & 6 \\ 0 & 6 & 3 & 0 \\ 7 & 3 & 1 & -5 \end{bmatrix}$$

Solution. This determinant could be computed as above by using elementary row operations to reduce A to row-echelon form, but we can put A in lower triangular form in one step by adding -3 times the first column to the fourth to obtain

$$\det(A) = \det \begin{bmatrix} 1 & 0 & 0 & 0 \\ 2 & 7 & 0 & 0 \\ 0 & 6 & 3 & 0 \\ 7 & 3 & 1 & -26 \end{bmatrix} = (1)(7)(3)(-26) = -546$$

This example points out the utility of keeping an eye open for column operations that can shorten computations. ▲

EXERCISE SET 2.2

1. Verify that $\det(A) = \det(A^T)$ for

(a) $A = \begin{bmatrix} -2 & 3 \\ 1 & 4 \end{bmatrix}$ (b) $\begin{bmatrix} 2 & -1 & 3 \\ 1 & 2 & 4 \\ 5 & -3 & 6 \end{bmatrix}$

2. Evaluate the following determinants by inspection.

(a) $\begin{vmatrix} 3 & -17 & 4 \\ 0 & 5 & 1 \\ 0 & 0 & -2 \end{vmatrix}$ (b) $\begin{vmatrix} \sqrt{2} & 0 & 0 & 0 \\ -8 & \sqrt{2} & 0 & 0 \\ 7 & 0 & -1 & 0 \\ 9 & 5 & 6 & 1 \end{vmatrix}$ (c) $\begin{vmatrix} -2 & 1 & 3 \\ 1 & -7 & 4 \\ -2 & 1 & 3 \end{vmatrix}$ (d) $\begin{vmatrix} 1 & -2 & 3 \\ 2 & -4 & 6 \\ 5 & -8 & 1 \end{vmatrix}$

3. Find the determinants of the following elementary matrices by inspection.

(a) $\begin{bmatrix} 1 & 0 & 0 & 0 \\ 0 & 1 & 0 & 0 \\ 0 & 0 & -5 & 0 \\ 0 & 0 & 0 & 1 \end{bmatrix}$ (b) $\begin{bmatrix} 1 & 0 & 0 & 0 \\ 0 & 0 & 1 & 0 \\ 0 & 1 & 0 & 0 \\ 0 & 0 & 0 & 1 \end{bmatrix}$ (c) $\begin{bmatrix} 1 & 0 & 0 & 0 \\ 0 & 1 & 0 & -9 \\ 0 & 0 & 1 & 0 \\ 0 & 0 & 0 & 1 \end{bmatrix}$

In Exercises 4–11 evaluate the determinant of the given matrix by reducing the matrix to row-echelon form.

4. $\begin{bmatrix} 3 & 6 & -9 \\ 0 & 0 & -2 \\ -2 & 1 & 5 \end{bmatrix}$ **5.** $\begin{bmatrix} 0 & 3 & 1 \\ 1 & 1 & 2 \\ 3 & 2 & 4 \end{bmatrix}$ **6.** $\begin{bmatrix} 1 & -3 & 0 \\ -2 & 4 & 1 \\ 5 & -2 & 2 \end{bmatrix}$ **7.** $\begin{bmatrix} 3 & -6 & 9 \\ -2 & 7 & -2 \\ 0 & 1 & 5 \end{bmatrix}$

8. $\begin{bmatrix} 1 & -2 & 3 & 1 \\ 5 & -9 & 6 & 3 \\ -1 & 2 & -6 & -2 \\ 2 & 8 & 6 & 1 \end{bmatrix}$

9. $\begin{bmatrix} 2 & 1 & 3 & 1 \\ 1 & 0 & 1 & 1 \\ 0 & 2 & 1 & 0 \\ 0 & 1 & 2 & 3 \end{bmatrix}$

10. $\begin{bmatrix} 0 & 1 & 1 & 1 \\ \frac{1}{2} & \frac{1}{2} & 1 & \frac{1}{2} \\ \frac{2}{3} & \frac{1}{3} & \frac{1}{3} & 0 \\ -\frac{1}{3} & \frac{2}{3} & 0 & 0 \end{bmatrix}$

11. $\begin{bmatrix} 1 & 3 & 1 & 5 & 3 \\ -2 & -7 & 0 & -4 & 2 \\ 0 & 0 & 1 & 0 & 1 \\ 0 & 0 & 2 & 1 & 1 \\ 0 & 0 & 0 & 1 & 1 \end{bmatrix}$

12. Given that $\begin{vmatrix} a & b & c \\ d & e & f \\ g & h & i \end{vmatrix} = -6$, find

(a) $\begin{vmatrix} d & e & f \\ g & h & i \\ a & b & c \end{vmatrix}$
(b) $\begin{vmatrix} 3a & 3b & 3c \\ -d & -e & -f \\ 4g & 4h & 4i \end{vmatrix}$
(c) $\begin{vmatrix} a+g & b+h & c+i \\ d & e & f \\ g & h & i \end{vmatrix}$
(d) $\begin{vmatrix} -3a & -3b & -3c \\ d & e & f \\ g-4d & h-4e & i-4f \end{vmatrix}$

13. Use row reduction to show that

$$\begin{vmatrix} 1 & 1 & 1 \\ a & b & c \\ a^2 & b^2 & c^2 \end{vmatrix} = (b-a)(c-a)(c-b)$$

14. Use an argument like that in the proof of Theorem 2.2.2 to show that

(a) $\det \begin{bmatrix} 0 & 0 & a_{13} \\ 0 & a_{22} & a_{23} \\ a_{31} & a_{32} & a_{33} \end{bmatrix} = -a_{13}a_{22}a_{31}$

(b) $\det \begin{bmatrix} 0 & 0 & 0 & a_{14} \\ 0 & 0 & a_{23} & a_{24} \\ 0 & a_{32} & a_{33} & a_{34} \\ a_{41} & a_{42} & a_{43} & a_{44} \end{bmatrix} = a_{14}a_{23}a_{32}a_{41}$

15. Prove the following special cases of Theorem 2.2.3.

(a) $\begin{vmatrix} ka_{11} & ka_{12} & ka_{13} \\ a_{21} & a_{22} & a_{23} \\ a_{31} & a_{32} & a_{33} \end{vmatrix} = k \begin{vmatrix} a_{11} & a_{12} & a_{13} \\ a_{21} & a_{22} & a_{23} \\ a_{31} & a_{32} & a_{33} \end{vmatrix}$

(b) $\begin{vmatrix} a_{21} & a_{22} & a_{23} \\ a_{11} & a_{12} & a_{13} \\ a_{31} & a_{32} & a_{33} \end{vmatrix} = - \begin{vmatrix} a_{11} & a_{12} & a_{13} \\ a_{21} & a_{22} & a_{23} \\ a_{31} & a_{32} & a_{33} \end{vmatrix}$

2.3 PROPERTIES OF THE DETERMINANT FUNCTION

In this section we shall develop some of the fundamental properties of the determinant function. Our work here will give us some further insight into the relationship between a square matrix and its determinant. One of the immediate consequences of this material will be an important determinant test for the invertibility of a matrix.

BASIC PROPERTIES OF DETERMINANTS

Suppose that A and B are $n \times n$ matrices and k is any scalar. We begin by considering possible relationships between $\det(A)$, $\det(B)$, and

$$\det(kA), \qquad \det(A+B), \qquad \text{and} \qquad \det(AB)$$

Since a common factor of any row of a matrix can be moved through the det sign, and since each of the n rows in kA has a common factor of k, we obtain

$$\det(kA) = k^n \det(A) \tag{1}$$

For example,

$$\begin{vmatrix} ka_{11} & ka_{12} & ka_{13} \\ ka_{21} & ka_{22} & ka_{23} \\ ka_{31} & ka_{32} & ka_{33} \end{vmatrix} = k^3 \begin{vmatrix} a_{11} & a_{12} & a_{13} \\ a_{21} & a_{22} & a_{23} \\ a_{31} & a_{32} & a_{33} \end{vmatrix}$$

Unfortunately, no simple relationship exists between the determinants $\det(A)$, $\det(B)$, and $\det(A + B)$ in general. In particular, we emphasize that $\det(A + B)$ is usually *not* equal to $\det(A) + \det(B)$. The following example illustrates this fact.

Example 1 Consider

$$A = \begin{bmatrix} 1 & 2 \\ 2 & 5 \end{bmatrix} \qquad B = \begin{bmatrix} 3 & 1 \\ 1 & 3 \end{bmatrix} \qquad A + B = \begin{bmatrix} 4 & 3 \\ 3 & 8 \end{bmatrix}$$

We have $\det(A) = 1$, $\det(B) = 8$, and $\det(A + B) = 23$; thus $\det(A + B) \neq \det(A) + \det(B)$. ▲

In spite of the negative tone of the preceding example, there is one important relationship concerning sums of determinants that is often useful. To obtain it, consider two 2×2 matrices that differ only in the second row:

$$A = \begin{bmatrix} a_{11} & a_{12} \\ a_{21} & a_{22} \end{bmatrix} \qquad \text{and} \qquad B = \begin{bmatrix} a_{11} & a_{12} \\ b_{21} & b_{22} \end{bmatrix}$$

We have

$$\det(A) + \det(B) = (a_{11}a_{22} - a_{12}a_{21}) + (a_{11}b_{22} - a_{12}b_{21})$$

$$= a_{11}(a_{22} + b_{22}) - a_{12}(a_{21} + b_{21})$$

$$= \det \begin{bmatrix} a_{11} & a_{12} \\ a_{21} + b_{21} & a_{22} + b_{22} \end{bmatrix}$$

Thus,

$$\det \begin{bmatrix} a_{11} & a_{12} \\ a_{21} & a_{22} \end{bmatrix} + \det \begin{bmatrix} a_{11} & a_{12} \\ b_{21} & b_{22} \end{bmatrix} = \det \begin{bmatrix} a_{11} & a_{12} \\ a_{21} + b_{21} & a_{22} + b_{22} \end{bmatrix}$$

This is a special case of the following general result.

> **Theorem 2.3.1.** *Let A, B, and C be $n \times n$ matrices that differ only in a single row, say the rth, and assume that the rth row of C can be obtained by adding corresponding entries in the rth rows of A and B. Then*
>
> $$det(C) = det(A) + det(B)$$
>
> *The same result holds for columns.*

Example 2   By evaluating the determinants, the reader can check that

$$\det \begin{bmatrix} 1 & 7 & 5 \\ 2 & 0 & 3 \\ 1+0 & 4+1 & 7+(-1) \end{bmatrix} = \det \begin{bmatrix} 1 & 7 & 5 \\ 2 & 0 & 3 \\ 1 & 4 & 7 \end{bmatrix} + \det \begin{bmatrix} 1 & 7 & 5 \\ 2 & 0 & 3 \\ 0 & 1 & -1 \end{bmatrix} \quad \blacktriangle$$

DETERMINANT OF A MATRIX PRODUCT

When one considers the complexity of the definitions of matrix multiplication and determinants, it would seem unlikely that any simple relationship should exist between them. This is what makes the elegant simplicity of the following result so surprising: We will show that if A and B are square matrices of the same size, then

$$\det(AB) = \det(A)\det(B) \tag{2}$$

The proof of this theorem is fairly intricate, so we will have to develop some preliminary results first. We begin with the special case of (2) in which A is an elementary matrix. Because this special case is only a prelude to (2), we call it a lemma.

Lemma 2.3.2.   *If B is an $n \times n$ matrix and E is an $n \times n$ elementary matrix, then*

$$det(EB) = det(E)\,det(B)$$

Proof.   We shall consider three cases, each depending on the row operation that produces E.

Case 1.   If E results from multiplying a row of I_n by k, then by Theorem 1.5.1 EB results from B by multiplying a row by k; so from Theorem 2.2.3a we have

$$\det(EB) = k\det(B)$$

But from Theorem 2.2.4a we have $\det(E) = k$, so

$$\det(EB) = \det(E)\det(B)$$

Cases 2 and 3.   The proofs of the cases where E results from interchanging two rows of I_n or from adding a multiple of one row to another follow the same pattern as Case 1 and are left as exercises. ▮

REMARK.   It follows by repeated applications of Lemma 2.3.2 that if B is an $n \times n$ matrix and $E_1, E_2, \ldots, E_r$ are $n \times n$ elementary matrices, then

$$\det(E_1 E_2 \cdots E_r B) = \det(E_1)\det(E_2)\cdots\det(E_r)\det(B) \tag{3}$$

For example,

$$\det(E_1 E_2 B) = \det(E_1)\det(E_2 B) = \det(E_1)\det(E_2)\det(B)$$

DETERMINANT TEST FOR INVERTIBILITY

The next theorem is one of the most fundamental in linear algebra; it provides an important criterion for invertibility in terms of determinants, and it will be used in proving (2).

Theorem 2.3.3. *A square matrix A is invertible if and only if det(A) ≠ 0.*

Proof. Let R be the reduced row-echelon form of A. As a preliminary step, we will show that $\det(A)$ and $\det(R)$ are both zero or both nonzero: Let $E_1, E_2, \ldots, E_r$ be the elementary matrices that correspond to the elementary row operations that produce R from A. Thus,

$$R = E_r \cdots E_2 E_1 A$$

and from (3)

$$\det(R) = \det(E_r) \cdots \det(E_2) \det(E_1) \det(A) \tag{4}$$

But from Theorem 2.2.4 the determinants of the elementary matrices are all nonzero. (Keep in mind that multiplying a row by zero is *not* an allowable elementary row operation, so $k \neq 0$ in this application of Theorem 2.2.4.) Thus, it follows from (4) that $\det(A)$ and $\det(R)$ are both zero or both nonzero. Now to the main body of the proof.

If A is invertible, then by Theorem 1.6.4 we have $R = I$, so $\det(R) = 1 \neq 0$ and consequently $\det(A) \neq 0$. Conversely, if $\det(A) \neq 0$, then $\det(R) \neq 0$, so R cannot have a row of zeros. It follows from Theorem 1.4.3 that $R = I$, so A is invertible by Theorem 1.6.4.

It follows from Theorems 2.3.3 and 2.2.5 that a square matrix with two proportional rows or columns is not invertible.

Example 3 Since the first and third rows of

$$A = \begin{bmatrix} 1 & 2 & 3 \\ 1 & 0 & 1 \\ 2 & 4 & 6 \end{bmatrix}$$

are proportional, $\det(A) = 0$. Thus, A is not invertible. ▲

We are now ready for the main result in this section.

Theorem 2.3.4. *If A and B are square matrices of the same size, then*

$$det(AB) = det(A)det(B)$$

Proof. We divide the proof into two cases that depend on whether or not A is invertible. If the matrix A is not invertible, then by Theorem 1.6.5 neither is the product AB. Thus, from Theorem 2.3.3, we have $\det(AB) = 0$ and $\det(A) = 0$, so it follows that $\det(AB) = \det(A) \det(B)$.

Now assume that A is invertible. By Theorem 1.6.4, the matrix A is expressible as a product of elementary matrices, say

$$A = E_1 E_2 \cdots E_r \tag{5}$$

so
$$AB = E_1 E_2 \cdots E_r B$$

Applying (3) to this equation yields

$$\det(AB) = \det(E_1)\det(E_2)\cdots\det(E_r)\det(B)$$

and applying (3) again yields

$$\det(AB) = \det(E_1 E_2 \cdots E_r)\det(B)$$

which, from (5), can be written as $\det(AB) = \det(A)\det(B)$.

Example 4   Consider the matrices

$$A = \begin{bmatrix} 3 & 1 \\ 2 & 1 \end{bmatrix} \qquad B = \begin{bmatrix} -1 & 3 \\ 5 & 8 \end{bmatrix} \qquad AB = \begin{bmatrix} 2 & 17 \\ 3 & 14 \end{bmatrix}$$

We leave it for the reader to verify that

$$\det(A) = 1 \qquad \det(B) = -23 \qquad \text{and} \qquad \det(AB) = -23$$

Thus, $\det(AB) = \det(A)\det(B)$ as guaranteed by Theorem 2.3.4. ▲

The following theorem gives a useful relationship between the determinant of an invertible matrix and the determinant of its inverse.

Theorem 2.3.5. *If A is invertible, then*

$$det(A^{-1}) = \frac{1}{det(A)}$$

Proof. Since $A^{-1}A = I$, it follows that $\det(A^{-1}A) = \det(I)$. Therefore, we must have $\det(A^{-1})\det(A) = 1$. Since $\det(A) \neq 0$, the proof can be completed by dividing through by $\det(A)$. ▮

LINEAR SYSTEMS OF THE FORM $A\mathbf{x} = \lambda\mathbf{x}$

Many applications of linear algebra are concerned with systems of n linear equations in n unknowns that are expressed in the form

$$A\mathbf{x} = \lambda\mathbf{x} \qquad (6)$$

where λ is a scalar. Such systems are really homogeneous linear systems in disguise, since (6) can be rewritten as $\lambda\mathbf{x} - A\mathbf{x} = \mathbf{0}$ or, by inserting an identity matrix and factoring, as

$$(\lambda I - A)\mathbf{x} = \mathbf{0} \qquad (7)$$

Here is an example.

Example 5 The linear system

$$x_1 + 3x_2 = \lambda x_1$$
$$4x_1 + 2x_2 = \lambda x_2$$

can be written in matrix form as

$$\begin{bmatrix} 1 & 3 \\ 4 & 2 \end{bmatrix} \begin{bmatrix} x_1 \\ x_2 \end{bmatrix} = \lambda \begin{bmatrix} x_1 \\ x_2 \end{bmatrix}$$

which is of form (6) with

$$A = \begin{bmatrix} 1 & 3 \\ 4 & 2 \end{bmatrix} \quad \text{and} \quad \mathbf{x} = \begin{bmatrix} x_1 \\ x_2 \end{bmatrix}$$

This system can be rewritten as

$$\lambda \begin{bmatrix} x_1 \\ x_2 \end{bmatrix} - \begin{bmatrix} 1 & 3 \\ 4 & 2 \end{bmatrix} \begin{bmatrix} x_1 \\ x_2 \end{bmatrix} = \begin{bmatrix} 0 \\ 0 \end{bmatrix}$$

or

$$\lambda \begin{bmatrix} 1 & 0 \\ 0 & 1 \end{bmatrix} \begin{bmatrix} x_1 \\ x_2 \end{bmatrix} - \begin{bmatrix} 1 & 3 \\ 4 & 2 \end{bmatrix} \begin{bmatrix} x_1 \\ x_2 \end{bmatrix} = \begin{bmatrix} 0 \\ 0 \end{bmatrix}$$

or

$$\begin{bmatrix} \lambda - 1 & -3 \\ -4 & \lambda - 2 \end{bmatrix} \begin{bmatrix} x_1 \\ x_2 \end{bmatrix} = \begin{bmatrix} 0 \\ 0 \end{bmatrix}$$

which is of form (7) with

$$\lambda I - A = \begin{bmatrix} \lambda - 1 & -3 \\ -4 & \lambda - 2 \end{bmatrix} \quad \blacktriangle$$

The primary problem of interest for linear systems of the form (7) is to determine those values of λ for which the system has a nontrivial solution; such a value of λ is called a ***characteristic value*** or an ***eigenvalue**** of A. If λ is an eigenvalue of A, then the nontrivial solutions of (7) are called the ***eigenvectors*** of A corresponding to λ.

It follows from Theorem 2.3.3 that the system $(\lambda I - A)\mathbf{x} = \mathbf{0}$ has a nontrivial solution if and only if

$$\det(\lambda I - A) = 0 \tag{8}$$

---

* The word *eigenvalue* is a mixture of German and English. The German prefix *eigen* can be translated as "proper," which stems from the older literature where eigenvalues were known as *proper values*; they were also called *latent roots*.

This is called the ***characteristic equation*** of A; the eigenvalues of A can be found by solving this equation for λ.

Eigenvalues and eigenvectors will be studied again in subsequent chapters, where we will discuss their geometric interpretation and develop their properties in more depth.

Example 6 Find the eigenvalues and corresponding eigenvectors of the matrix A in Example 5.

Solution. The characteristic equation of A is

$$\det(\lambda I - A) = \begin{vmatrix} \lambda - 1 & -3 \\ -4 & \lambda - 2 \end{vmatrix} = 0$$

or

$$\lambda^2 - 3\lambda - 10 = 0$$

The factored form of this equation is $(\lambda + 2)(\lambda - 5) = 0$, so the eigenvalues of A are $\lambda = -2$ and $\lambda = 5$.

By definition

$$\mathbf{x} = \begin{bmatrix} x_1 \\ x_2 \end{bmatrix}$$

is an eigenvector of A if and only if $\mathbf{x}$ is a nontrivial solution of $(\lambda I - A)\mathbf{x} = \mathbf{0}$; that is,

$$\begin{bmatrix} \lambda - 1 & -3 \\ -4 & \lambda - 2 \end{bmatrix} \begin{bmatrix} x_1 \\ x_2 \end{bmatrix} = \begin{bmatrix} 0 \\ 0 \end{bmatrix} \tag{9}$$

If $\lambda = -2$, then (9) becomes

$$\begin{bmatrix} -3 & -3 \\ -4 & -4 \end{bmatrix} \begin{bmatrix} x_1 \\ x_2 \end{bmatrix} = \begin{bmatrix} 0 \\ 0 \end{bmatrix}$$

Solving this system yields (verify)

$$x_1 = -t, \qquad x_2 = t$$

so the eigenvectors corresponding to $\lambda = -2$ are the nonzero solutions of the form

$$\mathbf{x} = \begin{bmatrix} x_1 \\ x_2 \end{bmatrix} = \begin{bmatrix} -t \\ t \end{bmatrix}$$

Again from (9), the eigenvectors of A corresponding to $\lambda = 5$ are the nontrivial solutions of

$$\begin{bmatrix} 4 & -3 \\ -4 & 3 \end{bmatrix} \begin{bmatrix} x_1 \\ x_2 \end{bmatrix} = \begin{bmatrix} 0 \\ 0 \end{bmatrix}$$

We leave it for the reader to solve this system and show that the eigenvectors of A corresponding to $\lambda = 5$ are the nonzero solutions of the form

$$\mathbf{x} = \begin{bmatrix} -\frac{3}{4}t \\ t \end{bmatrix} \quad \blacktriangle$$

SUMMARY

In Theorem 1.6.4 we listed five results that are equivalent to the invertibility of a matrix A. We conclude this section by merging Theorem 2.3.3 with that list to produce the following theorem that relates all of the major topics we have studied thus far.

> **Theorem 2.3.6.** *If A is an $n \times n$ matrix, then the following are equivalent.*
>
> (a) *A is invertible.*
> (b) *$A\mathbf{x} = \mathbf{0}$ has only the trivial solution.*
> (c) *The reduced row-echelon form of A is I_n.*
> (d) *A is expressible as a product of elementary matrices.*
> (e) *$A\mathbf{x} = \mathbf{b}$ is consistent for every $n \times 1$ matrix $\mathbf{b}$.*
> (f) *$A\mathbf{x} = \mathbf{b}$ has exactly one solution for every $n \times 1$ matrix $\mathbf{b}$.*
> (g) *$\det(A) \neq 0$.*

EXERCISE SET 2.3

1. Verify that $\det(kA) = k^n \det(A)$ for

(a) $A = \begin{bmatrix} -1 & 2 \\ 3 & 4 \end{bmatrix}$; $k = 2$ (b) $A = \begin{bmatrix} 2 & -1 & 3 \\ 3 & 2 & 1 \\ 1 & 4 & 5 \end{bmatrix}$; $k = -2$

2. Verify that $\det(AB) = \det(A)\det(B)$ for

$$A = \begin{bmatrix} 2 & 1 & 0 \\ 3 & 4 & 0 \\ 0 & 0 & 2 \end{bmatrix} \quad \text{and} \quad B = \begin{bmatrix} 1 & -1 & 3 \\ 7 & 1 & 2 \\ 5 & 0 & 1 \end{bmatrix}$$

3. By inspection, explain why $\det(A) = 0$.

$$A = \begin{bmatrix} -2 & 8 & 1 & 4 \\ 3 & 2 & 5 & 1 \\ 1 & 0 & 7 & 0 \\ 4 & -6 & 4 & -3 \end{bmatrix}$$

4. Use Theorem 2.3.3 to determine which of the following matrices are invertible.

(a) $\begin{bmatrix} 1 & 0 & -1 \\ 9 & -1 & 4 \\ 8 & 9 & -1 \end{bmatrix}$ (b) $\begin{bmatrix} 4 & 2 & 8 \\ -2 & 1 & -4 \\ 3 & 1 & 6 \end{bmatrix}$ (c) $\begin{bmatrix} \sqrt{2} & -\sqrt{7} & 0 \\ 3\sqrt{2} & -3\sqrt{7} & 0 \\ 5 & -9 & 0 \end{bmatrix}$ (d) $\begin{bmatrix} -3 & 0 & 1 \\ 5 & 0 & 6 \\ 8 & 0 & 3 \end{bmatrix}$

5. Let

$$A = \begin{bmatrix} a & b & c \\ d & e & f \\ g & h & i \end{bmatrix}$$

Assuming that $\det(A) = -7$, find

(a) $\det(3A)$ (b) $\det(A^{-1})$ (c) $\det(2A^{-1})$ (d) $\det((2A)^{-1})$ (e) $\det \begin{bmatrix} a & g & d \\ b & h & e \\ c & i & f \end{bmatrix}$

6. Without directly evaluating, show that $x = 0$ and $x = 2$ satisfy

$$\begin{vmatrix} x^2 & x & 2 \\ 2 & 1 & 1 \\ 0 & 0 & -5 \end{vmatrix} = 0$$

7. Without directly evaluating, show that

$$\det \begin{bmatrix} b+c & c+a & b+a \\ a & b & c \\ 1 & 1 & 1 \end{bmatrix} = 0$$

In Exercises 8–11 prove the identity without evaluating the determinants.

8. $\begin{vmatrix} a_1 & b_1 & a_1 + b_1 + c_1 \\ a_2 & b_2 & a_2 + b_2 + c_2 \\ a_3 & b_3 & a_3 + b_3 + c_3 \end{vmatrix} = \begin{vmatrix} a_1 & b_1 & c_1 \\ a_2 & b_2 & c_2 \\ a_3 & b_3 & c_3 \end{vmatrix}$

9. $\begin{vmatrix} a_1 + b_1 & a_1 - b_1 & c_1 \\ a_2 + b_2 & a_2 - b_2 & c_2 \\ a_3 + b_3 & a_3 - b_3 & c_3 \end{vmatrix} = -2 \begin{vmatrix} a_1 & b_1 & c_1 \\ a_2 & b_2 & c_2 \\ a_3 & b_3 & c_3 \end{vmatrix}$

10. $\begin{vmatrix} a_1 + b_1 t & a_2 + b_2 t & a_3 + b_3 t \\ a_1 t + b_1 & a_2 t + b_2 & a_3 t + b_3 \\ c_1 & c_2 & c_3 \end{vmatrix} = (1 - t^2) \begin{vmatrix} a_1 & a_2 & a_3 \\ b_1 & b_2 & b_3 \\ c_1 & c_2 & c_3 \end{vmatrix}$

11. $\begin{vmatrix} a_1 & b_1 + ta_1 & c_1 + rb_1 + sa_1 \\ a_2 & b_2 + ta_2 & c_2 + rb_2 + sa_2 \\ a_3 & b_3 + ta_3 & c_3 + rb_3 + sa_3 \end{vmatrix} = \begin{vmatrix} a_1 & a_2 & a_3 \\ b_1 & b_2 & b_3 \\ c_1 & c_2 & c_3 \end{vmatrix}$

12. For which value(s) of k does A fail to be invertible?

(a) $A = \begin{bmatrix} k - 3 & -2 \\ -2 & k - 2 \end{bmatrix}$ (b) $A = \begin{bmatrix} 1 & 2 & 4 \\ 3 & 1 & 6 \\ k & 3 & 2 \end{bmatrix}$

13. Use Theorem 2.3.3 to show that

$$\begin{bmatrix} \sin^2 \alpha & \sin^2 \beta & \sin^2 \gamma \\ \cos^2 \alpha & \cos^2 \beta & \cos^2 \gamma \\ 1 & 1 & 1 \end{bmatrix}$$

is not invertible for any values of α, β, and γ.

14. Express the following linear systems in the form $(\lambda I - A)\mathbf{x} = \mathbf{0}$.

(a) $\begin{aligned} x_1 + 2x_2 &= \lambda x_1 \\ 2x_1 + x_2 &= \lambda x_2 \end{aligned}$ (b) $\begin{aligned} 2x_1 + 3x_2 &= \lambda x_1 \\ 4x_1 + 3x_2 &= \lambda x_2 \end{aligned}$ (c) $\begin{aligned} 3x_1 + x_2 &= \lambda x_1 \\ -5x_1 - 3x_2 &= \lambda x_2 \end{aligned}$

15. For each of the systems in Exercise 14, find
(a) the characteristic equation
(b) the eigenvalues
(c) the eigenvectors corresponding to each of the eigenvalues

16. Let A and B be $n \times n$ matrices. Show that if A is invertible, then $\det(B) = \det(A^{-1}BA)$.

17. (a) Express

$$\begin{vmatrix} a_1 + b_1 & c_1 + d_1 \\ a_2 + b_2 & c_2 + d_2 \end{vmatrix}$$

as a sum of four determinants whose entries contain no sums.

   (b) Express

$$\begin{vmatrix} a_1 + b_1 & c_1 + d_1 & e_1 + f_1 \\ a_2 + b_2 & c_2 + d_2 & e_2 + f_2 \\ a_3 + b_3 & c_3 + d_3 & e_3 + f_3 \end{vmatrix}$$

as a sum of eight determinants whose entries contain no sums.

18. Prove that a square matrix A is invertible if and only if $A^T A$ is invertible.

19. Prove cases 2 and 3 of Lemma 2.3.2.

2.4 COFACTOR EXPANSION; CRAMER'S RULE

In this section we shall consider a method for evaluating determinants that is useful for hand computations and important theoretically. As a consequence of our work here, we will obtain a formula for the inverse of an invertible matrix as well as a formula for the solution to certain systems of linear equations in terms of determinants.

MINORS AND COFACTORS

Definition. If A is a square matrix, then the ***minor of entry*** a_{ij} is denoted by M_{ij} and is defined to be the determinant of the submatrix that remains after the ith row and jth column are deleted from A. The number $(-1)^{i+j}M_{ij}$ is denoted by C_{ij} and is called the ***cofactor of entry*** a_{ij}.

Example 1 Let

$$A = \begin{bmatrix} 3 & 1 & -4 \\ 2 & 5 & 6 \\ 1 & 4 & 8 \end{bmatrix}$$

The minor of entry a_{11} is

$$M_{11} = \begin{vmatrix} 3 & 1 & -4 \\ 2 & 5 & 6 \\ 1 & 4 & 8 \end{vmatrix} = \begin{vmatrix} 5 & 6 \\ 4 & 8 \end{vmatrix} = 16$$

The cofactor of a_{11} is

$$C_{11} = (-1)^{1+1}M_{11} = M_{11} = 16$$

Similarly, the minor of entry a_{32} is

$$M_{32} = \begin{vmatrix} 3 & 1 & -4 \\ 2 & 5 & 6 \\ 1 & 4 & 8 \end{vmatrix} = \begin{vmatrix} 3 & -4 \\ 2 & 6 \end{vmatrix} = 26$$

The cofactor of a_{32} is

$$C_{32} = (-1)^{3+2}M_{32} = -M_{32} = -26 \quad \blacktriangle$$

Notice that the cofactor and the minor of an element a_{ij} differ only in sign, that is, $C_{ij} = \pm M_{ij}$. A quick way for determining whether to use the + or − is to use the fact that the sign relating C_{ij} and M_{ij} is in the ith row and jth column of the "checkerboard" array

$$\begin{bmatrix} + & - & + & - & + & \cdots \\ - & + & - & + & - & \cdots \\ + & - & + & - & + & \cdots \\ - & + & - & + & - & \cdots \\ \vdots & \vdots & \vdots & \vdots & \vdots \end{bmatrix}$$

For example, $C_{11} = M_{11}$, $C_{21} = -M_{21}$, $C_{12} = -M_{12}$, $C_{22} = M_{22}$, and so on.

COFACTOR EXPANSIONS

Consider the general 3×3 matrix

$$A = \begin{bmatrix} a_{11} & a_{12} & a_{13} \\ a_{21} & a_{22} & a_{23} \\ a_{31} & a_{32} & a_{33} \end{bmatrix}$$

In Example 7 of Section 2.1 we showed that

$$\det(A) = a_{11}a_{22}a_{33} + a_{12}a_{23}a_{31} + a_{13}a_{21}a_{32}$$
$$- a_{13}a_{22}a_{31} - a_{12}a_{21}a_{33} - a_{11}a_{23}a_{32} \quad (1)$$

which can be rewritten as

$$\det(A) = a_{11}(a_{22}a_{33} - a_{23}a_{32}) + a_{21}(a_{13}a_{32} - a_{12}a_{33}) + a_{31}(a_{12}a_{23} - a_{13}a_{22})$$

Since the expressions in parentheses are just the cofactors C_{11}, C_{21}, and C_{31} (verify), we have

$$\det(A) = a_{11}C_{11} + a_{21}C_{21} + a_{31}C_{31} \quad (2)$$

Equation (2) shows that the determinant of A can be computed by multiplying the entries in the first column of A by their cofactors and adding the resulting products. This method of evaluating $\det(A)$ is called ***cofactor expansion*** along the first column of A.

Example 2   Let

$$A = \begin{bmatrix} 3 & 1 & 0 \\ -2 & -4 & 3 \\ 5 & 4 & -2 \end{bmatrix}$$

Evaluate $\det(A)$ by cofactor expansion along the first column of A.

Solution. From (2)

$$\det(A) = 3\begin{vmatrix} -4 & 3 \\ 4 & -2 \end{vmatrix} - (-2)\begin{vmatrix} 1 & 0 \\ 4 & -2 \end{vmatrix} + 5\begin{vmatrix} 1 & 0 \\ -4 & 3 \end{vmatrix}$$

$$= 3(-4) - (-2)(-2) + 5(3) = -1 \quad \blacktriangle$$

By rearranging the terms in (1) in various ways, it is possible to obtain other formulas like (2). There should be no trouble checking that all of the following are correct (see Exercise 28):

$$
\begin{aligned}
\det(A) &= a_{11}C_{11} + a_{12}C_{12} + a_{13}C_{13} \\
&= a_{11}C_{11} + a_{21}C_{21} + a_{31}C_{31} \\
&= a_{21}C_{21} + a_{22}C_{22} + a_{23}C_{23} \\
&= a_{12}C_{12} + a_{22}C_{22} + a_{32}C_{32} \\
&= a_{31}C_{31} + a_{32}C_{32} + a_{33}C_{33} \\
&= a_{13}C_{13} + a_{23}C_{23} + a_{33}C_{33}
\end{aligned}
\tag{3}
$$

Notice that in each equation the entries and cofactors all come from the same row or column. These equations are called the *cofactor expansions* of $\det(A)$.

The results we have just given for 3×3 matrices form a special case of the following general theorem, which we state without proof.

Theorem 2.4.1. *The determinant of an $n \times n$ matrix A can be computed by multiplying the entries in any row (or column) by their cofactors and adding the resulting products; that is, for each $1 \le i \le n$ and $1 \le j \le n$,*

$$\det(A) = a_{1j}C_{1j} + a_{2j}C_{2j} + \cdots + a_{nj}C_{nj}$$

(cofactor expansion along the jth column)

and

$$\det(A) = a_{i1}C_{i1} + a_{i2}C_{i2} + \cdots + a_{in}C_{in}$$

(cofactor expansion along the ith row)

Example 3 Let A be the matrix in Example 2. Evaluate $\det(A)$ by cofactor expansion along the first row.

Solution.

$$\det(A) = \begin{vmatrix} 3 & 1 & 0 \\ -2 & -4 & 3 \\ 5 & 4 & -2 \end{vmatrix} = 3\begin{vmatrix} -4 & 3 \\ 4 & -2 \end{vmatrix} - (1)\begin{vmatrix} -2 & 3 \\ 5 & -2 \end{vmatrix} + 0\begin{vmatrix} -2 & -4 \\ 5 & 4 \end{vmatrix}$$

$$= 3(-4) - (1)(-11) + 0 = -1$$

This agrees with the result obtained in Example 2. $\blacktriangle$

REMARK. In this example it was unnecessary to compute the last cofactor, since it was multiplied by zero. In general, the best strategy for evaluating a determinant by cofactor expansion is to expand along a row or column having the largest number of zeros.

Cofactor expansion and row or column operations can sometimes be used in combination to provide an effective method for evaluating determinants. The following example illustrates this idea.

Example 4   Evaluate $\det(A)$ where

$$A = \begin{bmatrix} 3 & 5 & -2 & 6 \\ 1 & 2 & -1 & 1 \\ 2 & 4 & 1 & 5 \\ 3 & 7 & 5 & 3 \end{bmatrix}$$

Solution. By adding suitable multiples of the second row to the remaining rows, we obtain

$$\det(A) = \begin{vmatrix} 0 & -1 & 1 & 3 \\ 1 & 2 & -1 & 1 \\ 0 & 0 & 3 & 3 \\ 0 & 1 & 8 & 0 \end{vmatrix}$$

$$= - \begin{vmatrix} -1 & 1 & 3 \\ 0 & 3 & 3 \\ 1 & 8 & 0 \end{vmatrix} \qquad \boxed{\begin{array}{l}\text{Cofactor expansion} \\ \text{along the first} \\ \text{column}\end{array}}$$

$$= - \begin{vmatrix} -1 & 1 & 3 \\ 0 & 3 & 3 \\ 0 & 9 & 3 \end{vmatrix} \qquad \boxed{\begin{array}{l}\text{We added the} \\ \text{first row to} \\ \text{the third row}\end{array}}$$

$$= -(-1)\begin{vmatrix} 3 & 3 \\ 9 & 3 \end{vmatrix} \qquad \boxed{\begin{array}{l}\text{Cofactor expansion} \\ \text{along the first} \\ \text{column}\end{array}}$$

$$= -18 \quad \blacktriangle$$

ADJOINT OF A MATRIX

In a cofactor expansion we compute $\det(A)$ by multiplying the entries in a row or column by their cofactors and adding the resulting products. It turns out that if one multiplies the entries in any row by the corresponding cofactors from a *different* row, the sum of these products is always zero. (This result also holds for columns.) Although we omit the general proof, the next example illustrates the idea of the proof in a special case.

Example 5 Let

$$A = \begin{bmatrix} a_{11} & a_{12} & a_{13} \\ a_{21} & a_{22} & a_{23} \\ a_{31} & a_{32} & a_{33} \end{bmatrix}$$

Consider the quantity

$$a_{11}C_{31} + a_{12}C_{32} + a_{13}C_{33}$$

that is formed by multiplying the entries in the first row by the cofactors of the corresponding entries in the third row and adding the resulting products. We now show that this quantity is equal to zero by the following trick. Construct a new matrix A' by replacing the third row of A with another copy of the first row. Thus,

$$A' = \begin{bmatrix} a_{11} & a_{12} & a_{13} \\ a_{21} & a_{22} & a_{23} \\ a_{11} & a_{12} & a_{13} \end{bmatrix}$$

Let $C'_{31}, C'_{32}, C'_{33}$ be the cofactors of the entries in the third row of A'. Since the first two rows of A and A' are the same, and since the computation of C_{31}, C_{32}, C_{33}, $C'_{31}, C'_{32},$ and C'_{33} involve only entries from the first two rows of A and A', it follows that

$$C_{31} = C'_{31}, \qquad C_{32} = C'_{32}, \qquad C_{33} = C'_{33}$$

Since A' has two identical rows,

$$\det(A') = 0 \tag{4}$$

On the other hand, evaluating $\det(A')$ by cofactor expansion along the third row gives

$$\det(A') = a_{11}C'_{31} + a_{12}C'_{32} + a_{13}C'_{33} = a_{11}C_{31} + a_{12}C_{32} + a_{13}C_{33} \tag{5}$$

From (4) and (5) we obtain

$$a_{11}C_{31} + a_{12}C_{32} + a_{13}C_{33} = 0 \quad \blacktriangle$$

Definition. If A is any $n \times n$ matrix and C_{ij} is the cofactor of a_{ij}, then the matrix

$$\begin{bmatrix} C_{11} & C_{12} & \cdots & C_{1n} \\ C_{21} & C_{22} & \cdots & C_{2n} \\ \vdots & \vdots & & \vdots \\ C_{n1} & C_{n2} & \cdots & C_{nn} \end{bmatrix}$$

is called the *matrix of cofactors from A*. The transpose of this matrix is called the *adjoint of A* and is denoted by $\text{adj}(A)$.

Example 6   Let

$$A = \begin{bmatrix} 3 & 2 & -1 \\ 1 & 6 & 3 \\ 2 & -4 & 0 \end{bmatrix}$$

The cofactors of A are

$$\begin{array}{ccc} C_{11} = 12 & C_{12} = 6 & C_{13} = -16 \\ C_{21} = 4 & C_{22} = 2 & C_{23} = 16 \\ C_{31} = 12 & C_{32} = -10 & C_{33} = 16 \end{array}$$

so that the matrix of cofactors is

$$\begin{bmatrix} 12 & 6 & -16 \\ 4 & 2 & 16 \\ 12 & -10 & 16 \end{bmatrix}$$

and the adjoint of A is

$$\text{adj}(A) = \begin{bmatrix} 12 & 4 & 12 \\ 6 & 2 & -10 \\ -16 & 16 & 16 \end{bmatrix} \quad \blacktriangle$$

We are now in a position to derive a formula for the inverse of an invertible matrix.

FORMULA FOR THE INVERSE OF A MATRIX

Theorem 2.4.2. *If A is an invertible matrix, then*

$$A^{-1} = \frac{1}{det(A)} adj(A) \tag{6}$$

Proof. We show first that

$$A \text{ adj}(A) = \det(A)I$$

Consider the product

$$A \text{ adj}(A) = \begin{bmatrix} a_{11} & a_{12} & \cdots & a_{1n} \\ a_{21} & a_{22} & \cdots & a_{2n} \\ \vdots & \vdots & & \vdots \\ a_{i1} & a_{i2} & \cdots & a_{in} \\ \vdots & \vdots & & \vdots \\ a_{n1} & a_{n2} & \cdots & a_{nn} \end{bmatrix} \begin{bmatrix} C_{11} & C_{21} & \cdots & C_{j1} & \cdots & C_{n1} \\ C_{12} & C_{22} & \cdots & C_{j2} & \cdots & C_{n2} \\ \vdots & \vdots & & \vdots & & \vdots \\ C_{1n} & C_{2n} & \cdots & C_{jn} & \cdots & C_{nn} \end{bmatrix}$$

The entry in the ith row and jth column of the product $A \text{ adj}(A)$ is

$$a_{i1}C_{j1} + a_{i2}C_{j2} + \cdots + a_{in}C_{jn} \tag{7}$$

(see the shaded lines above).

If $i = j$, then (7) is the cofactor expansion of $\det(A)$ along the ith row of A (Theorem 2.4.1), and if $i \neq j$, then the a's and the cofactors come from different rows of A, so the value of (7) is zero. Therefore,

$$A \, \mathrm{adj}(A) = \begin{bmatrix} \det(A) & 0 & \cdots & 0 \\ 0 & \det(A) & \cdots & 0 \\ \vdots & \vdots & & \vdots \\ 0 & 0 & \cdots & \det(A) \end{bmatrix} = \det(A)I \tag{8}$$

Since A is invertible, $\det(A) \neq 0$. Therefore, Equation (8) can be rewritten as

$$\frac{1}{\det(A)}[A \, \mathrm{adj}(A)] = I$$

or

$$A\left[\frac{1}{\det(A)} \mathrm{adj}(A)\right] = I$$

Multiplying both sides on the left by A^{-1} yields

$$A^{-1} = \frac{1}{\det(A)} \mathrm{adj}(A) \quad \blacksquare$$

Example 7 Use (6) to find the inverse of the matrix A in Example 6.

Solution. The reader can check that $\det(A) = 64$. Thus,

$$A^{-1} = \frac{1}{\det(A)} \mathrm{adj}(A) = \frac{1}{64}\begin{bmatrix} 12 & 4 & 12 \\ 6 & 2 & -10 \\ -16 & 16 & 16 \end{bmatrix}$$

$$= \begin{bmatrix} \frac{12}{64} & \frac{4}{64} & \frac{12}{64} \\ \frac{6}{64} & \frac{2}{64} & -\frac{10}{64} \\ -\frac{16}{64} & \frac{16}{64} & \frac{16}{64} \end{bmatrix} \quad \blacktriangle$$

APPLICATIONS OF THE ADJOINT FORMULA FOR THE INVERSE

Although the method in the preceding example is reasonable for inverting 3×3 matrices by hand, the inversion algorithm discussed in Section 1.5 is more efficient for larger matrices. It should be kept in mind, however, that the method of Section 1.5 is just a computational procedure, whereas Formula (6) is an actual formula for the inverse. As we shall now see, this formula is useful for deriving properties of the inverse.

In Section 1.7 we stated two results about inverses without proof.

- **Theorem 1.7.1c:** A triangular matrix is invertible if and only if its diagonal entries are all nonzero.
- **Theorem 1.7.1d:** The inverse of an invertible lower triangular matrix is lower triangular, and the inverse of an invertible upper triangular matrix is upper triangular.

We will now prove these results using the adjoint formula for the inverse.

Proof of Theorem 1.7.1c. Let $A = [a_{ij}]$ be a triangular matrix, so that its diagonal entries are

$$a_{11}, a_{22}, \ldots, a_{nn}$$

From Theorems 2.2.2 and 2.3.3, the matrix A is invertible if and only if

$$\det(A) = a_{11}a_{22} \cdots a_{nn} \neq 0$$

which is true if and only if the diagonal entries are all nonzero. ▊

We leave it as an exercise for the reader to use the adjoint formula for A^{-1} to show that if $A = [a_{ij}]$ is an invertible triangular matrix, then the successive diagonal entries of A^{-1} are

$$\frac{1}{a_{11}}, \frac{1}{a_{22}}, \ldots, \frac{1}{a_{nn}}$$

(See Example 3 of Section 1.7.)

Proof of Theorem 1.7.1d. We will prove the result for upper triangular matrices and leave the lower triangular case as an exercise. Assume that A is upper triangular and invertible. Since

$$A^{-1} = \frac{1}{\det(A)} \operatorname{adj}(A)$$

we can prove that A^{-1} is upper triangular by showing that $\operatorname{adj}(A)$ is upper triangular, or equivalently, that the matrix of cofactors is lower triangular. We can do this by showing that every cofactor C_{ij} with $i < j$ (i.e., above the main diagonal) is zero. Since

$$C_{ij} = (-1)^{i+j} M_{ij}$$

it suffices to show that each minor M_{ij} with $i < j$ is zero. For this purpose let B_{ij} be the matrix that results when the ith row and jth column of A are deleted, so that

$$M_{ij} = \det(B_{ij}) \tag{9}$$

From the assumption that $i < j$ it follows that B_{ij} is upper triangular (Exercise 32). Since A is upper triangular, its $(i + 1)$-st row begins with at least i zeros. But the ith row of B_{ij} is the $(i + 1)$-st row of A with the entry in the jth column removed. Since $i < j$, none of the first i zeros is removed by deleting the jth column; thus, the ith row of B_{ij} starts with at least i zeros, which implies that this row has a zero on the main diagonal. It now follows from Theorem 2.2.2 that $\det(B_{ij}) = 0$ and from (9) that $M_{ij} = 0$. ▊

CRAMER'S RULE

The next theorem provides a useful formula for the solution of certain linear systems of n equations in n unknowns. This formula, known as *Cramer's Rule** (see page 109), is of marginal interest for computational purposes, but it is useful for studying the mathematical properties of a solution without the need for solving the system.

Theorem 2.4.3 (*Cramer's Rule*). *If* $A\mathbf{x} = \mathbf{b}$ *is a system of n linear equations in n unknowns such that* $det(A) \neq 0$, *then the system has a unique solution. This solution is*

$$x_1 = \frac{det(A_1)}{det(A)}, \quad x_2 = \frac{det(A_2)}{det(A)}, \quad \ldots, \quad x_n = \frac{det(A_n)}{det(A)}$$

where A_j *is the matrix obtained by replacing the entries in the jth column of A by the entries in the matrix*

$$\mathbf{b} = \begin{bmatrix} b_1 \\ b_2 \\ \vdots \\ b_n \end{bmatrix}$$

Proof. If $det(A) \neq 0$, then A is invertible and, by Theorem 1.6.2, $\mathbf{x} = A^{-1}\mathbf{b}$ is the unique solution of $A\mathbf{x} = \mathbf{b}$. Therefore, by Theorem 2.4.2 we have

$$\mathbf{x} = A^{-1}\mathbf{b} = \frac{1}{det(A)} \text{adj}(A)\mathbf{b} = \frac{1}{det(A)} \begin{bmatrix} C_{11} & C_{21} & \cdots & C_{n1} \\ C_{12} & C_{22} & \cdots & C_{n2} \\ \vdots & \vdots & & \vdots \\ C_{1n} & C_{2n} & \cdots & C_{nn} \end{bmatrix} \begin{bmatrix} b_1 \\ b_2 \\ \vdots \\ b_n \end{bmatrix}$$

Multiplying the matrices out gives

$$\mathbf{x} = \frac{1}{det(A)} \begin{bmatrix} b_1 C_{11} + b_2 C_{21} + \cdots + b_n C_{n1} \\ b_1 C_{12} + b_2 C_{22} + \cdots + b_n C_{n2} \\ \vdots \quad \vdots \quad \vdots \\ b_1 C_{1n} + b_2 C_{2n} + \cdots + b_n C_{nn} \end{bmatrix}$$

---

* *Gabriel Cramer* (1704–1752), Swiss mathematician. Although Cramer does not rank with the great mathematicians of his time, his contributions as a disseminator of mathematical ideas have earned him a well-deserved place in the history of mathematics. Cramer traveled extensively and met many of the leading mathematicians of his day. These contacts and friendships led to extensive correspondence in which information about new mathematical discoveries was transmitted.

Cramer's most widely known work, *Introduction à l'analyse des lignes courbes algébriques* (1750), was a study and classification of algebraic curves; Cramer's Rule appeared in the appendix. Although the rule bears his name, variations of the basic idea were formulated earlier by various mathematicians. However, Cramer's superior notation helped clarify and popularize the technique.

Overwork combined with a fall from a carriage eventually led to his death in 1752. Cramer was apparently a good-natured and pleasant person, though he never married. His interests were broad. He wrote on philosophy of law and government and the history of mathematics. He served in public office, participated in artillery and fortifications activities for the government, instructed workers on techniques of cathedral repair, and undertook excavations of cathedral archives. Cramer received numerous honors for his activities.

The entry in the *j*th row of **x** is therefore

$$x_j = \frac{b_1 C_{1j} + b_2 C_{2j} + \cdots + b_n C_{nj}}{\det(A)} \tag{10}$$

Now let

$$A_j = \begin{bmatrix} a_{11} & a_{12} & \cdots & a_{1j-1} & b_1 & a_{1j+1} & \cdots & a_{1n} \\ a_{21} & a_{22} & \cdots & a_{2j-1} & b_2 & a_{2j+1} & \cdots & a_{2n} \\ \vdots & \vdots & & \vdots & \vdots & \vdots & & \vdots \\ a_{n1} & a_{n2} & \cdots & a_{nj-1} & b_n & a_{nj+1} & \cdots & a_{nn} \end{bmatrix}$$

Since A_j differs from A only in the *j*th column, it follows that the cofactors of entries $b_1, b_2, \ldots, b_n$ in A_j are the same as the cofactors of the corresponding entries in the *j*th column of A. The cofactor expansion of $\det(A_j)$ along the *j*th column is therefore

$$\det(A_j) = b_1 C_{1j} + b_2 C_{2j} + \cdots + b_n C_{nj}$$

Substituting this result in (10) gives

$$x_j = \frac{\det(A_j)}{\det(A)} \quad \blacksquare$$

Example 8   Use Cramer's Rule to solve

$$\begin{aligned} x_1 + \quad\quad\; + 2x_3 &= 6 \\ -3x_1 + 4x_2 + 6x_3 &= 30 \\ -x_1 - 2x_2 + 3x_3 &= 8 \end{aligned}$$

Solution.

$$A = \begin{bmatrix} 1 & 0 & 2 \\ -3 & 4 & 6 \\ -1 & -2 & 3 \end{bmatrix} \quad A_1 = \begin{bmatrix} 6 & 0 & 2 \\ 30 & 4 & 6 \\ 8 & -2 & 3 \end{bmatrix}$$

$$A_2 = \begin{bmatrix} 1 & 6 & 2 \\ -3 & 30 & 6 \\ -1 & 8 & 3 \end{bmatrix} \quad A_3 = \begin{bmatrix} 1 & 0 & 6 \\ -3 & 4 & 30 \\ -1 & -2 & 8 \end{bmatrix}$$

Therefore,

$$x_1 = \frac{\det(A_1)}{\det(A)} = \frac{-40}{44} = \frac{-10}{11}, \quad x_2 = \frac{\det(A_2)}{\det(A)} = \frac{72}{44} = \frac{18}{11},$$

$$x_3 = \frac{\det(A_3)}{\det(A)} = \frac{152}{44} = \frac{38}{11} \quad \blacktriangle$$

REMARK. To solve a system of n equations in n unknowns by Cramer's Rule, it is necessary to evaluate $n + 1$ determinants of $n \times n$ matrices. For systems with more than three equations, Gaussian elimination is far more efficient, since it is only necessary to reduce one $n \times (n + 1)$ augmented matrix. However, Cramer's Rule does give a formula for the solution if the determinant of the coefficient matrix is nonzero.

EXERCISE SET 2.4

1. Let

$$A = \begin{bmatrix} 1 & -2 & 3 \\ 6 & 7 & -1 \\ -3 & 1 & 4 \end{bmatrix}$$

(a) Find all the minors of A.   (b) Find all the cofactors.

2. Let

$$A = \begin{bmatrix} 4 & -1 & 1 & 6 \\ 0 & 0 & -3 & 3 \\ 4 & 1 & 0 & 14 \\ 4 & 1 & 3 & 2 \end{bmatrix}$$

Find
(a) M_{13} and C_{13} (b) M_{23} and C_{23} (c) M_{22} and C_{22} (d) M_{21} and C_{21}

3. Evaluate the determinant of the matrix in Exercise 1 by a cofactor expansion along
(a) the first row   (b) the first column   (c) the second row
(d) the second column   (e) the third row   (f) the third column

4. For the matrix in Exercise 1, find
(a) adj(A) (b) A^{-1} using Theorem 2.4.2

In Exercises 5–10 evaluate det(A) by a cofactor expansion along a row or column of your choice.

5. $A = \begin{bmatrix} -3 & 0 & 7 \\ 2 & 5 & 1 \\ -1 & 0 & 5 \end{bmatrix}$ **6.** $A = \begin{bmatrix} 3 & 3 & 1 \\ 1 & 0 & -4 \\ 1 & -3 & 5 \end{bmatrix}$ **7.** $A = \begin{bmatrix} 1 & k & k^2 \\ 1 & k & k^2 \\ 1 & k & k^2 \end{bmatrix}$

8. $A = \begin{bmatrix} k+1 & k-1 & 7 \\ 2 & k-3 & 4 \\ 5 & k+1 & k \end{bmatrix}$ **9.** $A = \begin{bmatrix} 3 & 3 & 0 & 5 \\ 2 & 2 & 0 & -2 \\ 4 & 1 & -3 & 0 \\ 2 & 10 & 3 & 2 \end{bmatrix}$ **10.** $A = \begin{bmatrix} 4 & 0 & 0 & 1 & 0 \\ 3 & 3 & 3 & -1 & 0 \\ 1 & 2 & 4 & 2 & 3 \\ 9 & 4 & 6 & 2 & 3 \\ 2 & 2 & 4 & 2 & 3 \end{bmatrix}$

In Exercises 11–14 find A^{-1} using Theorem 2.4.2.

11. $A = \begin{bmatrix} 2 & 5 & 5 \\ -1 & -1 & 0 \\ 2 & 4 & 3 \end{bmatrix}$ 　　**12.** $A = \begin{bmatrix} 2 & 0 & 3 \\ 0 & 3 & 2 \\ -2 & 0 & -4 \end{bmatrix}$

13. $A = \begin{bmatrix} 2 & -3 & 5 \\ 0 & 1 & -3 \\ 0 & 0 & 2 \end{bmatrix}$ 　　**14.** $A = \begin{bmatrix} 2 & 0 & 0 \\ 8 & 1 & 0 \\ -5 & 3 & 6 \end{bmatrix}$

15. Let

$$A = \begin{bmatrix} 1 & 3 & 1 & 1 \\ 2 & 5 & 2 & 2 \\ 1 & 3 & 8 & 9 \\ 1 & 3 & 2 & 2 \end{bmatrix}$$

(a) Evaluate A^{-1} using Theorem 2.4.2.
(b) Evaluate A^{-1} using the method of Example 4 in Section 1.5.
(c) Which method involves less computation?

In Exercises 16–21 solve by Cramer's Rule, where it applies.

16. $\begin{aligned} 7x_1 - 2x_2 &= 3 \\ 3x_1 + x_2 &= 5 \end{aligned}$ 　　**17.** $\begin{aligned} 4x + 5y &= 2 \\ 11x + y + 2z &= 3 \\ x + 5y + 2z &= 1 \end{aligned}$ 　　**18.** $\begin{aligned} x - 4y + z &= 6 \\ 4x - y + 2z &= -1 \\ 2x + 2y - 3z &= -20 \end{aligned}$

19. $\begin{aligned} x_1 - 3x_2 + x_3 &= 4 \\ 2x_1 - x_2 &= -2 \\ 4x_1 - 3x_3 &= 0 \end{aligned}$ 　　**20.** $\begin{aligned} -x_1 - 4x_2 + 2x_3 + x_4 &= -32 \\ 2x_1 - x_2 + 7x_3 + 9x_4 &= 14 \\ -x_1 + x_2 + 3x_3 + x_4 &= 11 \\ x_1 - 2x_2 + x_3 - 4x_4 &= -4 \end{aligned}$ 　　**21.** $\begin{aligned} 3x_1 - x_2 + x_3 &= 4 \\ -x_1 + 7x_2 - 2x_3 &= 1 \\ 2x_1 + 6x_2 - x_3 &= 5 \end{aligned}$

22. Show that the matrix

$$A = \begin{bmatrix} \cos\theta & \sin\theta & 0 \\ -\sin\theta & \cos\theta & 0 \\ 0 & 0 & 1 \end{bmatrix}$$

is invertible for all values of θ; then find A^{-1} using Theorem 2.4.2.

23. Use Cramer's Rule to solve for y without solving for x, z, and w.

$$\begin{aligned} 4x + y + z + w &= 6 \\ 3x + 7y - z + w &= 1 \\ 7x + 3y - 5z + 8w &= -3 \\ x + y + z + 2w &= 3 \end{aligned}$$

24. Let $Ax = b$ be the system in Exercise 23.
(a) Solve by Cramer's Rule.　　(b) Solve by Gauss–Jordan elimination.
(c) Which method involves fewer computations?

25. Prove that if $\det(A) = 1$ and all the entries in A are integers, then all the entries in A^{-1} are integers.

26. Let $A\mathbf{x} = \mathbf{b}$ be a system of n linear equations in n unknowns with integer coefficients and integer constants. Prove that if $\det(A) = 1$, the solution $\mathbf{x}$ has integer entries.

27. Prove that if A is an invertible lower triangular matrix, then A^{-1} is lower triangular.

28. Derive the first and last cofactor expansions listed in Formula (3).

29. Prove: The equation of the line through the distinct points (a_1, b_1) and (a_2, b_2) can be written as

$$\begin{vmatrix} x & y & 1 \\ a_1 & b_1 & 1 \\ a_2 & b_2 & 1 \end{vmatrix} = 0$$

expand) and get a linear equation

30. Prove: $(x_1, y_1), (x_2, y_2),$ and (x_3, y_3) are collinear points if and only if

$$\begin{vmatrix} x_1 & y_1 & 1 \\ x_2 & y_2 & 1 \\ x_3 & y_3 & 1 \end{vmatrix} = 0$$

31. Prove: The equation of the plane through the noncollinear points $(a_1, b_1, c_1), (a_2, b_2, c_2),$ and (a_3, b_3, c_3) can be written as

$$\begin{vmatrix} x & y & z & 1 \\ a_1 & b_1 & c_1 & 1 \\ a_2 & b_2 & c_2 & 1 \\ a_3 & b_3 & c_3 & 1 \end{vmatrix} = 0$$

32. Prove that if A is upper triangular and B_{ij} is the matrix that results when the ith row and jth column of A are deleted, then B_{ij} is upper triangular if $i < j$.

SUPPLEMENTARY EXERCISES

1. Use Cramer's Rule to solve for x' and y' in terms of x and y.

$$x = \tfrac{3}{5}x' - \tfrac{4}{5}y'$$
$$y = \tfrac{4}{5}x' + \tfrac{3}{5}y'$$

2. Use Cramer's Rule to solve for x' and y' in terms of x and y.

$$x = x' \cos \theta - y' \sin \theta$$
$$y = x' \sin \theta + y' \cos \theta$$

3. By examining the determinant of the coefficient matrix, show that the following system has a nontrivial solution if and only if $\alpha = \beta$.

$$x + y + \alpha z = 0$$
$$x + y + \beta z = 0$$
$$\alpha x + \beta y + z = 0$$

4. Let A be a 3×3 matrix, each of whose entries is 1 or 0. What is the largest possible value for $|A|$?

5. (a) For the triangle in Figure 1 below, use trigonometry to show that

$$b \cos \gamma + c \cos \beta = a$$
$$c \cos \alpha + a \cos \gamma = b$$
$$a \cos \beta + b \cos \alpha = c$$

and then apply Cramer's Rule to show that

$$\cos \alpha = \frac{b^2 + c^2 - a^2}{2bc}$$

(b) Use Cramer's Rule to obtain similar formulas for $\cos \beta$ and $\cos \gamma$.

Figure 1

6. Use determinants to show that for all real values of λ the only solution of

$$x - 2y = \lambda x$$
$$x - y = \lambda y$$

is $x = 0$, $y = 0$.

7. Prove: If A is invertible, then $\mathrm{adj}\,(A)$ is invertible and

$$[\mathrm{adj}\,(A)]^{-1} = \frac{1}{\det(A)} A = \mathrm{adj}\,(A^{-1})$$

8. Prove: If A is an $n \times n$ matrix, then $\det[\,\mathrm{adj}\,(A)] = [\det(A)]^{n-1}$.

9. **(For readers who have studied calculus.)** Show that if $f_1(x)$, $f_2(x)$, $g_1(x)$, and $g_2(x)$ are differentiable functions, and if

$$W = \begin{vmatrix} f_1(x) & f_2(x) \\ g_1(x) & g_2(x) \end{vmatrix}, \quad \text{then} \quad \frac{dW}{dx} = \begin{vmatrix} f_1'(x) & f_2'(x) \\ g_1(x) & g_2(x) \end{vmatrix} + \begin{vmatrix} f_1(x) & f_2(x) \\ g_1'(x) & g_2'(x) \end{vmatrix}$$

10. (a) In Figure 2 that follows, the area of the triangle ABC is expressible as

$$\text{area } ABC = \text{area } ADEC + \text{area } CEFB - \text{area } ADFB$$

Use this and the fact that the area of a trapezoid equals $\frac{1}{2}$ the altitude times the sum of the parallel sides to show that

$$\text{area } ABC = \frac{1}{2}\begin{vmatrix} x_1 & y_1 & 1 \\ x_2 & y_2 & 1 \\ x_3 & y_3 & 1 \end{vmatrix}$$

[*Note.* In the derivation of this formula, the vertices are labeled so that the triangle is traced counterclockwise proceeding from (x_1, y_1) to (x_2, y_2) to (x_3, y_3). For a clockwise orientation, the determinant above yields the *negative* of the area.]

(b) Use the result in (a) to find the area of the triangle with vertices $(3, 3)$, $(4, 0)$, $(-2, -1)$.

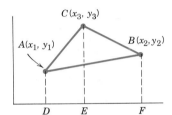

Figure 2

11. Prove: If the entries in each row of an $n \times n$ matrix A add up to zero, then the determinant of A is zero. [***Hint.*** Consider the product AX, where X is the $n \times 1$ matrix, each of whose entries is one.]

12. Let A be an $n \times n$ matrix and B the matrix that results when the rows of A are written in reverse order. How are $\det(A)$ and $\det(B)$ related?

13. How will A^{-1} be affected if
   (a) the ith and jth rows of A are interchanged;
   (b) the ith row of A is multiplied by a nonzero scalar, c;
   (c) c times the ith row of A is added to the jth row?

14. Let A be an $n \times n$ matrix. Suppose that B_1 is obtained by adding the same number t to each entry in the ith row of A and B_2 is obtained by subtracting t from each entry in the ith row of A. Show that $\det(A) = \frac{1}{2}[\det(B_1) + \det(B_2)]$.

15. Let

$$A = \begin{bmatrix} a_{11} & a_{12} & a_{13} \\ a_{21} & a_{22} & a_{23} \\ a_{31} & a_{32} & a_{33} \end{bmatrix}$$

   (a) Express $\det(\lambda I - A)$ as a polynomial $p(\lambda) = \lambda^3 + b\lambda^2 + c\lambda + d$.
   (b) Express the coefficients b and d in terms of determinants and traces.

16. Without directly evaluating the determinant, show that

$$\begin{vmatrix} \sin \alpha & \cos \alpha & \sin(\alpha + \delta) \\ \sin \beta & \cos \beta & \sin(\beta + \delta) \\ \sin \gamma & \cos \gamma & \sin(\gamma + \delta) \end{vmatrix} = 0$$

17. Use the fact that 21,375, 38,798, 34,162, 40,223, and 79,154 are all divisible by 19 to show that

$$\begin{vmatrix} 2 & 1 & 3 & 7 & 5 \\ 3 & 8 & 7 & 9 & 8 \\ 3 & 4 & 1 & 6 & 2 \\ 4 & 0 & 2 & 2 & 3 \\ 7 & 9 & 1 & 5 & 4 \end{vmatrix}$$

is divisible by 19 without directly evaluating the determinant.

18. Find the eigenvalues and corresponding eigenvectors for each of the following systems.

(a)
$$\begin{aligned} x_2 + 9x_3 &= \lambda x_1 \\ x_1 + 4x_2 - 7x_3 &= \lambda x_2 \\ x_1 \qquad\quad - 3x_3 &= \lambda x_3 \end{aligned}$$

(b)
$$\begin{aligned} x_2 + x_3 &= \lambda x_1 \\ x_1 \qquad - x_3 &= \lambda x_2 \\ x_1 + 5x_2 + 3x_3 &= \lambda x_3 \end{aligned}$$

CHAPTER **3**

VECTORS IN 2-SPACE AND 3-SPACE

Readers familiar with the contents of this chapter can go to Chapter 4 with no loss of continuity.

3.1 INTRODUCTION TO VECTORS (GEOMETRIC)

*Many physical quantities, such as area, length, mass, and temperature, are completely described once the magnitude of the quantity is given. Such quantities are called **scalars**. Other physical quantities, called **vectors**, are not completely determined until both a magnitude and a direction are specified. For example, wind movement is usually described by giving the speed and the direction, say 20 mph northeast. The wind speed and wind direction together form a vector quantity called the wind **velocity**. Other examples of vectors are force and displacement. In this section vectors in 2-space and 3-space will be introduced geometrically, arithmetic operations on vectors will be defined, and some basic properties of these operations will be established.*

GEOMETRIC VECTORS

Vectors can be represented geometrically as directed line segments or arrows in 2-space or 3-space; the direction of the arrow specifies the direction of the vector, and the length of the arrow describes its magnitude. The tail of the arrow is called the ***initial point*** of the vector, and the tip of the arrow the ***terminal point***. We shall denote vectors in lowercase boldface type (for instance, **a**, **k**, **v**, **w**, and **x**). When discussing vectors, we shall refer to numbers as *scalars*. All our scalars will be real numbers and will be denoted in lowercase italic type (for instance, a, k, v, w, and x).

If, as in Figure 1a, the initial point of a vector **v** is A and the terminal point is B, we write

$$\mathbf{v} = \overrightarrow{AB}$$

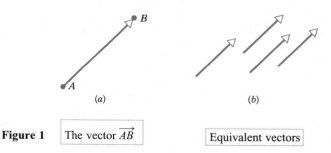

Figure 1   The vector $\overrightarrow{AB}$ Equivalent vectors

Vectors with the same length and same direction, such as those in Figure 1*b*, are called ***equivalent***. Since we want a vector to be determined solely by its length and direction, equivalent vectors are regarded as ***equal*** even though they may be located in different positions. If **v** and **w** are equivalent, we write

$$\mathbf{v} = \mathbf{w}$$

> **Definition.** If **v** and **w** are any two vectors, then the ***sum* v + w** is the vector determined as follows: Position the vector **w** so that its initial point coincides with the terminal point of **v**. The vector **v** + **w** is represented by the arrow from the initial point of **v** to the terminal point of **w** (Figure 2*a*).

In Figure 2*b* we have constructed two sums, **v** + **w** (blue arrows) and **w** + **v** (white arrows). It is evident that

$$\mathbf{v} + \mathbf{w} = \mathbf{w} + \mathbf{v}$$

and that the sum coincides with the diagonal of the parallelogram determined by **v** and **w** when these vectors are positioned so they have the same initial point.

The vector of length zero is called the ***zero vector*** and is denoted by **0**. We define

$$\mathbf{0} + \mathbf{v} = \mathbf{v} + \mathbf{0} = \mathbf{v}$$

for every vector **v**. Since there is no natural direction for the zero vector, we shall agree that it can be assigned any direction that is convenient for the problem being considered.

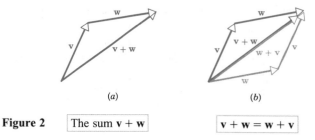

Figure 2   The sum **v** + **w**   **v** + **w** = **w** + **v**

If **v** is any nonzero vector, then −**v**, the *negative* of **v**, is defined to be the vector having the same magnitude as **v**, but oppositely directed (Figure 3).

Figure 3 | The negative of **v** has the same length as **v**, but is oppositely directed.

This vector has the property

$$\mathbf{v} + (-\mathbf{v}) = \mathbf{0}$$

(Why?) In addition, we define −**0** = **0**. Subtraction of vectors is defined as follows.

> **Definition.** If **v** and **w** are any two vectors, then *difference* of **w** from **v** is defined by
>
> $$\mathbf{v} - \mathbf{w} = \mathbf{v} + (-\mathbf{w})$$
>
> (Figure 4*a*).

To obtain the difference **v** − **w** without constructing −**w**, position **v** and **w** so their initial points coincide; the vector from the terminal point of **w** to the terminal point of **v** is then the vector **v** − **w** (Figure 4*b*).

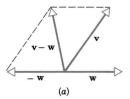

Figure 4　　　(a)　　　　　　(b)

> **Definition.** If **v** is a nonzero vector and k is a nonzero real number (scalar), then the *product* $k\mathbf{v}$ is defined to be the vector whose length is $|k|$ times the length of **v** and whose direction is the same as that of **v** if $k > 0$ and opposite to that of **v** if $k < 0$. We define $k\mathbf{v} = \mathbf{0}$ if $k = 0$ or **v** = **0**.

Figure 5 illustrates the relation between a vector **v** and the vectors $\frac{1}{2}\mathbf{v}$, $(-1)\mathbf{v}$, $2\mathbf{v}$, and $(-3)\mathbf{v}$. Note that the vector $(-1)\mathbf{v}$ has the same length as **v**, but is oppositely directed. Thus, $(-1)\mathbf{v}$ is just the negative of **v**; that is,

$$(-1)\mathbf{v} = -\mathbf{v}$$

Figure 5

A vector of the form $k\mathbf{v}$ is called a **scalar multiple** of $\mathbf{v}$. As evidenced by Figure 5, vectors that are scalar multiples of each other are parallel. Conversely, it can be shown that nonzero parallel vectors are scalar multiples of each other. We omit the proof.

VECTORS IN COORDINATE SYSTEMS

Problems involving vectors can often be simplified by introducing a rectangular coordinate system. For the moment we shall restrict the discussion to vectors in 2-space (the plane). Let $\mathbf{v}$ be any vector in the plane, and assume, as in Figure 6, that $\mathbf{v}$ has been positioned so its initial point is at the origin of a rectangular coordinate system. The coordinates (v_1, v_2) of the terminal point of $\mathbf{v}$ are called the **components of** $\mathbf{v}$, and we write

$$\mathbf{v} = (v_1, v_2)$$

If equivalent vectors, $\mathbf{v}$ and $\mathbf{w}$, are located so their initial points fall at the origin, then it is obvious that their terminal points must coincide (since the vectors have the same length and direction); thus, the vectors have the same components. Conversely, vectors with the same components are equivalent since they have the same length and same direction. In summary, two vectors

$$\mathbf{v} = (v_1, v_2) \qquad \text{and} \qquad \mathbf{w} = (w_1, w_2)$$

are equivalent if and only if

$$v_1 = w_1 \qquad \text{and} \qquad v_2 = w_2$$

Figure 6 | v_1 and v_2 are the components of $\mathbf{v}$.

The operations of vector addition and multiplication by scalars are easy to carry out in terms of components. As illustrated in Figure 7, if

$$\mathbf{v} = (v_1, v_2) \qquad \text{and} \qquad \mathbf{w} = (w_1, w_2)$$

then

$$\mathbf{v} + \mathbf{w} = (v_1 + w_1, v_2 + w_2) \tag{1}$$

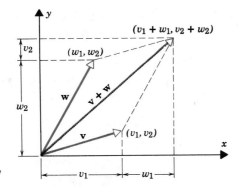

Figure 7

If $\mathbf{v} = (v_1, v_2)$ and k is any scalar, then by using a geometric argument involving similar triangles, it can be shown (Exercise 15) that

$$k\mathbf{v} = (kv_1, kv_2) \tag{2}$$

(Figure 8). Thus, for example, if $\mathbf{v} = (1, -2)$ and $\mathbf{w} = (7, 6)$, then

$$\mathbf{v} + \mathbf{w} = (1, -2) + (7, 6) = (1 + 7, -2 + 6) = (8, 4)$$

and

$$4\mathbf{v} = 4(1, -2) = (4(1), 4(-2)) = (4, -8)$$

Since $\mathbf{v} - \mathbf{w} = \mathbf{v} + (-1)\mathbf{w}$, it follows from Formulas (1) and (2) that

$$\mathbf{v} - \mathbf{w} = (v_1 - w_1, v_2 - w_2)$$

(Verify.)

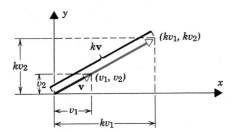

Figure 8

VECTORS IN 3-SPACE

Just as vectors in the plane can be described by pairs of real numbers, vectors in 3-space can be described by triples of real numbers by introducing a ***rectangular coordinate*** system. To construct such a coordinate system, select a point O, called the ***origin***, and choose three mutually perpendicular lines, called ***coordinate axes***, passing through the origin. Label these axes x, y, and z, and select a positive direction for each coordinate axis as well as a unit of length for measuring distances (Figure 9*a*). Each pair of coordinate axes determines a plane called a ***coordinate plane***. These are referred to as the ***xy-plane***, the ***xz-plane***, and the ***yz-plane***. To each point P in 3-space we assign a triple of numbers (x, y, z), called the ***coordinates of P***, as follows: Pass three planes through P parallel to the coordinate planes, and denote the points of intersections of these planes with the three coordinate axes by X, Y, and Z (Figure 9*b*).

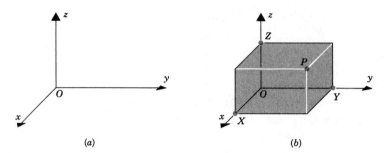

Figure 9 (*a*) (*b*)

The coordinates of P are defined to be the signed lengths

$$x = OX, \qquad y = OY, \qquad z = OZ$$

In Figure 10 we have constructed the points whose coordinates are $(4, 5, 6)$ and $(-3, 2, -4)$.

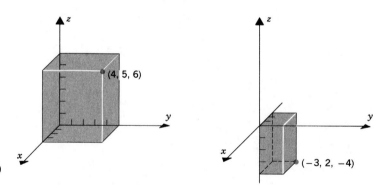

Figure 10

Rectangular coordinate systems in 3-space fall into two categories, ***left-handed*** and ***right-handed***. A right-handed system has the property that an ordinary screw pointed in the positive direction on the z-axis would be advanced if the positive x-axis is rotated 90° toward the positive y-axis (Figure 11*a*); the system is left-handed if the screw would be retracted (Figure 11*b*).

REMARK. In this book we shall use only right-handed coordinate systems.

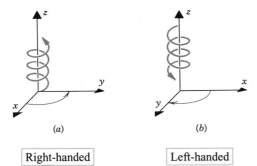

Figure 11   Right-handed   Left-handed

If, as in Figure 12, a vector **v** in 3-space is positioned so its initial point is at the origin of a rectangular coordinate system, then the coordinates of the terminal point are called the *components* of **v**, and we write

$$\mathbf{v} = (v_1, v_2, v_3)$$

If $\mathbf{v} = (v_1, v_2, v_3)$ and $\mathbf{w} = (w_1, w_2, w_3)$ are two vectors in 3-space, then arguments similar to those used for vectors in a plane can be used to establish the following results.

> **v** and **w** are equivalent if and only if $v_1 = w_1$, $v_2 = w_2$, and $v_3 = w_3$
>
> $\mathbf{v} + \mathbf{w} = (v_1 + w_1, v_2 + w_2, v_3 + w_3)$
>
> $k\mathbf{v} = (kv_1, kv_2, kv_3)$, where k is any scalar

Figure 12

Example 1   If $\mathbf{v} = (1, -3, 2)$ and $\mathbf{w} = (4, 2, 1)$, then

$$\mathbf{v} + \mathbf{w} = (5, -1, 3), \qquad 2\mathbf{v} = (2, -6, 4), \qquad -\mathbf{w} = (-4, -2, -1)$$
$$\mathbf{v} - \mathbf{w} = \mathbf{v} + (-\mathbf{w}) = (-3, -5, 1) \quad \text{▲}$$

Sometimes a vector is positioned so that its initial point is not at the origin. If the vector $\overrightarrow{P_1P_2}$ has initial point $P_1(x_1, y_1, z_1)$ and terminal point $P_2(x_2, y_2, z_2)$, then

$$\overrightarrow{P_1P_2} = (x_2 - x_1, y_2 - y_1, z_2 - z_1)$$

That is, the components of $\overrightarrow{P_1P_2}$ are obtained by subtracting the coordinates of the initial point from the coordinates of the terminal point. This may be seen using Figure 13: The vector $\overrightarrow{P_1P_2}$ is the difference of vectors $\overrightarrow{OP_2}$ and $\overrightarrow{OP_1}$, so

$$\overrightarrow{P_1P_2} = \overrightarrow{OP_2} - \overrightarrow{OP_1} = (x_2, y_2, z_2) - (x_1, y_1, z_1) = (x_2 - x_1, y_2 - y_1, z_2 - z_1)$$

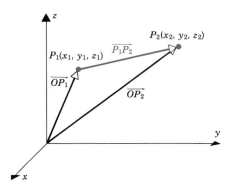

Figure 13

Example 2   The components of the vector $\mathbf{v} = \overrightarrow{P_1P_2}$ with initial point $P_1(2, -1, 4)$ and terminal point $P_2(7, 5, -8)$ are

$$\mathbf{v} = (7 - 2, 5 - (-1), (-8) - 4) = (5, 6, -12) \quad \triangle$$

In 2-space the vector with initial point $P_1(x_1, y_1)$ and terminal point $P_2(x_2, y_2)$ is

$$\overrightarrow{P_1P_2} = (x_2 - x_1, y_2 - y_1)$$

TRANSLATION OF AXES

The solutions to many problems can be simplified by translating the coordinate axes to obtain new axes parallel to the original ones.

In Figure 14*a* we have translated the axes of an xy-coordinate system to obtain an $x'y'$-coordinate system whose origin O' is at the point $(x, y) = (k, l)$. A point P in 2-space now has both (x, y) coordinates and (x', y') coordinates. To see how the two are related, consider the vector $\overrightarrow{O'P}$ (Figure 14*b*). In the xy-system its initial point is at

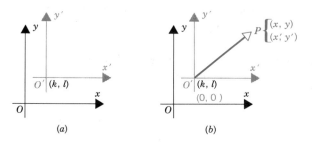

Figure 14   (a)   (b)

(k, l) and its terminal point is at (x, y), so $\overrightarrow{O'P} = (x - k, y - l)$. In the $x'y'$-system its initial point is at $(0, 0)$ and its terminal point is at (x', y'), so $\overrightarrow{O'P} = (x', y')$. Therefore,

$$x' = x - k \qquad y' = y - l$$

These formulas are called the ***translation equations***.

Example 3 Suppose that an xy-coordinate system is translated to obtain an $x'y'$-coordinate system whose origin has xy-coordinates $(k, l) = (4, 1)$.

(a) Find the $x'y'$-coordinates of the point with the xy-coordinates $P(2, 0)$.
(b) Find the xy-coordinates of the point with $x'y'$-coordinates $Q(-1, 5)$.

Solution (a). The translation equations are

$$x' = x - 4 \qquad y' = y - 1$$

so the $x'y'$-coordinates of $P(2, 0)$ are $x' = 2 - 4 = -2$ and $y' = 0 - 1 = -1$.

Solution (b). The translation equations in (a) can be rewritten as

$$x = x' + 4 \qquad y = y' + 1$$

so the xy-coordinates of Q are $x = -1 + 4 = 3$ and $y = 5 + 1 = 6$. ▲

In 3-space the translation equations are

$$x' = x - k \qquad y' = y - l \qquad z' = z - m$$

where (k, l, m) are the xyz-coordinates of the $x'y'z'$-origin.

EXERCISE SET 3.1

1. Draw a right-handed coordinate system and locate the points whose coordinates are
   (a) $(3, 4, 5)$ (b) $(-3, 4, 5)$ (c) $(3, -4, 5)$ (d) $(3, 4, -5)$
   (e) $(-3, -4, 5)$ (f) $(-3, 4, -5)$ (g) $(3, -4, -5)$ (h) $(-3, -4, -5)$
   (i) $(-3, 0, 0)$ (j) $(3, 0, 3)$ (k) $(0, 0, -3)$ (l) $(0, 3, 0)$

2. Sketch the following vectors with the initial points located at the origin:
   (a) $\mathbf{v}_1 = (3, 6)$ (b) $\mathbf{v}_2 = (-4, -8)$ (c) $\mathbf{v}_3 = (-4, -3)$ (d) $\mathbf{v}_4 = (5, -4)$ (e) $\mathbf{v}_5 = (3, 0)$
   (f) $\mathbf{v}_6 = (0, -7)$ (g) $\mathbf{v}_7 = (3, 4, 5)$ (h) $\mathbf{v}_8 = (3, 3, 0)$ (i) $\mathbf{v}_9 = (0, 0, -3)$

3. Find the components of the vector having initial point P_1 and terminal point P_2.
   (a) $P_1(4, 8)$, $P_2(3, 7)$ (b) $P_1(3, -5)$, $P_2(-4, -7)$ (c) $P_1(-5, 0)$, $P_2(-3, 1)$
   (d) $P_1(0, 0)$, $P_2(a, b)$ (e) $P_1(3, -7, 2)$, $P_2(-2, 5, -4)$ (f) $P_1(-1, 0, 2)$, $P_2(0, -1, 0)$
   (g) $P_1(a, b, c)$, $P_2(0, 0, 0)$ (h) $P_1(0, 0, 0)$, $P_2(a, b, c)$

4. Find a nonzero vector $\mathbf{u}$ with initial point $P(-1, 3, -5)$ such that
   (a) $\mathbf{u}$ has the same direction as $\mathbf{v} = (6, 7, -3)$ (b) $\mathbf{u}$ is oppositely directed to $\mathbf{v} = (6, 7, -3)$

5. Find a nonzero vector **u** with terminal point $Q(3, 0, -5)$ such that
   (a) **u** has the same direction as $\mathbf{v} = (4, -2, -1)$ (b) **u** is oppositely directed to $\mathbf{v} = (4, -2, -1)$

6. Let $\mathbf{u} = (-3, 1, 2)$, $\mathbf{v} = (4, 0, -8)$, and $\mathbf{w} = (6, -1, -4)$. Find the components of
   (a) $\mathbf{v} - \mathbf{w}$ (b) $6\mathbf{u} + 2\mathbf{v}$ (c) $-\mathbf{v} + \mathbf{u}$ (d) $5(\mathbf{v} - 4\mathbf{u})$ (e) $-3(\mathbf{v} - 8\mathbf{w})$ (f) $(2\mathbf{u} - 7\mathbf{w}) - (8\mathbf{v} + \mathbf{u})$

7. Let **u**, **v**, and **w** be the vectors in Exercise 6. Find the components of the vector **x** that satisfies $2\mathbf{u} - \mathbf{v} + \mathbf{x} = 7\mathbf{x} + \mathbf{w}$.

8. Let **u**, **v**, and **w** be the vectors in Exercise 6. Find scalars c_1, c_2, and c_3 such that

$$c_1\mathbf{u} + c_2\mathbf{v} + c_3\mathbf{w} = (2, 0, 4)$$

9. Show that there do not exist scalars c_1, c_2, and c_3 such that

$$c_1(-2, 9, 6) + c_2(-3, 2, 1) + c_3(1, 7, 5) = (0, 5, 4)$$

10. Find all scalars c_1, c_2, and c_3 such that

$$c_1(1, 2, 0) + c_2(2, 1, 1) + c_3(0, 3, 1) = (0, 0, 0)$$

11. Let P be the point $(2, 3, -2)$ and Q the point $(7, -4, 1)$.
   (a) Find the midpoint of the line segment connecting P and Q.
   (b) Find the point on the line segment connecting P and Q that is $\frac{3}{4}$ of the way from P to Q.

12. Suppose an xy-coordinate system is translated to obtain an $x'y'$-coordinate system whose origin O' has xy-coordinates $(2, -3)$.
   (a) Find the $x'y'$-coordinates of the point P whose xy-coordinates are $(7, 5)$.
   (b) Find the xy-coordinates of the point Q whose $x'y'$-coordinates are $(-3, 6)$.
   (c) Draw the xy and $x'y'$-coordinate axes and locate the points P and Q.

13. Suppose that an xyz-coordinate system is translated to obtain an $x'y'z'$-coordinate system. Let **v** be a vector whose components are $\mathbf{v} = (v_1, v_2, v_3)$ in the xyz-system. Show that **v** has the same components in the $x'y'z'$-system.

14. Find the components of **u**, **v**, $\mathbf{u} + \mathbf{v}$, and $\mathbf{u} - \mathbf{v}$ for the vectors shown in Figure 15.

Figure 15

15. Prove geometrically that if $\mathbf{v} = (v_1, v_2)$, then $k\mathbf{v} = (kv_1, kv_2)$. (Restrict the proof to the case $k > 0$ illustrated in Figure 8. The complete proof would involve various cases that depend on the sign of k and the quadrant in which the vector falls.)

3.2 NORM OF A VECTOR; VECTOR ARITHMETIC

In this section we shall establish the basic rules of vector arithmetic.

PROPERTIES OF VECTOR OPERATIONS

The following theorem lists the most important properties of vectors in 2-space and 3-space.

Theorem 3.2.1. *If* **u**, **v**, *and* **w** *are vectors in 2- or 3-space and k and l are scalars, then the following relationships hold.*

(*a*) $\mathbf{u} + \mathbf{v} = \mathbf{v} + \mathbf{u}$ (*b*) $(\mathbf{u} + \mathbf{v}) + \mathbf{w} = \mathbf{u} + (\mathbf{v} + \mathbf{w})$

(*c*) $\mathbf{u} + \mathbf{0} = \mathbf{0} + \mathbf{u} = \mathbf{u}$ (*d*) $\mathbf{u} + (-\mathbf{u}) = \mathbf{0}$

(*e*) $k(l\mathbf{u}) = (kl)\mathbf{u}$ (*f*) $k(\mathbf{u} + \mathbf{v}) = k\mathbf{u} + k\mathbf{v}$

(*g*) $(k + l)\mathbf{u} = k\mathbf{u} + l\mathbf{u}$ (*h*) $1\mathbf{u} = \mathbf{u}$

Before discussing the proof, we note that we have developed two approaches to vectors: *geometric*, in which vectors are represented by arrows or directed line segments, and *analytic*, in which vectors are represented by pairs or triples of numbers called components. As a consequence, the equations in Theorem 3.2.1 can be proved either geometrically or analytically. To illustrate, we shall prove part (*b*) both ways. The remaining proofs are left as exercises.

Proof of part (b) (analytic). We shall give the proof for vectors in 3-space; the proof for 2-space is similar. If $\mathbf{u} = (u_1, u_2, u_3)$, $\mathbf{v} = (v_1, v_2, v_3)$, and $\mathbf{w} = (w_1, w_2, w_3)$, then

$$
\begin{aligned}
(\mathbf{u} + \mathbf{v}) + \mathbf{w} &= [(u_1, u_2, u_3) + (v_1, v_2, v_3)] + (w_1, w_2, w_3) \\
&= (u_1 + v_1, u_2 + v_2, u_3 + v_3) + (w_1, w_2, w_3) \\
&= ([u_1 + v_1] + w_1, [u_2 + v_2] + w_2, [u_3 + v_3] + w_3) \\
&= (u_1 + [v_1 + w_1], u_2 + [v_2 + w_2], u_3 + [v_3 + w_3]) \\
&= (u_1, u_2, u_3) + (v_1 + w_1, v_2 + w_2, v_3 + w_3) \\
&= \mathbf{u} + (\mathbf{v} + \mathbf{w})
\end{aligned}
$$

Proof of part (b) (geometric). Let **u**, **v**, and **w** be represented by $\overrightarrow{PQ}$, $\overrightarrow{QR}$, and $\overrightarrow{RS}$ as shown in Figure 1. Then

$$\mathbf{v} + \mathbf{w} = \overrightarrow{QS} \quad \text{and} \quad \mathbf{u} + (\mathbf{v} + \mathbf{w}) = \overrightarrow{PS}$$

Also,

$$\mathbf{u} + \mathbf{v} = \overrightarrow{PR} \quad \text{and} \quad (\mathbf{u} + \mathbf{v}) + \mathbf{w} = \overrightarrow{PS}$$

Therefore,

$$\mathbf{u} + (\mathbf{v} + \mathbf{w}) = (\mathbf{u} + \mathbf{v}) + \mathbf{w} \quad \blacksquare$$

REMARK. In light of part (*b*) of this theorem, the symbol **u** + **v** + **w** is unambiguous since the same sum is obtained no matter where parentheses are inserted. Moreover, if the vectors **u**, **v**, and **w** are placed "tip to tail," then the sum **u** + **v** + **w** is the vector from the initial point of **u** to the terminal point of **w** (Figure 1).

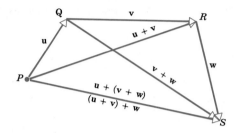

Figure 1   The vectors **u** + (**v** + **w**) and (**u** + **v**) + **w** are equal.

NORM OF A VECTOR

The ***length*** of a vector **u** is often called the ***norm*** of **u** and is denoted by $\|\mathbf{u}\|$. It follows from the theorem of Pythagoras that the norm of a vector $\mathbf{u} = (u_1, u_2)$ in 2-space is

$$\|\mathbf{u}\| = \sqrt{u_1^2 + u_2^2} \tag{1}$$

(Figure 2*a*). Let $\mathbf{u} = (u_1, u_2, u_3)$ be a vector in 3-space. Using Figure 2*b* and two applications of the theorem of Pythagoras, we obtain

$$\|\mathbf{u}\|^2 = (OR)^2 + (RP)^2$$
$$= (OQ)^2 + (OS)^2 + (RP)^2$$
$$= u_1^2 + u_2^2 + u_3^2$$

Thus,

$$\|\mathbf{u}\| = \sqrt{u_1^2 + u_2^2 + u_3^2} \tag{2}$$

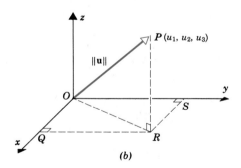

Figure 2   (*a*)   (*b*)

A vector of norm 1 is called a ***unit vector***.

If $P_1(x_1, y_1, z_1)$ and $P_2(x_2, y_2, z_2)$ are two points in 3-space, then the ***distance d*** between them is the norm of the vector $\overrightarrow{P_1P_2}$ (Figure 3). Since

$$\overrightarrow{P_1P_2} = (x_2 - x_1, y_2 - y_1, z_2 - z_1)$$

it follows from (2) that

$$d = \sqrt{(x_2 - x_1)^2 + (y_2 - y_1)^2 + (z_2 - z_1)^2} \tag{3}$$

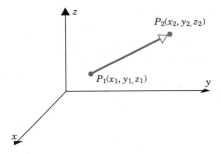

Figure 3 | The distance between P_1 and P_2 is the norm of the vector $\overrightarrow{P_1P_2}$.

Similarly, if $P_1(x_1, y_1)$ and $P_2(x_2, y_2)$ are points in 2-space, then the distance between them is given by

$$d = \sqrt{(x_2 - x_1)^2 + (y_2 - y_1)^2} \tag{4}$$

Example 1   The norm of the vector $\mathbf{u} = (-3, 2, 1)$ is
$$\|\mathbf{u}\| = \sqrt{(-3)^2 + (2)^2 + (1)^2} = \sqrt{14}$$
The distance d between the points $P_1(2, -1, -5)$ and $P_2(4, -3, 1)$ is
$$d = \sqrt{(4-2)^2 + (-3+1)^2 + (1+5)^2} = \sqrt{44} = 2\sqrt{11}$$

From the definition of the product $k\mathbf{u}$, the length of the vector $k\mathbf{u}$ is $|k|$ times the length of $\mathbf{u}$. Expressed as an equation, this statement says that

$$\|k\mathbf{u}\| = |k|\,\|\mathbf{u}\| \tag{5}$$

This useful formula is applicable in both 2-space and 3-space.

EXERCISE SET 3.2

1. Find the norm of **v**.
   (a) $\mathbf{v} = (4, -3)$ (b) $\mathbf{v} = (2, 3)$ (c) $\mathbf{v} = (-5, 0)$
   (d) $\mathbf{v} = (2, 2, 2)$ (e) $\mathbf{v} = (-7, 2, -1)$ (f) $\mathbf{v} = (0, 6, 0)$

2. Find the distance between P_1 and P_2.
   (a) $P_1(3, 4)$, $P_2(5, 7)$ (b) $P_1(-3, 6)$, $P_2(-1, -4)$
   (c) $P_1(7, -5, 1)$, $P_2(-7, -2, -1)$ (d) $P_1(3, 3, 3)$, $P_2(6, 0, 3)$

3. Let $\mathbf{u} = (2, -2, 3)$, $\mathbf{v} = (1, -3, 4)$, $\mathbf{w} = (3, 6, -4)$. In each part evaluate the expression.
   (a) $\|\mathbf{u} + \mathbf{v}\|$ (b) $\|\mathbf{u}\| + \|\mathbf{v}\|$ (c) $\|-2\mathbf{u}\| + 2\|\mathbf{u}\|$

   (d) $\|3\mathbf{u} - 5\mathbf{v} + \mathbf{w}\|$ (e) $\dfrac{1}{\|\mathbf{w}\|}\mathbf{w}$ (f) $\left\|\dfrac{1}{\|\mathbf{w}\|}\mathbf{w}\right\|$

4. Let $\mathbf{v} = (-1, 2, 5)$. Find all scalars k such that $\|k\mathbf{v}\| = 4$.

5. Let $\mathbf{u} = (7, -3, 1)$, $\mathbf{v} = (9, 6, 6)$, $\mathbf{w} = (2, 1, -8)$, $k = -2$, and $l = 5$. Verify that these vectors and scalars satisfy the stated equalities from Theorem 3.2.1.
   (a) part (b) (b) part (e) (c) part (f) (d) part (g)

6. (a) Show that if **v** is any nonzero vector, then

   $$\frac{1}{\|\mathbf{v}\|}\mathbf{v}$$

   is a unit vector.
   (b) Use the result in part (a) to find a unit vector that has the same direction as the vector $\mathbf{v} = (3, 4)$.
   (c) Use the result in part (a) to find a unit vector that is oppositely directed to the vector $\mathbf{v} = (-2, 3, -6)$.

7. (a) Show that the components of the vector $\mathbf{v} = (v_1, v_2)$ in Figure 4 are $v_1 = \|\mathbf{v}\| \cos \theta$ and $v_2 = \|\mathbf{v}\| \sin \theta$.
   (b) Let **u** and **v** be the vectors in Figure 5. Use the result in part (a) to find the components of $4\mathbf{u} - 5\mathbf{v}$.

Figure 4

Figure 5

8. Let $\mathbf{p}_0 = (x_0, y_0, z_0)$ and $\mathbf{p} = (x, y, z)$. Describe the set of all points (x, y, z) for which $\|\mathbf{p} - \mathbf{p}_0\| = 1$.

9. Prove geometrically that if $\mathbf{u}$ and $\mathbf{v}$ are vectors in 2- or 3-space, then $\|\mathbf{u} + \mathbf{v}\| \leq \|\mathbf{u}\| + \|\mathbf{v}\|$.

10. Prove parts (a), (c), and (e) of Theorem 3.2.1 analytically.

11. Prove parts (d), (g), and (h) of Theorem 3.2.1 analytically.

12. Prove part (f) of Theorem 3.2.1 geometrically.

3.3 DOT PRODUCT; PROJECTIONS

In this section we shall discuss a way of multiplying vectors in 2-space or 3-space, and we shall give some applications of this multiplication to geometry.

DOT PRODUCT OF VECTORS

Let $\mathbf{u}$ and $\mathbf{v}$ be two nonzero vectors in 2-space or 3-space, and assume these vectors have been positioned so their initial points coincide. By the *angle between* $\mathbf{u}$ *and* $\mathbf{v}$, we shall mean the angle θ determined by $\mathbf{u}$ and $\mathbf{v}$ that satisfies $0 \leq \theta \leq \pi$ (Figure 1).

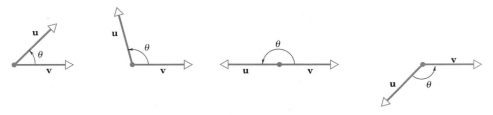

Figure 1

The angle θ between $\mathbf{u}$ and $\mathbf{v}$ satisfies $0° \leq \theta \leq \pi$.

Definition. If $\mathbf{u}$ and $\mathbf{v}$ are vectors in 2-space or 3-space and θ is the angle between $\mathbf{u}$ and $\mathbf{v}$, then the *dot product* or *Euclidean inner product* $\mathbf{u} \cdot \mathbf{v}$ is defined by

$$\mathbf{u} \cdot \mathbf{v} = \begin{cases} \|\mathbf{u}\| \, \|\mathbf{v}\| \cos \theta & \text{if } \mathbf{u} \neq \mathbf{0} \text{ and } \mathbf{v} \neq \mathbf{0} \\ 0 & \text{if } \mathbf{u} = \mathbf{0} \text{ or } \mathbf{v} = \mathbf{0} \end{cases} \tag{1}$$

Example 1 As shown in Figure 2, the angle between the vectors $\mathbf{u} = (0, 0, 1)$ and $\mathbf{v} = (0, 2, 2)$ is 45°. Thus,

$$\mathbf{u} \cdot \mathbf{v} = \|\mathbf{u}\| \, \|\mathbf{v}\| \cos \theta = (\sqrt{0^2 + 0^2 + 1^2})(\sqrt{0^2 + 2^2 + 2^2})\left(\frac{1}{\sqrt{2}}\right) = 2 \quad \blacktriangle$$

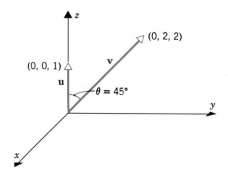

Figure 2

COMPONENT FORMULA FOR THE DOT PRODUCT

For purposes of computation, it is desirable to have a formula that expresses the dot product of two vectors in terms of the components of the vectors. We will derive such a formula for vectors in 3-space; the derivation for vectors in 2-space is similar.

Let $\mathbf{u} = (u_1, u_2, u_3)$ and $\mathbf{v} = (v_1, v_2, v_3)$ be two nonzero vectors. If, as shown in Figure 3, θ is the angle between $\mathbf{u}$ and $\mathbf{v}$, then the law of cosines yields

$$\|\overrightarrow{PQ}\|^2 = \|\mathbf{u}\|^2 + \|\mathbf{v}\|^2 - 2\|\mathbf{u}\|\,\|\mathbf{v}\|\cos\theta \tag{2}$$

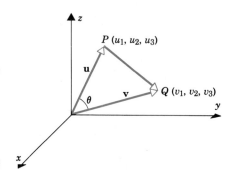

Figure 3

Since $\overrightarrow{PQ} = \mathbf{v} - \mathbf{u}$, we can rewrite (2) as

$$\|\mathbf{u}\|\,\|\mathbf{v}\|\cos\theta = \tfrac{1}{2}(\|\mathbf{u}\|^2 + \|\mathbf{v}\|^2 - \|\mathbf{v} - \mathbf{u}\|^2)$$

or

$$\mathbf{u}\cdot\mathbf{v} = \tfrac{1}{2}(\|\mathbf{u}\|^2 + \|\mathbf{v}\|^2 - \|\mathbf{v} - \mathbf{u}\|^2)$$

Substituting

$$\|\mathbf{u}\|^2 = u_1^2 + u_2^2 + u_3^2, \qquad \|\mathbf{v}\|^2 = v_1^2 + v_2^2 + v_3^2,$$

and

$$\|\mathbf{v} - \mathbf{u}\|^2 = (v_1 - u_1)^2 + (v_2 - u_2)^2 + (v_3 - u_3)^2$$

we obtain after simplifying

$$\mathbf{u} \cdot \mathbf{v} = u_1 v_1 + u_2 v_2 + u_3 v_3 \tag{3}$$

If $\mathbf{u} = (u_1, u_2)$ and $\mathbf{v} = (v_1, v_2)$ are two vectors in 2-space, then the corresponding formula is

$$\mathbf{u} \cdot \mathbf{v} = u_1 v_1 + u_2 v_2 \tag{4}$$

FINDING THE ANGLE BETWEEN VECTORS

If $\mathbf{u}$ and $\mathbf{v}$ are nonzero vectors, then Formula (1) can be written as

$$\cos \theta = \frac{\mathbf{u} \cdot \mathbf{v}}{\|\mathbf{u}\| \|\mathbf{v}\|} \tag{5}$$

Example 2 Consider the vectors

$$\mathbf{u} = (2, -1, 1) \quad \text{and} \quad \mathbf{v} = (1, 1, 2)$$

Find $\mathbf{u} \cdot \mathbf{v}$ and determine the angle θ between $\mathbf{u}$ and $\mathbf{v}$.

Solution.

$$\mathbf{u} \cdot \mathbf{v} = u_1 v_1 + u_2 v_2 + u_3 v_3 = (2)(1) + (-1)(1) + (1)(2) = 3$$

For the given vectors we have $\|\mathbf{u}\| = \|\mathbf{v}\| = \sqrt{6}$, so that from (5)

$$\cos \theta = \frac{\mathbf{u} \cdot \mathbf{v}}{\|\mathbf{u}\| \|\mathbf{v}\|} = \frac{3}{\sqrt{6}\sqrt{6}} = \frac{1}{2}$$

Thus, $\theta = 60°$. ▲

Example 3 Find the angle between a diagonal of a cube and one of its edges.

Solution. Let k be the length of an edge and introduce a coordinate system as shown in Figure 4.

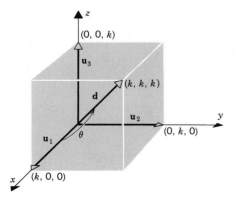

Figure 4

If we let $\mathbf{u}_1 = (k, 0, 0)$, $\mathbf{u}_2 = (0, k, 0)$, and $\mathbf{u}_3 = (0, 0, k)$, then the vector

$$\mathbf{d} = (k, k, k) = \mathbf{u}_1 + \mathbf{u}_2 + \mathbf{u}_3$$

is a diagonal of the cube. The angle θ between $\mathbf{d}$ and the edge $\mathbf{u}_1$ satisfies

$$\cos \theta = \frac{\mathbf{u}_1 \cdot \mathbf{d}}{\|\mathbf{u}_1\| \|\mathbf{d}\|} = \frac{k^2}{(k)(\sqrt{3k^2})} = \frac{1}{\sqrt{3}}$$

Thus,

$$\theta = \cos^{-1}\left(\frac{1}{\sqrt{3}}\right) \approx 54°44'$$

The following theorem shows how the dot product can be used to obtain information about the angle between two vectors; it also establishes an important relationship between the norm and the dot product.

Theorem 3.3.1. *Let $\mathbf{u}$ and $\mathbf{v}$ be vectors in 2- or 3-space.*

(a) $\mathbf{v} \cdot \mathbf{v} = \|\mathbf{v}\|^2;$ *that is,* $\|\mathbf{v}\| = (\mathbf{v} \cdot \mathbf{v})^{1/2}$
(b) *If the vectors $\mathbf{u}$ and $\mathbf{v}$ are nonzero and θ is the angle between them, then*

θ is acute	if and only if	$\mathbf{u} \cdot \mathbf{v} > 0$
θ is obtuse	if and only if	$\mathbf{u} \cdot \mathbf{v} < 0$
$\theta = \pi/2$	if and only if	$\mathbf{u} \cdot \mathbf{v} = 0$

Proof (a). Since the angle θ between $\mathbf{v}$ and $\mathbf{v}$ is 0, we have

$$\mathbf{v} \cdot \mathbf{v} = \|\mathbf{v}\| \|\mathbf{v}\| \cos \theta = \|\mathbf{v}\|^2 \cos 0 = \|\mathbf{v}\|^2$$

Proof (b). Since θ satisfies $0 \le \theta \le \pi$, it follows that: θ is acute if and only if $\cos \theta > 0$; θ is obtuse if and only if $\cos \theta < 0$; and $\theta = \pi/2$ if and only if $\cos \theta = 0$. But $\cos \theta$ has the same sign as $\mathbf{u} \cdot \mathbf{v}$ since $\mathbf{u} \cdot \mathbf{v} = \|\mathbf{u}\| \|\mathbf{v}\| \cos \theta$, $\|\mathbf{u}\| > 0$, and $\|\mathbf{v}\| > 0$. Thus, the result follows. ■

Example 4 If $\mathbf{u} = (1, -2, 3)$, $\mathbf{v} = (-3, 4, 2)$, and $\mathbf{w} = (3, 6, 3)$, then

$$\mathbf{u} \cdot \mathbf{v} = (1)(-3) + (-2)(4) + (3)(2) = -5$$
$$\mathbf{v} \cdot \mathbf{w} = (-3)(3) + (4)(6) + (2)(3) = 21$$
$$\mathbf{u} \cdot \mathbf{w} = (1)(3) + (-2)(6) + (3)(3) = 0$$

Therefore, $\mathbf{u}$ and $\mathbf{v}$ make an obtuse angle, $\mathbf{v}$ and $\mathbf{w}$ make an acute angle, and $\mathbf{u}$ and $\mathbf{w}$ are perpendicular.

ORTHOGONAL VECTORS

Perpendicular vectors are also called ***orthogonal*** vectors. In light of Theorem 3.3.1*b*, two *nonzero* vectors are orthogonal if and only if their dot product is zero. If we agree to consider $\mathbf{u}$ and $\mathbf{v}$ to be perpendicular when either or both of these vectors is $\mathbf{0}$, then we can state without exception that *two vectors $\mathbf{u}$ and $\mathbf{v}$ are orthogonal (perpendicu-*

lar) *if and only if* $\mathbf{u} \cdot \mathbf{v} = 0$. To indicate that $\mathbf{u}$ and $\mathbf{v}$ are orthogonal vectors we write $\mathbf{u} \perp \mathbf{v}$.

Example 5 Show that in 2-space the nonzero vector $\mathbf{n} = (a, b)$ is perpendicular to the line $ax + by + c = 0$.

Solution. Let $P_1(x_1, y_1)$ and $P_2(x_2, y_2)$ be distinct points on the line, so that

$$ax_1 + by_1 + c = 0$$
$$ax_2 + by_2 + c = 0 \tag{6}$$

Since the vector $\overrightarrow{P_1P_2} = (x_2 - x_1, y_2 - y_1)$ runs along the line (Figure 5), we need only show that $\mathbf{n}$ and $\overrightarrow{P_1P_2}$ are perpendicular. But on subtracting the equations in (6) we obtain

$$a(x_2 - x_1) + b(y_2 - y_1) = 0$$

which can be expressed in the form

$$(a, b) \cdot (x_2 - x_1, y_2 - y_1) = 0 \qquad \text{or} \qquad \mathbf{n} \cdot \overrightarrow{P_1P_2} = 0$$

Thus, $\mathbf{n}$ and $\overrightarrow{P_1P_2}$ are perpendicular. ▲

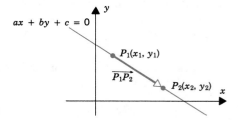

Figure 5

The following theorem lists the most important properties of the dot product. They are useful in calculations involving vectors.

Theorem 3.3.2. *If* $\mathbf{u}, \mathbf{v},$ *and* $\mathbf{w}$ *are vectors in 2- or 3-space and k is a scalar, then:*
(a) $\mathbf{u} \cdot \mathbf{v} = \mathbf{v} \cdot \mathbf{u}$
(b) $\mathbf{u} \cdot (\mathbf{v} + \mathbf{w}) = \mathbf{u} \cdot \mathbf{v} + \mathbf{u} \cdot \mathbf{w}$
(c) $k(\mathbf{u} \cdot \mathbf{v}) = (k\mathbf{u}) \cdot \mathbf{v} = \mathbf{u} \cdot (k\mathbf{v})$
(d) $\mathbf{v} \cdot \mathbf{v} > 0$ *if* $\mathbf{v} \neq \mathbf{0}$, *and* $\mathbf{v} \cdot \mathbf{v} = 0$ *if* $\mathbf{v} = \mathbf{0}$

Proof. We shall prove (c) for vectors in 3-space and leave the remaining proofs as exercises. Let $\mathbf{u} = (u_1, u_2, u_3)$ and $\mathbf{v} = (v_1, v_2, v_3)$; then

$$k(\mathbf{u} \cdot \mathbf{v}) = k(u_1v_1 + u_2v_2 + u_3v_3)$$
$$= (ku_1)v_1 + (ku_2)v_2 + (ku_3)v_3$$
$$= (k\mathbf{u}) \cdot \mathbf{v}$$

Similarly,

$$k(\mathbf{u} \cdot \mathbf{v}) = \mathbf{u} \cdot (k\mathbf{v})$$

ORTHOGONAL PROJECTIONS

In many applications it is of interest to "decompose" a vector $\mathbf{u}$ into a sum of two terms, one parallel to a specified nonzero vector $\mathbf{a}$ and the other perpendicular to $\mathbf{a}$. If $\mathbf{u}$ and $\mathbf{a}$ are positioned so their initial points coincide at a point Q, we can decompose the vector $\mathbf{u}$ as follows (Figure 6): Drop a perpendicular from the tip of $\mathbf{u}$ to the line through $\mathbf{a}$, and construct the vector $\mathbf{w}_1$ from Q to the foot of this perpendicular. Next form the difference

$$\mathbf{w}_2 = \mathbf{u} - \mathbf{w}_1$$

Figure 6

The vector $\mathbf{u}$ is the sum of $\mathbf{w}_1$ and $\mathbf{w}_2$, where $\mathbf{w}_1$ is parallel to $\mathbf{a}$ and $\mathbf{w}_2$ is perpendicular to $\mathbf{a}$.

As indicated in Figure 6, the vector $\mathbf{w}_1$ is parallel to $\mathbf{a}$, the vector $\mathbf{w}_2$ is perpendicular to $\mathbf{a}$, and

$$\mathbf{w}_1 + \mathbf{w}_2 = \mathbf{w}_1 + (\mathbf{u} - \mathbf{w}_1) = \mathbf{u}$$

The vector $\mathbf{w}_1$ is called the ***orthogonal projection of*** $\mathbf{u}$ ***on*** $\mathbf{a}$ or sometimes the ***vector component of*** $\mathbf{u}$ ***along*** $\mathbf{a}$. It is denoted by

$$\text{proj}_{\mathbf{a}} \mathbf{u} \tag{7}$$

The vector $\mathbf{w}_2$ is called the ***vector component of*** $\mathbf{u}$ ***orthogonal to*** $\mathbf{a}$. Since we have $\mathbf{w}_2 = \mathbf{u} - \mathbf{w}_1$, this vector can be written in notation (7) as

$$\mathbf{w}_2 = \mathbf{u} - \text{proj}_{\mathbf{a}} \mathbf{u}$$

The following theorem gives formulas for calculating the vectors $\text{proj}_{\mathbf{a}} \mathbf{u}$ and $\mathbf{u} - \text{proj}_{\mathbf{a}} \mathbf{u}$.

Theorem 3.3.3. *If* $\mathbf{u}$ *and* $\mathbf{a}$ *are vectors in 2-space or 3-space and if* $\mathbf{a} \neq \mathbf{0}$, *then*

$$\text{proj}_{\mathbf{a}} \mathbf{u} = \frac{\mathbf{u} \cdot \mathbf{a}}{\|\mathbf{a}\|^2} \mathbf{a} \qquad \textit{(vector component of } \mathbf{u} \textit{ along } \mathbf{a)}$$

$$\mathbf{u} - \text{proj}_{\mathbf{a}} \mathbf{u} = \mathbf{u} - \frac{\mathbf{u} \cdot \mathbf{a}}{\|\mathbf{a}\|^2} \mathbf{a} \qquad \textit{(vector component of } \mathbf{u} \textit{ orthogonal to } \mathbf{a)}$$

Proof. Let $\mathbf{w}_1 = \text{proj}_{\mathbf{a}}\,\mathbf{u}$ and $\mathbf{w}_2 = \mathbf{u} - \text{proj}_{\mathbf{a}}\,\mathbf{u}$. Since $\mathbf{w}_1$ is parallel to $\mathbf{a}$, it must be a scalar multiple of $\mathbf{a}$, so it can be written in the form $\mathbf{w}_1 = k\mathbf{a}$. Thus

$$\mathbf{u} = \mathbf{w}_1 + \mathbf{w}_2 = k\mathbf{a} + \mathbf{w}_2 \tag{8}$$

Taking the dot product of both sides of (8) with $\mathbf{a}$ and using Theorems 3.3.1a and 3.3.2 yields

$$\mathbf{u} \cdot \mathbf{a} = (k\mathbf{a} + \mathbf{w}_2) \cdot \mathbf{a} = k\|\mathbf{a}\|^2 + \mathbf{w}_2 \cdot \mathbf{a} \tag{9}$$

But $\mathbf{w}_2 \cdot \mathbf{a} = 0$ since $\mathbf{w}_2$ is perpendicular to $\mathbf{a}$; so (9) yields

$$k = \frac{\mathbf{u} \cdot \mathbf{a}}{\|\mathbf{a}\|^2}$$

Since $\text{proj}_{\mathbf{a}}\,\mathbf{u} = \mathbf{w}_1 = k\mathbf{a}$, we obtain

$$\text{proj}_{\mathbf{a}}\,\mathbf{u} = \frac{\mathbf{u} \cdot \mathbf{a}}{\|\mathbf{a}\|^2}\mathbf{a} \quad \blacksquare$$

Example 6 Let $\mathbf{u} = (2, -1, 3)$ and $\mathbf{a} = (4, -1, 2)$. Find the vector component of $\mathbf{u}$ along $\mathbf{a}$ and the vector component of $\mathbf{u}$ orthogonal to $\mathbf{a}$.

Solution.

$$\mathbf{u} \cdot \mathbf{a} = (2)(4) + (-1)(-1) + (3)(2) = 15$$
$$\|\mathbf{a}\|^2 = 4^2 + (-1)^2 + 2^2 = 21$$

Thus, the vector component of $\mathbf{u}$ along $\mathbf{a}$ is

$$\text{proj}_{\mathbf{a}}\,\mathbf{u} = \frac{\mathbf{u} \cdot \mathbf{a}}{\|\mathbf{a}\|^2}\mathbf{a} = \tfrac{15}{21}(4, -1, 2) = (\tfrac{20}{7}, -\tfrac{5}{7}, \tfrac{10}{7})$$

and the vector component of $\mathbf{u}$ orthogonal to $\mathbf{a}$ is

$$\mathbf{u} - \text{proj}_{\mathbf{a}}\,\mathbf{u} = (2, -1, 3) - (\tfrac{20}{7}, -\tfrac{5}{7}, \tfrac{10}{7}) = (-\tfrac{6}{7}, -\tfrac{2}{7}, \tfrac{11}{7})$$

As a check, the reader may wish to verify that the vectors $\mathbf{u} - \text{proj}_{\mathbf{a}}\,\mathbf{u}$ and $\mathbf{a}$ are perpendicular by showing that their dot product is zero. ▲

A formula for the length of the vector component of $\mathbf{u}$ along $\mathbf{a}$ can be obtained by writing

$$\|\text{proj}_{\mathbf{a}}\,\mathbf{u}\| = \left\|\frac{\mathbf{u} \cdot \mathbf{a}}{\|\mathbf{a}\|^2}\mathbf{a}\right\|$$

$$= \left|\frac{\mathbf{u} \cdot \mathbf{a}}{\|\mathbf{a}\|^2}\right|\|\mathbf{a}\| \qquad \boxed{\begin{array}{l}\text{Formula (5) of}\\\text{Section 3.2}\end{array}}$$

$$= \frac{|\mathbf{u} \cdot \mathbf{a}|}{\|\mathbf{a}\|^2}\|\mathbf{a}\| \qquad \boxed{\text{Since }\|\mathbf{a}\|^2 > 0}$$

which yields

$$\|\text{proj}_{\mathbf{a}}\,\mathbf{u}\| = \frac{|\mathbf{u}\cdot\mathbf{a}|}{\|\mathbf{a}\|} \tag{10}$$

If θ denotes the angle between $\mathbf{u}$ and $\mathbf{a}$, then $\mathbf{u}\cdot\mathbf{a} = \|\mathbf{u}\|\,\|\mathbf{a}\|\cos\theta$, so that (10) can also be written as

$$\|\text{proj}_{\mathbf{a}}\,\mathbf{u}\| = \|\mathbf{u}\|\,|\cos\theta| \tag{11}$$

(Verify.) A geometric interpretation of this result is given in Figure 7.

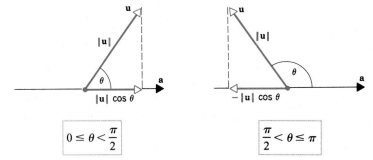

Figure 7

$$0 \le \theta < \frac{\pi}{2} \qquad\qquad \frac{\pi}{2} < \theta \le \pi$$

As an example, we will use vector methods to derive a formula for the distance from a point in the plane to a line.

Example 7 Find a formula for the distance D between the point $P_0(x_0, y_0)$ and the line $ax + by + c = 0$.

Solution. Let $Q(x_1, y_1)$ be any point on the line and position the vector

$$\mathbf{n} = (a, b)$$

so that its initial point is at Q.

By virtue of Example 5, the vector $\mathbf{n}$ is perpendicular to the line (Figure 8). As indicated in the figure, the distance D is equal to the length of the orthogonal projection of $\overrightarrow{QP}_0$ on $\mathbf{n}$; thus, from (10),

$$D = \|\text{proj}_{\mathbf{n}}\,\overrightarrow{QP}_0\| = \frac{|\overrightarrow{QP}_0\cdot\mathbf{n}|}{\|\mathbf{n}\|}$$

But

$$\overrightarrow{QP}_0 = (x_0 - x_1, y_0 - y_1)$$
$$\overrightarrow{QP}_0\cdot\mathbf{n} = a(x_0 - x_1) + b(y_0 - y_1)$$
$$\|\mathbf{n}\| = \sqrt{a^2 + b^2}$$

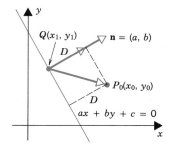

Figure 8

so that

$$D = \frac{|a(x_0 - x_1) + b(y_0 - y_1)|}{\sqrt{a^2 + b^2}} \tag{12}$$

Since the point $Q(x_1, y_1)$ lies on the line, its coordinates satisfy the equation of the line, so

$$ax_1 + by_1 + c = 0$$

or

$$c = -ax_1 - by_1$$

Substituting this expression in (12) yields the formula

$$D = \frac{|ax_0 + by_0 + c|}{\sqrt{a^2 + b^2}} \tag{13}$$

Example 8   It follows from Formula (13) that the distance D from the point $(1, -2)$ to the line $3x + 4y - 6 = 0$ is

$$D = \frac{|(3)(1) + 4(-2) - 6|}{\sqrt{3^2 + 4^2}} = \frac{|-11|}{\sqrt{25}} = \frac{11}{5}$$

EXERCISE SET 3.3

1. Find $\mathbf{u} \cdot \mathbf{v}$.
   (a) $\mathbf{u} = (2, 3), \mathbf{v} = (5, -7)$ (b) $\mathbf{u} = (-6, -2), \mathbf{v} = (4, 0)$
   (c) $\mathbf{u} = (1, -5, 4), \mathbf{v} = (3, 3, 3)$ (d) $\mathbf{u} = (-2, 2, 3), \mathbf{v} = (1, 7, -4)$

2. In each part of Exercise 1, find the cosine of the angle θ between $\mathbf{u}$ and $\mathbf{v}$.

3. Determine whether $\mathbf{u}$ and $\mathbf{v}$ make an acute angle, make an obtuse angle, or are orthogonal.
   (a) $\mathbf{u} = (6, 1, 4), \mathbf{v} = (2, 0, -3)$ (b) $\mathbf{u} = (0, 0, -1), \mathbf{v} = (1, 1, 1)$
   (c) $\mathbf{u} = (-6, 0, 4), \mathbf{v} = (3, 1, 6)$ (d) $\mathbf{u} = (2, 4, -8), \mathbf{v} = (5, 3, 7)$

4. Find the orthogonal projection of $\mathbf{u}$ on $\mathbf{a}$.
   (a) $\mathbf{u} = (6, 2), \mathbf{a} = (3, -9)$ (b) $\mathbf{u} = (-1, -2), \mathbf{a} = (-2, 3)$
   (c) $\mathbf{u} = (3, 1, -7), \mathbf{a} = (1, 0, 5)$ (d) $\mathbf{u} = (1, 0, 0), \mathbf{a} = (4, 3, 8)$

5. In each part of Exercise 4, find the vector component of **u** orthogonal to **a**.

6. In each part find $\|\text{proj}_{\mathbf{a}}\ \mathbf{u}\|$.
   (a) $\mathbf{u} = (1, -2)$, $\mathbf{a} = (-4, -3)$ (b) $\mathbf{u} = (5, 6)$, $\mathbf{a} = (2, -1)$
   (c) $\mathbf{u} = (3, 0, 4)$, $\mathbf{a} = (2, 3, 3)$ (d) $\mathbf{u} = (3, -2, 6)$, $\mathbf{a} = (1, 2, -7)$

7. Let $\mathbf{u} = (5, -2, 1)$, $\mathbf{v} = (1, 6, 3)$, and $k = -4$. Verify Theorem 3.3.2 for these quantities.

8. (a) Show that $\mathbf{v} = (a, b)$ and $\mathbf{w} = (-b, a)$ are orthogonal vectors.
   (b) Use the result in part (a) to find two vectors that are orthogonal to $\mathbf{v} = (2, -3)$.
   (c) Find two unit vectors that are orthogonal to $(-3, 4)$.

9. Let $\mathbf{u} = (3, 4)$, $\mathbf{v} = (5, -1)$, and $\mathbf{w} = (7, 1)$. Evaluate the expressions.
   (a) $\mathbf{u} \cdot (7\mathbf{v} + \mathbf{w})$ (b) $\|(\mathbf{u} \cdot \mathbf{w})\mathbf{w}\|$ (c) $\|\mathbf{u}\|(\mathbf{v} \cdot \mathbf{w})$ (d) $(\|\mathbf{u}\|\mathbf{v}) \cdot \mathbf{w}$

10. Explain why each of the following expressions makes no sense.
    (a) $\mathbf{u} \cdot (\mathbf{v} \cdot \mathbf{w})$ (b) $(\mathbf{u} \cdot \mathbf{v}) + \mathbf{w}$ (c) $\|\mathbf{u} \cdot \mathbf{v}\|$ (d) $k \cdot (\mathbf{u} + \mathbf{v})$

11. Use vectors to find the cosines of the interior angles of the triangle with vertices $(0, -1)$, $(1, -2)$, and $(4, 1)$.

12. Show that $A(3, 0, 2)$, $B(4, 3, 0)$, and $C(8, 1, -1)$ are vertices of a right triangle. At which vertex is the right angle?

13. Suppose that $\mathbf{a} \cdot \mathbf{b} = \mathbf{a} \cdot \mathbf{c}$ and $\mathbf{a} \neq \mathbf{0}$. Does it follow that $\mathbf{b} = \mathbf{c}$? Explain.

14. Let $\mathbf{p} = (2, k)$ and $\mathbf{q} = (3, 5)$. Find k such that
    (a) **p** and **q** are parallel
    (b) **p** and **q** are orthogonal
    (c) the angle between **p** and **q** is $\pi/3$
    (d) the angle between **p** and **q** is $\pi/4$

15. Use Formula (13) to calculate the distance between the point and the line.
    (a) $4x + 3y + 4 = 0$; $(-3, 1)$
    (b) $y = -4x + 2$; $(2, -5)$
    (c) $3x + y = 5$; $(1, 8)$

16. Establish the identity $\|\mathbf{u} + \mathbf{v}\|^2 + \|\mathbf{u} - \mathbf{v}\|^2 = 2\|\mathbf{u}\|^2 + 2\|\mathbf{v}\|^2$.

17. Establish the identity $\mathbf{u} \cdot \mathbf{v} = \frac{1}{4}\|\mathbf{u} + \mathbf{v}\|^2 - \frac{1}{4}\|\mathbf{u} - \mathbf{v}\|^2$.

18. Find the angle between a diagonal of a cube and one of its faces.

19. Let **i**, **j**, and **k** be unit vectors along the positive x, y, and z axes of a rectangular coordinate system in 3-space. If $\mathbf{v} = (a, b, c)$ is a nonzero vector, then the angles α, β, and γ between **v** and the vectors **i**, **j**, and **k**, respectively, are called the ***direction angles*** of **v** (Figure 9), and the numbers $\cos \alpha$, $\cos \beta$, and $\cos \gamma$ are called the ***direction cosines*** of **v**.
    (a) Show that $\cos \alpha = a/\|\mathbf{v}\|$.
    (b) Find $\cos \beta$ and $\cos \gamma$.
    (c) Show that $\mathbf{v}/\|\mathbf{v}\| = (\cos \alpha, \cos \beta, \cos \gamma)$.
    (d) Show that $\cos^2 \alpha + \cos^2 \beta + \cos^2 \gamma = 1$.

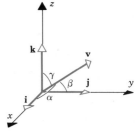

Figure 9

20. Use the result in Exercise 19 to estimate, to the nearest degree, the angles that a diagonal of a box with dimensions $10 \text{ cm} \times 15 \text{ cm} \times 25 \text{ cm}$ makes with edges of the box. [*Note.* A calculator or trigonometric table is needed.]

21. Referring to Exercise 19, show that v_1 and v_2 are perpendicular vectors in 3-space if and only if their direction cosines satisfy

$$\cos \alpha_1 \cos \alpha_2 + \cos \beta_1 \cos \beta_2 + \cos \gamma_1 \cos \gamma_2 = 0$$

22. Show that if v is orthogonal to both w_1 and w_2, then v is orthogonal to $k_1 w_1 + k_2 w_2$ for all scalars k_1 and k_2.

23. Let u and v be nonzero vectors in 2- or 3-space, and let $k = \|u\|$ and $l = \|v\|$. Show that the vector $w = lu + kv$ bisects the angle between u and v.

3.4 CROSS PRODUCT

In many applications of vectors to problems in geometry, physics, and engineering, it is of interest to construct a vector in 3-space that is perpendicular to two given vectors. In this section we shall introduce a type of vector multiplication that produces such a vector.

CROSS PRODUCT OF VECTORS

Definition. If $u = (u_1, u_2, u_3)$ and $v = (v_1, v_2, v_3)$ are vectors in 3-space, then the *cross product* $u \times v$ is the vector defined by

$$u \times v = (u_2 v_3 - u_3 v_2, \, u_3 v_1 - u_1 v_3, \, u_1 v_2 - u_2 v_1)$$

or in determinant notation

$$u \times v = \left(\begin{vmatrix} u_2 & u_3 \\ v_2 & v_3 \end{vmatrix}, \, -\begin{vmatrix} u_1 & u_3 \\ v_1 & v_3 \end{vmatrix}, \, \begin{vmatrix} u_1 & u_2 \\ v_1 & v_2 \end{vmatrix} \right) \tag{1}$$

REMARK. Instead of memorizing (1), you can obtain the components of $u \times v$ as follows:

- Form the 2×3 matrix

$$\begin{bmatrix} u_1 & u_2 & u_3 \\ v_1 & v_2 & v_3 \end{bmatrix}$$

whose first row contains the components of u and whose second row contains the components of v.

- To find the first component of $u \times v$, delete the first column and take the determinant; to find the second component, delete the second column and take the negative of the determinant; and to find the third component, delete the third column and take the determinant.

Example 1 Find $\mathbf{u} \times \mathbf{v}$, where $\mathbf{u} = (1, 2, -2)$ and $\mathbf{v} = (3, 0, 1)$.

Solution.

$$\begin{bmatrix} 1 & 2 & -2 \\ 3 & 0 & 1 \end{bmatrix}$$

$$\mathbf{u} \times \mathbf{v} = \left(\begin{vmatrix} 2 & -2 \\ 0 & 1 \end{vmatrix}, -\begin{vmatrix} 1 & -2 \\ 3 & 1 \end{vmatrix}, \begin{vmatrix} 1 & 2 \\ 3 & 0 \end{vmatrix} \right)$$

$$= (2, -7, -6) \quad \blacktriangle$$

There is an important difference between the dot product and cross product of two vectors—the dot product is a scalar and the cross product is a vector. The following theorem gives some important relationships between the dot product and cross product and also shows that $\mathbf{u} \times \mathbf{v}$ is orthogonal to both $\mathbf{u}$ and $\mathbf{v}$.

Theorem 3.4.1. *If* $\mathbf{u}$, $\mathbf{v}$, *and* $\mathbf{w}$ *are vectors in 3-space, then:*

(a) $\mathbf{u} \cdot (\mathbf{u} \times \mathbf{v}) = 0$ *($\mathbf{u} \times \mathbf{v}$ is orthogonal to $\mathbf{u}$)*

(b) $\mathbf{v} \cdot (\mathbf{u} \times \mathbf{v}) = 0$ *($\mathbf{u} \times \mathbf{v}$ is orthogonal to $\mathbf{v}$)*

(c) $\|\mathbf{u} \times \mathbf{v}\|^2 = \|\mathbf{u}\|^2 \|\mathbf{v}\|^2 - (\mathbf{u} \cdot \mathbf{v})^2$ *(Lagrange's* identity)*

(d) $\mathbf{u} \times (\mathbf{v} \times \mathbf{w}) = (\mathbf{u} \cdot \mathbf{w})\mathbf{v} - (\mathbf{u} \cdot \mathbf{v})\mathbf{w}$ *(relationship between cross and dot products)*

(e) $(\mathbf{u} \times \mathbf{v}) \times \mathbf{w} = (\mathbf{u} \cdot \mathbf{w})\mathbf{v} - (\mathbf{v} \cdot \mathbf{w})\mathbf{u}$ *(relationship between cross and dot products)*

Proof (a). Let $\mathbf{u} = (u_1, u_2, u_3)$ and $\mathbf{v} = (v_1, v_2, v_3)$. Then

$$\mathbf{u} \cdot (\mathbf{u} \times \mathbf{v}) = (u_1, u_2, u_3) \cdot (u_2 v_3 - u_3 v_2, u_3 v_1 - u_1 v_3, u_1 v_2 - u_2 v_1)$$

$$= u_1(u_2 v_3 - u_3 v_2) + u_2(u_3 v_1 - u_1 v_3) + u_3(u_1 v_2 - u_2 v_1)$$

$$= 0$$

**Joseph Louis Lagrange* (1736–1813), French–Italian mathematician and astronomer. Lagrange, the son of a public official, was born in Turin, Italy. (Baptismal records list his name as Giuseppe Lodovico Lagrangia.) Although his father wanted him to be a lawyer, Lagrange was attracted to mathematics and astronomy after reading a memoir by the astronomer Halley. At age 16 he began to study mathematics on his own and by age 19 was appointed to a professorship at the Royal Artillery School in Turin. The following year he solved some famous problems using new methods that eventually blossomed into a branch of mathematics called calculus of variations. These methods and Lagrange's applications of them to problems in celestial mechanics were so monumental that by age 25 he was regarded by many of his contemporaries as the greatest living mathematician. One of Lagrange's most famous works is a memoir, *Mécanique Analytique*, in which he reduced the theory of mechanics to a few general formulas from which all other necessary equations could be derived.

It is an interesting historical fact that Lagrange's father speculated unsuccessfully in several financial ventures, so his family was forced to live quite modestly. Lagrange himself stated that if his family had money, he would not have made mathematics his vocation.

Napoleon was a great admirer of Lagrange and showered him with honors—count, senator, and Legion of Honor. In spite of his fame, Lagrange was always a shy and modest man. On his death, he was buried with honor in the Pantheon.

Proof (b). Similar to (*a*).

Proof (c). Since

$$\|\mathbf{u} \times \mathbf{v}\|^2 = (u_2 v_3 - u_3 v_2)^2 + (u_3 v_1 - u_1 v_3)^2 + (u_1 v_2 - u_2 v_1)^2 \qquad (2)$$

and

$$\|\mathbf{u}\|^2 \|\mathbf{v}\|^2 - (\mathbf{u} \cdot \mathbf{v})^2 = (u_1^2 + u_2^2 + u_3^2)(v_1^2 + v_2^2 + v_3^2) - (u_1 v_1 + u_2 v_2 + u_3 v_3)^2 \qquad (3)$$

The proof can be completed by "multiplying out" the right sides of (2) and (3) and verifying their equality.

Proof (d) and (e). See Exercises 26 and 27. ▌

Example 2 Consider the vectors

$$\mathbf{u} = (1, 2, -2) \qquad \text{and} \qquad \mathbf{v} = (3, 0, 1)$$

In Example 1 we showed that

$$\mathbf{u} \times \mathbf{v} = (2, -7, -6)$$

Since

$$\mathbf{u} \cdot (\mathbf{u} \times \mathbf{v}) = (1)(2) + (2)(-7) + (-2)(-6) = 0$$

and

$$\mathbf{v} \cdot (\mathbf{u} \times \mathbf{v}) = (3)(2) + (0)(-7) + (1)(-6) = 0$$

$\mathbf{u} \times \mathbf{v}$ is orthogonal to both $\mathbf{u}$ and $\mathbf{v}$ as guaranteed by Theorem 3.4.1. ▲

The main arithmetic properties of the cross product are listed in the next theorem.

Theorem 3.4.2. *If* $\mathbf{u}$, $\mathbf{v}$, *and* $\mathbf{w}$ *are any vectors in 3-space and* k *is any scalar, then:*
(*a*) $\mathbf{u} \times \mathbf{v} = -(\mathbf{v} \times \mathbf{u})$
(*b*) $\mathbf{u} \times (\mathbf{v} + \mathbf{w}) = (\mathbf{u} \times \mathbf{v}) + (\mathbf{u} \times \mathbf{w})$
(*c*) $(\mathbf{u} + \mathbf{v}) \times \mathbf{w} = (\mathbf{u} \times \mathbf{w}) + (\mathbf{v} \times \mathbf{w})$
(*d*) $k(\mathbf{u} \times \mathbf{v}) = (k\mathbf{u}) \times \mathbf{v} = \mathbf{u} \times (k\mathbf{v})$
(*e*) $\mathbf{u} \times \mathbf{0} = \mathbf{0} \times \mathbf{u} = \mathbf{0}$
(*f*) $\mathbf{u} \times \mathbf{u} = \mathbf{0}$

The proofs follow immediately from Formula (1) and properties of determinants; for example, (*a*) can be proved as follows:

Proof (a). Interchanging $\mathbf{u}$ and $\mathbf{v}$ in (1) interchanges the rows of the three determinants on the right side of (1) and hence changes the sign of each component in the cross product. Thus, $\mathbf{u} \times \mathbf{v} = -(\mathbf{v} \times \mathbf{u})$. ▌

The proofs of the remaining parts are left as exercises.

Example 3   Consider the vectors

$$\mathbf{i} = (1, 0, 0) \qquad \mathbf{j} = (0, 1, 0) \qquad \mathbf{k} = (0, 0, 1)$$

These vectors each have length 1 and lie along the coordinate axes (Figure 1). They are called the ***standard unit vectors*** in 3-space. Every vector $\mathbf{v} = (v_1, v_2, v_3)$ in 3-space is expressible in terms of $\mathbf{i}$, $\mathbf{j}$, and $\mathbf{k}$ since we can write

$$\mathbf{v} = (v_1, v_2, v_3) = v_1(1, 0, 0) + v_2(0, 1, 0) + v_3(0, 0, 1) = v_1\mathbf{i} + v_2\mathbf{j} + v_3\mathbf{k}$$

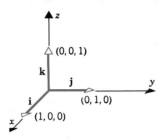

Figure 1   The standard unit vectors

For example,

$$(2, -3, 4) = 2\mathbf{i} - 3\mathbf{j} + 4\mathbf{k}$$

From (1) we obtain

$$\mathbf{i} \times \mathbf{j} = \left(\begin{vmatrix} 0 & 0 \\ 1 & 0 \end{vmatrix}, \ -\begin{vmatrix} 1 & 0 \\ 0 & 0 \end{vmatrix}, \ \begin{vmatrix} 1 & 0 \\ 0 & 1 \end{vmatrix} \right) = (0, 0, 1) = \mathbf{k} \quad \blacktriangle$$

The reader should have no trouble obtaining the following results:

$$\mathbf{i} \times \mathbf{i} = \mathbf{j} \times \mathbf{j} = \mathbf{k} \times \mathbf{k} = \mathbf{0}$$
$$\mathbf{i} \times \mathbf{j} = \mathbf{k}, \qquad \mathbf{j} \times \mathbf{k} = \mathbf{i}, \qquad \mathbf{k} \times \mathbf{i} = \mathbf{j}$$
$$\mathbf{j} \times \mathbf{i} = -\mathbf{k}, \qquad \mathbf{k} \times \mathbf{j} = -\mathbf{i}, \qquad \mathbf{i} \times \mathbf{k} = -\mathbf{j}$$

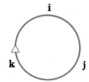

Figure 2

Figure 2 is helpful for remembering these results. Referring to this diagram, the cross product of two consecutive vectors going clockwise is the next vector around, and the cross product of two consecutive vectors going counterclockwise is the negative of the next vector around.

DETERMINANT FORMULA FOR CROSS PRODUCT

It is also worth noting that a cross product can be represented symbolically in the form of a 3×3 determinant:

$$\mathbf{u} \times \mathbf{v} = \begin{vmatrix} \mathbf{i} & \mathbf{j} & \mathbf{k} \\ u_1 & u_2 & u_3 \\ v_1 & v_2 & v_3 \end{vmatrix} = \begin{vmatrix} u_2 & u_3 \\ v_2 & v_3 \end{vmatrix}\mathbf{i} - \begin{vmatrix} u_1 & u_3 \\ v_1 & v_3 \end{vmatrix}\mathbf{j} + \begin{vmatrix} u_1 & u_2 \\ v_1 & v_2 \end{vmatrix}\mathbf{k} \qquad (4)$$

For example, if $\mathbf{u} = (1, 2, -2)$ and $\mathbf{v} = (3, 0, 1)$, then

$$\mathbf{u} \times \mathbf{v} = \begin{vmatrix} \mathbf{i} & \mathbf{j} & \mathbf{k} \\ 1 & 2 & -2 \\ 3 & 0 & 1 \end{vmatrix} = 2\mathbf{i} - 7\mathbf{j} - 6\mathbf{k}$$

which agrees with the result obtained in Example 1.

Warning. It is not true in general that $\mathbf{u} \times (\mathbf{v} \times \mathbf{w}) = (\mathbf{u} \times \mathbf{v}) \times \mathbf{w}$. For example,

$$\mathbf{i} \times (\mathbf{j} \times \mathbf{j}) = \mathbf{i} \times \mathbf{0} = \mathbf{0}$$

and

$$(\mathbf{i} \times \mathbf{j}) \times \mathbf{j} = \mathbf{k} \times \mathbf{j} = -\mathbf{i}$$

so that

$$\mathbf{i} \times (\mathbf{j} \times \mathbf{j}) \neq (\mathbf{i} \times \mathbf{j}) \times \mathbf{j}$$

We know from Theorem 3.4.1 that $\mathbf{u} \times \mathbf{v}$ is orthogonal to both $\mathbf{u}$ and $\mathbf{v}$. If $\mathbf{u}$ and $\mathbf{v}$ are nonzero vectors, it can be shown that the direction of $\mathbf{u} \times \mathbf{v}$ can be determined using the following "right-hand rule"* (Figure 3): Let θ be the angle between $\mathbf{u}$ and $\mathbf{v}$, and suppose $\mathbf{u}$ is rotated through the angle θ until it coincides with $\mathbf{v}$. If the fingers of the right hand are cupped so they point in the direction of rotation, then the thumb indicates (roughly) the direction of $\mathbf{u} \times \mathbf{v}$.

Figure 3

The reader may find it instructive to practice this rule with the products

$$\mathbf{i} \times \mathbf{j} = \mathbf{k} \qquad \mathbf{j} \times \mathbf{k} = \mathbf{i} \qquad \mathbf{k} \times \mathbf{i} = \mathbf{j}$$

GEOMETRIC INTERPRETATION OF THE CROSS PRODUCT

If $\mathbf{u}$ and $\mathbf{v}$ are vectors in 3-space, then the norm of $\mathbf{u} \times \mathbf{v}$ has a useful geometric interpretation. Lagrange's identity, given in Theorem 3.4.1, states that

$$\|\mathbf{u} \times \mathbf{v}\|^2 = \|\mathbf{u}\|^2 \|\mathbf{v}\|^2 - (\mathbf{u} \cdot \mathbf{v})^2 \tag{5}$$

---

* Recall that we agreed to consider only right-handed coordinate systems in this text. Had we used left-handed systems instead, a "left-hand rule" would apply here.

If θ denotes the angle between $\mathbf{u}$ and $\mathbf{v}$, then $\mathbf{u} \cdot \mathbf{v} = \|\mathbf{u}\| \|\mathbf{v}\| \cos\theta$, so that (5) can be rewritten as

$$\|\mathbf{u} \times \mathbf{v}\|^2 = \|\mathbf{u}\|^2 \|\mathbf{v}\|^2 - \|\mathbf{u}\|^2 \|\mathbf{v}\|^2 \cos^2\theta$$
$$= \|\mathbf{u}\|^2 \|\mathbf{v}\|^2 (1 - \cos^2\theta)$$
$$= \|\mathbf{u}\|^2 \|\mathbf{v}\|^2 \sin^2\theta$$

Since $0 \le \theta \le \pi$, it follows that $\sin\theta \ge 0$, so this can be rewritten as,

$$\|\mathbf{u} \times \mathbf{v}\| = \|\mathbf{u}\| \|\mathbf{v}\| \sin\theta \qquad (6)$$

But $\|\mathbf{v}\| \sin\theta$ is the altitude of the parallelogram determined by $\mathbf{u}$ and $\mathbf{v}$ (Figure 4). Thus,

Figure 4

from (6), the area A of this parallelogram is given by

$$A = (\text{base})(\text{altitude}) = \|\mathbf{u}\| \|\mathbf{v}\| \sin\theta = \|\mathbf{u} \times \mathbf{v}\|$$

This result is even correct if $\mathbf{u}$ and $\mathbf{v}$ are collinear, since the parallelogram determined by $\mathbf{u}$ and $\mathbf{v}$ has zero area and from (6) we have $\mathbf{u} \times \mathbf{v} = \mathbf{0}$ because $\theta = 0$ in this case. Thus we have the following theorem.

Theorem 3.4.3. *If $\mathbf{u}$ and $\mathbf{v}$ are vectors in 3-space, then $\|\mathbf{u} \times \mathbf{v}\|$ is equal to the area of the parallelogram determined by $\mathbf{u}$ and $\mathbf{v}$.*

Example 4 Find the area of the triangle determined by the points $P_1(2, 2, 0)$, $P_2(-1, 0, 2)$, and $P_3(0, 4, 3)$.

Solution. The area A of the triangle is $\frac{1}{2}$ the area of the parallelogram determined by the vectors $\overrightarrow{P_1P_2}$ and $\overrightarrow{P_1P_3}$ (Figure 5). Using the method discussed in Example 2 of Section 3.1, $\overrightarrow{P_1P_2} = (-3, -2, 2)$ and $\overrightarrow{P_1P_3} = (-2, 2, 3)$. It follows that

$$\overrightarrow{P_1P_2} \times \overrightarrow{P_1P_3} = (-10, 5, -10)$$

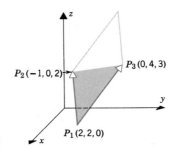

Figure 5

and consequently

$$A = \tfrac{1}{2}\|\overrightarrow{P_1P_2} \times \overrightarrow{P_1P_3}\| = \tfrac{1}{2}(15) = \tfrac{15}{2} \quad \blacktriangle$$

SCALAR TRIPLE PRODUCT

> **Definition.** If **u**, **v**, and **w** are vectors in 3-space, then
> $$\mathbf{u} \cdot (\mathbf{v} \times \mathbf{w})$$
> is called the *scalar triple product* of **u**, **v**, and **w**.

The scalar triple product of $\mathbf{u} = (u_1, u_2, u_3)$, $\mathbf{v} = (v_1, v_2, v_3)$, and $\mathbf{w} = (w_1, w_2, w_3)$ can be calculated from the formula

$$\mathbf{u} \cdot (\mathbf{v} \times \mathbf{w}) = \begin{vmatrix} u_1 & u_2 & u_3 \\ v_1 & v_2 & v_3 \\ w_1 & w_2 & w_3 \end{vmatrix} \tag{7}$$

This follows from Formula (4) since

$$\mathbf{u} \cdot (\mathbf{v} \times \mathbf{w}) = \mathbf{u} \cdot \left(\begin{vmatrix} v_2 & v_3 \\ w_2 & w_3 \end{vmatrix} \mathbf{i} - \begin{vmatrix} v_1 & v_3 \\ w_1 & w_3 \end{vmatrix} \mathbf{j} + \begin{vmatrix} v_1 & v_2 \\ w_1 & w_2 \end{vmatrix} \mathbf{k} \right)$$

$$= \begin{vmatrix} v_2 & v_3 \\ w_2 & w_3 \end{vmatrix} u_1 - \begin{vmatrix} v_1 & v_3 \\ w_1 & w_3 \end{vmatrix} u_2 + \begin{vmatrix} v_1 & v_2 \\ w_1 & w_2 \end{vmatrix} u_3$$

$$= \begin{vmatrix} u_1 & u_2 & u_3 \\ v_1 & v_2 & v_3 \\ w_1 & w_2 & w_3 \end{vmatrix}$$

Example 5 Calculate the scalar triple product $\mathbf{u} \cdot (\mathbf{v} \times \mathbf{w})$ of the vectors

$$\mathbf{u} = 3\mathbf{i} - 2\mathbf{j} - 5\mathbf{k}, \qquad \mathbf{v} = \mathbf{i} + 4\mathbf{j} - 4\mathbf{k}, \qquad \mathbf{w} = 3\mathbf{j} + 2\mathbf{k}$$

Solution. From (7)

$$\mathbf{u} \cdot (\mathbf{v} \times \mathbf{w}) = \begin{vmatrix} 3 & -2 & -5 \\ 1 & 4 & -4 \\ 0 & 3 & 2 \end{vmatrix}$$

$$= 3 \begin{vmatrix} 4 & -4 \\ 3 & 2 \end{vmatrix} - (-2) \begin{vmatrix} 1 & -4 \\ 0 & 2 \end{vmatrix} + (-5) \begin{vmatrix} 1 & 4 \\ 0 & 3 \end{vmatrix}$$

$$= 60 + 4 - 15 = 49 \quad \blacktriangle$$

REMARK. The symbol $(\mathbf{u} \cdot \mathbf{v}) \times \mathbf{w}$ makes no sense since we cannot form the cross product of a scalar and a vector. Thus, no ambiguity arises if we write $\mathbf{u} \cdot \mathbf{v} \times \mathbf{w}$ rather than $\mathbf{u} \cdot (\mathbf{v} \times \mathbf{w})$. However, for clarity we shall usually keep the parentheses.

It follows from (7) that

$$\mathbf{u} \cdot (\mathbf{v} \times \mathbf{w}) = \mathbf{w} \cdot (\mathbf{u} \times \mathbf{v}) = \mathbf{v} \cdot (\mathbf{w} \times \mathbf{u})$$

since the 3×3 determinants that represent these products can be obtained from one another by *two* row interchanges. (Verify.) These relationships can be remembered by moving the vectors $\mathbf{u}$, $\mathbf{v}$, and $\mathbf{w}$ clockwise around the vertices of the triangle in Figure 6.

Figure 6

GEOMETRIC INTERPRETATION OF DETERMINANTS

The next theorem provides a useful geometric interpretation of 2×2 and 3×3 determinants.

Theorem 3.4.4

(a) *The absolute value of the determinant*

$$\det \begin{bmatrix} u_1 & u_2 \\ v_1 & v_2 \end{bmatrix}$$

is equal to the area of the parallelogram in 2-space determined by the vectors $\mathbf{u} = (u_1, u_2)$ *and* $\mathbf{v} = (v_1, v_2)$. *(See Figure 7a.)*

(b) *The absolute value of the determinant*

$$\det \begin{bmatrix} u_1 & u_2 & u_3 \\ v_1 & v_2 & v_3 \\ w_1 & w_2 & w_3 \end{bmatrix}$$

is equal to the volume of the parallelepiped in 3-space determined by the vectors $\mathbf{u} = (u_1, u_2, u_3)$, $\mathbf{v} = (v_1, v_2, v_3)$, *and* $\mathbf{w} = (w_1, w_2, w_3)$. *(See Figure 7b.)*

Proof (a). The key to the proof is to use Theorem 3.4.3. However, that theorem applies to vectors in 3-space, whereas $\mathbf{u} = (u_1, u_2)$ and $\mathbf{v} = (v_1, v_2)$ are vectors in 2-space. To circumvent this "dimension problem," we shall view $\mathbf{u}$ and $\mathbf{v}$ as vectors in the xy-plane

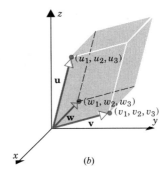

Figure 7 (a) (b)

of an *xyz*-coordinate system (Figure 8*a*), in which case these vectors are expressed as $\mathbf{u} = (u_1, u_2, 0)$ and $\mathbf{v} = (v_1, v_2, 0)$. Thus,

$$\mathbf{u} \times \mathbf{v} = \begin{vmatrix} \mathbf{i} & \mathbf{j} & \mathbf{k} \\ u_1 & u_2 & 0 \\ v_1 & v_2 & 0 \end{vmatrix} = \begin{vmatrix} u_1 & u_2 \\ v_1 & v_2 \end{vmatrix} \mathbf{k} = \det \begin{bmatrix} u_1 & u_2 \\ v_1 & v_2 \end{bmatrix} \mathbf{k}$$

It now follows from Theorem 3.4.3 and the fact that $\|\mathbf{k}\| = 1$ that the area A of the parallelogram determined by $\mathbf{u}$ and $\mathbf{v}$ is

$$A = \|\mathbf{u} \times \mathbf{v}\| = \left\| \det \begin{bmatrix} u_1 & u_2 \\ v_1 & v_2 \end{bmatrix} \mathbf{k} \right\| = \left| \det \begin{bmatrix} u_1 & u_2 \\ v_1 & v_2 \end{bmatrix} \right| \|\mathbf{k}\| = \left| \det \begin{bmatrix} u_1 & u_2 \\ v_1 & v_2 \end{bmatrix} \right|$$

which completes the proof.

Proof (b). As shown in Figure 8*b*, take the base of the parallelepiped determined by $\mathbf{u}$, $\mathbf{v}$, and $\mathbf{w}$ to be the parallelogram determined by $\mathbf{v}$ and $\mathbf{w}$. It follows from Theorem 3.4.3 that the area of the base is $\|\mathbf{v} \times \mathbf{w}\|$ and, as illustrated in Figure 8*b*, the height h

Figure 8 (a) (b)

of the parallelepiped is the length of the orthogonal projection of $\mathbf{u}$ on $\mathbf{v} \times \mathbf{w}$. Therefore, by Formula (10) of Section 3.3,

$$h = \|\text{proj}_{\mathbf{v} \times \mathbf{w}}\, \mathbf{u}\| = \frac{|\mathbf{u} \cdot (\mathbf{v} \times \mathbf{w})|}{\|\mathbf{v} \times \mathbf{w}\|}$$

It follows that the volume V of the parallelepiped is

$$V = (\text{area of base}) \cdot \text{height} = \|\mathbf{v} \times \mathbf{w}\| \frac{|\mathbf{u} \cdot (\mathbf{v} \times \mathbf{w})|}{\|\mathbf{v} \times \mathbf{w}\|} = |\mathbf{u} \cdot (\mathbf{v} \times \mathbf{w})|$$

so from (7)

$$V = \left| \det \begin{bmatrix} u_1 & u_2 & u_3 \\ v_1 & v_2 & v_3 \\ w_1 & w_2 & w_3 \end{bmatrix} \right|$$

which completes the proof. ▮

REMARK. If V denotes the volume of the parallelepiped determined by vectors $\mathbf{u}$, $\mathbf{v}$, and $\mathbf{w}$, then it follows from Theorem 3.4.4 and Formula (7) that

$$V = \begin{bmatrix} \text{volume of parallelepiped} \\ \text{determined by } \mathbf{u}, \mathbf{v}, \text{ and } \mathbf{w} \end{bmatrix} = |\mathbf{u} \cdot (\mathbf{v} \times \mathbf{w})| \tag{8}$$

From this and Theorem 3.3.1b, we can conclude that

$$\mathbf{u} \cdot (\mathbf{v} \times \mathbf{w}) = \pm V$$

where the $+$ or $-$ results depending on whether $\mathbf{u}$ makes an acute or obtuse angle with $(\mathbf{v} \times \mathbf{w})$.

Formula (8) leads to a useful test for ascertaining whether three given vectors lie in the same plane. Since three vectors not in the same plane determine a parallelepiped of positive volume, it follows from (8) that $|\mathbf{u} \cdot (\mathbf{v} \times \mathbf{w})| = 0$ if and only if the vectors $\mathbf{u}$, $\mathbf{v}$, and $\mathbf{w}$ lie in the same plane. Thus, we have the following result.

Theorem 3.4.5. *If the vectors* $\mathbf{u} = (u_1, u_2, u_3)$, $\mathbf{v} = (v_1, v_2, v_3)$, *and* $\mathbf{w} = (w_1, w_2, w_3)$ *have the same initial point, then they lie in the same plane if and only if*

$$\mathbf{u} \cdot (\mathbf{v} \times \mathbf{w}) = \begin{vmatrix} u_1 & u_2 & u_3 \\ v_1 & v_2 & v_3 \\ w_1 & w_2 & w_3 \end{vmatrix} = 0$$

**INDEPENDENCE
OF CROSS
PRODUCT AND
COORDINATES**

Initially, we defined a vector to be a directed line segment or arrow in 2-space or 3-space; coordinate systems and components were introduced later in order to simplify computations with vectors. Thus, a vector has a "mathematical existence" regardless of whether a coordinate system has been introduced. Further, the components of a vector are not determined by the vector alone; they depend as well on the coordinate system chosen. For example, in Figure 9 we have indicated a fixed vector **v** in the plane and two different coordinate systems. In the xy-coordinate system the components of **v** are $(1, 1)$, and in the $x'y'$-system they are $(\sqrt{2}, 0)$.

This raises an important question about our definition of cross product. Since we defined the cross product $\mathbf{u} \times \mathbf{v}$ in terms of the components of **u** and **v**, and since these components depend on the coordinate system chosen, it seems possible that two *fixed* vectors **u** and **v** might have different cross products in different coordinate systems. Fortunately, this is not the case. To see that this is so, we need only recall that

- $\mathbf{u} \times \mathbf{v}$ is perpendicular to both **u** and **v**.
- The orientation of $\mathbf{u} \times \mathbf{v}$ is determined by the right-hand rule.
- $\|\mathbf{u} \times \mathbf{v}\| = \|\mathbf{u}\| \|\mathbf{v}\| \sin \theta$.

These three properties completely determine the vector $\mathbf{u} \times \mathbf{v}$: the first and second properties determine the direction, and the third property determines the length. Since these properties of $\mathbf{u} \times \mathbf{v}$ depend only on the lengths and relative positions of **u** and **v** and not on the particular right-hand coordinate system being used, the vector $\mathbf{u} \times \mathbf{v}$ will remain unchanged if a different right-hand coordinate system is introduced. Thus, we say that the definition of $\mathbf{u} \times \mathbf{v}$ is ***coordinate free***. This result is of importance to physicists and engineers who often work with many coordinate systems in the same problem.

Figure 9

Example 6 Consider two perpendicular vectors **u** and **v**, each of length 1 (as shown in Figure 10*a*). If we introduce an xyz-coordinate system as shown in Figure 10*b*, then

$$\mathbf{u} = (1, 0, 0) = \mathbf{i} \quad \text{and} \quad \mathbf{v} = (0, 1, 0) = \mathbf{j}$$

so that

$$\mathbf{u} \times \mathbf{v} = \mathbf{i} \times \mathbf{j} = \mathbf{k} = (0, 0, 1)$$

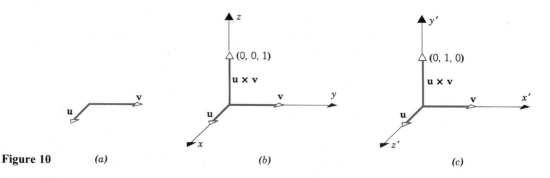

Figure 10   (*a*)   (*b*)   (*c*)

However, if we introduce an $x'y'z'$-coordinate system as shown in Figure 10*c*, then

$$\mathbf{u} = (0, 0, 1) = \mathbf{k} \quad \text{and} \quad \mathbf{v} = (1, 0, 0) = \mathbf{i}$$

so that

$$\mathbf{u} \times \mathbf{v} = \mathbf{k} \times \mathbf{i} = \mathbf{j} = (0, 1, 0)$$

But it is clear from Figures 10*b* and 10*c* that the vector $(0, 0, 1)$ in the xyz-system is the same as the vector $(0, 1, 0)$ in the $x'y'z'$-system. Thus, we obtain the same vector $\mathbf{u} \times \mathbf{v}$ whether we compute with coordinates from the xyz-system or with coordinates from the $x'y'z'$-system. ▲

EXERCISE SET 3.4

1. Let $\mathbf{u} = (3, 2, -1)$, $\mathbf{v} = (0, 2, -3)$, and $\mathbf{w} = (2, 6, 7)$. Compute
   (a) $\mathbf{v} \times \mathbf{w}$ (b) $\mathbf{u} \times (\mathbf{v} \times \mathbf{w})$ (c) $(\mathbf{u} \times \mathbf{v}) \times \mathbf{w}$
   (d) $(\mathbf{u} \times \mathbf{v}) \times (\mathbf{v} \times \mathbf{w})$ (e) $\mathbf{u} \times (\mathbf{v} - 2\mathbf{w})$ (f) $(\mathbf{u} \times \mathbf{v}) - 2\mathbf{w}$

2. Find a vector that is orthogonal to both $\mathbf{u}$ and $\mathbf{v}$.
   (a) $\mathbf{u} = (-6, 4, 2)$, $\mathbf{v} = (3, 1, 5)$ (b) $\mathbf{u} = (-2, 1, 5)$, $\mathbf{v} = (3, 0, -3)$

3. Find the area of the parallelogram determined by $\mathbf{u}$ and $\mathbf{v}$.
   (a) $\mathbf{u} = (1, -1, 2)$, $\mathbf{v} = (0, 3, 1)$ (b) $\mathbf{u} = (2, 3, 0)$, $\mathbf{v} = (-1, 2, -2)$
   (c) $\mathbf{u} = (3, -1, 4)$, $\mathbf{v} = (6, -2, 8)$

4. Find the area of the triangle having vertices P, Q, and R.
   (a) $P(2, 6, -1)$, $Q(1, 1, 1)$, $R(4, 6, 2)$ (b) $P(1, -1, 2)$, $Q(0, 3, 4)$, $R(6, 1, 8)$

5. Verify Theorem 3.4.1 for the vectors $\mathbf{u} = (4, 2, 1)$ and $\mathbf{v} = (-3, 2, 7)$.

6. Verify Theorem 3.4.2 for $\mathbf{u} = (5, -1, 2)$, $\mathbf{v} = (6, 0, -2)$, $\mathbf{w} = (1, 2, -1)$, and $k = -5$.

7. What is wrong with the expression $\mathbf{u} \times \mathbf{v} \times \mathbf{w}$?

8. Find the scalar triple product $\mathbf{u} \cdot (\mathbf{v} \times \mathbf{w})$.
   (a) $\mathbf{u} = (-1, 2, 4)$, $\mathbf{v} = (3, 4, -2)$, $\mathbf{w} = (-1, 2, 5)$
   (b) $\mathbf{u} = (3, -1, 6)$, $\mathbf{v} = (2, 4, 3)$, $\mathbf{w} = (5, -1, 2)$

9. Suppose that $\mathbf{u} \cdot (\mathbf{v} \times \mathbf{w}) = 3$. Find
   (a) $\mathbf{u} \cdot (\mathbf{w} \times \mathbf{v})$ (b) $(\mathbf{v} \times \mathbf{w}) \cdot \mathbf{u}$ (c) $\mathbf{w} \cdot (\mathbf{u} \times \mathbf{v})$ (d) $\mathbf{v} \cdot (\mathbf{u} \times \mathbf{w})$ (e) $(\mathbf{u} \times \mathbf{w}) \cdot \mathbf{v}$ (f) $\mathbf{v} \cdot (\mathbf{w} \times \mathbf{w})$

10. Find the volume of the parallelepiped with sides $\mathbf{u}$, $\mathbf{v}$, and $\mathbf{w}$.
    (a) $\mathbf{u} = (2, -6, 2)$, $\mathbf{v} = (0, 4, -2)$, $\mathbf{w} = (2, 2, -4)$ (b) $\mathbf{u} = (3, 1, 2)$, $\mathbf{v} = (4, 5, 1)$, $\mathbf{w} = (1, 2, 4)$

11. Determine whether $\mathbf{u}$, $\mathbf{v}$, and $\mathbf{w}$ lie in the same plane when positioned so that their initial points coincide.
    (a) $\mathbf{u} = (-1, -2, 1)$, $\mathbf{v} = (3, 0, -2)$, $\mathbf{w} = (5, -4, 0)$
    (b) $\mathbf{u} = (5, -2, 1)$, $\mathbf{v} = (4, -1, 1)$, $\mathbf{w} = (1, -1, 0)$
    (c) $\mathbf{u} = (4, -8, 1)$, $\mathbf{v} = (2, 1, -2)$, $\mathbf{w} = (3, -4, 12)$

12. Find all unit vectors parallel to the yz-plane that are perpendicular to the vector $(3, -1, 2)$.

13. Find all unit vectors in the plane determined by $\mathbf{u} = (3, 0, 1)$ and $\mathbf{v} = (1, -1, 1)$ that are perpendicular to the vector $\mathbf{w} = (1, 2, 0)$.

14. Let $\mathbf{a} = (a_1, a_2, a_3)$, $\mathbf{b} = (b_1, b_2, b_3)$, $\mathbf{c} = (c_1, c_2, c_3)$, and $\mathbf{d} = (d_1, d_2, d_3)$. Show that

    $$(\mathbf{a} + \mathbf{d}) \cdot (\mathbf{b} \times \mathbf{c}) = \mathbf{a} \cdot (\mathbf{b} \times \mathbf{c}) + \mathbf{d} \cdot (\mathbf{b} \times \mathbf{c})$$

15. Simplify $(\mathbf{u} + \mathbf{v}) \times (\mathbf{u} - \mathbf{v})$.

16. Use the cross product to find the sine of the angle between the vectors $\mathbf{u} = (2, 3, -6)$ and $\mathbf{v} = (2, 3, 6)$.

17. (a) Find the area of the triangle having vertices $A(1, 0, 1)$, $B(0, 2, 3)$, and $C(2, 1, 0)$.
    (b) Use the result of part (a) to find the length of the altitude from vertex C to side AB.

18. Show that if $\mathbf{u}$ is a vector from any point on a line to a point P not on the line, and $\mathbf{v}$ is a vector parallel to the line, then the distance between P and the line is given by $\|\mathbf{u} \times \mathbf{v}\|/\|\mathbf{v}\|$.

19. Use the result of Exercise 18 to find the distance between the point P and the line through the points A and B:
    (a) $P(-3, 1, 2)$, $A(1, 1, 0)$, $B(-2, 3, -4)$ (b) $P(4, 3, 0)$, $A(2, 1, -3)$, $B(0, 2, -1)$

20. Prove: If θ is the angle between $\mathbf{u}$ and $\mathbf{v}$ and $\mathbf{u} \cdot \mathbf{v} \neq 0$, then $\tan \theta = \|\mathbf{u} \times \mathbf{v}\|/(\mathbf{u} \cdot \mathbf{v})$.

21. Consider the parallelepiped with sides $\mathbf{u} = (3, 2, 1)$, $\mathbf{v} = (1, 1, 2)$, and $\mathbf{w} = (1, 3, 3)$.
    (a) Find the area of the face determined by $\mathbf{u}$ and $\mathbf{w}$.
    (b) Find the angle between $\mathbf{u}$ and the plane containing the face determined by $\mathbf{v}$ and $\mathbf{w}$.
    [*Note*. The *angle between a vector and a plane* is defined to be the angle θ between the vector and that normal for which $0 \leq \theta \leq \pi/2$.]

22. Find a vector $\mathbf{n}$ perpendicular to the plane determined by the points $A(0, -2, 1)$, $B(1, -1, -2)$, and $C(-1, 1, 0)$. [See the note in Exercise 21.]

23. Let $\mathbf{m}$ and $\mathbf{n}$ be vectors whose components in the xyz-system of Figure 10 are $\mathbf{m} = (0, 0, 1)$ and $\mathbf{n} = (0, 1, 0)$.
    (a) Find the components of $\mathbf{m}$ and $\mathbf{n}$ in the $x'y'z'$-system of Figure 10.
    (b) Compute $\mathbf{m} \times \mathbf{n}$ using the components in the xyz-system.
    (c) Compute $\mathbf{m} \times \mathbf{n}$ using the components in the $x'y'z'$-system.
    (d) Show that the vectors obtained in (b) and (c) are the same.

24. Prove the following identities.
    (a) $(\mathbf{u} + k\mathbf{v}) \times \mathbf{v} = \mathbf{u} \times \mathbf{v}$ (b) $\mathbf{u} \cdot (\mathbf{v} \times \mathbf{z}) = -(\mathbf{u} \times \mathbf{z}) \cdot \mathbf{v}$

25. Let **u**, **v**, and **w** be nonzero vectors in 3-space with the same initial point, but such that no two of them are collinear. Show that
(a) $\mathbf{u} \times (\mathbf{v} \times \mathbf{w})$ lies in the plane determined by **v** and **w**
(b) $(\mathbf{u} \times \mathbf{v}) \times \mathbf{w}$ lies in the plane determined by **u** and **v**

26. Prove part (*d*) of Theorem 3.4.1. [***Hint.*** First prove the result in the case where $\mathbf{w} = \mathbf{i} = (1, 0, 0)$, then when $\mathbf{w} = \mathbf{j} = (0, 1, 0)$, and then when $\mathbf{w} = \mathbf{k} = (0, 0, 1)$. Finally prove it for an arbitrary vector $\mathbf{w} = (w_1, w_2, w_3)$ by writing $\mathbf{w} = w_1\mathbf{i} + w_2\mathbf{j} + w_3\mathbf{k}$.]

27. Prove part (*e*) of Theorem 3.4.1. [***Hint.*** Apply part (*a*) of Theorem 3.4.2 to the result in part (*d*) of Theorem 3.4.1.]

28. Let $\mathbf{u} = (1, 3, -1)$, $\mathbf{v} = (1, 1, 2)$, and $\mathbf{w} = (3, -1, 2)$. Calculate $\mathbf{u} \times (\mathbf{v} \times \mathbf{w})$ using Exercise 26; then check your result by calculating directly.

29. Prove: If **a**, **b**, **c**, and **d** lie in the same plane, then $(\mathbf{a} \times \mathbf{b}) \times (\mathbf{c} \times \mathbf{d}) = \mathbf{0}$.

30. It is a theorem of solid geometry that the volume of a tetrahedron is $\frac{1}{3}$(area of base)·(height). Use this result to prove that the volume of a tetrahedron whose sides are the vectors **a**, **b**, and **c** is $\frac{1}{6}|\mathbf{a} \cdot (\mathbf{b} \times \mathbf{c})|$ (Figure 11).

Figure 11

31. Use the result of Exercise 30 to find the volume of the tetrahedron with vertices P, Q, R, and S.
(a) $P(-1, 2, 0)$, $Q(2, 1, -3)$, $R(1, 0, 1)$, $S(3, -2, 3)$
(b) $P(0, 0, 0)$, $Q(1, 2, -1)$, $R(3, 4, 0)$, $S(-1, -3, 4)$

32. Prove parts (*a*) and (*b*) of Theorem 3.4.2.

33. Prove parts (*c*) and (*d*) of Theorem 3.4.2.

34. Prove parts (*e*) and (*f*) of Theorem 3.4.2.

3.5 LINES AND PLANES IN 3-SPACE

In this section we shall use vectors to derive equations of lines and planes in 3-space, and we shall use these equations to solve some basic geometric problems.

PLANES IN 3-SPACE

In plane analytic geometry a line can be specified by giving its slope and one of its points. Similarly, a plane in 3-space can be specified by giving its inclination and specifying one of its points. A convenient method for describing the inclination is to specify a nonzero vector (called a ***normal***) that is perpendicular to the plane.

Suppose that we want the equation of the plane passing through the point $P_0(x_0, y_0, z_0)$ and having the nonzero vector $\mathbf{n} = (a, b, c)$ as a normal. It is evident from Figure 1 that the plane consists precisely of those points $P(x, y, z)$ for which the vector $\overrightarrow{P_0P}$ is orthogonal to $\mathbf{n}$, that is,

$$\mathbf{n} \cdot \overrightarrow{P_0P} = 0 \tag{1}$$

Since $\overrightarrow{P_0P} = (x - x_0, y - y_0, z - z_0)$, Equation (1) can be written as

$$a(x - x_0) + b(y - y_0) + c(z - z_0) = 0 \tag{2}$$

We call this the *point-normal* form of the equation of a plane.

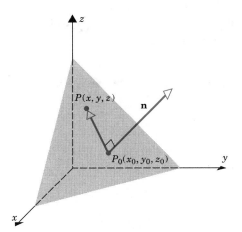

Figure 1

Example 1   Find an equation of the plane passing through the point $(3, -1, 7)$ and perpendicular to the vector $\mathbf{n} = (4, 2, -5)$.

Solution.   From (2) a point-normal form is

$$4(x - 3) + 2(y + 1) - 5(z - 7) = 0 \quad \blacktriangle$$

By multiplying out and collecting terms, (2) can be rewritten in the form

$$ax + by + cz + d = 0$$

where a, b, c, and d are constants, and a, b, and c are not all zero. For example, the equation in Example 1 can be rewritten as

$$4x + 2y - 5z + 25 = 0$$

As the next theorem shows, every equation of the form $ax + by + cz + d = 0$ represents a plane in 3-space.

> **Theorem 3.5.1.** *If a, b, c, and d are constants and a, b, and c are not all zero, then the graph of the equation*
>
> $$ax + by + cz + d = 0 \tag{3}$$
>
> *is a plane having the vector* $\mathbf{n} = (a, b, c)$ *as a normal.*

Equation (3) is a linear equation in x, y, and z; it is called the **general form** of the equation of a plane.

Proof. By hypothesis, the coefficients a, b, and c are not all zero. Assume, for the moment, that $a \neq 0$. Then the equation $ax + by + cz + d = 0$ can be rewritten in the form $a(x + (d/a)) + by + cz = 0$. But this is a point-normal form of the plane passing through the point $(-d/a, 0, 0)$ and having $\mathbf{n} = (a, b, c)$ as a normal.

If $a = 0$, then either $b \neq 0$ or $c \neq 0$. A straightforward modification of the above argument will handle these other cases. ▨

Just as the solutions of a system of linear equations

$$ax + by = k_1$$
$$cx + dy = k_2$$

correspond to points of intersection of the lines $ax + by = k_1$ and $cx + dy = k_2$ in the xy-plane, so the solutions of a system

$$ax + by + cz = k_1$$
$$dx + ey + fz = k_2 \tag{4}$$
$$gx + hy + iz = k_3$$

correspond to the points of intersection of the three planes $ax + by + cz = k_1$, $dx + ey + fz = k_2$, and $gx + hy + iz = k_3$.

In Figure 2 we have illustrated some of the geometric possibilities that occur when (4) has zero, one, or infinitely many solutions.

Example 2 Find the equation of the plane passing through the points $P_1(1, 2, -1)$, $P_2(2, 3, 1)$, and $P_3(3, -1, 2)$.

Solution. Since the three points lie in the plane, their coordinates must satisfy the general equation $ax + by + cz + d = 0$ of the plane. Thus,

$$a + 2b - c + d = 0$$
$$2a + 3b + c + d = 0$$
$$3a - b + 2c + d = 0$$

Solving this system gives

$$a = -\tfrac{9}{16}t, \qquad b = -\tfrac{1}{16}t, \qquad c = \tfrac{5}{16}t, \qquad d = t$$

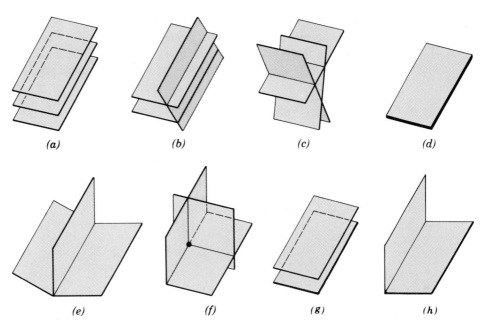

Figure 2	(*a*) No solutions (3 parallel planes). (*b*) No solutions (2 planes parallel). (*c*) No solutions (3 planes with no common intersection). (*d*) Infinitely many solutions (3 coincident planes). (*e*) Infinitely many solutions (3 planes intersecting in a line). (*f*) One solution (3 planes intersecting at a point). (*g*) No solutions (2 coincident planes parallel to a third plane). (*h*) Infinitely many solutions (2 coincident planes intersecting a third plane).

Letting $t = -16$, for example, yields the desired equation

$$9x + y - 5z - 16 = 0$$

We note that any other choice of t gives a multiple of this equation, so that any value of $t \neq 0$ would also give a valid equation of the plane.

Alternative Solution. Since $P_1(1, 2, -1)$, $P_2(2, 3, 1)$, and $P_3(3, -1, 2)$ lie in the plane, the vectors $\overrightarrow{P_1P_2} = (1, 1, 2)$ and $\overrightarrow{P_1P_3} = (2, -3, 3)$ are parallel to the plane. Therefore, $\overrightarrow{P_1P_2} \times \overrightarrow{P_1P_3} = (9, 1, -5)$ is normal to the plane, since it is perpendicular to both $\overrightarrow{P_1P_2}$ and $\overrightarrow{P_1P_3}$. From this and the fact that P_1 lies in the plane, a point-normal form for the equation of the plane is

$$9(x - 1) + (y - 2) - 5(z + 1) = 0$$

or

$$9x + y - 5z - 16 = 0 \quad \blacktriangle$$

VECTOR FORM OF THE EQUATION OF A PLANE

Vector notation provides a useful alternative way of writing the point-normal form of the equation of a plane: Referring to Figure 3, let $\mathbf{r} = (x, y, z)$ be the vector from the origin to the point $P(x, y, z)$, let $\mathbf{r}_0 = (x_0, y_0, z_0)$ be the vector from the origin to the point $P_0(x_0, y_0, z_0)$, and let $\mathbf{n} = (a, b, c)$ be a vector normal to the plane (Figure 3).

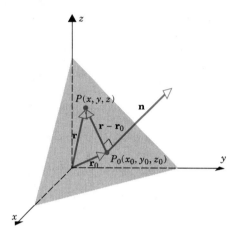

Figure 3

Then $\overrightarrow{P_0 P} = \mathbf{r} - \mathbf{r}_0$, so Formula (1) can be rewritten as

$$\mathbf{n} \cdot (\mathbf{r} - \mathbf{r}_0) = 0 \tag{5}$$

This is called the ***vector form of the equation of a plane***.

Example 3   The equation

$$(-1, 2, 5) \cdot (x - 6, y - 3, z + 4) = 0$$

is the vector equation of the plane that passes through the point $(6, 3, -4)$ and is perpendicular to the vector $\mathbf{n} = (-1, 2, 5)$. ▲

LINES IN 3-SPACE

We shall now show how to obtain equations for lines in 3-space. Suppose that l is the line in 3-space through the point $P_0(x_0, y_0, z_0)$ and parallel to the nonzero vector $\mathbf{v} = (a, b, c)$. It is clear (Figure 4) that l consists precisely of those points $P(x, y, z)$ for which the vector $\overrightarrow{P_0 P}$ is parallel to $\mathbf{v}$, that is, for which there is a scalar t such that

$$\overrightarrow{P_0 P} = t\mathbf{v} \tag{6}$$

In terms of components, (6) can be written as

$$(x - x_0, y - y_0, z - z_0) = (ta, tb, tc)$$

from which it follows that $x - x_0 = ta$, $y - y_0 = tb$, and $z - z_0 = tc$, so that

$$x = x_0 + ta, \qquad y = y_0 + tb, \qquad z = z_0 + tc$$

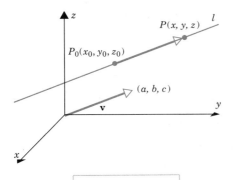

Figure 4

$\overrightarrow{P_0P}$ is parallel to **v**.

As the parameter t varies from $-\infty$ to $+\infty$ the point $P(x, y, z)$ traces out the line l. The equations

$$x = x_0 + ta, \quad y = y_0 + tb, \quad z = z_0 + tc \qquad (-\infty < t < +\infty) \tag{7}$$

are called *parametric equations* for l.

Example 4   The line through the point $(1, 2, -3)$ and parallel to the vector $\mathbf{v} = (4, 5, -7)$ has parametric equations

$$x = 1 + 4t, \quad y = 2 + 5t, \quad z = -3 - 7t \qquad (-\infty < t < +\infty) \;▲$$

Example 5

(a)   Find parametric equations for the line l passing through the points $P_1(2, 4, -1)$ and $P_2(5, 0, 7)$.
(b)   Where does the line intersect the xy-plane?

Solution (a).   Since the vector $\overrightarrow{P_1P_2} = (3, -4, 8)$ is parallel to l and $P_1(2, 4, -1)$ lies on l, the line l is given by

$$x = 2 + 3t, \quad y = 4 - 4t, \quad z = -1 + 8t \qquad (-\infty < t < +\infty)$$

Solution (b).   The line intersects the xy-plane at the point where $z = -1 + 8t = 0$, that is, where $t = \frac{1}{8}$. Substituting this value of t in the parametric equations for l yields as the point of intersection

$$(x, y, z) = (\tfrac{19}{8}, \tfrac{7}{2}, 0) \;▲$$

Example 6   Find parametric equations for the line of intersection of the planes

$$3x + 2y - 4z - 6 = 0 \qquad \text{and} \qquad x - 3y - 2z - 4 = 0$$

Solution. The line of intersection consists of all points (x, y, z) that satisfy the two equations in the system

$$3x + 2y - 4z = 6$$
$$x - 3y - 2z = 4$$

Solving this system gives

$$x = \tfrac{26}{11} + \tfrac{16}{11}t, \qquad y = -\tfrac{6}{11} - \tfrac{2}{11}t, \qquad z = t$$

The parametric equations for l are therefore

$$x = \tfrac{26}{11} + \tfrac{16}{11}t, \quad y = -\tfrac{6}{11} - \tfrac{2}{11}t, \quad z = t \qquad (-\infty < t < +\infty) \quad \blacktriangle$$

VECTOR FORM OF THE EQUATION OF A LINE

Vector notation provides a useful alternative way of writing the parametric equations of a line: Referring to Figure 3, let $\mathbf{r} = (x, y, z)$ be the vector from the origin to the point $P(x, y, z)$, let $\mathbf{r}_0 = (x_0, y_0, z_0)$ be the vector from the origin to the point $P_0(x_0, y_0, z_0)$, and let $\mathbf{v} = (a, b, c)$ be a vector parallel to the line (Figure 5). Then $\overrightarrow{P_0 P} = \mathbf{r} - \mathbf{r}_0$, so Formula (6) can be rewritten as

$$\mathbf{r} - \mathbf{r}_0 = t\mathbf{v}$$

Taking into account the range of t-values, this can be rewritten as

$$\mathbf{r} = \mathbf{r}_0 + t\mathbf{v} \qquad (-\infty < t < +\infty) \tag{8}$$

This is called the ***vector form of the equation of a line*** in 3-space.

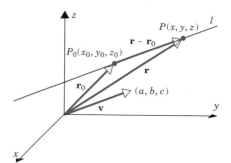

Figure 5

Example 7   The equation

$$(x, y, z) = (-2, 0, 3) + t(4, -7, 1) \qquad (-\infty < t < +\infty)$$

is the vector equation of the line through the point $(-2, 0, 3)$ that is parallel to the vector $\mathbf{v} = (4, -7, 1)$. $\blacktriangle$

SOME PROBLEMS INVOLVING DISTANCE

We conclude this section by discussing two basic "distance problems" in 3-space:

Problems

(a) Find the distance between a point and a plane.
(b) Find the distance between two parallel planes.

The two problems are related. If we can find the distance between a point and a plane, then we can find the distance between parallel planes by computing the distance between either one of the planes and an arbitrary point P_0 in the other (Figure 6).

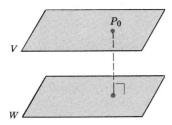

Figure 6 | The distance between the parallel planes V and W is equal to the distance between P_0 and W.

Theorem 3.5.2. *The distance D between a point $P_0(x_0, y_0, z_0)$ and the plane $ax + by + cz + d = 0$ is*

$$D = \frac{|ax_0 + by_0 + cz_0 + d|}{\sqrt{a^2 + b^2 + c^2}} \tag{9}$$

Proof. Let $Q(x_1, y_1, z_1)$ be any point in the plane. Position the normal $\mathbf{n} = (a, b, c)$ so that its initial point is at Q. As illustrated in Figure 7, the distance D is equal to the length of the orthogonal projection of $\overrightarrow{QP_0}$ on $\mathbf{n}$. Thus, from (10) of Section 3.3,

$$D = \|\text{proj}_{\mathbf{n}} \overrightarrow{QP_0}\| = \frac{|\overrightarrow{QP_0} \cdot \mathbf{n}|}{\|\mathbf{n}\|}$$

But

$$\overrightarrow{QP_0} = (x_0 - x_1, y_0 - y_1, z_0 - z_1)$$
$$\overrightarrow{QP_0} \cdot \mathbf{n} = a(x_0 - x_1) + b(y_0 - y_1) + c(z_0 - z_1)$$
$$\|\mathbf{n}\| = \sqrt{a^2 + b^2 + c^2}$$

Figure 7

Thus,

$$D = \frac{|a(x_0 - x_1) + b(y_0 - y_1) + c(z_0 - z_1)|}{\sqrt{a^2 + b^2 + c^2}} \qquad (10)$$

Since the point $Q(x_1, y_1, z_1)$ lies in the plane, its coordinates satisfy the equation of the plane; thus

$$ax_1 + by_1 + cz_1 + d = 0$$

or

$$d = -ax_1 - by_1 - cz_1$$

Substituting this expression in (10) yields (9).

REMARK. Note the similarity between (9) and the formula for the distance between a point and a line in 2-space [(13) of Section 3.3].

Example 8   Find the distance D between the point $(1, -4, -3)$ and the plane $2x - 3y + 6z = -1$.

Solution.   To apply (9), we first rewrite the equation of the plane in the form

$$2x - 3y + 6z + 1 = 0$$

Then

$$D = \frac{|(2)(1) + (-3)(-4) + 6(-3) + 1|}{\sqrt{2^2 + (-3)^2 + 6^2}} = \frac{|-3|}{7} = \frac{3}{7} \quad \blacktriangle$$

Given two planes, either they intersect, in which case we can ask for their line of intersection, as in Example 6, or they are parallel, in which case we can ask for the distance between them. The following example illustrates the latter problem.

Example 9   The planes

$$x + 2y - 2z = 3 \qquad \text{and} \qquad 2x + 4y - 4z = 7$$

are parallel since their normals, $(1, 2, -2)$ and $(2, 4, -4)$, are parallel vectors. Find the distance between these planes.

Solution. To find the distance D between the planes, we may select an arbitrary point in one of the planes and compute its distance to the other plane. By setting $y = z = 0$ in the equation $x + 2y - 2z = 3$, we obtain the point $P_0(3, 0, 0)$ in this plane. From (9), the distance between P_0 and the plane $2x + 4y - 4z = 7$ is

$$D = \frac{|(2)(3) + 4(0) + (-4)(0) - 7|}{\sqrt{2^2 + 4^2 + (-4)^2}} = \frac{1}{6} \quad \blacktriangle$$

EXERCISE SET 3.5

1. Find a point-normal form of the equation of the plane passing through P and having $\mathbf{n}$ as a normal.
   (a) $P(-1, 3, -2)$; $\mathbf{n} = (-2, 1, -1)$ (b) $P(1, 1, 4)$; $\mathbf{n} = (1, 9, 8)$
   (c) $P(2, 0, 0)$; $\mathbf{n} = (0, 0, 2)$ (d) $P(0, 0, 0)$; $\mathbf{n} = (1, 2, 3)$

2. Write the equations of the planes in Exercise 1 in general form.

3. Find a point-normal form.
   (a) $-3x + 7y + 2z = 10$ (b) $x - 4z = 0$

4. Find an equation for the plane passing through the given points.
   (a) $P(-4, -1, -1)$, $Q(-2, 0, 1)$, $R(-1, -2, -3)$ (b) $P(5, 4, 3)$, $Q(4, 3, 1)$, $R(1, 5, 4)$

5. Determine whether the planes are parallel.
   (a) $4x - y + 2z = 5$ and $7x - 3y + 4z = 8$
   (b) $x - 4y - 3z - 2 = 0$ and $3x - 12y - 9z - 7 = 0$
   (c) $2y = 8x - 4z + 5$ and $x = \frac{1}{2}z + \frac{1}{4}y$

6. Determine whether the line and plane are parallel.
   (a) $x = -5 - 4t$, $y = 1 - t$, $z = 3 + 2t$; $x + 2y + 3z - 9 = 0$
   (b) $x = 3t$, $y = 1 + 2t$, $z = 2 - t$; $4x - y + 2z = 1$

7. Determine whether the planes are perpendicular.
   (a) $3x - y + z - 4 = 0$, $x + 2z = -1$ (b) $x - 2y + 3z = 4$, $-2x + 5y + 4z = -1$

8. Determine whether the line and plane are perpendicular.
   (a) $x = -2 - 4t$, $y = 3 - 2t$, $z = 1 + 2t$; $2x + y - z = 5$
   (b) $x = 2 + t$, $y = 1 - t$, $z = 5 + 3t$; $6x + 6y - 7 = 0$

9. Find parametric equations for the line passing through P and parallel to $\mathbf{n}$.
   (a) $P(3, -1, 2)$, $\mathbf{n} = (2, 1, 3)$ (b) $P(-2, 3, -3)$; $\mathbf{n} = (6, -6, -2)$
   (c) $P(2, 2, 6)$; $\mathbf{n} = (0, 1, 0)$ (d) $P(0, 0, 0)$; $\mathbf{n} = (1, -2, 3)$

10. Find parametric equations for the line passing through the given points.
    (a) $(5, -2, 4)$, $(7, 2, -4)$ (b) $(0, 0, 0)$, $(2, -1, -3)$

11. Find parametric equations for the line of intersection of the given planes.
    (a) $7x - 2y + 3z = -2$ and $-3x + y + 2z + 5 = 0$ (b) $2x + 3y - 5z = 0$ and $y = 0$

12. Find the vector form of the equation of the plane that passes through P_0 and has normal $\mathbf{n}$.
    (a) $P_0(-1, 2, 4)$; $\mathbf{n} = (-2, 4, 1)$ (b) $P_0(2, 0, -5)$; $\mathbf{n} = (-1, 4, 3)$
    (c) $P_0(5, -2, 1)$; $\mathbf{n} = (-1, 0, 0)$ (d) $P_0(0, 0, 0)$; $\mathbf{n} = (a, b, c)$

13. Determine whether the planes are parallel.
 (a) $(-1, 2, 4) \cdot (x - 5, y + 3, z - 7) = 0$; $(2, -4, -8) \cdot (x + 3, y + 5, z - 9) = 0$
 (b) $(3, 0, -1) \cdot (x + 1, y - 2, z - 3) = 0$; $(-1, 0, 3) \cdot (x + 1, y - z, z - 3) = 0$

14. Determine whether the planes are perpendicular.
 (a) $(-2, 1, 4) \cdot (x - 1, y, z + 3) = 0$; $(1, -2, 1) \cdot (x + 3, y - 5, z) = 0$
 (b) $(3, 0, -2) \cdot (x + 4, y - 7, z + 1) = 0$; $(1, 1, 1) \cdot (x, y, z) = 0$

15. Find the vector form of the equation of the line through P_0 and parallel to **v**.
 (a) $P_0(-1, 2, 3)$; $\mathbf{v} = (7, -1, 5)$ (b) $P_0(2, 0, -1)$; $\mathbf{v} = (1, 1, 1)$
 (c) $P_0(2, -4, 1)$; $\mathbf{v} = (0, 0, -2)$ (d) $P_0(0, 0, 0)$; $\mathbf{v} = (a, b, c)$

16. Show that the line

$$x = 0, \quad y = t, \quad z = t \qquad (-\infty < t < +\infty)$$

 (a) lies in the plane $6x + 4y - 4z = 0$
 (b) is parallel to and below the plane $5x - 3y + 3z = 1$
 (c) is parallel to and above the plane $6x + 2y - 2z = 3$

17. Find an equation for the plane through $(-2, 1, 7)$ that is perpendicular to the line
 $x - 4 = 2t, y + 2 = 3t, z = -5t$.

18. Find an equation of
 (a) the xy-plane   (b) the xz-plane   (c) the yz-plane

19. Find an equation of the plane that contains the point (x_0, y_0, z_0) and is
 (a) parallel to the xy-plane   (b) parallel to the yz-plane   (c) parallel to the xz-plane

20. Find an equation for the plane that passes through the origin and is parallel to the plane
 $7x + 4y - 2z + 3 = 0$.

21. Find an equation for the plane that passes through the point $(3, -6, 7)$ and is parallel to the
 plane $5x - 2y + z - 5 = 0$.

22. Find the point of intersection of the line

$$x - 9 = -5t, \quad y + 1 = -t, \quad z - 3 = t \qquad (-\infty < t < +\infty)$$

 and the plane $2x - 3y + 4z + 7 = 0$.

23. Find an equation for the plane that contains the line $x = -1 + 3t, y = 5 + 2t, z = 2 - t$ and
 is perpendicular to the plane $2x - 4y + 2z = 9$.

24. Find an equation for the plane that passes through $(2, 4, -1)$ and contains the line of
 intersection of the planes $x - y - 4z = 2$ and $-2x + y + 2z = 3$.

25. Show that the points $(-1, -2, -3)$, $(-2, 0, 1)$, $(-4, -1, -1)$, and $(2, 0, 1)$ lie in the
 same plane.

26. Find parametric equations for the line through $(-2, 5, 0)$ that is parallel to the planes
 $2x + y - 4z = 0$ and $-x + 2y + 3z + 1 = 0$.

27. Find an equation for the plane through $(-2, 1, 5)$ that is perpendicular to the planes
 $4x - 2y + 2z = -1$ and $3x + 3y - 6z = 5$.

28. Find an equation for the plane through $(2, -1, 4)$ that is perpendicular to the line of intersection of the planes $4x + 2y + 2z = -1$ and $3x + 6y + 3z = 7$.

29. Find an equation for the plane that is perpendicular to the plane $8x - 2y + 6z = 1$ and passes through the points $P_1(-1, 2, 5)$ and $P_2(2, 1, 4)$.

30. Show that the lines

$$x = 3 - 2t, \quad y = 4 + t, \quad z = 1 - t \quad (-\infty < t < +\infty)$$

and

$$x = 5 + 2t, \quad y = 1 - t, \quad z = 7 + t \quad (-\infty < t < +\infty)$$

are parallel, and find an equation for the plane they determine.

31. Find an equation for the plane that contains the point $(1, -1, 2)$ and the line $x = t$, $y = t + 1$, $z = -3 + 2t$.

32. Find an equation for the plane that contains the line $x = 1 + t$, $y = 3t$, $z = 2t$ and is parallel to the line of intersection of the planes $-x + 2y + z = 0$ and $x + z + 1 = 0$.

33. Find an equation for the plane, each of whose points is equidistant from $(-1, -4, -2)$ and $(0, -2, 2)$.

34. Show that the line

$$x - 5 = -t, \quad y + 3 = 2t, \quad z + 1 = -5t \quad (-\infty < t < +\infty)$$

is parallel to the plane $-3x + y + z - 9 = 0$.

35. Show that the lines

$$x - 3 = 4t, \quad y - 4 = t, \quad z - 1 = 0 \quad (-\infty < t < +\infty)$$

and

$$x + 1 = 12t, \quad y - 7 = 6t, \quad z - 5 = 3t \quad (-\infty < t < +\infty)$$

intersect, and find the point of intersection.

36. Find an equation for the plane containing the lines in Exercise 35.

37. Find parametric equations for the line of intersection of the planes
(a) $-3x + 2y + z = -5$ and $7x + 3y - 2z = -2$
(b) $5x - 7y + 2z = 0$ and $y = 0$

38. Show that the plane whose intercepts with the coordinate axes are $x = a$, $y = b$, and $z = c$ has equation

$$\frac{x}{a} + \frac{y}{b} + \frac{z}{c} = 1$$

provided a, b, and c are nonzero.

39. Find the distance between the point and the plane.
(a) $(3, 1, -2); \ x + 2y - 2z = 4$
(b) $(-1, 2, 1); \ 2x + 3y - 4z = 1$
(c) $(0, 3, -2); \ x - y - z = 3$

40. Find the distance between the given parallel planes.
(a) $3x - 4y + z = 1$ and $6x - 8y + 2z = 3$
(b) $-4x + y - 3z = 0$ and $8x - 2y + 6z = 0$
(c) $2x - y + z = 1$ and $2x - y + z = -1$

41. Show that if a, b, and c are all nonzero, then the line

$$x = x_0 + at, \quad y = y_0 + bt, \quad z = z_0 + ct \qquad (-\infty < t < +\infty)$$

consists of all points (x, y, z) that satisfy

$$\frac{x - x_0}{a} = \frac{y - y_0}{b} = \frac{z - z_0}{c}$$

These are called ***symmetric equations*** for the line.

42. Find symmetric equations for the lines in parts (a) and (b) of Exercise 9. [***Note.*** See Exercise 41 for the terminology.]

43. In each part find equations for two planes whose intersection is the given line.
(a) $x = 7 - 4t, \quad y = -5 - 2t, \quad z = 5 + t \qquad (-\infty < t < +\infty)$
(b) $x = 4t, \quad y = 2t, \quad z = 7t \qquad (-\infty < t < +\infty)$
[***Hint.*** Each equality in the symmetric equations of a line represents a plane containing the line. See Exercise 41 for terminology.]

44. Two intersecting planes in 3-space determine two angles of intersection, an acute angle $(0 \le \theta \le 90°)$ and its supplement $180° - \theta$ (Figure 8a). If $\mathbf{n}_1$ and $\mathbf{n}_2$ are nonzero normals to the planes, then the angle between $\mathbf{n}_1$ and $\mathbf{n}_2$ is θ or $180° - \theta$, depending on the directions of the normals (Figure 8b). In each part find the acute angle of intersection of the planes to the nearest degree.
(a) $x = 0$ and $2x - y + z - 4 = 0$
(b) $x + 2y - 2z = 5$ and $6x - 3y + 2z = 8$
[***Note.*** A calculator is needed.]

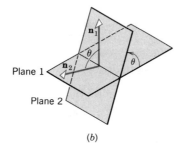

(a)   (b)   **Figure 8**

45. Find the acute angle of intersection between the plane $x - y - 3z = 5$ and the line $x = 2 - t, y = 2t, z = 3t - 1$ to the nearest degree. [***Hint.*** See Exercise 44.]

CHAPTER **4**

EUCLIDEAN VECTOR SPACES

4.1 EUCLIDEAN *n*-SPACE

The idea of using pairs of numbers to locate points in the plane and triples of numbers to locate points in 3-space was first clearly spelled out in the mid-seventeenth century. By the latter part of the nineteenth century mathematicians and physicists began to realize that there was no need to stop with triples. It was recognized that quadruples of numbers (a_1, a_2, a_3, a_4) could be regarded as points in "4-dimensional" space, quintuples $(a_1, a_2, \ldots, a_5)$ as points in "5-dimensional" space, and so on. Although our geometric visualization does not extend beyond 3-space, it is nevertheless possible to extend many familiar ideas beyond 3-space by working with analytic or numerical properties of points and vectors rather than the geometric properties. In this section we shall make these ideas more precise.

VECTORS IN
n-SPACE

> **Definition.** If n is a positive integer, then an ***ordered n-tuple*** is a sequence of n real numbers $(a_1, a_2, \ldots, a_n)$. The set of all ordered n-tuples is called ***n-space*** and is denoted by R^n.

When $n = 2$ or 3, it is usual to use the terms ***ordered pair*** and ***ordered triple***, respectively, rather than ordered 2-tuple and 3-tuple. When $n = 1$, each ordered n-tuple consists of one real number, and so R^1 may be viewed as the set of real numbers. It is usual to write R rather than R^1 for this set.

It might have occurred to the reader in the study of 3-space that the symbol (a_1, a_2, a_3) has two different geometric interpretations: it can be interpreted as a point, in which case a_1, a_2, and a_3 are the coordinates (Figure 1*a*), or it can be interpreted as a vector, in which case a_1, a_2, and a_3 are the components (Figure 1*b*). It follows, therefore, that an ordered n-tuple $(a_1, a_2, \ldots, a_n)$ can be viewed either as a "gener-

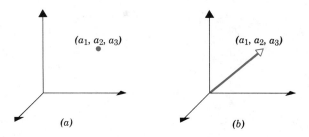

(a)   *(b)*

Figure 1   The ordered triple (a_1, a_2, a_3) can be interpreted geometrically as a point or a vector.

alized point'' or a ''generalized vector''—the distinction is mathematically unimportant. Thus, we can describe the 5-tuple $(-2, 4, 0, 1, 6)$ either as a point in R^5 or a vector in R^5.

Definition. Two vectors $\mathbf{u} = (u_1, u_2, \ldots, u_n)$ and $\mathbf{v} = (v_1, v_2, \ldots, v_n)$ in R^n are called *equal* if

$$u_1 = v_1, u_2 = v_2, \ldots, u_n = v_n$$

The *sum* $\mathbf{u} + \mathbf{v}$ is defined by

$$\mathbf{u} + \mathbf{v} = (u_1 + v_1, u_2 + v_2, \ldots, u_n + v_n)$$

and if k is any scalar, the *scalar multiple* $k\mathbf{u}$ is defined by

$$k\mathbf{u} = (ku_1, ku_2, \ldots, ku_n)$$

The operations of addition and scalar multiplication in this definition are called the *standard operations* on R^n.

The *zero vector* in R^n is denoted by $\mathbf{0}$ and is defined to be the vector

$$\mathbf{0} = (0, 0, \ldots, 0)$$

If $\mathbf{u} = (u_1, u_2, \ldots, u_n)$ is any vector in R^n, then the *negative* (or *additive inverse*) of $\mathbf{u}$ is denoted by $-\mathbf{u}$ and is defined by

$$-\mathbf{u} = (-u_1, -u_2, \ldots, -u_n)$$

The *difference* of vectors in R^n is defined by

$$\mathbf{v} - \mathbf{u} = \mathbf{v} + (-\mathbf{u})$$

or in terms of components

$$\mathbf{v} - \mathbf{u} = (v_1 - u_1, v_2 - u_2, \ldots, v_n - u_n)$$

PROPERTIES OF VECTOR OPERATIONS IN *n*-SPACE

The most important arithmetic properties of addition and scalar multiplication of vectors in R^n are listed in the following theorem. The proofs are all easy and are left as exercises.

Theorem 4.1.1. *If* $\mathbf{u} = (u_1, u_2, \ldots, u_n)$, $\mathbf{v} = (v_1, v_2, \ldots, v_n)$, *and* $\mathbf{w} = (w_1, w_2, \ldots, w_n)$ *are vectors in R^n and k and l are scalars, then:*

(*a*) $\mathbf{u} + \mathbf{v} = \mathbf{v} + \mathbf{u}$
(*b*) $\mathbf{u} + (\mathbf{v} + \mathbf{w}) = (\mathbf{u} + \mathbf{v}) + \mathbf{w}$
(*c*) $\mathbf{u} + \mathbf{0} = \mathbf{0} + \mathbf{u} = \mathbf{u}$
(*d*) $\mathbf{u} + (-\mathbf{u}) = \mathbf{0}$; *that is,* $\mathbf{u} - \mathbf{u} = \mathbf{0}$
(*e*) $k(l\mathbf{u}) = (kl)\mathbf{u}$
(*f*) $k(\mathbf{u} + \mathbf{v}) = k\mathbf{u} + k\mathbf{v}$
(*g*) $(k + l)\mathbf{u} = k\mathbf{u} + l\mathbf{u}$
(*h*) $1\mathbf{u} = \mathbf{u}$

This theorem enables us to manipulate vectors in R^n without expressing the vectors in terms of components. For example, to solve the vector equation $\mathbf{x} + \mathbf{u} = \mathbf{v}$ for $\mathbf{x}$, we can add $-\mathbf{u}$ to both sides and proceed as follows:

$$(\mathbf{x} + \mathbf{u}) + (-\mathbf{u}) = \mathbf{v} + (-\mathbf{u})$$
$$\mathbf{x} + (\mathbf{u} - \mathbf{u}) = \mathbf{v} - \mathbf{u}$$
$$\mathbf{x} + \mathbf{0} = \mathbf{v} - \mathbf{u}$$
$$\mathbf{x} = \mathbf{v} - \mathbf{u}$$

The reader will find it instructive to name the parts of Theorem 4.1.1 that justify the last three steps in this computation.

EUCLIDEAN *n*-SPACE

To extend the notions of distance, norm, and angle to R^n, we begin with the following generalization of the dot product on R^2 and R^3 [Formula (3) of Section 3.3].

Definition. If $\mathbf{u} = (u_1, u_2, \ldots, u_n)$ and $\mathbf{v} = (v_1, v_2, \ldots, v_n)$ are any vectors in R^n, then the *Euclidean inner product* $\mathbf{u} \cdot \mathbf{v}$ is defined by

$$\mathbf{u} \cdot \mathbf{v} = u_1 v_1 + u_2 v_2 + \cdots + u_n v_n$$

Observe that when $n = 2$ or 3, the Euclidean inner product is the ordinary dot product.

Example 1 The Euclidean inner product of the vectors

$$\mathbf{u} = (-1, 3, 5, 7) \qquad \text{and} \qquad \mathbf{v} = (5, -4, 7, 0)$$

in R^4 is

$$\mathbf{u} \cdot \mathbf{v} = (-1)(5) + (3)(-4) + (5)(7) + (7)(0) = 18 \quad \blacktriangle$$

Since so many of the familiar ideas from 2-space and 3-space carry over to *n*-space, it is common to refer to R^n with the operations of addition, scalar multiplication, and the Euclidean inner product as ***Euclidean n-space***.

The four main arithmetic properties of the Euclidean inner product are listed in the next theorem.

Theorem 4.1.2. *If* **u**, **v**, *and* **w** *are vectors in R^n and k is any scalar, then:*

(*a*) $\mathbf{u} \cdot \mathbf{v} = \mathbf{v} \cdot \mathbf{u}$

(*b*) $(\mathbf{u} + \mathbf{v}) \cdot \mathbf{w} = \mathbf{u} \cdot \mathbf{w} + \mathbf{v} \cdot \mathbf{w}$

(*c*) $(k\mathbf{u}) \cdot \mathbf{v} = k(\mathbf{u} \cdot \mathbf{v})$

(*d*) $\mathbf{v} \cdot \mathbf{v} \geq 0$. *Further,* $\mathbf{v} \cdot \mathbf{v} = 0$ *if and only if* $\mathbf{v} = \mathbf{0}$.

We shall prove parts (*b*) and (*d*) and leave proofs of the rest as exercises.

Proof (b). Let $\mathbf{u} = (u_1, u_2, \ldots, u_n)$, $\mathbf{v} = (v_1, v_2, \ldots, v_n)$, and $\mathbf{w} = (w_1, w_2, \ldots, w_n)$. Then

$$
\begin{aligned}
(\mathbf{u} + \mathbf{v}) \cdot \mathbf{w} &= (u_1 + v_1, u_2 + v_2, \ldots, u_n + v_n) \cdot (w_1, w_2, \ldots, w_n) \\
&= (u_1 + v_1)w_1 + (u_2 + v_2)w_2 + \cdots + (u_n + v_n)w_n \\
&= (u_1 w_1 + u_2 w_2 + \cdots + u_n w_n) + (v_1 w_1 + v_2 w_2 + \cdots + v_n w_n) \\
&= \mathbf{u} \cdot \mathbf{w} + \mathbf{v} \cdot \mathbf{w}
\end{aligned}
$$

Proof (d). We have $\mathbf{v} \cdot \mathbf{v} = v_1^2 + v_2^2 + \cdots + v_n^2 \geq 0$. Further, equality holds if and only if $v_1 = v_2 = \cdots = v_n = 0$, that is, if and only if $\mathbf{v} = \mathbf{0}$. ∎

Example 2 Theorem 4.1.2 allows us to perform computations with Euclidean inner products in much the same way that we perform them with ordinary arithmetic products. For example,

$$
\begin{aligned}
(3\mathbf{u} + 2\mathbf{v}) \cdot (4\mathbf{u} + \mathbf{v}) &= (3\mathbf{u}) \cdot (4\mathbf{u} + \mathbf{v}) + (2\mathbf{v}) \cdot (4\mathbf{u} + \mathbf{v}) \\
&= (3\mathbf{u}) \cdot (4\mathbf{u}) + (3\mathbf{u}) \cdot \mathbf{v} + (2\mathbf{v}) \cdot (4\mathbf{u}) + (2\mathbf{v}) \cdot \mathbf{v} \\
&= 12(\mathbf{u} \cdot \mathbf{u}) + 3(\mathbf{u} \cdot \mathbf{v}) + 8(\mathbf{v} \cdot \mathbf{u}) + 2(\mathbf{v} \cdot \mathbf{v}) \\
&= 12(\mathbf{u} \cdot \mathbf{u}) + 11(\mathbf{u} \cdot \mathbf{v}) + 2(\mathbf{v} \cdot \mathbf{v})
\end{aligned}
$$

The reader should determine which parts of Theorem 4.1.2 were used in each step. ▲

NORM AND DISTANCE IN EUCLIDEAN *n*-SPACE

By analogy with the familiar formulas in R^2 and R^3, we define the ***Euclidean norm*** (or ***Euclidean length***) of a vector $\mathbf{u} = (u_1, u_2, \ldots, u_n)$ in R^n by

$$
\|\mathbf{u}\| = (\mathbf{u} \cdot \mathbf{u})^{1/2} = \sqrt{u_1^2 + u_2^2 + \cdots + u_n^2} \tag{1}
$$

[Compare this formula to Formulas (1) and (2) in Section 3.2.]

Similarly, the ***Euclidean distance*** between the points $\mathbf{u} = (u_1, u_2, \ldots, u_n)$ and $\mathbf{v} = (v_1, v_2, \ldots, v_n)$ in R^n is defined by

$$d(\mathbf{u}, \mathbf{v}) = \|\mathbf{u} - \mathbf{v}\| = \sqrt{(u_1 - v_1)^2 + (u_2 - v_2)^2 + \cdots + (u_n - v_n)^2} \qquad (2)$$

[See Formulas (3) and (4) of Section 3.2.]

Example 3 If $\mathbf{u} = (1, 3, -2, 7)$ and $\mathbf{v} = (0, 7, 2, 2)$, then in the Euclidean space R^4

$$\|\mathbf{u}\| = \sqrt{(1)^2 + (3)^2 + (-2)^2 + (7)^2} = \sqrt{63} = 3\sqrt{7}$$

and

$$d(\mathbf{u}, \mathbf{v}) = \sqrt{(1 - 0)^2 + (3 - 7)^2 + (-2 - 2)^2 + (7 - 2)^2} = \sqrt{58} \quad \blacktriangle$$

The following theorem provides one of the most important inequalities in linear algebra, the **Cauchy–Schwarz*** inequality.

Theorem 4.1.3 (*Cauchy–Schwarz Inequality in R^n*). *If*

$$\mathbf{u} = (u_1, u_2 \ldots, u_n) \qquad and \qquad \mathbf{v} = (v_1, v_2, \ldots, v_n)$$

are vectors in R^n, then

$$|\mathbf{u} \cdot \mathbf{v}| \leq \|\mathbf{u}\| \, \|\mathbf{v}\| \qquad (3)$$

or expressed in terms of components,

$$|u_1 v_1 + u_2 v_2 + \cdots + u_n v_n| \leq (u_1^2 + u_2^2 + \cdots + u_n^2)^{1/2}(v_1^2 + v_2^2 + \cdots + v_n^2)^{1/2} \qquad (4)$$

We omit the proof at this time, since a more general version of this theorem will be proved later in the text. However, for vectors in R^2 and R^3, this result is a simple consequence of Formula (1) of Section 3.3: If **u** and **v** are nonzero vectors in R^2 or R^3, then

$$|\mathbf{u} \cdot \mathbf{v}| = |\,\|\mathbf{u}\|\,\|\mathbf{v}\| \, \cos \theta| = \|\mathbf{u}\|\,\|\mathbf{v}\|\,|\cos\theta| \leq \|\mathbf{u}\|\,\|\mathbf{v}\| \tag{5}$$

and if either $\mathbf{u} = \mathbf{0}$ or $\mathbf{v} = \mathbf{0}$, then both sides of (3) are zero, so the inequality holds in this case as well.

The next two theorems list the basic properties of length and distance in Euclidean n-space.

Theorem 4.1.4. *If* **u** *and* **v** *are vectors in* R^n *and* k *is any scalar, then:*

(*a*) $\|\mathbf{u}\| \geq 0$
(*b*) $\|\mathbf{u}\| = 0$ *if and only if* $\mathbf{u} = \mathbf{0}$
(*c*) $\|k\mathbf{u}\| = |k|\,\|\mathbf{u}\|$
(*d*) $\|\mathbf{u} + \mathbf{v}\| \leq \|\mathbf{u}\| + \|\mathbf{v}\|$ (*Triangle inequality*) $\rangle$ losh at this

---

We shall prove (*c*) and (*d*) and leave (*a*) and (*b*) as exercises.

Proof (*c*). If $\mathbf{u} = (u_1, u_2, \ldots, u_n)$, then $k\mathbf{u} = (ku_1, ku_2, \ldots, ku_n)$, so

$$\|k\mathbf{u}\| = \sqrt{(ku_1)^2 + (ku_2)^2 + \cdots + (ku_n)^2}$$
$$= |k|\sqrt{u_1^2 + u_2^2 + \cdots + u_n^2}$$
$$= |k|\,\|\mathbf{u}\|$$

Proof (*d*)

$$\|\mathbf{u} + \mathbf{v}\|^2 = (\mathbf{u} + \mathbf{v}) \cdot (\mathbf{u} + \mathbf{v}) = (\mathbf{u} \cdot \mathbf{u}) + 2(\mathbf{u} \cdot \mathbf{v}) + (\mathbf{v} \cdot \mathbf{v})$$
$$= \|\mathbf{u}\|^2 + 2(\mathbf{u} \cdot \mathbf{v}) + \|\mathbf{v}\|^2$$
$$\leq \|\mathbf{u}\|^2 + 2|\mathbf{u} \cdot \mathbf{v}| + \|\mathbf{v}\|^2 \qquad \boxed{\text{Property of absolute value}}$$
$$\leq \|\mathbf{u}\|^2 + 2\|\mathbf{u}\|\,\|\mathbf{v}\| + \|\mathbf{v}\|^2 \qquad \boxed{\text{Cauchy–Schwarz inequality}}$$
$$= (\|\mathbf{u}\| + \|\mathbf{v}\|)^2$$

The result now follows on taking square roots of both sides. ▌

Part (*c*) of this theorem states that multiplying a vector by a scalar k multiplies the length of that vector by a factor of $|k|$ (Figure 2*a*). Part (*d*) of this theorem is known as the ***triangle inequality*** because it generalizes the familiar result from Euclidean geometry which states that the sum of two sides of a triangle is at least as large as the third side (Figure 2*b*).

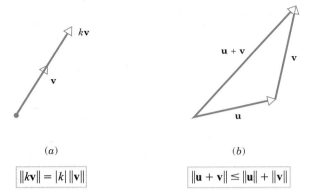

(a)

(b)

Figure 2 $\boxed{\|k\mathbf{v}\| = |k|\,\|\mathbf{v}\|}$ $\boxed{\|\mathbf{u} + \mathbf{v}\| \leq \|\mathbf{u}\| + \|\mathbf{v}\|}$

> **Theorem 4.1.5.** *If* **u**, **v**, *and* **w** *are vectors in* R^n *and k is any scalar, then:*
> (a) $d(\mathbf{u}, \mathbf{v}) \geq 0$
> (b) $d(\mathbf{u}, \mathbf{v}) = 0$ *if and only if* $\mathbf{u} = \mathbf{v}$
> (c) $d(\mathbf{u}, \mathbf{v}) = d(\mathbf{v}, \mathbf{u})$
> (d) $d(\mathbf{u}, \mathbf{v}) \leq d(\mathbf{u}, \mathbf{w}) + d(\mathbf{w}, \mathbf{v})$ *(Triangle inequality)*

The results in this theorem are immediate consequences of Theorem 4.1.4. We shall prove part (*d*) and leave the remaining parts as exercises.

Proof (*d*). From (2) and part (*d*) of Theorem 4.1.4 we have

$$d(\mathbf{u}, \mathbf{v}) = \|\mathbf{u} - \mathbf{v}\| = \|(\mathbf{u} - \mathbf{w}) + (\mathbf{w} - \mathbf{v})\|$$
$$\leq \|\mathbf{u} - \mathbf{w}\| + \|\mathbf{w} - \mathbf{v}\| = d(\mathbf{u}, \mathbf{w}) + d(\mathbf{w}, \mathbf{v})$$

Part (*d*) of this theorem, which is also called the *triangle inequality*, generalizes the familiar result from Euclidean geometry which states that the shortest distance between two points is a straight line (Figure 3).

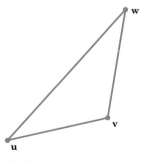

Figure 3 $\boxed{d(\mathbf{u}, \mathbf{w}) \leq d(\mathbf{u}, \mathbf{v}) + d(\mathbf{v}, \mathbf{w})}$

Formula (1) expresses the norm of a vector in terms of a dot product. The following useful theorem expresses the dot product in terms of norms.

Theorem 4.1.6. *If* **u** *and* **v** *are vectors in* R^n *with the Euclidean inner product, then*

$$\mathbf{u} \cdot \mathbf{v} = \tfrac{1}{4}\|\mathbf{u} + \mathbf{v}\|^2 - \tfrac{1}{4}\|\mathbf{u} - \mathbf{v}\|^2 \tag{6}$$

Proof

$$\|\mathbf{u} + \mathbf{v}\|^2 = (\mathbf{u} + \mathbf{v}) \cdot (\mathbf{u} + \mathbf{v}) = \|\mathbf{u}\|^2 + 2(\mathbf{u} \cdot \mathbf{v}) + \|\mathbf{v}\|^2$$
$$\|\mathbf{u} - \mathbf{v}\|^2 = (\mathbf{u} - \mathbf{v}) \cdot (\mathbf{u} - \mathbf{v}) = \|\mathbf{u}\|^2 - 2(\mathbf{u} \cdot \mathbf{v}) + \|\mathbf{v}\|^2$$

from which (6) follows by simple algebra. ▮

Some numerical problems that use this theorem are given in the exercises.

ORTHOGONALITY Recall that in the Euclidean spaces R^2 and R^3 two vectors **u** and **v** are defined to be *orthogonal* (perpendicular) if $\mathbf{u} \cdot \mathbf{v} = 0$ (Section 3.3). Motivated by this, we make the following definition.

Definition. Two vectors **u** and **v** in R^n are called *orthogonal* if $\mathbf{u} \cdot \mathbf{v} = 0$.

Example 4 In the Euclidean space R^4 the vectors

$$\mathbf{u} = (-2, 3, 1, 4) \qquad \text{and} \qquad \mathbf{v} = (1, 2, 0, -1)$$

are orthogonal, since

$$\mathbf{u} \cdot \mathbf{v} = (-2)(1) + (3)(2) + (1)(0) + (4)(-1) = 0 \;\; \triangle$$

Properties of orthogonal vectors will be discussed in more detail later in the text, but we note at this point that many of the familiar properties of orthogonal vectors in the Euclidean spaces R^2 and R^3 continue to hold in the Euclidean space R^n. For example, if **u** and **v** are orthogonal vectors in R^2 or R^3, then **u**, **v**, and $\mathbf{u} + \mathbf{v}$ form the sides of a right triangle (Figure 4); thus, by the Theorem of Pythagoras

$$\|\mathbf{u} + \mathbf{v}\|^2 = \|\mathbf{u}\|^2 + \|\mathbf{v}\|^2$$

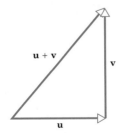

Figure 4

The following theorem shows that this result extends to R^n.

Theorem 4.1.7 (*Theorem of Pythagoras for R^n*). *If* **u** *and* **v** *are orthogonal vectors in R^n with the Euclidean inner product, then*

$$\|\mathbf{u} + \mathbf{v}\|^2 = \|\mathbf{u}\|^2 + \|\mathbf{v}\|^2$$

Proof

$$\|\mathbf{u} + \mathbf{v}\|^2 = (\mathbf{u} + \mathbf{v}) \cdot (\mathbf{u} + \mathbf{v}) = \|\mathbf{u}\|^2 + 2(\mathbf{u} \cdot \mathbf{v}) + \|\mathbf{u}\|^2 = \|\mathbf{u}\|^2 + \|\mathbf{v}\|^2 \quad \blacksquare$$

ALTERNATIVE NOTATIONS FOR VECTORS IN R^n

A vector $\mathbf{u} = (u_1, u_2, \ldots, u_n)$ in R^n can also be written in matrix notation as a row matrix or a column matrix:

$$\mathbf{u} = \begin{bmatrix} u_1 \\ u_2 \\ \vdots \\ u_n \end{bmatrix} \quad \text{or} \quad \mathbf{u} = [\, u_1 \quad u_2 \quad \cdots \quad u_n \,]$$

This is justified because the matrix operations

$$\mathbf{u} + \mathbf{v} = \begin{bmatrix} u_1 \\ u_2 \\ \vdots \\ u_n \end{bmatrix} + \begin{bmatrix} v_1 \\ v_2 \\ \vdots \\ v_n \end{bmatrix} = \begin{bmatrix} u_1 + v_1 \\ u_2 + v_2 \\ \vdots \\ u_n + v_n \end{bmatrix}, \quad k\mathbf{u} = k\begin{bmatrix} u_1 \\ u_2 \\ \vdots \\ u_n \end{bmatrix} = \begin{bmatrix} ku_1 \\ ku_2 \\ \vdots \\ ku_n \end{bmatrix}$$

or

$$\mathbf{u} + \mathbf{v} = [\, u_1 \quad u_2 \quad \cdots \quad u_n \,] + [\, v_1 \quad v_2 \quad \cdots \quad v_n \,]$$

$$= [\, u_1 + v_1 \quad u_2 + v_2 \quad \cdots \quad u_n + v_n \,]$$

$$k\mathbf{u} = k[\, u_1 \quad u_2 \quad \cdots \quad u_n \,] = [\, ku_1 \quad ku_2 \quad \cdots \quad ku_n \,]$$

produce the same results as the vector operations

$$\mathbf{u} + \mathbf{v} = (u_1, u_2, \ldots, u_n) + (v_1, v_2, \ldots, v_n) = (u_1 + v_1, u_2 + v_2, \ldots, u_n + v_n)$$

$$k\mathbf{u} = k(u_1, u_2, \ldots, u_n) = (ku_1, ku_2, \ldots, ku_n)$$

The only difference is the form in which the vectors are written.

A MATRIX FORMULA FOR THE DOT PRODUCT

If we use column matrix notation for the vectors

$$\mathbf{u} = \begin{bmatrix} u_1 \\ u_2 \\ \vdots \\ u_n \end{bmatrix} \quad \text{and} \quad \mathbf{v} = \begin{bmatrix} v_1 \\ v_2 \\ \vdots \\ v_n \end{bmatrix}$$

and omit the brackets on 1×1 matrices, then it follows that

$$\mathbf{v}^T\mathbf{u} = [\, v_1 \quad v_2 \quad \cdots \quad v_n \,] \begin{bmatrix} u_1 \\ u_2 \\ \vdots \\ u_n \end{bmatrix} = [\, u_1v_1 + u_2v_2 + \cdots + u_nv_n \,] = [\, \mathbf{u} \cdot \mathbf{v} \,] = \mathbf{u} \cdot \mathbf{v}$$

Thus, for vectors in column matrix notation we have the following formula for the Euclidean inner product:

$$\mathbf{v}^T\mathbf{u} = \mathbf{u} \cdot \mathbf{v} \tag{7}$$

For example, if

$$\mathbf{u} = \begin{bmatrix} -1 \\ 3 \\ 5 \\ 7 \end{bmatrix} \quad \text{and} \quad \mathbf{v} = \begin{bmatrix} 5 \\ -4 \\ 7 \\ 0 \end{bmatrix}$$

then

$$\mathbf{u} \cdot \mathbf{v} = \mathbf{v}^T\mathbf{u} = [\, 5 \quad -4 \quad 7 \quad 0 \,] \begin{bmatrix} -1 \\ 3 \\ 5 \\ 7 \end{bmatrix} = [\, 18 \,] = 18$$

If A is an $n \times n$ matrix, then it follows from Formula (7) and properties of the transpose that

$$A\mathbf{u} \cdot \mathbf{v} = \mathbf{v}^T(A\mathbf{u}) = (\mathbf{v}^T A)\mathbf{u} = (A^T\mathbf{v})^T\mathbf{u} = \mathbf{u} \cdot A^T\mathbf{v}$$
$$\mathbf{u} \cdot A\mathbf{v} = (A\mathbf{v})^T\mathbf{u} = (\mathbf{v}^T A^T)\mathbf{u} = \mathbf{v}^T(A^T\mathbf{u}) = A^T\mathbf{u} \cdot \mathbf{v}$$

The resulting formulas

$$A\mathbf{u} \cdot \mathbf{v} = \mathbf{u} \cdot A^T\mathbf{v} \tag{8}$$

$$\mathbf{u} \cdot A\mathbf{v} = A^T\mathbf{u} \cdot \mathbf{v} \tag{9}$$

provide an important link between multiplication by an $n \times n$ matrix A and multiplication by A^T.

Example 5 Suppose that

$$A = \begin{bmatrix} 1 & -2 & 3 \\ 2 & 4 & 1 \\ -1 & 0 & 1 \end{bmatrix}, \quad \mathbf{u} = \begin{bmatrix} -1 \\ 2 \\ 4 \end{bmatrix}, \quad \mathbf{v} = \begin{bmatrix} -2 \\ 0 \\ 5 \end{bmatrix}$$

Then

$$Au = \begin{bmatrix} 1 & -2 & 3 \\ 2 & 4 & 1 \\ -1 & 0 & 1 \end{bmatrix}\begin{bmatrix} -1 \\ 2 \\ 4 \end{bmatrix} = \begin{bmatrix} 7 \\ 10 \\ 5 \end{bmatrix}$$

$$A^T v = \begin{bmatrix} 1 & 2 & -1 \\ -2 & 4 & 0 \\ 3 & 1 & 1 \end{bmatrix}\begin{bmatrix} -2 \\ 0 \\ 5 \end{bmatrix} = \begin{bmatrix} -7 \\ 4 \\ -1 \end{bmatrix}$$

from which we obtain

$$Au \cdot v = 7(-2) + 10(0) + 5(5) = 11$$
$$u \cdot A^T v = (-1)(-7) + 2(4) + 4(-1) = 11$$

Thus, $Au \cdot v = u \cdot A^T v$ as guaranteed by Formula (8). We leave it for the reader to verify that (9) also holds. ▲

A DOT PRODUCT VIEW OF MATRIX MULTIPLICATION

Dot products provide another way of thinking about matrix multiplication. Recall that if $A = [a_{ij}]$ is an $m \times r$ matrix and $B = [b_{ij}]$ is an $r \times n$ matrix, then the ijth entry of AB is

$$a_{i1}b_{1j} + a_{i2}b_{2j} + \cdots + a_{ir}b_{rj}$$

which is the dot product of the ith row vector of A

$$[a_{i1} \quad a_{i2} \quad \cdots \quad a_{ir}]$$

and the jth column vector of B

$$\begin{bmatrix} b_{ij} \\ b_{2j} \\ \vdots \\ b_{rj} \end{bmatrix}$$

Thus, if the row vectors of A are $r_1, r_2, \ldots, r_m$ and the column vectors of B are $c_1, c_2, \ldots, c_n$, then the matrix product AB can be expressed as

$$AB = \begin{bmatrix} r_1 \cdot c_1 & r_1 \cdot c_2 & \cdots & r_1 \cdot c_n \\ r_2 \cdot c_1 & r_2 \cdot c_2 & \cdots & r_2 \cdot c_n \\ \vdots & \vdots & & \vdots \\ r_m \cdot c_1 & r_m \cdot c_2 & \cdots & r_m \cdot c_n \end{bmatrix} \tag{10}$$

In particular, a linear system $A\mathbf{x} = \mathbf{b}$ can be expressed in dot product form as

$$\begin{bmatrix} \mathbf{r}_1 \cdot \mathbf{x} \\ \mathbf{r}_2 \cdot \mathbf{x} \\ \vdots \\ \mathbf{r}_m \cdot \mathbf{x} \end{bmatrix} = \begin{bmatrix} b_1 \\ b_2 \\ \vdots \\ b_m \end{bmatrix} \qquad (11)$$

where $\mathbf{r}_1, \mathbf{r}_2, \ldots, \mathbf{r}_m$, are the row vectors of A, and $b_1, b_2, \ldots, b_m$ are the entries of $\mathbf{b}$.

Example 6 The following is an example of a linear system expressed in dot product form (11).

System	**Dot Product Form**

$$\begin{aligned} 3x_1 - 4x_2 + x_3 &= 1 \\ 2x_1 - 7x_2 - 4x_3 &= 5 \\ x_1 + 5x_2 - 8x_3 &= 0 \end{aligned} \qquad \begin{bmatrix} (3, -4, 1) \cdot (x_1, x_2, x_3) \\ (2, -7, -4) \cdot (x_1, x_2, x_3) \\ (1, 5, -8) \cdot (x_1, x_2, x_3) \end{bmatrix} = \begin{bmatrix} 1 \\ 5 \\ 0 \end{bmatrix} \quad \blacktriangle$$

EXERCISE SET 4.1

1. Let $\mathbf{u} = (-3, 2, 1, 0)$, $\mathbf{v} = (4, 7, -3, 2)$, and $\mathbf{w} = (5, -2, 8, 1)$. Find
   (a) $\mathbf{v} - \mathbf{w}$ (b) $2\mathbf{u} + 7\mathbf{v}$ (c) $-\mathbf{u} + (\mathbf{v} - 4\mathbf{w})$
   (d) $6(\mathbf{u} - 3\mathbf{v})$ (e) $-\mathbf{v} - \mathbf{w}$ (f) $(6\mathbf{v} - \mathbf{w}) - (4\mathbf{u} + \mathbf{v})$

2. Let $\mathbf{u}$, $\mathbf{v}$, and $\mathbf{w}$ be the vectors in Exercise 1. Find the vector $\mathbf{x}$ that satisfies $5\mathbf{x} - 2\mathbf{v} = 2(\mathbf{w} - 5\mathbf{x})$.

3. Let $\mathbf{u}_1 = (-1, 3, 2, 0)$, $\mathbf{u}_2 = (2, 0, 4, -1)$, $\mathbf{u}_3 = (7, 1, 1, 4)$, and $\mathbf{u}_4 = (6, 3, 1, 2)$. Find scalars c_1, c_2, c_3, and c_4 such that $c_1\mathbf{u}_1 + c_2\mathbf{u}_2 + c_3\mathbf{u}_3 + c_4\mathbf{u}_4 = (0, 5, 6, -3)$.

4. Show that there do not exist scalars c_1, c_2, and c_3 such that
$$c_1(1, 0, 1, 0) + c_2(1, 0, -2, 1) + c_3(2, 0, 1, 2) = (1, -2, 2, 3)$$

5. In each part compute the Euclidean norm of the vector.
   (a) $(-2, 5)$ (b) $(1, 2, -2)$ (c) $(3, 4, 0, -12)$ (d) $(-2, 1, 1, -3, 4)$

6. Let $\mathbf{u} = (4, 1, 2, 3)$, $\mathbf{v} = (0, 3, 8, -2)$, and $\mathbf{w} = (3, 1, 2, 2)$. Evaluate the expression.

   (a) $\|\mathbf{u} + \mathbf{v}\|$ (b) $\|\mathbf{u}\| + \|\mathbf{v}\|$ (c) $\|-2\mathbf{u}\| + 2\|\mathbf{u}\|$

   (d) $\|3\mathbf{u} - 5\mathbf{v} + \mathbf{w}\|$ (e) $\dfrac{1}{\|\mathbf{w}\|}\mathbf{w}$ (f) $\left\|\dfrac{1}{\|\mathbf{w}\|}\mathbf{w}\right\|$

7. Show that if $\mathbf{v}$ is a nonzero vector in R^n, then $(1/\|\mathbf{v}\|)\mathbf{v}$ has Euclidean norm 1.

8. Let $\mathbf{v} = (-2, 3, 0, 6)$. Find all scalars k such that $\|k\mathbf{v}\| = 5$.

9. Find the Euclidean inner product $\mathbf{u} \cdot \mathbf{v}$.
   (a) $\mathbf{u} = (2, 5)$, $\mathbf{v} = (-4, 3)$ (b) $\mathbf{u} = (4, 8, 2)$, $\mathbf{v} = (0, 1, 3)$
   (c) $\mathbf{u} = (3, 1, 4, -5)$, $\mathbf{v} = (2, 2, -4, -3)$ (d) $\mathbf{u} = (-1, 1, 0, 4, -3)$, $\mathbf{v} = (-2, -2, 0, 2, -1)$

10. (a) Find two vectors in R^2 with Euclidean norm 1 whose Euclidean inner product with $(3, -1)$ is zero.

(b) Show that there are infinitely many vectors in R^3 with Euclidean norm 1 whose Euclidean inner product with $(1, -3, 5)$ is zero.

11. Find the Euclidean distance between **u** and **v**.
    (a) $\mathbf{u} = (1, -2)$, $\mathbf{v} = (2, 1)$
    (b) $\mathbf{u} = (2, -2, 2)$, $\mathbf{v} = (0, 4, -2)$
    (c) $\mathbf{u} = (0, -2, -1, 1)$, $\mathbf{v} = (-3, 2, 4, 4)$
    (d) $\mathbf{u} = (3, -3, -2, 0, -3)$, $\mathbf{v} = (-4, 1, -1, 5, 0)$

12. Verify parts (b), (e), (f), and (g) of Theorem 4.1.1 for $\mathbf{u} = (2, 0, -3, 1)$, $\mathbf{v} = (4, 0, 3, 5)$, $\mathbf{w} = (1, 6, 2, -1)$, $k = 5$, and $l = -3$.

13. Verify parts (b) and (c) of Theorem 4.1.2 for the values of **u**, **v**, **w**, and k in Exercise 12.

14. In each part determine whether the given vectors are orthogonal.
    (a) $\mathbf{u} = (-1, 3, 2)$, $\mathbf{v} = (4, 2, -1)$
    (b) $\mathbf{u} = (-2, -2, -2)$, $\mathbf{v} = (1, 1, 1)$
    (c) $\mathbf{u} = (u_1, u_2, u_3)$, $\mathbf{v} = (0, 0, 0)$
    (d) $\mathbf{u} = (-4, 6, -10, 1)$, $\mathbf{v} = (2, 1, -2, 9)$
    (e) $\mathbf{u} = (0, 3, -2, 1)$, $\mathbf{v} = (5, 2, -1, 0)$
    (f) $\mathbf{u} = (a, b)$, $\mathbf{v} = (-b, a)$

15. For which values of k are **u** and **v** orthogonal?
    (a) $\mathbf{u} = (2, 1, 3)$, $\mathbf{v} = (1, 7, k)$
    (b) $\mathbf{u} = (k, k, 1)$, $\mathbf{v} = (k, 5, 6)$

16. Find two vectors of norm 1 that are orthogonal to the three vectors $\mathbf{u} = (2, 1, -4, 0)$, $\mathbf{v} = (-1, -1, 2, 2)$, and $\mathbf{w} = (3, 2, 5, 4)$.

17. In each part verify that the Cauchy–Schwarz inequality holds.
    (a) $\mathbf{u} = (3, 2)$, $\mathbf{v} = (4, -1)$
    (b) $\mathbf{u} = (-3, 1, 0)$, $\mathbf{v} = (2, -1, 3)$
    (c) $\mathbf{u} = (-4, 2, 1)$, $\mathbf{v} = (8, -4, -2)$
    (d) $\mathbf{u} = (0, -2, 2, 1)$, $\mathbf{v} = (-1, -1, 1, 1)$

18. In each part verify that Formulas (8) and (9) hold.

    (a) $A = \begin{bmatrix} 2 & -1 \\ 3 & 4 \end{bmatrix}$, $\mathbf{u} = \begin{bmatrix} 3 \\ 1 \end{bmatrix}$, $\mathbf{v} = \begin{bmatrix} -2 \\ 6 \end{bmatrix}$

    (b) $A = \begin{bmatrix} -1 & 2 & 4 \\ 3 & 1 & 0 \\ 5 & -2 & 3 \end{bmatrix}$, $\mathbf{u} = \begin{bmatrix} -1 \\ 2 \\ 5 \end{bmatrix}$, $\mathbf{v} = \begin{bmatrix} 0 \\ 2 \\ -4 \end{bmatrix}$

19. Solve the following linear system for x_1, x_2, and x_3.

    $$(1, -1, 4) \cdot (x_1, x_2, x_3) = 10$$
    $$(3, 2, 0) \cdot (x_1, x_2, x_3) = 1$$
    $$(4, -5, -1) \cdot (x_1, x_2, x_3) = 7$$

20. Find $\mathbf{u} \cdot \mathbf{v}$ given that $\|\mathbf{u} + \mathbf{v}\| = 1$ and $\|\mathbf{u} - \mathbf{v}\| = 5$.

21. Use Theorem 4.1.7 to show that **u** and **v** are orthogonal vectors in R^n if $\|\mathbf{u} + \mathbf{v}\| = \|\mathbf{u} - \mathbf{v}\|$. Interpret this result geometrically in R^2.

22. Prove the identity

    $$\|\mathbf{u} + \mathbf{v}\|^2 + \|\mathbf{u} - \mathbf{v}\|^2 = 2\|\mathbf{u}\|^2 + 2\|\mathbf{v}\|^2$$

    for vectors in R^n. Interpret this result geometrically in R^2.

23. Prove that if **u** and **v** are orthogonal vectors in R^n such that $\|\mathbf{u}\| = 1$ and $\|\mathbf{v}\| = 1$, then $d(\mathbf{u}, \mathbf{v}) = \sqrt{2}$. Interpret this result geometrically in R^2.

24. Prove the following generalization of Theorem 4.1.7. If $v_1, v_2, \ldots, v_r$ are pairwise orthogonal vectors in R^n, then

$$\|v_1 + v_2 + \cdots + v_r\|^2 = \|v_1\|^2 + \|v_2\|^2 + \cdots + \|v_r\|^2$$

25. Prove: If u and v are $n \times 1$ matrices and A is an invertible $n \times n$ matrix, then

$$[v^T A^T A u]^2 \leq (u^T A^T A u)(v^T A^T A v)$$

26. Use the Cauchy–Schwarz inequality to prove that for all real values of a, b, and θ,

$$[a \cos \theta + b \sin \theta]^2 \leq a^2 + b^2$$

27. Prove: If u, v, and w are vectors in R^n and k is any scalar, then
  (a) $u \cdot (kv) = k(u \cdot v)$ (b) $u \cdot (v + w) = u \cdot v + u \cdot w$

28. Prove parts (*a*) through (*d*) of Theorem 4.1.1.

29. Prove parts (*e*) through (*h*) of Theorem 4.1.1.

30. Prove parts (*a*) and (*c*) of Theorem 4.1.2.

31. Prove parts (*a*) and (*b*) of Theorem 4.1.4.

32. Prove parts (*a*), (*b*), and (*c*) of Theorem 4.1.5.

33. Suppose that $a_1, a_2, \ldots, a_n$ are positive real numbers. In R^2, the vectors $v_1 = (a_1, 0)$ and $v_2 = (0, a_2)$ determine a rectangle of area $A = a_1 a_2$ (Figure 5a), and in R^3 the vectors $v_1 = (a_1, 0, 0)$, $v_2 = (0, a_2, 0)$, and $v_3 = (0, 0, a_3)$ determine a box of volume $V = a_1 a_2 a_3$ (Figure 5b). The area A and the volume V are sometimes called the **Euclidean measure** of the rectangle and box, respectively.
  (a) How would you define the Euclidean measure of the "box" in R^n that is determined by the vectors

$$v_1 = (a_1, 0, 0, \ldots, 0),\ v_2 = (0, a_2, 0, \ldots, 0),\ \ldots,\ v_n = (0, 0, 0, \ldots, a_n)?$$

  (b) How would you define the Euclidean length of the "diagonal" of the box in part (a)?

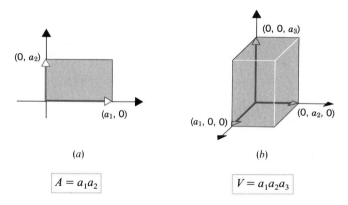

(a)

$A = a_1 a_2$

(b)

$V = a_1 a_2 a_3$

Figure 5

4.2 LINEAR TRANSFORMATIONS FROM R^n TO R^m

In this section we shall begin the study of functions of the form $\mathbf{w} = F(\mathbf{x})$, where the independent variable $\mathbf{x}$ is a vector in R^n and the dependent variable $\mathbf{w}$ is a vector in R^m. We shall concentrate on a special class of such functions called "linear transformations." Linear transformations are fundamental in the study of linear algebra and have many important applications in physics, engineering, social sciences, and various branches of mathematics.

FUNCTIONS FROM R^n TO R

Recall that a ***function*** is a rule f that associates with each element in a set A one and only one element in a set B. If f associates the element b with the element a, then we write $b = f(a)$ and say that b is the ***image*** of a under f or that $f(a)$ is the ***value*** of f at a. The set A is called the ***domain*** of f and the set B is called the ***codomain*** of f. The subset of B consisting of all possible values for f as a varies over A is called the ***range*** of f. For the most common functions, A and B are sets of real numbers, in which case f is called a ***real-valued function of a real variable***. Other common functions occur when B is a set of real numbers and A is a set of vectors in R^2, R^3, or more generally, in R^n. Some examples are shown in Table 1.

TABLE 1

Formula	Example	Classification	Description
$f(x)$	$f(x) = x^2$	Real-valued function of a real variable	Function from R to R
$f(x, y)$	$f(x, y) = x^2 + y^2$	Real-valued function of two real variables	Function from R^2 to R
$f(x, y, z)$	$f(x, y, z) = x^2 + y^2 + z^2$	Real-valued function of three real variables	Function from R^3 to R
$f(x_1, x_2, \ldots, x_n)$	$f(x_1, x_2, \ldots, x_n) = x_1^2 + x_2^2 + \cdots + x_n^2$	Real-valued function of n real variables	Function from R^n to R

Two functions f_1 and f_2 are regarded to be ***equal***, written $f_1 = f_2$, if they have the same domain and $f_1(a) = f_2(a)$ for all a in the domain.

FUNCTIONS FROM R^n TO R^m

If the domain of a function f is R^n and the codomain is R^m (m and n possibly the same), then f is called a ***transformation*** from R^n to R^m, and we say that f ***maps*** R^n into R^m. We denote this by writing $f : R^n \to R^m$. The functions in Table 1 are transformations for which $m = 1$. In the special case where $m = n$ the transformation $f : R^n \to R^n$ is called an ***operator*** on R^n. The first entry in Table 1 is an operator on R.

To illustrate one important way in which transformations can arise, suppose that $f_1, f_2, \ldots, f_m$ are real-valued functions of n real variables, say

$$
\begin{aligned}
w_1 &= f_1(x_1, x_2, \ldots, x_n) \\
w_2 &= f_2(x_1, x_2, \ldots, x_n) \\
&\ \ \vdots \qquad\qquad \vdots \\
w_m &= f_m(x_1, x_2, \ldots, x_n)
\end{aligned}
\tag{1}
$$

These m equations assign a unique point $(w_1, w_2, \ldots, w_m)$ in R^m to each point $(x_1, x_2, \ldots, x_n)$ in R^n and thus define a transformation from R^n to R^m. If we denote this transformation by T, then $T: R^n \rightarrow R^m$ and

$$
T(x_1, x_2, \ldots, x_n) = (w_1, w_2, \ldots, w_m)
$$

Example 1   The equations

$$
\begin{aligned}
w_1 &= x_1 + x_2 \\
w_2 &= 3x_1 x_2 \\
w_3 &= x_1^2 - x_2^2
\end{aligned}
$$

define a transformation $T: R^2 \rightarrow R^3$. With this transformation the image of the point (x_1, x_2) is

$$
T(x_1, x_2) = (x_1 + x_2, 3x_1 x_2, x_1^2 - x_2^2)
$$

Thus, for example,

$$
T(1, -2) = (-1, -6, -3) \ \ \blacktriangle
$$

LINEAR TRANSFORMATIONS FROM R^n TO R^m

In the special case where the equations in (1) are linear, the transformation $T: R^n \rightarrow R^m$ defined by those equations is called a ***linear transformation*** (or a ***linear operator*** if $m = n$). Thus, a linear transformation $T: R^n \rightarrow R^m$ is defined by equations of the form

$$
\begin{aligned}
w_1 &= a_{11}x_1 + a_{12}x_2 + \cdots + a_{1n}x_n \\
w_2 &= a_{21}x_1 + a_{22}x_2 + \cdots + a_{2n}x_n \\
&\ \ \vdots \qquad \vdots \qquad \vdots \qquad\qquad \vdots \\
w_m &= a_{m1}x_1 + a_{m2}x_2 + \cdots + a_{mn}x_n
\end{aligned}
\tag{2}
$$

or in matrix notation

$$
\begin{bmatrix} w_1 \\ w_2 \\ \vdots \\ w_m \end{bmatrix} = \begin{bmatrix} a_{11} & a_{12} & \cdots & a_{1n} \\ a_{21} & a_{22} & \cdots & a_{2n} \\ \vdots & \vdots & & \vdots \\ a_{m1} & a_{m2} & \cdots & a_{mn} \end{bmatrix} \begin{bmatrix} x_1 \\ x_2 \\ \vdots \\ x_n \end{bmatrix}
\tag{3}
$$

or more briefly by

$$
\mathbf{w} = A\mathbf{x}
\tag{4}
$$

The matrix $A = [a_{ij}]$ is called the ***standard matrix*** for the linear transformation T, and T is called ***multiplication by A***.

Example 2 The linear transformation $T: R^4 \to R^3$ defined by the equations

$$\begin{aligned} w_1 &= 2x_1 - 3x_2 + x_3 - 5x_4 \\ w_2 &= 4x_1 + x_2 - 2x_3 + x_4 \\ w_3 &= 5x_1 - x_2 + 4x_3 \end{aligned} \tag{5}$$

can be expressed in matrix form as

$$\begin{bmatrix} w_1 \\ w_2 \\ w_3 \end{bmatrix} = \begin{bmatrix} 2 & -3 & 1 & -5 \\ 4 & 1 & -2 & 1 \\ 5 & -1 & 4 & 0 \end{bmatrix} \begin{bmatrix} x_1 \\ x_2 \\ x_3 \\ x_4 \end{bmatrix} \tag{6}$$

so the standard matrix for T is

$$A = \begin{bmatrix} 2 & -3 & 1 & -5 \\ 4 & 1 & -2 & 1 \\ 5 & -1 & 4 & 0 \end{bmatrix}$$

The image of a point (x_1, x_2, x_3, x_4) can be computed directly from the defining equations (5) or from (6) by matrix multiplication. For example, if $(x_1, x_2, x_3, x_4) = (1, -3, 0, 2)$, then substituting in (5) yields

$$w_1 = 1, \qquad w_2 = 3, \qquad w_3 = 8$$

(verify) or alternatively from (6)

$$\begin{bmatrix} w_1 \\ w_2 \\ w_3 \end{bmatrix} = \begin{bmatrix} 2 & -3 & 1 & -5 \\ 4 & 1 & -2 & 1 \\ 5 & -1 & 4 & 0 \end{bmatrix} \begin{bmatrix} 1 \\ -3 \\ 0 \\ 2 \end{bmatrix} = \begin{bmatrix} 1 \\ 3 \\ 8 \end{bmatrix} \ \blacktriangle$$

SOME NOTATIONAL MATTERS

If $T: R^n \to R^m$ is multiplication by A, and if it is important to emphasize that A is the standard matrix for T, we shall denote the linear transformation $T: R^n \to R^m$ by $T_A: R^n \to R^m$. Thus,

$$T_A(\mathbf{x}) = A\mathbf{x} \tag{7}$$

It is understood in this equation that the vector $\mathbf{x}$ in R^n is expressed as a column matrix.

Sometimes it is awkward to introduce a new letter to denote the standard matrix for a linear transformation $T: R^n \to R^m$. In such cases we will denote the standard matrix for T by the symbol $[\,T\,]$. With this notation equation (7) would take the form

$$T(\mathbf{x}) = [\,T\,]\mathbf{x} \tag{8}$$

Occasionally, the two notations for a standard matrix will be mixed, in which case we have the relationship

$$[\,T_A\,] = A \tag{9}$$

REMARK. Amidst all of this notation it is important to keep in mind that we have established a correspondence between $m \times n$ matrices and linear transformations from R^n to R^m: To each matrix A there corresponds a linear transformation T_A (multiplication by A), and to each linear transformation $T: R^n \rightarrow R^m$, there corresponds an $m \times n$ matrix $[T]$ (the standard matrix for T).

GEOMETRY OF LINEAR TRANSFORMA- TIONS

Depending on whether n-tuples are regarded as points or vectors, the geometric effect of an operator $T: R^n \rightarrow R^n$ is to transform each point (or vector) in R^n into some new point (or vector) (Figure 1).

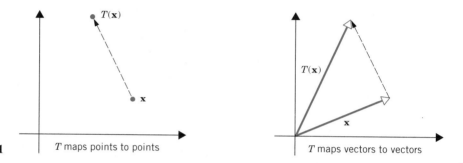

Figure 1 T maps points to points T maps vectors to vectors

Example 3   If 0 is the $m \times n$ zero matrix and $\mathbf{0}$ is the zero vector in R^n, then for every vector $\mathbf{x}$ in R^n

$$T_0(\mathbf{x}) = 0\mathbf{x} = \mathbf{0}$$

so multiplication by zero maps every vector in R^n into the zero vector in R^m. We call T_0 the **zero transformation** from R^n to R^m. Sometimes the zero transformation is denoted by 0. Although this is the same notation used for the zero matrix, the appropriate interpretation will usually be clear from the context. ▲

Example 4   If I is the $n \times n$ identity matrix, then for every vector $\mathbf{x}$ in R^n

$$T_I(\mathbf{x}) = I\mathbf{x} = \mathbf{x}$$

so multiplication by I maps every vector in R^n into itself. We call T_I the **identity operator** on R^n. Sometimes the identity operator is denoted by I. Although this is the same notation used for the identity matrix, the appropriate interpretation will usually be clear from the context. ▲

Among the most important linear operations on R^2 and R^3 are those that produce reflections, projections, and rotations. We shall now discuss such operators.

REFLECTION OPERATORS

Consider the operator $T: R^2 \rightarrow R^2$ that maps each vector into its symmetric image about the y-axis (Figure 2).

If we let $\mathbf{w} = T(\mathbf{x})$, then the equations relating the components of $\mathbf{x}$ and $\mathbf{w}$ are

$$\begin{aligned} w_1 &= -x = -x + 0y \\ w_2 &= y = 0x + y \end{aligned} \qquad (10)$$

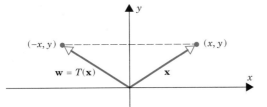

Figure 2

or in matrix form,

$$\begin{bmatrix} w_1 \\ w_2 \end{bmatrix} = \begin{bmatrix} -1 & 0 \\ 0 & 1 \end{bmatrix} \begin{bmatrix} x \\ y \end{bmatrix} \tag{11}$$

Since the equations in (10) are linear, T is a linear operator and from (11) the standard matrix for T is

$$[T] = \begin{bmatrix} -1 & 0 \\ 0 & 1 \end{bmatrix}$$

In general, operators on R^2 and R^3 that map each vector into its symmetric image about some line or plane are called **reflection operators**. Such operators are linear. Tables 2 and 3 list some of the common reflection operators.

TABLE 2

Operator	Illustration	Equations	Standard Matrix
Reflection about the y-axis		$w_1 = -x$ $w_2 = y$	$\begin{bmatrix} -1 & 0 \\ 0 & 1 \end{bmatrix}$
Reflection about the x-axis		$w_1 = x$ $w_2 = -y$	$\begin{bmatrix} 1 & 0 \\ 0 & -1 \end{bmatrix}$
Reflection about the line $y = x$		$w_1 = y$ $w_2 = x$	$\begin{bmatrix} 0 & 1 \\ 1 & 0 \end{bmatrix}$

TABLE 3

Operator	Illustration	Equations	Standard Matrix
Reflection about the xy-plane	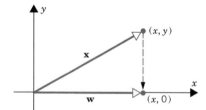	$\begin{aligned} w_1 &= x \\ w_2 &= y \\ w_3 &= -z \end{aligned}$	$\begin{bmatrix} 1 & 0 & 0 \\ 0 & 1 & 0 \\ 0 & 0 & -1 \end{bmatrix}$
Reflection about the xz-plane		$\begin{aligned} w_1 &= x \\ w_2 &= -y \\ w_3 &= z \end{aligned}$	$\begin{bmatrix} 1 & 0 & 0 \\ 0 & -1 & 0 \\ 0 & 0 & 1 \end{bmatrix}$
Reflection about the yz-plane		$\begin{aligned} w_1 &= -x \\ w_2 &= y \\ w_3 &= z \end{aligned}$	$\begin{bmatrix} -1 & 0 & 0 \\ 0 & 1 & 0 \\ 0 & 0 & 1 \end{bmatrix}$

PROJECTION OPERATORS

Consider the operator $T: R^2 \rightarrow R^2$ that maps each vector into its orthogonal projection on the x-axis (Figure 3).

Figure 3

The equations relating the components of $\mathbf{x}$ and $\mathbf{w} = T(\mathbf{x})$ are

$$\begin{aligned} w_1 &= x = \; x + 0y \\ w_2 &= 0 = 0x + 0y \end{aligned} \tag{12}$$

or in matrix form,

$$\begin{bmatrix} w_1 \\ w_2 \end{bmatrix} = \begin{bmatrix} 1 & 0 \\ 0 & 0 \end{bmatrix} \begin{bmatrix} x \\ y \end{bmatrix} \tag{13}$$

The equations in (12) are linear, so T is a linear operator and from (13) the standard matrix for T is

$$[T] = \begin{bmatrix} 1 & 0 \\ 0 & 0 \end{bmatrix}$$

In general, a *projection operator* (or more precisely an *orthogonal projection operator*) on R^2 or R^3 is any operator that maps each vector into its orthogonal projection on a line or plane through the origin. It can be shown that such operators are linear. Some of the basic projection operators on R^2 and R^3 are listed in Tables 4 and 5.

TABLE 4

Operator	Illustration	Equations	Standard Matrix
Orthogonal projection on the x-axis		$w_1 = x$ $w_2 = 0$	$\begin{bmatrix} 1 & 0 \\ 0 & 0 \end{bmatrix}$
Orthogonal projection on the y-axis		$w_1 = 0$ $w_2 = y$	$\begin{bmatrix} 0 & 0 \\ 0 & 1 \end{bmatrix}$

TABLE 5

Operator	Illustration	Equations	Standard Matrix
Orthogonal projection on the xy-plane		$w_1 = x$ $w_2 = y$ $w_3 = 0$	$\begin{bmatrix} 1 & 0 & 0 \\ 0 & 1 & 0 \\ 0 & 0 & 0 \end{bmatrix}$
Orthogonal projection on the xz-plane		$w_1 = x$ $w_2 = 0$ $w_3 = z$	$\begin{bmatrix} 1 & 0 & 0 \\ 0 & 0 & 0 \\ 0 & 0 & 1 \end{bmatrix}$
Orthogonal projection on the yz-plane		$w_1 = 0$ $w_2 = y$ $w_3 = z$	$\begin{bmatrix} 0 & 0 & 0 \\ 0 & 1 & 0 \\ 0 & 0 & 1 \end{bmatrix}$

ROTATION OPERATORS

An operator that rotates each vector in R^2 through a fixed angle θ is called a ***rotation operator*** on R^2. Table 6 lists the two basic rotation operators on R^2. To show how these results are derived, consider the rotation operator that rotates each vector counterclockwise through a fixed positive angle θ. To find equations relating $\mathbf{x}$ and $\mathbf{w} = T(\mathbf{x})$, let ϕ be the angle from the positive x-axis to $\mathbf{x}$, and let r be the common length of $\mathbf{x}$ and $\mathbf{w}$ (Figure 4).

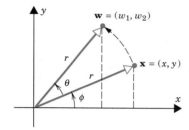

Figure 4

Then from basic trigonometry

$$x = r \cos \phi, \qquad y = r \sin \phi \tag{14}$$

and

$$w_1 = r \cos(\theta + \phi), \qquad w_2 = r \sin(\theta + \phi) \tag{15}$$

Using trigonometric identities on (15) yields

$$w_1 = r \cos \theta \cos \phi - r \sin \theta \sin \phi$$
$$w_2 = r \sin \theta \cos \phi + r \cos \theta \sin \phi$$

and substituting (14) yields

$$\begin{aligned} w_1 &= x \cos \theta - y \sin \theta \\ w_2 &= x \sin \theta + y \cos \theta \end{aligned} \tag{16}$$

The equations in (16) are linear, so T is a linear operator; moreover, it follows from these equations that the standard matrix for T is

$$[T] = \begin{bmatrix} \cos \theta & -\sin \theta \\ \sin \theta & \cos \theta \end{bmatrix}$$

TABLE 6

Operator	Illustration	Equations	Standard Matrix
Rotation through an angle θ		$w_1 = x \cos \theta - y \sin \theta$ $w_2 = x \sin \theta + y \cos \theta$	$\begin{bmatrix} \cos \theta & -\sin \theta \\ \sin \theta & \cos \theta \end{bmatrix}$

Example 5 If each vector in R^2 is rotated through an angle of $\pi/6 = (30°)$, then the image **w** of a vector

$$\mathbf{x} = \begin{bmatrix} x \\ y \end{bmatrix}$$

is

$$\mathbf{w} = \begin{bmatrix} \cos \pi/6 & -\sin \pi/6 \\ \sin \pi/6 & \cos \pi/6 \end{bmatrix} \begin{bmatrix} x \\ y \end{bmatrix} = \begin{bmatrix} \sqrt{3}/2 & -1/2 \\ 1/2 & \sqrt{3}/2 \end{bmatrix} \begin{bmatrix} x \\ y \end{bmatrix} = \begin{bmatrix} \dfrac{\sqrt{3}}{2}x - \dfrac{1}{2}y \\ \dfrac{1}{2}x + \dfrac{\sqrt{3}}{2}y \end{bmatrix}$$

For example, the image of the vector

$$\mathbf{x} = \begin{bmatrix} 1 \\ 1 \end{bmatrix}$$

is

$$\mathbf{w} = \begin{bmatrix} \dfrac{\sqrt{3}-1}{2} \\ \dfrac{1+\sqrt{3}}{2} \end{bmatrix}$$

A rotation of vectors in R^3 is usually described in relation to a ray emanating from the origin, called the ***axis of rotation***. As a vector revolves around the axis of rotation it sweeps out some portion of a cone (Figure 5*a*). The ***angle of rotation***, which is measured in the base of the cone, is described as "clockwise" or "counterclockwise" in relation to a viewpoint that is along the axis of rotation *looking toward the origin*. For example, in Figure 5*a* the vector **w** results from rotating the vector **x** counterclockwise around the axis *l* through an angle θ. As in R^2, angles are *positive* if they are generated by counterclockwise rotations and *negative* if they are generated by clockwise rotations.

The most common way of describing a general axis of rotation is to specify a nonzero vector **u** that runs along the axis of rotation and has its initial point at the origin. The counterclockwise direction for a rotation about the axis can then be determined by a "right-hand rule" (Figure 5*b*): If the thumb of the right hand points in the direction of **u**, then the cupped fingers point in a counterclockwise direction.

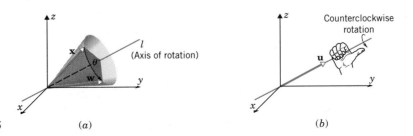

Figure 5   (*a*)   (*b*)

A ***rotation operator*** on R^3 is a linear operator that rotates each vector in R^3 about some rotation axis through a fixed angle θ. In Table 7 we have described the rotation operators on R^3 whose axes of rotation are the positive coordinate axes. For each of these rotations one of the components is unchanged by the rotation, and the relationships between the other components can be derived by the same procedure used to derive (16). For example, in the rotation about the z-axis, the z-components of $\mathbf{x}$ and $\mathbf{w} = T(\mathbf{x})$ are the same, and the x- and y-components are related as in (16). This yields the rotation equations shown in the last row of Table 7.

TABLE 7

Operator	Illustration	Equations	Standard Matrix
Counterclockwise rotation about the positive x-axis through an angle θ		$w_1 = x$ $w_2 = y \cos\theta - z \sin\theta$ $w_3 = y \sin\theta + z \cos\theta$	$\begin{bmatrix} 1 & 0 & 0 \\ 0 & \cos\theta & -\sin\theta \\ 0 & \sin\theta & \cos\theta \end{bmatrix}$
Counterclockwise rotation about the positive y-axis through an angle θ		$w_1 = x \cos\theta + z \sin\theta$ $w_2 = y$ $w_3 = -x \sin\theta + z \cos\theta$	$\begin{bmatrix} \cos\theta & 0 & \sin\theta \\ 0 & 1 & 0 \\ -\sin\theta & 0 & \cos\theta \end{bmatrix}$
Counterclockwise rotation about the positive z-axis through an angle θ		$w_1 = x \cos\theta - y \sin\theta$ $w_2 = x \sin\theta + y \cos\theta$ $w_3 = z$	$\begin{bmatrix} \cos\theta & -\sin\theta & 0 \\ \sin\theta & \cos\theta & 0 \\ 0 & 0 & 1 \end{bmatrix}$

For completeness, we note that the standard matrix for a counterclockwise rotation through an angle θ about an axis in R^3, which is determined by an arbitrary *unit vector* $\mathbf{u} = (a, b, c)$ that has its initial point at the origin, is

$$\begin{bmatrix} a^2(1 - \cos\theta) + \cos\theta & ab(1 - \cos\theta) - c\sin\theta & ac(1 - \cos\theta) + b\sin\theta \\ ab(1 - \cos\theta) + c\sin\theta & b^2(1 - \cos\theta) + \cos\theta & bc(1 - \cos\theta) - a\sin\theta \\ ac(1 - \cos\theta) - b\sin\theta & bc(1 - \cos\theta) + a\sin\theta & c^2(1 - \cos\theta) + \cos\theta \end{bmatrix}$$

(17)

The derivation can be found in the book *Principles of Interactive Computer Graphics*, by W. M. Newman and R. F. Sproull, New York, McGraw-Hill, 1979. The reader may find it instructive to derive the results in Table 7 as special cases of this more general result.

DILATION AND CONTRACTION OPERATORS

If k is a nonnegative scalar, then the operator $T(\mathbf{x}) = k\mathbf{x}$ on R^2 or R^3 is called a ***contraction with factor k*** if $0 \le k \le 1$ and a ***dilation with factor k*** if $k \ge 1$. The geometric effect of a contraction is to compress each vector by a factor of k (Figure 6a), and the effect of a dilation is to stretch each vector by a factor of k (Figure 6b). A contraction compresses R^2 or R^3 uniformly toward the origin from all directions, and a dilation stretches R^2 or R^3 uniformly away from the origin in all directions.

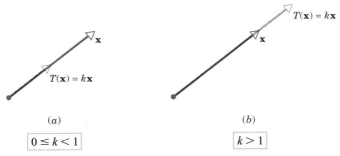

(a)   (b)

Figure 6 $\boxed{0 \le k < 1}$ $\boxed{k > 1}$

The most extreme contraction occurs when $k = 0$, in which case $T(\mathbf{x}) = k\mathbf{x}$ reduces to the zero operator $T(\mathbf{x}) = \mathbf{0}$, which compresses every vector into a single point (the origin). If $k = 1$, then $T(\mathbf{x}) = k\mathbf{x}$ reduces to the identity operator $T(\mathbf{x}) = \mathbf{x}$, which leaves each vector unchanged; this can be regarded as either a contraction or a dilation. Tables 8 and 9 list the dilation and contraction operators on R^2 and R^3.

TABLE 8

Operator	Illustration	Equations	Standard Matrix
Contraction with factor k on R^2 $(0 \le k \le 1)$		$w_1 = kx$ $w_2 = ky$	$\begin{bmatrix} k & 0 \\ 0 & k \end{bmatrix}$
Dilation with factor k on R^2 $(k \ge 1)$		$w_1 = kx$ $w_2 = ky$	

TABLE 9

Operator	Illustration	Equations	Standard Matrix
Contraction with factor k on R^3 $(0 \leq k \leq 1)$		$w_1 = kx$ $w_2 = ky$ $w_3 = kz$	
Dilation with factor k on R^3 $(k \geq 1)$		$w_1 = kx$ $w_2 = ky$ $w_3 = kz$	$\begin{bmatrix} k & 0 & 0 \\ 0 & k & 0 \\ 0 & 0 & k \end{bmatrix}$

COMPOSITIONS OF LINEAR TRANSFORMATIONS

If $T_A: R^n \to R^k$ and $T_B: R^k \to R^m$ are linear transformations, then for each $\mathbf{x}$ in R^n one can first compute $T_A(\mathbf{x})$, which is a vector in R^k, and then one can compute $T_B(T_A(\mathbf{x}))$, which is a vector in R^m. Thus, the application of T_A followed by T_B produces a transformation from R^n to R^m. This transformation is called the *composition of T_B with T_A* and is denoted by $T_B \circ T_A$ (read "T_B circle T_A"). Thus,

$$(T_B \circ T_A)(\mathbf{x}) = T_B(T_A(\mathbf{x})) \tag{18}$$

The composition $T_B \circ T_A$ is linear since

$$(T_B \circ T_A)(\mathbf{x}) = T_B(T_A(\mathbf{x})) = B(A\mathbf{x}) = (BA)\mathbf{x} \tag{19}$$

so $T_B \circ T_A$ is multiplication by BA, which is a linear transformation. Formula (19) also tells us that the standard matrix for $T_B \circ T_A$ is BA. This is expressed by the formula

$$T_B \circ T_A = T_{BA} \tag{20}$$

REMARK. Formula (20) captures an important idea: *Multiplying matrices is equivalent to composing the corresponding linear transformations in the right-to-left order of the factors.*

There is an alternative form of Formula (20): If $T_1: R^n \to R^k$ and $T_2: R^k \to R^m$ are linear transformations, then because the standard matrix for the composition $T_2 \circ T_1$ is the product of the standard matrices of T_2 and T_1, we have

$$[T_2 \circ T_1] = [T_2][T_1] \tag{21}$$

Example 6   Let $T_1 : R^2 \to R^2$ and $T_2 : R^2 \to R^2$ be the linear operators that rotate vectors through the angles θ_1 and θ_2, respectively. Thus, the operation

$$(T_2 \circ T_1)(\mathbf{x}) = T_2(T_1(\mathbf{x}))$$

first rotates $\mathbf{x}$ through the angle θ_1, then rotates $T_1(\mathbf{x})$ through the angle θ_2. It follows that the net effect of $T_2 \circ T_1$ is to rotate each vector in R^2 through the angle $\theta_1 + \theta_2$ (Figure 7).

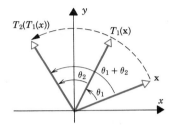

Figure 7

Thus, the standard matrices for these linear operators are

$$[\,T_1\,] = \begin{bmatrix} \cos\theta_1 & -\sin\theta_1 \\ \sin\theta_1 & \cos\theta_1 \end{bmatrix}, \qquad [\,T_2\,] = \begin{bmatrix} \cos\theta_2 & -\sin\theta_2 \\ \sin\theta_2 & \cos\theta_2 \end{bmatrix},$$

$$[\,T_2 \circ T_1\,] = \begin{bmatrix} \cos(\theta_1 + \theta_2) & -\sin(\theta_1 + \theta_2) \\ \sin(\theta_1 + \theta_2) & \cos(\theta_1 + \theta_2) \end{bmatrix}$$

These matrices should satisfy (21). With the help of some basic trigonometric identities, we can show that this is so as follows:

$$[\,T_2\,][\,T_1\,] = \begin{bmatrix} \cos\theta_2 & -\sin\theta_2 \\ \sin\theta_2 & \cos\theta_2 \end{bmatrix} \begin{bmatrix} \cos\theta_1 & -\sin\theta_1 \\ \sin\theta_1 & \cos\theta_1 \end{bmatrix}$$

$$= \begin{bmatrix} \cos\theta_2\cos\theta_1 - \sin\theta_2\sin\theta_1 & -(\cos\theta_2\sin\theta_1 + \sin\theta_2\cos\theta_1) \\ \sin\theta_2\cos\theta_1 + \cos\theta_2\sin\theta_1 & -\sin\theta_2\sin\theta_1 + \cos\theta_2\cos\theta_1 \end{bmatrix}$$

$$= \begin{bmatrix} \cos(\theta_1 + \theta_2) & -\sin(\theta_1 + \theta_2) \\ \sin(\theta_1 + \theta_2) & \cos(\theta_1 + \theta_2) \end{bmatrix}$$

$$= [\,T_2 \circ T_1\,]$$

REMARK.   In general, the order in which linear transformations are composed matters. This is to be expected, since the composition of two linear transformations corresponds to the multiplication of their standard matrices, and we know that the order in which matrices are multiplied makes a difference.

Example 7   Let $T_1 : R^2 \to R^2$ be the reflection operator about the line $y = x$, and let $T_2 : R^2 \to R^2$ be the orthogonal projection on the y-axis. Figure 8 illustrates graphically that $T_1 \circ T_2$ and $T_2 \circ T_1$ have different effects on a vector $\mathbf{x}$. This same conclusion can be reached by showing that the standard matrices for T_1 and T_2 do not commute:

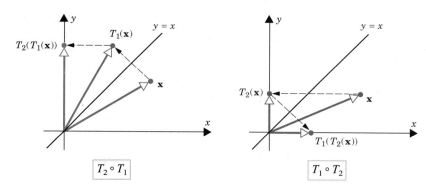

Figure 8

$$[\, T_1 \circ T_2 \,] = [\, T_1 \,][\, T_2 \,] = \begin{bmatrix} 0 & 1 \\ 1 & 0 \end{bmatrix} \begin{bmatrix} 0 & 0 \\ 0 & 1 \end{bmatrix} = \begin{bmatrix} 0 & 1 \\ 0 & 0 \end{bmatrix}$$

$$[\, T_2 \circ T_1 \,] = [\, T_2 \,][\, T_1 \,] = \begin{bmatrix} 0 & 0 \\ 0 & 1 \end{bmatrix} \begin{bmatrix} 0 & 1 \\ 1 & 0 \end{bmatrix} = \begin{bmatrix} 0 & 0 \\ 1 & 0 \end{bmatrix}$$

so $[\, T_2 \circ T_1 \,] \neq [\, T_1 \circ T_2 \,]$. ▲

Example 8   Let $T_1 : R^2 \to R^2$ be the reflection about the y-axis, and let $T_2 : R^2 \to R^2$ be the reflection about the x-axis. In this case $T_1 \circ T_2$ and $T_2 \circ T_1$ are the same; both map each vector $\mathbf{x} = (x, y)$ into its negative $-\mathbf{x} = (-x, -y)$ (Figure 9):

$$(T_1 \circ T_2)(x, y) = T_1(x, -y) = (-x, -y)$$
$$(T_2 \circ T_1)(x, y) = T_2(-x, y) = (-x, -y)$$

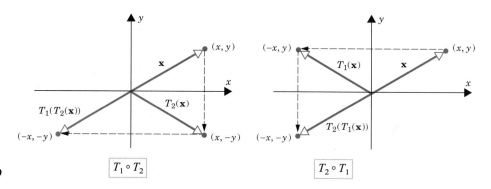

Figure 9

The equality of $T_1 \circ T_2$ and $T_2 \circ T_1$ can also be deduced by showing that the standard matrices for T_1 and T_2 commute:

$$[\, T_1 \circ T_2 \,] = [\, T_1 \,][\, T_2 \,] = \begin{bmatrix} -1 & 0 \\ 0 & 1 \end{bmatrix} \begin{bmatrix} 1 & 0 \\ 0 & -1 \end{bmatrix} = \begin{bmatrix} -1 & 0 \\ 0 & -1 \end{bmatrix}$$

$$[\, T_2 \circ T_1 \,] = [\, T_2 \,][\, T_1 \,] = \begin{bmatrix} 1 & 0 \\ 0 & -1 \end{bmatrix} \begin{bmatrix} -1 & 0 \\ 0 & 1 \end{bmatrix} = \begin{bmatrix} -1 & 0 \\ 0 & -1 \end{bmatrix}$$

The operator $T(\mathbf{x}) = -\mathbf{x}$ on R^2 or R^3 is called the **reflection about the origin**. As the computations above show, the standard matrix for this operator on R^2 is

$$[T] = \begin{bmatrix} -1 & 0 \\ 0 & -1 \end{bmatrix} \quad \triangle$$

COMPOSITIONS OF THREE OR MORE LINEAR TRANSFORMA- TIONS

Compositions can be defined for three or more linear transformations. For example, consider the linear transformations

$$T_1 : R^n \to R^k, \qquad T_2 : R^k \to R^l, \qquad T_3 : R^l \to R^m$$

We define the composition $(T_3 \circ T_2 \circ T_1) : R^n \to R^m$ by

$$(T_3 \circ T_2 \circ T_1)(\mathbf{x}) = T_3(T_2(T_1(\mathbf{x})))$$

It can be shown that this composition is a linear transformation and that the standard matrix for $T_3 \circ T_2 \circ T_1$ is related to the standard matrices for T_1, T_2, and T_3 by

$$[\, T_3 \circ T_2 \circ T_1 \,] = [\, T_3 \,][\, T_2 \,][\, T_1 \,] \tag{22}$$

which is a generalization of (21). If the standard matrices for T_1, T_2, and T_3 are denoted by A, B, and C, respectively, then we also have the following generalization of (20):

$$T_C \circ T_B \circ T_A = T_{CBA} \tag{23}$$

Example 9 Find the standard matrix for the linear operator $T : R^3 \to R^3$ that first rotates a vector clockwise about the z-axis through an angle θ, then reflects the resulting vector about the yz-plane, and then projects that vector orthogonally onto the xy-plane.

Solution. The linear transformation T can be expressed as the composition

$$T = T_C \circ T_B \circ T_A = T_{CBA}$$

where T_A is the rotation about the z-axis, T_B is the reflection about the yz-plane, and T_C is the orthogonal projection on the xy-plane. From Tables 3, 5, and 7 the standard matrices for these linear transformations are

$$[T_A] = \begin{bmatrix} \cos\theta & -\sin\theta & 0 \\ \sin\theta & \cos\theta & 0 \\ 0 & 0 & 1 \end{bmatrix}, \qquad [T_B] = \begin{bmatrix} -1 & 0 & 0 \\ 0 & 1 & 0 \\ 0 & 0 & 1 \end{bmatrix}, \qquad [T_C] = \begin{bmatrix} 1 & 0 & 0 \\ 0 & 1 & 0 \\ 0 & 0 & 0 \end{bmatrix}$$

Thus, from (22) the standard matrix for T is

$$CBA = \begin{bmatrix} \cos\theta & -\sin\theta & 0 \\ \sin\theta & \cos\theta & 0 \\ 0 & 0 & 1 \end{bmatrix} \begin{bmatrix} 1 & 0 & 0 \\ 0 & 1 & 0 \\ 0 & 0 & 0 \end{bmatrix} \begin{bmatrix} -1 & 0 & 0 \\ 0 & 1 & 0 \\ 0 & 0 & 1 \end{bmatrix}$$

$$= \begin{bmatrix} -\cos\theta & \sin\theta & 0 \\ \sin\theta & \cos\theta & 0 \\ 0 & 0 & 0 \end{bmatrix} \quad \triangle$$

EXERCISE SET 4.2

1. Find the domain and codomain of the transformation defined by the equations, and determine whether the transformation is linear.

(a) $w_1 = 3x_1 - 2x_2 + 4x_3$
$\quad w_2 = 5x_1 - 8x_2 + x_3$

(b) $w_1 = 2x_1x_2 - x_2$
$\quad w_2 = x_1 + 3x_1x_2$
$\quad w_3 = x_1 + x_2$

(c) $w_1 = 5x_1 - x_2 + x_3$
$\quad w_2 = -x_1 + x_2 + 7x_3$
$\quad w_3 = 2x_1 - 4x_2 - x_3$

(d) $w_1 = x_1^2 - 3x_2 + x_3 - 2x_4$
$\quad w_2 = 3x_1 - 4x_2 - x_3^2 + x_4$

2. Find the standard matrix for the linear transformation defined by the equations.

(a) $w_1 = 2x_1 - 3x_2 + x_4$
$\quad w_2 = 3x_1 + 5x_2 - x_4$

(b) $w_1 = 7x_1 + 2x_2 - 8x_3$
$\quad w_2 = \quad - x_2 + 5x_3$
$\quad w_3 = 4x_1 + 7x_2 - x_3$

(c) $w_1 = -x_1 + x_2$
$\quad w_2 = 3x_1 - 2x_2$
$\quad w_3 = 5x_1 - 7x_2$

(d) $w_1 = x_1$
$\quad w_2 = x_1 + x_2$
$\quad w_3 = x_1 + x_2 + x_3$
$\quad w_4 = x_1 + x_2 + x_3 + x_4$

3. Find the standard matrix for the linear transformation $T: R^3 \to R^3$ given by

$$w_1 = 3x_1 + 5x_2 - x_3$$
$$w_2 = 4x_1 - x_2 + x_3$$
$$w_3 = 3x_1 + 2x_2 - x_3$$

and calculate $T(-1, 2, 4)$ by directly substituting in the equations and by matrix multiplication.

4. Find the standard matrix for the linear operator T defined by the formula.

(a) $T(x_1, x_2) = (2x_1 - x_2, x_1 + x_2)$
(b) $T(x_1, x_2) = (x_1, x_2)$
(c) $T(x_1, x_2, x_3) = (x_1 + 2x_2 + x_3, x_1 + 5x_2, x_3)$
(d) $T(x_1, x_2, x_3) = (4x_1, 7x_2, -8x_3)$

5. Find the standard matrix for the linear transformation T defined by the formula.

(a) $T(x_1, x_2) = (x_2, -x_1, x_1 + 3x_2, x_1 - x_2)$
(b) $T(x_1, x_2, x_3, x_4) = (7x_1 + 2x_2 - x_3 + x_4, x_2 + x_3, -x_1)$
(c) $T(x_1, x_2, x_3) = (0, 0, 0, 0, 0)$
(d) $T(x_1, x_2, x_3, x_4) = (x_4, x_1, x_3, x_2, x_1 - x_3)$

6. In each part the standard matrix $[T]$ of a linear transformation T is given. Use it to find $T(x)$. [Express the answers in matrix form.]

(a) $[T] = \begin{bmatrix} 1 & 2 \\ 3 & 4 \end{bmatrix}$; $\mathbf{x} = \begin{bmatrix} 3 \\ -2 \end{bmatrix}$

(b) $[T] = \begin{bmatrix} -1 & 2 & 0 \\ 3 & 1 & 5 \end{bmatrix}$; $\mathbf{x} = \begin{bmatrix} -1 \\ 1 \\ 3 \end{bmatrix}$

(c) $[T] = \begin{bmatrix} -2 & 1 & 4 \\ 3 & 5 & 7 \\ 6 & 0 & -1 \end{bmatrix}$; $\mathbf{x} = \begin{bmatrix} x_1 \\ x_2 \\ x_3 \end{bmatrix}$

(d) $[T] = \begin{bmatrix} -1 & 1 \\ 2 & 4 \\ 7 & 8 \end{bmatrix}$; $\mathbf{x} = \begin{bmatrix} x_1 \\ x_2 \end{bmatrix}$

7. In each part use the standard matrix for T to find $T(x)$; then check the result by calculating $T(\mathbf{x})$ directly.

(a) $T(x_1, x_2) = (-x_1 + x_2, x_2)$; $\mathbf{x} = (-1, 4)$
(b) $T(x_1, x_2, x_3) = (2x_1 - x_2 + x_3, x_2 + x_3, 0)$; $\mathbf{x} = (2, 1, -3)$

8. Use matrix multiplication to find the reflection of $(-1, 2)$ about
   (a) the *x*-axis   (b) the *y*-axis   (c) the line $y = x$

9. Use matrix multiplication to find the reflection of $(2, -5, 3)$ about
   (a) the *xy*-plane   (b) the *xz*-plane   (c) the *yz*-plane

10. Use matrix multiplication to find the orthogonal projection of $(2, -5)$ on
    (a) the *x*-axis   (b) the *y*-axis

11. Use matrix multiplication to find the orthogonal projection of $(-2, 1, 3)$ on
    (a) the *xy*-plane   (b) the *xz*-plane   (c) the *yz*-plane

12. Use matrix multiplication to find the image of the vector $(3, -4)$ when it is rotated through an angle of
    (a) $\theta = 30°$ (b) $\theta = -60°$ (c) $\theta = 45°$ (d) $\theta = 90°$

13. Use matrix multiplication to find the image of the vector $(-2, 1, 2)$ if it is rotated
    (a) counterclockwise 30° about the *x*-axis
    (b) counterclockwise 45° about the *y*-axis
    (c) counterclockwise 90° about the *z*-axis

14. Find the standard matrix for the linear operator that rotates a vector in R^3 clockwise through an angle of $-60°$ about
    (a) the *x*-axis   (b) the *y*-axis   (c) the *z*-axis

15. Use matrix multiplication to find the image of the vector $(-2, 1, 2)$ if it is rotated
    (a) clockwise $-30°$ about the *x*-axis
    (b) clockwise $-45°$ about the *y*-axis
    (c) clockwise $-90°$ about the *z*-axis

16. Find the standard matrix for the stated composition of linear operators on R^2.
    (a) A counterclockwise rotation of 90°, followed by a reflection about the line $y = x$.
    (b) An orthogonal projection on the *y*-axis, followed by a contraction with factor $k = \frac{1}{2}$.
    (c) A reflection about the *x*-axis, followed by a dilation with factor $k = 3$.

17. Find the standard matrix for the stated composition of linear operators on R^2.
    (a) A counterclockwise rotation of 60°, followed by an orthogonal projection on the *x*-axis, followed by a reflection about the line $y = x$.
    (b) A dilation with factor $k = 2$, followed by a counterclockwise rotation of 45°, followed by a reflection about the *y*-axis.
    (c) A counterclockwise rotation of 15°, followed by a counterclockwise rotation of 105°, followed by a counterclockwise rotation of 60°.

18. Find the standard matrix for the stated composition of linear operators on R^3.
    (a) A reflection about the *yz*-plane, followed by an orthogonal projection on the *xz*-plane.
    (b) A counterclockwise rotation of 45° about the *y*-axis, followed by a dilation with factor $k = \sqrt{2}$.
    (c) An orthogonal projection on the *xy*-plane, followed by a reflection about the *yz*-plane.

19. Find the standard matrix for the stated composition of linear operators on R^3.
    (a) A counterclockwise rotation of 30° about the *x*-axis, followed by a counterclockwise rotation of 30° about the *z*-axis, followed by a contraction with factor $k = \frac{1}{4}$.
    (b) A reflection about the *xy*-plane, followed by a reflection about the *xz*-plane, followed by an orthogonal projection on the *yz*-plane.
    (c) A counterclockwise rotation of 270° about the *x*-axis, followed by a counterclockwise rotation of 90° about the *y*-axis, followed by a rotation of 180° about the *z*-axis.

20. Determine whether $T_1 \circ T_2 = T_2 \circ T_1$.

(a) $T_1 : R^2 \to R^2$ is the orthogonal projection on the x-axis and $T_2 : R^2 \to R^2$ is the orthogonal projection on the y-axis.

(b) $T_1 : R^2 \to R^2$ is the counterclockwise rotation through an angle θ_1 and $T_2 : R^2 \to R^2$ is the counterclockwise rotation through an angle θ_2.

(c) $T_1 : R^2 \to R^2$ is the reflection about the x-axis and $T_2 : R^2 \to R^2$ is the reflection about the y-axis.

(d) $T_1 : R^2 \to R^2$ is the orthogonal projection on the x-axis and $T_2 : R^2 \to R^2$ is the counterclockwise rotation through an angle θ.

21. Determine whether $T_1 \circ T_2 = T_2 \circ T_1$.

(a) $T_1 : R^3 \to R^3$ is a dilation by a factor k and $T_2 : R^3 \to R^3$ is the counterclockwise rotation about the z-axis through an angle θ.

(b) $T_1 : R^3 \to R^3$ is the rotation about the x-axis through an angle θ_1 and $T_2 : R^3 \to R^3$ is the rotation about the z-axis through an angle θ_2.

22. In R^3 the ***orthogonal projections*** on the x-axis, y-axis, and z-axis are defined by

$$T_1(x,y,z) = (x,0,0), \qquad T_2(x,y,z) = (0,y,0), \qquad T_3(x,y,z) = (0,0,z)$$

respectively.

(a) Show that the orthogonal projections on the coordinate axes are linear operators and find their standard matrices.

(b) Show that if $T : R^3 \to R^3$ is an orthogonal projection on one of the coordinate axes, then for every vector $\mathbf{x}$ in R^3, the vectors $T(\mathbf{x})$ and $\mathbf{x} - T(\mathbf{x})$ are orthogonal vectors.

(c) Make a sketch showing $\mathbf{x}$ and $\mathbf{x} - T(\mathbf{x})$ in the case where T is the orthogonal projection on the x-axis.

23. Derive the standard matrices for the counterclockwise rotations about the x-axis, y-axis, and z-axis in R^3 from Formula (17).

24. Use Formula (17) to find the standard matrix for a rotation $90°$ counterclockwise about the axis determined by the vector $\mathbf{v} = (1, 1, 1)$. [***Note.*** Formula (17) requires that the vector defining the axis of rotation have length 1.]

25. Verify Formula (21) for the given linear transformations.

(a) $T_1(x_1, x_2) = (x_1 + x_2, x_1 - x_2)$ and $T_2(x_1, x_2) = (3x_1, 2x_1 + 4x_2)$

(b) $T_1(x_1, x_2) = (4x_1, -2x_1 + x_2, -x_1 - 3x_2)$ and $T_2(x_1, x_2, x_3) = (x_1 + 2x_2 - x_3, 4x_1 - x_3)$

(c) $T_1(x_1, x_2, x_3) = (-x_1 + x_2, -x_2 + x_3, -x_3 + x_1)$ and $T_2(x_1, x_2, x_3) = (-2x_1, 3x_3, -4x_2)$

26. It can be proved that if A is a 2×2 matrix with $\det(A) = 1$ and such that the column vectors of A are orthogonal and have length 1, then multiplication by A is a counterclockwise rotation through some angle θ. Verify that

$$A = \begin{bmatrix} -1/\sqrt{2} & -1/\sqrt{2} \\ 1/\sqrt{2} & -1/\sqrt{2} \end{bmatrix}$$

satisfies the stated conditions and find the angle of rotation.

27. The result stated in Exercise 26 is also true in R^3: It can be proved that if A is a 3×3 matrix with $\det(A) = 1$ and such that the column vectors of A are pairwise orthogonal and have length 1, then multiplication by A is a counterclockwise rotation about some axis of rotation through some angle θ. Use Formula (17) to show that if A satisfies the stated conditions, then the angle of rotation satisfies the equation

$$\cos\theta = \frac{\mathrm{tr}(A) - 1}{2}$$

28. Let A be a 3×3 matrix satisfying the conditions stated in Exercise 27. It can be shown that if **x** is *any* vector in R^3, then the vector

$$\mathbf{u} = A\mathbf{x} + A^T\mathbf{x} + [1 - \mathrm{tr}(A)]\mathbf{x}$$

determines an axis of rotation when **u** is positioned with its initial point at the origin. [See *The Axis of Rotation: Analysis, Algebra, Geometry,* by Dan Kalman, *Mathematics Magazine,* Vol. 62, No. 4, Oct. 1989.]

(a) Show that multiplication by

$$A = \begin{bmatrix} \frac{1}{9} & -\frac{4}{9} & \frac{8}{9} \\ \frac{8}{9} & \frac{4}{9} & \frac{1}{9} \\ -\frac{4}{9} & \frac{7}{9} & \frac{4}{9} \end{bmatrix}$$

is a rotation.

(b) Find a vector of length 1 that defines an axis for the rotation.

(c) Use the result in Exercise 27 to find the counterclockwise angle of rotation about the axis obtained in part (b).

4.3 PROPERTIES OF LINEAR TRANSFORMATIONS FROM R^n TO R^m

In this section we shall investigate the relationship between the invertibility of a matrix and properties of the corresponding matrix transformation. We shall also obtain a characterization of linear transformations from R^n to R^m that will form the basis for more general linear transformations that will be discussed in later sections, and we shall discuss some geometric properties of eigenvectors.

ONE-TO-ONE LINEAR TRANSFORMATIONS

Linear transformations that map distinct vectors (or points) into distinct vectors (or points) are of special importance. One example of such a transformation is the linear operator $T: R^2 \to R^2$ that rotates each vector through an angle θ. It is obvious geometrically that if **u** and **v** are distinct vectors in R^2, then so are the rotated vectors $T(\mathbf{u})$ and $T(\mathbf{v})$ (Figure 1).

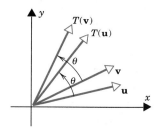

Figure 1   Distinct vectors **u** and **v** are rotated into distinct vectors $T(\mathbf{u})$ and $T(\mathbf{v})$.

In contrast, if $T: R^3 \rightarrow R^3$ is the orthogonal projection of R^3 on the *xy*-plane, then distinct points on the same vertical line get mapped into the same point in the *xy*-plane (Figure 2).

Figure 2 | The distinct points P and Q are mapped into the same point M.

Definition. A linear transformation $T: R^n \rightarrow R^m$ is said to be ***one-to-one*** if T maps distinct vectors (points) in R^n into distinct vectors (points) in R^m.

REMARK. It follows from this definition that for each vector **w** in the range of a one-to-one linear transformation T, there is exactly one vector **x** such that $T(\mathbf{x}) = \mathbf{w}$.

Example 1 In the terminology of the preceding definition, the rotation operator of Figure 1 is one-to-one, but the orthogonal projection operator of Figure 2 is not. ▲

Let A be an $n \times n$ matrix, and let $T_A: R^n \rightarrow R^n$ be multiplication by A. We shall now investigate relationships between the invertibility of A and properties of T_A.

Recall from Theorem 2.3.6 (with **w** in place of **b**) that the following are equivalent:

- A is invertible.
- $A\mathbf{x} = \mathbf{w}$ is consistent for every $n \times 1$ matrix **w**.
- $A\mathbf{x} = \mathbf{w}$ has exactly one solution for every $n \times 1$ matrix **w**.

However, the last of these statements is actually stronger than necessary. One can show that the following are equivalent (Exercise 24):

- A is invertible.
- $A\mathbf{x} = \mathbf{w}$ is consistent for every $n \times 1$ matrix **w**.
- $A\mathbf{x} = \mathbf{w}$ has exactly one solution when the system is consistent.

Translating these into the corresponding statements about the linear operator T_A, we deduce that the following are equivalent:

- A is invertible.
- For every vector **w** in R^n, there is some vector **x** in R^n such that $T_A(\mathbf{x}) = \mathbf{w}$. Stated another way, the range of T_A is all of R^n.
- For every vector **w** in the range of T_A, there is exactly one vector **x** in R^n such that $T_A(\mathbf{x}) = \mathbf{w}$. Stated another way, T_A is one-to-one.

In summary, we have established the following theorem about linear operators on R^n.

> **Theorem 4.3.1.** *if A is an $n \times n$ matrix and $T_A : R^n \to R^n$ is multiplication by A, then the following statements are equivalent.*
>
> (a) *A is invertible.*
> (b) *The range of T_A is R^n.*
> (c) *T_A is one-to-one.*

Example 2   In Example 1 we observed that the rotation operator $T : R^2 \to R^2$ illustrated in Figure 1 is one-to-one. It follows from Theorem 4.3.1 that the range of T must be all of R^2 and that the standard matrix for T must be invertible. To show that the range of T is all of R^2 we must show that every vector $\mathbf{w}$ in R^2 is the image of some vector $\mathbf{x}$ under T. But this is clearly so, since the vector $\mathbf{x}$ obtained by rotating $\mathbf{w}$ through the angle $-\theta$ maps into $\mathbf{w}$ when rotated through the angle θ. Moreover, from Table 6 of Section 4.2, the standard matrix for T is

$$[\,T\,] = \begin{bmatrix} \cos\theta & -\sin\theta \\ \sin\theta & \cos\theta \end{bmatrix}$$

which is invertible, since

$$\det[\,T\,] = \begin{vmatrix} \cos\theta & -\sin\theta \\ \sin\theta & \cos\theta \end{vmatrix} = \cos^2\theta + \sin^2\theta = 1 \neq 0 \quad \blacktriangle$$

Example 3   In Example 1 we observed that the projection operator $T : R^3 \to R^3$ illustrated in Figure 2 is not one-to-one. It follows from Theorem 4.3.1 that the range of T is *not* all of R^3 and the standard matrix for T is not invertible. To show that the range of T is not all of R^3 we must find a vector $\mathbf{w}$ in R^3 that is not the image of any vector $\mathbf{x}$ under T. But any vector $\mathbf{w}$ outside of the xy-plane has this property, since all images under T lie in the xy-plane. Moreover, from Table 5 of Section 4.2, the standard matrix for T is

$$[\,T\,] = \begin{bmatrix} 1 & 0 & 0 \\ 0 & 1 & 0 \\ 0 & 0 & 0 \end{bmatrix}$$

which is not invertible, since $\det[\,T\,] = 0$. $\blacktriangle$

INVERSE OF A ONE-TO-ONE LINEAR OPERATOR

If $T_A : R^n \to R^n$ is a one-to-one linear operator, then from Theorem 4.3.1 the matrix A is invertible. Thus, $T_{A^{-1}} : R^n \to R^n$ is itself a linear operator; it is called the *inverse of T_A*. The linear operators T_A and $T_{A^{-1}}$ cancel the effect of one another in the sense that for all $\mathbf{x}$ in R^n

$$T_A(T_{A^{-1}}(\mathbf{x})) = AA^{-1}\mathbf{x} = I\mathbf{x} = \mathbf{x}$$
$$T_{A^{-1}}(T_A(\mathbf{x})) = A^{-1}A\mathbf{x} = I\mathbf{x} = \mathbf{x}$$

or equivalently,

$$T_A \circ T_{A^{-1}} = T_{AA^{-1}} = T_I$$
$$T_{A^{-1}} \circ T_A = T_{A^{-1}A} = T_I$$

From a more geometric viewpoint, if **w** is the image of **x** under T_A, then $T_{A^{-1}}$ maps **w** back into **x**, since

$$T_{A^{-1}}(\mathbf{w}) = T_{A^{-1}}(T_A(\mathbf{x})) = \mathbf{x}$$

(Figure 3).

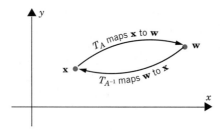

Figure 3

Before turning to an example, it will be helpful to touch on a notational matter. When a one-to-one linear operator on R^n is written as $T: R^n \rightarrow R^n$ (rather than $T_A: R^n \rightarrow R^n$), then the inverse of the operator T is denoted by T^{-1} (rather than $T_{A^{-1}}$). Since the standard matrix for T^{-1} is the inverse of the standard matrix for T, we have

$$[T^{-1}] = [T]^{-1} \tag{1}$$

Example 4 Let $T: R^2 \rightarrow R^2$ be the operator that rotates each vector in R^2 through the angle θ; so from Table 6 of Section 4.2

$$[T] = \begin{bmatrix} \cos\theta & -\sin\theta \\ \sin\theta & \cos\theta \end{bmatrix} \tag{2}$$

It is evident geometrically that to undo the effect of T one must rotate each vector in R^2 through the angle $-\theta$. But this is exactly what the operator T^{-1} does, since the standard matrix for T^{-1} is

$$[T^{-1}] = [T]^{-1} = \begin{bmatrix} \cos\theta & \sin\theta \\ -\sin\theta & \cos\theta \end{bmatrix} = \begin{bmatrix} \cos(-\theta) & -\sin(-\theta) \\ \sin(-\theta) & \cos(-\theta) \end{bmatrix}$$

(verify), which is identical to (2) except that θ is replaced by $-\theta$. ▲

Example 5 Show that the linear operator $T: R^2 \rightarrow R^2$ defined by the equations

$$w_1 = 2x_1 + x_2$$
$$w_2 = 3x_1 + 4x_2$$

is one-to-one, and find $T^{-1}(w_1, w_2)$.

Solution. The matrix form of these equations is

$$\begin{bmatrix} w_1 \\ w_2 \end{bmatrix} = \begin{bmatrix} 2 & 1 \\ 3 & 4 \end{bmatrix} \begin{bmatrix} x_1 \\ x_2 \end{bmatrix}$$

so the standard matrix for T is

$$[T] = \begin{bmatrix} 2 & 1 \\ 3 & 4 \end{bmatrix}$$

This matrix is invertible (so T is one-to-one) and the standard matrix for T^{-1} is

$$[T^{-1}] = [T]^{-1} = \begin{bmatrix} \frac{4}{5} & -\frac{1}{5} \\ -\frac{3}{5} & \frac{2}{5} \end{bmatrix}$$

Thus,

$$[T^{-1}]\begin{bmatrix} w_1 \\ w_2 \end{bmatrix} = \begin{bmatrix} \frac{4}{5} & -\frac{1}{5} \\ -\frac{3}{5} & \frac{2}{5} \end{bmatrix}\begin{bmatrix} w_1 \\ w_2 \end{bmatrix} = \begin{bmatrix} \frac{4}{5}w_1 - \frac{1}{5}w_2 \\ -\frac{3}{5}w_1 + \frac{2}{5}w_2 \end{bmatrix}$$

from which we conclude that

$$T^{-1}(w_1, w_2) = (\tfrac{4}{5}w_1 - \tfrac{1}{5}w_2, \ -\tfrac{3}{5}w_1 + \tfrac{2}{5}w_2) \quad \blacktriangle$$

LINEARITY PROPERTIES

In the preceding section we defined a transformation $T: R^n \to R^m$ to be linear if the equations relating $\mathbf{x}$ and $\mathbf{w} = T(\mathbf{x})$ are linear equations. The following theorem provides an alternative characterization of linearity. This theorem is fundamental and will be the basis for extending the concept of a linear transformation to more general settings later in this text.

Theorem 4.3.2. *A transformation $T: R^n \to R^m$ is linear if and only if the following relationships hold for all vectors $\mathbf{u}$ and $\mathbf{v}$ in R^n and every scalar c.*

(a) $T(\mathbf{u} + \mathbf{v}) = T(\mathbf{u}) + T(\mathbf{v})$
(b) $T(c\mathbf{u}) = cT(\mathbf{u})$

Proof. Assume first that T is a linear transformation, and let A be the standard matrix for T. It follows from the basic arithmetic properties of matrices that

$$T(\mathbf{u} + \mathbf{v}) = A(\mathbf{u} + \mathbf{v}) = A\mathbf{u} + A\mathbf{v} = T(\mathbf{u}) + T(\mathbf{v})$$

and

$$T(c\mathbf{u}) = A(c\mathbf{u}) = c(A\mathbf{u}) = cT(\mathbf{u})$$

Conversely, assume that properties (*a*) and (*b*) hold for the transformation *T*. We can prove that *T* is linear by finding a matrix *A* with the property that

$$T(\mathbf{x}) = A\mathbf{x} \tag{3}$$

for all vectors **x** in R^n. This will show that *T* is multiplication by *A* and therefore linear. But before we can produce this matrix we need to observe that property (*a*) can be extended to three or more terms; for example, if **u**, **v**, and **w** are any vectors in R^n, then by first grouping **v** and **w** and applying property (a) we obtain

$$T(\mathbf{u} + \mathbf{v} + \mathbf{w}) = T(\mathbf{u} + (\mathbf{v} + \mathbf{w})) = T(\mathbf{u}) + T(\mathbf{v} + \mathbf{w}) = T(\mathbf{u}) + T(\mathbf{v}) + T(\mathbf{w})$$

More generally, for any vectors $\mathbf{v}_1, \mathbf{v}_2, \ldots, \mathbf{v}_k$ in R^n we have

$$T(\mathbf{v}_1 + \mathbf{v}_2 + \cdots + \mathbf{v}_k) = T(\mathbf{v}_1) + T(\mathbf{v}_2) + \cdots + T(\mathbf{v}_k)$$

Now, to find the matrix *A* let $\mathbf{e}_1, \mathbf{e}_2, \ldots, \mathbf{e}_n$ be the vectors

$$\mathbf{e}_1 = \begin{bmatrix} 1 \\ 0 \\ 0 \\ \vdots \\ 0 \end{bmatrix}, \quad \mathbf{e}_2 = \begin{bmatrix} 0 \\ 1 \\ 0 \\ \vdots \\ 0 \end{bmatrix}, \quad \ldots, \quad \mathbf{e}_n = \begin{bmatrix} 0 \\ 0 \\ 0 \\ \vdots \\ 1 \end{bmatrix} \tag{4}$$

and let *A* be the matrix whose successive column vectors are $T(\mathbf{e}_1), T(\mathbf{e}_2), \ldots, T(\mathbf{e}_n)$; that is,

$$A = [\, T(\mathbf{e}_1) \; \vdots \; T(\mathbf{e}_2) \; \vdots \; \cdots \; \vdots \; T(\mathbf{e}_n) \,] \tag{5}$$

If

$$\mathbf{x} = \begin{bmatrix} x_1 \\ x_2 \\ \vdots \\ x_n \end{bmatrix}$$

is any vector in R^n, then as discussed in Section 1.3, the product *A***x** is a linear combination of the column vectors of *A* with coefficients from **x**, so that

$$
\begin{aligned}
A\mathbf{x} &= x_1 T(\mathbf{e}_1) + x_2 T(\mathbf{e}_2) + \cdots + x_n T(\mathbf{e}_n) \\
&= T(x_1 \mathbf{e}_1) + T(x_2 \mathbf{e}_2) + \cdots + T(x_n \mathbf{e}_n) \qquad \boxed{\text{Property } (b)} \\
&= T(x_1 \mathbf{e}_1 + x_2 \mathbf{e}_2 + \cdots + x_n \mathbf{e}_n) \qquad \boxed{\text{Property } (a) \text{ for } n \text{ terms}} \\
&= T(\mathbf{x})
\end{aligned}
$$

which completes the proof. ▊

Expression (5) is important in its own right, since it provides an explicit formula for the standard matrix of a linear operator $T: R^n \to R^m$ in terms of the images of the vectors $\mathbf{e}_1, \mathbf{e}_2, \ldots, \mathbf{e}_n$ under *T*. For reasons that will be discussed later, the vectors

$\mathbf{e}_1, \mathbf{e}_2, \ldots, \mathbf{e}_n$ in (4) are called the **standard basis** vectors for R^n. In R^2 and R^3 these are the vectors of length 1 along the coordinate axes (Figure 4).

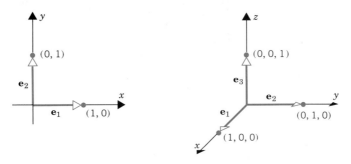

Figure 4 | Standard basis for R^2 | Standard basis for R^3

Because of its importance we shall state (5) as a theorem for future reference.

Theorem 4.3.3. *If $T: R^n \to R^m$ is a linear transformation, and $\mathbf{e}_1, \mathbf{e}_2, \ldots, \mathbf{e}_n$ are the standard basis vectors for R^n, then the standard matrix for T is*

$$[T] = [T(\mathbf{e}_1) \mid T(\mathbf{e}_2) \mid \cdots \mid T(\mathbf{e}_n)] \tag{6}$$

Formula (6) is a powerful tool for finding standard matrices and analyzing the geometric effect of a linear transformation. For example, suppose that $T: R^3 \to R^3$ is the orthogonal projection on the xy-plane. Referring to Figure 4, it is evident geometrically that

$$T(\mathbf{e}_1) = \mathbf{e}_1 = \begin{bmatrix} 1 \\ 0 \\ 0 \end{bmatrix}, \qquad T(\mathbf{e}_2) = \mathbf{e}_2 = \begin{bmatrix} 0 \\ 1 \\ 0 \end{bmatrix}, \qquad T(\mathbf{e}_3) = \mathbf{0} = \begin{bmatrix} 0 \\ 0 \\ 0 \end{bmatrix}$$

so by (6)

$$[T] = \begin{bmatrix} 1 & 0 & 0 \\ 0 & 1 & 0 \\ 0 & 0 & 0 \end{bmatrix}$$

which agrees with the result in Table 5.

Using (6) another way, suppose that $T_A: R^3 \to R^2$ is multiplication by

$$A = \begin{bmatrix} -1 & 2 & 1 \\ 3 & 0 & 6 \end{bmatrix}$$

The images of the standard basis vectors can be read directly from the columns of the matrix A:

$$T_A\left(\begin{bmatrix} 1 \\ 0 \\ 0 \end{bmatrix} \right) = \begin{bmatrix} -1 \\ 3 \end{bmatrix}, \qquad T_A\left(\begin{bmatrix} 0 \\ 1 \\ 0 \end{bmatrix} \right) = \begin{bmatrix} 2 \\ 0 \end{bmatrix}, \qquad T_A\left(\begin{bmatrix} 0 \\ 0 \\ 1 \end{bmatrix} \right) = \begin{bmatrix} 1 \\ 6 \end{bmatrix}$$

Example 6 Let l be the line in the xy-plane that passes through the origin and makes an angle θ with the positive x-axis, where $0 \le \theta < \pi$. As illustrated in Figure 5a, let $T: R^2 \to R^2$ be a linear operator that maps each vector into its orthogonal projection on l.

(a) Find the standard matrix for T.
(b) Find the orthogonal projection of the vector $\mathbf{x} = (1, 5)$ onto the line through the origin that makes an angle of $\theta = \pi/6$ with the positive x-axis.

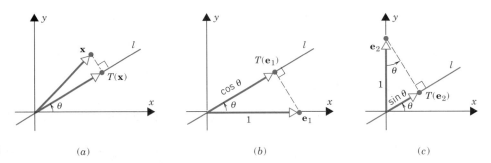

Figure 5   (a)   (b)   (c)

Solution (a). From (6)

$$[\,T\,] = [\,T(\mathbf{e}_1) \mid T(\mathbf{e}_2)\,]$$

where $\mathbf{e}_1$ and $\mathbf{e}_2$ are the standard basis vectors for R^2. We consider the case where $0 \le \theta \le \pi/2$; the case where $\pi/2 < \theta < \pi$ is similar. Referring to Figure 5b, we have $\|T(\mathbf{e}_1)\| = \cos\,\theta$, so

$$T(\mathbf{e}_1) = \begin{bmatrix} \|T(\mathbf{e}_1)\| \cos\theta \\ \|T(\mathbf{e}_1)\| \sin\theta \end{bmatrix} = \begin{bmatrix} \cos^2\theta \\ \sin\theta\cos\theta \end{bmatrix}$$

and referring to Figure 5c, we have $\|T(\mathbf{e}_2)\| = \sin\,\theta$, so

$$T(\mathbf{e}_2) = \begin{bmatrix} \|T(\mathbf{e}_2)\| \cos\theta \\ \|T(\mathbf{e}_2)\| \sin\theta \end{bmatrix} = \begin{bmatrix} \sin\theta\cos\theta \\ \sin^2\theta \end{bmatrix}$$

Thus, the standard matrix for T is

$$[\,T\,] = \begin{bmatrix} \cos^2\theta & \sin\theta\cos\theta \\ \sin\theta\cos\theta & \sin^2\theta \end{bmatrix}$$

Solution (b). Since $\sin \pi/6 = 1/2$ and $\cos \pi/6 = \sqrt{3}/2$, it follows from part (a) that the standard matrix for this projection operator is

$$[\,T\,] = \begin{bmatrix} 3/4 & \sqrt{3}/4 \\ \sqrt{3}/4 & 1/4 \end{bmatrix}$$

Thus,

$$T\left(\begin{bmatrix} 1 \\ 5 \end{bmatrix}\right) = \begin{bmatrix} 3/4 & \sqrt{3}/4 \\ \sqrt{3}/4 & 1/4 \end{bmatrix}\begin{bmatrix} 1 \\ 5 \end{bmatrix} = \begin{bmatrix} \dfrac{3 + 5\sqrt{3}}{4} \\ \dfrac{\sqrt{3} + 5}{4} \end{bmatrix}$$

or in horizontal notation

$$T(1,5) = \left(\frac{3 + 5\sqrt{3}}{4}, \frac{\sqrt{3} + 5}{4}\right) \quad \blacktriangle$$

GEOMETRIC INTERPRETATION OF EIGENVECTORS

Recall from Section 2.3 that if A is an $n \times n$ matrix, then λ is called an *eigenvalue* of A if there is a nonzero vector $\mathbf{x}$ such that

$$A\mathbf{x} = \lambda\mathbf{x} \qquad \text{or equivalently} \qquad (\lambda I - A)\mathbf{x} = \mathbf{0}$$

The nonzero vectors $\mathbf{x}$ satisfying this equation are called the *eigenvectors* of A corresponding to λ.

Eigenvalues and eigenvectors can also be defined for linear operators on R^n; the definitions parallel those for matrices.

Definition. If $T: R^n \to R^n$ is a linear operator, then a scalar λ is called an *eigenvalue of T* if there is a nonzero $\mathbf{x}$ in R^n such that

$$T(\mathbf{x}) = \lambda\mathbf{x} \qquad\qquad (7)$$

Those nonzero vectors $\mathbf{x}$ that satisfy this equation are called the *eigenvectors of T corresponding to λ*.

Observe that if A is the standard matrix for T, then (7) can be written as

$$A\mathbf{x} = \lambda\mathbf{x}$$

from which it follows that

- The eigenvalues of T are precisely the eigenvalues of its standard matrix A.
- $\mathbf{x}$ is an eigenvector of T corresponding to λ if and only if $\mathbf{x}$ is an eigenvector of A corresponding to λ.

If λ is an eigenvalue of A and $\mathbf{x}$ is a corresponding eigenvector, then $A\mathbf{x} = \lambda\mathbf{x}$, so multiplication by A maps $\mathbf{x}$ into a scalar multiple of itself. In R^2 and R^3, this means that *multiplication by A maps each eigenvector $\mathbf{x}$ into a vector that lies on the same line as $\mathbf{x}$* (Figure 6).

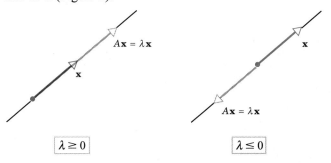

Figure 6

Recall from Section 4.2 that if $\lambda \geq 0$, then the linear operator $A\mathbf{x} = \lambda\mathbf{x}$ compresses $\mathbf{x}$ by a factor of λ if $0 \leq \lambda \leq 1$ or stretches $\mathbf{x}$ by a factor of λ if $\lambda \geq 1$. If $\lambda < 0$, then $A\mathbf{x} = \lambda\mathbf{x}$ reverses the direction of $\mathbf{x}$, and compresses the reversed vector by a factor of $|\lambda|$ if $0 \leq |\lambda| \leq 1$ or stretches the reversed vector by a factor of $|\lambda|$ if $|\lambda| \geq 1$ (Figure 7).

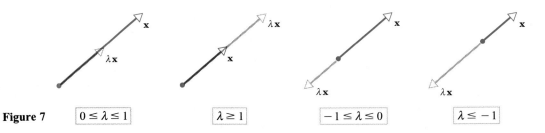

Figure 7 $\boxed{0 \leq \lambda \leq 1}$ $\boxed{\lambda \geq 1}$ $\boxed{-1 \leq \lambda \leq 0}$ $\boxed{\lambda \leq -1}$

Example 7   Let $T: R^2 \rightarrow R^2$ be the linear operator that rotates each vector through an angle θ. It is evident geometrically that unless θ is a multiple of π, then T does not map any nonzero vector $\mathbf{x}$ onto the same line as $\mathbf{x}$; consequently, T has no real eigenvalues. But if θ *is* a multiple of π, then every nonzero vector $\mathbf{x}$ is mapped onto the same line as $\mathbf{x}$, so *every* nonzero vector is an eigenvector of T. Let us verify these geometric observations algebraically. The standard matrix for T is

$$A = \begin{bmatrix} \cos\theta & -\sin\theta \\ \sin\theta & \cos\theta \end{bmatrix}$$

As discussed in Section 2.3, the eigenvalues of this matrix are the solutions of the characteristic equation

$$\det(\lambda I - A) = \begin{vmatrix} \lambda - \cos\theta & \sin\theta \\ -\sin\theta & \lambda - \cos\theta \end{vmatrix} = 0$$

that is,

$$(\lambda - \cos\theta)^2 + \sin^2\theta = 0 \tag{8}$$

But if θ is not a multiple of π, then $\sin^2\theta > 0$, so this equation has no real solution for λ and consequently A has no real eigenvectors.* If θ *is* a multiple of π, then $\sin\theta = 0$ and either $\cos\theta = 1$ or $\cos\theta = -1$, depending on the particular multiple of π. In the case where $\sin\theta = 0$ and $\cos\theta = 1$, the characteristic equation (8) becomes $(\lambda - 1)^2 = 0$, so $\lambda = 1$ is the only eigenvalue of A. In this case the matrix A is

$$A = \begin{bmatrix} 1 & 0 \\ 0 & 1 \end{bmatrix} = I$$

---

* There are applications that require complex scalars and vectors with complex components. In such cases, complex eigenvalues and eigenvectors with complex components are allowed. However, these have no direct geometric significance here. In later chapters we will discuss such eigenvalues and eigenvectors, but until explicitly stated otherwise, it will be assumed that only real eigenvalues and eigenvectors with real components are to be considered.

Thus, for all $\mathbf{x}$ in R^2

$$T(\mathbf{x}) = A\mathbf{x} = I\mathbf{x} = \mathbf{x}$$

so T maps every vector to itself, and hence to the same line. In the case where $\sin \theta = 0$ and $\cos \theta = -1$, the characteristic equation (8) becomes $(\lambda + 1)^2 = 0$, so that $\lambda = -1$ is the only eigenvalue of A. In this case the matrix A is

$$A = \begin{bmatrix} -1 & 0 \\ 0 & -1 \end{bmatrix} = -I$$

Thus, for all $\mathbf{x}$ in R^2

$$T(\mathbf{x}) = A\mathbf{x} = -I\mathbf{x} = -\mathbf{x}$$

so T maps every vector to its negative, and hence to the same line as x. ▲

Example 8 Let $T: R^3 \rightarrow R^3$ be the orthogonal projection on the xy-plane. Vectors in the xy-plane are mapped into themselves under T, so each nonzero vector in the xy-plane is an eigenvector corresponding to the eigenvalue $\lambda = 1$. Every vector $\mathbf{x}$ along the z-axis is mapped into $\mathbf{0}$ under T, which is on the same line as $\mathbf{x}$, so every nonzero vector on the z-axis is an eigenvector corresponding to the eigenvalue $\lambda = 0$. Vectors not in the xy-plane or along the z-axis are not mapped into scalar multiples of themselves, so there are no other eigenvectors or eigenvalues.

To verify these geometric observations algebraically, recall from Table 5 of Section 4.2 that the standard matrix for T is

$$A = \begin{bmatrix} 1 & 0 & 0 \\ 0 & 1 & 0 \\ 0 & 0 & 0 \end{bmatrix}$$

The characteristic equation of A is

$$\det(\lambda I - A) = \begin{vmatrix} \lambda - 1 & 0 & 0 \\ 0 & \lambda - 1 & 0 \\ 0 & 0 & \lambda \end{vmatrix} = 0$$

or

$$(\lambda - 1)^2 \lambda = 0$$

which has the solutions $\lambda = 0$ and $\lambda = 1$ anticipated above.

As discussed in Section 2.3, the eigenvectors of the matrix A corresponding to an eigenvalue λ are the nonzero solutions of

$$\begin{bmatrix} \lambda - 1 & 0 & 0 \\ 0 & \lambda - 1 & 0 \\ 0 & 0 & \lambda \end{bmatrix} \begin{bmatrix} x_1 \\ x_2 \\ x_3 \end{bmatrix} = \begin{bmatrix} 0 \\ 0 \\ 0 \end{bmatrix} \tag{9}$$

If $\lambda = 0$, this system is

$$\begin{bmatrix} -1 & 0 & 0 \\ 0 & -1 & 0 \\ 0 & 0 & 0 \end{bmatrix} \begin{bmatrix} x_1 \\ x_2 \\ x_3 \end{bmatrix} = \begin{bmatrix} 0 \\ 0 \\ 0 \end{bmatrix}$$

which has the solutions $x_1 = 0$, $x_2 = 0$, $x_3 = t$ (verify), or in matrix form

$$\begin{bmatrix} x_1 \\ x_2 \\ x_3 \end{bmatrix} = \begin{bmatrix} 0 \\ 0 \\ t \end{bmatrix}$$

As anticipated, these are the vectors along the z-axis. If $\lambda = 1$, then system (9) is

$$\begin{bmatrix} 0 & 0 & 0 \\ 0 & 0 & 0 \\ 0 & 0 & 1 \end{bmatrix} \begin{bmatrix} x_1 \\ x_2 \\ x_3 \end{bmatrix} = \begin{bmatrix} 0 \\ 0 \\ 0 \end{bmatrix}$$

which has the solutions $x_1 = s$, $x_2 = t$, $x_3 = 0$ (verify), or in matrix form,

$$\begin{bmatrix} x_1 \\ x_2 \\ x_3 \end{bmatrix} = \begin{bmatrix} s \\ t \\ 0 \end{bmatrix}$$

As anticipated, these are the vectors in the xy-plane. ▲

SUMMARY

In Theorem 2.3.6 we listed six results that are equivalent to the invertibility of a matrix A. We conclude this section by merging Theorem 4.3.1 with that list to produce the following theorem that relates all of the major topics we have studied thus far.

> **Theorem 4.3.4.** *If A is an $n \times n$ matrix, and if $T_A : R^n \to R^n$ is multiplication by A, then the following are equivalent.*
>
> (a) *A is invertible.*
> (b) *$A\mathbf{x} = \mathbf{0}$ has only the trivial solution.*
> (c) *The reduced row-echelon form of A is I_n.*
> (d) *A is expressible as a product of elementary matrices.*
> (e) *$A\mathbf{x} = \mathbf{b}$ is consistent for every $n \times 1$ matrix $\mathbf{b}$.*
> (f) *$A\mathbf{x} = \mathbf{b}$ has exactly one solution for every $n \times 1$ matrix $\mathbf{b}$.*
> (g) *$\det(A) \neq 0$.*
> (h) *The range of T_A is R^n.*
> (i) *T_A is one-to-one.*

EXERCISE SET 4.3

1. By inspection, determine whether the linear operator is one-to-one.
   (a) the orthogonal projection on the x-axis in R^2 (b) the reflection about the y-axis in R^2
   (c) the reflection about the line $y = x$ in R^2 (d) a contraction with factor $k > 0$ in R^2
   (e) a rotation about the z-axis in R^3 (f) a reflection about the xy-plane in R^3
   (g) a dilation with factor $k > 0$ in R^3

2. Find the standard matrix for the linear operator defined by the equations and use Theorem 4.3.1 to determine whether the operator is one-to-one.

   (a) $\begin{aligned} w_1 &= 8x_1 + 4x_2 \\ w_2 &= 2x_1 + x_2 \end{aligned}$
   (b) $\begin{aligned} w_1 &= 2x_1 - 3x_2 \\ w_2 &= 5x_1 + x_2 \end{aligned}$
   (c) $\begin{aligned} w_1 &= -x_1 + 3x_2 + 2x_3 \\ w_2 &= 2x_1 + 4x_3 \\ w_3 &= x_1 + 3x_2 + 6x_3 \end{aligned}$
   (d) $\begin{aligned} w_1 &= x_1 + 2x_2 + 3x_3 \\ w_2 &= 2x_1 + 5x_2 + 3x_3 \\ w_3 &= x_1 + 8x_3 \end{aligned}$

3. Show that the range of the linear operator defined by the equations

   $$w_1 = 4x_1 - 2x_2$$
   $$w_2 = 2x_1 - x_2$$

   is not all of R^2, and find a vector that is not in the range.

4. Show that the range of the linear operator defined by the equations

   $$w_1 = x_1 - 2x_2 + x_3$$
   $$w_2 = 5x_1 - x_2 + 3x_3$$
   $$w_3 = 4x_1 + x_2 + 2x_3$$

   is not R^3, and find a vector that is not in the range.

5. Determine whether the linear operator $T: R^2 \to R^2$ defined by the equations is one-to-one; if so, find the standard matrix for the inverse operator, and find $T^{-1}(w_1, w_2)$.

   (a) $\begin{aligned} w_1 &= x_1 + 2x_2 \\ w_2 &= -x_1 + x_2 \end{aligned}$
   (b) $\begin{aligned} w_1 &= 4x_1 - 6x_2 \\ w_2 &= -2x_1 + 3x_2 \end{aligned}$
   (c) $\begin{aligned} w_1 &= -x_2 \\ w_2 &= -x_1 \end{aligned}$
   (d) $\begin{aligned} w_1 &= 3x_1 \\ w_2 &= -5x_1 \end{aligned}$

6. Determine whether the linear operator $T: R^3 \to R^3$ defined by the equations is one-to-one; if so, find the standard matrix for the inverse operator, and find $T^{-1}(w_1, w_2, w_3)$.

   (a) $\begin{aligned} w_1 &= x_1 - 2x_2 + 2x_3 \\ w_2 &= 2x_1 + x_2 + x_3 \\ w_3 &= x_1 + x_2 \end{aligned}$
   (b) $\begin{aligned} w_1 &= x_1 - 3x_2 + 4x_3 \\ w_2 &= -x_1 + x_2 + x_3 \\ w_3 &= -2x_2 + 5x_3 \end{aligned}$

   (c) $\begin{aligned} w_1 &= x_1 + 4x_2 - x_3 \\ w_2 &= 2x_1 + 7x_2 + x_3 \\ w_3 &= x_1 + 3x_2 \end{aligned}$
   (d) $\begin{aligned} w_1 &= x_1 + 2x_2 + x_3 \\ w_2 &= -2x_1 + x_2 + 4x_3 \\ w_3 &= 7x_1 + 4x_2 - 5x_3 \end{aligned}$

7. By inspection, determine the inverse of the given one-to-one linear operator.
   (a) the reflection about the x-axis in R^2
   (b) the rotation through an angle of $\pi/4$ in R^2
   (c) the dilation by a factor of 3 in R^2
   (d) the reflection about the yz-plane in R^3
   (e) the contraction by a factor of $\frac{1}{5}$ in R^3

In Exercises 8 and 9 use Theorem 4.3.2 to determine whether $T: R^2 \to R^2$ is a linear operator.

8. (a) $T(x, y) = (2x, y)$ (b) $T(x, y) = (x^2, y)$ (c) $T(x, y) = (-y, x)$ (d) $T(x, y) = (x, 0)$

9. (a) $T(x, y) = (2x + y, x - y)$ (b) $T(x, y) = (x + 1, y)$
    (c) $T(x, y) = (y, y)$ (d) $T(x, y) = (\sqrt[3]{x}, \sqrt[3]{y})$

In Exercises 10 and 11 use Theorem 4.3.2 to determine whether $T: R^3 \to R^2$ is a linear transformation.

10. (a) $T(x, y, z) = (x, x + y + z)$ (b) $T(x, y, z) = (1, 1)$

11. (a) $T(x, y, z) = (0, 0)$ (b) $T(x, y, z) = (3x - 4y, 2x - 5z)$

12. In each part use Theorem 4.3.3 to find the standard matrix for the linear operator from the images of the standard basis vectors.
(a) the reflection operators on R^2 in Table 2 of Section 4.2
(b) the reflection operators on R^3 in Table 3 of Section 4.2
(c) the projection operators on R^2 in Table 4 of Section 4.2
(d) the projection operators on R^3 in Table 5 of Section 4.2
(e) the rotation operators on R^2 in Table 6 of Section 4.2
(f) the dilation and contraction operators on R^3 in Table 9 of Section 4.2

13. Use Theorem 4.3.3 to find the standard matrix for $T: R^2 \to R^2$ from the images of the standard basis vectors.
(a) $T: R^2 \to R^2$ projects a vector orthogonally onto the x-axis and then reflects that vector about the y-axis.
(b) $T: R^2 \to R^2$ reflects a vector about the line $y = x$ and then reflects that vector about the x-axis.
(c) $T: R^2 \to R^2$ dilates a vector by a factor of 3, then reflects that vector about the line $y = x$, and then projects that vector orthogonally onto the y-axis.

14. Use Theorem 4.3.3 to find the standard matrix for $T: R^3 \to R^3$ from the images of the standard basis vectors.
(a) $T: R^3 \to R^3$ reflects a vector about the xz-plane and then contracts that vector by a factor of $\frac{1}{5}$.
(b) $T: R^3 \to R^3$ projects a vector orthogonally onto the xz-plane and then projects that vector orthogonally onto the xy-plane.
(c) $T: R^3 \to R^3$ reflects a vector about the xy-plane, then reflects that vector about the xz-plane, and then reflects that vector about the yz-plane.

15. Let $T_A: R^3 \to R^3$ be multiplication by

$$A = \begin{bmatrix} -1 & 3 & 0 \\ 2 & 1 & 2 \\ 4 & 5 & -3 \end{bmatrix}$$

and let e_1, e_2, and e_3 be the standard basis vectors for R^3. Find the following vectors by inspection.
(a) $T_A(e_1)$, $T_A(e_2)$, and $T_A(e_3)$ (b) $T_A(e_1 + e_2 + e_3)$ (c) $T_A(7e_3)$

16. Determine whether multiplication by A is a one-to-one linear transformation.

(a) $A = \begin{bmatrix} 1 & -1 \\ 2 & 0 \\ 3 & -4 \end{bmatrix}$ (b) $A = \begin{bmatrix} 1 & 2 & 3 \\ -1 & 0 & -4 \end{bmatrix}$

17. Use the result in Example 6 to find the orthogonal projection of $\mathbf{x}$ onto the line through the origin that makes an angle θ with the positive x-axis.

(a) $\mathbf{x} = (-1, 2)$; $\theta = 45°$ (b) $\mathbf{x} = (1, 0)$; $\theta = 30°$ (c) $\mathbf{x} = (1, 5)$; $\theta = 120°$

18. Use the type of argument given in Example 8 to find the eigenvalues and corresponding eigenvectors of T. Check your conclusions by calculating the eigenvalues and corresponding eigenvectors from the standard matrix for T.

(a) $T:R^2 \rightarrow R^2$ is the reflection about the x-axis.
(b) $T:R^2 \rightarrow R^2$ is the reflection about the line $y = x$.
(c) $T:R^2 \rightarrow R^2$ is the orthogonal projection on the x-axis.
(d) $T:R^2 \rightarrow R^2$ is the contraction by a factor of $\frac{1}{2}$.

19. Follow the directions of Exercise 18.

(a) $T:R^3 \rightarrow R^3$ is the reflection about the yz-plane.
(b) $T:R^3 \rightarrow R^3$ is the orthogonal projection on the xz-plane.
(c) $T:R^3 \rightarrow R^3$ is the dilation by a factor of 2.
(d) $T:R^3 \rightarrow R^3$ is a counterclockwise rotation of $45°$ about the z-axis.

20. (a) Is a composition of one-to-one linear transformations one-to-one? Justify your conclusion.
(b) Can the composition of a one-to-one linear transformation and a linear transformation that is not one-to-one be one-to-one? Justify your conclusion.

21. Show that $T(x, y) = (0, 0)$ defines a linear operator on R^2 but $T(x, y) = (1, 1)$ does not.

22. Prove that if $T:R^n \rightarrow R^m$ is a linear transformation, then $T(\mathbf{0}) = \mathbf{0}$; that is, T maps the zero vector in R^n into the zero vector in R^m.

23. Let l be the line in the xy-plane that passes through the origin and makes an angle θ with the positive x-axis, where $0 \le \theta < \pi$. Let $T:R^2 \rightarrow R^2$ be the linear operator that reflects each vector about l (Figure 8).

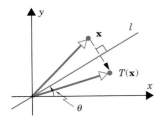

Figure 8

(a) Use the method of Example 6 to find the standard matrix for T.
(b) Find the reflection of the vector $\mathbf{x} = (1, 5)$ about the line l through the origin that makes an angle of $\theta = 30°$ with the positive x-axis.

24. Prove: An $n \times n$ matrix A is invertible if and only if the linear system $A\mathbf{x} = \mathbf{w}$ has exactly one solution for every vector $\mathbf{w}$ in R^n for which the system is consistent.

CHAPTER **5**

GENERAL VECTOR SPACES

5.1 REAL VECTOR SPACES

In this section we shall generalize the concept of a vector still further. We shall state a set of axioms which, if satisfied by a class of objects, will entitle those objects to be called "vectors." The axioms will be chosen by abstracting the most important properties of vectors in R^n; as a consequence, vectors in R^n will automatically satisfy these axioms. Thus, our new concept of a vector will include our old vectors and many new kinds of vectors as well. These new types of vectors will include, among other things, various kinds of matrices and functions. Our work in this section is not an idle exercise in theoretical mathematics; it will provide a powerful tool for extending our geometric visualization to a wide variety of important mathematical problems where geometric intuition would not otherwise be available. Briefly stated, the idea is this: We can visualize vectors in R^2 and R^3 geometrically as arrows, which enables us to draw physical or mental pictures to help solve problems. Because the axioms that we will be using to create our new kinds of vectors will be based on properties of vectors in R^2 and R^3, these new vectors will have many of the familiar properties of vectors in R^2 and R^3. Consequently, when we want to solve a problem involving our new kinds of vectors, say matrices or functions, we may be able to get a foothold on the problem by visualizing geometrically what the corresponding problem would be like in R^2 and R^3.

VECTOR SPACE AXIOMS

Definition. Let V be an arbitrary nonempty set of objects on which two operations are defined, addition and multiplication by scalars (numbers). By **addition** we mean a rule for associating with each pair of objects **u** and **v** in V an object **u** + **v**, called the **sum** of **u** and **v**; by **scalar multiplication** we mean a rule for associating with each scalar k and each object **u** in V an object $k\mathbf{u}$, called the **scalar multiple** of **u** by k. If the following axioms are satisfied by all objects **u**, **v**, **w** in V and all scalars k and l, then we call V a **vector space** and we call the objects in V **vectors**.

(1) If **u** and **v** are objects in V, then $\mathbf{u} + \mathbf{v}$ is in V.

(2) $\mathbf{u} + \mathbf{v} = \mathbf{v} + \mathbf{u}$

(3) $\mathbf{u} + (\mathbf{v} + \mathbf{w}) = (\mathbf{u} + \mathbf{v}) + \mathbf{w}$

(4) There is an object **0** in V, called a ***zero vector*** for V, such that $\mathbf{0} + \mathbf{u} = \mathbf{u} + \mathbf{0} = \mathbf{u}$ for all **u** in V.

(5) For each **u** in V, there is an object $-\mathbf{u}$ in V, called a ***negative*** of **u**, such that $\mathbf{u} + (-\mathbf{u}) = (-\mathbf{u}) + \mathbf{u} = \mathbf{0}$.

(6) If k is any scalar and **u** is any object in V, then $k\mathbf{u}$ is in V.

(7) $k(\mathbf{u} + \mathbf{v}) = k\mathbf{u} + k\mathbf{v}$

(8) $(k + l)\mathbf{u} = k\mathbf{u} + l\mathbf{u}$

(9) $k(l\mathbf{u}) = (kl)(\mathbf{u})$

(10) $1\mathbf{u} = \mathbf{u}$

REMARK. Depending on the application, scalars may be real numbers or complex numbers. Vector spaces in which the scalars are complex numbers are called ***complex vector spaces***, and those in which the scalars must be real are called ***real vector spaces***. In Chapter 10 we shall discuss complex vector spaces; until then, *all of our scalars will be real numbers.*

The reader should keep in mind that the definition of a vector space specifies neither the nature of the vectors nor the operations. Any kind of object can be a vector, and the operations of addition and scalar multiplication may not have any relationship or similarity to the standard vector operations on R^n. The only requirement is that the ten vector space axioms be satisfied. Some authors use the notations $\oplus$ and $\odot$ for vector addition and scalar multiplication to distinguish these operations from addition and multiplication of real numbers; we will not use this convention, however.

EXAMPLES OF VECTOR SPACES

The following examples will illustrate the variety of possible vector spaces. In each example we will specify a nonempty set V and two operations: addition and scalar multiplication; then we shall verify that the ten vector space axioms are satisfied, thereby entitling V, with the specified operations, to be called a vector space.

Example 1 The set $V = R^n$ with the standard operations of addition and scalar multiplication defined in Section 4.1 is a vector space. Axioms 1 and 6 follow from the definitions of the standard operations on R^n; the remaining axioms follow from Theorem 4.1.1. ▲

The three most important special cases of R^n are R (the real numbers), R^2 (the vectors in the plane), and R^3 (the vectors in 3-space).

Example 2 Show that the set V of all 2×2 matrices with real entries is a vector space if vector addition is defined to be matrix addition and vector scalar multiplication is defined to be matrix scalar multiplication.

Solution. In this example we will find it convenient to verify the axioms in the following order: 1, 6, 2, 3, 7, 8, 9, 4, 5, and 10. Let

$$\mathbf{u} = \begin{bmatrix} u_{11} & u_{12} \\ u_{21} & u_{22} \end{bmatrix} \quad \text{and} \quad \mathbf{v} = \begin{bmatrix} v_{11} & v_{12} \\ v_{21} & v_{22} \end{bmatrix}$$

To prove Axiom 1, we must show that $\mathbf{u} + \mathbf{v}$ is an object in V; that is, we must show that $\mathbf{u} + \mathbf{v}$ is a 2×2 matrix. But this follows from the definition of matrix addition, since

$$\mathbf{u} + \mathbf{v} = \begin{bmatrix} u_{11} & u_{12} \\ u_{21} & u_{22} \end{bmatrix} + \begin{bmatrix} v_{11} & v_{12} \\ v_{21} & v_{22} \end{bmatrix} = \begin{bmatrix} u_{11} + v_{11} & u_{12} + v_{12} \\ u_{21} + v_{21} & u_{22} + v_{22} \end{bmatrix}$$

Similarly, Axiom 6 holds because for any real number k we have

$$k\mathbf{u} = k\begin{bmatrix} u_{11} & u_{12} \\ u_{21} & u_{22} \end{bmatrix} = \begin{bmatrix} ku_{11} & ku_{12} \\ ku_{21} & ku_{22} \end{bmatrix}$$

so that $k\mathbf{u}$ is a 2×2 matrix and consequently is an object in V.

Axiom 2 follows from Theorem 1.4.1a since

$$\mathbf{u} + \mathbf{v} = \begin{bmatrix} u_{11} & u_{12} \\ u_{21} & u_{22} \end{bmatrix} + \begin{bmatrix} v_{11} & v_{12} \\ v_{21} & v_{22} \end{bmatrix} = \begin{bmatrix} v_{11} & v_{12} \\ v_{21} & v_{22} \end{bmatrix} + \begin{bmatrix} u_{11} & u_{12} \\ u_{21} & u_{22} \end{bmatrix} = \mathbf{v} + \mathbf{u}$$

Similarly, Axiom 3 follows for part (b) of that theorem; and Axioms 7, 8, and 9 follow from parts (h), (j), and (l), respectively.

To prove Axiom 4, we must find an object $\mathbf{0}$ in V such that $\mathbf{0} + \mathbf{u} = \mathbf{u} + \mathbf{0} = \mathbf{u}$ for all $\mathbf{u}$ in V. This can be done by defining $\mathbf{0}$ to be

$$\mathbf{0} = \begin{bmatrix} 0 & 0 \\ 0 & 0 \end{bmatrix}$$

With this definition

$$\mathbf{0} + \mathbf{u} = \begin{bmatrix} 0 & 0 \\ 0 & 0 \end{bmatrix} + \begin{bmatrix} u_{11} & u_{12} \\ u_{21} & u_{22} \end{bmatrix} = \begin{bmatrix} u_{11} & u_{12} \\ u_{21} & u_{22} \end{bmatrix} = \mathbf{u}$$

and similarly $\mathbf{u} + \mathbf{0} = \mathbf{u}$. To prove Axiom 5 we must show that each object $\mathbf{u}$ in V has a negative $-\mathbf{u}$ such that $\mathbf{u} + (-\mathbf{u}) = \mathbf{0}$ and $(-\mathbf{u}) + \mathbf{u} = \mathbf{0}$. This can be done by defining the negative of $\mathbf{u}$ to be

$$-\mathbf{u} = \begin{bmatrix} -u_{11} & -u_{12} \\ -u_{21} & -u_{22} \end{bmatrix}$$

With this definition

$$\mathbf{u} + (-\mathbf{u}) = \begin{bmatrix} u_{11} & u_{12} \\ u_{21} & u_{22} \end{bmatrix} + \begin{bmatrix} -u_{11} & -u_{12} \\ -u_{21} & -u_{22} \end{bmatrix} = \begin{bmatrix} 0 & 0 \\ 0 & 0 \end{bmatrix} = \mathbf{0}$$

and similarly $(-\mathbf{u}) + \mathbf{u} = \mathbf{0}$. Finally, Axiom 10 is a simple computation:

$$1\mathbf{u} = 1\begin{bmatrix} u_{11} & u_{12} \\ u_{21} & u_{22} \end{bmatrix} = \begin{bmatrix} u_{11} & u_{12} \\ u_{21} & u_{22} \end{bmatrix} = \mathbf{u} \quad \blacktriangle$$

Example 3 Example 2 is a special case of a more general class of vector spaces. The arguments in that example can be adapted to show that the set V of all $m \times n$ matrices with real entries, together with the operations of matrix addition and scalar multiplication, is a vector space. The $m \times n$ zero matrix is the zero vector **0**, and if **u** is the $m \times n$ matrix U, then the matrix $-U$ is the negative $-\mathbf{u}$ of the vector **u**. We shall denote this vector space by the symbol M_{mn}. ▲

Example 4 Let V be the set of real-valued functions defined on the entire real line $(-\infty, \infty)$. If $\mathbf{f} = f(x)$ and $\mathbf{g} = g(x)$ are two such functions and k is any real number, define the sum function $\mathbf{f} + \mathbf{g}$ and the scalar multiple $k\mathbf{f}$ by

$$(\mathbf{f} + \mathbf{g})(x) = f(x) + g(x)$$
$$(k\mathbf{f})(x) = kf(x)$$

In other words, the value of the function $\mathbf{f} + \mathbf{g}$ at x is obtained by adding together the values of **f** and **g** at x (Figure 1a). Similarly, the value of $k\mathbf{f}$ at x is k times the value of **f** at x (Figure 1b). In the exercises we shall ask you to show that V is a vector space with respect to these operations. This vector space is denoted by $F(-\infty, \infty)$. If **f** and **g** are vectors in this space, then to say that $\mathbf{f} = \mathbf{g}$ is equivalent to saying that $f(x) = g(x)$ for all x in the interval $(-\infty, \infty)$.

The vector **0** in $F(-\infty, \infty)$ is the constant function that is identically zero for all values of x. The graph of this function is the line that coincides with the x-axis. The negative of a vector **f** is the function $-\mathbf{f} = -f(x)$. Geometrically, the graph of $-\mathbf{f}$ is the reflection of the graph of **f** across the x-axis (Figure 1c). ▲

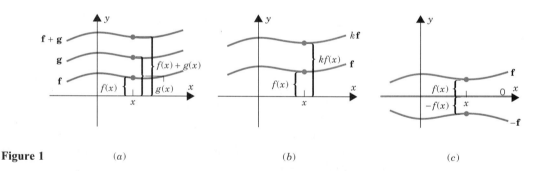

Figure 1 (a) (b) (c)

REMARK. In the preceding example we focused attention on the interval $(-\infty, \infty)$. Had we restricted our attention to some closed interval $[a, b]$ or some open interval (a, b), the functions defined on those intervals with the operations stated in the example would also have produced vector spaces. Those vector spaces are denoted by $F[a, b]$ and $F(a, b)$, respectively.

Example 5 Let $V = R^2$ and define addition and scalar multiplication operations as follows: If $\mathbf{u} = (u_1, u_2)$ and $\mathbf{v} = (v_1, v_2)$, then define

$$\mathbf{u} + \mathbf{v} = (u_1 + v_1, u_2 + v_2)$$

and if k is any real number, then define

$$k\mathbf{u} = (ku_1, 0)$$

For example, if $\mathbf{u} = (2, 4)$, $\mathbf{v} = (-3, 5)$, and $k = 7$, then

$$\mathbf{u} + \mathbf{v} = (2 + (-3), 4 + 5) = (-1, 9)$$
$$k\mathbf{u} = 7\mathbf{u} = (7 \cdot 2, 0) = (14, 0)$$

The addition operation is the standard addition operation on R^2, but the scalar multiplication operation is not the standard scalar multiplication. In the exercises we will ask you to show that the first nine vector space axioms are satisfied; however, there are values of $\mathbf{u}$ for which Axiom 10 fails to hold. For example, if $\mathbf{u} = (u_1, u_2)$ is such that $u_2 \neq 0$, then

$$1\mathbf{u} = 1(u_1, u_2) = (1 \cdot u_1, 0) = (u_1, 0) \neq \mathbf{u}$$

Thus, V is not a vector space with the stated operations. ▲

Example 6 Let V be any plane through the origin in R^3. We shall show that the points in V form a vector space under the standard addition and scalar multiplication operations for vectors in R^3. From Example 1, we know that R^3 itself is a vector space under these operations. Thus, Axioms 2, 3, 7, 8, 9, and 10 hold for all points in R^3 and consequently for all points in the plane V. We therefore need only show that Axioms 1, 4, 5, and 6 are satisfied.

Since the plane V passes through the origin, it has an equation of the form

$$ax + by + cz = 0 \tag{1}$$

(Theorem 3.5.1). Thus, if $\mathbf{u} = (u_1, u_2, u_3)$ and $\mathbf{v} = (v_1, v_2, v_3)$ are points in V, then $au_1 + bu_2 + cu_3 = 0$ and $av_1 + bv_2 + cv_3 = 0$. Adding these equations gives

$$a(u_1 + v_1) + b(u_2 + v_2) + c(u_3 + v_3) = 0$$

This equality tells us that the coordinates of the point

$$\mathbf{u} + \mathbf{v} = (u_1 + v_1, u_2 + v_2, u_3 + v_3)$$

satisfy (1); thus, $\mathbf{u} + \mathbf{v}$ lies in the plane V. This proves that Axiom 1 is satisfied. The verifications of Axioms 4 and 6 are left as exercises; however, we shall prove Axiom 5. Multiplying $au_1 + bu_2 + cu_3 = 0$ through by -1 gives

$$a(-u_1) + b(-u_2) + c(-u_3) = 0$$

Thus, $-\mathbf{u} = (-u_1, -u_2, -u_3)$ lies in V. This establishes Axiom 5. ▲

Example 7 Let V consist of a single object, which we denote by $\mathbf{0}$, and define

$$\mathbf{0} + \mathbf{0} = \mathbf{0}$$
$$k\mathbf{0} = \mathbf{0}$$

for all scalars k. It is easy to check that all the vector space axioms are satisfied. We call this the ***zero vector space***. ▲

SOME PROPERTIES OF VECTORS

As we progress, we shall add more examples of vector spaces to our list. We conclude this section with a theorem that gives a useful list of vector properties.

> **Theorem 5.1.1.** *Let V be a vector space, **u** a vector in V, and k a scalar; then:*
> (a) $0\mathbf{u} = \mathbf{0}$
> (b) $k\mathbf{0} = \mathbf{0}$
> (c) $(-1)\mathbf{u} = -\mathbf{u}$
> (d) *If* $k\mathbf{u} = \mathbf{0}$, *then* $k = 0$ *or* $\mathbf{u} = \mathbf{0}$.

We shall prove parts (a) and (c) and leave proofs of the remaining parts as exercises.

Proof (a). We can write

$$0\mathbf{u} + 0\mathbf{u} = (0 + 0)\mathbf{u} \qquad [\text{Axiom 8}]$$
$$= 0\mathbf{u} \qquad [\text{Property of the number 0}]$$

By Axiom 5 the vector $0\mathbf{u}$ has a negative, $-0\mathbf{u}$. Adding this negative to both sides above yields

$$[0\mathbf{u} + 0\mathbf{u}] + (-0\mathbf{u}) = 0\mathbf{u} + (-0\mathbf{u})$$

or

$$0\mathbf{u} + [0\mathbf{u} + (-0\mathbf{u})] = 0\mathbf{u} + (-0\mathbf{u}) \qquad [\text{Axiom 3}]$$
$$0\mathbf{u} + \mathbf{0} = \mathbf{0} \qquad [\text{Axiom 5}]$$
$$0\mathbf{u} = \mathbf{0} \qquad [\text{Axiom 4}]$$

Proof (c). To show $(-1)\mathbf{u} = -\mathbf{u}$, we must demonstrate that $\mathbf{u} + (-1)\mathbf{u} = \mathbf{0}$. To see this, observe that

$$\mathbf{u} + (-1)\mathbf{u} = 1\mathbf{u} + (-1)\mathbf{u} \qquad [\text{Axiom 10}]$$
$$= (1 + (-1))\mathbf{u} \qquad [\text{Axiom 8}]$$
$$= 0\mathbf{u} \qquad [\text{Property of numbers}]$$
$$= \mathbf{0} \qquad [\text{Part (a) above}]$$

EXERCISE SET 5.1

In Exercises 1–13 a set of objects is given together with operations of addition and scalar multiplication. Determine which sets are vector spaces under the given operations. For those that are not, list all axioms that fail to hold.

1. The set of all triples of real numbers (x, y, z) with the operations

$$(x, y, z) + (x', y', z') = (x + x', y + y', z + z') \text{ and } k(x, y, z) = (kx, y, z)$$

2. The set of all triples of real numbers (x, y, z) with the operations

$$(x, y, z) + (x', y', z') = (x + x', y + y', z + z') \quad \text{and} \quad k(x, y, z) = (0, 0, 0)$$

3. The set of all pairs of real numbers (x, y) with the operations

$$(x, y) + (x', y') = (x + x', y + y') \quad \text{and} \quad k(x, y) = (2kx, 2ky)$$

4. The set of all real numbers x with the standard operations of addition and multiplication.

5. The set of all pairs of real numbers of the form $(x, 0)$ with the standard operations on R^2.

6. The set of all pairs of real numbers of the form (x, y), where $x \geq 0$, with the standard operations on R^2.

7. The set of all n-tuples of real numbers of the form $(x, x, \ldots, x)$ with the standard operations on R^n.

8. The set of all pairs of real numbers (x, y) with the operations

$$(x, y) + (x', y') = (x + x' + 1, y + y' + 1) \quad \text{and} \quad k(x, y) = (kx, ky)$$

9. The set of all 2×2 matrices of the form

$$\begin{bmatrix} a & 1 \\ 1 & b \end{bmatrix}$$

with matrix addition and scalar multiplication.

10. The set of all 2×2 matrices of the form

$$\begin{bmatrix} a & 0 \\ 0 & b \end{bmatrix}$$

with matrix addition and scalar multiplication.

11. The set of all real-valued functions f defined everywhere on the real line and such that $f(1) = 0$, with the operations defined in Example 4.

12. The set of all 2×2 matrices of the form

$$\begin{bmatrix} a & a + b \\ a + b & b \end{bmatrix}$$

with matrix addition and scalar multiplication.

13. The set whose only element is the moon. The operations are moon + moon = moon and $k(\text{moon}) = \text{moon}$, where k is a real number.

14. Prove that a line passing through the origin in R^3 is a vector space under the standard operations on R^3.

15. Show that the set of all positive real numbers with the operations

$$x + y = xy \quad \text{and} \quad kx = x^k$$

is a vector space.

16. Complete the unfinished details in Example 4.

17. Complete the unfinished details in Example 6.

18. Prove part (b) of Theorem 5.1.1.

19. Prove part (d) of Theorem 5.1.1.

20. Prove that a vector space cannot have more than one zero vector.

21. Prove that a vector has exactly one negative.

22. Show that the first nine vector space axioms are satisfied if $V = R^2$ has the addition and scalar multiplication operations defined in Example 5.

5.2 SUBSPACES

It is possible for one vector space to be contained within a larger vector space. For example, we showed in the preceding section that planes through the origin are vector spaces that are contained within the larger vector space R^3. In this section we shall study this important concept in more detail.

DEFINITION OF A SUBSPACE

> **Definition.** A subset W of a vector space V is called a ***subspace*** of V if W is itself a vector space under the addition and scalar multiplication defined on V.

In general, one must verify the ten vector space axioms to show that a set W with addition and scalar multiplication forms a vector space. However, if W is part of a larger set V that is already known to be a vector space, then certain axioms need not be verified for W because they are "inherited" from V. For example, there is no need to check that $\mathbf{u} + \mathbf{v} = \mathbf{v} + \mathbf{u}$ (Axiom 2) for W because this holds for all vectors in V and consequently for all vectors in W. Other axioms inherited by W from V are 3, 7, 8, 9, and 10. Thus, to show that a set W is a subspace of a vector space V, we need only verify Axioms 1, 4, 5, and 6. The following theorem shows that even Axioms 4 and 5 can be dispensed with.

> **Theorem 5.2.1.** *If W is a set of one or more vectors from a vector space V, then W is a subspace of V if and only if the following conditions hold.*
>
> *(a) If $\mathbf{u}$ and $\mathbf{v}$ are vectors in W, then $\mathbf{u} + \mathbf{v}$ is in W.*
> *(b) If k is any scalar and $\mathbf{u}$ is any vector in W, then $k\mathbf{u}$ is in W.*

Proof. If W is a subspace of V, then all the vector space axioms are satisfied; in particular, Axioms 1 and 6 hold. But these are precisely conditions (*a*) and (*b*).

Conversely, assume conditions (*a*) and (*b*) hold. Since these conditions are vector space Axioms 1 and 6, we need only show that W satisfies the remaining eight axioms. Axioms 2, 3, 7, 8, 9, and 10 are automatically satisfied by the vectors in W since they are satisfied by all vectors in V. Therefore, to complete the proof, we need only verify that Axioms 4 and 5 are satisfied by vectors in W.

Let $\mathbf{u}$ be any vector in W. By condition (*b*), $k\mathbf{u}$ is in W for every scalar k. Setting $k = 0$, it follows from Theorem 5.1.1 that $0\mathbf{u} = \mathbf{0}$ is in W, and setting $k = -1$, it follows that $(-1)\mathbf{u} = -\mathbf{u}$ is in W. ▮

REMARK. A set W of one or more vectors from a vector space V is said to be ***closed under addition*** if condition (*a*) in Theorem 5.2.1 holds and ***closed under scalar multiplication*** if condition (*b*) holds. Thus, Theorem 5.2.1 states that W is a subspace of V *if and only if W is closed under addition and closed under scalar multiplication.*

EXAMPLES OF SUBSPACES

Example 1   In Example 6 of Section 5.1 we verified the ten vector space axioms to show that the points in a plane through the origin of R^3 form a subspace of R^3. In light of Theorem 5.2.1 we can see that much of that work was unnecessary; it would have been sufficient to verify that the plane is closed under addition and scalar multiplication (Axioms 1 and 6). In Section 5.1 we verified those two axioms algebraically; however,

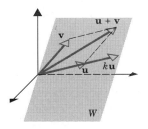

they can also be proved geometrically as follows: Let W be any plane through the origin and let **u** and **v** be any vectors in W. Then **u** + **v** must lie in W because it is the diagonal of the parallelogram determined by **u** and **v** (Figure 1), and k**u** must lie in W for any scalar k because k**u** lies on a line through **u**. Thus, W is closed under addition and scalar multiplication, so it is a subspace of R^3. ▲

The vectors **u** + **v** and k**u** both lie in the same plane as **u** and **v**.

Figure 1

Example 2 Show that a line through the origin of R^3 is a subspace of R^3.

Solution. Let W be a line through the origin of R^3. It is evident geometrically that the sum of two vectors on this line also lies on the line and that a scalar multiple of a vector on the line is on the line as well (Figure 2). Thus, W is closed under addition and scalar multiplication, so it is a subspace of R^3. In the exercises we will ask you to prove this result algebraically using parametric equations for the line. ▲

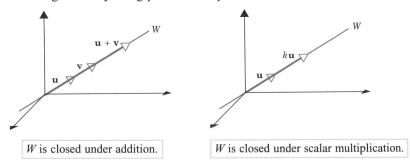

Figure 2

W is closed under addition.

W is closed under scalar multiplication.

Example 3 Let W be the set of all points (x, y) in R^2 such that $x \geq 0$ and $y \geq 0$. These are the points in the first quadrant. The set W is *not* a subspace of R^2 since it is not closed under scalar multiplication. For example, **v** = (1, 1) lies in W, but its negative (-1)**v** = $-$**v** = $(-1, -1)$ does not (Figure 3). ▲

Every nonzero vector space V has at least two subspaces: V itself is a subspace, and the set $\{\mathbf{0}\}$ consisting of just the zero vector in V is a subspace called the *zero subspace*. Combining this with Examples 1 and 2, we obtain the following list of subspaces of R^2 and R^3:

W is closed under scalar multiplication.

Figure 3

Subspaces of R^2

- $\{\mathbf{0}\}$
- Lines through the origin
- R^2

Subspaces of R^3

- $\{\mathbf{0}\}$
- Lines through the origin
- Planes through the origin
- R^3

Later, we will show that these are the only subspaces of R^2 and R^3.

Example 4 From Theorem 1.7.2, the sum of two symmetric matrices is symmetric, and a scalar multiple of a symmetric matrix is symmetric. Thus, the set of $n \times n$ symmetric matrices is a subspace of the vector space M_{nn} of all $n \times n$ matrices. Similarly, the set of $n \times n$ upper triangular matrices, the set of $n \times n$ lower triangular matrices, and the set of $n \times n$ diagonal matrices all form subspaces of M_{nn}, since each of these sets is closed under addition and scalar multiplication. ▲

Example 5 Let n be a positive integer, and let W consist of all functions expressible in the form

$$p(x) = a_0 + a_1 x + \cdots + a_n x^n \tag{1}$$

where $a_0, \ldots, a_n$ are real numbers. Thus, W consists of the zero function together with all real polynomials of degree n or less. The set W is a subspace of the vector space of all real-valued functions discussed in Example 4 of the preceding section. To see this, let $\mathbf{p}$ and $\mathbf{q}$ be the polynomials

$$p(x) = a_0 + a_1 x + \cdots + a_n x^n$$

and

$$q(x) = b_0 + b_1 x + \cdots + b_n x^n$$

Then

$$(\mathbf{p} + \mathbf{q})(x) = p(x) + q(x) = (a_0 + b_0) + (a_1 + b_1)x + \cdots + (a_n + b_n)x^n$$

and

$$(k\mathbf{p})(x) = kp(x) = (ka_0) + (ka_1)x + \cdots + (ka_n)x^n$$

These functions have the form given in (1), so $\mathbf{p} + \mathbf{q}$ and $k\mathbf{p}$ lie in W. We shall denote the vector space W in this example by the symbol P_n. ▲

Example 6 (***For readers who have studied calculus.***) Recall from calculus that if $\mathbf{f}$ and $\mathbf{g}$ are continuous functions on the interval $(-\infty, \infty)$ and k is a constant, then $\mathbf{f} + \mathbf{g}$ and $k\mathbf{f}$ are also continuous. Thus, the continuous functions on the interval $(-\infty, \infty)$ form a subspace of $F(-\infty, \infty)$, since they are closed under addition and scalar multiplication. We denote this subspace by $C(-\infty, \infty)$. Similarly, if $\mathbf{f}$ and $\mathbf{g}$ are differentiable functions, then so are $\mathbf{f} + \mathbf{g}$ and $k\mathbf{f}$. Thus, the functions with continuous first derivatives on $(-\infty, \infty)$ form a subspace of $F(-\infty, \infty)$. We denote this subspace by $C^1(-\infty, \infty)$, where the superscript 1 is used to emphasize the *first* derivative. However, it is a theorem of calculus that every differentiable function is continuous, so $C^1(-\infty, \infty)$ is actually a subspace of $C(-\infty, \infty)$.

To take this a step further, for each positive integer m, the functions with continuous mth derivatives on $(-\infty, \infty)$ form a subspace of $C^1(-\infty, \infty)$ as do the functions that have continuous derivatives of all orders. We denote the subspace of functions with continuous mth derivatives on $(-\infty, \infty)$ by $C^m(-\infty, \infty)$, and we denote the subspace of functions that have continuous derivatives of all orders on $(-\infty, \infty)$ by $C^\infty(-\infty, \infty)$. Finally, it is a theorem of calculus that polynomials have continuous derivatives of all orders, so P_n is a subspace of $C^\infty(-\infty, \infty)$. The hierarchy of subspaces discussed in this example is pictured in Figure 4. ▲

REMARK. In the preceding example we focused on the interval $(-\infty, \infty)$. Had we focused on a closed interval $[a, b]$, then the subspaces corresponding to those defined in the example would be denoted by $C[a, b]$, $C^m[a, b]$, and $C^\infty[a, b]$. Similarly, on an open interval (a, b) they would be denoted by $C(a, b)$, $C^m(a, b)$, and $C^\infty(a, b)$.

Figure 4

**SOLUTION
SPACES OF
HOMOGENEOUS
SYSTEMS**

If $A\mathbf{x} = \mathbf{b}$ is a system of linear equations, then each vector $\mathbf{x}$ that satisfies this equation is called a **solution vector** of the system. The following theorem shows that the solution vectors of a *homogeneous* linear system form a vector space, which we shall call the **solution space** of the system.

Theorem 5.2.2. *If $A\mathbf{x} = \mathbf{0}$ is a homogeneous linear system of m equations in n unknowns, then the set of solution vectors is a subspace of R^n.*

Proof. Let W be the set of solution vectors. There is at least one vector in W, namely $\mathbf{0}$. To show that W is closed under addition and scalar multiplication, we must show that if $\mathbf{x}$ and $\mathbf{x}'$ are any solution vectors and k is any scalar, then $\mathbf{x} + \mathbf{x}'$ and $k\mathbf{x}$ are also solution vectors. But if $\mathbf{x}$ and $\mathbf{x}'$ are solution vectors, then

$$A\mathbf{x} = \mathbf{0} \qquad \text{and} \qquad A\mathbf{x}' = \mathbf{0}$$

from which it follows that

$$A(\mathbf{x} + \mathbf{x}') = A\mathbf{x} + A\mathbf{x}' = \mathbf{0} + \mathbf{0} = \mathbf{0}$$

and

$$A(k\mathbf{x}) = kA\mathbf{x} = k\mathbf{0} = \mathbf{0}$$

which proves that $\mathbf{x} + \mathbf{x}'$ and $k\mathbf{x}$ are solution vectors. ▨

Example 7 Consider the linear systems

(a) $\begin{bmatrix} 1 & -2 & 3 \\ 2 & -4 & 6 \\ 3 & -6 & 9 \end{bmatrix} \begin{bmatrix} x \\ y \\ z \end{bmatrix} = \begin{bmatrix} 0 \\ 0 \\ 0 \end{bmatrix}$ (b) $\begin{bmatrix} 1 & -2 & 3 \\ -3 & 7 & -8 \\ -2 & 4 & -6 \end{bmatrix} \begin{bmatrix} x \\ y \\ z \end{bmatrix} = \begin{bmatrix} 0 \\ 0 \\ 0 \end{bmatrix}$

(c) $\begin{bmatrix} 1 & -2 & 3 \\ -3 & 7 & -8 \\ 4 & 1 & 2 \end{bmatrix} \begin{bmatrix} x \\ y \\ z \end{bmatrix} = \begin{bmatrix} 0 \\ 0 \\ 0 \end{bmatrix}$ (d) $\begin{bmatrix} 0 & 0 & 0 \\ 0 & 0 & 0 \\ 0 & 0 & 0 \end{bmatrix} \begin{bmatrix} x \\ y \\ z \end{bmatrix} = \begin{bmatrix} 0 \\ 0 \\ 0 \end{bmatrix}$

Each of these systems has three unknowns, so the solutions form subspaces of R^3. Geometrically, this means that each solution space must be a line through the origin, a plane through the origin, the origin only, or all of R^3. We shall now verify that this is so (leaving it to the reader to solve the systems).

Solution.

(*a*) The solutions are

$$x = 2s - 3t, \qquad y = s, \qquad z = t$$

from which it follows that

$$x = 2y - 3z \qquad \text{or} \qquad x - 2y + 3z = 0$$

This is the equation of the plane through the origin with $\mathbf{n} = (1, -2, 3)$ as a normal vector.

(*b*) The solutions are

$$x = -5t, \qquad y = -t, \qquad z = t$$

which are parametric equations for the line through the origin parallel to the vector $\mathbf{v} = (-5, -1, 1)$.

(*c*) The solution is $x = 0$, $y = 0$, $z = 0$, so the solution space is the origin only, that is, $\{\mathbf{0}\}$.

(*d*) The solutions are

$$x = r, \qquad y = s, \qquad z = t$$

where r, s, and t have arbitrary values, so the solution space is all of R^3. ▲

LINEAR COMBINATIONS OF VECTORS

In Section 1.3 we introduced the concept of a linear combination of column vectors. The following definition extends this idea to more general vectors.

> **Definition.** A vector $\mathbf{w}$ is called a *linear combination* of the vectors $\mathbf{v}_1, \mathbf{v}_2, \ldots, \mathbf{v}_r$ if it can be expressed in the form
>
> $$\mathbf{w} = k_1\mathbf{v}_1 + k_2\mathbf{v}_2 + \cdots + k_r\mathbf{v}_r$$
>
> where $k_1, k_2, \ldots, k_r$ are scalars.

REMARK. If $r = 1$, then the equation in the preceding definition reduces to $\mathbf{w} = k_1\mathbf{v}_1$; that is, $\mathbf{w}$ is a linear combination of a single vector $\mathbf{v}_1$ if it is a scalar multiple of $\mathbf{v}_1$.

Example 8 Every vector $\mathbf{v} = (a, b, c)$ in R^3 is expressible as a linear combination of the standard basis vectors

$$\mathbf{i} = (1, 0, 0), \qquad \mathbf{j} = (0, 1, 0), \qquad \mathbf{k} = (0, 0, 1)$$

since

$$\mathbf{v} = (a, b, c) = a(1, 0, 0) + b(0, 1, 0) + c(0, 0, 1) = a\mathbf{i} + b\mathbf{j} + c\mathbf{k} \quad ▲$$

Example 9 Consider the vectors $\mathbf{u} = (1, 2, -1)$ and $\mathbf{v} = (6, 4, 2)$ in R^3. Show that $\mathbf{w} = (9, 2, 7)$ is a linear combination of $\mathbf{u}$ and $\mathbf{v}$ and that $\mathbf{w}' = (4, -1, 8)$ is *not* a linear combination of $\mathbf{u}$ and $\mathbf{v}$.

Solution. In order for $\mathbf{w}$ to be a linear combination of $\mathbf{u}$ and $\mathbf{v}$, there must be scalars k_1 and k_2 such that $\mathbf{w} = k_1\mathbf{u} + k_2\mathbf{v}$; that is,

$$(9, 2, 7) = k_1(1, 2, -1) + k_2(6, 4, 2)$$

or

$$(9, 2, 7) = (k_1 + 6k_2, 2k_1 + 4k_2, -k_1 + 2k_2)$$

Equating corresponding components gives

$$k_1 + 6k_2 = 9$$
$$2k_1 + 4k_2 = 2$$
$$-k_1 + 2k_2 = 7$$

Solving this system yields $k_1 = -3$, $k_2 = 2$, so

$$\mathbf{w} = -3\mathbf{u} + 2\mathbf{v}$$

Similarly, for $\mathbf{w}'$ to be a linear combination of $\mathbf{u}$ and $\mathbf{v}$, there must be scalars k_1 and k_2 such that $\mathbf{w}' = k_1\mathbf{u} + k_2\mathbf{v}$; that is,

$$(4, -1, 8) = k_1(1, 2, -1) + k_2(6, 4, 2)$$

or

$$(4, -1, 8) = (k_1 + 6k_2, 2k_1 + 4k_2, -k_1 + 2k_2)$$

Equating corresponding components gives

$$k_1 + 6k_2 = \quad 4$$
$$2k_1 + 4k_2 = -1$$
$$-k_1 + 2k_2 = \quad 8$$

This system of equations is inconsistent (verify), so no such scalars k_1 and k_2 exist. Consequently, $\mathbf{w}'$ is not a linear combination of $\mathbf{u}$ and $\mathbf{v}$. ▲

SPANNING

If $\mathbf{v}_1, \mathbf{v}_2, \ldots, \mathbf{v}_r$ are vectors in a vector space V, then generally some vectors in V may be linear combinations of $\mathbf{v}_1, \mathbf{v}_2, \ldots, \mathbf{v}_r$ and others may not. The following theorem shows that if we construct a set W consisting of all those vectors that are expressible as linear combinations of $\mathbf{v}_1, \mathbf{v}_2, \ldots, \mathbf{v}_r$, then W forms a subspace of V.

Theorem 5.2.3. *If $\mathbf{v}_1, \mathbf{v}_2, \ldots, \mathbf{v}_r$ are vectors in a vector space V, then:*

(a) The set W of all linear combinations of $\mathbf{v}_1, \mathbf{v}_2, \ldots, \mathbf{v}_r$ is a subspace of V.

(b) W is the smallest subspace of V that contains $\mathbf{v}_1, \mathbf{v}_2, \ldots, \mathbf{v}_r$ in the sense that every other subspace of V that contains $\mathbf{v}_1, \mathbf{v}_2, \ldots, \mathbf{v}_r$ must contain W.

Proof (a). To show that W is a subspace of V, we must prove it is closed under addition and scalar multiplication. There is at least one vector in W, namely, $\mathbf{0}$, since $\mathbf{0} = 0\mathbf{v}_1 + 0\mathbf{v}_2 + \cdots + 0\mathbf{v}_r$. If $\mathbf{u}$ and $\mathbf{v}$ are vectors in W, then

$$\mathbf{u} = c_1\mathbf{v}_1 + c_2\mathbf{v}_2 + \cdots + c_r\mathbf{v}_r$$

and

$$\mathbf{v} = k_1\mathbf{v}_1 + k_2\mathbf{v}_2 + \cdots + k_r\mathbf{v}_r$$

where $c_1, c_2, \ldots, c_r, k_1, k_2, \ldots, k_r$ are scalars. Therefore,

$$\mathbf{u} + \mathbf{v} = (c_1 + k_1)\mathbf{v}_1 + (c_2 + k_2)\mathbf{v}_2 + \cdots + (c_r + k_r)\mathbf{v}_r$$

and, for any scalar k,

$$k\mathbf{u} = (kc_1)\mathbf{v}_1 + (kc_2)\mathbf{v}_2 + \cdots + (kc_r)\mathbf{v}_r$$

Thus, $\mathbf{u} + \mathbf{v}$ and $k\mathbf{u}$ are linear combinations of $\mathbf{v}_1, \mathbf{v}_2, \ldots, \mathbf{v}_r$ and consequently lie in W. Therefore, W is closed under addition and scalar multiplication.

Proof (b). Each vector $\mathbf{v}_i$ is a linear combination of $\mathbf{v}_1, \mathbf{v}_2, \ldots, \mathbf{v}_r$ since we can write

$$\mathbf{v}_i = 0\mathbf{v}_1 + 0\mathbf{v}_2 + \cdots + 1\mathbf{v}_i + \cdots + 0\mathbf{v}_r$$

Therefore, the subspace W contains each of the vectors $\mathbf{v}_1, \mathbf{v}_2, \ldots, \mathbf{v}_r$. Let W' be any other subspace that contains $\mathbf{v}_1, \mathbf{v}_2, \ldots, \mathbf{v}_r$. Since W' is closed under addition and scalar multiplication, it must contain all linear combinations of $\mathbf{v}_1, \mathbf{v}_2, \ldots, \mathbf{v}_r$. Thus, W' contains each vector of W. ▮

We make the following definition.

Definition. If $S = \{\mathbf{v}_1, \mathbf{v}_2, \ldots, \mathbf{v}_r\}$ is a set of vectors in a vector space V, then the subspace W of V consisting of all linear combinations of the vectors in S is called the *space spanned* by $\mathbf{v}_1, \mathbf{v}_2, \ldots, \mathbf{v}_r$, and we say that the vectors $\mathbf{v}_1, \mathbf{v}_2, \ldots, \mathbf{v}_r$ *span* W. To indicate that W is the space spanned by the vectors in the set $S = \{\mathbf{v}_1, \mathbf{v}_2, \ldots, \mathbf{v}_r\}$ we write

$$W = \text{span}(S) \qquad \text{or} \qquad W = \text{span}\{\mathbf{v}_1, \mathbf{v}_2, \ldots, \mathbf{v}_r\}$$

Example 10 If $\mathbf{v}_1$ and $\mathbf{v}_2$ are noncollinear vectors in R^3 with initial points at the origin, then span $\{\mathbf{v}_1, \mathbf{v}_2\}$, which consists of all linear combinations $k_1\mathbf{v}_1 + k_2\mathbf{v}_2$, is the plane determined by $\mathbf{v}_1$ and $\mathbf{v}_2$ (Figure 5a). Similarly, if $\mathbf{v}$ is a nonzero vector in R^2 or R^3, then span $\{\mathbf{v}\}$, which is the set of all scalar multiples $k\mathbf{v}$, is the line determined by $\mathbf{v}$ (Figure 5b). ▲

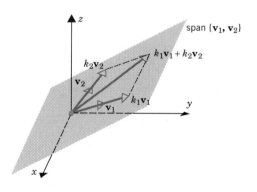

Span $\{\mathbf{v}_1, \mathbf{v}_2\}$ is the plane through the origin determined by $\mathbf{v}_1$ and $\mathbf{v}_2$.

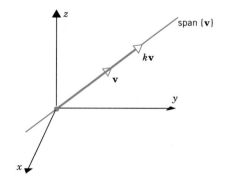

Span $\{\mathbf{v}\}$ is the line through the origin determined by $\mathbf{v}$.

Figure 5

Example 11   The polynomials $1, x, x^2, \ldots, x^n$ span the vector space P_n defined in Example 5 since each polynomial $\mathbf{p}$ in P_n can be written as

$$\mathbf{p} = a_0 + a_1 x + \cdots + a_n x^n$$

which is a linear combination of $1, x, x^2, \ldots, x^n$. We can denote this by writing

$$P_n = \text{span}\{1, x, x^2, \ldots, x^n\} \quad \blacktriangle$$

Example 12   Determine whether $\mathbf{v}_1 = (1, 1, 2)$, $\mathbf{v}_2 = (1, 0, 1)$, and $\mathbf{v}_3 = (2, 1, 3)$ span the vector space R^3.

Solution.   We must determine whether an arbitrary vector $\mathbf{b} = (b_1, b_2, b_3)$ in R^3 can be expressed as a linear combination

$$\mathbf{b} = k_1 \mathbf{v}_1 + k_2 \mathbf{v}_2 + k_3 \mathbf{v}_3$$

of the vectors $\mathbf{v}_1$, $\mathbf{v}_2$, and $\mathbf{v}_3$. Expressing this equation in terms of components gives

$$(b_1, b_2, b_3) = k_1(1, 1, 2) + k_2(1, 0, 1) + k_3(2, 1, 3)$$

or

$$(b_1, b_2, b_3) = (k_1 + k_2 + 2k_3, \, k_1 + k_3, \, 2k_1 + k_2 + 3k_3)$$

or

$$\begin{aligned}
k_1 + k_2 + 2k_3 &= b_1 \\
k_1 \quad\;\; + k_3 &= b_2 \\
2k_1 + k_2 + 3k_3 &= b_3
\end{aligned}$$

The problem thus reduces to determining whether this system is consistent for all values of b_1, b_2, and b_3. By parts (*a*) and (*e*) of Theorem 4.3.4, this system is consistent for all b_1, b_2, and b_3 if and only if the coefficient matrix

$$A = \begin{bmatrix} 1 & 1 & 2 \\ 1 & 0 & 1 \\ 2 & 1 & 3 \end{bmatrix}$$

is invertible. But $\det(A) = 0$ (verify), so that A is not invertible; consequently, $\mathbf{v}_1$, $\mathbf{v}_2$, and $\mathbf{v}_3$ do not span R^3. $\blacktriangle$

Spanning sets are not unique. For example, any two noncollinear vectors that lie in the plane shown in Figure 5 will span that same plane, and any nonzero vector on the line in that figure will span the same line. We leave the proof of the following useful theorem as an exercise.

Theorem 5.2.4.   *If $S = \{\mathbf{v}_1, \mathbf{v}_2, \ldots, \mathbf{v}_r\}$ and $S' = \{\mathbf{w}_1, \mathbf{w}_2, \ldots, \mathbf{w}_k\}$ are two sets of vectors in a vector space V, then*

$$\text{span}\{\mathbf{v}_1, \mathbf{v}_2, \ldots, \mathbf{v}_r\} = \text{span}\{\mathbf{w}_1, \mathbf{w}_2, \ldots, \mathbf{w}_k\}$$

if and only if each vector in S is a linear combination of those in S', and conversely each vector in S' is a linear combination of those in S.

EXERCISE SET 5.2

1. Use Theorem 5.2.1 to determine which of the following are subspaces of R^3.
   (a) all vectors of the form $(a, 0, 0)$
   (b) all vectors of the form $(a, 1, 1)$
   (c) all vectors of the form (a, b, c), where $b = a + c$
   (d) all vectors of the form (a, b, c), where $b = a + c + 1$

2. Use Theorem 5.2.1 to determine which of the following are subspaces of M_{22}.
   (a) all 2×2 matrices with integer entries
   (b) all matrices

$$\begin{bmatrix} a & b \\ c & d \end{bmatrix}$$

   where $a + b + c + d = 0$
   (c) all 2×2 matrices A such that $\det(A) = 0$

3. Use Theorem 5.2.1 to determine which of the following are subspaces of P_3.
   (a) all polynomials $a_0 + a_1 x + a_2 x^2 + a_3 x^3$ for which $a_0 = 0$
   (b) all polynomials $a_0 + a_1 x + a_2 x^2 + a_3 x^3$ for which $a_0 + a_1 + a_2 + a_3 = 0$
   (c) all polynomials $a_0 + a_1 x + a_2 x^2 + a_3 x^3$ for which a_0, a_1, a_2, and a_3 are integers
   (d) all polynomials of the form $a_0 + a_1 x$, where a_0 and a_1 are real numbers

4. Use Theorem 5.2.1 to determine which of the following are subspaces of the space of $F(-\infty, \infty)$.
   (a) all f such that $f(x) \leq 0$ for all x (b) all f such that $f(0) = 0$
   (c) all f such that $f(0) = 2$ (d) all constant functions
   (e) all f of the form $k_1 + k_2 \sin x$, where k_1 and k_2 are real numbers

5. Use Theorem 5.2.1 to determine which of the following are subspaces of M_{nn}.
   (a) all $n \times n$ matrices A such that $\text{tr}(A) = 0$
   (b) all $n \times n$ matrices A such that $A^T = -A$
   (c) all $n \times n$ matrices A such that the linear system $A\mathbf{x} = \mathbf{0}$ has only the trivial solution

6. Determine whether the solution space of the system $A\mathbf{x} = \mathbf{0}$ is a line through the origin, a plane through the origin, or the origin only. If it is a plane, find an equation for it; if it is a line, find parametric equations for it.

(a) $A = \begin{bmatrix} -1 & 1 & 1 \\ 3 & -1 & 0 \\ 2 & -4 & -5 \end{bmatrix}$ (b) $A = \begin{bmatrix} 1 & -2 & 3 \\ -3 & 6 & 9 \\ -2 & 4 & -6 \end{bmatrix}$ (c) $A = \begin{bmatrix} 1 & 2 & 3 \\ 2 & 5 & 3 \\ 1 & 0 & 8 \end{bmatrix}$

(d) $A = \begin{bmatrix} 1 & 2 & -6 \\ 1 & 4 & 4 \\ 3 & 10 & 6 \end{bmatrix}$ (e) $A = \begin{bmatrix} 1 & -1 & 1 \\ 2 & -1 & 4 \\ 3 & 1 & 11 \end{bmatrix}$ (f) $A = \begin{bmatrix} 1 & -3 & 1 \\ 2 & -6 & 2 \\ 3 & -9 & 3 \end{bmatrix}$

7. Which of the following are linear combinations of $\mathbf{u} = (0, -2, 2)$ and $\mathbf{v} = (1, 3, -1)$?
   (a) $(2, 2, 2)$ (b) $(3, 1, 5)$ (c) $(0, 4, 5)$ (d) $(0, 0, 0)$

8. Express the following as linear combinations of $\mathbf{u} = (2, 1, 4)$, $\mathbf{v} = (1, -1, 3)$, and $\mathbf{w} = (3, 2, 5)$.
   (a) $(-9, -7, -15)$ (b) $(6, 11, 6)$ (c) $(0, 0, 0)$ (d) $(7, 8, 9)$

9. Express the following as linear combinations of $p_1 = 2 + x + 4x^2$, $p_2 = 1 - x + 3x^2$, and $p_3 = 3 + 2x + 5x^2$.
   (a) $-9 - 7x - 15x^2$ (b) $6 + 11x + 6x^2$ (c) 0 (d) $7 + 8x + 9x^2$

10. Which of the following are linear combinations of

$$A = \begin{bmatrix} 4 & 0 \\ -2 & -2 \end{bmatrix}, \quad B = \begin{bmatrix} 1 & -1 \\ 2 & 3 \end{bmatrix}, \quad C = \begin{bmatrix} 0 & 2 \\ 1 & 4 \end{bmatrix}?$$

   (a) $\begin{bmatrix} 6 & -8 \\ -1 & -8 \end{bmatrix}$ (b) $\begin{bmatrix} 0 & 0 \\ 0 & 0 \end{bmatrix}$ (c) $\begin{bmatrix} 6 & 0 \\ 3 & 8 \end{bmatrix}$ (d) $\begin{bmatrix} -1 & 5 \\ 7 & 1 \end{bmatrix}$

11. In each part determine whether the given vectors span R^3.
    (a) $v_1 = (2, 2, 2)$, $v_2 = (0, 0, 3)$, $v_3 = (0, 1, 1)$
    (b) $v_1 = (2, -1, 3)$, $v_2 = (4, 1, 2)$, $v_3 = (8, -1, 8)$
    (c) $v_1 = (3, 1, 4)$, $v_2 = (2, -3, 5)$, $v_3 = (5, -2, 9)$, $v_4 = (1, 4, -1)$
    (d) $v_1 = (1, 2, 6)$, $v_2 = (3, 4, 1)$, $v_3 = (4, 3, 1)$, $v_4 = (3, 3, 1)$

12. Let $f = \cos^2 x$ and $g = \sin^2 x$. Which of the following lie in the space spanned by f and g?
    (a) $\cos 2x$ (b) $3 + x^2$ (c) 1 (d) $\sin x$ (e) 0

13. Determine whether the following polynomials span P_2.

$$p_1 = 1 - x + 2x^2, \qquad p_2 = 3 + x,$$
$$p_3 = 5 - x + 4x^2, \qquad p_4 = -2 - 2x + 2x^2$$

14. Let $v_1 = (2, 1, 0, 3)$, $v_2 = (3, -1, 5, 2)$, and $v_3 = (-1, 0, 2, 1)$. Which of the following vectors are in span$\{v_1, v_2, v_3\}$?
    (a) $(2, 3, -7, 3)$ (b) $(0, 0, 0, 0)$ (c) $(1, 1, 1, 1)$ (d) $(-4, 6, -13, 4)$

15. Find an equation for the plane spanned by the vectors $u = (-1, 1, 1)$ and $v = (3, 4, 4)$.

16. Find parametric equations for the line spanned by the vector $u = (3, -2, 5)$.

17. Show that the solution vectors of a consistent nonhomogeneous system of m linear equations in n unknowns do not form a subspace of R^n.

18. Prove Theorem 5.2.4.

19. Use Theorem 5.2.4 to show that

$$v_1 = (1, 6, 4), \qquad v_2 = (2, 4, -1), \qquad v_3 = (-1, 2, 5),$$

   and

$$w_1 = (1, -2, -5), \qquad w_2 = (0, 8, 9)$$

   span the same subspace of R^3.

20. A line L through the origin in R^3 can be represented by parametric equations of the form $x = at$, $y = bt$, and $z = ct$. Use these equations to show that L is a subspace of R^3; that is, if $v_1 = (x_1, y_1, z_1)$ and $v_2 = (x_2, y_2, z_2)$ are points on L and k is any real number, then kv_1 and $v_1 + v_2$ are also points on L.

21. (*For readers who have studied calculus.*) Show that the following sets of functions are subspaces of $F(-\infty, \infty)$.
    (a) all everywhere continuous functions   (b) all everywhere differentiable functions
    (c) all everywhere differentiable functions that satisfy $f' + 2f = 0$

22. (*For readers who have studied calculus.*) Show that the set of continuous functions $\mathbf{f} = f(x)$ on $[a, b]$ such that

$$\int_a^b f(x)\, dx = 0$$

is a subspace of $C[a, b]$.

5.3 LINEAR INDEPENDENCE

In the preceding section we learned that a set of vectors $S = \{\mathbf{v}_1, \mathbf{v}_2, \ldots, \mathbf{v}_r\}$ spans a given vector space V if every vector in V is expressible as a linear combination of the vectors in S. In general, there may be more than one way to express a vector in V as a linear combination of vectors in a spanning set. In this section we shall study conditions under which each vector in V is expressible as a linear combination of the spanning vectors in exactly one way. Spanning sets with this property play a fundamental role in the study of vector spaces.

DEFINITION OF LINEAR INDEPENDENCE

Definition. If $S = \{\mathbf{v}_1, \mathbf{v}_2, \ldots, \mathbf{v}_r\}$ is a nonempty set of vectors, then the vector equation

$$k_1\mathbf{v}_1 + k_2\mathbf{v}_2 + \cdots + k_r\mathbf{v}_r = \mathbf{0}$$

has at least one solution, namely

$$k_1 = 0, \quad k_2 = 0, \quad \ldots, \quad k_r = 0$$

If this is the only solution, then S is called a ***linearly independent*** set. If there are other solutions, then S is called a ***linearly dependent*** set.

Example 1 If $\mathbf{v}_1 = (2, -1, 0, 3)$, $\mathbf{v}_2 = (1, 2, 5, -1)$, and $\mathbf{v}_3 = (7, -1, 5, 8)$, then the set of vectors $S = \{\mathbf{v}_1, \mathbf{v}_2, \mathbf{v}_3\}$ is linearly dependent, since $3\mathbf{v}_1 + \mathbf{v}_2 - \mathbf{v}_3 = \mathbf{0}$. ▲

Example 2 The polynomials

$$\mathbf{p}_1 = 1 - x, \quad \mathbf{p}_2 = 5 + 3x - 2x^2, \quad \text{and} \quad \mathbf{p}_3 = 1 + 3x - x^2$$

form a linearly dependent set in P_2 since $3\mathbf{p}_1 - \mathbf{p}_2 + 2\mathbf{p}_3 = \mathbf{0}$. ▲

Example 3 Consider the vectors $\mathbf{i} = (1, 0, 0)$, $\mathbf{j} = (0, 1, 0)$, and $\mathbf{k} = (0, 0, 1)$ in R^3. In terms of components the vector equation

$$k_1\mathbf{i} + k_2\mathbf{j} + k_3\mathbf{k} = \mathbf{0}$$

becomes

$$k_1(1, 0, 0) + k_2(0, 1, 0) + k_3(0, 0, 1) = (0, 0, 0)$$

or equivalently,

$$(k_1, k_2, k_3) = (0, 0, 0)$$

This implies that $k_1 = 0$, $k_2 = 0$, and $k_3 = 0$, so the set $S = \{\mathbf{i}, \mathbf{j}, \mathbf{k}\}$ is linearly independent. A similar argument can be used to show that the vectors

$$\mathbf{e}_1 = (1, 0, 0, \ldots, 0), \quad \mathbf{e}_2 = (0, 1, 0, \ldots, 0), \quad \ldots, \quad \mathbf{e}_n = (0, 0, 0, \ldots, 1)$$

form a linearly independent set in R^n. ▲

Example 4 Determine whether the vectors

$$\mathbf{v}_1 = (1, -2, 3), \qquad \mathbf{v}_2 = (5, 6, -1), \qquad \mathbf{v}_3 = (3, 2, 1)$$

form a linearly dependent set or a linearly independent set.

Solution. In terms of components the vector equation

$$k_1 \mathbf{v}_1 + k_2 \mathbf{v}_2 + k_3 \mathbf{v}_3 = \mathbf{0}$$

becomes

$$k_1(1, -2, 3) + k_2(5, 6, -1) + k_3(3, 2, 1) = (0, 0, 0)$$

or equivalently,

$$(k_1 + 5k_2 + 3k_3, -2k_1 + 6k_2 + 2k_3, 3k_1 - k_2 + k_3) = (0, 0, 0)$$

Equating corresponding components gives

$$k_1 + 5k_2 + 3k_3 = 0$$
$$-2k_1 + 6k_2 + 2k_3 = 0$$
$$3k_1 - k_2 + k_3 = 0$$

Thus, $\mathbf{v}_1$, $\mathbf{v}_2$, and $\mathbf{v}_3$ form a linearly dependent set if this system has a nontrivial solution, or a linearly independent set if it has only the trivial solution. Solving this system yields

$$k_1 = -\tfrac{1}{2}t, \qquad k_2 = -\tfrac{1}{2}t, \qquad k_3 = t$$

Thus, the system has nontrivial solutions and $\mathbf{v}_1$, $\mathbf{v}_2$, and $\mathbf{v}_3$ form a linearly dependent set. Alternatively, we could show the existence of nontrivial solutions without solving the system by showing that the coefficient matrix has determinant zero and consequently is not invertible (verify). ▲

Example 5 Show that the polynomials

$$1, x, x^2, \ldots, x^n$$

form a linearly independent set of vectors in P_n.

Solution. Let

$$\mathbf{p}_0 = 1, \quad \mathbf{p}_1 = x, \quad \mathbf{p}_2 = x^2, \quad \ldots, \quad \mathbf{p}_n = x^n$$

and assume that some linear combination of these polynomials is zero, say

$$a_0\mathbf{p}_0 + a_1\mathbf{p}_1 + a_2\mathbf{p}_2 + \cdots + a_n\mathbf{p}_n = \mathbf{0}$$

or equivalently,

$$a_0 + a_1x + a_2x^2 + \cdots + a_nx^n = 0 \qquad \text{for all } x \text{ in } (-\infty, \infty) \tag{1}$$

We must show that

$$a_0 = a_1 = a_2 = \cdots = a_n = 0$$

To see that this is so, recall from algebra that a *nonzero* polynomial of degree n has at most n distinct roots. But this implies that $a_0 = a_1 = a_2 = \cdots = a_n = 0$; otherwise, it would follow from (1) that $a_0 + a_1x + a_2x^2 + \cdots + a_nx^n$ is a nonzero polynomial with infinitely many roots. ▲

The term "linearly dependent" suggests that the vectors "depend" on each other in some way. The following theorem shows that this is in fact the case.

> **Theorem 5.3.1.** *A set S with two or more vectors is:*
>
> (a) *Linearly dependent if and only if at least one of the vectors in S is expressible as a linear combination of the other vectors in S.*
>
> (b) *Linearly independent if and only if no vector in S is expressible as a linear combination of the other vectors in S.*

We shall prove part (a) and leave the proof of part (b) as an exercise.

Proof (a). Let $S = \{\mathbf{v}_1, \mathbf{v}_2, \ldots, \mathbf{v}_r\}$ be a set with two or more vectors. If we assume that S is linearly dependent, then there are scalars $k_1, k_2, \ldots, k_r$, not all zero, such that

$$k_1\mathbf{v}_1 + k_2\mathbf{v}_2 + \cdots + k_r\mathbf{v}_r = \mathbf{0} \tag{2}$$

To be specific, suppose that $k_1 \neq 0$. Then (1) can be rewritten as

$$\mathbf{v}_1 = \left(-\frac{k_2}{k_1}\right)\mathbf{v}_2 + \cdots + \left(-\frac{k_r}{k_1}\right)\mathbf{v}_r$$

which expresses $\mathbf{v}_1$ as a linear combination of the other vectors in S. Similarly, if $k_j \neq 0$ in (2) for some $j = 2, 3, \ldots, r$, then $\mathbf{v}_j$ is expressible as a linear combination of the other vectors in S.

Conversely, let us assume that at least one of the vectors in S is expressible as a linear combination of the other vectors. To be specific, suppose that

$$\mathbf{v}_1 = c_2\mathbf{v}_2 + c_3\mathbf{v}_3 + \cdots + c_r\mathbf{v}_r$$

so

$$\mathbf{v}_1 - c_2\mathbf{v}_2 - c_3\mathbf{v}_3 - \cdots - c_r\mathbf{v}_r = \mathbf{0}$$

It follows that S is linearly dependent since the equation

$$k_1\mathbf{v}_1 + k_2\mathbf{v}_2 + \cdots + k_r\mathbf{v}_r = \mathbf{0}$$

is satisfied by

$$k_1 = 1, \quad k_2 = -c_2, \quad \ldots, \quad k_r = -c_r$$

which are not all zero. The proof in the case where some vector other than $\mathbf{v}_1$ is expressible as a linear combination of the other vectors in S is similar. ▮

Example 6 In Example 1 we saw that the vectors

$$\mathbf{v}_1 = (2, -1, 0, 3), \qquad \mathbf{v}_2 = (1, 2, 5, -1), \qquad \text{and} \qquad \mathbf{v}_3 = (7, -1, 5, 8)$$

form a linearly dependent set. It follows from Theorem 5.3.1 that at least one of these vectors is expressible as a linear combination of the other two. In this example each vector is expressible as a linear combination of the other two since it follows from the equation $3\mathbf{v}_1 + \mathbf{v}_2 - \mathbf{v}_3 = \mathbf{0}$ (see Example 1) that

$$\mathbf{v}_1 = -\tfrac{1}{3}\mathbf{v}_2 + \tfrac{1}{3}\mathbf{v}_3, \qquad \mathbf{v}_2 = -3\mathbf{v}_1 + \mathbf{v}_3, \qquad \text{and} \qquad \mathbf{v}_3 = 3\mathbf{v}_1 + \mathbf{v}_2 \quad ▲$$

Example 7 In Example 3 we saw that the vectors $\mathbf{i} = (1, 0, 0)$, $\mathbf{j} = (0, 1, 0)$, and $\mathbf{k} = (0, 0, 1)$ form a linearly independent set. Thus, it follows from Theorem 5.3.1 that none of these vectors is expressible as a linear combination of the other two. To see directly that this is so, suppose that $\mathbf{k}$ is expressible as

$$\mathbf{k} = k_1\mathbf{i} + k_2\mathbf{j}$$

Then, in terms of components,

$$(0, 0, 1) = k_1(1, 0, 0) + k_2(0, 1, 0)$$

or

$$(0, 0, 1) = (k_1, k_2, 0)$$

But this equation is not satisfied by any values of k_1 and k_2, so $\mathbf{k}$ cannot be expressed as a linear combination of $\mathbf{i}$ and $\mathbf{j}$. Similarly, $\mathbf{i}$ is not expressible as a linear combination of $\mathbf{j}$ and $\mathbf{k}$, and $\mathbf{j}$ is not expressible as a linear combination of $\mathbf{i}$ and $\mathbf{k}$. ▲

The following theorem gives two simple facts about linear independence that are important to know.

Theorem 5.3.2

(*a*) *A finite set of vectors that contains the zero vector is linearly dependent.*

(*b*) *A set with exactly two vectors is linearly independent if and only if neither vector is a scalar multiple of the other.*

We shall prove part (*a*) and leave the proof of part (*b*) as an exercise.

Proof (a). For any vectors $\mathbf{v}_1, \mathbf{v}_2, \ldots, \mathbf{v}_r$, the set $S = \{\mathbf{v}_1, \mathbf{v}_2, \ldots, \mathbf{v}_r, \mathbf{0}\}$ is linearly dependent since the equation

$$0\mathbf{v}_1 + 0\mathbf{v}_2 + \cdots + 0\mathbf{v}_r + 1(\mathbf{0}) = \mathbf{0}$$

expresses $\mathbf{0}$ as a linear combination of the vectors in S with coefficients that are not all zero. ▊

Example 8 The functions $\mathbf{f}_1 = x$ and $\mathbf{f}_2 = \sin x$ form a linearly independent set of vectors in $F(-\infty, \infty)$, since neither function is a constant multiple of the other. ▲

GEOMETRIC INTERPRETATION OF LINEAR INDEPENDENCE

Linear independence has some useful geometric interpretations in R^2 and R^3:

- In R^2 or R^3, a set of two vectors is linearly independent if and only if the vectors do not lie on the same line when they are placed with their initial points at the origin (Figure 1).

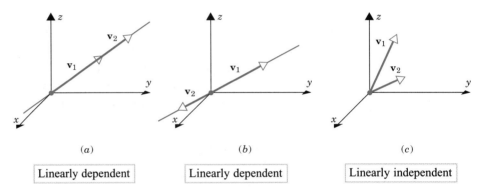

(a)	(b)	(c)

Figure 1   Linearly dependent   Linearly dependent   Linearly independent

- In R^3, a set of three vectors is linearly independent if and only if the vectors do not lie in the same plane when they are placed with their initial points at the origin (Figure 2).

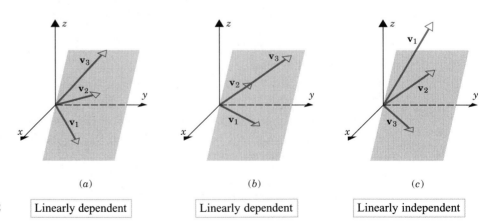

(a)	(b)	(c)

Figure 2   Linearly dependent   Linearly dependent   Linearly independent

The first result follows from the fact that two vectors are linearly independent if and only if neither vector is a scalar multiple of the other. Geometrically, this is equivalent to stating that the vectors do not lie on the same line when they are positioned with their initial points at the origin.

The second result follows from the fact that three vectors are linearly independent if and only if none of the vectors is a linear combination of the other two. Geometrically, this is equivalent to stating that none of the vectors lies in the same plane as the other two, or alternatively, that the three vectors do not lie in a common plane when they are positioned with their initial points at the origin (why?).

The next theorem shows that a linearly independent set in R^n can contain at most n vectors.

Theorem 5.3.3. *Let $S = \{\mathbf{v}_1, \mathbf{v}_2, \ldots, \mathbf{v}_r\}$ be a set of vectors in R^n. If $r > n$, then S is linearly dependent.*

Proof. Suppose that

$$\mathbf{v}_1 = (v_{11}, v_{12}, \ldots, v_{1n})$$
$$\mathbf{v}_2 = (v_{21}, v_{22}, \ldots, v_{2n})$$
$$\vdots \qquad \vdots$$
$$\mathbf{v}_r = (v_{r1}, v_{r2}, \ldots, v_{rn})$$

Consider the equation

$$k_1\mathbf{v}_1 + k_2\mathbf{v}_2 + \cdots + k_r\mathbf{v}_r = \mathbf{0}$$

If, as illustrated in Example 4, we express both sides of this equation in terms of components and then equate corresponding components, we obtain the system

$$v_{11}k_1 + v_{21}k_2 + \cdots + v_{r1}k_r = 0$$
$$v_{12}k_1 + v_{22}k_2 + \cdots + v_{r2}k_r = 0$$
$$\vdots \qquad \vdots \qquad \qquad \vdots \qquad \vdots$$
$$v_{1n}k_1 + v_{2n}k_2 + \cdots + v_{rn}k_r = 0$$

This is a homogeneous system of n equations in the r unknowns $k_1, \ldots, k_r$. Since $r > n$, it follows from Theorem 1.2.1 that the system has nontrivial solutions. Therefore, $S = \{\mathbf{v}_1, \mathbf{v}_2, \ldots, \mathbf{v}_r\}$ is a linearly dependent set. ▨

REMARK. The preceding theorem tells us that a set in R^2 with more than two vectors is linearly dependent, and a set in R^3 with more than three vectors is linearly dependent.

FOR READERS WHO HAVE STUDIED CALCULUS

LINEAR INDEPENDENCE OF FUNCTIONS

Sometimes linear dependence of functions can be deduced from known identities. For example, the functions

$$\mathbf{f}_1 = \sin^2 x, \qquad \mathbf{f}_2 = \cos^2 x, \qquad \text{and} \qquad \mathbf{f}_3 = 5$$

form a linearly dependent set in $F(-\infty, \infty)$, since the equation

$$5\mathbf{f}_1 + 5\mathbf{f}_2 - \mathbf{f}_3 = 5\sin^2 x + 5\cos^2 x - 5 = 5(\sin^2 x + \cos^2 x) - 5 = \mathbf{0}$$

expresses $\mathbf{0}$ as a linear combination of $\mathbf{f}_1$, $\mathbf{f}_2$, and $\mathbf{f}_3$ with coefficients that are not all zero. However, it is only in special situations that such identities can be applied. Although there is no general method that can be used to establish linear independence or linear dependence of functions in $F(-\infty, \infty)$, we shall now develop a theorem that can sometimes be used to show that a given set of functions is linearly independent.

If $\mathbf{f}_1 = f_1(x)$, $\mathbf{f}_2 = f_2(x)$, ..., $\mathbf{f}_n = f_n(x)$ are $n-1$ times differentiable functions on the interval $(-\infty, \infty)$, then the determinant of

$$W(x) = \begin{bmatrix} f_1(x) & f_2(x) & \cdots & f_n(x) \\ f_1'(x) & f_2'(x) & \cdots & f_n'(x) \\ \vdots & \vdots & & \vdots \\ f_1^{(n-1)}(x) & f_2^{(n-1)}(x) & \cdots & f_n^{(n-1)}(x) \end{bmatrix}$$

is called the **_Wronskian_**[*] of $f_1, f_2, \ldots, f_n$. As we shall now show, this determinant is useful for ascertaining whether the functions $\mathbf{f}_1, \mathbf{f}_2, \ldots, \mathbf{f}_n$ form a linearly independent set of vectors in the vector space $C^{(n-1)}(-\infty, \infty)$.

Suppose, for the moment, that $\mathbf{f}_1, \mathbf{f}_2, \ldots, \mathbf{f}_n$ are linearly dependent vectors in $C^{(n-1)}(-\infty, \infty)$. Then there exist scalars $k_1, k_2, \ldots, k_n$, *not all zero*, such that

$$k_1 f_1(x) + k_2 f_2(x) + \cdots + k_n f_n(x) = 0$$

for all x in the interval $(-\infty, \infty)$. Combining this equation with the equations obtained by $n-1$ successive differentiations yields

$$
\begin{aligned}
k_1 f_1(x) &+ k_2 f_2(x) &+ \cdots + k_n f_n(x) &= 0 \\
k_1 f_1'(x) &+ k_2 f_2'(x) &+ \cdots + k_n f_n'(x) &= 0 \\
&\vdots &\vdots \\
k_1 f_1^{(n-1)}(x) &+ k_2 f_2^{(n-1)}(x) &+ \cdots + k_n f_n^{(n-1)}(x) &= 0
\end{aligned}
$$

Thus, the linear dependence of $\mathbf{f}_1, \mathbf{f}_2, \ldots, \mathbf{f}_n$ implies that the linear system

---

[*]*Józef Maria Hoëne-Wrónski* (1776–1853). Polish–French mathematician and philosopher. Wrónski received his early education in Poznán and Warsaw. He served as an artillery officer in the Prussian army in a national uprising in 1794, was taken prisoner by the Russian army, and on his release studied philosophy at various German universities. He became a French citizen in 1800 and eventually settled in Paris, where he did research in analysis leading to some controversial mathematical papers and relatedly to a famous court trial over financial matters. Several years thereafter, his proposed research on the determination of longitude at sea was rebuffed by the British Board of Longitude and Wrónski turned to studies in Messianic philosophy. In the 1830s he investigated the feasibility of caterpillar vehicles to compete with trains, with no luck, and spent his last years in poverty. Much of his mathematical work was fraught with errors and imprecision, but it often contained valuable isolated results and ideas. Some writers attribute this lifelong pattern of argumentation to psychopathic tendencies and an exaggeration of the importance of his own work.

$$\begin{bmatrix} f_1(x) & f_2(x) & \cdots & f_n(x) \\ f_1'(x) & f_2'(x) & \cdots & f_n'(x) \\ \vdots & \vdots & & \vdots \\ f_1^{(n-1)}(x) & f_2^{(n-1)}(x) & \cdots & f_n^{(n-1)}(x) \end{bmatrix} \begin{bmatrix} k_1 \\ k_2 \\ \vdots \\ k_n \end{bmatrix} \begin{bmatrix} 0 \\ 0 \\ \vdots \\ 0 \end{bmatrix}$$

has a nontrivial solution for *every* x in the interval $(-\infty, \infty)$. This implies in turn that for every x in $(-\infty, \infty)$ the coefficient matrix is not invertible, or equivalently, that its determinant (the Wronskian) is zero for every x in $(-\infty, \infty)$. Thus, if the Wronskian is *not* identically zero on $(-\infty, \infty)$, then the functions $\mathbf{f}_1, \mathbf{f}_2, \ldots, \mathbf{f}_n$ must be linearly independent vectors in $C^{(n-1)}(-\infty, \infty)$. This is the content of the following theorem.

> **Theorem 5.3.4.** *If the functions* $\mathbf{f}_1, \mathbf{f}_2, \ldots, \mathbf{f}_n$ *have* $n-1$ *continuous derivatives on the interval* $(-\infty, \infty)$, *and if the Wronskian of these functions is not identically zero on* $(-\infty, \infty)$, *then these functions form a linearly independent set of vectors in* $C^{(n-1)}(-\infty, \infty)$.

Example 9 Show that $\mathbf{f}_1 = x$ and $\mathbf{f}_2 = \sin x$ form a linearly independent set of vectors in $C^1(-\infty, \infty)$.

Solution. In Example 8 we showed that these vectors form a linearly independent set by noting that neither vector is a scalar multiple of the other. However, for illustrative purposes, we shall obtain this same result using Theorem 5.3.4. The Wronskian is

$$W(x) = \begin{vmatrix} x & \sin x \\ 1 & \cos x \end{vmatrix} = x \cos x - \sin x$$

This function does not have value zero for all x in the interval $(-\infty, \infty)$(verify), so $\mathbf{f}_1$ and $\mathbf{f}_2$ form a linearly independent set. ▲

Example 10 Show that $\mathbf{f}_1 = 1$, $\mathbf{f}_2 = e^x$, and $\mathbf{f}_3 = e^{2x}$ form a linearly independent set of vectors in $C^2(-\infty, \infty)$.

Solution. The Wronskian is

$$W(x) = \begin{vmatrix} 1 & e^x & e^{2x} \\ 0 & e^x & 2e^{2x} \\ 0 & e^x & 4e^{2x} \end{vmatrix} = 2e^{3x}$$

This function does not have value zero for all x (in fact, for any x) in the interval $(-\infty, \infty)$, so $\mathbf{f}_1, \mathbf{f}_2$, and $\mathbf{f}_3$ form a linearly independent set. ▲

REMARK. The converse of Theorem 5.3.4 is false. If the Wronskian of $\mathbf{f}_1, \mathbf{f}_2, \ldots, \mathbf{f}_n$ is identically zero on $(-\infty, \infty)$, then no conclusion can be reached about the linear independence of $\{\mathbf{f}_1, \mathbf{f}_2, \ldots, \mathbf{f}_n\}$; this set of vectors may be linearly independent or linearly dependent. We omit the details.

EXERCISE SET 5.3

1. Explain why the following are linearly dependent sets of vectors. (Solve this problem by inspection.)
   (a) $\mathbf{u}_1 = (-1, 2, 4)$ and $\mathbf{u}_2 = (5, -10, -20)$ in R^3 (b) $\mathbf{u}_1 = (3, -1)$, $\mathbf{u}_2 = (4, 5)$, $\mathbf{u}_3 = (-4, 7)$ in R^2
   (c) $\mathbf{p}_1 = 3 - 2x + x^2$ and $\mathbf{p}_2 = 6 - 4x + 2x^2$ in P_2 (d) $A = \begin{bmatrix} -3 & 4 \\ 2 & 0 \end{bmatrix}$ and $B = \begin{bmatrix} 3 & -4 \\ -2 & 0 \end{bmatrix}$ in M_{22}

2. Which of the following sets of vectors in R^3 are linearly dependent?
   (a) $(4, -1, 2)$, $(-4, 10, 2)$ (b) $(-3, 0, 4)$, $(5, -1, 2)$, $(1, 1, 3)$
   (c) $(8, -1, 3)$, $(4, 0, 1)$ (d) $(-2, 0, 1)$, $(3, 2, 5)$, $(6, -1, 1)$, $(7, 0, -2)$

3. Which of the following sets of vectors in R^4 are linearly dependent?
   (a) $(3, 8, 7, -3)$, $(1, 5, 3, -1)$, $(2, -1, 2, 6)$, $(1, 4, 0, 3)$
   (b) $(0, 0, 2, 2)$, $(3, 3, 0, 0)$, $(1, 1, 0, -1)$
   (c) $(0, 3, -3, -6)$, $(-2, 0, 0, -6)$, $(0, -4, -2, -2)$, $(0, -8, 4, -4)$
   (d) $(3, 0, -3, 6)$, $(0, 2, 3, 1)$, $(0, -2, -2, 0)$, $(-2, 1, 2, 1)$

4. Which of the following sets of vectors in P_2 are linearly dependent?
   (a) $2 - x + 4x^2$, $3 + 6x + 2x^2$, $2 + 10x - 4x^2$ (b) $3 + x + x^2$, $2 - x + 5x^2$, $4 - 3x^2$
   (c) $6 - x^2$, $1 + x + 4x^2$ (d) $1 + 3x + 3x^2$, $x + 4x^2$, $5 + 6x + 3x^2$, $7 + 2x - x^2$

5. Assume that $\mathbf{v}_1$, $\mathbf{v}_2$, and $\mathbf{v}_3$ are vectors in R^3 that have their initial points at the origin. In each part determine whether the three vectors lie in a plane.
   (a) $\mathbf{v}_1 = (2, -2, 0)$, $\mathbf{v}_2 = (6, 1, 4)$, $\mathbf{v}_3 = (2, 0, -4)$ (b) $\mathbf{v}_1 = (-6, 7, 2)$, $\mathbf{v}_2 = (3, 2, 4)$, $\mathbf{v}_3 = (4, -1, 2)$

6. Assume that $\mathbf{v}_1$, $\mathbf{v}_2$, and $\mathbf{v}_3$ are vectors in R^3 that have their initial points at the origin. In each part determine whether the three vectors lie on the same line.
   (a) $\mathbf{v}_1 = (-1, 2, 3)$, $\mathbf{v}_2 = (2, -4, -6)$, $\mathbf{v}_3 = (-3, 6, 0)$ (b) $\mathbf{v}_1 = (2, -1, 4)$, $\mathbf{v}_2 = (4, 2, 3)$, $\mathbf{v}_3 = (2, 7, -6)$
   (c) $\mathbf{v}_1 = (4, 6, 8)$, $\mathbf{v}_2 = (2, 3, 4)$, $\mathbf{v}_3 = (-2, -3, -4)$

7. (a) Show that the vectors $\mathbf{v}_1 = (0, 3, 1, -1)$, $\mathbf{v}_2 = (6, 0, 5, 1)$, and $\mathbf{v}_3 = (4, -7, 1, 3)$ form a linearly dependent set in R^4.
   (b) Express each vector as a linear combination of the other two.

8. For which real values of λ do the following vectors form a linearly dependent set in R^3?
   $$\mathbf{v}_1 = (\lambda, -\tfrac{1}{2}, -\tfrac{1}{2}), \qquad \mathbf{v}_2 = (-\tfrac{1}{2}, \lambda, -\tfrac{1}{2}), \qquad \mathbf{v}_3 = (-\tfrac{1}{2}, -\tfrac{1}{2}, \lambda)$$

9. Show that if $\{\mathbf{v}_1, \mathbf{v}_2, \mathbf{v}_3\}$ is a linearly independent set of vectors, then so are $\{\mathbf{v}_1, \mathbf{v}_2\}$, $\{\mathbf{v}_1, \mathbf{v}_3\}$, $\{\mathbf{v}_2, \mathbf{v}_3\}$, $\{\mathbf{v}_1\}$, $\{\mathbf{v}_2\}$, and $\{\mathbf{v}_3\}$.

10. Show that if $S = \{\mathbf{v}_1, \mathbf{v}_2, \ldots, \mathbf{v}_r\}$ is a linearly independent set of vectors, then so is every nonempty subset of S.

11. Show that if $\{\mathbf{v}_1, \mathbf{v}_2, \mathbf{v}_3\}$ is a linearly dependent set of vectors in a vector space V, and $\mathbf{v}_4$ is any vector in V, then $\{\mathbf{v}_1, \mathbf{v}_2, \mathbf{v}_3, \mathbf{v}_4\}$ is also linearly dependent.

12. Show that if $\{\mathbf{v}_1, \mathbf{v}_2, \ldots, \mathbf{v}_r\}$ is a linearly dependent set of vectors in a vector space V, and if $\mathbf{v}_{r+1}, \ldots, \mathbf{v}_n$ are any vectors in V, then $\{\mathbf{v}_1, \mathbf{v}_2, \ldots, \mathbf{v}_{r+1}, \ldots, \mathbf{v}_n\}$ is also linearly dependent.

13. Show that every set with more than three vectors from P_2 is linearly dependent.

14. Show that if $\{\mathbf{v}_1, \mathbf{v}_2\}$ is linearly independent and $\mathbf{v}_3$ does not lie in span$\{\mathbf{v}_1, \mathbf{v}_2\}$, then $\{\mathbf{v}_1, \mathbf{v}_2, \mathbf{v}_3\}$ is linearly independent.

15. Prove: For any vectors **u**, **v**, and **w**, the vectors **u** − **v**, **v** − **w**, and **w** − **u** form a linearly dependent set.

16. Prove: The space spanned by two vectors in R^3 is either a line through the origin, a plane through the origin, or the origin itself.

17. Under what conditions is a set with one vector linearly independent?

18. Are the vectors $\mathbf{v}_1$, $\mathbf{v}_2$, and $\mathbf{v}_3$ in Figure 3a linearly independent? What about those in Figure 3b? Explain.

 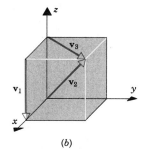

(a)   (b)   **Figure 3**

19. Use appropriate identities, where required, to determine which of the following sets of vectors in $F(-\infty, \infty)$ are linearly dependent.
    (a) 6, $3 \sin^2 x$, $2 \cos^2 x$ (b) x, $\cos x$ (c) 1, $\sin x$, $\sin 2x$
    (d) $\cos 2x$, $\sin^2 x$, $\cos^2 x$ (e) $(3 - x)^2$, $x^2 - 6x$, 5   (f) 0, $\cos^3 \pi x$, $\sin^5 3\pi x$

20. (*For readers who have studied calculus.*) Use the Wronskian to show that the following sets of vectors are linearly independent.
    (a) 1, x, e^x (b) $\sin x$, $\cos x$, $x \sin x$ (c) e^x, xe^x, $x^2 e^x$ (d) 1, x, x^2

21. Use part (a) of Theorem 5.3.1 to prove part (b).

22. Prove part (b) of Theorem 5.3.2.

5.4 BASIS AND DIMENSION

We usually think of a line as being one-dimensional, a plane as two-dimensional, and space around us as three-dimensional. It is the primary purpose of this section to make this intuitive notion of dimension precise.

NON-RECTANGULAR COORDINATE SYSTEMS

In plane analytic geometry we learned to associate a pair of coordinates (a, b) with a point P in the plane by projecting P onto a pair of perpendicular coordinate axes (Figure 1a). By this process, each point in the plane is assigned a unique set of coordinates and, conversely, each pair of coordinates is associated with a unique point in the plane. We describe this by saying that the coordinate system establishes a ***one-to-one correspondence*** between points in the plane and ordered pairs of real numbers. Although perpendicular coordinate axes are the most common, any two nonparallel lines can be used to define a coordinate system in the plane. For example, in Figure 1*b*, we have attached a pair of coordinates (a, b) to the point P by projecting P parallel to the nonperpendicular coordinate axes. Similarly, in 3-space any three noncoplanar coordinate axes can be used to define a coordinate system (Figure 1*c*).

Figure 1

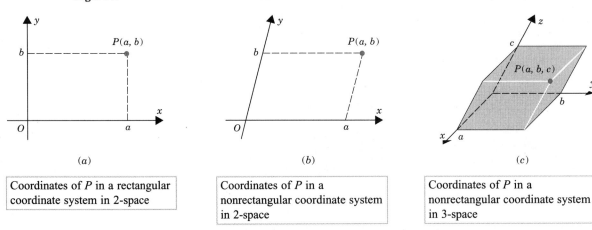

(a)

Coordinates of P in a rectangular coordinate system in 2-space

(b)

Coordinates of P in a nonrectangular coordinate system in 2-space

(c)

Coordinates of P in a nonrectangular coordinate system in 3-space

Our first objective in this section is to extend the concept of a coordinate system to general vector spaces. As a start, it will be helpful to reformulate the notion of a coordinate system in 2-space or 3-space using vectors rather than coordinate axes to specify the coordinate system. This can be done by replacing each coordinate axis with a vector of length 1 that points in the positive direction of the axis. In Figure 2*a*, for example, $\mathbf{v}_1$ and $\mathbf{v}_2$ are such vectors. As illustrated in that figure, if P is any point in the plane, the vector $\overrightarrow{OP}$ can be written as a linear combination of $\mathbf{v}_1$ and $\mathbf{v}_2$ by projecting P parallel to $\mathbf{v}_1$ and $\mathbf{v}_2$ to make $\overrightarrow{OP}$ the diagonal of a parallelogram determined by vectors $a\mathbf{v}_1$ and $b\mathbf{v}_2$:

$$\overrightarrow{OP} = a\mathbf{v}_1 + b\mathbf{v}_2$$

It is evident that the numbers a and b in this vector formula are precisely the coordinates of P in the coordinate system of Figure 1*b*. Similarly, the coordinates (a, b, c) of the point P in Figure 1*c* can be obtained by expressing $\overrightarrow{OP}$ as a linear combination of the vectors shown in Figure 2*b*.

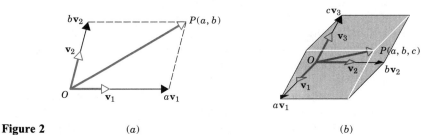

Figure 2 (a) (b)

Informally stated, vectors that specify a coordinate system are called "basis vectors" for that system. Although we used basis vectors of length 1 in the preceding discussion, we shall see in a moment that this is not essential—nonzero vectors of any length will suffice.

The scales of measurement along the coordinate axes are essential ingredients of any coordinate system. Usually, one tries to use the same scale on each axis and have the integer points on the axes spaced 1 unit of distance apart. However, this is not always practical or appropriate: Unequal scales, or scales in which the integer points are more or less than 1 unit apart, may be required to fit a particular graph on a printed page or to represent physical quantities with diverse units in the same coordinate system (time in seconds on one axis and temperature in hundreds of degrees on another, for example). When a coordinate system is specified by a set of basis vectors, then the lengths of those vectors correspond to the distances between successive integer points on the coordinate axes (Figure 3). Thus, it is the directions of the basis vectors that define the positive directions of the coordinate axes and the lengths of the basis vectors that establish the scales of measurement.

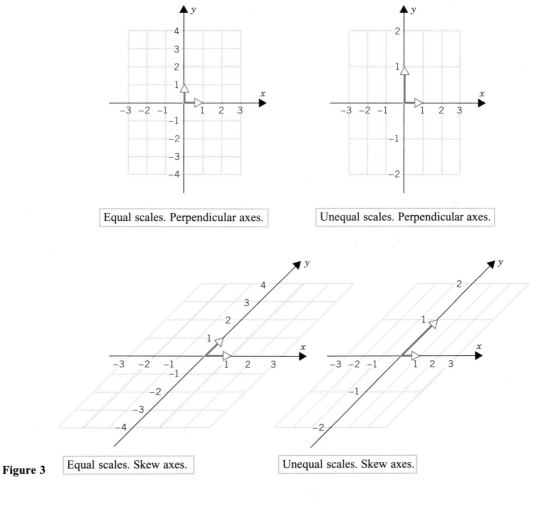

Equal scales. Perpendicular axes.

Unequal scales. Perpendicular axes.

Equal scales. Skew axes.

Unequal scales. Skew axes.

Figure 3

The following key definition will make the preceding ideas more precise and enable us to extend the concept of a coordinate system to general vector spaces.

BASIS FOR A VECTOR SPACE

Definition. If V is any vector space and $S = \{\mathbf{v}_1, \mathbf{v}_2, \ldots, \mathbf{v}_n\}$ is a set of vectors in V, then S is called a ***basis*** for V if the following two conditions hold:

(a) S is linearly independent.
(b) S spans V.

A basis is the vector space generalization of a coordinate system in 2-space and 3-space. The following theorem will help us to see why this is so.

Theorem 5.4.1. *If* $S = \{\mathbf{v}_1, \mathbf{v}_2, \ldots, \mathbf{v}_n\}$ *is a basis for a vector space* V, *then every vector* $\mathbf{v}$ *in* V *can be expressed in the form* $\mathbf{v} = c_1\mathbf{v}_1 + c_2\mathbf{v}_2 + \cdots + c_n\mathbf{v}_n$ *in exactly one way.*

Proof. Since S spans V, it follows from the definition of a spanning set that every vector in V is expressible as a linear combination of the vectors in S. To see that there is only *one* way to express a vector as a linear combination of the vectors in S, suppose that some vector $\mathbf{v}$ can be written as

$$\mathbf{v} = c_1\mathbf{v}_1 + c_2\mathbf{v}_2 + \cdots + c_n\mathbf{v}_n$$

and also as

$$\mathbf{v} = k_1\mathbf{v}_1 + k_2\mathbf{v}_2 + \cdots + k_n\mathbf{v}_n$$

Subtracting the second equation from the first gives

$$\mathbf{0} = (c_1 - k_1)\mathbf{v}_1 + (c_2 - k_2)\mathbf{v}_2 + \cdots + (c_n - k_n)\mathbf{v}_n$$

Since the right side of this equation is a linear combination of vectors in S, the linear independence of S implies that

$$c_1 - k_1 = 0, \qquad c_2 - k_2 = 0, \quad \ldots, \quad c_n - k_n = 0$$

that is,

$$c_1 = k_1, \quad c_2 = k_2, \quad \ldots, \quad c_n = k_n$$

Thus, the two expressions for $\mathbf{v}$ are the same. ∎

COORDINATES RELATIVE TO A BASIS

If $S = \{\mathbf{v}_1, \mathbf{v}_2, \ldots, \mathbf{v}_n\}$ is a basis for a vector space V, and

$$\mathbf{v} = c_1\mathbf{v}_1 + c_2\mathbf{v}_2 + \cdots + c_n\mathbf{v}_n$$

is the expression for a vector $\mathbf{v}$ in terms of the basis S, then the scalars $c_1, c_2, \ldots, c_n$ are called the ***coordinates*** of $\mathbf{v}$ relative to the basis S. The vector $(c_1, c_2, \ldots, c_n)$ in R^n constructed from these coordinates is called the ***coordinate vector of v relative to S***; it is denoted by

$$(\mathbf{v})_S = (c_1, c_2, \ldots, c_n)$$

REMARK. It should be noted that coordinate vectors depend not only on the basis S but also on the order in which the basis vectors are written; a change in the order of the basis vectors results in a corresponding change of order for the entries in the coordinate vectors.

Example 1 In Example 3 of the preceding section we showed that if

$$\mathbf{i} = (1, 0, 0), \qquad \mathbf{j} = (0, 1, 0), \qquad \text{and} \qquad \mathbf{k} = (0, 0, 1)$$

then $S = \{\mathbf{i}, \mathbf{j}, \mathbf{k}\}$ is a linearly independent set in R^3. This set also spans R^3 since any vector $\mathbf{v} = (a, b, c)$ in R^3 can be written as

$$\mathbf{v} = (a, b, c) = a(1, 0, 0) + b(0, 1, 0) + c(0, 0, 1) = a\mathbf{i} + b\mathbf{j} + c\mathbf{k} \qquad (1)$$

Thus, S is a basis for R^3; it is called the ***standard basis*** for R^3. Looking at the coefficients of $\mathbf{i}$, $\mathbf{j}$, and $\mathbf{k}$ in (1), it follows that the coordinates of $\mathbf{v}$ relative to the standard basis are a, b, and c, so

$$(\mathbf{v})_S = (a, b, c)$$

Comparing this result to (1) we see that

$$\mathbf{v} = (\mathbf{v})_S$$

This equation states that the components of a vector $\mathbf{v}$ relative to a rectangular *xyz*-coordinate system and the coordinates of $\mathbf{v}$ relative to the standard basis are the same; thus, the coordinate system and the basis produce precisely the same one-to-one correspondence between points in 3-space and ordered triples of real numbers (Figure 4).

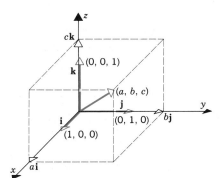

Figure 4

The results in the preceding example are a special case of those in the next example.

STANDARD BASIS FOR R^n **Example 2** In Example 3 of the preceding section we showed that if

$$\mathbf{e}_1 = (1, 0, 0, \ldots, 0), \quad \mathbf{e}_2 = (0, 1, 0, \ldots, 0), \quad \ldots, \quad \mathbf{e}_n = (0, 0, 0, \ldots, 1)$$

then

$$S = \{\mathbf{e}_1, \mathbf{e}_2, \ldots, \mathbf{e}_n\}$$

is a linearly independent set in R^n. This set also spans R^n since any vector $\mathbf{v} = (v_1, v_2, \ldots, v_n)$ in R^n can be written as

$$\mathbf{v} = v_1\mathbf{e}_1 + v_2\mathbf{e}_2 + \cdots + v_n\mathbf{e}_n \tag{2}$$

Thus, S is a basis for R^n; it is called the ***standard basis for R^n***. It follows from (2) that the coordinates of $\mathbf{v} = (v_1, v_2, \ldots, v_n)$ relative to the standard basis are $v_1, v_2, \ldots, v_n$, so

$$(\mathbf{v})_S = (v_1, v_2, \ldots, v_n)$$

As in Example 1, we have

$$\mathbf{v} = (\mathbf{v})_S$$

so a vector $\mathbf{v}$ and its coordinate vector relative to the standard basis for R^n are the same. ▲

REMARK. We will see in a subsequent example that a vector and its coordinate vector will not be the same; the equality that we observed in the two preceding examples is a special situation that occurs only with the standard basis for R^n.

REMARK. In R^2 and R^3, the standard basis vectors are commonly denoted by $\mathbf{i}$, $\mathbf{j}$, and $\mathbf{k}$, rather than by $\mathbf{e}_1$, $\mathbf{e}_2$, and $\mathbf{e}_3$. We shall use both notations, depending on the particular situation.

Example 3   Let $\mathbf{v}_1 = (1, 2, 1)$, $\mathbf{v}_2 = (2, 9, 0)$, and $\mathbf{v}_3 = (3, 3, 4)$. Show that the set $S = \{\mathbf{v}_1, \mathbf{v}_2, \mathbf{v}_3\}$ is a basis for R^3.

Solution. To show that the set S spans R^3, we must show that an arbitrary vector $\mathbf{b} = (b_1, b_2, b_3)$ can be expressed as a linear combination

$$\mathbf{b} = c_1\mathbf{v}_1 + c_2\mathbf{v}_2 + c_3\mathbf{v}_3$$

of the vectors in S. Expressing this equation in terms of components gives

$$(b_1, b_2, b_3) = c_1(1, 2, 1) + c_2(2, 9, 0) + c_3(3, 3, 4)$$

or

$$(b_1, b_2, b_3) = (c_1 + 2c_2 + 3c_3, \ 2c_1 + 9c_2 + 3c_3, \ c_1 + 4c_3)$$

or on equating corresponding components

$$\begin{aligned} c_1 + 2c_2 + 3c_3 &= b_1 \\ 2c_1 + 9c_2 + 3c_3 &= b_2 \\ c_1 \qquad\quad + 4c_3 &= b_3 \end{aligned} \tag{3}$$

Thus, to show that S spans R^3, we must demonstrate that system (3) has a solution for all choices of $\mathbf{b} = (b_1, b_2, b_3)$.

To prove that S is linearly independent, we must show that the only solution of

$$c_1\mathbf{v}_1 + c_2\mathbf{v}_2 + c_3\mathbf{v}_3 = \mathbf{0} \tag{4}$$

is $c_1 = c_2 = c_3 = 0$. As above, if (4) is expressed in terms of components, the verification of independence reduces to showing that the homogeneous system

$$c_1 + 2c_2 + 3c_3 = 0$$
$$2c_1 + 9c_2 + 3c_3 = 0 \qquad (5)$$
$$c_1 \qquad + 4c_3 = 0$$

has only the trivial solution. Observe that systems (3) and (5) have the same coefficient matrix. Thus, by parts (*a*), (*b*), and (*g*) of Theorem 4.3.4, we can simultaneously prove that S is linearly independent and spans R^3 by demonstrating that in systems (3) and (5) the matrix of coefficients

$$A = \begin{bmatrix} 1 & 2 & 3 \\ 2 & 9 & 3 \\ 1 & 0 & 4 \end{bmatrix}$$

has a nonzero determinant. But

$$\det(A) = \begin{vmatrix} 1 & 2 & 3 \\ 2 & 9 & 3 \\ 1 & 0 & 4 \end{vmatrix} = -1$$

so S is a basis for R^3. ▲

Example 4   Let $S = \{\mathbf{v}_1, \mathbf{v}_2, \mathbf{v}_3\}$ be the basis for R^3 in the preceding example.

(a)   Find the coordinate vector of $\mathbf{v} = (5, -1, 9)$ with respect to S.
(b)   Find the vector $\mathbf{v}$ in R^3 whose coordinate vector with respect to the basis S is $(\mathbf{v})_S = (-1, 3, 2)$.

Solution (a). We must find scalars c_1, c_2, c_3 such that

$$\mathbf{v} = c_1\mathbf{v}_1 + c_2\mathbf{v}_2 + c_3\mathbf{v}_3$$

or, in terms of components,

$$(5, -1, 9) = c_1(1, 2, 1) + c_2(2, 9, 0) + c_3(3, 3, 4)$$

Equating corresponding components gives

$$c_1 + 2c_2 + 3c_3 = \quad 5$$
$$2c_1 + 9c_2 + 3c_3 = -1$$
$$c_1 \qquad + 4c_3 = \quad 9$$

Solving this system, we obtain $c_1 = 1$, $c_2 = -1$, $c_3 = 2$ (verify). Therefore,

$$(\mathbf{v})_S = (1, -1, 2)$$

Solution (b). Using the definition of the coordinate vector $(\mathbf{v})_S$, we obtain

$$\mathbf{v} = (-1)\mathbf{v}_1 + 3\mathbf{v}_2 + 2\mathbf{v}_3$$
$$= (-1)(1, 2, 1) + 3(2, 9, 0) + 2(3, 3, 4) = (11, 31, 7) \quad ▲$$

Example 5

(a) Show that $S = \{1, x, x^2, \ldots, x^n\}$ is a basis for the vector space P_n of polynomials of the form $a_0 + a_1 x + \cdots + a_n x^n$.

(b) Find the coordinate vector of the polynomial $\mathbf{p} = a_0 + a_1 x + a_2 x^2$ relative to the basis $S = \{1, x, x^2\}$ for P_2.

Solution (a). We showed that S spans P_n in Example 11 of Section 5.2, and we showed that S is a linearly independent set in Example 5 of Section 5.3. Thus, S is a basis for P_n; it is called the **standard basis for P_n**.

Solution (b). The coordinates of $\mathbf{p} = a_0 + a_1 x + a_2 x^2$ are the scalar coefficients of the basis vectors 1, x, and x^2, so $(\mathbf{p})_S = (a_0, a_1, a_2)$. ▲

Example 6 Let

$$M_1 = \begin{bmatrix} 1 & 0 \\ 0 & 0 \end{bmatrix}, \qquad M_2 = \begin{bmatrix} 0 & 1 \\ 0 & 0 \end{bmatrix}, \qquad M_3 = \begin{bmatrix} 0 & 0 \\ 1 & 0 \end{bmatrix}, \qquad M_4 = \begin{bmatrix} 0 & 0 \\ 0 & 1 \end{bmatrix}$$

The set $S = \{M_1, M_2, M_3, M_4\}$ is a basis for the vector space M_{22} of 2×2 matrices. To see that S spans M_{22}, note that an arbitrary vector (matrix)

$$\begin{bmatrix} a & b \\ c & d \end{bmatrix}$$

can be written as

$$\begin{bmatrix} a & b \\ c & d \end{bmatrix} = a \begin{bmatrix} 1 & 0 \\ 0 & 0 \end{bmatrix} + b \begin{bmatrix} 0 & 1 \\ 0 & 0 \end{bmatrix} + c \begin{bmatrix} 0 & 0 \\ 1 & 0 \end{bmatrix} + d \begin{bmatrix} 0 & 0 \\ 0 & 1 \end{bmatrix}$$
$$= aM_1 + bM_2 + cM_3 + dM_4$$

To see that S is linearly independent, assume that

$$aM_1 + bM_2 + cM_3 + dM_4 = 0$$

That is,

$$a \begin{bmatrix} 1 & 0 \\ 0 & 0 \end{bmatrix} + b \begin{bmatrix} 0 & 1 \\ 0 & 0 \end{bmatrix} + c \begin{bmatrix} 0 & 0 \\ 1 & 0 \end{bmatrix} + d \begin{bmatrix} 0 & 0 \\ 0 & 1 \end{bmatrix} = \begin{bmatrix} 0 & 0 \\ 0 & 0 \end{bmatrix}$$

It follows that

$$\begin{bmatrix} a & b \\ c & d \end{bmatrix} = \begin{bmatrix} 0 & 0 \\ 0 & 0 \end{bmatrix}$$

Thus, $a = b = c = d = 0$, so S is linearly independent. The basis S in this example is called the *standard basis for M_{22}*. More generally, the **standard basis for M_{mn}** consists of the mn different matrices with a single 1 and zeros for the remaining entries. ▲

Example 7 If $S = \{\mathbf{v}_1, \mathbf{v}_2, \ldots, \mathbf{v}_r\}$ is a *linearly independent* set in a vector space V, then S is a basis for the subspace $\text{span}(S)$ since the set S spans $\text{span}(S)$ by definition of $\text{span}(S)$. ▲

DIMENSION

> **Definition.** A nonzero vector space V is called **finite-dimensional** if it contains a finite set of vectors $\{\mathbf{v}_1, \mathbf{v}_2, \ldots, \mathbf{v}_n\}$ that forms a basis. If no such set exists, V is called **infinite-dimensional**. In addition, we shall regard the zero vector space to be finite-dimensional.

Example 8 By Examples 2, 5, and 6, the vector spaces R^n, P^n, and M_{mn} are finite-dimensional. The vector spaces $F(-\infty, \infty)$, $C(-\infty, \infty)$, $C^m(-\infty, \infty)$, and $C^\infty(-\infty, \infty)$ are infinite-dimensional (Exercise 23). ▲

The next theorem will provide the key to the concept of dimension.

> **Theorem 5.4.2.** *If V is a finite-dimensional vector space and $\{\mathbf{v}_1, \mathbf{v}_2, \ldots, \mathbf{v}_n\}$ is any basis, then:*
>
> (a) *Every set with more than n vectors is linearly dependent.*
> (b) *No set with fewer than n vectors spans V.*

Proof (a) Let $S' = \{\mathbf{w}_1, \mathbf{w}_2, \ldots, \mathbf{w}_m\}$ be any set of m vectors in V, where $m > n$. We want to show that S' is linearly dependent. Since $S = \{\mathbf{v}_1, \mathbf{v}_2, \ldots, \mathbf{v}_n\}$ is a basis, each $\mathbf{w}_i$ can be expressed as a linear combination of the vectors in S, say

$$
\begin{aligned}
\mathbf{w}_1 &= a_{11}\mathbf{v}_1 + a_{21}\mathbf{v}_2 + \cdots + a_{n1}\mathbf{v}_n \\
\mathbf{w}_2 &= a_{12}\mathbf{v}_1 + a_{22}\mathbf{v}_2 + \cdots + a_{n2}\mathbf{v}_n \\
&\;\;\vdots \qquad \vdots \qquad \vdots \qquad \vdots \\
\mathbf{w}_m &= a_{1m}\mathbf{v}_1 + a_{2m}\mathbf{v}_2 + \cdots + a_{nm}\mathbf{v}_n
\end{aligned}
\tag{6}
$$

To show that S' is linearly dependent, we must find scalars $k_1, k_2, \ldots, k_m$, not all zero, such that

$$
k_1\mathbf{w}_1 + k_2\mathbf{w}_2 + \cdots + k_m\mathbf{w}_m = \mathbf{0}
\tag{7}
$$

Using the equations in (6), we can rewrite (7) as

$$
\begin{aligned}
(k_1 a_{11} + k_2 a_{12} + \cdots + k_m a_{1m})\mathbf{v}_1 & \\
+ (k_1 a_{21} + k_2 a_{22} + \cdots + k_m a_{2m})\mathbf{v}_2 & \\
\ddots \qquad\qquad & \\
+ (k_1 a_{n1} + k_2 a_{n2} + \cdots + k_m a_{nm})\mathbf{v}_n &= \mathbf{0}
\end{aligned}
$$

Thus, from the linear independence of S, the problem of proving that S' is a linearly dependent set reduces to showing there are scalars $k_1, k_2, \ldots, k_m$, not all zero, that satisfy

$$
\begin{aligned}
a_{11}k_1 + a_{12}k_2 + \cdots + a_{1m}k_m &= 0 \\
a_{21}k_1 + a_{22}k_2 + \cdots + a_{2m}k_m &= 0 \\
\;\;\vdots \qquad \vdots \qquad\qquad \vdots \qquad \vdots & \\
a_{n1}k_1 + a_{n2}k_2 + \cdots + a_{nm}k_m &= 0
\end{aligned}
\tag{8}
$$

But (8) has more unknowns than equations, so the proof is complete, since Theorem 1.2.1 guarantees the existence of nontrivial solutions.

Proof (b). Let $S' = \{\mathbf{w}_1, \mathbf{w}_2, \ldots, \mathbf{w}_m\}$ by any set of m vectors in V, where $m < n$. We want to show that S' does not span V. The proof will be by contradiction: We will show that assuming S' spans V leads to a contradiction of the linear independence of $\{\mathbf{v}_1, \mathbf{v}_2, \ldots, \mathbf{v}_n\}$.

If S' spans V, then every vector in V is a linear combination of the vectors in S'. In particular, each basis vector $\mathbf{v}_i$ is a linear combination of the vectors in S', say

$$
\begin{aligned}
\mathbf{v}_1 &= a_{11}\mathbf{w}_1 + a_{21}\mathbf{w}_2 + \cdots + a_{m1}\mathbf{w}_m \\
\mathbf{v}_2 &= a_{12}\mathbf{w}_1 + a_{22}\mathbf{w}_2 + \cdots + a_{m2}\mathbf{w}_m \\
&\;\;\vdots \qquad\quad \vdots \qquad\qquad\quad \vdots \\
\mathbf{v}_n &= a_{1n}\mathbf{w}_1 + a_{2n}\mathbf{w}_2 + \cdots + a_{mn}\mathbf{w}_m
\end{aligned}
\tag{9}
$$

To obtain our contradiction we will show that there are scalars $k_1, k_2, \ldots, k_n$, not all zero, such that

$$
k_1\mathbf{v}_1 + k_2\mathbf{v}_2 + \cdots + k_n\mathbf{v}_n = \mathbf{0}
\tag{10}
$$

But observe that (9) and (10) have the same form as (6) and (7) except that m and n are interchanged and the $\mathbf{w}$'s and $\mathbf{v}$'s are interchanged. Thus, the computations that led to (8) now yield

$$
\begin{aligned}
a_{11}k_1 + a_{12}k_2 + \cdots + a_{1n}k_n &= 0 \\
a_{21}k_1 + a_{22}k_2 + \cdots + a_{2n}k_n &= 0 \\
\vdots \qquad \vdots \qquad\qquad \vdots \qquad\;\; \vdots \\
a_{m1}k_1 + a_{m2}k_2 + \cdots + a_{mn}k_n &= 0
\end{aligned}
$$

This linear system has more unknowns than equations, and hence has nontrivial solutions by Theorem 1.2.1. ▨

It follows from the preceding theorem that if $S = \{\mathbf{v}_1, \mathbf{v}_2, \ldots, \mathbf{v}_n\}$ is any basis for a vector space V, then all sets in V that simultaneously span V and are linearly independent must have precisely n vectors. Thus, all bases for V must have the same number of vectors as the arbitrary basis S. This yields the following result, which is one of the most important in linear algebra.

> **Theorem 5.4.3.** *All bases for a finite-dimensional vector space have the same number of vectors.*

To see how this theorem relates to the concept of "dimension," recall that the standard basis for R^n has n vectors (Example 2). Thus, Theorem 5.4.3 implies that all bases for R^n have n vectors. In particular, every basis for R^3 has three vectors, every basis for R^2 has two vectors, and every basis for R^1 ($= R$) has one vector. Intuitively, R^3 is three-dimensional, R^2 (a plane) is two-dimensional, and R (a line) is one-dimensional. Thus, for familiar vector spaces, the number of vectors in a basis is the same as the dimension. This suggests the following definition.

Definition. The **_dimension_** of a finite-dimensional vector space V, denoted by $\dim(V)$, is defined to be the number of vectors in a basis for V. In addition, we define the zero vector space to have dimension zero.

REMARK. From here on we shall follow a common convention of regarding the empty set to be a basis for the zero vector space. This is consistent with the preceding definition, since the empty set has no vectors and the zero vector space has dimension zero.

Example 9

$$\dim(R^n) = n \qquad \text{[The standard basis has } n \text{ vectors (Example 2).]}$$
$$\dim(P_n) = n + 1 \qquad \text{[The standard basis has } n + 1 \text{ vectors (Example 5).]}$$
$$\dim(M_{mn}) = mn \qquad \text{[The standard basis has } mn \text{ vectors (Example 6).]}$$

Example 10   Determine a basis for and the dimension of the solution space of the homogeneous system

$$
\begin{array}{rcl}
2x_1 + 2x_2 - x_3 \qquad\quad + x_5 &=& 0 \\
-x_1 - x_2 + 2x_3 - 3x_4 + x_5 &=& 0 \\
x_1 + x_2 - 2x_3 \qquad\quad - x_5 &=& 0 \\
x_3 + x_4 + x_5 &=& 0
\end{array}
$$

Solution. In Example 6 of Section 1.2 it was shown that the general solution of the given system is

$$x_1 = -s - t, \qquad x_2 = s, \qquad x_3 = -t, \qquad x_4 = 0, \qquad x_5 = t$$

Therefore, the solution vectors can be written as

$$
\begin{bmatrix} x_1 \\ x_2 \\ x_3 \\ x_4 \\ x_5 \end{bmatrix}
=
\begin{bmatrix} -s - t \\ s \\ -t \\ 0 \\ t \end{bmatrix}
=
\begin{bmatrix} -s \\ s \\ 0 \\ 0 \\ 0 \end{bmatrix}
+
\begin{bmatrix} -t \\ 0 \\ -t \\ 0 \\ t \end{bmatrix}
= s
\begin{bmatrix} -1 \\ 1 \\ 0 \\ 0 \\ 0 \end{bmatrix}
+ t
\begin{bmatrix} -1 \\ 0 \\ -1 \\ 0 \\ 1 \end{bmatrix}
$$

which shows that the vectors

$$
\mathbf{v}_1 =
\begin{bmatrix} -1 \\ 1 \\ 0 \\ 0 \\ 0 \end{bmatrix}
\qquad \text{and} \qquad
\mathbf{v}_2 =
\begin{bmatrix} -1 \\ 0 \\ -1 \\ 0 \\ 1 \end{bmatrix}
$$

span the solution space. Since they are also linearly independent (verify), $\{\mathbf{v}_1, \mathbf{v}_2\}$ is a basis, and the solution space is two-dimensional.

SOME FUNDAMENTAL THEOREMS

We shall devote the remainder of this section to a series of theorems that reveal the subtle interrelationships among the concepts of spanning, linear independence, basis, and dimension. These theorems are not idle exercises in mathematical theory—they are essential to the understanding of vector spaces, and many practical applications of linear algebra build on them.

The following theorem, which we call the *Plus/Minus Theorem* (our own name), establishes two basic principles on which most of the theorems to follow will rely.

Theorem 5.4.4 (*Plus/Minus Theorem*). *Let S be a nonempty set of vectors in a vector space V.*

(a) *If S is a linearly independent set, and if **v** is a vector in V that is outside of span(S), then the set that results by inserting **v** into S is still linearly independent.*

(b) *If **v** is a vector in S that is expressible as a linear combination of other vectors in S, and if S − {**v**} denotes the set obtained by removing **v** from S, then S and S − {**v**} span the same space; that is,*

$$span(S) = span(S - \{\mathbf{v}\})$$

We shall defer the proof to the end of the section, so that we may move more immediately to the consequences of the theorem. However, the theorem can be visualized in R^3 as follows:

(a) A set S of two linearly independent vectors in R^3 spans a plane through the origin. If we enlarge S by inserting any vector **v** outside of this plane (Figure 5*a*), then the resulting set of three vectors is still linearly independent since none of the three vectors lies in the same plane as the other two.

(b) If S is a set of three noncollinear vectors in R^3 that lie in a common plane through the origin (Figure 5*b*), then the three vectors span the plane. However, if we remove from S any vector **v** that is a linear combination of the other two, the remaining set of two vectors still spans the plane.

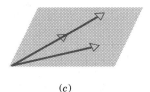

(a)	(b)	(c)
None of the three vectors lies in the same plane as the other two.	Any of the vectors can be removed, and the remaining two will still span the plane.	Either of the collinear vectors can be removed, and the remaining two will still span the plane.

Figure 5

In general, to show that a set of vectors $\{\mathbf{v}_1, \mathbf{v}_2, \ldots, \mathbf{v}_n\}$ is a basis for a vector space V, we must show that the vectors are linearly independent and span V. However, if we happen to know that V has dimension n (so that $\{\mathbf{v}_1, \mathbf{v}_2, \ldots, \mathbf{v}_n\}$ contains the right number of vectors for a basis), then it suffices to check *either* linear independence *or* spanning—the remaining condition will hold automatically. This is the content of the following theorem.

Theorem 5.4.5. *If V is an n-dimensional vector space, and if S is a set in V with exactly n vectors, then S is a basis for V if either S spans V or S is linearly independent.*

Proof. Assume that S has exactly n vectors and spans V. To prove that S is a basis we must show that S is a linearly independent set. But if this is not so, then some vector $\mathbf{v}$ in S is a linear combination of the remaining vectors. If we remove this vector from S, then it follows from the Plus/Minus Theorem (5.4.4b) that the remaining set of $n-1$ vectors still spans V. But this is impossible, since it follows from Theorem 5.4.2b that no set with fewer than n vectors can span an n-dimensional vector space. Thus, S is linearly independent.

Assume that S has exactly n vectors and is a linearly independent set. To prove that S is a basis we must show that S spans V. But if this is not so, then there is some vector $\mathbf{v}$ in V that is not in span(S). If we insert this vector into S, then it follows from the Plus/Minus Theorem (5.4.4a) that this set of $n+1$ vectors is still linearly independent. But this is impossible, since it follows from Theorem 5.4.2a that no set with more than n vectors in an n-dimensional vector space can be linearly independent. Thus, S spans V. ▨

Example 11

(a) Show that $\mathbf{v}_1 = (-3, 7)$ and $\mathbf{v}_2 = (5, 5)$ form a basis for R^2 by inspection.
(b) Show that $\mathbf{v}_1 = (2, 0, -1)$, $\mathbf{v}_2 = (4, 0, 7)$, $\mathbf{v}_3 = (-1, 1, 4)$ form a basis for R^3 by inspection.

Solution (a). Since neither vector is a scalar multiple of the other, the two vectors form a linearly independent set in the two-dimensional space R^2, and hence form a basis by Theorem 5.4.5.

Solution (b). The vectors $\mathbf{v}_1$ and $\mathbf{v}_2$ form a linearly independent set in the xz-plane (why?). The vector $\mathbf{v}_3$ is outside of the xz-plane, so the set $\{\mathbf{v}_1, \mathbf{v}_2, \mathbf{v}_3\}$ is also linearly independent. Since R^3 is three-dimensional, Theorem 5.4.5 implies that $\{\mathbf{v}_1, \mathbf{v}_2, \mathbf{v}_3\}$ is a basis for R^3. ▲

The following theorem shows that for a finite-dimensional vector space V every set that spans V contains a basis for V within it, and every linearly independent set in V is part of some basis for V.

Theorem 5.4.6. *Let S be a finite set of vectors in a finite-dimensional vector space V.*

(a) *If S spans V but is not a basis for V, then S can be reduced to a basis for V by removing appropriate vectors from S.*
(b) *If S is a linearly independent set that is not already a basis for V, then S can be enlarged to a basis for V by inserting appropriate vectors into S.*

Proof (a). If S is a set of vectors that spans V but is not a basis for V, then S is a linearly dependent set. Thus, some vector $\mathbf{v}$ in S is expressible as a linear combination of the other vectors in S. By the Plus/Minus Theorem (5.4.4b), we can remove $\mathbf{v}$ from S, and the resulting set S' will still span V. If S' is linearly independent, then S' is a basis for V, and we are done. If S' is linearly dependent, then we can remove some appropriate vector from S' to produce a set S'' that still spans V. We can continue removing vectors in this way until we finally arrive at a set of vectors in S that is linearly independent and spans V. This subset of S is a basis for V.

Proof (b) Suppose that $\dim(V) = n$. If S is a linearly independent set that is not already a basis for V, then S fails to span V, and there is some vector $\mathbf{v}$ in V that is not in $\mathrm{span}(S)$. By the Plus/Minus Theorem (5.4.4a), we can insert $\mathbf{v}$ into S, and the resulting set S' will still be linearly independent. If S' spans V, then S' is a basis for V, and we are done. If S' does not span V, then we can insert an appropriate vector into S' to produce a set S'' that is still linearly independent. We can continue inserting vectors in this way until we reach a set with n linearly independent vectors in V. This set will be a basis for V by Theorem 5.4.5. ▮

In the next section we shall give numerical examples that illustrate this theorem.

It can be proved (Exercise 29) that any subspace of a finite-dimensional vector space is finite dimensional. We conclude this section with a theorem which shows that the dimension of a subspace of a finite-dimensional vector space V cannot exceed the dimension of V itself and that the only way a subspace can have the same dimension as V is if the subspace is the entire vector space V. Figure 6 illustrates this idea in R^3. In that figure observe that successively larger subspaces increase in dimension.

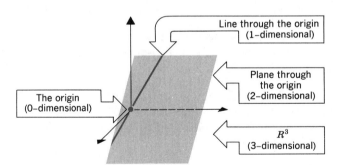

Figure 6

Theorem 5.4.7. *If W is a subspace of a finite-dimensional vector space V, then $\dim(W) \leq \dim(V)$; moreover, if $\dim(W) = \dim(V)$, then $W = V$.*

Proof. Let $S = \{\mathbf{w}_1, \mathbf{w}_2, \ldots, \mathbf{w}_m\}$ be a basis for W. Either S is also a basis for V or it is not. If it is, then $\dim(W) = \dim(V) = m$. If it is not, then by Theorem 5.4.6b, vectors can be added to the linearly independent set S to make it into a basis for V, so $\dim(W) < \dim(V)$. Thus, $\dim(W) \leq \dim(V)$ in all cases. If $\dim(W) = \dim(V)$, then S

is a set of m linearly independent vectors in the m-dimensional vector space V; hence, S is a basis for V by Theorem 5.4.5. This implies that $W = V$ (why?). ▮

ADDITIONAL PROOFS

Proof of Theorem 5.4.4a. Assume that $S = \{\mathbf{v}_1, \mathbf{v}_2, \ldots, \mathbf{v}_r\}$ is a linearly independent set of vectors in V, and $\mathbf{v}$ is a vector in V outside of span(S). To show that $S' = (\mathbf{v}_1, \mathbf{v}_2, \ldots, \mathbf{v}_r, \mathbf{v}\}$ is a linearly independent set, we must show that the only scalars that satisfy

$$k_1\mathbf{v}_1 + k_2\mathbf{v}_2 + \cdots + k_r\mathbf{v}_r + k_{r+1}\mathbf{v} = \mathbf{0} \tag{11}$$

are $k_1 = k_2 = \cdots = k_r = k_{r+1} = 0$. But we must have $k_{r+1} = 0$; otherwise, we could solve (11) for $\mathbf{v}$ as a linear combination of $\mathbf{v}_1, \mathbf{v}_2, \ldots, \mathbf{v}_r$, contradicting the assumption that $\mathbf{v}$ is outside of span(S). Thus, (11) simplifies to

$$k_1\mathbf{v}_1 + k_2\mathbf{v}_2 + \cdots + k_r\mathbf{v}_r = \mathbf{0} \tag{12}$$

which, by the linear independence of $\{\mathbf{v}_1, \mathbf{v}_2, \ldots, \mathbf{v}_r\}$, implies that

$$k_1 = k_2 = \cdots = k_r = 0.$$

Proof of Theorem 5.4.4b. Assume that $S = \{\mathbf{v}_1, \mathbf{v}_2, \ldots, \mathbf{v}_r\}$ is a set of vectors in V, and to be specific suppose that $\mathbf{v}_r$ is a linear combination of $\mathbf{v}_1, \mathbf{v}_2, \ldots, \mathbf{v}_{r-1}$, say

$$\mathbf{v}_r = c_1\mathbf{v}_1 + c_2\mathbf{v}_2 + \cdots + c_{r-1}\mathbf{v}_{r-1} \tag{13}$$

We want to show that if $\mathbf{v}_r$ is removed from S, then the remaining set of vectors $\{\mathbf{v}_1, \mathbf{v}_2, \ldots, \mathbf{v}_{r-1}\}$ still spans span(S); that is, we must show that every vector $\mathbf{w}$ in span(S) is expressible as a linear combination of $\{\mathbf{v}_1, \mathbf{v}_2, \ldots, \mathbf{v}_{r-1}\}$. But if $\mathbf{w}$ is in span(S), then $\mathbf{w}$ is expressible in the form

$$\mathbf{w} = k_1\mathbf{v}_1 + k_2\mathbf{v}_2 + \cdots + k_{r-1}\mathbf{v}_{r-1} + k_r\mathbf{v}_r$$

or on substituting (13),

$$\mathbf{w} = k_1\mathbf{v}_1 + k_2\mathbf{v}_2 + \cdots + k_{r-1}\mathbf{v}_{r-1} + k_r(c_1\mathbf{v}_1 + c_2\mathbf{v}_2 + \cdots + c_{r-1}\mathbf{v}_{r-1})$$

which expresses $\mathbf{w}$ as a linear combination of $\mathbf{v}_1, \mathbf{v}_2, \ldots, \mathbf{v}_{r-1}$. ▮

EXERCISE SET 5.4

1. Explain why the following sets of vectors are *not* bases for the indicated vector spaces. (Solve this problem by inspection.)
   (a) $\mathbf{u}_1 = (1, 2)$, $\mathbf{u}_2 = (0, 3)$, $\mathbf{u}_3 = (2, 7)$ for R^2
   (b) $\mathbf{u}_1 = (-1, 3, 2)$, $\mathbf{u}_2 = (6, 1, 1)$ for R^3
   (c) $\mathbf{p}_1 = 1 + x + x^2$, $\mathbf{p}_2 = x - 1$ for P_2
   (d) $A = \begin{bmatrix} 1 & 1 \\ 2 & 3 \end{bmatrix}$, $B = \begin{bmatrix} 6 & 0 \\ -1 & 4 \end{bmatrix}$, $C = \begin{bmatrix} 3 & 0 \\ 1 & 7 \end{bmatrix}$, $D = \begin{bmatrix} 5 & 1 \\ 4 & 2 \end{bmatrix}$, $E = \begin{bmatrix} 7 & 1 \\ 2 & 9 \end{bmatrix}$ for M_{22}

2. Which of the following sets of vectors are bases for R^2?
   (a) $(2, 1), (3, 0)$ (b) $(4, 1), (-7, -8)$ (c) $(0, 0), (1, 3)$ (d) $(3, 9), (-4, -12)$

3. Which of the following sets of vectors are bases for R^3?
   (a) $(1, 0, 0), (2, 2, 0), (3, 3, 3)$　　　(b) $(3, 1, -4), (2, 5, 6), (1, 4, 8)$
   (c) $(2, -3, 1), (4, 1, 1), (0, -7, 1)$　　(d) $(1, 6, 4), (2, 4, -1), (-1, 2, 5)$

4. Which of the following sets of vectors are bases for P_2?
   (a) $1 - 3x + 2x^2, 1 + x + 4x^2, 1 - 7x$　　(b) $4 + 6x + x^2, -1 + 4x + 2x^2, 5 + 2x - x^2$
   (c) $1 + x + x^2, x + x^2, x^2$　　　　　　(d) $-4 + x + 3x^2, 6 + 5x + 2x^2, 8 + 4x + x^2$

5. Show that the following set of vectors is a basis for M_{22}.

$$\begin{bmatrix} 3 & 6 \\ 3 & -6 \end{bmatrix}, \quad \begin{bmatrix} 0 & -1 \\ -1 & 0 \end{bmatrix}, \quad \begin{bmatrix} 0 & -8 \\ -12 & -4 \end{bmatrix}, \quad \begin{bmatrix} 1 & 0 \\ -1 & 2 \end{bmatrix}$$

6. Let V be the space spanned by $v_1 = \cos^2 x$, $v_2 = \sin^2 x$, $v_3 = \cos 2x$.
   (a) Show that $S = \{v_1, v_2, v_3\}$ is not a basis for V.　　(b) Find a basis for V.

7. Find the coordinate vector of w relative to the basis $S = \{u_1, u_2\}$ for R^2.
   (a) $u_1 = (1, 0)$, $u_2 = (0, 1)$; $w = (3, -7)$　　(b) $u_1 = (2, -4)$, $u_2 = (3, 8)$; $w = (1, 1)$
   (c) $u_1 = (1, 1)$, $u_2 = (0, 2)$; $w = (a, b)$

8. Find the coordinate vector of v relative to the basis $S = \{v_1, v_2, v_3\}$.
   (a) $v = (2, -1, 3)$; $v_1 = (1, 0, 0)$, $v_2 = (2, 2, 0)$, $v_3 = (3, 3, 3)$
   (b) $v = (5, -12, 3)$; $v_1 = (1, 2, 3)$, $v_2 = (-4, 5, 6)$, $v_3 = (7, -8, 9)$

9. Find the coordinate vector of p relative to the basis $S = \{p_1, p_2, p_3\}$.
   (a) $p = 4 - 3x + x^2$; $p_1 = 1$, $p_2 = x$, $p_3 = x^2$
   (b) $p = 2 - x + x^2$; $p_1 = 1 + x$, $p_2 = 1 + x^2$, $p_3 = x + x^2$

10. Find the coordinate vector of A relative to the basis $S = \{A_1, A_2, A_3, A_4\}$.

$$A = \begin{bmatrix} 2 & 0 \\ -1 & 3 \end{bmatrix}, \quad A_1 = \begin{bmatrix} -1 & 1 \\ 0 & 0 \end{bmatrix}, \quad A_2 = \begin{bmatrix} 1 & 1 \\ 0 & 0 \end{bmatrix}, \quad A_3 = \begin{bmatrix} 0 & 0 \\ 1 & 0 \end{bmatrix}, \quad A_4 = \begin{bmatrix} 0 & 0 \\ 0 & 1 \end{bmatrix}$$

In Exercises 11–16 determine the dimension of and a basis for the solution space of the system.

11.　$\begin{aligned} x_1 + x_2 - x_3 &= 0 \\ -2x_1 - x_2 + 2x_3 &= 0 \\ -x_1 \quad\;\; + x_3 &= 0 \end{aligned}$　
12.　$\begin{aligned} 3x_1 + x_2 + x_3 + x_4 &= 0 \\ 5x_1 - x_2 + x_3 - x_4 &= 0 \end{aligned}$　
13.　$\begin{aligned} x_1 - 4x_2 + 3x_3 - x_4 &= 0 \\ 2x_1 - 8x_2 + 6x_3 - 2x_4 &= 0 \end{aligned}$

14.　$\begin{aligned} x_1 - 3x_2 + x_3 &= 0 \\ 2x_1 - 6x_2 + 2x_3 &= 0 \\ 3x_1 - 9x_2 + 3x_3 &= 0 \end{aligned}$　
15.　$\begin{aligned} 2x_1 + x_2 + 3x_3 &= 0 \\ x_1 \quad\;\; + 5x_3 &= 0 \\ x_2 + x_3 &= 0 \end{aligned}$　
16.　$\begin{aligned} x + y + z &= 0 \\ 3x + 2y - 2z &= 0 \\ 4x + 3y - z &= 0 \\ 6x + 5y + z &= 0 \end{aligned}$

17. Determine bases for the following subspaces of R^3.
    (a) the plane $3x - 2y + 5z = 0$　　　(b) the plane $x - y = 0$
    (c) the line $x = 2t, y = -t, z = 4t$　　(d) all vectors of the form (a, b, c), where $b = a + c$

18. Determine the dimensions of the following subspaces of R^4.
    (a) all vectors of the form $(a, b, c, 0)$
    (b) all vectors of the form (a, b, c, d), where $d = a + b$ and $c = a - b$
    (c) all vectors of the form (a, b, c, d), where $a = b = c = d$

19. Determine the dimension of the subspace of P_3 consisting of all polynomials $a_0 + a_1 x + a_2 x^2 + a_3 x^3$ for which $a_0 = 0$.

20. Find a standard basis vector that can be added to the set $\{v_1, v_2\}$ to produce a basis for R^3.
    (a) $v_1 = (-1, 2, 3)$, $v_2 = (1, -2, -2)$　　(b) $v_1 = (1, -1, 0)$, $v_2 = (3, 1, -2)$

21. Find standard basis vectors that can be added to the set $\{v_1, v_2\}$ to produce a basis for R^4.

$$v_1 = (1, -4, 2, -3), \qquad v_2 = (-3, 8, -4, 6)$$

22. Let $\{v_1, v_2, v_3\}$ be a basis for a vector space V. Show that $\{u_1, u_2, u_3\}$ is also a basis, where $u_1 = v_1$, $u_2 = v_1 + v_2$, and $u_3 = v_1 + v_2 + v_3$.

23. (a) Show that for every positive integer n, one can find $n + 1$ linearly independent vectors in $F(-\infty, \infty)$. [**Hint.** Look for polynomials.]
    (b) Use the result in part (a) to prove that $F(-\infty, \infty)$ is infinite-dimensional.
    (c) Prove that $C(-\infty, \infty)$, $C^m(-\infty, \infty)$, and $C^\infty(-\infty, \infty)$ are infinite-dimensional vector spaces.

24. Let S be a basis for an n-dimensional vector space V. Show that if $v_1, v_2, \ldots, v_r$ form a linearly independent set of vectors in V, then the coordinate vectors $(v_1)_S, (v_2)_S, \ldots, (v_r)_S$ form a linearly independent set in R^n, and conversely.

25. Using the notation from Exercise 24, show that if $v_1, v_2, \ldots, v_r$ span V, then the coordinate vectors $(v_1)_S, (v_2)_S, \ldots, (v_r)_S$ span R^n, and conversely.

26. Find a basis for the subspace of P_2 spanned by the given vectors.
    (a) $-1 + x - 2x^2$, $3 + 3x + 6x^2$, 9 (b) $1 + x$, x^2, $-2 + 2x^2$, $-3x$
    (c) $1 + x - 3x^2$, $2 + 2x - 6x^2$, $3 + 3x - 9x^2$
    [**Hint.** Let S be the standard basis for P_2 and work with the coordinate vectors relative to S; note Exercises 24, 25.]

27. Figure 7 shows a rectangular xy-coordinate system and an $x'y'$-coordinate system with skewed axes. Assuming that 1-unit scales are used on all the axes, find the $x'y'$-coordinates of the points whose xy-coordinates are given.
    (a) $(1, 1)$ (b) $(1, 0)$ (c) $(0, 1)$ (d) (a, b)

Figure 7

28. Figure 8 shows a rectangular xy-coordinate system determined by the unit basis vectors $\mathbf{i}$ and $\mathbf{j}$ and an $x'y'$-coordinate system determined by unit basis vectors u_1 and u_2. Find the $x'y'$-coordinates of the points whose xy-coordinates are given.
    (a) $(\sqrt{3}, 1)$ (b) $(1, 0)$ (c) $(0, 1)$ (d) (a, b)

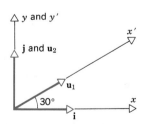

Figure 8

29. Prove: Any subspace of a finite-dimensional vector space is finite dimensional.

5.5 ROW SPACE, COLUMN SPACE, AND NULLSPACE

In this section we shall study three important vector spaces associated with matrices. Our work here will provide us with a deeper understanding of the relationships between the solutions of a linear system and properties of its coefficient matrix.

We begin with some definitions.

ROW AND COLUMN VECTORS

Definition. For an $m \times n$ matrix

$$A = \begin{bmatrix} a_{11} & a_{12} & \cdots & a_{1n} \\ a_{21} & a_{22} & \cdots & a_{2n} \\ \vdots & \vdots & & \vdots \\ a_{m1} & a_{m2} & \cdots & a_{mn} \end{bmatrix}$$

the vectors

$$\mathbf{r}_1 = [a_{11} \quad a_{12} \quad \cdots \quad a_{1n}]$$
$$\mathbf{r}_2 = [a_{21} \quad a_{22} \quad \cdots \quad a_{2n}]$$
$$\vdots$$
$$\mathbf{r}_m = [a_{m1} \quad a_{m2} \quad \cdots \quad a_{mn}]$$

in R^n formed from the rows of A are called the **row vectors** of A, and the vectors

$$\mathbf{c}_1 = \begin{bmatrix} a_{11} \\ a_{21} \\ \vdots \\ a_{m1} \end{bmatrix}, \quad \mathbf{c}_2 = \begin{bmatrix} a_{12} \\ a_{22} \\ \vdots \\ a_{m2} \end{bmatrix}, \quad \ldots, \quad \mathbf{c}_n = \begin{bmatrix} a_{1n} \\ a_{2n} \\ \vdots \\ a_{mn} \end{bmatrix}$$

in R^m formed from the columns of A are called the **column vectors** of A.

Example 1 Let

$$A = \begin{bmatrix} 2 & 1 & 0 \\ 3 & -1 & 4 \end{bmatrix}$$

The row vectors of A are

$$\mathbf{r}_1 = [2 \quad 1 \quad 0] \quad \text{and} \quad \mathbf{r}_2 = [3 \quad -1 \quad 4]$$

and the column vectors of A are

$$\mathbf{c}_1 = \begin{bmatrix} 2 \\ 3 \end{bmatrix}, \quad \mathbf{c}_2 = \begin{bmatrix} 1 \\ -1 \end{bmatrix}, \quad \text{and} \quad \mathbf{c}_3 = \begin{bmatrix} 0 \\ 4 \end{bmatrix} \quad \blacktriangle$$

The following definition defines three important vector spaces associated with a matrix.

ROW SPACE,
COLUMN
SPACE, AND
NULLSPACE

Definition. If A is an $m \times n$ matrix, then the subspace of R^n spanned by the row vectors of A is called the ***row space*** of A, and the subspace of R^m spanned by the column vectors is called the ***column space*** of A. The solution space of the homogeneous system of equations $A\mathbf{x} = \mathbf{0}$, which is a subspace of R^n, is called the ***nullspace*** of A.

In this section and the next we shall be concerned with the following general questions:

- What relationships exist between the solutions of a linear system $A\mathbf{x} = \mathbf{b}$ and the row space, column space, and nullspace of the coefficient matrix A?
- What relationships exist between the row space, column space, and nullspace of a matrix?

To investigate the first of these questions, suppose that

$$A = \begin{bmatrix} a_{11} & a_{12} & \cdots & a_{1n} \\ a_{21} & a_{22} & \cdots & a_{2n} \\ \vdots & \vdots & & \vdots \\ a_{m1} & a_{m2} & \cdots & a_{mn} \end{bmatrix} \quad \text{and} \quad \mathbf{x} = \begin{bmatrix} x_1 \\ x_2 \\ \vdots \\ x_n \end{bmatrix}$$

It follows from Formula (7) of Section 1.3 that if $\mathbf{c}_1, \mathbf{c}_2, \ldots, \mathbf{c}_n$ denote the column vectors of A, then the product $A\mathbf{x}$ can be expressed as a linear combination of these column vectors with coefficients from $\mathbf{x}$; that is,

$$A\mathbf{x} = x_1\mathbf{c}_1 + x_2\mathbf{c}_2 + \cdots + x_n\mathbf{c}_n \tag{1}$$

Thus, a linear system, $A\mathbf{x} = \mathbf{b}$, of m equations in n unknowns can be written as

$$x_1\mathbf{c}_1 + x_2\mathbf{c}_2 + \cdots + x_n\mathbf{c}_n = \mathbf{b} \tag{2}$$

from which we conclude that $A\mathbf{x} = \mathbf{b}$ is consistent if and only if $\mathbf{b}$ is expressible as a linear combination of the column vectors of A or, equivalently, if and only if $\mathbf{b}$ is in the column space of A. This yields the following theorem.

Theorem 5.5.1. *A system of linear equations $A\mathbf{x} = \mathbf{b}$ is consistent if and only if $\mathbf{b}$ is in the column space of A.*

Example 2 Let $A\mathbf{x} = \mathbf{b}$ be the linear system

$$\begin{bmatrix} -1 & 3 & 2 \\ 1 & 2 & -3 \\ 2 & 1 & -2 \end{bmatrix} \begin{bmatrix} x_1 \\ x_2 \\ x_3 \end{bmatrix} = \begin{bmatrix} 1 \\ -9 \\ -3 \end{bmatrix}$$

Show that **b** is in the column space of A, and express **b** as a linear combination of the column vectors of A.

Solution. Solving the system by Gaussian elimination yields (verify)

$$x_1 = 2, \qquad x_2 = -1, \qquad x_3 = 3$$

Since the system is consistent, **b** is in the column space of A. Moreover, from (2) and the solution obtained, it follows that

$$2\begin{bmatrix} -1 \\ 1 \\ 2 \end{bmatrix} - \begin{bmatrix} 3 \\ 2 \\ 1 \end{bmatrix} + 3\begin{bmatrix} 2 \\ -3 \\ -2 \end{bmatrix} = \begin{bmatrix} 1 \\ -9 \\ -3 \end{bmatrix}$$

RELATIONSHIP BETWEEN SOLUTIONS OF $A\mathbf{x} = \mathbf{0}$ AND $A\mathbf{x} = \mathbf{b}$

The next theorem establishes a fundamental relationship between the solutions of a nonhomogeneous linear system $A\mathbf{x} = \mathbf{b}$ and those of the corresponding homogeneous linear system $A\mathbf{x} = \mathbf{0}$ with the same coefficient matrix.

> **Theorem 5.5.2.** *If $\mathbf{x}_0$ denotes any single solution of a consistent nonhomogeneous linear system $A\mathbf{x} = \mathbf{b}$, and if $\mathbf{v}_1, \mathbf{v}_2, \ldots, \mathbf{v}_k$ form a basis for the nullspace of A, that is, the solution space of the homogeneous system $A\mathbf{x} = \mathbf{0}$, then every solution of $A\mathbf{x} = \mathbf{b}$ can be expressed in the form*
>
> $$\mathbf{x} = \mathbf{x}_0 + c_1\mathbf{v}_1 + c_2\mathbf{v}_2 + \cdots + c_k\mathbf{v}_k \qquad (3)$$
>
> *and, conversely, for all choices of scalars $c_1, c_2, \ldots, c_k$, the vector $\mathbf{x}$ in this formula is a solution of $A\mathbf{x} = \mathbf{b}$.*

Proof. Assume that $\mathbf{x}_0$ is any fixed solution of $A\mathbf{x} = \mathbf{b}$ and that $\mathbf{x}$ is an arbitrary solution. Then

$$A\mathbf{x}_0 = \mathbf{b} \qquad \text{and} \qquad A\mathbf{x} = \mathbf{b}$$

Subtracting these equations yields

$$A\mathbf{x} - A\mathbf{x}_0 = \mathbf{0}$$

or

$$A(\mathbf{x} - \mathbf{x}_0) = \mathbf{0}$$

which shows that $\mathbf{x} - \mathbf{x}_0$ is a solution of the homogeneous system $A\mathbf{x} = \mathbf{0}$. Since $\mathbf{v}_1, \mathbf{v}_2, \ldots, \mathbf{v}_k$ is a basis for the solution space of this system, we can express $\mathbf{x} - \mathbf{x}_0$ as a linear combination of these vectors, say

$$\mathbf{x} - \mathbf{x}_0 = c_1\mathbf{v}_1 + c_2\mathbf{v}_2 + \cdots + c_k\mathbf{v}_k$$

Thus,

$$\mathbf{x} = \mathbf{x}_0 + c_1\mathbf{v}_1 + c_2\mathbf{v}_2 + \cdots + c_k\mathbf{v}_k$$

which proves the first part of the theorem. Conversely, for all choices of the scalars c_1, $c_2, \ldots, c_k$ in (3) we have

$$A\mathbf{x} = A(\mathbf{x}_0 + c_1\mathbf{v}_1 + c_2\mathbf{v}_2 + \cdots + c_k\mathbf{v}_k)$$

or

$$A\mathbf{x} = A\mathbf{x}_0 + c_1(A\mathbf{v}_1) + c_2(A\mathbf{v}_2) + \cdots + c_k(A\mathbf{v}_k)$$

But $\mathbf{x}_0$ is a solution of the nonhomogeneous system and $\mathbf{v}_1, \mathbf{v}_2, \ldots, \mathbf{v}_k$ are solutions of the homogeneous system, so the last equation implies that

$$A\mathbf{x} = \mathbf{b} + \mathbf{0} + \mathbf{0} + \cdots + \mathbf{0} = \mathbf{b}$$

which shows that $\mathbf{x}$ is a solution of $A\mathbf{x} = \mathbf{b}$. ▮

REMARK. There is some terminology associated with Formula (3). The vector $\mathbf{x}_0$ is called a ***particular solution*** of $A\mathbf{x} = \mathbf{b}$. The expression $\mathbf{x}_0 + c_1\mathbf{v}_1 + c_2\mathbf{v}_2 + \cdots + c_k\mathbf{v}_k$ is called the ***general solution*** of $A\mathbf{x} = \mathbf{b}$, and the expression $c_1\mathbf{v}_1 + c_2\mathbf{v}_2 + \cdots + c_k\mathbf{v}_k$ is called the ***general solution*** of $A\mathbf{x} = \mathbf{0}$. With this terminology, Formula (3) states, *the general solution of $A\mathbf{x} = \mathbf{b}$ is the sum of any particular solution of $A\mathbf{x} = \mathbf{b}$ and the general solution of $A\mathbf{x} = \mathbf{0}$.*

For linear systems with two or three unknowns, Theorem 5.5.2 has a nice geometric interpretation in R^2 and R^3. For example, consider the case where $A\mathbf{x} = \mathbf{0}$ and $A\mathbf{x} = \mathbf{b}$ are linear systems with two unknowns. The solutions of $A\mathbf{x} = \mathbf{0}$ form a subspace of R^2, and hence constitute a line through the origin, the origin only, or all of R^2. From Theorem 5.5.2, the solutions of $A\mathbf{x} = \mathbf{b}$ can be obtained by adding any particular solution of $A\mathbf{x} = \mathbf{b}$, say $\mathbf{x}_0$, to the solutions of $A\mathbf{x} = \mathbf{0}$. Assuming that $\mathbf{x}_0$ is positioned with its initial point at the origin, this has the geometric effect of translating the solution space of $A\mathbf{x} = \mathbf{0}$ so that the point at the origin is moved to the tip of $\mathbf{x}_0$ (Figure 1). This means that the solution vectors of $A\mathbf{x} = \mathbf{b}$ form a line through the tip of $\mathbf{x}_0$, the point at the tip of $\mathbf{x}_0$, or all of R^2. (Can you visualize the last case?) Similarly, for linear systems with three unknowns, the solutions of $A\mathbf{x} = \mathbf{b}$ constitute a plane through the tip of any particular solution $\mathbf{x}_0$, a line through the tip of $\mathbf{x}_0$, the point at the tip of $\mathbf{x}_0$, or all of R^3.

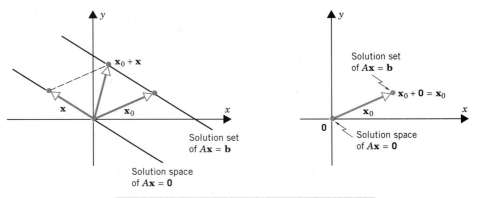

Figure 1

Adding $\mathbf{x}_0$ to each vector $\mathbf{x}$ in the solution space of $A\mathbf{x} = \mathbf{0}$ translates the solution space.

Example 3 In Example 3 of Section 1.2 we solved the nonhomogeneous linear system

$$
\begin{array}{rcl}
x_1 + 3x_2 - 2x_3 \phantom{{}+2x_4} + 2x_5 \phantom{{}-3x_6} &=& 0 \\
2x_1 + 6x_2 - 5x_3 - 2x_4 + 4x_5 - 3x_6 &=& -1 \\
5x_3 + 10x_4 \phantom{{}+4x_5} + 15x_6 &=& 5 \\
2x_1 + 6x_2 \phantom{{}-5x_3} + 8x_4 + 4x_5 + 18x_6 &=& 6
\end{array}
\tag{4}
$$

and obtained

$$
x_1 = -3r - 4s - 2t, \quad x_2 = r, \quad x_3 = -2s, \quad x_4 = s, \quad x_5 = t, \quad x_6 = \tfrac{1}{3}
$$

This result can be written in vector form as

$$
\begin{bmatrix} x_1 \\ x_2 \\ x_3 \\ x_4 \\ x_5 \\ x_6 \end{bmatrix}
=
\begin{bmatrix} -3r - 4s - 2t \\ r \\ -2s \\ s \\ t \\ \tfrac{1}{3} \end{bmatrix}
=
\begin{bmatrix} 0 \\ 0 \\ 0 \\ 0 \\ 0 \\ \tfrac{1}{3} \end{bmatrix}
+ r
\begin{bmatrix} -3 \\ 1 \\ 0 \\ 0 \\ 0 \\ 0 \end{bmatrix}
+ s
\begin{bmatrix} -4 \\ 0 \\ -2 \\ 1 \\ 0 \\ 0 \end{bmatrix}
+ t
\begin{bmatrix} -2 \\ 0 \\ 0 \\ 0 \\ 1 \\ 0 \end{bmatrix}
$$

which is the general solution of (4). Comparing this to (3), the vector

$$
\mathbf{x}_0 =
\begin{bmatrix} 0 \\ 0 \\ 0 \\ 0 \\ 0 \\ \tfrac{1}{3} \end{bmatrix}
$$

is a particular solution of (4) and

$$
\mathbf{x} = r
\begin{bmatrix} -3 \\ 1 \\ 0 \\ 0 \\ 0 \\ 0 \end{bmatrix}
+ s
\begin{bmatrix} -4 \\ 0 \\ -2 \\ 1 \\ 0 \\ 0 \end{bmatrix}
+ t
\begin{bmatrix} -2 \\ 0 \\ 0 \\ 0 \\ 1 \\ 0 \end{bmatrix}
$$

is the general solution of the homogeneous system

$$
\begin{array}{rcl}
x_1 + 3x_2 - 2x_3 \phantom{{}+2x_4} + 2x_5 \phantom{{}-3x_6} &=& 0 \\
2x_1 + 6x_2 - 5x_3 - 2x_4 + 4x_5 - 3x_6 &=& 0 \\
5x_3 + 10x_4 \phantom{{}+4x_5} + 15x_6 &=& 0 \\
2x_1 + 6x_2 \phantom{{}-5x_3} + 8x_4 + 4x_5 + 18x_6 &=& 0
\end{array}
$$

(verify). ▲

BASES FOR ROW SPACES, COLUMN SPACES, AND NULLSPACES

We first developed elementary row operations for the purpose of solving linear systems, and we know from that work that performing an elementary row operation on an augmented matrix does not change the solution set of the corresponding linear system. It follows that applying an elementary row operation to a matrix A does not change the solution set of the corresponding linear system $A\mathbf{x} = \mathbf{0}$, or, stated another way, it does not change the nullspace of A. Thus, we have the following theorem.

> **Theorem 5.5.3.** *Elementary row operations do not change the nullspace of a matrix.*

Example 4 Find a basis for the nullspace of

$$A = \begin{bmatrix} 2 & 2 & -1 & 0 & 1 \\ -1 & -1 & 2 & -3 & 1 \\ 1 & 1 & -2 & 0 & -1 \\ 0 & 0 & 1 & 1 & 1 \end{bmatrix}$$

Solution. The nullspace of A is the solution space of the homogeneous system

$$\begin{aligned} 2x_1 + 2x_2 - \ x_3 \qquad\quad + x_5 &= 0 \\ -x_1 - \ x_2 + 2x_3 - 3x_4 + x_5 &= 0 \\ x_1 + \ x_2 - 2x_3 \qquad\quad - x_5 &= 0 \\ x_3 + \ x_4 + x_5 &= 0 \end{aligned}$$

In Example 10 of Section 5.4 we showed that the vectors

$$\mathbf{v}_1 = \begin{bmatrix} -1 \\ 1 \\ 0 \\ 0 \\ 0 \end{bmatrix} \quad \text{and} \quad \mathbf{v}_2 = \begin{bmatrix} -1 \\ 0 \\ -1 \\ 0 \\ 1 \end{bmatrix}$$

form a basis for this space. ▲

The following theorem is a companion to Theorem 5.5.3.

> **Theorem 5.5.4.** *Elementary row operations do not change the row space of a matrix.*

Proof. Suppose that the row vectors of a matrix A are $\mathbf{r}_1, \mathbf{r}_2, \ldots, \mathbf{r}_m$, and let B be obtained from A by performing an elementary row operation. We shall show that every vector in the row space of B is also in the row space of A, and conversely that every vector in the row space of A is in the row space of B. We can then conclude that A and B have the same row space.

Consider the possibilities: If the row operation is a row interchange, then B and A have the same row vectors and consequently have the same row space. If the row operation is multiplication of a row by a nonzero scalar or the addition of a multiple of one row to another, then the row vectors $\mathbf{r}_1', \mathbf{r}_2', \ldots, \mathbf{r}_m'$ of B are linear combinations of $\mathbf{r}_1, \mathbf{r}_2, \ldots, \mathbf{r}_m$; thus, they lie in the row space of A. Since a vector space is closed under addition and scalar multiplication, all linear combinations of $\mathbf{r}_1', \mathbf{r}_2', \ldots, \mathbf{r}_m'$ will also lie in the row space of A. Therefore, each vector in the row space of B is in the row space of A.

Since B is obtained from A by performing a row operation, A can be obtained from B by performing the inverse operation (Section 1.5). Thus, the argument above shows that the row space of A is contained in the row space of B.

In light of Theorems 5.5.3 and 5.5.4, one might anticipate that elementary row operations should not change the column space of a matrix. However, this is *not* so— elementary row operations can change the column space. For example, consider the matrix

$$A = \begin{bmatrix} 1 & 3 \\ 2 & 6 \end{bmatrix}$$

The second column is a scalar multiple of the first, so the column space of A consists of all scalar multiples of the first column vector. However, if we add -2 times the first row of A to the second row, we obtain

$$B = \begin{bmatrix} 1 & 3 \\ 0 & 0 \end{bmatrix}$$

Here again the second column is a scalar multiple of the first, so the column space of B consists of all scalar multiples of the first column vector. This is not the same as the column space of A.

Although elementary row operations can change the column space of a matrix, we shall show that whatever relationships of linear independence or linear dependence that exist among the column vectors prior to a row operation will also hold for the corresponding columns of the matrix that results from that operation. To make this more precise, suppose that a matrix B results by performing an elementary row operation on an $m \times n$ matrix A. By Theorem 5.5.3 the two homogeneous linear systems

$$A\mathbf{x} = \mathbf{0} \quad \text{and} \quad B\mathbf{x} = \mathbf{0}$$

have the same solution set. Thus, the first system has a nontrivial solution if and only if the same is true of the second. But if the column vectors of A and B, respectively, are

$$\mathbf{c}_1, \mathbf{c}_2, \ldots, \mathbf{c}_n \quad \text{and} \quad \mathbf{c}_1', \mathbf{c}_2', \ldots, \mathbf{c}_n'$$

then from (2) the two systems can be rewritten as

$$x_1\mathbf{c}_1 + x_2\mathbf{c}_2 + \cdots + x_n\mathbf{c}_n = \mathbf{0} \tag{5}$$

and

$$x_1\mathbf{c}_1' + x_2\mathbf{c}_2' + \cdots + x_n\mathbf{c}_n' = \mathbf{0} \tag{6}$$

Thus, (5) has a nontrivial solution for $x_1, x_2, \ldots, x_n$ if and only if the same is true of (6). This implies that the column vectors of A are linearly independent if and only if the same is true of B. Although we shall omit the proof, this conclusion also applies to any subset of the column vectors. Thus, we have the following result.

Theorem 5.5.5. *If A and B are row equivalent matrices, then:*

(a) A given set of column vectors of A is linearly independent if and only if the corresponding column vectors of B are linearly independent.

(b) A given set of column vectors of A forms a basis for the column space of A if and only if the corresponding column vectors of B form a basis for the column space of B.

The following theorem makes it possible to find bases for the row and column spaces of a matrix in row-echelon form by inspection.

Theorem 5.5.6. *If a matrix R is in row-echelon form, then the row vectors with the leading 1's (i.e., the nonzero row vectors) form a basis for the row space of R, and the column vectors with the leading 1's of the row vectors form a basis for the column space of R.*

Since this result is virtually self-evident when one looks at numerical examples, we shall omit the proof; the proof involves little more than an analysis of the positions of the 0's and 1's of R.

Example 5 The matrix

$$R = \begin{bmatrix} 1 & -2 & 5 & 0 & 3 \\ 0 & 1 & 3 & 0 & 0 \\ 0 & 0 & 0 & 1 & 0 \\ 0 & 0 & 0 & 0 & 0 \end{bmatrix}$$

is in row-echelon form. From Theorem 5.5.6 the vectors

$$\mathbf{r}_1 = [\, 1 \quad -2 \quad 5 \quad 0 \quad 3 \,]$$
$$\mathbf{r}_2 = [\, 0 \quad 1 \quad 3 \quad 0 \quad 0 \,]$$
$$\mathbf{r}_3 = [\, 0 \quad 0 \quad 0 \quad 1 \quad 0 \,]$$

form a basis for the row space of R, and the vectors

$$\mathbf{c}_1 = \begin{bmatrix} 1 \\ 0 \\ 0 \\ 0 \end{bmatrix}, \quad \mathbf{c}_2 = \begin{bmatrix} -2 \\ 1 \\ 0 \\ 0 \end{bmatrix}, \quad \mathbf{c}_4 = \begin{bmatrix} 0 \\ 0 \\ 1 \\ 0 \end{bmatrix}$$

form a basis for the column space of R.

Example 6   Find bases for the row and column spaces of

$$A = \begin{bmatrix} 1 & -3 & 4 & -2 & 5 & 4 \\ 2 & -6 & 9 & -1 & 8 & 2 \\ 2 & -6 & 9 & -1 & 9 & 7 \\ -1 & 3 & -4 & 2 & -5 & -4 \end{bmatrix}$$

Solution.   Since elementary row operations do not change the row space of a matrix, we can find a basis for the row space of A by finding a basis for the row space of any row-echelon form of A. Reducing A to row-echelon form we obtain (verify)

$$R = \begin{bmatrix} 1 & -3 & 4 & -2 & 5 & 4 \\ 0 & 0 & 1 & 3 & -2 & -6 \\ 0 & 0 & 0 & 0 & 1 & 5 \\ 0 & 0 & 0 & 0 & 0 & 0 \end{bmatrix}$$

By Theorem 5.5.6 the nonzero row vectors of R form a basis for the row space of R, and hence form a basis for the row space of A. These basis vectors are

$$\mathbf{r}_1 = [\, 1 \quad -3 \quad 4 \quad -2 \quad 5 \quad 4 \,]$$
$$\mathbf{r}_2 = [\, 0 \quad 0 \quad 1 \quad 3 \quad -2 \quad -6 \,]$$
$$\mathbf{r}_3 = [\, 0 \quad 0 \quad 0 \quad 0 \quad 1 \quad 5 \,]$$

Keeping in mind that A and R may have different column spaces, we cannot find a basis for the column space of A *directly* from the column vectors of R. However, it follows from Theorem 5.5.5b that if we can find a set of column vectors of R that forms a basis for the column space of R, then the *corresponding* column vectors of A will form a basis for the column space of A.

The first, third, and fifth columns of R contain the leading 1's of the row vectors, so

$$\mathbf{c}'_1 = \begin{bmatrix} 1 \\ 0 \\ 0 \\ 0 \end{bmatrix}, \quad \mathbf{c}'_3 = \begin{bmatrix} 4 \\ 1 \\ 0 \\ 0 \end{bmatrix}, \quad \mathbf{c}'_5 = \begin{bmatrix} 5 \\ -2 \\ 1 \\ 0 \end{bmatrix}$$

form a basis for the column space of R; thus the corresponding column vectors of A, namely,

$$\mathbf{c}_1 = \begin{bmatrix} 1 \\ 2 \\ 2 \\ -1 \end{bmatrix}, \quad \mathbf{c}_3 = \begin{bmatrix} 4 \\ 9 \\ 9 \\ -4 \end{bmatrix}, \quad \mathbf{c}_5 = \begin{bmatrix} 5 \\ 8 \\ 9 \\ -5 \end{bmatrix}$$

form a basis for the column space of A.   ▲

Example 7 Find a basis for the space spanned by the vectors

$$\mathbf{v}_1 = (1, -2, 0, 0, 3), \quad \mathbf{v}_2 = (2, -5, -3, -2, 6), \quad \mathbf{v}_3 = (0, 5, 15, 10, 0),$$
$$\mathbf{v}_4 = (2, 6, 18, 8, 6)$$

Solution. Except for a variation in notation, the space spanned by these vectors is the row space of the matrix

$$\begin{bmatrix} 1 & -2 & 0 & 0 & 3 \\ 2 & -5 & -3 & -2 & 6 \\ 0 & 5 & 15 & 10 & 0 \\ 2 & 6 & 18 & 8 & 6 \end{bmatrix}$$

Reducing this matrix to row-echelon form we obtain

$$\begin{bmatrix} 1 & -2 & 0 & 0 & 3 \\ 0 & 1 & 3 & 2 & 0 \\ 0 & 0 & 1 & 1 & 0 \\ 0 & 0 & 0 & 0 & 0 \end{bmatrix}$$

The nonzero row vectors in this matrix are

$$\mathbf{w}_1 = (1, -2, 0, 0, 3), \quad \mathbf{w}_2 = (0, 1, 3, 2, 0), \quad \mathbf{w}_3 = (0, 0, 1, 1, 0)$$

These vectors form a basis for the row space and consequently form a basis for the subspace of R^5 spanned by $\mathbf{v}_1$, $\mathbf{v}_2$, $\mathbf{v}_3$, and $\mathbf{v}_4$. ▲

Observe that in Example 6 the basis vectors obtained for the column space of A consisted of column vectors of A, but the basis vectors obtained for the row space of A were not all row vectors of A. The following example illustrates a procedure for finding a basis for the row space of a matrix A that consists entirely of row vectors of A.

Example 8 Find a basis for the row space of

$$A = \begin{bmatrix} 1 & -2 & 0 & 0 & 3 \\ 2 & -5 & -3 & -2 & 6 \\ 0 & 5 & 15 & 10 & 0 \\ 2 & 6 & 18 & 8 & 6 \end{bmatrix}$$

consisting entirely of row vectors from A.

Solution. We will transpose A, thereby converting the row space of A into the column space of A^T; then we will use the method of Example 6 to find a basis for the column space of A^T; and then we will transpose again to convert column vectors back to row vectors. Transposing A yields

$$A^T = \begin{bmatrix} 1 & 2 & 0 & 2 \\ -2 & -5 & 5 & 6 \\ 0 & -3 & 15 & 18 \\ 0 & -2 & 10 & 8 \\ 3 & 6 & 0 & 6 \end{bmatrix}$$

Reducing this matrix to row-echelon form yields

$$\begin{bmatrix} 1 & 2 & 0 & 2 \\ 0 & 1 & -5 & -10 \\ 0 & 0 & 0 & 1 \\ 0 & 0 & 0 & 0 \\ 0 & 0 & 0 & 0 \end{bmatrix}$$

The first, second, and fourth columns contain the leading 1's, so the corresponding column vectors in A^T form a basis for the column space of A^T; these are

$$\mathbf{c}_1 = \begin{bmatrix} 1 \\ -2 \\ 0 \\ 0 \\ 3 \end{bmatrix}, \qquad \mathbf{c}_2 = \begin{bmatrix} 2 \\ -5 \\ -3 \\ -2 \\ 6 \end{bmatrix}, \qquad \text{and} \qquad \mathbf{c}_4 = \begin{bmatrix} 2 \\ 6 \\ 18 \\ 8 \\ 6 \end{bmatrix}$$

Transposing again and adjusting the notation appropriately yields the basis vectors

$$\mathbf{r}_1 = [\,1 \quad -2 \quad 0 \quad 0 \quad 3\,], \qquad \mathbf{r}_2 = [\,2 \quad -5 \quad -3 \quad -2 \quad 6\,],$$

and

$$\mathbf{r}_4 = [\,2 \quad 6 \quad 18 \quad 8 \quad 6\,]$$

for the row space of A. ▲

We know from Theorem 5.5.5 that elementary row operations do not alter relationships of linear independence and linear dependence among the column vectors; however, Formulas (5) and (6) imply an even deeper result. Because these formulas actually have *the same scalar coefficients* $x_1, x_2, \ldots, x_n$, it follows that elementary row operations do not alter the *formulas* (linear combinations) that relate linearly dependent column vectors. We omit the formal proof.

Example 9

(a) Find a subset of the vectors

$$\mathbf{v}_1 = (1, -2, 0, 3), \qquad \mathbf{v}_2 = (2, -5, -3, 6),$$
$$\mathbf{v}_3 = (0, 1, 3, 0), \qquad \mathbf{v}_4 = (2, -1, 4, -7), \qquad \mathbf{v}_5 = (5, -8, 1, 2)$$

that forms a basis for the space spanned by these vectors.

(b) Express the vectors not in the basis as a linear combination of the basis vectors.

Solution (a). We begin by constructing a matrix that has $\mathbf{v}_1, \mathbf{v}_2, \ldots, \mathbf{v}_5$ as its column vectors:

$$\begin{bmatrix} 1 & 2 & 0 & 2 & 5 \\ -2 & -5 & 1 & -1 & -8 \\ 0 & -3 & 3 & 4 & 1 \\ 3 & 6 & 0 & -7 & 2 \end{bmatrix} \qquad (7)$$

$$\begin{array}{ccccc} \uparrow & \uparrow & \uparrow & \uparrow & \uparrow \\ \mathbf{v}_1 & \mathbf{v}_2 & \mathbf{v}_3 & \mathbf{v}_4 & \mathbf{v}_5 \end{array}$$

The first part of our problem can be solved by finding a basis for the column space of this matrix. Reducing the matrix to *reduced* row-echelon form and denoting the column vectors of the resulting matrix by $\mathbf{w}_1, \mathbf{w}_2, \mathbf{w}_3, \mathbf{w}_4$, and $\mathbf{w}_5$ yields

$$\begin{bmatrix} 1 & 0 & 2 & 0 & 1 \\ 0 & 1 & -1 & 0 & 1 \\ 0 & 0 & 0 & 1 & 1 \\ 0 & 0 & 0 & 0 & 0 \end{bmatrix} \qquad (8)$$

$$\begin{array}{ccccc} \uparrow & \uparrow & \uparrow & \uparrow & \uparrow \\ \mathbf{w}_1 & \mathbf{w}_2 & \mathbf{w}_3 & \mathbf{w}_4 & \mathbf{w}_5 \end{array}$$

The leading 1's occur in columns 1, 2, and 4, so that by Theorem 5.5.6

$$\{\mathbf{w}_1, \mathbf{w}_2, \mathbf{w}_4\}$$

is a basis for the column space of (8) and consequently

$$\{\mathbf{v}_1, \mathbf{v}_2, \mathbf{v}_4\}$$

is a basis for the column space of (7).

Solution (b). We shall start by expressing $\mathbf{w}_3$ and $\mathbf{w}_5$ as linear combinations of the basis vectors $\mathbf{w}_1, \mathbf{w}_2, \mathbf{w}_4$. The simplest way of doing this is to express $\mathbf{w}_3$ and $\mathbf{w}_5$ in terms of basis vectors with smaller subscripts. Thus, we shall express $\mathbf{w}_3$ as a linear combination of $\mathbf{w}_1$ and $\mathbf{w}_2$, and we shall express $\mathbf{w}_5$ as a linear combination of $\mathbf{w}_1, \mathbf{w}_2$, and $\mathbf{w}_4$. By inspection of (8), these linear combinations are

$$\mathbf{w}_3 = 2\mathbf{w}_1 - \mathbf{w}_2$$
$$\mathbf{w}_5 = \mathbf{w}_1 + \mathbf{w}_2 + \mathbf{w}_4$$

We call these the ***dependency equations***. The corresponding relationships in (7) are

$$\mathbf{v}_3 = 2\mathbf{v}_1 - \mathbf{v}_2$$
$$\mathbf{v}_5 = \mathbf{v}_1 + \mathbf{v}_2 + \mathbf{v}_4$$

The procedure illustrated in the preceding example is sufficiently important that we shall summarize the steps:

Given a set of vectors $S = \{\mathbf{v}_1, \mathbf{v}_2, \ldots, \mathbf{v}_k\}$ in R^n, the following procedure produces a subset of these vectors that forms a basis for span(S) and expresses those vectors of S that are not in the basis as linear combinations of the basis vectors.

Step 1. Form the matrix A having $\mathbf{v}_1, \mathbf{v}_2, \ldots, \mathbf{v}_k$ as its column vectors.

Step 2. Reduce the matrix A to its reduced row-echelon form R, and let $\mathbf{w}_1,$ $\mathbf{w}_2, \ldots, \mathbf{w}_k$ be the column vectors of R.

Step 3. Identify the columns that contain the leading 1's in R. The corresponding column vectors of A are the basis vectors for span (S).

Step 4. Express each column vector of R that does *not* contain a leading 1 as a linear combination of preceding column vectors that do contain leading 1's. (You will be able to do this by inspection.) This yields a set of dependency equations involving the column vectors of R. The corresponding equations for the column vectors of A express the vectors not in the basis as linear combinations of the basis vectors.

EXERCISE SET 5.5

1. List the row vectors and column vectors of the matrix

$$\begin{bmatrix} 2 & -1 & 0 & 1 \\ 3 & 5 & 7 & -1 \\ 1 & 4 & 2 & 7 \end{bmatrix}$$

2. Express the product $A\mathbf{x}$ as a linear combination of the column vectors of A.

(a) $\begin{bmatrix} 2 & 3 \\ -1 & 4 \end{bmatrix}\begin{bmatrix} 1 \\ 2 \end{bmatrix}$

(b) $\begin{bmatrix} 4 & 0 & -1 \\ 3 & 6 & 2 \\ 0 & -1 & 4 \end{bmatrix}\begin{bmatrix} -2 \\ 3 \\ 5 \end{bmatrix}$

(c) $\begin{bmatrix} -3 & 6 & 2 \\ 5 & -4 & 0 \\ 2 & 3 & -1 \\ 1 & 8 & 3 \end{bmatrix}\begin{bmatrix} -1 \\ 2 \\ 5 \end{bmatrix}$

(d) $\begin{bmatrix} 2 & 1 & 5 \\ 6 & 3 & -8 \end{bmatrix}\begin{bmatrix} 3 \\ 0 \\ -5 \end{bmatrix}$

3. Determine whether $\mathbf{b}$ is in the column space of A, and if so, express $\mathbf{b}$ as a linear combination of the column vectors of A.

(a) $A = \begin{bmatrix} 1 & 3 \\ 4 & -6 \end{bmatrix}$; $\mathbf{b} = \begin{bmatrix} -2 \\ 10 \end{bmatrix}$

(b) $A = \begin{bmatrix} 1 & 1 & 2 \\ 1 & 0 & 1 \\ 2 & 1 & 3 \end{bmatrix}$; $\mathbf{b} = \begin{bmatrix} -1 \\ 0 \\ 2 \end{bmatrix}$

(c) $A = \begin{bmatrix} 1 & -1 & 1 \\ 9 & 3 & 1 \\ 1 & 1 & 1 \end{bmatrix}$; $\mathbf{b} = \begin{bmatrix} 5 \\ 1 \\ -1 \end{bmatrix}$

(d) $A = \begin{bmatrix} 1 & -1 & 1 \\ 1 & 1 & -1 \\ -1 & -1 & 1 \end{bmatrix}$; $\mathbf{b} = \begin{bmatrix} 2 \\ 0 \\ 0 \end{bmatrix}$

(e) $A = \begin{bmatrix} 1 & 2 & 0 & 1 \\ 0 & 1 & 2 & 1 \\ 1 & 2 & 1 & 3 \\ 0 & 1 & 2 & 2 \end{bmatrix}$; $\mathbf{b} = \begin{bmatrix} 4 \\ 3 \\ 5 \\ 7 \end{bmatrix}$

4. Suppose that $x_1 = -1$, $x_2 = 2$, $x_3 = 4$, $x_4 = -3$ is a solution of a nonhomogeneous linear system $A\mathbf{x} = \mathbf{b}$ and that the solution set of the homogeneous system $A\mathbf{x} = \mathbf{0}$ is given by the formulas

$$x_1 = -3r + 4s, \quad x_2 = r - s, \quad x_3 = r, \quad x_4 = s$$

(a) Find the vector form of the general solution of $A\mathbf{x} = \mathbf{0}$.

(b) Find the vector form of the general solution of $A\mathbf{x} = \mathbf{b}$.

5. Find the vector form of the general solution of the given linear system $A\mathbf{x} = \mathbf{b}$; then use that result to find the vector form of the general solution of $A\mathbf{x} = \mathbf{0}$.

(a) $\begin{aligned} x_1 - 3x_2 &= 1 \\ 2x_1 - 6x_2 &= 2 \end{aligned}$

(b) $\begin{aligned} x_1 + x_2 + 2x_3 &= 5 \\ x_1 \quad\;\; + x_3 &= -2 \\ 2x_1 + x_2 + 3x_3 &= 3 \end{aligned}$

(c) $\begin{aligned} x_1 - 2x_2 + x_3 + 2x_4 &= -1 \\ 2x_1 - 4x_2 + 2x_3 + 4x_4 &= -2 \\ -x_1 + 2x_2 - x_3 - 2x_4 &= 1 \\ 3x_1 - 6x_2 + 3x_3 + 6x_4 &= -3 \end{aligned}$

(d) $\begin{aligned} x_1 + 2x_2 - 3x_3 + x_4 &= 4 \\ -2x_1 + x_2 + 2x_3 + x_4 &= -1 \\ -x_1 + 3x_2 - x_3 + 2x_4 &= 3 \\ 4x_1 - 7x_2 \quad\;\; - 5x_4 &= -5 \end{aligned}$

6. Find a basis for the nullspace of A.

(a) $A = \begin{bmatrix} 1 & -1 & 3 \\ 5 & -4 & -4 \\ 7 & -6 & 2 \end{bmatrix}$

(b) $A = \begin{bmatrix} 2 & 0 & -1 \\ 4 & 0 & -2 \\ 0 & 0 & 0 \end{bmatrix}$

(c) $A = \begin{bmatrix} 1 & 4 & 5 & 2 \\ 2 & 1 & 3 & 0 \\ -1 & 3 & 2 & 2 \end{bmatrix}$

(d) $A = \begin{bmatrix} 1 & 4 & 5 & 6 & 9 \\ 3 & -2 & 1 & 4 & -1 \\ -1 & 0 & -1 & -2 & -1 \\ 2 & 3 & 5 & 7 & 8 \end{bmatrix}$

(e) $A = \begin{bmatrix} 1 & -3 & 2 & 2 & 1 \\ 0 & 3 & 6 & 0 & -3 \\ 2 & -3 & -2 & 4 & 4 \\ 3 & -6 & 0 & 6 & 5 \\ -2 & 9 & 2 & -4 & -5 \end{bmatrix}$

7. In each part, a matrix in row-echelon form is given. By inspection, find bases for the row and column spaces of A.

(a) $\begin{bmatrix} 1 & 0 & 2 \\ 0 & 0 & 1 \\ 0 & 0 & 0 \end{bmatrix}$

(b) $\begin{bmatrix} 1 & -3 & 0 & 0 \\ 0 & 1 & 0 & 0 \\ 0 & 0 & 0 & 0 \\ 0 & 0 & 0 & 0 \end{bmatrix}$

(c) $\begin{bmatrix} 1 & 2 & 4 & 5 \\ 0 & 1 & -3 & 0 \\ 0 & 0 & 1 & -3 \\ 0 & 0 & 0 & 1 \\ 0 & 0 & 0 & 0 \end{bmatrix}$

(d) $\begin{bmatrix} 1 & 2 & -1 & 5 \\ 0 & 1 & 4 & 3 \\ 0 & 0 & 1 & -7 \\ 0 & 0 & 0 & 1 \end{bmatrix}$

8. For the matrices in Exercise 6, find a basis for the row space of A by reducing the matrix to row-echelon form.

9. For the matrices in Exercise 6, find a basis for the column space of A.

10. For the matrices in Exercise 6, find a basis for the row space of A consisting entirely of row vectors of A.

11. Find a basis for the subspace of R^4 spanned by the given vectors.

(a) $(1, 1, -4, -3)$, $(2, 0, 2, -2)$, $(2, -1, 3, 2)$ (b) $(-1, 1, -2, 0)$, $(3, 3, 6, 0)$, $(9, 0, 0, 3)$

(c) $(1, 1, 0, 0)$, $(0, 0, 1, 1)$, $(-2, 0, 2, 2)$, $(0, -3, 0, 3)$

12. Find a subset of the vectors that forms a basis for the space spanned by the vectors; then express each vector that is not in the basis as a linear combination of the basis vectors.

(a) $\mathbf{v}_1 = (1, 0, 1, 1)$, $\mathbf{v}_2 = (-3, 3, 7, 1)$, $\mathbf{v}_3 = (-1, 3, 9, 3)$, $\mathbf{v}_4 = (-5, 3, 5, -1)$

(b) $\mathbf{v}_1 = (1, -2, 0, 3)$, $\mathbf{v}_2 = (2, -4, 0, 6)$, $\mathbf{v}_3 = (-1, 1, 2, 0)$, $\mathbf{v}_4 = (0, -1, 2, 3)$

(c) $\mathbf{v}_1 = (1, -1, 5, 2)$, $\mathbf{v}_2 = (-2, 3, 1, 0)$, $\mathbf{v}_3 = (4, -5, 9, 4)$, $\mathbf{v}_4 = (0, 4, 2, -3)$, $\mathbf{v}_5 = (-7, 18, 2, -8)$

13. Prove that the row vectors of an $n \times n$ invertible matrix A form a basis for R^n.

14. (a) Let

$$A = \begin{bmatrix} 0 & 1 & 0 \\ 1 & 0 & 0 \\ 0 & 0 & 0 \end{bmatrix}$$

and consider a rectangular *xyz*-coordinate system in 3-space. Show that the nullspace of A consists of all points on the *z*-axis, and the column space consists of all points in the *xy*-plane (Figure 2).

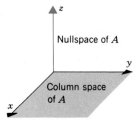

Figure 2

(b) Find a 3×3 matrix whose nullspace is the *x*-axis and whose column space is the *yz*-plane.

5.6 RANK AND NULLITY

In the preceding section we investigated the relationships between systems of linear equations and the row space, column space, and nullspace of the coefficient matrix. In this section we shall be concerned with relationships between the dimensions of the row space, column space, and nullspace of a matrix and its transpose. The results we will obtain are fundamental and will provide deeper insights into linear systems and linear transformations.

THE FOUR FUNDAMENTAL MATRIX SPACES

If we consider a matrix A and its transpose A^T together, then there are six vector spaces of interest:

row space of A	row space of A^T
column space of A	column space of A^T
nullspace of A	nullspace of A^T

However, transposing a matrix converts row vectors into column vectors and column vectors into row vectors, so that, except for a difference in notation, the row space of A^T is the same as the column space of A, and the column space of A^T is the same as the row space of A. This leaves four vector spaces of interest:

row space of A	column space of A
nullspace of A	nullspace of A^T

These are known as the **fundamental matrix spaces** associated with A. If A is an $m \times n$ matrix, then the row space of A and nullspace of A are subspaces of R^n and the column space of A and the nullspace of A^T are subspaces of R^m. Our primary goal in this section is to establish relationships between the dimensions of these four vector spaces.

ROW SPACE AND COLUMN SPACE HAVE EQUAL DIMENSIONS

In Example 6 of Section 5.5, we found that the row space and column space of the matrix

$$A = \begin{bmatrix} 1 & -3 & 4 & -2 & 5 & 4 \\ 2 & -6 & 9 & -1 & 8 & 2 \\ 2 & -6 & 9 & -1 & 9 & 7 \\ -1 & 3 & -4 & 2 & -5 & -4 \end{bmatrix}$$

each have three basis vectors; that is, both are three-dimensional. It is not accidental that these dimensions are the same; it is a consequence of the following general result.

Theorem 5.6.1. *If A is any matrix, then the row space and column space of A have the same dimension.*

Proof. Let R be the reduced row-echelon form of A. It follows from Theorem 5.5.4 that

$$\dim(\text{row space of } A) = \dim(\text{row space of } R)$$

and it follows from Theorem 5.5.5b that

$$\dim(\text{column space of } A) = \dim(\text{column space of } R)$$

Thus, the proof will be complete if we can show that the row space and column space of R have the same dimension. But the dimension of the row space of R is the number of nonzero rows and the dimension of the column space of R is the number of columns that contain leading 1's (Theorem 5.5.6). However, the nonzero rows are precisely the rows in which the leading 1's occur, so the number of leading 1's and the number of nonzero rows is the same. This shows that the row space and column space of R have the same dimension. ▮

The dimensions of the row space, column space, and nullspace of a matrix are such important numbers that there is some notation and terminology associated with them.

RANK AND NULLITY

Definition. The common dimension of the row space and column space of a matrix A is called the **rank** of A and is denoted by rank(A); the dimension of the nullspace of A is called the **nullity** of A and is denoted by nullity(A).

Example 1 Find the rank and nullity of the matrix

$$A = \begin{bmatrix} -1 & 2 & 0 & 4 & 5 & -3 \\ 3 & -7 & 2 & 0 & 1 & 4 \\ 2 & -5 & 2 & 4 & 6 & 1 \\ 4 & -9 & 2 & -4 & -4 & 7 \end{bmatrix}$$

Solution. The reduced row-echelon form of A is

$$\begin{bmatrix} 1 & 0 & -4 & -28 & -37 & 13 \\ 0 & 1 & -2 & -12 & -16 & 5 \\ 0 & 0 & 0 & 0 & 0 & 0 \\ 0 & 0 & 0 & 0 & 0 & 0 \end{bmatrix} \qquad (1)$$

(verify). Since there are two nonzero rows (or equivalently, two leading 1's), the row space and column space are both two-dimensional, so rank$(A) = 2$. To find the nullity of A, we must find the dimension of the solution space of the linear system $A\mathbf{x} = \mathbf{0}$. This system can be solved by reducing the augmented matrix to reduced row-echelon form. The resulting matrix will be identical to (1), except with an additional last column of zeros, and the corresponding system of equations will be

$$x_1 - 4x_3 - 28x_4 - 37x_5 + 13x_6 = 0$$
$$x_2 - 2x_3 - 12x_4 - 16x_5 + 5x_6 = 0$$

or, on solving for the leading variables,

$$x_1 = 4x_3 + 28x_4 + 37x_5 - 13x_6$$
$$x_2 = 2x_3 + 12x_4 + 16x_5 - 5x_6 \qquad (2)$$

It follows that the general solution of the system is

$$x_1 = 4r + 28s + 37t - 13u$$
$$x_2 = 2r + 12s + 16t - 5u$$
$$x_3 = r$$
$$x_4 = s$$
$$x_5 = t$$
$$x_6 = u$$

or equivalently,

$$\begin{bmatrix} x_1 \\ x_2 \\ x_3 \\ x_4 \\ x_5 \\ x_6 \end{bmatrix} = r \begin{bmatrix} 4 \\ 2 \\ 1 \\ 0 \\ 0 \\ 0 \end{bmatrix} + s \begin{bmatrix} 28 \\ 12 \\ 0 \\ 1 \\ 0 \\ 0 \end{bmatrix} + t \begin{bmatrix} 37 \\ 16 \\ 0 \\ 0 \\ 1 \\ 0 \end{bmatrix} + u \begin{bmatrix} -13 \\ -5 \\ 0 \\ 0 \\ 0 \\ 1 \end{bmatrix} \qquad (3)$$

The four vectors on the right side of (3) form a basis for the solution space, so nullity$(A) = 4$. ▲

The following theorem shows that a matrix and its transpose have the same rank.

Theorem 5.6.2. *If A is any matrix, then rank$(A) =$ rank(A^T).*

Proof

$$\text{rank}(A) = \dim(\text{row space of } A) = \dim(\text{column space of } A^T) = \text{rank}(A^T). \blacksquare$$

The following theorem establishes an important relationship between the rank and nullity of a matrix.

DIMENSION THEOREM

Theorem 5.6.3 (*Dimension Theorem for Matrices*). *If A is a matrix with n columns, then*

$$rank(A) + nullity(A) = n \qquad (4)$$

Proof. Since A has n columns, the homogeneous linear system $A\mathbf{x} = \mathbf{0}$ has n unknowns (variables). These fall into two categories: the leading variables and the free variables. Thus,

$$\begin{bmatrix} \text{number of leading} \\ \text{variables} \end{bmatrix} + \begin{bmatrix} \text{number of free} \\ \text{variables} \end{bmatrix} = n$$

But the number of leading variables is the same as the number of leading 1's in the reduced row-echelon form of A, and this is the rank of A. Thus,

$$\text{rank}(A) + \begin{bmatrix} \text{number of free} \\ \text{variables} \end{bmatrix} = n$$

The number of free variables is equal to the nullity of A. This is so because the nullity of A is the dimension of the solution space of $A\mathbf{x} = \mathbf{0}$, which is the same as the number of parameters in the general solution [see (3), for example], which is the same as the number of free variables. Thus,

$$\text{rank}(A) + \text{nullity}(A) = n \quad \blacksquare$$

The proof of the preceding theorem contains two results that are of importance in their own right.

Theorem 5.6.4. *If A is an $m \times n$ matrix, then:*

(a) rank$(A) =$ *the number of leading variables in the solution of $A\mathbf{x} = \mathbf{0}$.*
(b) nullity$(A) =$ *the number of parameters in the solution of $A\mathbf{x} = \mathbf{0}$.*

Example 2 The matrix

$$A = \begin{bmatrix} -1 & 2 & 0 & 4 & 5 & -3 \\ 3 & -7 & 2 & 0 & 1 & 4 \\ 2 & -5 & 2 & 4 & 6 & 1 \\ 4 & -9 & 2 & -4 & -4 & 7 \end{bmatrix}$$

has 6 columns, so

$$\text{rank}(A) + \text{nullity}(A) = 6$$

This is consistent with Example 1, where we showed that rank$(A) = 2$ and nullity$(A) = 4$. ▲

Example 3 Find the number of parameters in the solution set of $A\mathbf{x} = \mathbf{0}$ if A is a 5×7 matrix of rank 3.

Solution. From (4)

$$\text{nullity}(A) = n - \text{rank}(A) = 7 - 3 = 4$$

Thus, there are four parameters. ▲

Suppose now that A is an $m \times n$ matrix of rank r; it follows from Theorem 5.6.2 that A^T is an $n \times m$ matrix of rank r. Applying Theorem 5.6.3 to A and A^T yields

$$\text{nullity}(A) = n - r, \qquad \text{nullity}(A^T) = m - r$$

from which we deduce the following table relating the dimensions of the four fundamental spaces of a matrix A of rank r.

Fundamental Space	Dimension
Row space of A	r
Column space of A	r
Nullspace of A	$n - r$
Nullspace of A^T	$m - r$

MAXIMUM VALUE FOR RANK

If A is an $m \times n$ matrix, then the row vectors lie in R^n and the column vectors lie in R^m. This implies that the row space of A is at most n-dimensional and that the column space is at most m-dimensional. Since the row and column spaces have the same dimension (the rank of A), we must conclude that if $m \neq n$, then the rank of A is at most the smaller of the values of m and n. We denote this by writing

$$\text{rank}(A) \leq \min(m, n) \tag{5}$$

where $\min(m, n)$ denotes the smaller of the numbers m and n if $m \neq n$ or their common value if $m = n$.

Example 4 If A is a 7×4 matrix, then the rank of A is at most 4 and, consequently, the seven row vectors must be linearly dependent. If A is a 4×7 matrix, then again the rank of A is at most 4 and, consequently, the seven column vectors must be linearly dependent. ▲

LINEAR SYSTEMS OF *m* EQUATIONS IN *n* UNKNOWNS

In earlier sections we obtained a wide range of theorems concerning linear systems of n equations in n unknowns. (See Theorem 4.3.4.) We shall now turn our attention to linear systems of m equations in n unknowns in which m and n need not be the same.

The following theorem provides conditions under which a linear system of m equations in n unknowns is guaranteed to be consistent.

Theorem 5.6.5 (*The Consistency Theorem*). *If $Ax = b$ is a linear system of m equations in n unknowns, then the following are equivalent.*

(a) $Ax = b$ is consistent.
(b) b is in the column space of A.
(c) The coefficient matrix A and the augmented matrix $[A \mid b]$ have the same rank.

Proof. It suffices to prove the two equivalences $(a) \Leftrightarrow (b)$ and $(b) \Leftrightarrow (c)$, since it will then follow as a matter of logic that $(a) \Leftrightarrow (c)$.

$(a) \Leftrightarrow (b)$. See Theorem 5.5.1.

$(b) \Rightarrow (c)$. We will show that if b is in the column space of A, then the column spaces of A and $[A \mid b]$ are actually the same, from which it will follow that these two matrices have the same rank.

By definition, the column space of a matrix is the space spanned by its column vectors, so the column spaces of A and $[A \mid b]$ can be expressed as

$$\text{span}\{c_1, c_2, \ldots, c_n\} \quad \text{and} \quad \text{span}\{c_1, c_2, \ldots, c_n, b\}$$

respectively. If b is in the column space of A, then each vector in the set $\{c_1, c_2, \ldots, c_n, b\}$ is a linear combination of the vectors in $\{c_1, c_2, \ldots, c_n\}$ and conversely (why?). Thus, from Theorem 5.2.4 the column spaces of A and $[A \mid b]$ are the same.

$(c) \Rightarrow (b)$. Assume that A and $[A \mid b]$ have the same rank r. By Theorem 5.4.4, there is some subset of the column vectors of A that forms a basis for the column space of A. Suppose that those column vectors are

$$c_1', c_2', \ldots, c_r'$$

These r basis vectors also belong to the r-dimensional column space of $[A \mid b]$; hence they also form a basis for the column space of $[A \mid b]$ by Theorem 5.4.6a. This means that b is expressible as a linear combination of $c_1', c_2', \ldots, c_r'$, and consequently b lies in the column space of A. ▮

It is not hard to visualize why this theorem is true if one views the rank of a matrix as the number of nonzero rows in its reduced row-echelon form. For example, the augmented matrix for the system

$$
\begin{aligned}
x_1 - 2x_2 - 3x_3 + 2x_4 &= -4 \\
-3x_1 + 7x_2 - x_3 + x_4 &= -3 \\
2x_1 - 5x_2 + 4x_3 - 3x_4 &= 7 \\
-3x_1 + 6x_2 + 9x_3 - 6x_4 &= -1
\end{aligned}
$$

is

$$
\left[\begin{array}{cccc|c}
1 & -2 & -3 & 2 & -4 \\
-3 & 7 & -1 & 1 & -3 \\
2 & -5 & 4 & -3 & 7 \\
-3 & 6 & 9 & -6 & -1
\end{array}\right]
$$

which has the following reduced row-echelon form (verify):

$$
\left[\begin{array}{cccc|c}
1 & 0 & -23 & 16 & 0 \\
0 & 1 & -10 & 7 & 0 \\
0 & 0 & 0 & 0 & 1 \\
0 & 0 & 0 & 0 & 0
\end{array}\right]
$$

Because of the row

$$0 \quad 0 \quad 0 \quad 0 \quad 1$$

we see that the system is inconsistent. However, it is also because of this row that the reduced row-echelon form of the augmented matrix has fewer zero rows than the reduced row-echelon form of the coefficient matrix. This forces the coefficient matrix and the augmented matrix for the system to have different ranks.

The Consistency Theorem is concerned with conditions under which a linear system $A\mathbf{x} = \mathbf{b}$ is consistent for a *specific* vector $\mathbf{b}$. The following theorem is concerned with conditions under which a linear system is consistent for *all possible* choices of $\mathbf{b}$.

Theorem 5.6.6. *If $A\mathbf{x} = \mathbf{b}$ is a linear system of m equations in n unknowns, then the following are equivalent.*

(a) $A\mathbf{x} = \mathbf{b}$ *is consistent for every $m \times 1$ matrix $\mathbf{b}$.*
(b) *The column vectors of A span R^m.*
(c) $rank(A) = m$.

Proof. It suffices to prove the two equivalences $(a) \Leftrightarrow (b)$ and $(a) \Leftrightarrow (c)$, since it will then follow as a matter of logic that $(b) \Leftrightarrow (c)$.

$(a) \Leftrightarrow (b)$. From Formula (2) of Section 5.5, the system $A\mathbf{x} = \mathbf{b}$ can be expressed as

$$x_1\mathbf{c}_1 + x_2\mathbf{c}_2 + \cdots + x_n\mathbf{c}_n = \mathbf{b}$$

from which we can conclude that $A\mathbf{x} = \mathbf{b}$ is consistent for every $m \times 1$ matrix $\mathbf{b}$ if and only if every such $\mathbf{b}$ is expressible as a linear combination of the column vectors $\mathbf{c}_1$, $\mathbf{c}_2, \ldots, \mathbf{c}_n$, or equivalently, if and only if these column vectors span R^m.

$(a) \Rightarrow (c)$. From the assumption that $A\mathbf{x} = \mathbf{b}$ is consistent for every $m \times 1$ matrix $\mathbf{b}$, and from parts (a) and (b) of the Consistency Theorem (5.6.5), it follows that every vector $\mathbf{b}$ in R^m lies in the column space of A; that is, the column space of A is all of R^m. Thus, $\text{rank}(A) = \dim(R^m) = m$.

$(c) \Rightarrow (a)$. From the assumption that $\text{rank}(A) = m$, it follows that the column space of A is a subspace of R^m of dimension m, and hence must be all of R^m by part (b) of Theorem 5.4.7. It now follows from parts (a) and (b) of the Consistency Theorem (5.6.5) that $A\mathbf{x} = \mathbf{b}$ is consistent for every vector $\mathbf{b}$ in R^m, since every such $\mathbf{b}$ is in the column space of A. ▮

A linear system with more equations than unknowns is called an ***overdetermined linear system***. If $A\mathbf{x} = \mathbf{b}$ is an overdetermined linear system of m equations in n unknowns (so that $m > n$), then the column vectors of A cannot span R^m (why?); it follows from the last theorem that *for a fixed $m \times n$ matrix A with $m > n$ the overdetermined linear system $A\mathbf{x} = \mathbf{b}$ cannot be consistent for every possible $\mathbf{b}$.*

Example 5 The linear system

$$
\begin{aligned}
x_1 - 2x_2 &= b_1 \\
x_1 - x_2 &= b_2 \\
x_1 + x_2 &= b_3 \\
x_1 + 2x_2 &= b_4 \\
x_1 + 3x_2 &= b_5
\end{aligned}
$$

is overdetermined, so it cannot be consistent for all possible values of b_1, b_2, b_3, b_4, and b_5. Exact conditions under which the system is consistent can be obtained by solving the linear system by Gauss–Jordan elimination. We leave it for the reader to show that the reduced row-echelon form of the augmented matrix is

$$
\begin{bmatrix}
1 & 0 & 2b_2 - b_1 \\
0 & 1 & b_2 - b_1 \\
0 & 0 & b_3 - 3b_2 + 2b_1 \\
0 & 0 & b_4 - 4b_2 + 3b_1 \\
0 & 0 & b_5 - 5b_2 + 4b_1
\end{bmatrix}
$$

Thus, the system is consistent if and only if b_1, b_2, b_3, b_4, and b_5 satisfy the conditions

$$
\begin{aligned}
2b_1 - 3b_2 + b_3 \quad\quad &= 0 \\
3b_1 - 4b_2 \quad\quad + b_4 \quad &= 0 \\
4b_1 - 5b_2 \quad\quad\quad\quad + b_5 &= 0
\end{aligned}
$$

or, on solving this homogeneous linear system,

$$
b_1 = 5r - 4s, \quad\quad b_2 = 4r - 3s, \quad\quad b_3 = 2r - s, \quad\quad b_4 = r, \quad\quad b_5 = s
$$

where r and s are arbitrary. ▲

In Formula (3) of Theorem 5.5.2, the scalars $c_1, c_2, \ldots, c_k$ are the arbitrary parameters in the general solutions of both $A\mathbf{x} = \mathbf{b}$ and $A\mathbf{x} = \mathbf{0}$. Thus, these two systems have the same number of parameters in their general solutions. Moreover, it follows from part (*b*) of Theorem 5.6.4 that the number of such parameters is nullity(A). This fact and the Dimension Theorem for Matrices (5.6.3) yields the following theorem.

Theorem 5.6.7. *If $A\mathbf{x} = \mathbf{b}$ is a consistent linear system of m equations in n unknowns, and if A has rank r, then the general solution of the system contains $n - r$ parameters.*

Example 6   If A is a 5×7 matrix with rank 4, and if $A\mathbf{x} = \mathbf{b}$ is a consistent linear system, then the general solution of the system contains $7 - 4 = 3$ parameters. ▲

In earlier sections we obtained a wide range of conditions under which a homogeneous linear system $A\mathbf{x} = \mathbf{0}$ of n equations in n unknowns is guaranteed to have only the trivial solution. (See Theorem 4.3.4.) The following theorem obtains some corresponding results for systems of m equations in n unknowns, where m and n may differ.

Theorem 5.6.8. *If A is an $m \times n$ matrix, then the following are equivalent.*

(*a*) *$A\mathbf{x} = \mathbf{0}$ has only the trivial solution.*
(*b*) *The column vectors of A are linearly independent.*
(*c*) *$A\mathbf{x} = \mathbf{b}$ has at most one solution (none or one) for every $m \times 1$ matrix $\mathbf{b}$.*

Proof. If suffices to prove the two equivalences (*a*)⇔(*b*) and (*a*)⇔(*c*), since it will then follow as a matter of logic that (*b*)⇔(*c*).

(*a*)⇔(*b*). If $\mathbf{c}_1, \mathbf{c}_2, \ldots, \mathbf{c}_n$ are the column vectors of A, then the linear system $A\mathbf{x} = \mathbf{0}$ can be written as

$$x_1\mathbf{c}_1 + x_2\mathbf{c}_2 + \cdots + x_n\mathbf{c}_n = \mathbf{0} \tag{6}$$

If $\mathbf{c}_1, \mathbf{c}_2, \ldots, \mathbf{c}_n$ are linearly independent, then this equation is satisfied only by $x_1 = x_2 = \cdots = x_n = 0$, which means that $A\mathbf{x} = \mathbf{0}$ has only the trivial solution. Conversely, if $A\mathbf{x} = \mathbf{0}$ has only the trivial solution, then (6) is satisfied only by $x_1 = x_2 = \cdots = x_n = 0$, which means that $\mathbf{c}_1, \mathbf{c}_2, \ldots, \mathbf{c}_n$ are linearly independent.

(*a*)⇒(*c*). Assume that $A\mathbf{x} = \mathbf{0}$ has only the trivial solution. Either $A\mathbf{x} = \mathbf{b}$ is consistent or it is not. If it is not consistent, then there are no solutions of $A\mathbf{x} = \mathbf{b}$, and we are done. If $A\mathbf{x} = \mathbf{b}$ is consistent, let $\mathbf{x}_0$ be any solution. From the remark following Theorem 5.5.2 and the fact that $A\mathbf{x} = \mathbf{0}$ has only the trivial solution, we conclude that the general solution of $A\mathbf{x} = \mathbf{b}$ is $\mathbf{x}_0 + \mathbf{0} = \mathbf{x}_0$. Thus, the only solution of $A\mathbf{x} = \mathbf{b}$ is $\mathbf{x}_0$.

(*c*)⇒(*a*). Assume that $A\mathbf{x} = \mathbf{b}$ has at most one solution for every $m \times 1$ matrix $\mathbf{b}$. Then, in particular, $A\mathbf{x} = \mathbf{0}$ has at most one solution. Thus, $A\mathbf{x} = \mathbf{0}$ has only the trivial solution. ▌

A linear system with more unknowns than equations is called an ***underdetermined linear system***. If $A\mathbf{x} = \mathbf{b}$ is a consistent underdetermined linear system of m equations in n unknowns (so that $m < n$), then it follows from Theorem 5.6.7 that the general solution has at least one parameter (why?); hence *a consistent underdetermined linear system must have infinitely many solutions*. Moreover, if $A\mathbf{x} = \mathbf{b}$ is any underdetermined linear system, then the column vectors of A cannot be linearly independent (why?); it follows from Theorem 5.6.3 that *for a fixed $m \times n$ matrix A with $m < n$, the underdetermined linear system $A\mathbf{x} = \mathbf{b}$ has infinitely many solutions for some $\mathbf{b}$*.

REMARK. It also follows from Theorem 5.6.3 that an underdetermined homogeneous linear system has infinitely many solutions; but this was already proved in Chapter 1 (Theorem 1.2.1).

Example 7 If A is a 5×7 matrix, then for every 7×1 matrix $\mathbf{b}$, the linear system $A\mathbf{x} = \mathbf{b}$ is underdetermined. Thus, $A\mathbf{x} = \mathbf{b}$ must be consistent for some $\mathbf{b}$, and for each such $\mathbf{b}$ the general solution must have $7 - r$ parameters, where r is the rank of A. ▲

SUMMARY

In Theorem 4.3.4 we listed eight results that are equivalent to the invertibility of a matrix A. We conclude this section by adding eight more results to that list to produce the following theorem that relates all of the major topics we have studied thus far.

Theorem 5.6.9. *If A is an $n \times n$ matrix, and if $T_A : R^n \rightarrow R^n$ is multiplication by A, then the following are equivalent.*

(*a*) *A is invertible.*
(*b*) *$A\mathbf{x} = \mathbf{0}$ has only the trivial solution.*
(*c*) *The reduced row-echelon form of A is I_n.*
(*d*) *A is expressible as a product of elementary matrices.*
(*e*) *$A\mathbf{x} = \mathbf{b}$ is consistent for every $n \times 1$ matrix $\mathbf{b}$.*
(*f*) *$A\mathbf{x} = \mathbf{b}$ has exactly one solution for every $n \times 1$ matrix $\mathbf{b}$.*
(*g*) *$\det(A) \neq 0$.*
(*h*) *The range of T_A is R^n.*
(*i*) *T_A is one-to-one.*
(*j*) *The column vectors of A are linearly independent.*
(*k*) *The row vectors of A are linearly independent.*
(*l*) *The column vectors of A span R^n.*
(*m*) *The row vectors of A span R^n.*
(*n*) *The column vectors of A form a basis for R^n.*
(*o*) *The row vectors of A form a basis for R^n.*
(*p*) *A has rank n.*
(*q*) *A has nullity 0.*

Proof. We already know from Theorem 4.3.4 that statements (*a*) through (*i*) are equivalent. To complete the proof we will show that (*j*) through (*q*) are equivalent to (*b*) by proving the sequence of implications $(b) \Rightarrow (j) \Rightarrow (k) \Rightarrow (l) \Rightarrow (m) \Rightarrow (n) \Rightarrow (o) \Rightarrow (p) \Rightarrow (q) \Rightarrow (b)$.

$(b) \Rightarrow (j)$. If $A\mathbf{x} = \mathbf{0}$ has only the trivial solution, then by Theorem 5.6.8 the column vectors of A are linearly independent.

$(j) \Rightarrow (k) \Rightarrow (l) \Rightarrow (m) \Rightarrow (n) \Rightarrow (o)$. This follows from Theorem 5.4.5 and the fact that R^n is an n-dimensional vector space. (The details are left as an exercise.)

$(o) \Rightarrow (p)$. If the n row vectors of A form a basis for R^n, then the row space of A is n-dimensional and A has rank n.

$(p) \Rightarrow (q)$. This follows from the Dimension Theorem (5.6.3).

$(q) \Rightarrow (b)$. If A has nullity 0, then the solution space of $A\mathbf{x} = \mathbf{0}$ has dimension 0, which means that it contains only the zero vector. Hence, $A\mathbf{x} = \mathbf{0}$ has only the trivial solution. ▊

EXERCISE SET 5.6

1. Verify that $\text{rank}(A) = \text{rank}(A^T)$.

$$A = \begin{bmatrix} 1 & 2 & 4 & 0 \\ -3 & 1 & 5 & 2 \\ -2 & 3 & 9 & 2 \end{bmatrix}$$

2. Find the rank and nullity of the matrix; then verify that the values obtained satisfy Formula (4) of the dimension theorem.

(a) $A = \begin{bmatrix} 1 & -1 & 3 \\ 5 & -4 & -4 \\ 7 & -6 & 2 \end{bmatrix}$ (b) $A = \begin{bmatrix} 2 & 0 & -1 \\ 4 & 0 & -2 \\ 0 & 0 & 0 \end{bmatrix}$ (c) $A = \begin{bmatrix} 1 & 4 & 5 & 2 \\ 2 & 1 & 3 & 0 \\ -1 & 3 & 2 & 2 \end{bmatrix}$

(d) $A = \begin{bmatrix} 1 & 4 & 5 & 6 & 9 \\ 3 & -2 & 1 & 4 & -1 \\ -1 & 0 & -1 & -2 & -1 \\ 2 & 3 & 5 & 7 & 8 \end{bmatrix}$ (e) $A = \begin{bmatrix} 1 & -3 & 2 & 2 & 1 \\ 0 & 3 & 6 & 0 & -3 \\ 2 & -3 & -2 & 4 & 4 \\ 3 & -6 & 0 & 6 & 5 \\ -2 & 9 & 2 & -4 & -5 \end{bmatrix}$

3. In each part of Exercise 2 use the results obtained to find the number of leading variables and the number of parameters in the solution of $A\mathbf{x} = \mathbf{0}$ without solving the system.

4. In each part use the information in the table to find the dimension of the row space of A, the column space of A, the nullspace of A, and the nullspace of A^T.

	(a)	(b)	(c)	(d)	(e)	(f)	(g)
Size of A	3×3	3×3	3×3	5×9	9×5	4×4	6×2
Rank (A)	3	2	1	2	2	0	2

5. In each part, find the largest possible value for the rank of A and the smallest possible value for the nullity of A.
   (a) A is 4×4 (b) A is 3×5 (c) A is 5×3

6. If A is an $m \times n$ matrix, what is the largest possible value for its rank and the smallest possible value for its nullity? [**Hint.** See Exercise 5.]

7. In each part, use the information in the table to determine whether the linear system $A\mathbf{x} = \mathbf{b}$ is consistent, If so, state the number of parameters in its general solution.

	(a)	(b)	(c)	(d)	(e)	(f)	(g)
Size of A	3×3	3×3	3×3	5×9	5×9	4×4	6×2
Rank (A)	3	2	1	2	2	0	2
Rank $[A \mid \mathbf{b}]$	3	3	1	2	3	0	2

8. For each of the matrices in Exercise 7, find the nullity of A, and determine the number of parameters in the general solution of the homogeneous linear system $A\mathbf{x} = \mathbf{0}$.

9. What conditions must be satisfied by b_1, b_2, b_3, b_4, and b_5 for the overdetermined linear system

$$\begin{aligned} x_1 - 3x_2 &= b_1 \\ x_1 - 2x_2 &= b_2 \\ x_1 + x_2 &= b_3 \\ x_1 - 4x_2 &= b_4 \\ x_1 + 5x_2 &= b_5 \end{aligned}$$

to be consistent?

10. Let

$$A = \begin{bmatrix} a_{11} & a_{12} & a_{13} \\ a_{21} & a_{22} & a_{23} \end{bmatrix}$$

Show that A has rank 2 if and only if one or more of the determinants

$$\begin{vmatrix} a_{11} & a_{12} \\ a_{21} & a_{22} \end{vmatrix}, \qquad \begin{vmatrix} a_{11} & a_{13} \\ a_{21} & a_{23} \end{vmatrix}, \qquad \begin{vmatrix} a_{12} & a_{13} \\ a_{22} & a_{23} \end{vmatrix}$$

is nonzero.

11. Suppose that A is a 3×3 matrix whose nullspace is a line through the origin in 3-space. Can the row or column space of A also be a line through the origin? Explain.

12. Discuss how the rank of A varies with t.

(a) $A = \begin{bmatrix} 1 & 1 & t \\ 1 & t & 1 \\ t & 1 & 1 \end{bmatrix}$ (b) $A = \begin{bmatrix} t & 3 & -1 \\ 3 & 6 & -2 \\ -1 & -3 & t \end{bmatrix}$

13. Are there values of r and s for which the rank of

$$\begin{bmatrix} 1 & 0 & 0 \\ 0 & r-2 & 2 \\ 0 & s-1 & r+2 \\ 0 & 0 & 3 \end{bmatrix}$$

is one or two? If so, find those values.

14. Suppose that A is a 3×3 matrix whose column space is a plane through the origin in 3-space. Can the nullspace be a plane through the origin? Can the row space? Explain.

15. (a) Prove: If A is a 3×5 matrix, then the column vectors of A are linearly dependent.
   (b) Prove: If A is a 5×3 matrix, then the row vectors of A are linearly dependent.

16. Prove: If A is a matrix that is not square, then either the row vectors of A or the column vectors of A are linearly dependent. [**Hint.** See Exercise 15.]

17. Use the result in Exercise 10 to show that the set of points (x, y, z) in R^3 for which the matrix

$$\begin{bmatrix} x & y & z \\ 1 & x & y \end{bmatrix}$$

has rank 1 is the curve with parametric equations $x = t$, $y = t^2$, $z = t^3$.

18. Prove: If $k \neq 0$, then A and kA have the same rank.

SUPPLEMENTARY EXERCISES

1. In each part, the solution space is a subspace of R^3 and so must be a line through the origin, a plane through the origin, all of R^3, or the origin only. For each system, determine which is the case. If the subspace is a plane, find an equation for it, and if it is a line, find parametric equations.

(a) $0x + 0y + 0z = 0$

(b) $2x - 3y + z = 0$
   $6x - 9y + 3z = 0$
   $-4x + 6y - 2z = 0$

(c) $x - 2y + 7z = 0$
   $-4x + 8y + 5z = 0$
   $2x - 4y + 3z = 0$

(d) $x + 4y + 8z = 0$
   $2x + 5y + 6z = 0$
   $3x + y - 4z = 0$

2. For what values of s is the solution space of

$$x_1 + x_2 + sx_3 = 0$$
$$x_1 + sx_2 + x_3 = 0$$
$$sx_1 + x_2 + x_3 = 0$$

a line through the origin, a plane through the origin, the origin only, or all of R^3?

3. (a) Express $(4a, a - b, a + 2b)$ as a linear combination of $(4, 1, 1)$ and $(0, -1, 2)$.

(b) Express $(3a + b + 3c, -a + 4b - c, 2a + b + 2c)$ as a linear combination of $(3, -1, 2)$ and $(1, 4, 1)$.

(c) Express $(2a - b + 4c, 3a - c, 4b + c)$ as a linear combination of three nonzero vectors.

4. Let W be the space spanned by $\mathbf{f} = \sin x$ and $\mathbf{g} = \cos x$.

(a) Show that for any value of θ, $\mathbf{f}_1 = \sin(x + \theta)$ and $\mathbf{g}_1 = \cos(x + \theta)$ are vectors in W.

(b) Show that $\mathbf{f}_1$ and $\mathbf{g}_1$ form a basis for W.

5. (a) Express $\mathbf{v} = (1, 1)$ as a linear combination of $\mathbf{v}_1 = (1, -1)$, $\mathbf{v}_2 = (3, 0)$, $\mathbf{v}_3 = (2, 1)$ in two different ways.

(b) Show that this does not violate Theorem 5.4.1.

6. Let A be an $n \times n$ matrix, and let $\mathbf{v}_1, \mathbf{v}_2, \ldots, \mathbf{v}_n$ be linearly independent vectors in R^n expressed as $n \times 1$ matrices. What must be true about A for $A\mathbf{v}_1, A\mathbf{v}_2, \ldots, A\mathbf{v}_n$ to be linearly independent?

7. Must a basis for P_n contain a polynomial of degree k for each $k = 0, 1, 2, \ldots, n$? Justify your answer.

8. For purposes of this problem, let us define a "checkerboard matrix" to be a square matrix $A = [\,a_{ij}\,]$ such that

$$a_{ij} = \begin{cases} 1 & \text{if } i + j \text{ is even} \\ 0 & \text{if } i + j \text{ is odd} \end{cases}$$

Find the rank and nullity of the following checkerboard matrices:

(a) the 3×3 checkerboard matrix   (b) the 4×4 checkerboard matrix   (c) the $n \times n$ checkerboard matrix

9. For purposes of this exercise, let us define an "X-matrix" to be a square matrix with an odd number of rows and columns that has 0's everywhere except on the two diagonals, where it has 1's. Find the rank and nullity of the following X-matrices:

(a) $\begin{bmatrix} 1 & 0 & 1 \\ 0 & 1 & 0 \\ 1 & 0 & 1 \end{bmatrix}$ (b) $\begin{bmatrix} 1 & 0 & 0 & 0 & 1 \\ 0 & 1 & 0 & 1 & 0 \\ 0 & 0 & 1 & 0 & 0 \\ 0 & 1 & 0 & 1 & 0 \\ 1 & 0 & 0 & 0 & 1 \end{bmatrix}$ (c) the X-matrix of size $(2n + 1) \times (2n + 1)$

10. In each part, show that the set of polynomials is a subspace of P_n and find a basis for it.

(a) all polynomials in P_n such that $p(-x) = p(x)$ (b) all polynomials in P_n such that $p(0) = 0$

11. (*For readers who have studied calculus.*) Show that the set of all polynomials in P_n that have a horizontal tangent at $x = 0$ is a subspace of P_n. Find a basis for this subspace.

12. In advanced linear algebra one proves the following determinant criterion for rank: *The rank of a matrix A is r if and only if A has some $r \times r$ submatrix with a nonzero determinant, and all square submatrices of larger size have determinant zero.* (A submatrix of A is any matrix obtained by deleting rows or columns of A. The matrix A itself is also considered to be a submatrix of A.) In each part use this criterion to find the rank of the matrix.

(a) $\begin{bmatrix} 1 & 2 & 0 \\ 2 & 4 & -1 \end{bmatrix}$ (b) $\begin{bmatrix} 1 & 2 & 3 \\ 2 & 4 & 6 \end{bmatrix}$ (c) $\begin{bmatrix} 1 & 0 & 1 \\ 2 & -1 & 3 \\ 3 & -1 & 4 \end{bmatrix}$ (d) $\begin{bmatrix} 1 & -1 & 2 & 0 \\ 3 & 1 & 0 & 0 \\ -1 & 2 & 4 & 0 \end{bmatrix}$

13. Use the result in Exercise 12 to find the possible ranks for matrices of the form

$$\begin{bmatrix} 0 & 0 & 0 & 0 & 0 & a_{16} \\ 0 & 0 & 0 & 0 & 0 & a_{26} \\ 0 & 0 & 0 & 0 & 0 & a_{36} \\ 0 & 0 & 0 & 0 & 0 & a_{46} \\ a_{51} & a_{52} & a_{53} & a_{54} & a_{55} & a_{56} \end{bmatrix}$$

14. Prove: If S is a basis for a vector space V, then for any vectors $\mathbf{u}$ and $\mathbf{v}$ in V and any scalar k the following relationships hold:

(a) $(\mathbf{u} + \mathbf{v})_S = (\mathbf{u})_S + (\mathbf{v})_S$ (b) $(k\mathbf{u})_S = k(\mathbf{u})_S$

CHAPTER **6**

INNER PRODUCT SPACES

6.1 INNER PRODUCTS

In Section 4.1 we defined the Euclidean inner product on R^n and used it to extend the concepts of length and distance to Euclidean n-space. In this section we shall use the most important properties of the Euclidean inner product as axioms for defining a general concept of an inner product; then we will show how inner products can be used to define notions of length and distance in vector spaces other than R^n.

GENERAL INNER PRODUCTS

In Section 4.1 we denoted the Euclidean inner product of two vectors in R^n by $\mathbf{u} \cdot \mathbf{v}$. It will be convenient in this section to introduce the alternative notation $\langle \mathbf{u}, \mathbf{v} \rangle$ for this inner product. With this notation, the fundamental properties of the Euclidean inner product that were listed in Theorem 4.1.2 are precisely the axioms in the following definition.

Definition. An *inner product* on a real vector space V is a function that associates a real number $\langle \mathbf{u}, \mathbf{v} \rangle$ with each pair of vectors $\mathbf{u}$ and $\mathbf{v}$ in V in such a way that the following axioms are satisfied for all vectors $\mathbf{u}$, $\mathbf{v}$, and $\mathbf{w}$ in V and all scalars k.

(1) $\langle \mathbf{u}, \mathbf{v} \rangle = \langle \mathbf{v}, \mathbf{u} \rangle$ [Symmetry axiom]
(2) $\langle \mathbf{u} + \mathbf{v}, \mathbf{w} \rangle = \langle \mathbf{u}, \mathbf{w} \rangle + \langle \mathbf{v}, \mathbf{w} \rangle$ [Additivity axiom]
(3) $\langle k\mathbf{u}, \mathbf{v} \rangle = k\langle \mathbf{u}, \mathbf{v} \rangle$ [Homogeneity axiom]
(4) $\langle \mathbf{v}, \mathbf{v} \rangle \geq 0$ [Positivity axiom]
    where $\langle \mathbf{v}, \mathbf{v} \rangle = 0$
    if and only if $\mathbf{v} = \mathbf{0}$

A real vector space with an inner product is called a *real inner product space*.

REMARK. In Chapter 10 we shall study complex inner products, that is, inner products whose values are complex numbers. Until that time we shall use the term "inner product space" to mean "real inner product space."

Because the inner product axioms are based on properties of the Euclidean inner product, the Euclidean inner product automatically satisfies these axioms; this is the content of the following example.

Example 1 If $\mathbf{u} = (u_1, u_2, \ldots, u_n)$ and $\mathbf{v} = (v_1, v_2, \ldots, v_n)$ are vectors in R^n, then the formula

$$\langle \mathbf{u}, \mathbf{v} \rangle = \mathbf{u} \cdot \mathbf{v} = u_1 v_1 + u_2 v_2 + \cdots + u_n v_n$$

defines $\langle \mathbf{u}, \mathbf{v} \rangle$ to be the Euclidean inner product on R^n. The four inner product axioms hold by Theorem 4.1.2. ▲

The Euclidean inner product is the most important inner product on R^n. However, there are various applications in which it is desirable to modify the Euclidean inner product by *weighting* its terms differently. More precisely, if

$$w_1, w_2, \ldots, w_n$$

are *positive* real numbers, which we shall call **weights**, and if $\mathbf{u} = (u_1, u_2, \ldots, u_n)$ and $\mathbf{v} = (v_1, v_2, \ldots, v_n)$ are vectors in R^n, then it can be shown (Exercise 26) that the formula

$$\langle \mathbf{u}, \mathbf{v} \rangle = w_1 u_1 v_1 + w_2 u_2 v_2 + \cdots + w_n u_n v_n \tag{1}$$

defines an inner product on R^n; it is called the **weighted Euclidean inner product with weights** $w_1, w_2, \ldots, w_n$.

To illustrate one way in which a weighted Euclidean inner product can arise, suppose that some physical experiment can produce any of n possible numerical values

$$x_1, x_2, \ldots, x_n$$

and that m repetitions of the experiment yields these values with various frequencies; that is, x_1 occurs f_1 times, x_2 occurs f_2 times, and so forth. Since there is a total of m repetitions of the experiment,

$$f_1 + f_2 + \cdots + f_n = m$$

Thus, the **arithmetic average** or **mean** of the observed numerical values (denoted by $\bar{x}$) is

$$\bar{x} = \frac{f_1 x_1 + f_2 x_2 + \cdots + f_n x_n}{f_1 + f_2 + \cdots + f_n} = \frac{1}{m}(f_1 x_1 + f_2 x_2 + \cdots + f_n x_n) \tag{2}$$

If we let

$$\mathbf{f} = (f_1, f_2, \ldots, f_n)$$
$$\mathbf{x} = (x_1, x_2, \ldots, x_n)$$
$$w_1 = w_2 = \cdots = w_n = 1/m$$

then (2) can be expressed as the weighted inner product

$$\bar{x} = \langle \mathbf{f}, \mathbf{x} \rangle = w_1 f_1 x_1 + w_2 f_2 x_2 + \cdots + w_n f_n x_n$$

REMARK. It will always be assumed that R^n has the Euclidean inner product unless some other inner product is specifically specified. As defined in Section 4.1, we refer to R^n with the Euclidean inner product as *Euclidean n-space*.

Example 2 Let $\mathbf{u} = (u_1, u_2)$ and $\mathbf{v} = (v_1, v_2)$ be vectors in R^2. Verify that the weighted Euclidean inner product

$$\langle \mathbf{u}, \mathbf{v} \rangle = 3u_1 v_1 + 2u_2 v_2$$

satisfies the four inner product axioms.

Solution. Note first that if $\mathbf{u}$ and $\mathbf{v}$ are interchanged in this equation, the right side remains the same. Therefore,

$$\langle \mathbf{u}, \mathbf{v} \rangle = \langle \mathbf{v}, \mathbf{u} \rangle$$

If $\mathbf{w} = (w_1, w_2)$, then

$$\begin{aligned}
\langle \mathbf{u} + \mathbf{v}, \mathbf{w} \rangle &= 3(u_1 + v_1)w_1 + 2(u_2 + v_2)w_2 \\
&= (3u_1 w_1 + 2u_2 w_2) + (3v_1 w_1 + 2v_2 w_2) \\
&= \langle \mathbf{u}, \mathbf{w} \rangle + \langle \mathbf{v}, \mathbf{w} \rangle
\end{aligned}$$

which establishes the second axiom.

Next,

$$\begin{aligned}
\langle k\mathbf{u}, \mathbf{v} \rangle &= 3(ku_1)v_1 + 2(ku_2)v_2 \\
&= k(3u_1 v_1 + 2u_2 v_2) \\
&= k\langle \mathbf{u}, \mathbf{v} \rangle
\end{aligned}$$

which establishes the third axiom.

Finally,

$$\langle \mathbf{v}, \mathbf{v} \rangle = 3v_1 v_1 + 2v_2 v_2 = 3v_1^2 + 2v_2^2$$

Obviously, $\langle \mathbf{v}, \mathbf{v} \rangle = 3v_1^2 + 2v_2^2 \geq 0$. Further, $\langle \mathbf{v}, \mathbf{v} \rangle = 3v_1^2 + 2v_2^2 = 0$ if and only if $v_1 = v_2 = 0$, that is, if and only if $\mathbf{v} = (v_1, v_2) = \mathbf{0}$. Thus, the fourth axiom is satisfied. ▲

LENGTH AND DISTANCE IN INNER PRODUCT SPACES

Before discussing more examples of inner products, we shall pause to explain how inner products are used to introduce notions of length and distance in inner product spaces. Recall that in Euclidean *n*-space the Euclidean length of a vector $\mathbf{u} = (u_1, u_2, \ldots, u_n)$ can be expressed in terms of the Euclidean inner product as

$$\|\mathbf{u}\| = (\mathbf{u} \cdot \mathbf{u})^{1/2}$$

and the Euclidean distance between two arbitrary points $\mathbf{u} = (u_1, u_2, \ldots, u_n)$ and $\mathbf{v} = (v_1, v_2, \ldots, v_n)$ can be expressed as

$$d(\mathbf{u}, \mathbf{v}) = \|\mathbf{u} - \mathbf{v}\| = [(\mathbf{u} - \mathbf{v}) \cdot (\mathbf{u} - \mathbf{v})]^{1/2}$$

[see Formulas (1) and (2) of Section 4.1]. Motivated by these formulas, we make the following definition.

Definition. If V is an inner product space, then the *norm* (or *length*) of a vector $\mathbf{u}$ in V is denoted by $\|\mathbf{u}\|$ and is defined by

$$\|\mathbf{u}\| = \langle \mathbf{u}, \mathbf{u} \rangle^{1/2}$$

The *distance* between two points (vectors) $\mathbf{u}$ and $\mathbf{v}$ is denoted by $d(\mathbf{u}, \mathbf{v})$ and is defined by

$$d(\mathbf{u}, \mathbf{v}) = \|\mathbf{u} - \mathbf{v}\|$$

Example 3   If $\mathbf{u} = (u_1, u_2, \ldots, u_n)$ and $\mathbf{v} = (v_1, v_2, \ldots, v_n)$ are vectors in R^n with the Euclidean inner product, then

$$\|\mathbf{u}\| = \langle \mathbf{u}, \mathbf{u} \rangle^{1/2} = (\mathbf{u} \cdot \mathbf{u})^{1/2} = \sqrt{u_1^2 + u_2^2 + \cdots + u_n^2}$$

and

$$d(\mathbf{u}, \mathbf{v}) = \|\mathbf{u} - \mathbf{v}\| = \langle \mathbf{u} - \mathbf{v}, \mathbf{u} - \mathbf{v} \rangle^{1/2} = [(\mathbf{u} - \mathbf{v}) \cdot (\mathbf{u} - \mathbf{v})]^{1/2}$$
$$= \sqrt{(u_1 - v_1)^2 + (u_2 - v_2)^2 + \cdots + (u_n - v_n)^2}$$

Observe that these are simply the standard formulas for the Euclidean norm and distance discussed in Section 4.1 [see Formulas (1) and (2) in that section]. ▲

Example 4   It is important to keep in mind that norm and distance depend on the inner product being used. If the inner product is changed, then the norms and distances between vectors also change. For example, for the vectors $\mathbf{u} = (1, 0)$ and $\mathbf{v} = (0, 1)$ in R^2 with the Euclidean inner product, we have

$$\|\mathbf{u}\| = \sqrt{1^2 + 0^2} = 1$$

and

$$d(\mathbf{u}, \mathbf{v}) = \|\mathbf{u} - \mathbf{v}\| = \|(1, -1)\| = \sqrt{1^2 + (-1)^2} = \sqrt{2}$$

However, if we change to the weighted Euclidean inner product

$$\langle \mathbf{u}, \mathbf{v} \rangle = 3u_1v_1 + 2u_2v_2$$

then we obtain

$$\|\mathbf{u}\| = \langle \mathbf{u}, \mathbf{u} \rangle^{1/2} = [3(1)(1) + 2(0)(0)]^{1/2} = \sqrt{3}$$

and

$$d(\mathbf{u}, \mathbf{v}) = \|\mathbf{u} - \mathbf{v}\| = \langle (1, -1), (1, -1) \rangle^{1/2}$$
$$= [3(1)(1) + 2(-1)(-1)]^{1/2} = \sqrt{5} \quad ▲$$

UNIT CIRCLES AND SPHERES IN INNER PRODUCT SPACES

If V is an inner product space, then the set of points in V that satisfy

$$\|\mathbf{u}\| = 1$$

is called the *unit sphere* or sometimes the *unit circle* in V. In R^2 and R^3 these are the points that lie 1 unit away from the origin.

Example 5

(a) Sketch the unit circle in an xy-coordinate system in R^2 using the Euclidean inner product $\langle \mathbf{u}, \mathbf{v} \rangle = u_1v_1 + u_2v_2$.
(b) Sketch the unit circle in an xy-coordinate system in R^2 using the weighted Euclidean inner product $\langle \mathbf{u}, \mathbf{v} \rangle = \frac{1}{9}u_1v_1 + \frac{1}{4}u_2v_2$.

Solution (a). If $\mathbf{u} = (x, y)$, then $\|\mathbf{u}\| = \langle \mathbf{u}, \mathbf{u} \rangle^{1/2} = \sqrt{x^2 + y^2}$, so the equation of the unit circle is $\sqrt{x^2 + y^2} = 1$, or on squaring both sides,

$$x^2 + y^2 = 1$$

As expected, the graph of this equation is a circle of radius 1 centered at the origin (Figure 1a).

(a)

(b)

Figure 1

| The unit circle with the Euclidean norm $\|\mathbf{u}\| = \sqrt{x^2 + y^2}$ | The unit circle with the norm $\|\mathbf{u}\| = \sqrt{\frac{1}{9}x^2 + \frac{1}{4}y^2}$ |

Solution (b). If $\mathbf{u} = (x, y)$, then $\|\mathbf{u}\| = \langle \mathbf{u}, \mathbf{u} \rangle^{1/2} = \sqrt{\frac{1}{9}x^2 + \frac{1}{4}y^2}$, so the equation of the unit circle is $\sqrt{\frac{1}{9}x^2 + \frac{1}{4}y^2} = 1$, or on squaring both sides,

$$\frac{x^2}{9} + \frac{y^2}{4} = 1$$

The graph of this equation is the ellipse shown in Figure 1b.

It would be reasonable for you to feel uncomfortable with the results in the last example. For even though our definitions of length and distance reduce to the standard definitions when applied to R^2 with the Euclidean inner product, it does require a stretch of the imagination to think of the unit "circle" as having an elliptical shape. However, even though nonstandard inner products distort familiar spaces and lead to strange values for lengths and distances, many of the basic theorems of Euclidean geometry continue to apply in these unusual spaces. For example, it is a basic fact in Euclidean geometry that the sum of the lengths of two sides of a triangle is at least as large as the length of the third side (Figure 2a). We shall see later that this familiar result holds in all inner product spaces, regardless of how unusual the inner product might be. As another example, recall the theorem from Euclidean geometry which states that the sum of the squares of the diagonals of a parallelogram is equal to the sum of the squares of the four sides (Figure 2b). This result also holds in all inner product spaces, regardless of the inner product (Exercise 20).

(a)

(b)

Figure 2 $\quad \boxed{\|\mathbf{u} + \mathbf{v}\| \leq \|\mathbf{u}\| + \|\mathbf{v}\|} \qquad \boxed{\|\mathbf{u} + \mathbf{v}\|^2 + \|\mathbf{u} - \mathbf{v}\|^2 = 2(\|\mathbf{u}\|^2 + \|\mathbf{v}\|^2)}$

INNER PRODUCTS GENERATED BY MATRICES

The Euclidean inner product and the weighted Euclidean inner products are special cases of a general class of inner products on R^n, which we shall now describe. Let

$$\mathbf{u} = \begin{bmatrix} u_1 \\ u_2 \\ \vdots \\ u_n \end{bmatrix} \quad \text{and} \quad \mathbf{v} = \begin{bmatrix} v_1 \\ v_2 \\ \vdots \\ v_n \end{bmatrix}$$

be vectors in R^n (expressed as $n \times 1$ matrices), and let A be an *invertible* $n \times n$ matrix. It can be shown (Exercise 30) that if $\mathbf{u} \cdot \mathbf{v}$ is the Euclidean inner product on R^n, then the formula

$$\langle \mathbf{u}, \mathbf{v} \rangle = A\mathbf{u} \cdot A\mathbf{v} \tag{3}$$

defines an inner product; it is called the ***inner product on R^n generated by A***.

Recalling that the Euclidean inner product $\mathbf{u} \cdot \mathbf{v}$ can be written as the matrix product $\mathbf{v}^T\mathbf{u}$ [see (7) in Section 4.1], it follows that (3) can be written in the alternative form

$$\langle \mathbf{u}, \mathbf{v} \rangle = (A\mathbf{v})^T A\mathbf{u}$$

or equivalently,

$$\langle \mathbf{u}, \mathbf{v} \rangle = \mathbf{v}^T A^T A\mathbf{u} \tag{4}$$

Example 6   The inner product on R^n generated by the $n \times n$ identity matrix is the Euclidean inner product, since substituting $A = I$ in (3) yields

$$\langle \mathbf{u}, \mathbf{v} \rangle = I\mathbf{u} \cdot I\mathbf{v} = \mathbf{u} \cdot \mathbf{v}$$

The weighted Euclidean inner product $\langle \mathbf{u}, \mathbf{v} \rangle = 3u_1v_1 + 2u_2v_2$ discussed in Example 2 is the inner product on R^2 generated by

$$A = \begin{bmatrix} \sqrt{3} & 0 \\ 0 & \sqrt{2} \end{bmatrix}$$

because substituting this matrix in (4) yields

$$\langle \mathbf{u}, \mathbf{v} \rangle = [v_1 \quad v_2] \begin{bmatrix} \sqrt{3} & 0 \\ 0 & \sqrt{2} \end{bmatrix} \begin{bmatrix} \sqrt{3} & 0 \\ 0 & \sqrt{2} \end{bmatrix} \begin{bmatrix} u_1 \\ u_2 \end{bmatrix}$$

$$= [v_1 \quad v_2] \begin{bmatrix} 3 & 0 \\ 0 & 2 \end{bmatrix} \begin{bmatrix} u_1 \\ u_2 \end{bmatrix}$$

$$= 3u_1v_1 + 2u_2v_2$$

In general, the weighted Euclidean inner product

$$\langle \mathbf{u}, \mathbf{v} \rangle = w_1 u_1 v_1 + w_2 u_2 v_2 + \cdots + w_n u_n v_n$$

is the inner product on R^n generated by

$$A = \begin{bmatrix} \sqrt{w_1} & 0 & 0 & \cdots & 0 \\ 0 & \sqrt{w_2} & 0 & \cdots & 0 \\ \vdots & \vdots & \vdots & & \vdots \\ 0 & 0 & 0 & \cdots & \sqrt{w_n} \end{bmatrix} \tag{5}$$

(verify). ▲

In the following examples we shall describe some inner products on vector spaces other than R^n.

Example 7   If

$$U = \begin{bmatrix} u_1 & u_2 \\ u_3 & u_4 \end{bmatrix} \quad \text{and} \quad V = \begin{bmatrix} v_1 & v_2 \\ v_3 & v_4 \end{bmatrix}$$

are any two 2×2 matrices, then the following formula defines an inner product on M_{22} (verify):

$$\langle U, V \rangle = u_1 v_1 + u_2 v_2 + u_3 v_3 + u_4 v_4$$

For example, if

$$U = \begin{bmatrix} 1 & 2 \\ 3 & 4 \end{bmatrix} \quad \text{and} \quad V = \begin{bmatrix} -1 & 0 \\ 3 & 2 \end{bmatrix}$$

then

$$\langle U, V \rangle = 1(-1) + 2(0) + 3(3) + 4(2) = 16$$

The norm of a matrix U relative to this inner product is

$$\|U\| = \langle U, U \rangle^{1/2} = \sqrt{u_1^2 + u_2^2 + u_3^2 + u_4^2}$$

and the unit sphere in this space consists of all 2×2 matrices U whose entries satisfy the equation $\|U\| = 1$, which on squaring yields

$$u_1^2 + u_2^2 + u_3^2 + u_4^2 = 1 \quad ▲$$

Example 8 If

$$\mathbf{p} = a_0 + a_1 x + a_2 x^2 \quad \text{and} \quad \mathbf{q} = b_0 + b_1 x + b_2 x^2$$

are any two vectors in P_2, then the following formula defines an inner product on P_2 (verify):

$$\langle \mathbf{p}, \mathbf{q} \rangle = a_0 b_0 + a_1 b_1 + a_2 b_2$$

The norm of the polynomial $\mathbf{p}$ relative to this inner product is

$$\|\mathbf{p}\| = \langle \mathbf{p}, \mathbf{p} \rangle^{1/2} = \sqrt{a_0^2 + a_1^2 + a_2^2}$$

and the unit sphere in this space consists of all polynomials $\mathbf{p}$ in P_2 whose coefficients satisfy the equation $\|\mathbf{p}\| = 1$, which on squaring yields

$$a_0^2 + a_1^2 + a_2^2 = 1 \quad \blacktriangle$$

Example 9 (*For readers who have studied calculus.*) Let $\mathbf{f} = f(x)$ and $\mathbf{g} = g(x)$ be two continuous functions in $C[a, b]$ and define

$$\langle \mathbf{f}, \mathbf{g} \rangle = \int_a^b f(x)g(x)\,dx \qquad (6)$$

We shall show that this formula defines an inner product on $C[a, b]$ by verifying the four inner product axioms for functions $\mathbf{f} = f(x)$, $\mathbf{g} = g(x)$, and $\mathbf{s} = s(x)$ in $C[a, b]$:

(1) $\langle \mathbf{f}, \mathbf{g} \rangle = \int_a^b f(x)g(x)\,dx = \int_a^b g(x)f(x)\,dx = \langle \mathbf{g}, \mathbf{f} \rangle$

which proves that Axiom 1 holds.

(2) $\langle \mathbf{f} + \mathbf{g}, \mathbf{s} \rangle = \int_a^b (f(x) + g(x))s(x)\,dx$

$$= \int_a^b f(x)s(x)\,dx + \int_a^b g(x)s(x)\,dx$$

$$= \langle \mathbf{f}, \mathbf{s} \rangle + \langle \mathbf{g}, \mathbf{s} \rangle$$

which proves that Axiom 2 holds.

(3) $\langle k\mathbf{f}, \mathbf{g} \rangle = \int_a^b kf(x)g(x)\,dx = k \int_a^b f(x)g(x)\,dx = k\langle \mathbf{f}, \mathbf{g} \rangle$

which proves that Axiom 3 holds.

(4) If $\mathbf{f} = f(x)$ is any function in $C[a, b]$, then $f^2(x) \geq 0$ for all x in $[a, b]$; therefore,

$$\langle \mathbf{f}, \mathbf{f} \rangle = \int_a^b f^2(x)\,dx \geq 0$$

Further, because $f^2(x) \geq 0$ and $\mathbf{f} = f(x)$ is continuous on $[a, b]$, it follows that $\int_a^b f^2(x)\,dx = 0$ if and only if $f(x) = 0$ for all x in $[a, b]$. Therefore, we have $\langle \mathbf{f}, \mathbf{f} \rangle = \int_a^b f^2(x)\,dx = 0$ if and only if $\mathbf{f} = \mathbf{0}$. This proves that Axiom 4 holds. $\blacktriangle$

Example 10 (*For readers who have studied calculus.*)   If $C[a, b]$ has the inner product defined in the preceding example, then the norm of a function $\mathbf{f} = f(x)$ relative to this inner product is

$$\|\mathbf{f}\| = \langle \mathbf{f}, \mathbf{f} \rangle^{1/2} = \sqrt{\int_a^b f^2(x)\, dx} \tag{7}$$

and the unit sphere in this space consists of all functions $\mathbf{f}$ in $C[a, b]$ that satisfy the equation $\|\mathbf{f}\| = 1$, which on squaring yields

$$\int_a^b f^2(x)\, dx = 1 \quad \blacktriangle$$

REMARK (*For readers who have studied calculus*). Since polynomials are continuous functions on $(-\infty, \infty)$, they are continuous on any closed interval $[a, b]$. Thus, for all such intervals the vector space P_n is a subspace of $C[a, b]$, and Formula (6) defines an inner product on P_n.

REMARK (*For readers who have studied calculus*). Recall from calculus that the arc length of a curve $y = f(x)$ over an interval $[a, b]$ is given by the formula

$$L = \int_a^b \sqrt{1 + [f'(x)]^2}\, dx \tag{8}$$

Do not confuse this concept of arc length with $\|\mathbf{f}\|$, which is the length (norm) of $\mathbf{f}$ when $\mathbf{f}$ is viewed as a vector in $C[a, b]$. Formulas (7) and (8) are quite different.

SOME PROPERTIES OF INNER PRODUCTS

The following theorem lists some basic algebraic properties of inner products.

Theorem 6.1.1. *If* $\mathbf{u}$, $\mathbf{v}$, *and* $\mathbf{w}$ *are vectors in a real inner product space, and* k *is any scalar, then:*

(a) $\langle \mathbf{0}, \mathbf{v} \rangle = \langle \mathbf{v}, \mathbf{0} \rangle = 0$
(b) $\langle \mathbf{u}, \mathbf{v} + \mathbf{w} \rangle = \langle \mathbf{u}, \mathbf{v} \rangle + \langle \mathbf{u}, \mathbf{w} \rangle$
(c) $\langle \mathbf{u}, k\mathbf{v} \rangle = k\langle \mathbf{u}, \mathbf{v} \rangle$
(d) $\langle \mathbf{u} - \mathbf{v}, \mathbf{w} \rangle = \langle \mathbf{u}, \mathbf{w} \rangle - \langle \mathbf{v}, \mathbf{w} \rangle$
(e) $\langle \mathbf{u}, \mathbf{v} - \mathbf{w} \rangle = \langle \mathbf{u}, \mathbf{v} \rangle - \langle \mathbf{u}, \mathbf{w} \rangle$

Proof. We shall prove part (b) and leave the proofs of the remaining parts as exercises.

$$
\begin{aligned}
\langle \mathbf{u}, \mathbf{v} + \mathbf{w} \rangle &= \langle \mathbf{v} + \mathbf{w}, \mathbf{u} \rangle && \text{[By symmetry]} \\
&= \langle \mathbf{v}, \mathbf{u} \rangle + \langle \mathbf{w}, \mathbf{u} \rangle && \text{[By additivity]} \\
&= \langle \mathbf{u}, \mathbf{v} \rangle + \langle \mathbf{u}, \mathbf{w} \rangle && \text{[By symmetry]} \quad\blacksquare
\end{aligned}
$$

The following example illustrates how Theorem 6.1.1 and the defining properties of inner products can be used to perform algebraic computations with inner products. As you read through the example, you will find it instructive to justify the steps.

Example 11

$$\langle \mathbf{u} - 2\mathbf{v}, 3\mathbf{u} + 4\mathbf{v} \rangle = \langle \mathbf{u}, 3\mathbf{u} + 4\mathbf{v} \rangle - \langle 2\mathbf{v}, 3\mathbf{u} + 4\mathbf{v} \rangle$$
$$= \langle \mathbf{u}, 3\mathbf{u} \rangle + \langle \mathbf{u}, 4\mathbf{v} \rangle - \langle 2\mathbf{v}, 3\mathbf{u} \rangle - \langle 2\mathbf{v}, 4\mathbf{v} \rangle$$
$$= 3\langle \mathbf{u}, \mathbf{u} \rangle + 4\langle \mathbf{u}, \mathbf{v} \rangle - 6\langle \mathbf{v}, \mathbf{u} \rangle - 8\langle \mathbf{v}, \mathbf{v} \rangle$$
$$= 3\|\mathbf{u}\|^2 + 4\langle \mathbf{u}, \mathbf{v} \rangle - 6\langle \mathbf{u}, \mathbf{v} \rangle - 8\|\mathbf{v}\|^2$$
$$= 3\|\mathbf{u}\|^2 - 2\langle \mathbf{u}, \mathbf{v} \rangle - 8\|\mathbf{v}\|^2 \quad \blacktriangle$$

Since Theorem 6.1.1 is a general result, it is guaranteed to hold for *all* real inner product spaces. This is the real power of the axiomatic development of vector spaces and inner products—a single theorem proves a multitude of results at once. For example, we are guaranteed without any further proof that the five properties given in Theorem 6.1.1 are true for the inner product on R^n generated by any matrix A [Formula (3)]. For example, let us check part (*b*) of Theorem 6.1.1 for this inner product:

$$\langle \mathbf{u}, \mathbf{v} + \mathbf{w} \rangle = (\mathbf{v} + \mathbf{w})^T A^T A \mathbf{u}$$
$$= (\mathbf{v}^T + \mathbf{w}^T) A^T A \mathbf{u} \qquad [\text{Property of transpose}]$$
$$= (\mathbf{v}^T A^T A \mathbf{u}) + (\mathbf{w}^T A^T A \mathbf{u}) \qquad [\text{Property of matrix multiplication}]$$
$$= \langle \mathbf{u}, \mathbf{v} \rangle + \langle \mathbf{u}, \mathbf{w} \rangle$$

The reader will find it instructive to check the remaining parts of Theorem 6.1.1 for this inner product.

EXERCISE SET 6.1

1. Let $\langle \mathbf{u}, \mathbf{v} \rangle$ be the Euclidean inner product on R^2, and let $\mathbf{u} = (3, -2)$, $\mathbf{v} = (4, 5)$, $\mathbf{w} = (-1, 6)$, and $k = -4$. Find
   (a) $\langle \mathbf{u}, \mathbf{v} \rangle = \langle \mathbf{v}, \mathbf{u} \rangle$
   (b) $\langle \mathbf{u} + \mathbf{v}, \mathbf{w} \rangle = \langle \mathbf{u}, \mathbf{w} \rangle + \langle \mathbf{v}, \mathbf{w} \rangle$
   (c) $\langle \mathbf{u}, \mathbf{v} + \mathbf{w} \rangle = \langle \mathbf{u}, \mathbf{v} \rangle + \langle \mathbf{u}, \mathbf{w} \rangle$
   (d) $\langle k\mathbf{u}, \mathbf{v} \rangle = k\langle \mathbf{u}, \mathbf{v} \rangle = \langle \mathbf{u}, k\mathbf{v} \rangle$
   (e) $\langle \mathbf{0}, \mathbf{v} \rangle = \langle \mathbf{v}, \mathbf{0} \rangle = 0$

2. Repeat Exercise 1 for the weighted Euclidean inner product $\langle \mathbf{u}, \mathbf{v} \rangle = 4u_1 v_1 + 5u_2 v_2$.

3. Compute $\langle \mathbf{u}, \mathbf{v} \rangle$ using the inner product in Example 7.

   (a) $\mathbf{u} = \begin{bmatrix} 3 & -2 \\ 4 & 8 \end{bmatrix}$, $\mathbf{v} = \begin{bmatrix} -1 & 3 \\ 1 & 1 \end{bmatrix}$
   (b) $\mathbf{u} = \begin{bmatrix} 1 & 2 \\ -3 & 5 \end{bmatrix}$, $\mathbf{v} = \begin{bmatrix} 4 & 6 \\ 0 & 8 \end{bmatrix}$

4. Compute $\langle \mathbf{p}, \mathbf{q} \rangle$ using the inner product in Example 8.
   (a) $\mathbf{p} = -2 + x + 3x^2$, $\mathbf{q} = 4 - 7x^2$
   (b) $\mathbf{p} = -5 + 2x + x^2$, $\mathbf{q} = 3 + 2x - 4x^2$

5. (a) Use Formula (3) to show that $\langle \mathbf{u}, \mathbf{v} \rangle = 9u_1 v_1 + 4u_2 v_2$ is the inner product on R^2 generated by

   $$A = \begin{bmatrix} 3 & 0 \\ 0 & 2 \end{bmatrix}$$

   (b) Use the inner product in (a) to compute $\langle \mathbf{u}, \mathbf{v} \rangle$ if $\mathbf{u} = (-3, 2)$ and $\mathbf{v} = (1, 7)$.

6. (a) Use Formula (3) to show that $\langle \mathbf{u}, \mathbf{v} \rangle = 5u_1v_1 - u_1v_2 - u_2v_1 + 10u_2v_2$ is the inner product on R^2 generated by

$$A = \begin{bmatrix} 2 & 1 \\ -1 & 3 \end{bmatrix}$$

(b) Use the inner product in (a) to compute $\langle \mathbf{u}, \mathbf{v} \rangle$ if $\mathbf{u} = (0, -3)$ and $\mathbf{v} = (6, 2)$.

7. Let $\mathbf{u} = (u_1, u_2)$ and $\mathbf{v} = (v_1, v_2)$. In each part, the given expression is an inner product on R^2. Find a matrix that generates it.
   (a) $\langle \mathbf{u}, \mathbf{v} \rangle = 3u_1v_1 + 5u_2v_2$ (b) $\langle \mathbf{u}, \mathbf{v} \rangle = 4u_1v_1 + 6u_2v_2$

8. Let $\mathbf{u} = (u_1, u_2)$ and $\mathbf{v} = (v_1, v_2)$. Show that the following are inner products on R^2 by verifying that the inner product axioms hold.
   (a) $\langle \mathbf{u}, \mathbf{v} \rangle = 3u_1v_1 + 5u_2v_2$ (b) $\langle \mathbf{u}, \mathbf{v} \rangle = 4u_1v_1 + u_2v_1 + u_1v_2 + 4u_2v_2$

9. Let $\mathbf{u} = (u_1, u_2, u_3)$ and $\mathbf{v} = (v_1, v_2, v_3)$. Determine which of the following are inner products on R^3. For those that are not, list the axioms that do not hold.
   (a) $\langle \mathbf{u}, \mathbf{v} \rangle = u_1v_1 + u_3v_3$ (b) $\langle \mathbf{u}, \mathbf{v} \rangle = u_1^2v_1^2 + u_2^2v_2^2 + u_3^2v_3^2$
   (c) $\langle \mathbf{u}, \mathbf{v} \rangle = 2u_1v_1 + u_2v_2 + 4u_3v_3$ (d) $\langle \mathbf{u}, \mathbf{v} \rangle = u_1v_1 - u_2v_2 + u_3v_3$

10. In each part use the given inner product on R^2 to find $\|\mathbf{w}\|$, where $\mathbf{w} = (-1, 3)$.
   (a) the Euclidean inner product
   (b) the weighted Euclidean inner product $\langle \mathbf{u}, \mathbf{v} \rangle = 3u_1v_1 + 2u_2v_2$, where $\mathbf{u} = (u_1, u_2)$ and $\mathbf{v} = (v_1, v_2)$
   (c) the inner product generated by the matrix

$$A = \begin{bmatrix} 1 & 2 \\ -1 & 3 \end{bmatrix}$$

11. Use the inner products in Exercise 10 to find $d(\mathbf{u}, \mathbf{v})$ for $\mathbf{u} = (-1, 2)$ and $\mathbf{v} = (2, 5)$.

12. Let P_2 have the inner product in Example 8. In each part find $\|\mathbf{p}\|$.
   (a) $\mathbf{p} = -2 + 3x + 2x^2$ (b) $\mathbf{p} = 4 - 3x^2$

13. Let M_{22} have the inner product in Example 7. In each part find $\|A\|$.
   (a) $A = \begin{bmatrix} -2 & 5 \\ 3 & 6 \end{bmatrix}$ (b) $A = \begin{bmatrix} 0 & 0 \\ 0 & 0 \end{bmatrix}$

14. Let P_2 have the inner product in Example 8. Find $d(\mathbf{p}, \mathbf{q})$.
$$\mathbf{p} = 3 - x + x^2, \qquad \mathbf{q} = 2 + 5x^2$$

15. Let M_{22} have the inner product in Example 7. Find $d(A, B)$.
   (a) $A = \begin{bmatrix} 2 & 6 \\ 9 & 4 \end{bmatrix}$, $B = \begin{bmatrix} -4 & 7 \\ 1 & 6 \end{bmatrix}$ (b) $A = \begin{bmatrix} -2 & 4 \\ 1 & 0 \end{bmatrix}$, $B = \begin{bmatrix} -5 & 1 \\ 6 & 2 \end{bmatrix}$

16. Suppose that $\mathbf{u}$, $\mathbf{v}$, and $\mathbf{w}$ are vectors such that
$$\langle \mathbf{u}, \mathbf{v} \rangle = 2, \quad \langle \mathbf{v}, \mathbf{w} \rangle = -3, \quad \langle \mathbf{u}, \mathbf{w} \rangle = 5, \quad \|\mathbf{u}\| = 1, \quad \|\mathbf{v}\| = 2, \quad \|\mathbf{w}\| = 7$$

Evaluate the given expression.
   (a) $\langle \mathbf{u} + \mathbf{v}, \mathbf{v} + \mathbf{w} \rangle$ (b) $\langle 2\mathbf{v} - \mathbf{w}, 3\mathbf{u} + 2\mathbf{w} \rangle$ (c) $\langle \mathbf{u} - \mathbf{v} - 2\mathbf{w}, 4\mathbf{u} + \mathbf{v} \rangle$
   (d) $\|\mathbf{u} + \mathbf{v}\|$ (e) $\|2\mathbf{w} - \mathbf{v}\|$ (f) $\|\mathbf{u} - 2\mathbf{v} + 4\mathbf{w}\|$

17. (*For readers who have studied calculus.*) Let the vector space P_2 have the inner product

$$\langle \mathbf{p}, \mathbf{q} \rangle = \int_{-1}^{1} p(x)q(x)\, dx$$

(a) Find $\|\mathbf{p}\|$ for $\mathbf{p} = 1$, $\mathbf{p} = x$, and $\mathbf{p} = x^2$. (b) Find $d(\mathbf{p}, \mathbf{q})$ if $\mathbf{p} = 1$ and $\mathbf{q} = x$.

18. Sketch the unit circle in R^2 using the given inner product.
   (a) $\langle \mathbf{u}, \mathbf{v} \rangle = \frac{1}{4}u_1v_1 + \frac{1}{16}u_2v_2$ (b) $\langle \mathbf{u}, \mathbf{v} \rangle = 2u_1v_1 + u_2v_2$

19. Find a weighted Euclidean inner product on R^2 for which the unit circle is the ellipse shown in Figure 3.

Figure 3

20. Show that the following identity holds for vectors in any inner product space.

$$\|\mathbf{u} + \mathbf{v}\|^2 + \|\mathbf{u} - \mathbf{v}\|^2 = 2\|\mathbf{u}\|^2 + 2\|\mathbf{v}\|^2$$

21. Show that the following identity holds for vectors in any inner product space.

$$\langle \mathbf{u}, \mathbf{v} \rangle = \frac{1}{4}\|\mathbf{u} + \mathbf{v}\|^2 - \frac{1}{4}\|\mathbf{u} - \mathbf{v}\|^2$$

22. Let $U = \begin{bmatrix} u_1 & u_2 \\ u_3 & u_4 \end{bmatrix}$ and $V = \begin{bmatrix} v_1 & v_2 \\ v_3 & v_4 \end{bmatrix}$.

   Show that $\langle U, V \rangle = u_1v_1 + u_2v_3 + u_3v_2 + u_4v_4$ is *not* an inner product on M_{22}.

23. Let $\mathbf{p} = p(x)$ and $\mathbf{q} = q(x)$ be polynomials in P_2. Show that

$$\langle \mathbf{p}, \mathbf{q} \rangle = p(0)q(0) + p(\tfrac{1}{2})q(\tfrac{1}{2}) + p(1)q(1)$$

   is an inner product on P_2.

24. Prove: If $\langle \mathbf{u}, \mathbf{v} \rangle$ is the Euclidean inner product on R^n, and if A is an $n \times n$ matrix, then

$$\langle \mathbf{u}, A\mathbf{v} \rangle = \langle A^T\mathbf{u}, \mathbf{v} \rangle$$

   [*Hint.* Use the fact that $\langle \mathbf{u}, \mathbf{v} \rangle = \mathbf{u} \cdot \mathbf{v} = \mathbf{v}^T\mathbf{u}$.]

25. Verify the result in Exercise 24 for the Euclidean inner product on R^3 and

$$\mathbf{u} = \begin{bmatrix} -1 \\ 2 \\ 4 \end{bmatrix}, \quad \mathbf{v} = \begin{bmatrix} 3 \\ 0 \\ -2 \end{bmatrix}, \quad A = \begin{bmatrix} 1 & -2 & 1 \\ 3 & 4 & 0 \\ 5 & -1 & 2 \end{bmatrix}$$

26. Let $\mathbf{u} = (u_1, u_2, \ldots, u_n)$ and $\mathbf{v} = (v_1, v_2, \ldots, v_n)$. Show that

$$\langle \mathbf{u}, \mathbf{v} \rangle = w_1u_1v_1 + w_2u_2v_2 + \cdots + w_nu_nv_n$$

   is an inner product on R^n if $w_1, w_2, \ldots, w_n$ are positive real numbers.

27. (**For readers who have studied calculus.**) Use the inner product

$$\langle \mathbf{p}, \mathbf{q} \rangle = \int_{-1}^{1} p(x)q(x)\, dx$$

to compute $\langle \mathbf{p}, \mathbf{q} \rangle$ for the vectors $\mathbf{p} = p(x)$ and $\mathbf{q} = q(x)$ in P_3.
(a) $\mathbf{p} = 1 - x + x^2 + 5x^3 \qquad \mathbf{q} = x - 3x^2$
(b) $\mathbf{p} = x - 5x^3 \qquad\qquad \mathbf{q} = 2 + 8x^2$

28. (**For readers who have studied calculus.**) In each part use the inner product

$$\langle \mathbf{f}, \mathbf{g} \rangle = \int_{0}^{1} f(x)g(x)\, dx$$

to compute $\langle \mathbf{f}, \mathbf{g} \rangle$ for the vectors $\mathbf{f} = f(x)$ and $\mathbf{g} = g(x)$ in $C[0, 1]$.

(a) $\mathbf{f} = \cos 2\pi x, \quad \mathbf{g} = \sin 2\pi x$ \qquad (b) $\mathbf{f} = x, \quad \mathbf{g} = e^x$ \qquad (c) $\mathbf{f} = \tan \dfrac{\pi}{4}x, \quad \mathbf{g} = 1$

29. Show that the inner product in Example 7 can be written as $\langle U, V \rangle = \mathrm{tr}(U^T V)$.

30. Prove that Formula (3) defines an inner product on R^n. [**Hint.** Use the alternative version of Formula (3) given by (4).]

31. Show that matrix (5) generates the weighted Euclidean inner product
$\langle \mathbf{u}, \mathbf{v} \rangle = w_1 u_1 v_1 + w_2 u_2 v_2 + \cdots + w_n u_n v_n$ on R^n.

32. Prove parts (*a*) and (*d*) of Theorem 6.1.1.

33. Prove parts (*c*) and (*e*) of Theorem 6.1.1.

6.2 ANGLE AND ORTHOGONALITY IN INNER PRODUCT SPACES

In this section we shall define the notion of an angle between two vectors in an inner product space, and we shall use this concept to obtain some basic relations between vectors in an inner product, including a fundamental geometric relationship between the nullspace and column space of a matrix.

CAUCHY–SCHWARZ INEQUALITY

Recall from Formula (1) of Section 3.3 that if $\mathbf{u}$ and $\mathbf{v}$ are nonzero vectors in R^2 or R^3 and θ is the angle between them, then

$$\mathbf{u} \cdot \mathbf{v} = \|\mathbf{u}\|\,\|\mathbf{v}\| \cos \theta \tag{1}$$

or alternatively,

$$\cos \theta = \frac{\mathbf{u} \cdot \mathbf{v}}{\|\mathbf{u}\|\,\|\mathbf{v}\|} \tag{2}$$

Our first goal in this section is to define the concept of an angle between two vectors in a general inner product space. For such a definition to be reasonable, we would want it to be consistent with Formula (2) when it is applied to the special case of R^2 and R^3 with the Euclidean inner product. Thus, we will want our definition of the angle θ between two nonzero vectors in an inner product space to satisfy the relationship

$$\cos\theta = \frac{\langle \mathbf{u}, \mathbf{v} \rangle}{\|\mathbf{u}\|\,\|\mathbf{v}\|} \tag{3}$$

However, because $|\cos\theta| \leq 1$, there would be no hope of satisfying (3) unless we were assured that every pair of nonzero vectors in an inner product space satisfies the inequality

$$\left| \frac{\langle \mathbf{u}, \mathbf{v} \rangle}{\|\mathbf{u}\|\,\|\mathbf{v}\|} \right| \leq 1$$

Fortunately, we will be able to prove that this is the case by using the following generalization of the Cauchy–Schwarz inequality (see Theorem 4.1.3).

Theorem 6.2.1 (*Cauchy–Schwarz Inequality*). *If* $\mathbf{u}$ *and* $\mathbf{v}$ *are vectors in a real inner product space, then*

$$|\langle \mathbf{u}, \mathbf{v} \rangle| \leq \|\mathbf{u}\|\,\|\mathbf{v}\| \tag{4}$$

Proof. We warn the reader in advance that the proof presented here depends on a clever trick that is not easy to motivate. If $\mathbf{u} = \mathbf{0}$, then $\langle \mathbf{u}, \mathbf{v} \rangle = \langle \mathbf{u}, \mathbf{u} \rangle = 0$, so that the two sides of (4) are equal. Assume now that $\mathbf{u} \neq \mathbf{0}$. Let $a = \langle \mathbf{u}, \mathbf{u} \rangle$, $b = 2\langle \mathbf{u}, \mathbf{v} \rangle$, $c = \langle \mathbf{v}, \mathbf{v} \rangle$, and let t be any real number. By the positivity axiom, the inner product of any vector with itself is always nonnegative. Therefore,

$$0 \leq \langle (t\mathbf{u} + \mathbf{v}), (t\mathbf{u} + \mathbf{v}) \rangle = \langle \mathbf{u}, \mathbf{u} \rangle t^2 + 2\langle \mathbf{u}, \mathbf{v} \rangle t + \langle \mathbf{v}, \mathbf{v} \rangle$$
$$= at^2 + bt + c$$

This inequality implies that the quadratic polynomial $at^2 + bt + c$ has either no real roots or a repeated real root. Therefore, its discriminant must satisfy the inequality $b^2 - 4ac \leq 0$. Expressing the coefficients a, b, and c in terms of the vectors $\mathbf{u}$ and $\mathbf{v}$ gives $4\langle \mathbf{u}, \mathbf{v} \rangle^2 - 4\langle \mathbf{u}, \mathbf{u} \rangle\langle \mathbf{v}, \mathbf{v} \rangle \leq 0$, or equivalently,

$$\langle \mathbf{u}, \mathbf{v} \rangle^2 \leq \langle \mathbf{u}, \mathbf{u} \rangle\langle \mathbf{v}, \mathbf{v} \rangle$$

Taking square roots of both sides and using the fact that $\langle \mathbf{u}, \mathbf{u} \rangle$ and $\langle \mathbf{v}, \mathbf{v} \rangle$ are nonnegative yields

$$|\langle \mathbf{u}, \mathbf{v} \rangle| \leq \langle \mathbf{u}, \mathbf{u} \rangle^{1/2}\langle \mathbf{v}, \mathbf{v} \rangle^{1/2}$$

or equivalently,

$$|\langle \mathbf{u}, \mathbf{v} \rangle| \leq \|\mathbf{u}\|\,\|\mathbf{v}\|$$

which completes the proof. ▮

For reference, we note that the Cauchy–Schwarz inequality can be written in the following two alternative forms:

$$\langle \mathbf{u}, \mathbf{v} \rangle^2 \leq \langle \mathbf{u}, \mathbf{u} \rangle \langle \mathbf{v}, \mathbf{v} \rangle \tag{5}$$

$$\langle \mathbf{u}, \mathbf{v} \rangle^2 \leq \|\mathbf{u}\|^2 \|\mathbf{v}\|^2 \tag{6}$$

The first of these formulas was obtained in the proof of Theorem 6.2.1, and the second is derived from the first using the fact that $\|\mathbf{u}\|^2 = \langle \mathbf{u}, \mathbf{u} \rangle$ and $\|\mathbf{v}\|^2 = \langle \mathbf{v}, \mathbf{v} \rangle$.

Example 1   The Cauchy–Schwarz inequality for R^n (Theorem 4.1.3) follows as a special case of Theorem 6.2.1 by taking $\langle \mathbf{u}, \mathbf{v} \rangle$ to be the Euclidean inner product $\mathbf{u} \cdot \mathbf{v}$. ▲

PROPERTIES OF LENGTH AND DISTANCE IN INNER PRODUCT SPACES

The next two theorems show that the basic properties of length and distance that were established in Theorems 4.1.4 and 4.1.5 for vectors in Euclidean n-space continue to hold in general inner product spaces. This is strong evidence that our definitions of inner product, length, and distance are well chosen.

Theorem 6.2.2. *If* $\mathbf{u}$ *and* $\mathbf{v}$ *are vectors in an inner product space V, and if k is any scalar, then:*

(a) $\|\mathbf{u}\| \geq 0$
(b) $\|\mathbf{u}\| = 0$ *if and only if* $\mathbf{u} = \mathbf{0}$
(c) $\|k\mathbf{u}\| = |k| \|\mathbf{u}\|$
(d) $\|\mathbf{u} + \mathbf{v}\| \leq \|\mathbf{u}\| + \|\mathbf{v}\|$ *(Triangle inequality)*

Theorem 6.2.3. *If* $\mathbf{u}$, $\mathbf{v}$, *and* $\mathbf{w}$ *are vectors in an inner product space V, and if k is any scalar, then:*

(a) $d(\mathbf{u}, \mathbf{v}) \geq 0$
(b) $d(\mathbf{u}, \mathbf{v}) = 0$ *if and only if* $\mathbf{u} = \mathbf{v}$
(c) $d(\mathbf{u}, \mathbf{v}) = d(\mathbf{v}, \mathbf{u})$
(d) $d(\mathbf{u}, \mathbf{v}) \leq d(\mathbf{u}, \mathbf{w}) + d(\mathbf{w}, \mathbf{v})$ *(Triangle inequality)*

We shall prove part (d) of Theorem 6.2.2 and leave the remaining parts of Theorems 6.2.2 and 6.2.3 as exercises.

Proof of Theorem 6.2.2d.   By definition

$$\begin{aligned}
\|\mathbf{u} + \mathbf{v}\|^2 &= \langle \mathbf{u} + \mathbf{v}, \mathbf{u} + \mathbf{v} \rangle \\
&= \langle \mathbf{u}, \mathbf{u} \rangle + 2\langle \mathbf{u}, \mathbf{v} \rangle + \langle \mathbf{v}, \mathbf{v} \rangle \\
&\leq \langle \mathbf{u}, \mathbf{u} \rangle + 2|\langle \mathbf{u}, \mathbf{v} \rangle| + \langle \mathbf{v}, \mathbf{v} \rangle \qquad [\text{Property of absolute value}] \\
&\leq \langle \mathbf{u}, \mathbf{u} \rangle + 2\|\mathbf{u}\|\|\mathbf{v}\| + \langle \mathbf{v}, \mathbf{v} \rangle \qquad [\text{By} (4)] \\
&= \|\mathbf{u}\|^2 + 2\|\mathbf{u}\|\|\mathbf{v}\| + \|\mathbf{v}\|^2 \\
&= (\|\mathbf{u}\| + \|\mathbf{v}\|)^2
\end{aligned}$$

Taking square roots gives

$$\|\mathbf{u} + \mathbf{v}\| \le \|\mathbf{u}\| + \|\mathbf{v}\| \quad \blacksquare$$

ANGLE BETWEEN VECTORS

We shall now show how the Cauchy–Schwarz inequality can be used to define angles in general inner product spaces. Suppose that $\mathbf{u}$ and $\mathbf{v}$ are nonzero vectors in an inner product space V. If we divide both sides of Formula (6) by $\|\mathbf{u}\|^2 \|\mathbf{v}\|^2$, we obtain

$$\left[\frac{\langle \mathbf{u}, \mathbf{v} \rangle}{\|\mathbf{u}\| \|\mathbf{v}\|} \right]^2 \le 1$$

or equivalently,

$$-1 \le \frac{\langle \mathbf{u}, \mathbf{v} \rangle}{\|\mathbf{u}\| \|\mathbf{v}\|} \le 1 \tag{7}$$

Now if θ is an angle whose radian measure varies from 0 to π, then $\cos \theta$ assumes every value between -1 and 1 inclusive exactly once (Figure 1).

Figure 1

Thus, from (7) there is a unique angle θ such that

$$\cos \theta = \frac{\langle \mathbf{u}, \mathbf{v} \rangle}{\|\mathbf{u}\| \|\mathbf{v}\|} \quad \text{and} \quad 0 \le \theta \le \pi \tag{8}$$

We define θ to be the ***angle between u and v***. Observe that in R^2 or R^3 with the Euclidean inner product, (8) agrees with the usual formula for the cosine of the angle between two nonzero vectors [Formula (2)].

Example 2 Let R^4 have the Euclidean inner product. Find the cosine of the angle θ between the vectors $\mathbf{u} = (4, 3, 1, -2)$ and $\mathbf{v} = (-2, 1, 2, 3)$.

Solution. We leave it for the reader to verify that

$$\|\mathbf{u}\| = \sqrt{30}, \qquad \|\mathbf{v}\| = \sqrt{18}, \qquad \text{and} \qquad \langle \mathbf{u}, \mathbf{v} \rangle = -9$$

so that

$$\cos \theta = \frac{\langle \mathbf{u}, \mathbf{v} \rangle}{\|\mathbf{u}\| \|\mathbf{v}\|} = -\frac{9}{\sqrt{30}\sqrt{18}} = -\frac{3}{2\sqrt{15}} \quad \blacktriangle$$

ORTHOGONALITY

Example 2 is primarily a mathematical exercise, for there is relatively little need to find angles between vectors, except in R^2 and R^3 with the Euclidean inner product. However, a problem of major importance in all inner product spaces is to determine whether two vectors are *orthogonal*—that is, whether the angle between them is $\theta = \pi/2$.

It follows from (8) that if **u** and **v** are *nonzero* vectors in an inner product space and θ is the angle between them, then $\cos\theta = 0$ if and only if $\langle \mathbf{u}, \mathbf{v} \rangle = 0$. Equivalently, for nonzero vectors we have $\theta = \pi/2$ if and only if $\langle \mathbf{u}, \mathbf{v} \rangle = 0$. If we agree to consider the angle between **u** and **v** to be $\pi/2$ when either or both of these vectors is **0**, then we can state without exception that the angle between **u** and **v** is $\pi/2$ if and only if $\langle \mathbf{u}, \mathbf{v} \rangle = 0$. This suggests the following definition.

Definition. Two vectors **u** and **v** in an inner product space are called **orthogonal** if $\langle \mathbf{u}, \mathbf{v} \rangle = 0$.

Observe that in the special case where $\langle \mathbf{u}, \mathbf{v} \rangle = \mathbf{u} \cdot \mathbf{v}$ is the Euclidean inner product on R^n, this definition reduces to the definition of orthogonality in Euclidean n-space given in Section 4.1. We also emphasize that orthogonality depends on the inner product; two vectors can be orthogonal with respect to one inner product but not another.

Example 3 If M_{22} has the inner product of Example 7 in the preceding section, then the matrices

$$U = \begin{bmatrix} 1 & 0 \\ 1 & 1 \end{bmatrix} \quad \text{and} \quad V = \begin{bmatrix} 0 & 2 \\ 0 & 0 \end{bmatrix}$$

are orthogonal, since

$$\langle U, V \rangle = 1(0) + 0(2) + 1(0) + 1(0) = 0 \quad \triangle$$

Example 4 (*For readers who have studied calculus.*) Let P_2 have the inner product

$$\langle \mathbf{p}, \mathbf{q} \rangle = \int_{-1}^{1} p(x)q(x)\,dx$$

and let

$$\mathbf{p} = x, \qquad \mathbf{q} = x^2$$

Then

$$\|\mathbf{p}\| = \langle \mathbf{p}, \mathbf{p} \rangle^{1/2} = \left[\int_{-1}^{1} xx\,dx \right]^{1/2} = \left[\int_{-1}^{1} x^2\,dx \right]^{1/2} = \sqrt{\frac{2}{3}}$$

$$\|\mathbf{q}\| = \langle \mathbf{q}, \mathbf{q} \rangle^{1/2} = \left[\int_{-1}^{1} x^2 x^2\,dx \right]^{1/2} = \left[\int_{-1}^{1} x^4\,dx \right]^{1/2} = \sqrt{\frac{2}{5}}$$

$$\langle \mathbf{p}, \mathbf{q} \rangle = \int_{-1}^{1} xx^2\,dx = \int_{-1}^{1} x^3\,dx = 0$$

Because $\langle \mathbf{p}, \mathbf{q} \rangle = 0$, the vectors $\mathbf{p} = x$ and $\mathbf{q} = x^2$ are orthogonal relative to the given inner product. ▲

In Section 4.1 we proved the Theorem of Pythagoras for vectors in Euclidean n-space. The following theorem extends this result to vectors in any inner product space.

Theorem 6.2.4 (*Generalized Theorem of Pythagoras*). *If* $\mathbf{u}$ *and* $\mathbf{v}$ *are orthogonal vectors in an inner product space, then*

$$\|\mathbf{u} + \mathbf{v}\|^2 = \|\mathbf{u}\|^2 + \|\mathbf{v}\|^2$$

Proof. The orthogonality of $\mathbf{u}$ and $\mathbf{v}$ implies that $\langle \mathbf{u}, \mathbf{v} \rangle = 0$, so

$$\|\mathbf{u} + \mathbf{v}\|^2 = \langle (\mathbf{u} + \mathbf{v}), (\mathbf{u} + \mathbf{v}) \rangle = \|\mathbf{u}\|^2 + 2\langle \mathbf{u}, \mathbf{v} \rangle + \|\mathbf{v}\|^2$$
$$= \|\mathbf{u}\|^2 + \|\mathbf{v}\|^2$$

Example 5 (*For readers who have studied calculus.*) In Example 4 we showed that $\mathbf{p} = x$ and $\mathbf{q} = x^2$ are orthogonal relative to the inner product

$$\langle \mathbf{p}, \mathbf{q} \rangle = \int_{-1}^{1} p(x)q(x)\, dx$$

on P_2. It follows from the Theorem of Pythagoras that

$$\|\mathbf{p} + \mathbf{q}\|^2 = \|\mathbf{p}\|^2 + \|\mathbf{q}\|^2$$

Thus, from the computations in Example 4 we have

$$\|\mathbf{p} + \mathbf{q}\|^2 = \left(\sqrt{\frac{2}{3}} \right)^2 + \left(\sqrt{\frac{2}{5}} \right)^2 = \frac{2}{3} + \frac{2}{5} = \frac{16}{15}$$

We can check this result by direct integration:

$$\|\mathbf{p} + \mathbf{q}\|^2 = \langle \mathbf{p} + \mathbf{q}, \mathbf{p} + \mathbf{q} \rangle = \int_{-1}^{1} (x + x^2)(x + x^2)\, dx$$

$$= \int_{-1}^{1} x^2\, dx + 2 \int_{-1}^{1} x^3\, dx + \int_{-1}^{1} x^4\, dx = \frac{2}{3} + 0 + \frac{2}{5} = \frac{16}{15} \;▲$$

ORTHOGONAL COMPLEMENTS

If V is a plane through the origin of R^3 with the Euclidean inner product, then the set of all vectors that are orthogonal to every vector in V forms the line L through the origin that is perpendicular to V (Figure 2). In the language of linear algebra we say that the line and the plane are *orthogonal complements* of one another. The following definition extends this concept to general inner product spaces.

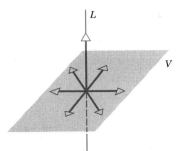

Every vector in L is orthogonal
to every vector in V.

Figure 2

Definition. Let W be a subspace of an inner product space V. A vector **u** in V is
said to be ***orthogonal to W*** if it is orthogonal to every vector in W, and the set of
all vectors in V that are orthogonal to W is called the ***orthogonal complement
of W***.

Recall from geometry that the symbol $\perp$ is used to indicate perpendicularity. In
linear algebra the orthogonal complement of a subspace W is denoted by $W^{\perp}$ (read "W
perp"). The following theorem lists the basic properties of orthogonal complements.

Theorem 6.2.5. *If W is a subspace of a finite-dimensional inner product space V,
then:*

(a) $W^{\perp}$ *is a subspace of V.*
*(b) The only vector common to W and $W^{\perp}$ is **0**.*
(c) The orthogonal complement of $W^{\perp}$ is W; that is, $(W^{\perp})^{\perp} = W$.

We shall prove part (a) and leave (b) as an exercise for this section (Exercise 25). The
proof of (c) requires results covered later in this chapter, so its proof is left for the
exercises at the end of the chapter.

Proof (a). Note first that $\langle \mathbf{0}, \mathbf{w} \rangle = 0$ for every vector **w** in W, so $W^{\perp}$ contains at least
the zero vector. We want to show that $W^{\perp}$ is closed under addition and scalar multi-
plication; that is, we want to show that the sum of two vectors in $W^{\perp}$ is orthogonal to
every vector in W and that any scalar multiple of a vector in $W^{\perp}$ is orthogonal to every
vector in W. Let **u** and **v** be any vectors in $W^{\perp}$, let k be any scalar, and let **w** be any
vector in W. Then from the definition of $W^{\perp}$ we have $\langle \mathbf{u}, \mathbf{w} \rangle = 0$ and $\langle \mathbf{v}, \mathbf{w} \rangle = 0$. Using
basic properties of the inner product we have

$$\langle \mathbf{u} + \mathbf{v}, \mathbf{w} \rangle = \langle \mathbf{u}, \mathbf{w} \rangle + \langle \mathbf{v}, \mathbf{w} \rangle = 0 + 0 = 0$$
$$\langle k\mathbf{u}, \mathbf{w} \rangle = k\langle \mathbf{u}, \mathbf{w} \rangle = k(0) = 0$$

which proves that $\mathbf{u} + \mathbf{v}$ and $k\mathbf{u}$ are in $W^{\perp}$. ▇

REMARK. Because W and $W^{\perp}$ are orthogonal complements of one another by part (c)
of the preceding theorem, we shall say that W and $W^{\perp}$ are ***orthogonal complements***.

**A GEOMETRIC
LINK BETWEEN
NULLSPACE AND
ROW SPACE**

The following fundamental theorem provides a geometric link between the nullspace and row space of a matrix.

Theorem 6.2.6. *If A is an $m \times n$ matrix, then:*

(a) *The nullspace of A and the row space of A are orthogonal complements in R^n with respect to the Euclidean inner product.*

(b) *The nullspace of A^T and the column space of A are orthogonal complements in R^m with respect to the Euclidean inner product.*

Proof (a). We want to show that the orthogonal complement of the row space of A is the nullspace of A. To do this we must show that if a vector $\mathbf{v}$ is orthogonal to every vector in the row space, then $A\mathbf{v} = \mathbf{0}$, and conversely, if $A\mathbf{v} = \mathbf{0}$, then $\mathbf{v}$ is orthogonal to every vector in the row space.

Assume first that $\mathbf{v}$ is orthogonal to every vector in the row space of A. Then in particular $\mathbf{v}$ is orthogonal to the row vectors $\mathbf{r}_1, \mathbf{r}_2, \ldots, \mathbf{r}_n$ of A; that is,

$$\mathbf{r}_1 \cdot \mathbf{v} = \mathbf{r}_2 \cdot \mathbf{v} = \cdots = \mathbf{r}_n \cdot \mathbf{v} = 0 \tag{9}$$

But by Formula (11) of Section 4.1, the linear system $A\mathbf{x} = \mathbf{0}$ can be expressed in dot product notation as

$$\begin{bmatrix} \mathbf{r}_1 \cdot \mathbf{x} \\ \mathbf{r}_2 \cdot \mathbf{x} \\ \vdots \\ \mathbf{r}_n \cdot \mathbf{x} \end{bmatrix} = \begin{bmatrix} 0 \\ 0 \\ \vdots \\ 0 \end{bmatrix} \tag{10}$$

so from (9), $\mathbf{v}$ is a solution of this system and hence lies in the nullspace of A.

Conversely, assume that $\mathbf{v}$ is a vector in the nullspace of A, so that $A\mathbf{v} = \mathbf{0}$. It follows from (10) that

$$\mathbf{r}_1 \cdot \mathbf{v} = \mathbf{r}_2 \cdot \mathbf{v} = \cdots = \mathbf{r}_n \cdot \mathbf{v} = 0$$

But if $\mathbf{r}$ is any vector in the row space of A, then $\mathbf{r}$ is expressible as a linear combination of the row vectors of A, say

$$\mathbf{r} = c_1 \mathbf{r}_1 + c_2 \mathbf{r}_2 + \cdots + c_n \mathbf{r}_n$$

Thus,

$$\begin{aligned} \mathbf{r} \cdot \mathbf{v} &= (c_1 \mathbf{r}_1 + c_2 \mathbf{r}_2 + \cdots + c_n \mathbf{r}_n) \cdot \mathbf{v} \\ &= c_1(\mathbf{r}_1 \cdot \mathbf{v}) + c_2(\mathbf{r}_2 \cdot \mathbf{v}) + \cdots + c_n(\mathbf{r}_n \cdot \mathbf{v}) \\ &= 0 + 0 + \cdots + 0 = 0 \end{aligned}$$

which proves that $\mathbf{v}$ is orthogonal to every vector in the row space of A.

Proof (b). Since the column space of A is the row space of A^T (except for a difference in notation), the proof follows by applying the result in part (*a*) to A^T. ▮

The following example shows how Theorem 6.2.6 can be used to find a basis for the orthogonal complement of a subspace of Euclidean n-space.

Example 6 Let W be the subspace of R^5 spanned by the vectors

$$\mathbf{w}_1 = (2, 2, -1, 0, 1), \qquad \mathbf{w}_2 = (-1, -1, 2, -3, 1),$$
$$\mathbf{w}_3 = (1, 1, -2, 0, -1), \qquad \mathbf{w}_4 = (0, 0, 1, 1, 1)$$

Find a basis for the orthogonal complement of W.

Solution. The space W spanned by $\mathbf{w}_1$, $\mathbf{w}_2$, $\mathbf{w}_3$, and $\mathbf{w}_4$ is the same as the row space of the matrix

$$A = \begin{bmatrix} 2 & 2 & -1 & 0 & 1 \\ -1 & -1 & 2 & -3 & 1 \\ 1 & 1 & -2 & 0 & -1 \\ 0 & 0 & 1 & 1 & 1 \end{bmatrix}$$

and by part (*a*) of Theorem 6.2.6 the nullspace of A is the orthogonal complement of W. In Example 4 of Section 5.5 we showed that

$$\mathbf{v}_1 = \begin{bmatrix} -1 \\ 1 \\ 0 \\ 0 \\ 0 \end{bmatrix} \quad \text{and} \quad \mathbf{v}_2 = \begin{bmatrix} -1 \\ 0 \\ -1 \\ 0 \\ 1 \end{bmatrix}$$

form a basis for this nullspace. Expressing these vectors in the same notation as $\mathbf{w}_1$, $\mathbf{w}_2$, $\mathbf{w}_3$, and $\mathbf{w}_4$, we conclude that the vectors

$$\mathbf{v}_1 = (-1, 1, 0, 0, 0) \quad \text{and} \quad \mathbf{v}_2 = (-1, 0, -1, 0, 1)$$

form a basis for the orthogonal complement of W. As a check, the reader may want to verify that $\mathbf{v}_1$ and $\mathbf{v}_2$ are orthogonal to $\mathbf{w}_1$, $\mathbf{w}_2$, $\mathbf{w}_3$, and $\mathbf{w}_4$ by calculating the necessary dot products. ▲

SUMMARY

We leave it as an exercise for the reader to show that in any inner product space V, the zero space $\{\mathbf{0}\}$ and the entire space V are orthogonal complements. Thus, if A is an $n \times n$ matrix, to say that $A\mathbf{x} = \mathbf{0}$ has only the trivial solution is equivalent to saying that the orthogonal complement of the nullspace of A is all of R^n, or equivalently, that the rowspace of A is all of R^n. This enables us to add two new results to the seventeen listed in Theorem 5.6.9.

Theorem 6.2.7. *If A is an $n \times n$ matrix, and if $T_A : R^n \rightarrow R^n$ is multiplication by A, then the following are equivalent.*

(a) *A is invertible.*

(b) *$A\mathbf{x} = \mathbf{0}$ has only the trivial solution.*

(c) *The reduced row-echelon form of A is I_n.*

(d) *A is expressible as a product of elementary matrices.*

(e) *$A\mathbf{x} = \mathbf{b}$ is consistent for every $n \times 1$ matrix $\mathbf{b}$.*

(f) *$A\mathbf{x} = \mathbf{b}$ has exactly one solution for every $n \times 1$ matrix $\mathbf{b}$.*

(g) *$det(A) \neq 0$.*

(h) *The range of T_A is R^n.*

(i) *T_A is one-to-one.*

(j) *The column vectors of A are linearly independent.*

(k) *The row vectors of A are linearly independent.*

(l) *The column vectors of A span R^n.*

(m) *The row vectors of A span R^n.*

(n) *The column vectors of A form a basis for R^n.*

(o) *The row vectors of A form a basis for R^n.*

(p) *A has rank n.*

(q) *A has nullity 0.*

(r) *The orthogonal complement of the nullspace of A is R^n.*

(s) *The orthogonal complement of the row space of A is $\{\mathbf{0}\}$.*

This theorem relates all of the major topics we have studied thus far.

EXERCISE SET 6.2

1. In each part determine whether the given vectors are orthogonal with respect to the Euclidean inner product.

(a) $\mathbf{u} = (-1, 3, 2)$, $\mathbf{v} = (4, 2, -1)$ (b) $\mathbf{u} = (-2, -2, -2)$, $\mathbf{v} = (1, 1, 1)$

(c) $\mathbf{u} = (u_1, u_2, u_3)$, $\mathbf{v} = (0, 0, 0)$ (d) $\mathbf{u} = (-4, 6, -10, 1)$, $\mathbf{v} = (2, 1, -2, 9)$

(e) $\mathbf{u} = (0, 3, -2, 1)$, $\mathbf{v} = (5, 2, -1, 0)$ (f) $\mathbf{u} = (a, b)$, $\mathbf{v} = (-b, a)$

2. Let R^4 have the Euclidean inner product, and let $\mathbf{u} = (-1, 1, 0, 2)$. Determine whether the vector $\mathbf{u}$ is orthogonal to the set of vectors $W = \{\mathbf{w}_1, \mathbf{w}_2, \mathbf{w}_3\}$, where $\mathbf{w}_1 = (0, 0, 0, 0)$, $\mathbf{w}_2 = (1, -1, 3, 0)$, and $\mathbf{w}_3 = (4, 0, 9, 2)$.

3. Let R^2, R^3, and R^4 have the Euclidean inner product. In each part find the cosine of the angle between $\mathbf{u}$ and $\mathbf{v}$.

(a) $\mathbf{u} = (1, -3)$, $\mathbf{v} = (2, 4)$ (b) $\mathbf{u} = (-1, 0)$, $\mathbf{v} = (3, 8)$

(c) $\mathbf{u} = (-1, 5, 2)$, $\mathbf{v} = (2, 4, -9)$ (d) $\mathbf{u} = (4, 1, 8)$, $\mathbf{v} = (1, 0, -3)$

(e) $\mathbf{u} = (1, 0, 1, 0)$, $\mathbf{v} = (-3, -3, -3, -3)$ (f) $\mathbf{u} = (2, 1, 7, -1)$, $\mathbf{v} = (4, 0, 0, 0)$

4. Let P_2 have the inner product in Example 8 of Section 6.1. Find the cosine of the angle between **p** and **q**.

(a) $\mathbf{p} = -1 + 5x + 2x^2$, $\mathbf{q} = 2 + 4x - 9x^2$ (b) $\mathbf{p} = x - x^2$, $\mathbf{q} = 7 + 3x + 3x^2$

5. Show that $\mathbf{p} = 1 - x + 2x^2$ and $\mathbf{q} = 2x + x^2$ are orthogonal with respect to the inner product in Exercise 4.

6. Let M_{22} have the inner product in Example 7 of Section 6.1. Find the cosine of the angle between A and B.

(a) $A = \begin{bmatrix} 2 & 6 \\ 1 & -3 \end{bmatrix}$, $B = \begin{bmatrix} 3 & 2 \\ 1 & 0 \end{bmatrix}$ (b) $A = \begin{bmatrix} 2 & 4 \\ -1 & 3 \end{bmatrix}$, $B = \begin{bmatrix} -3 & 1 \\ 4 & 2 \end{bmatrix}$

7. Let

$$A = \begin{bmatrix} 2 & 1 \\ -1 & 3 \end{bmatrix}$$

Which of the following matrices are orthogonal to A with respect to the inner product in Exercise 6?

(a) $\begin{bmatrix} -3 & 0 \\ 0 & 2 \end{bmatrix}$ (b) $\begin{bmatrix} 1 & 1 \\ 0 & -1 \end{bmatrix}$ (c) $\begin{bmatrix} 0 & 0 \\ 0 & 0 \end{bmatrix}$ (d) $\begin{bmatrix} 2 & 1 \\ 5 & 2 \end{bmatrix}$

8. Let R^3 have the Euclidean inner product. For which values of k are **u** and **v** orthogonal?

(a) $\mathbf{u} = (2, 1, 3)$, $\mathbf{v} = (1, 7, k)$ (b) $\mathbf{u} = (k, k, 1)$, $\mathbf{v} = (k, 5, 6)$

9. Let R^4 have the Euclidean inner product. Find two vectors of norm 1 that are orthogonal to the three vectors $\mathbf{u} = (2, 1, -4, 0)$, $\mathbf{v} = (-1, -1, 2, 2)$, and $\mathbf{w} = (3, 2, 5, 4)$.

10. In each part verify that the Cauchy–Schwarz inequality holds for the given vectors using the Euclidean inner product.

(a) $\mathbf{u} = (3, 2)$, $\mathbf{v} = (4, -1)$ (b) $\mathbf{u} = (-3, 1, 0)$, $\mathbf{v} = (2, -1, 3)$

(c) $\mathbf{u} = (-4, 2, 1)$, $\mathbf{v} = (8, -4, -2)$ (d) $\mathbf{u} = (0, -2, 2, 1)$, $\mathbf{v} = (-1, -1, 1, 1)$

11. In each part verify that the Cauchy–Schwarz inequality holds for the given vectors.

(a) $\mathbf{u} = (-2, 1)$ and $\mathbf{v} = (1, 0)$ using the inner product of Example 2 of Section 6.1

(b) $U = \begin{bmatrix} -1 & 2 \\ 6 & 1 \end{bmatrix}$ and $V = \begin{bmatrix} 1 & 0 \\ 3 & 3 \end{bmatrix}$

using the inner product in Example 7 of Section 6.1

(c) $\mathbf{p} = -1 + 2x + x^2$ and $\mathbf{q} = 2 - 4x^2$ using the inner product given in Example 8 of Section 6.1

12. Let W be the line in R^2 with equation $y = 2x$. Find an equation for $W^\perp$.

13. (a) Let W be the plane in R^3 with equation $x - 2y - 3z = 0$. Find parametric equations for $W^\perp$.

(b) Let W be the line in R^3 with parametric equations

$$x = 2t, \quad y = -5t, \quad z = 4t \qquad (-\infty < t < \infty)$$

Find an equation for $W^\perp$.

14. Let

$$A = \begin{bmatrix} 1 & 2 & -1 & 2 \\ 3 & 5 & 0 & 4 \\ 1 & 1 & 2 & 0 \end{bmatrix}$$

(a) Find bases for the row space and nullspace of A.

(b) Verify that every vector in the row space is orthogonal to every vector in the nullspace (as guaranteed by Theorem 6.2.6a).

15. Let A be the matrix in Exercise 14.

(a) Find bases for the column space of A and nullspace of A^T.

(b) Verify that every vector in the column space of A is orthogonal to every vector in the nullspace of A^T (as guaranteed by Theorem 6.2.6b).

16. Find a basis for the orthogonal complement of the subspace of R^n spanned by the vectors.

(a) $\mathbf{v}_1 = (1, -1, 3)$, $\mathbf{v}_2 = (5, -4, -4)$, $\mathbf{v}_3 = (7, -6, 2)$

(b) $\mathbf{v}_1 = (2, 0, -1)$, $\mathbf{v}_2 = (4, 0, -2)$

(c) $\mathbf{v}_1 = (1, 4, 5, 2)$, $\mathbf{v}_2 = (2, 1, 3, 0)$, $\mathbf{v}_3 = (-1, 3, 2, 2)$

(d) $\mathbf{v}_1 = (1, 4, 5, 6, 9)$, $\mathbf{v}_2 = (3, -2, 1, 4, -1)$, $\mathbf{v}_3 = (-1, 0, -1, -2, -1)$,
    $\mathbf{v}_4 = (2, 3, 5, 7, 8)$

17. Let V be an inner product space. Show that if $\mathbf{u}$ and $\mathbf{v}$ are orthogonal vectors in V such that $\|\mathbf{u}\| = \|\mathbf{v}\| = 1$, then $\|\mathbf{u} - \mathbf{v}\| = \sqrt{2}$.

18. Let V be an inner product space. Show that if $\mathbf{w}$ is orthogonal to both $\mathbf{u}_1$ and $\mathbf{u}_2$, it is orthogonal to $k_1\mathbf{u}_1 + k_2\mathbf{u}_2$ for all scalars k_1 and k_2. Interpret this result geometrically in the case where V is R^3 with the Euclidean inner product.

19. Let V be an inner product space. Show that if $\mathbf{w}$ is orthogonal to each of the vectors $\mathbf{u}_1, \mathbf{u}_2$, ..., $\mathbf{u}_r$, then it is orthogonal to every vector in span$\{\mathbf{u}_1, \mathbf{u}_2, \ldots, \mathbf{u}_r\}$.

20. Let $\{\mathbf{v}_1, \mathbf{v}_2, \ldots, \mathbf{v}_r\}$ be a basis for an inner product space V. Show that the zero vector is the only vector in V that is orthogonal to all of the basis vectors.

21. Let $\{\mathbf{w}_1, \mathbf{w}_2, \ldots, \mathbf{w}_k\}$ be a basis for a subspace W of V. Show that $W^\perp$ consists of all vectors in V that are orthogonal to every basis vector.

22. Prove the following generalization of Theorem 6.2.4. If $\mathbf{v}_1, \mathbf{v}_2, \ldots, \mathbf{v}_r$ are pairwise orthogonal vectors in an inner product space V, then

$$\|\mathbf{v}_1 + \mathbf{v}_2 + \cdots + \mathbf{v}_r\|^2 = \|\mathbf{v}_1\|^2 + \|\mathbf{v}_2\|^2 + \cdots + \|\mathbf{v}_r\|^2$$

23. Prove the following parts of Theorem 6.2.2:

(a) part (*a*)   (b) part (*b*)   (c) part (*c*)

24. Prove the following parts of Theorem 6.2.3:

(a) part (*a*)   (b) part (*b*)   (c) part (*c*)   (d) part (*d*)

25. Prove part (b) of Theorem 6.2.5.

26. Prove: If $\mathbf{u}$ and $\mathbf{v}$ are $n \times 1$ matrices and A is an invertible $n \times n$ matrix, then

$$[\mathbf{v}^T A^T A \mathbf{u}]^2 \le (\mathbf{u}^T A^T A \mathbf{u})(\mathbf{v}^T A^T A \mathbf{v})$$

27. Use the Cauchy–Schwarz inequality to prove that for all real values of a, b, and θ,

$$[a \cos\theta + b \sin\theta]^2 \le a^2 + b^2$$

28. Prove: If $w_1, w_2, \ldots, w_n$ are positive real numbers and if $\mathbf{u} = (u_1, u_2, \ldots, u_n)$ and $\mathbf{v} = (v_1, v_2, \ldots, v_n)$ are any two vectors R^n, then

$$|w_1 u_1 v_1 + w_2 u_2 v_2 + \cdots + w_n u_n v_n| \le (w_1 u_1^2 + w_2 u_2^2 + \cdots + w_n u_n^2)^{1/2}(w_1 v_1^2 + w_2 v_2^2 + \cdots + w_n v_n^2)^{1/2}$$

29. Show that equality holds in the Cauchy–Schwarz inequality if and only if $\mathbf{u}$ and $\mathbf{v}$ are linearly dependent.

30. (*For readers who have studied calculus.*) Let $C[0, \pi]$ have the inner product

$$\langle \mathbf{f}, \mathbf{g} \rangle = \int_0^\pi f(x)g(x)\, dx$$

and let $\mathbf{f}_n = \cos nx \ (n = 0, 1, 2, \ldots)$. Show that if $k \ne l$, then $\mathbf{f}_k$ and $\mathbf{f}_l$ are orthogonal with respect to the given inner product.

31. (*For readers who have studied calculus.*) Let $f(x)$ and $g(x)$ be continuous functions on $[0, 1]$. Prove:

(a) $\left[\int_0^1 f(x)g(x)\, dx \right]^2 \le \left[\int_0^1 f^2(x)\, dx \right]\left[\int_0^1 g^2(x)\, dx \right]$

(b) $\left[\int_0^1 [\, f(x) + g(x)\,]^2\, dx \right]^{1/2} \le \left[\int_0^1 f^2(x)\, dx \right]^{1/2} + \left[\int_0^1 g^2(x)\, dx \right]^{1/2}$

[*Hint.* Use the Cauchy–Schwarz inequality.]

32. Use vector methods to prove that a triangle that is inscribed in a circle so that it has a diameter for a side must be a right triangle. [*Hint.* Express the vectors $\overrightarrow{AB}$ and $\overrightarrow{BC}$ in Figure 3 in terms of $\mathbf{u}$ and $\mathbf{v}$.]

Figure 3

33. With respect to the Euclidean inner product, the vectors $\mathbf{u} = (1, \sqrt{3})$ and $\mathbf{v} = (-1, \sqrt{3})$ have norm 2, and the angle between them is 60° (Figure 4). Find a weighted Euclidean inner product with respect to which $\mathbf{u}$ and $\mathbf{v}$ are orthogonal unit vectors.

Figure 4

6.3 ORTHONORMAL BASES; GRAM–SCHMIDT PROCESS; *QR*-DECOMPOSITION

In many problems involving vector spaces, the problem solver is free to choose any basis for the vector space that seems appropriate. In inner product spaces the solution of a problem is often greatly simplified by choosing a basis in which the vectors are orthogonal to one another. In this section we shall show how such bases can be obtained.

ORTHOGONAL AND ORTHONORMAL BASES

> **Definition.** A set of vectors in an inner product space is called an ***orthogonal set*** if all pairs of distinct vectors in the set are orthogonal. An orthogonal set in which each vector has norm 1 is called ***orthonormal***.

Example 1 Let

$$\mathbf{u}_1 = (0, 1, 0), \qquad \mathbf{u}_2 = (1, 0, 1), \qquad \mathbf{u}_3 = (1, 0, -1)$$

and assume that R^3 has the Euclidean inner product. It follows that the set of vectors $S = \{\mathbf{u}_1, \mathbf{u}_2, \mathbf{u}_3\}$ is orthogonal since $\langle \mathbf{u}_1, \mathbf{u}_2 \rangle = \langle \mathbf{u}_1, \mathbf{u}_3 \rangle = \langle \mathbf{u}_2, \mathbf{u}_3 \rangle = 0$. ▲

If $\mathbf{v}$ is a nonzero vector in an inner product space, then by part (*c*) of Theorem 6.2.2 the vector

$$\frac{1}{\|\mathbf{v}\|} \mathbf{v}$$

has norm 1, since

$$\left\| \frac{1}{\|\mathbf{v}\|} \mathbf{v} \right\| = \left| \frac{1}{\|\mathbf{v}\|} \right| \|\mathbf{v}\| = \frac{1}{\|\mathbf{v}\|} \|\mathbf{v}\| = 1$$

The process of multiplying a nonzero vector $\mathbf{v}$ by the reciprocal of its length to obtain a vector of norm 1 is called ***normalizing*** $\mathbf{v}$. An orthogonal set of *nonzero* vectors can always be converted to an orthonormal set by normalizing each of its vectors.

Example 2 The Euclidean norms of the vectors in Example 1 are

$$\|\mathbf{u}_1\| = 1, \qquad \|\mathbf{u}_2\| = \sqrt{2}, \qquad \|\mathbf{u}_3\| = \sqrt{2}$$

Consequently, normalizing $\mathbf{u}_1$, $\mathbf{u}_2$, and $\mathbf{u}_3$ yields

$$\mathbf{v}_1 = \frac{\mathbf{u}_1}{\|\mathbf{u}_1\|} = (0, 1, 0), \qquad \mathbf{v}_2 = \frac{\mathbf{u}_2}{\|\mathbf{u}_2\|} = \left(\frac{1}{\sqrt{2}}, 0, \frac{1}{\sqrt{2}} \right),$$

$$\mathbf{v}_3 = \frac{\mathbf{u}_3}{\|\mathbf{u}_3\|} = \left(\frac{1}{\sqrt{2}}, 0, -\frac{1}{\sqrt{2}} \right)$$

We leave it for you to verify that the set $S = \{\mathbf{v}_1, \mathbf{v}_2, \mathbf{v}_3\}$ is orthonormal by showing that

$$\langle \mathbf{v}_1, \mathbf{v}_2 \rangle = \langle \mathbf{v}_1, \mathbf{v}_3 \rangle = \langle \mathbf{v}_2, \mathbf{v}_3 \rangle = 0$$
$$\|\mathbf{v}_1\| = \|\mathbf{v}_2\| = \|\mathbf{v}_3\| = 1$$

In an inner product space, a basis consisting of orthonormal vectors is called an **orthonormal basis**, and a basis consisting of orthogonal vectors is called an **orthogonal basis**. A familiar example of an orthonormal basis is the standard basis for R^3 with the Euclidean inner product:

$$\mathbf{i} = (1, 0, 0), \qquad \mathbf{j} = (0, 1, 0), \qquad \mathbf{k} = (0, 0, 1)$$

This is the basis that is associated with rectangular coordinate systems (Figure 4 of Section 5.4). More generally, in R^n with the Euclidean inner product, the standard basis

$$\mathbf{e}_1 = (1, 0, 0, \ldots, 0), \quad \mathbf{e}_2 = (0, 1, 0, \ldots, 0), \quad \ldots, \quad \mathbf{e}_n = (0, 0, 0, \ldots, 1)$$

is orthonormal.

COORDINATES RELATIVE TO ORTHONORMAL BASES

The interest in finding orthonormal bases for inner product spaces is motivated in part by the following theorem, which shows that it is exceptionally simple to express a vector in terms of an orthonormal basis.

> **Theorem 6.3.1.** *If* $S = \{\mathbf{v}_1, \mathbf{v}_2, \ldots, \mathbf{v}_n\}$ *is an orthonormal basis for an inner product space* V, *and* $\mathbf{u}$ *is any vector in* V, *then*
>
> $$\mathbf{u} = \langle \mathbf{u}, \mathbf{v}_1 \rangle \mathbf{v}_1 + \langle \mathbf{u}, \mathbf{v}_2 \rangle \mathbf{v}_2 + \cdots + \langle \mathbf{u}, \mathbf{v}_n \rangle \mathbf{v}_n$$

Proof. Since $S = \{\mathbf{v}_1, \mathbf{v}_2, \ldots, \mathbf{v}_n\}$ is a basis, a vector $\mathbf{u}$ can be expressed in the form

$$\mathbf{u} = k_1 \mathbf{v}_1 + k_2 \mathbf{v}_2 + \cdots + k_n \mathbf{v}_n$$

We shall complete the proof by showing that $k_i = \langle \mathbf{u}, \mathbf{v}_i \rangle$ for $i = 1, 2, \ldots, n$. For each vector $\mathbf{v}_i$ in S we have

$$\langle \mathbf{u}, \mathbf{v}_i \rangle = \langle k_1 \mathbf{v}_1 + k_2 \mathbf{v}_2 + \cdots + k_n \mathbf{v}_n, \mathbf{v}_i \rangle$$
$$= k_1 \langle \mathbf{v}_1, \mathbf{v}_i \rangle + k_2 \langle \mathbf{v}_2, \mathbf{v}_i \rangle + \cdots + k_n \langle \mathbf{v}_n, \mathbf{v}_i \rangle$$

Since $S = \{\mathbf{v}_1, \mathbf{v}_2, \ldots, \mathbf{v}_n\}$ is an orthonormal set, we have

$$\langle \mathbf{v}_i, \mathbf{v}_i \rangle = \|\mathbf{v}_i\|^2 = 1 \qquad \text{and} \qquad \langle \mathbf{v}_j, \mathbf{v}_i \rangle = 0 \qquad \text{if } j \neq i$$

Therefore, the above expression for $\langle \mathbf{u}, \mathbf{v}_i \rangle$ simplifies to

$$\langle \mathbf{u}, \mathbf{v}_i \rangle = k_i$$

Using the terminology and notation introduced in Section 5.4, the scalars

$$\langle \mathbf{u}, \mathbf{v}_1 \rangle, \langle \mathbf{u}, \mathbf{v}_2 \rangle, \ldots, \langle \mathbf{u}, \mathbf{v}_n \rangle$$

in Theorem 6.3.1 are the coordinates of **u** relative to the orthonormal basis $S = \{\mathbf{v}_1, \mathbf{v}_2, \ldots, \mathbf{v}_n\}$ and

$$(\mathbf{u})_S = (\langle \mathbf{u}, \mathbf{v}_1 \rangle, \langle \mathbf{u}, \mathbf{v}_2 \rangle, \ldots, \langle \mathbf{u}, \mathbf{v}_n \rangle)$$

is the coordinate vector of **u** relative to this basis.

Example 3   Let

$$\mathbf{v}_1 = (0, 1, 0), \qquad \mathbf{v}_2 = (-\tfrac{4}{5}, 0, \tfrac{3}{5}), \qquad \mathbf{v}_3 = (\tfrac{3}{5}, 0, \tfrac{4}{5})$$

It is easy to check that $S = \{\mathbf{v}_1, \mathbf{v}_2, \mathbf{v}_3\}$ is an orthonormal basis for R^3 with the Euclidean inner product. Express the vector $\mathbf{u} = (1, 1, 1)$ as a linear combination of the vectors in S, and find the coordinate vector $(\mathbf{u})_S$.

Solution.

$$\langle \mathbf{u}, \mathbf{v}_1 \rangle = 1, \quad \langle \mathbf{u}, \mathbf{v}_2 \rangle = -\tfrac{1}{5}, \quad \text{and} \quad \langle \mathbf{u}, \mathbf{v}_3 \rangle = \tfrac{7}{5}$$

Therefore, by Theorem 6.3.1 we have

$$\mathbf{u} = \mathbf{v}_1 - \tfrac{1}{5}\mathbf{v}_2 + \tfrac{7}{5}\mathbf{v}_3$$

that is,

$$(1, 1, 1) = (0, 1, 0) - \tfrac{1}{5}(-\tfrac{4}{5}, 0, \tfrac{3}{5}) + \tfrac{7}{5}(\tfrac{3}{5}, 0, \tfrac{4}{5})$$

The coordinate vector of **u** relative to S is

$$(\mathbf{u})_S = (\langle \mathbf{u}, \mathbf{v}_1 \rangle, \langle \mathbf{u}, \mathbf{v}_2 \rangle, \langle \mathbf{u}, \mathbf{v}_3 \rangle) = (1, -\tfrac{1}{5}, \tfrac{7}{5}) \quad \blacktriangle$$

REMARK. The usefulness of Theorem 6.3.1 should be evident from this example if it is kept in mind that for nonorthonormal bases, it is usually necessary to solve a system of equations in order to express a vector in terms of the basis.

Orthonormal bases for inner product spaces are convenient because, as the following theorem shows, many familiar formulas hold for such bases.

Theorem 6.3.2.   *If S is an orthonormal basis for an n-dimensional inner product space, and if*

$$(\mathbf{u})_S = (u_1, u_2, \ldots, u_n) \qquad and \qquad (\mathbf{v})_S = (v_1, v_2, \ldots, v_n)$$

then:

(a) $\|\mathbf{u}\| = \sqrt{u_1^2 + u_2^2 + \cdots + u_n^2}$

(b) $d(\mathbf{u}, \mathbf{v}) = \sqrt{(u_1 - v_1)^2 + (u_2 - v_2)^2 + \cdots + (u_n - v_n)^2}$

(c) $\langle \mathbf{u}, \mathbf{v} \rangle = u_1 v_1 + u_2 v_2 + \cdots + u_n v_n$

The proof is left for the exercises.

REMARK. Observe that the right side of the equality in part (*a*) is the norm of the coordinate vector $(\mathbf{u})_S$ with respect to the Euclidean inner product on R^n, and the right side of the equality in part (*c*) is the Euclidean inner product of $(\mathbf{u})_S$ and $(\mathbf{v})_S$. Thus, by working with orthonormal bases, the computation of general norms and inner products can be reduced to the computation of Euclidean norms and inner products of the coordinate vectors.

Example 4 If R^3 has the Euclidean inner product, then the norm of the vector $\mathbf{u} = (1, 1, 1)$ is

$$\|\mathbf{u}\| = (\mathbf{u} \cdot \mathbf{u})^{1/2} = \sqrt{1^2 + 1^2 + 1^2} = \sqrt{3}$$

However, if we let R^3 have the orthonormal basis S in the last example, then we know from that example that the coordinate vector of $\mathbf{u}$ relative to S is

$$(\mathbf{u})_S = (1, -\tfrac{1}{5}, \tfrac{7}{5})$$

The norm of $\mathbf{u}$ can also be calculated from this vector using part (*a*) of Theorem 6.3.2. This yields

$$\|\mathbf{u}\| = \sqrt{1^2 + (-\tfrac{1}{5})^2 + (\tfrac{7}{5})^2} = \sqrt{\tfrac{75}{25}} = \sqrt{3} \ \text{▲}$$

COORDINATES RELATIVE TO ORTHOGONAL BASES

If $S = \{\mathbf{v}_1, \mathbf{v}_2, \ldots, \mathbf{v}_n\}$ is an *orthogonal* basis for a vector space V, then normalizing each of these vectors yields the orthonormal basis

$$S' = \left\{ \frac{\mathbf{v}_1}{\|\mathbf{v}_1\|}, \frac{\mathbf{v}_2}{\|\mathbf{v}_2\|}, \ldots, \frac{\mathbf{v}_n}{\|\mathbf{v}_n\|} \right\}$$

Thus, if $\mathbf{u}$ is any vector in V, it follows from Theorem 6.3.1 that

$$\mathbf{u} = \left\langle \mathbf{u}, \frac{\mathbf{v}_1}{\|\mathbf{v}_1\|} \right\rangle \frac{\mathbf{v}_1}{\|\mathbf{v}_1\|} + \left\langle \mathbf{u}, \frac{\mathbf{v}_2}{\|\mathbf{v}_2\|} \right\rangle \frac{\mathbf{v}_2}{\|\mathbf{v}_2\|} + \cdots + \left\langle \mathbf{u}, \frac{\mathbf{v}_n}{\|\mathbf{v}_n\|} \right\rangle \frac{\mathbf{v}_n}{\|\mathbf{v}_n\|}$$

which by part (*c*) of Theorem 6.1.1 can be rewritten as

$$\mathbf{u} = \frac{\langle \mathbf{u}, \mathbf{v}_1 \rangle}{\|\mathbf{v}_1\|^2} \mathbf{v}_1 + \frac{\langle \mathbf{u}, \mathbf{v}_2 \rangle}{\|\mathbf{v}_2\|^2} \mathbf{v}_2 + \cdots + \frac{\langle \mathbf{u}, \mathbf{v}_n \rangle}{\|\mathbf{v}_n\|^2} \mathbf{v}_n \tag{1}$$

This formula expresses $\mathbf{u}$ as a linear combination of the vectors in the orthogonal basis S. Some problems requiring the use of this formula are given in the exercises.

It is self-evident that if $\mathbf{v}_1$, $\mathbf{v}_2$, and $\mathbf{v}_3$ are three nonzero mutually perpendicular vectors in R^3, then none of these vectors lies in the same plane as the other two; that is, the vectors are linearly independent. The following theorem generalizes this result.

Theorem 6.3.3. *If $S = \{\mathbf{v}_1, \mathbf{v}_2, \ldots, \mathbf{v}_n\}$ is an orthogonal set of nonzero vectors in an inner product space, then S is linearly independent.*

Proof. Assume that

$$k_1\mathbf{v}_1 + k_2\mathbf{v}_2 + \cdots + k_n\mathbf{v}_n = \mathbf{0} \tag{2}$$

To demonstrate that $S = \{\mathbf{v}_1, \mathbf{v}_2, \ldots, \mathbf{v}_n\}$ is linearly independent, we must prove that $k_1 = k_2 = \cdots = k_n = 0$.

For each $\mathbf{v}_i$ in S, it follows from (2) that

$$\langle k_1\mathbf{v}_1 + k_2\mathbf{v}_2 + \cdots + k_n\mathbf{v}_n, \mathbf{v}_i \rangle = \langle \mathbf{0}, \mathbf{v}_i \rangle = 0$$

or equivalently,

$$k_1\langle \mathbf{v}_1, \mathbf{v}_i \rangle + k_2\langle \mathbf{v}_2, \mathbf{v}_i \rangle + \cdots + k_n\langle \mathbf{v}_n, \mathbf{v}_i \rangle = 0$$

From the orthogonality of S it follows that $\langle \mathbf{v}_j, \mathbf{v}_i \rangle = 0$ when $j \neq i$, so that this equation reduces to

$$k_i\langle \mathbf{v}_i, \mathbf{v}_i \rangle = 0$$

Since the vectors in S are assumed to be nonzero, $\langle \mathbf{v}_i, \mathbf{v}_i \rangle \neq 0$ by the positivity axiom for inner products. Therefore, $k_i = 0$. Since the subscript i is arbitrary, we have $k_1 = k_2 = \cdots = k_n = 0$; thus, S is linearly independent. ∎

Example 5 In Example 2 we showed that the vectors

$$\mathbf{v}_1 = (0, 1, 0), \qquad \mathbf{v}_2 = \left(\frac{1}{\sqrt{2}}, 0, \frac{1}{\sqrt{2}} \right), \qquad \text{and} \qquad \mathbf{v}_3 = \left(\frac{1}{\sqrt{2}}, 0, -\frac{1}{\sqrt{2}} \right)$$

form an orthonormal set with respect to the Euclidean inner product on R^3. By Theorem 6.3.3, these vectors form a linearly independent set, and since R^3 is three-dimensional, $S = \{\mathbf{v}_1, \mathbf{v}_2, \mathbf{v}_3\}$ is an orthonormal basis for R^3 by Theorem 5.4.6a. ▲

ORTHOGONAL PROJECTIONS

We shall now develop some results that will help us to construct orthogonal and orthonormal bases for inner product spaces.

In R^2 or R^3 with the Euclidean inner product, it is evident geometrically that if W is a line or a plane through the origin, then each vector $\mathbf{u}$ in the space can be expressed as a sum

$$\mathbf{u} = \mathbf{w}_1 + \mathbf{w}_2$$

where $\mathbf{w}_1$ is in W and $\mathbf{w}_2$ is perpendicular to W (Figure 1). This result is a special case of the following general theorem whose proof is given at the end of this section.

Figure 1

> **Theorem 6.3.4** (***Projection Theorem***). *If W is a finite-dimensional subspace of an inner product space V, then every vector $\mathbf{u}$ in V can be expressed in exactly one way as*
>
> $$\mathbf{u} = \mathbf{w}_1 + \mathbf{w}_2 \tag{3}$$
>
> *where $\mathbf{w}_1$ is in W and $\mathbf{w}_2$ is in $W^{\perp}$.*

The vector $\mathbf{w}_1$ in the preceding theorem is called the ***orthogonal projection of*** $\mathbf{u}$ ***on*** W and is denoted by $\text{proj}_W \mathbf{u}$. The vector $\mathbf{w}_2$ is called the ***component of*** $\mathbf{u}$ ***orthogonal to*** W and is denoted by $\text{proj}_{W^{\perp}} \mathbf{u}$. Thus, Formula (3) in the Projection Theorem can be expressed as

$$\mathbf{u} = \text{proj}_W \mathbf{u} + \text{proj}_{W^{\perp}} \mathbf{u} \tag{4}$$

Since $\mathbf{w}_2 = \mathbf{u} - \mathbf{w}_1$, it follows that

$$\text{proj}_{W^{\perp}} \mathbf{u} = \mathbf{u} - \text{proj}_W \mathbf{u}$$

so Formula (4) can also be written as

$$\mathbf{u} = \text{proj}_W \mathbf{u} + (\mathbf{u} - \text{proj}_W \mathbf{u}) \tag{5}$$

(Figure 2).

Figure 2

The following theorem, whose proof is requested in the exercises, provides formulas for calculating orthogonal projections.

> **Theorem 6.3.5.** *Let W be a finite-dimensional subspace of an inner product space V.*
>
> (*a*) *If $\{\mathbf{v}_1, \mathbf{v}_2, \ldots, \mathbf{v}_r\}$ is an orthonormal basis for W, and $\mathbf{u}$ is any vector in V, then*
>
> $$\text{proj}_W \mathbf{u} = \langle \mathbf{u}, \mathbf{v}_1 \rangle \mathbf{v}_1 + \langle \mathbf{u}, \mathbf{v}_2 \rangle \mathbf{v}_2 + \cdots + \langle \mathbf{u}, \mathbf{v}_r \rangle \mathbf{v}_r \tag{6}$$
>
> (*b*) *If $\{\mathbf{v}_1, \mathbf{v}_2, \ldots, \mathbf{v}_r\}$ is an orthogonal basis for W, and $\mathbf{u}$ is any vector in V, then*
>
> $$\text{proj}_W \mathbf{u} = \frac{\langle \mathbf{u}, \mathbf{v}_1 \rangle}{\|\mathbf{v}_1\|^2} \mathbf{v}_1 + \frac{\langle \mathbf{u}, \mathbf{v}_2 \rangle}{\|\mathbf{v}_2\|^2} \mathbf{v}_2 + \cdots + \frac{\langle \mathbf{u}, \mathbf{v}_r \rangle}{\|\mathbf{v}_r\|^2} \mathbf{v}_r \tag{7}$$

Example 6 Let R^3 have the Euclidean inner product, and let W be the subspace spanned by the orthonormal vectors $\mathbf{v}_1 = (0, 1, 0)$ and $\mathbf{v}_2 = (-\frac{4}{5}, 0, \frac{3}{5})$. From (6) the orthogonal projection of $\mathbf{u} = (1, 1, 1)$ on W is

$$\text{proj}_W \mathbf{u} = \langle \mathbf{u}, \mathbf{v}_1 \rangle \mathbf{v}_1 + \langle \mathbf{u}, \mathbf{v}_2 \rangle \mathbf{v}_2$$
$$= (1)(0, 1, 0) + (-\tfrac{1}{5})(-\tfrac{4}{5}, 0, \tfrac{3}{5})$$
$$= (\tfrac{4}{25}, 1, -\tfrac{3}{25})$$

The component of $\mathbf{u}$ orthogonal to W is

$$\text{proj}_{W\perp} \mathbf{u} = \mathbf{u} - \text{proj}_W \mathbf{u} = (1, 1, 1) - (\tfrac{4}{25}, 1, -\tfrac{3}{25}) = (\tfrac{21}{25}, 0, \tfrac{28}{25})$$

Observe that $\text{proj}_{W\perp} \mathbf{u}$ is orthogonal to both $\mathbf{v}_1$ and $\mathbf{v}_2$ so that this vector is orthogonal to each vector in the space W spanned by $\mathbf{v}_1$ and $\mathbf{v}_2$ as it should be. ▲

FINDING ORTHOGONAL AND ORTHONORMAL BASES

We have seen that orthonormal bases enjoy a variety of useful properties. Our next theorem, which is the main result in this section, shows that every nonzero finite-dimensional vector space has an orthonormal basis. The proof of this result is extremely important, since it provides an algorithm, or method, for converting an arbitrary basis into an orthonormal basis.

> **Theorem 6.3.6.** *Every nonzero finite-dimensional inner product space has an orthonormal basis.*

Proof. Let V be any nonzero finite-dimensional inner product space, and let $\{\mathbf{u}_1, \mathbf{u}_2, \ldots, \mathbf{u}_n\}$ be any basis for V. It suffices to show that V has an orthogonal basis, since the vectors in the orthogonal basis can be normalized to produce an orthonormal basis for V. The following sequence of steps will produce an orthogonal basis $\{\mathbf{v}_1, \mathbf{v}_2, \ldots, \mathbf{v}_n\}$ for V.

Step 1. Let $\mathbf{v}_1 = \mathbf{u}_1$.

Step 2. As illustrated in Figure 3, we can obtain a vector $\mathbf{v}_2$ that is orthogonal to $\mathbf{v}_1$ by computing the component of $\mathbf{u}_2$ that is orthogonal to the space W_1 spanned by $\mathbf{v}_1$. We use Formula (7):

$$\mathbf{v}_2 = \mathbf{u}_2 - \text{proj}_{W_1} \mathbf{u}_2 = \mathbf{u}_2 - \frac{\langle \mathbf{u}_2, \mathbf{v}_1 \rangle}{\|\mathbf{v}_1\|^2} \mathbf{v}_1$$

Of course, if $\mathbf{v}_2 = \mathbf{0}$, then $\mathbf{v}_2$ is not a basis vector. But this cannot happen, since it would then follow from the preceding formula for $\mathbf{v}_2$ that

$$\mathbf{u}_2 = \frac{\langle \mathbf{u}_2, \mathbf{v}_1 \rangle}{\|\mathbf{v}_1\|^2} \mathbf{v}_1 = \frac{\langle \mathbf{u}_2, \mathbf{v}_1 \rangle}{\|\mathbf{u}_1\|^2} \mathbf{u}_1$$

which says that $\mathbf{u}_2$ is a multiple of $\mathbf{u}_1$, contradicting the linear independence of the basis $S = \{\mathbf{u}_1, \mathbf{u}_2, \ldots, \mathbf{u}_n\}$.

Step 3. To construct a vector $\mathbf{v}_3$ that is orthogonal to both $\mathbf{v}_1$ and $\mathbf{v}_2$, we compute the component of $\mathbf{u}_3$ orthogonal to the space W_2 spanned by $\mathbf{v}_1$ and $\mathbf{v}_2$ (Figure 4). From (7)

$$\mathbf{v}_3 = \mathbf{u}_3 - \text{proj}_{W_2}\, \mathbf{u}_3 = \mathbf{u}_3 - \frac{\langle \mathbf{u}_3, \mathbf{v}_1 \rangle}{\|\mathbf{v}_1\|^2}\mathbf{v}_1 - \frac{\langle \mathbf{u}_3, \mathbf{v}_2 \rangle}{\|\mathbf{v}_2\|^2}\mathbf{v}_2$$

As in Step 2, the linear independence of $\{\mathbf{u}_1, \mathbf{u}_2, \ldots, \mathbf{u}_n\}$ ensures that $\mathbf{v}_3 \neq \mathbf{0}$. We leave the details as an exercise.

Step 4. To determine a vector $\mathbf{v}_4$ that is orthogonal to $\mathbf{v}_1$, $\mathbf{v}_2$, and $\mathbf{v}_3$, we compute the component of $\mathbf{u}_4$ orthogonal to the space W_3 spanned by $\mathbf{v}_1$, $\mathbf{v}_2$, and $\mathbf{v}_3$. From (7)

$$\mathbf{v}_4 = \mathbf{u}_4 - \text{proj}_{W_3}\, \mathbf{u}_4 = \mathbf{u}_4 - \frac{\langle \mathbf{u}_4, \mathbf{v}_1 \rangle}{\|\mathbf{v}_1\|^2}\mathbf{v}_1 - \frac{\langle \mathbf{u}_4, \mathbf{v}_2 \rangle}{\|\mathbf{v}_2\|^2}\mathbf{v}_2 - \frac{\langle \mathbf{u}_4, \mathbf{v}_3 \rangle}{\|\mathbf{v}_3\|^2}\mathbf{v}_3$$

Figure 3

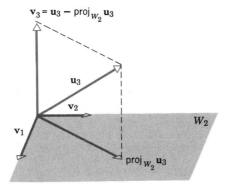

Figure 4

Continuing in this way, we will obtain, after n steps, an orthogonal set of vectors, $\{\mathbf{v}_1, \mathbf{v}_2, \ldots, \mathbf{v}_n\}$. Since V is n-dimensional and every orthogonal set is linearly independent, the set $\{\mathbf{v}_1, \mathbf{v}_2, \ldots, \mathbf{v}_n\}$ is an orthogonal basis for V.

he preceding step-by-step construction for converting an arbitrary basis into an orthogonal basis is called the **Gram–Schmidt* process**.

Example 7 Consider the vector space R^3 with the Euclidean inner product. Apply the Gram–Schmidt process to transform the basis vectors $\mathbf{u}_1 = (1, 1, 1)$, $\mathbf{u}_2 = (0, 1, 1)$, $\mathbf{u}_3 = (0, 0, 1)$ into an orthogonal basis $\{\mathbf{v}_1, \mathbf{v}_2, \mathbf{v}_3\}$; then normalize the orthogonal basis vectors to obtain an orthonormal basis $\{\mathbf{q}_1, \mathbf{q}_2, \mathbf{q}_3\}$.

Solution.

Step 1. $\quad \mathbf{v}_1 = \mathbf{u}_1 = (1, 1, 1)$

Step 2. $\quad \mathbf{v}_2 = \mathbf{u}_2 - \mathrm{proj}_{W_1} \mathbf{u}_2 = \mathbf{u}_2 - \dfrac{\langle \mathbf{u}_2, \mathbf{v}_1 \rangle}{\|\mathbf{v}_1\|^2} \mathbf{v}_1$

$\qquad = (0, 1, 1) - \dfrac{2}{3}(1, 1, 1) = \left(-\dfrac{2}{3}, \dfrac{1}{3}, \dfrac{1}{3} \right)$

Step 3. $\quad \mathbf{v}_3 = \mathbf{u}_3 - \mathrm{proj}_{W_2} \mathbf{u}_3 = \mathbf{u}_3 - \dfrac{\langle \mathbf{u}_3, \mathbf{v}_1 \rangle}{\|\mathbf{v}_1\|^2} \mathbf{v}_1 - \dfrac{\langle \mathbf{u}_3, \mathbf{v}_2 \rangle}{\|\mathbf{v}_2\|^2} \mathbf{v}_2$

$\qquad = (0, 0, 1) - \dfrac{1}{3}(1, 1, 1) - \dfrac{1/3}{2/3}\left(-\dfrac{2}{3}, \dfrac{1}{3}, \dfrac{1}{3} \right)$

$\qquad = \left(0, -\dfrac{1}{2}, \dfrac{1}{2} \right)$

---

* *Jörgen Pederson Gram* (1850–1916) was a Danish actuary. Gram's early education was at village schools supplemented by private tutoring. After graduating from high school he obtained a master's degree in mathematics with specialization in the newly developing modern algebra. Gram then took a position as an actuary for the Hafnia Life Insurance Company, where he developed mathematical foundations of accident insurance for the company Skjold. He served on the Board of Directors of Hafnia and directed Skjold until 1910, at which time he became director of the Danish Insurance Board. During his employ as an actuary, he earned a Ph.D. based on his dissertation "On Series Development Utilizing the Least Squares Method." It was in this thesis that his contributions to the Gram–Schmidt process were first formulated. Gram eventually became interested in abstract number theory and won a gold medal from the Royal Danish Society of Sciences and Letters for his contributions to that field. However, he also had a lifelong interest in the interplay between theoretical and applied mathematics that led to four treatises on Danish forest management. Gram was killed one evening in a bicycle collision on the way to a meeting of the Royal Danish Society.

Erhardt Schmidt (1876–1959) was a German mathematician. Schmidt received his doctoral degree from Göttingen University in 1905, where he studied under one of the giants of mathematics, David Hilbert. He eventually went to teach at Berlin University in 1917, where he stayed for the rest of his life. Schmidt made important contributions to a variety of mathematical fields, but is most noteworthy for fashioning many of Hilbert's diverse ideas into a general concept (called a Hilbert space), which is fundamental in the study of infinite-dimensional vector spaces. Schmidt first described the process that bears his name in a paper on integral equations published in 1907.

Thus,

$$\mathbf{v}_1 = (1,\ 1,\ 1), \qquad \mathbf{v}_2 = \left(-\frac{2}{3}, \frac{1}{3}, \frac{1}{3}\right), \qquad \mathbf{v}_3 = \left(0,\ -\frac{1}{2}, \frac{1}{2}\right)$$

form an orthogonal basis for R^3. The norms of these vectors are

$$\|\mathbf{v}_1\| = \sqrt{3}, \qquad \|\mathbf{v}_2\| = \frac{\sqrt{6}}{3}, \qquad \|\mathbf{v}_3\| = \frac{1}{\sqrt{2}}$$

so an orthonormal basis for R^3 is

$$\mathbf{q}_1 = \frac{\mathbf{v}_1}{\|\mathbf{v}_1\|} = \left(\frac{1}{\sqrt{3}}, \frac{1}{\sqrt{3}}, \frac{1}{\sqrt{3}}\right), \qquad \mathbf{q}_2 = \frac{\mathbf{v}_2}{\|\mathbf{v}_2\|} = \left(-\frac{2}{\sqrt{6}}, \frac{1}{\sqrt{6}}, \frac{1}{\sqrt{6}}\right),$$

$$\mathbf{q}_3 = \frac{\mathbf{v}_3}{\|\mathbf{v}_3\|} = \left(0,\ -\frac{1}{\sqrt{2}}, \frac{1}{\sqrt{2}}\right)$$

REMARK. In the preceding example we used the Gram–Schmidt process to produce an orthogonal basis; then, after the entire orthogonal basis was obtained, we normalized to obtain an orthonormal basis. Alternatively, one can normalize each orthogonal basis vector as soon as it is obtained, thereby generating the orthonormal basis step by step. However, this method has the slight disadvantage of producing more square roots to manipulate.

The Gram–Schmidt process with subsequent normalization not only converts an arbitrary basis $\{\mathbf{u}_1,\ \mathbf{u}_2,\ \dots,\ \mathbf{u}_n\}$ into an orthonormal basis $\{\mathbf{q}_1,\ \mathbf{q}_2,\ \dots,\ \mathbf{q}_n\}$, but it does it in such a way that for $k \geq 2$ the following relationships hold:

- $\{\mathbf{q}_1,\ \mathbf{q}_2, \dots,\ \mathbf{q}_k\}$ is an orthonormal basis for the space spanned by $\{\mathbf{u}_1,\ \mathbf{u}_2,\ \dots,\ \mathbf{u}_k\}$.
- $\mathbf{q}_k$ is orthogonal to $\{\mathbf{u}_1,\ \mathbf{u}_2,\ \dots,\ \mathbf{u}_{k-1}\}$.

We omit the proofs, but these facts should become evident after some thoughtful examination of the proof of Theorem 6.3.6.

QR-DECOMPOSITION   *skip*

We pose the following problem.

> **Problem.** If A is an $m \times n$ matrix with linearly independent column vectors, and if Q is the matrix with orthonormal column vectors that results from applying the Gram–Schmidt process to the column vectors of A, what relationship, if any, exists between A and Q?

To solve this problem, suppose that the column vectors of A are $\mathbf{u}_1,\ \mathbf{u}_2,\ \dots,\ \mathbf{u}_n$ and the orthonormal column vectors of Q are $\mathbf{q}_1,\ \mathbf{q}_2,\ \dots,\ \mathbf{q}_n$; thus,

$$A = [\ \mathbf{u}_1 \ \vdots \ \mathbf{u}_2 \ \vdots \ \cdots \ \vdots \ \mathbf{u}_n\ ] \qquad \text{and} \qquad Q = [\ \mathbf{q}_1 \ \vdots \ \mathbf{q}_2 \ \vdots \ \cdots \ \vdots \ \mathbf{q}_n\ ]$$

It follows from Theorem 6.3.1 that $\mathbf{u}_1, \mathbf{u}_2, \ldots, \mathbf{u}_n$ are expressible in terms of $\mathbf{q}_1, \mathbf{q}_2, \ldots, \mathbf{q}_n$ as

$$\mathbf{u}_1 = \langle \mathbf{u}_1, \mathbf{q}_1 \rangle \mathbf{q}_1 + \langle \mathbf{u}_1, \mathbf{q}_2 \rangle \mathbf{q}_2 + \cdots + \langle \mathbf{u}_1, \mathbf{q}_n \rangle \mathbf{q}_n$$
$$\mathbf{u}_2 = \langle \mathbf{u}_2, \mathbf{q}_1 \rangle \mathbf{q}_1 + \langle \mathbf{u}_2, \mathbf{q}_2 \rangle \mathbf{q}_2 + \cdots + \langle \mathbf{u}_2, \mathbf{q}_n \rangle \mathbf{q}_n$$
$$\vdots \qquad \vdots \qquad \vdots \qquad \vdots$$
$$\mathbf{u}_n = \langle \mathbf{u}_n, \mathbf{q}_1 \rangle \mathbf{q}_1 + \langle \mathbf{u}_n, \mathbf{q}_2 \rangle \mathbf{q}_2 + \cdots + \langle \mathbf{u}_n, \mathbf{q}_n \rangle \mathbf{q}_n$$

Recalling from Section 1.3 that the jth column vector of a matrix product is a linear combination of the column vectors of the first factor with coefficients coming from the jth column of the second factor, it follows that these relationships can be expressed in matrix form as

$$[\, \mathbf{u}_1 \mid \mathbf{u}_2 \mid \cdots \mid \mathbf{u}_n \,] = [\, \mathbf{q}_1 \mid \mathbf{q}_2 \mid \cdots \mid \mathbf{q}_n \,] \begin{bmatrix} \langle \mathbf{u}_1, \mathbf{q}_1 \rangle & \langle \mathbf{u}_2, \mathbf{q}_1 \rangle & \cdots & \langle \mathbf{u}_n, \mathbf{q}_1 \rangle \\ \langle \mathbf{u}_1, \mathbf{q}_2 \rangle & \langle \mathbf{u}_2, \mathbf{q}_2 \rangle & \cdots & \langle \mathbf{u}_n, \mathbf{q}_2 \rangle \\ \vdots & \vdots & & \vdots \\ \langle \mathbf{u}_1, \mathbf{q}_n \rangle & \langle \mathbf{u}_2, \mathbf{q}_n \rangle & \cdots & \langle \mathbf{u}_n, \mathbf{q}_n \rangle \end{bmatrix}$$

or more briefly as

$$A = QR \tag{8}$$

However, it is a property of the Gram–Schmidt process that for $j \geq 2$, the vector $\mathbf{q}_j$ is orthogonal to $\mathbf{u}_1, \mathbf{u}_2, \ldots, \mathbf{u}_{j-1}$; thus, all entries below the main diagonal of R are zero,

$$R = \begin{bmatrix} \langle \mathbf{u}_1, \mathbf{q}_1 \rangle & \langle \mathbf{u}_2, \mathbf{q}_1 \rangle & \cdots & \langle \mathbf{u}_n, \mathbf{q}_1 \rangle \\ 0 & \langle \mathbf{u}_2, \mathbf{q}_2 \rangle & \cdots & \langle \mathbf{u}_n, \mathbf{q}_2 \rangle \\ \vdots & \vdots & & \vdots \\ 0 & 0 & \cdots & \langle \mathbf{u}_n, \mathbf{q}_n \rangle \end{bmatrix} \tag{9}$$

We leave it as an exercise to show that the diagonal entries of R are nonzero, so that R is invertible. Thus, (8) is a factorization of A into the product of a matrix Q with orthonormal column vectors and an invertible upper triangular matrix R. We call (8) the **QR-decomposition of A**. In summary, we have the following theorem.

Theorem 6.3.7 (**QR-Decomposition**). *If A is an $m \times n$ matrix with linearly independent column vectors, then A can be factored as*

$$A = QR$$

where Q is an $m \times n$ matrix with orthonormal column vectors, and R is an $n \times n$ invertible upper triangular matrix.

REMARK. Recall from Theorem 6.2.7 that if A is an $n \times n$ matrix, then the invertibility of A is equivalent to linear independence of the column vectors; thus, every invertible matrix has a QR-decomposition.

Example 8   Find the QR-decomposition of

$$A = \begin{bmatrix} 1 & 0 & 0 \\ 1 & 1 & 0 \\ 1 & 1 & 1 \end{bmatrix}$$

Solution. The column vectors of A are

$$\mathbf{u}_1 = \begin{bmatrix} 1 \\ 1 \\ 1 \end{bmatrix}, \quad \mathbf{u}_2 = \begin{bmatrix} 0 \\ 1 \\ 1 \end{bmatrix}, \quad \mathbf{u}_3 = \begin{bmatrix} 0 \\ 0 \\ 1 \end{bmatrix}$$

Applying the Gram–Schmidt process with subsequent normalization to these column vectors yields the orthonormal vectors (see Example 7)

$$\mathbf{q}_1 = \begin{bmatrix} 1/\sqrt{3} \\ 1/\sqrt{3} \\ 1/\sqrt{3} \end{bmatrix}, \quad \mathbf{q}_2 = \begin{bmatrix} -2/\sqrt{6} \\ 1/\sqrt{6} \\ 1/\sqrt{6} \end{bmatrix}, \quad \mathbf{q}_3 = \begin{bmatrix} 0 \\ -1/\sqrt{2} \\ 1/\sqrt{2} \end{bmatrix}$$

and from (9) the matrix R is

$$R = \begin{bmatrix} \langle \mathbf{u}_1, \mathbf{q}_1 \rangle & \langle \mathbf{u}_2, \mathbf{q}_1 \rangle & \langle \mathbf{u}_3, \mathbf{q}_1 \rangle \\ 0 & \langle \mathbf{u}_2, \mathbf{q}_2 \rangle & \langle \mathbf{u}_3, \mathbf{q}_2 \rangle \\ 0 & 0 & \langle \mathbf{u}_3, \mathbf{q}_3 \rangle \end{bmatrix} = \begin{bmatrix} 3/\sqrt{3} & 2/\sqrt{3} & 1/\sqrt{3} \\ 0 & 2/\sqrt{6} & 1/\sqrt{6} \\ 0 & 0 & 1/\sqrt{2} \end{bmatrix}$$

Thus, the QR-decomposition of A is

$$\underbrace{\begin{bmatrix} 1 & 0 & 0 \\ 1 & 1 & 0 \\ 1 & 1 & 1 \end{bmatrix}}_{A} = \underbrace{\begin{bmatrix} 1/\sqrt{3} & -2/\sqrt{6} & 0 \\ 1/\sqrt{3} & 1/\sqrt{6} & -1/\sqrt{2} \\ 1/\sqrt{3} & 1/\sqrt{6} & 1/\sqrt{2} \end{bmatrix}}_{Q} \underbrace{\begin{bmatrix} 3/\sqrt{3} & 2/\sqrt{3} & 1/\sqrt{3} \\ 0 & 2/\sqrt{6} & 1/\sqrt{6} \\ 0 & 0 & 1/\sqrt{2} \end{bmatrix}}_{R}$$

THE ROLE OF THE QR-DECOMPOSITION IN LINEAR ALGEBRA

In recent years the QR-decomposition has assumed growing importance as the mathematical foundation for a wide variety of practical numerical algorithms, including a widely used algorithm for computing eigenvalues of large matrices. Such algorithms are discussed in textbooks that deal with advanced numerical methods of linear algebra.

ADDITIONAL PROOF

Proof of Theorem 6.3.4. There are two parts to the proof. First we must find vectors $\mathbf{w}_1$ and $\mathbf{w}_2$ with the stated properties, and then we must show that these are the only such vectors.

By the Gram–Schmidt process there is an orthonormal basis $\{\mathbf{v}_1, \mathbf{v}_2, \ldots, \mathbf{v}_n\}$ for W. Let

$$\mathbf{w}_1 = \langle \mathbf{u}, \mathbf{v}_1 \rangle \mathbf{v}_1 + \langle \mathbf{u}, \mathbf{v}_2 \rangle \mathbf{v}_2 + \cdots + \langle \mathbf{u}, \mathbf{v}_n \rangle \mathbf{v}_n \tag{10}$$

and

$$\mathbf{w}_2 = \mathbf{u} - \mathbf{w}_1 \tag{11}$$

It follows that $\mathbf{w}_1 + \mathbf{w}_2 = \mathbf{w}_1 + (\mathbf{u} - \mathbf{w}_1) = \mathbf{u}$, so it remains to show that $\mathbf{w}_1$ is in W and $\mathbf{w}_2$ is orthogonal to W. But $\mathbf{w}_1$ lies in W because it is a linear combination of the basis vectors for W. To show that $\mathbf{w}_2$ is orthogonal to W, we must show that $\langle \mathbf{w}_2, \mathbf{w} \rangle = 0$ for every vector $\mathbf{w}$ in W. But if $\mathbf{w}$ is any vector in W, it can be expressed as a linear combination

$$\mathbf{w} = k_1\mathbf{v}_1 + k_2\mathbf{v}_2 + \cdots + k_n\mathbf{v}_n$$

of the basis vectors $\mathbf{v}_1, \mathbf{v}_2, \ldots, \mathbf{v}_n$. Thus,

$$\langle \mathbf{w}_2, \mathbf{w} \rangle = \langle \mathbf{u} - \mathbf{w}_1, \mathbf{w} \rangle = \langle \mathbf{u}, \mathbf{w} \rangle - \langle \mathbf{w}_1, \mathbf{w} \rangle \tag{12}$$

But

$$\langle \mathbf{u}, \mathbf{w} \rangle = \langle \mathbf{u}, k_1\mathbf{v}_1 + k_2\mathbf{v}_2 + \cdots + k_n\mathbf{v}_n \rangle$$
$$= k_1\langle \mathbf{u}, \mathbf{v}_1 \rangle + k_2\langle \mathbf{u}, \mathbf{v}_2 \rangle + \cdots + k_n\langle \mathbf{u}, \mathbf{v}_n \rangle$$

and by part (*c*) of Theorem 6.3.2

$$\langle \mathbf{w}_1, \mathbf{w} \rangle = \langle \mathbf{u}, \mathbf{v}_1 \rangle k_1 + \langle \mathbf{u}, \mathbf{v}_2 \rangle k_2 + \cdots + \langle \mathbf{u}, \mathbf{v}_n \rangle k_n$$

Thus, $\langle \mathbf{u}, \mathbf{w} \rangle$ and $\langle \mathbf{w}_1, \mathbf{w} \rangle$ are equal, so (12) yields $\langle \mathbf{w}_2, \mathbf{w} \rangle = 0$, which is what we want to show.

To see that (10) and (11) are the only vectors with the properties stated in the theorem, suppose that we can also write

$$\mathbf{u} = \mathbf{w}_1' + \mathbf{w}_2' \tag{13}$$

where $\mathbf{w}_1'$ is in W and $\mathbf{w}_2'$ is orthogonal to W. If we subtract from (13) the equation

$$\mathbf{u} = \mathbf{w}_1 + \mathbf{w}_2$$

we obtain

$$\mathbf{0} = (\mathbf{w}_1' - \mathbf{w}_1) + (\mathbf{w}_2' - \mathbf{w}_2)$$

or

$$\mathbf{w}_1 - \mathbf{w}_1' = \mathbf{w}_2' - \mathbf{w}_2 \tag{14}$$

Since $\mathbf{w}_2$ and $\mathbf{w}_2'$ are orthogonal to W, their difference is also orthogonal to W, since for any vector $\mathbf{w}$ in W we can write

$$\langle \mathbf{w}, \mathbf{w}_2' - \mathbf{w}_2 \rangle = \langle \mathbf{w}, \mathbf{w}_2' \rangle - \langle \mathbf{w}, \mathbf{w}_2 \rangle = 0 - 0 = 0$$

But $\mathbf{w}_2' - \mathbf{w}_2$ is itself a vector in W, since from (14) it is the difference of the two vectors $\mathbf{w}_1$ and $\mathbf{w}_1'$ which lie in the subspace W. Thus, $\mathbf{w}_2' - \mathbf{w}_2$ must be orthogonal to itself; that is,

$$\langle \mathbf{w}_2' - \mathbf{w}_2, \mathbf{w}_2' - \mathbf{w}_2 \rangle = 0$$

But this implies that $\mathbf{w}_2' - \mathbf{w}_2 = 0$ by Axiom 4 for inner products. Thus, $\mathbf{w}_2' = \mathbf{w}_2$, and by (14), $\mathbf{w}_1' = \mathbf{w}_1$.

EXERCISE SET 6.3

1. Which of the following sets of vectors are orthogonal with respect to the Euclidean inner product on R^2?
   (a) $(0, 1)$, $(2, 0)$ (b) $(-1/\sqrt{2},\ 1/\sqrt{2})$, $(1/\sqrt{2},\ 1/\sqrt{2})$
   (c) $(-1/\sqrt{2},\ -1/\sqrt{2})$, $(1/\sqrt{2},\ 1/\sqrt{2})$ (d) $(0, 0)$, $(0, 1)$

2. Which of the sets in Exercise 1 are orthonormal with respect to the Euclidean inner product on R^2?

3. Which of the following sets of vectors are orthogonal with respect to the Euclidean inner product on R^3?
   (a) $\left(\dfrac{1}{\sqrt{2}}, 0, \dfrac{1}{\sqrt{2}}\right)$, $\left(\dfrac{1}{\sqrt{3}}, \dfrac{1}{\sqrt{3}}, -\dfrac{1}{\sqrt{3}}\right)$, $\left(-\dfrac{1}{\sqrt{2}}, 0, \dfrac{1}{\sqrt{2}}\right)$ (b) $\left(\dfrac{2}{3}, -\dfrac{2}{3}, \dfrac{1}{3}\right)$, $\left(\dfrac{2}{3}, \dfrac{1}{3}, -\dfrac{2}{3}\right)$, $\left(\dfrac{1}{3}, \dfrac{2}{3}, \dfrac{2}{3}\right)$

   (c) $(1, 0, 0)$, $\left(0, \dfrac{1}{\sqrt{2}}, \dfrac{1}{\sqrt{2}}\right)$, $(0, 0, 1)$ (d) $\left(\dfrac{1}{\sqrt{6}}, \dfrac{1}{\sqrt{6}}, -\dfrac{2}{\sqrt{6}}\right)$, $\left(\dfrac{1}{\sqrt{2}}, -\dfrac{1}{\sqrt{2}}, 0\right)$

4. Which of the sets in Exercise 3 are orthonormal with respect to the Euclidean inner product on R^3?

5. Which of the following sets of polynomials are orthonormal with respect to the inner product on P_2 discussed in Example 8 of Section 6.1?
   (a) $\frac{2}{3} - \frac{2}{3}x + \frac{1}{3}x^2$, $\frac{2}{3} + \frac{1}{3}x - \frac{2}{3}x^2$, $\frac{1}{3} + \frac{2}{3}x + \frac{2}{3}x^2$ (b) 1, $\dfrac{1}{\sqrt{2}}x + \dfrac{1}{\sqrt{2}}x^2$, x^2

6. Which of the following sets of matrices are orthonormal with respect to the inner product on M_{22} discussed in Example 7 of Section 6.1?
   (a) $\begin{bmatrix} 1 & 0 \\ 0 & 0 \end{bmatrix}$, $\begin{bmatrix} 0 & \frac{2}{3} \\ \frac{1}{3} & -\frac{2}{3} \end{bmatrix}$, $\begin{bmatrix} 0 & \frac{2}{3} \\ -\frac{2}{3} & \frac{1}{3} \end{bmatrix}$, $\begin{bmatrix} 0 & \frac{1}{3} \\ \frac{2}{3} & \frac{2}{3} \end{bmatrix}$
   (b) $\begin{bmatrix} 1 & 0 \\ 0 & 0 \end{bmatrix}$, $\begin{bmatrix} 0 & 1 \\ 0 & 0 \end{bmatrix}$, $\begin{bmatrix} 0 & 0 \\ 1 & 1 \end{bmatrix}$, $\begin{bmatrix} 0 & 0 \\ 1 & -1 \end{bmatrix}$

7. Verify that the given set of vectors is orthogonal with respect to the Euclidean inner product; then convert it to an orthonormal set by normalizing the vectors.
   (a) $(-1, 2)$, $(6, 3)$ (b) $(1, 0, -1)$, $(2, 0, 2)$, $(0, 5, 0)$ (c) $(\frac{1}{5}, \frac{1}{5}, \frac{1}{5})$, $(-\frac{1}{2}, \frac{1}{2}, 0)$, $(\frac{1}{3}, \frac{1}{3}, -\frac{2}{3})$

8. Let $\mathbf{x} = \left(\dfrac{1}{\sqrt{5}}, -\dfrac{1}{\sqrt{5}}\right)$ and $\mathbf{y} = \left(\dfrac{2}{\sqrt{30}}, \dfrac{3}{\sqrt{30}}\right)$.

   Show that $\{\mathbf{x}, \mathbf{y}\}$ is orthonormal if R^2 has the inner product $\langle \mathbf{u}, \mathbf{v} \rangle = 3u_1 v_1 + 2u_2 v_2$, but is not orthonormal if R^2 has the Euclidean inner product.

9. Verify that the vectors $\mathbf{v}_1 = (-\frac{3}{5}, \frac{4}{5}, 0)$, $\mathbf{v}_2 = (\frac{4}{5}, \frac{3}{5}, 0)$, $\mathbf{v}_3 = (0, 0, 1)$ form an orthonormal basis for R^3 with the Euclidean inner product; then use Theorem 6.3.1 to express each of the following as linear combinations of $\mathbf{v}_1$, $\mathbf{v}_2$, and $\mathbf{v}_3$.
   (a) $(1, -1, 2)$ (b) $(3, -7, 4)$ (c) $(\frac{1}{7}, -\frac{3}{7}, \frac{5}{7})$

10. Verify that the vectors

   $$\mathbf{v}_1 = (1, -1, 2, -1), \quad \mathbf{v}_2 = (-2, 2, 3, 2), \quad \mathbf{v}_3 = (1, 2, 0, -1), \quad \mathbf{v}_4 = (1, 0, 0, 1)$$

form an orthogonal basis for R^4 with the Euclidean inner product; then use Formula (1) to express each of the following as linear combinations of v_1, v_2, v_3, and v_4.
(a) $(1, 1, 1, 1)$ (b) $(\sqrt{2}, -3\sqrt{2}, 5\sqrt{2}, -\sqrt{2})$ (c) $(-\frac{1}{3}, \frac{2}{3}, -\frac{1}{3}, \frac{4}{3})$

11. In each part an orthonormal basis relative to the Euclidean inner product is given. Use Theorem 6.3.1 to find the coordinate vector of **w** with respect to that basis.

(a) $\mathbf{w} = (3, 7)$; $\mathbf{u}_1 = \left(\dfrac{1}{\sqrt{2}}, -\dfrac{1}{\sqrt{2}}\right)$, $\mathbf{u}_2 = \left(\dfrac{1}{\sqrt{2}}, \dfrac{1}{\sqrt{2}}\right)$

(b) $\mathbf{w} = (-1, 0, 2)$; $\mathbf{u}_1 = (\frac{2}{3}, -\frac{2}{3}, \frac{1}{3})$, $\mathbf{u}_2 = (\frac{2}{3}, \frac{1}{3}, -\frac{2}{3})$, $\mathbf{u}_3 = (\frac{1}{3}, \frac{2}{3}, \frac{2}{3})$

12. Let R^2 have the Euclidean inner product, and let $S = \{\mathbf{w}_1, \mathbf{w}_2\}$ be the orthonormal basis with $\mathbf{w}_1 = (\frac{3}{5}, -\frac{4}{5})$, $\mathbf{w}_2 = (\frac{4}{5}, \frac{3}{5})$.
(a) Find the vectors **u** and **v** that have coordinate vectors $(\mathbf{u})_S = (1, 1)$ and $(\mathbf{v})_S = (-1, 4)$.
(b) Compute $\|\mathbf{u}\|$, $d(\mathbf{u}, \mathbf{v})$, and $\langle \mathbf{u}, \mathbf{v} \rangle$ by applying Theorem 6.3.2 to the coordinate vectors $(\mathbf{u})_S$ and $(\mathbf{v})_S$; then check the results by performing the computations directly on **u** and **v**.

13. Let R^3 have the Euclidean inner product, and let $S = \{\mathbf{w}_1, \mathbf{w}_2, \mathbf{w}_3\}$ be the orthonormal basis with $\mathbf{w}_1 = (0, -\frac{3}{5}, \frac{4}{5})$, $\mathbf{w}_2 = (1, 0, 0)$, and $\mathbf{w}_3 = (0, \frac{4}{5}, \frac{3}{5})$.
(a) Find the vectors **u**, **v**, and **w** that have the coordinate vectors $(\mathbf{u})_S = (-2, 1, 2)$, $(\mathbf{v})_S = (3, 0, -2)$, and $(\mathbf{w})_S = (5, -4, 1)$.
(b) Compute $\|\mathbf{v}\|$, $d(\mathbf{u}, \mathbf{w})$, and $\langle \mathbf{w}, \mathbf{v} \rangle$ by applying Theorem 6.3.2 to the coordinate vectors $(\mathbf{u})_S$, $(\mathbf{v})_S$, and $(\mathbf{w})_S$; then check the results by performing the computations directly on **u** and **v**.

14. In each part, S represents some orthonormal basis for a four-dimensional inner product space. Use the given information to find $\|\mathbf{u}\|$, $\|\mathbf{v} - \mathbf{w}\|$, $\|\mathbf{v} + \mathbf{w}\|$, and $\langle \mathbf{v}, \mathbf{w} \rangle$.
(a) $(\mathbf{u})_S = (-1, 2, 1, 3)$, $(\mathbf{v})_S = (0, -3, 1, 5)$, $(\mathbf{w})_S = (-2, -4, 3, 1)$
(b) $(\mathbf{u})_S = (0, 0, -1, -1)$, $(\mathbf{v})_S = (5, 5, -2, -2)$, $(\mathbf{w})_S = (3, 0, -3, 0)$

15. (a) Show that the vectors $\mathbf{v}_1 = (1, -2, 3, -4)$, $\mathbf{v}_2 = (2, 1, -4, -3)$, $\mathbf{v}_3 = (-3, 4, 1, -2)$, and $\mathbf{v}_4 = (4, 3, 2, 1)$ form an orthogonal basis for R^4 with the Euclidean inner product.
(b) Use (1) to express $\mathbf{u} = (-1, 2, 3, 7)$ as a linear combination of the vectors in (a).

16. Let R^2 have the Euclidean inner product. Use the Gram–Schmidt process to transform the basis $\{\mathbf{u}_1, \mathbf{u}_2\}$ into an orthonormal basis.
(a) $\mathbf{u}_1 = (1, -3)$, $\mathbf{u}_2 = (2, 2)$ (b) $\mathbf{u}_1 = (1, 0)$, $\mathbf{u}_2 = (3, -5)$

17. Let R^3 have the Euclidean inner product. Use the Gram–Schmidt process to transform the basis $\{\mathbf{u}_1, \mathbf{u}_2, \mathbf{u}_3\}$ into an orthonormal basis.
(a) $\mathbf{u}_1 = (1, 1, 1)$, $\mathbf{u}_2 = (-1, 1, 0)$, $\mathbf{u}_3 = (1, 2, 1)$
(b) $\mathbf{u}_1 = (1, 0, 0)$, $\mathbf{u}_2 = (3, 7, -2)$, $\mathbf{u}_3 = (0, 4, 1)$

18. Let R^4 have the Euclidean inner product. Use the Gram–Schmidt process to transform the basis $\{\mathbf{u}_1, \mathbf{u}_2, \mathbf{u}_3, \mathbf{u}_4\}$ into an orthonormal basis.

$$\mathbf{u}_1 = (0, 2, 1, 0), \quad \mathbf{u}_2 = (1, -1, 0, 0), \quad \mathbf{u}_3 = (1, 2, 0, -1), \quad \mathbf{u}_4 = (1, 0, 0, 1)$$

19. Let R^3 have the Euclidean inner product. Find an orthonormal basis for the subspace spanned by $(0, 1, 2)$, $(-1, 0, 1)$, $(-1, 1, 3)$.

20. Let R^3 have the inner product $\langle \mathbf{u}, \mathbf{v} \rangle = u_1 v_1 + 2u_2 v_2 + 3u_3 v_3$. Use the Gram–Schmidt process to transform $\mathbf{u}_1 = (1, 1, 1)$, $\mathbf{u}_2 = (1, 1, 0)$, $\mathbf{u}_3 = (1, 0, 0)$ into an orthonormal basis.

21. The subspace of R^3 spanned by the vectors $\mathbf{u}_1 = (\frac{4}{5}, 0, -\frac{3}{5})$ and $\mathbf{u}_2 = (0, 1, 0)$ is a plane passing through the origin. Express $\mathbf{w} = (1, 2, 3)$ in the form $\mathbf{w} = \mathbf{w}_1 + \mathbf{w}_2$, where $\mathbf{w}_1$ lies in the plane and $\mathbf{w}_2$ is perpendicular to the plane.

22. Repeat Exercise 21 with $\mathbf{u}_1 = (1, 1, 1)$ and $\mathbf{u}_2 = (2, 0, -1)$.

23. Let R^4 have the Euclidean inner product. Express $\mathbf{w} = (-1, 2, 6, 0)$ in the form $\mathbf{w} = \mathbf{w}_1 + \mathbf{w}_2$, where $\mathbf{w}_1$ is in the space W spanned by $\mathbf{u}_1 = (-1, 0, 1, 2)$ and $\mathbf{u}_2 = (0, 1, 0, 1)$, and $\mathbf{w}_2$ is orthogonal to W.

24. Find the *QR*-decomposition of the matrix.

(a) $\begin{bmatrix} 1 & -1 \\ 2 & 3 \end{bmatrix}$
  (b) $\begin{bmatrix} 1 & 2 \\ 0 & 1 \\ 1 & 4 \end{bmatrix}$
  (c) $\begin{bmatrix} 1 & 1 \\ -2 & 1 \\ 2 & 1 \end{bmatrix}$

(d) $\begin{bmatrix} 1 & 0 & 2 \\ 0 & 1 & 1 \\ 1 & 2 & 0 \end{bmatrix}$
  (e) $\begin{bmatrix} 1 & 2 & 1 \\ 1 & 1 & 1 \\ 0 & 3 & 1 \end{bmatrix}$
  (f) $\begin{bmatrix} 1 & 0 & 1 \\ -1 & 1 & 1 \\ 1 & 0 & 1 \\ -1 & 1 & 1 \end{bmatrix}$

25. Let $\{\mathbf{v}_1, \mathbf{v}_2, \mathbf{v}_3\}$ be an orthonormal basis for an inner product space V. Show that if $\mathbf{w}$ is a vector in V, then $\|\mathbf{w}\|^2 = \langle \mathbf{w}, \mathbf{v}_1 \rangle^2 + \langle \mathbf{w}, \mathbf{v}_2 \rangle^2 + \langle \mathbf{w}, \mathbf{v}_3 \rangle^2$.

26. Let $\{\mathbf{v}_1, \mathbf{v}_2, \ldots, \mathbf{v}_n\}$ be an orthonormal basis for an inner product space V. Show that if $\mathbf{w}$ is a vector in V, then $\|\mathbf{w}\|^2 = \langle \mathbf{w}, \mathbf{v}_1 \rangle^2 + \langle \mathbf{w}, \mathbf{v}_2 \rangle^2 + \cdots + \langle \mathbf{w}, \mathbf{v}_n \rangle^2$.

27. In Step 3 of the proof of Theorem 6.3.6, it was stated that "the linear independence of $\{\mathbf{u}_1, \mathbf{u}_2, \ldots, \mathbf{u}_n\}$ ensures that $\mathbf{v}_3 \neq \mathbf{0}$." Prove this statement.

28. Prove that the diagonal entries of R in Formula (9) are nonzero.

29. (*For readers who have studied calculus.*) Let the vector space P_2 have the inner product

$$\langle \mathbf{p}, \mathbf{q} \rangle = \int_{-1}^{1} p(x)q(x)\, dx$$

Apply the Gram–Schmidt process to transform the standard basis $S = \{1, x, x^2\}$ into an orthonormal basis. (The polynomials in the resulting basis are called the first three ***normalized Legendre polynomials***.)

30. (*For readers who have studied calculus.*) Use Theorem 6.3.1 to express the following as linear combinations of the first three normalized Legendre polynomials (Exercise 29).
  (a) $1 + x + 4x^2$ (b) $2 - 7x^2$ (c) $4 + 3x$

31. (*For readers who have studied calculus.*) Let P_2 have the inner product

$$\langle \mathbf{p}, \mathbf{q} \rangle = \int_{0}^{1} p(x)q(x)\, dx$$

Apply the Gram–Schmidt process to transform the standard basis $S = \{1, x, x^2\}$ into an orthonormal basis.

32. Prove Theorem 6.3.5.

33. Prove Theorem 6.3.2*a*.

34. Prove Theorem 6.3.2*b*.

35. Prove Theorem 6.3.2*c*.

6.4 BEST APPROXIMATION; LEAST SQUARES

In this section we shall show how orthogonal projections can be used to solve certain approximation problems. The results obtained in this section have a wide variety of applications in both mathematics and science.

ORTHOGONAL PROJECTIONS VIEWED AS APPROXIMATIONS

If P is a point in ordinary 3-space and W is a plane through the origin, then the point Q in W closest to P is obtained by dropping a perpendicular from P to W (Figure 1*a*). Therefore, if we let $\mathbf{u} = \overrightarrow{OP}$, the distance between P and W is given by

$$\|\mathbf{u} - \text{proj}_W \, \mathbf{u}\|$$

In other words, among all vectors $\mathbf{w}$ in W the vector $\mathbf{w} = \text{proj}_W \, \mathbf{u}$ minimizes the distance $\|\mathbf{u} - \mathbf{w}\|$ (Figure 1*b*).

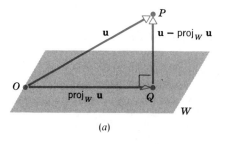

(*a*)	(*b*)

Figure 1 Q is the point in W closest to P. $\|\mathbf{u} - \mathbf{w}\|$ is minimized by $\mathbf{w} = \text{proj}_W \, \mathbf{u}$.

There is another way of thinking about this idea. View $\mathbf{u}$ as a fixed vector that we would like to approximate by a vector in W. Any such approximation $\mathbf{w}$ will result in an "error vector,"

$$\mathbf{u} - \mathbf{w}$$

which, unless $\mathbf{u}$ is in W, cannot be made equal to $\mathbf{0}$. However, by choosing

$$\mathbf{w} = \text{proj}_W \, \mathbf{u}$$

we can make the length of the error vector

$$\|\mathbf{u} - \mathbf{w}\| = \|\mathbf{u} - \text{proj}_W \, \mathbf{u}\|$$

as small as possible. Thus, we can describe $\text{proj}_W \, \mathbf{u}$ as the "best approximation" to $\mathbf{u}$ by vectors in W. The following theorem will make these intuitive ideas precise.

Theorem 6.4.1 (***Best Approximation Theorem***). *If W is a finite-dimensional subspace of an inner product space V, and if $\mathbf{u}$ is a vector in V, then $\text{proj}_W \, \mathbf{u}$ is the **best approximation** to $\mathbf{u}$ from W in the sense that*

$$\|\mathbf{u} - \text{proj}_W \, \mathbf{u}\| < \|\mathbf{u} - \mathbf{w}\|$$

for every vector $\mathbf{w}$ in W that is different from $\text{proj}_W \, \mathbf{u}$.

Proof. For every vector **w** in W we can write

$$\mathbf{u} - \mathbf{w} = (\mathbf{u} - \text{proj}_W \mathbf{u}) + (\text{proj}_W \mathbf{u} - \mathbf{w}) \tag{1}$$

But $\text{proj}_W \mathbf{u} - \mathbf{w}$, being a difference of vectors in W, is in W; and $\mathbf{u} - \text{proj}_W \mathbf{u}$ is orthogonal to W, so that the two terms on the right side of (1) are orthogonal. Thus, by the Theorem of Pythagoras (6.2.4),

$$\|\mathbf{u} - \mathbf{w}\|^2 = \|\mathbf{u} - \text{proj}_W \mathbf{u}\|^2 + \|\text{proj}_W \mathbf{u} - \mathbf{w}\|^2$$

If $\mathbf{w} \neq \text{proj}_W \mathbf{u}$, then the second term in this sum will be positive, so that

$$\|\mathbf{u} - \mathbf{w}\|^2 > \|\mathbf{u} - \text{proj}_W \mathbf{u}\|^2$$

or equivalently,

$$\|\mathbf{u} - \mathbf{w}\| > \|\mathbf{u} - \text{proj}_W \mathbf{u}\| \quad \blacksquare$$

Applications of this theorem will be given later in the text.

**LEAST SQUARES
SOLUTIONS OF
LINEAR SYSTEMS**

Up to now we have been concerned primarily with consistent systems of linear equations. However, inconsistent linear systems are also important in physical applications. It is a common situation that some physical problem leads to a linear system $A\mathbf{x} = \mathbf{b}$ that should be consistent on theoretical grounds but fails to be so because "measurement errors" in the entries of A and **b** perturb the system enough to cause inconsistency. In such situations one looks for a value of **x** that comes "as close as possible" to being a solution in the sense that it minimizes the value of $\|A\mathbf{x} - \mathbf{b}\|$ with respect to the Euclidean inner product. The quantity $\|A\mathbf{x} - \mathbf{b}\|$ can be viewed as a measure of the "error" that results from regarding **x** as an approximate solution of the linear system $A\mathbf{x} = \mathbf{b}$. If the system is consistent and **x** is an exact solution, then the error is zero, since $\|A\mathbf{x} - \mathbf{b}\| = \|\mathbf{0}\| = 0$. In general, the larger the value of $\|A\mathbf{x} - \mathbf{b}\|$, the more poorly **x** approximates a solution of the system.

> **Least Squares Problem.** Given a linear system $A\mathbf{x} = \mathbf{b}$ of m equations in n unknowns, find a vector **x**, if possible, that minimizes $\|A\mathbf{x} - \mathbf{b}\|$ with respect to the Euclidean inner product on R^m. Such a vector is called a ***least squares solution*** of $A\mathbf{x} = \mathbf{b}$.

REMARK. To understand the origin of the term *least squares*, let $\mathbf{e} = A\mathbf{x} - \mathbf{b}$, which we can view as the error vector that results from the approximation **x**. If $\mathbf{e} = (e_1, e_2, \ldots, e_m)$, then a least squares solution minimizes $\|\mathbf{e}\| = (e_1^2 + e_2^2 + \cdots + e_m^2)^{1/2}$; hence it also minimizes $\|\mathbf{e}\|^2 = e_1^2 + e_2^2 + \cdots + e_m^2$—thus, the term *least squares*.

To solve the least squares problem, let W be the column space of A. For each $n \times 1$ matrix **x**, the product $A\mathbf{x}$ is a linear combination of the column vectors of A. Thus, as **x** varies over R^n the vector $A\mathbf{x}$ varies over all possible linear combinations of the column vectors of A; that is, $A\mathbf{x}$ varies over the entire column space W. Geometrically, solving the least squares problem amounts to finding a vector **x** in R^n such that $A\mathbf{x}$ is the closest vector in W to **b** (Figure 2).

Figure 2 | A least squares solution **x** produces the vector $A\mathbf{x}$ in W closest to **b**.

It follows from the Best Approximation Theorem (6.4.1) that the closest vector in W to **b** is the orthogonal projection of **b** on W. Thus, for a vector **x** to be a least squares solution of $A\mathbf{x} = \mathbf{b}$, this vector must satisfy

$$A\mathbf{x} = \text{proj}_W \mathbf{b} \tag{2}$$

One could attempt to find least squares solutions of $A\mathbf{x} = \mathbf{b}$ by first calculating the vector $\text{proj}_W \mathbf{b}$ and then solving (2); however, there is a better approach: It follows from the Projection Theorem (6.3.4) and Formula (5) of Section 6.3 that

$$\mathbf{b} - A\mathbf{x} = \mathbf{b} - \text{proj}_W \mathbf{b}$$

is orthogonal to W. But W is the column space of A, so it follows from Theorem 6.2.6 that $\mathbf{b} - A\mathbf{x}$ lies in the nullspace of A^T. Therefore, a least squares solution of $A\mathbf{x} = \mathbf{b}$ must satisfy

$$A^T(\mathbf{b} - A\mathbf{x}) = 0$$

or equivalently,

$$A^T A\mathbf{x} = A^T \mathbf{b} \tag{3}$$

This is called the ***normal system*** associated with $A\mathbf{x} = \mathbf{b}$, and the individual equations are called the ***normal equations*** associated with $A\mathbf{x} = \mathbf{b}$. Thus, the problem of finding a least squares solution of $A\mathbf{x} = \mathbf{b}$ has been reduced to the problem of finding an exact solution of the associated normal system.

Note the following observations about the normal system:

- The normal system involves n equations in n unknowns (verify).

- The normal system is consistent, since it is satisfied by a least squares solution of $A\mathbf{x} = \mathbf{b}$.

- The normal system may have infinitely many solutions, in which case all of its solutions are least squares solutions of $A\mathbf{x} = \mathbf{b}$.

From these observations and Formula (2) we have the following theorem.

Theorem 6.4.2. *For any linear system* $A\mathbf{x} = \mathbf{b}$, *the associated normal system*

$$A^T A\mathbf{x} = A^T\mathbf{b}$$

is consistent, and all solutions of the normal system are least squares solutions of $A\mathbf{x} = \mathbf{b}$. *Moreover, if* W *is the column space of* A, *and* $\mathbf{x}$ *is any least squares solution of* $A\mathbf{x} = \mathbf{b}$, *then the orthogonal projection of* $\mathbf{b}$ *on* W *is*

$$\text{proj}_W\,\mathbf{b} = A\mathbf{x}$$

UNIQUENESS OF LEAST SQUARES SOLUTIONS

Before we examine some numerical examples, we shall establish conditions under which a linear system is guaranteed to have a unique least squares solution. We shall need the following theorem.

Theorem 6.4.3. *If* A *is an* $m \times n$ *matrix, then the following are equivalent.*
(*a*) *A has linearly independent column vectors.*
(*b*) $A^T A$ *is invertible.*

Proof. We shall prove that $(a) \Rightarrow (b)$ and $(b) \Rightarrow (a)$.

$(a) \Rightarrow (b)$: Assume that A has linearly independent column vectors. The matrix $A^T A$ has size $n \times n$, so we can prove that this matrix is invertible by showing that the linear system $A^T A\mathbf{x} = \mathbf{0}$ has only the trivial solution. But if $\mathbf{x}$ is any solution of this system, then $A\mathbf{x}$ is in the nullspace of A^T and also in the column space of A. By Theorem 6.2.6 these spaces are orthogonal complements, so part (*b*) of Theorem 6.2.5 implies that $A\mathbf{x} = \mathbf{0}$. But A has linearly independent column vectors, so $\mathbf{x} = \mathbf{0}$ by Theorem 5.6.8.

$(b) \Rightarrow (a)$: Assume that $A^T A$ is invertible. To prove that A has linearly independent column vectors, it suffices to prove that $A\mathbf{x} = \mathbf{0}$ has only the trivial solution by Theorem 5.6.8. But if $\mathbf{x}$ is any solution of $A\mathbf{x} = \mathbf{0}$, then $A^T A\mathbf{x} = A^T\mathbf{0} = \mathbf{0}$, so $\mathbf{x} = \mathbf{0}$ from the invertibility of $A^T A$. ▮

The next theorem is a direct consequence of Theorems 6.4.2 and 6.4.3. We omit the details.

Theorem 6.4.4 (*Uniqueness of Least Squares Solutions*). *If* A *is an* $m \times n$ *matrix with linearly independent column vectors, then for every* $n \times 1$ *matrix* $\mathbf{b}$, *the linear system* $A\mathbf{x} = \mathbf{b}$ *has a unique least squares solution. This solution is given by*

$$\mathbf{x} = (A^T A)^{-1} A^T\mathbf{b} \tag{4}$$

Moreover, if W *is the column space of* A, *then the orthogonal projection of* $\mathbf{b}$ *on* W *is*

$$\text{proj}_W\,\mathbf{b} = A\mathbf{x} = A(A^T A)^{-1} A^T\mathbf{b} \tag{5}$$

REMARK. Formulas (4) and (5) have various theoretical applications, but they are not efficient for numerical calculations. Least squares solutions of $A\mathbf{x} = \mathbf{b}$ are best computed by using Gaussian elimination or Gauss–Jordan elimination to solve the normal equations, and the orthogonal projection of $\mathbf{b}$ on the column space of A is best obtained by computing $A\mathbf{x}$, where $\mathbf{x}$ is the least squares solution of $A\mathbf{x} = \mathbf{b}$.

Example 1   Find the least squares solution of the linear system $A\mathbf{x} = \mathbf{b}$ given by

$$x_1 - x_2 = 4$$
$$3x_1 + 2x_2 = 1$$
$$-2x_1 + 4x_2 = 3$$

and find the orthogonal projection of $\mathbf{b}$ on the column space of A.

Solution. Here

$$A = \begin{bmatrix} 1 & -1 \\ 3 & 2 \\ -2 & 4 \end{bmatrix} \quad \text{and} \quad \mathbf{b} = \begin{bmatrix} 4 \\ 1 \\ 3 \end{bmatrix}$$

Observe that A has linearly independent column vectors, so we know in advance that there is a unique least squares solution. We have

$$A^T A = \begin{bmatrix} 1 & 3 & -2 \\ -1 & 2 & 4 \end{bmatrix} \begin{bmatrix} 1 & -1 \\ 3 & 2 \\ -2 & 4 \end{bmatrix} = \begin{bmatrix} 14 & -3 \\ -3 & 21 \end{bmatrix}$$

$$A^T \mathbf{b} = \begin{bmatrix} 1 & 3 & -2 \\ -1 & 2 & 4 \end{bmatrix} \begin{bmatrix} 4 \\ 1 \\ 3 \end{bmatrix} = \begin{bmatrix} 1 \\ 10 \end{bmatrix}$$

so the normal system $A^T A \mathbf{x} = A^T \mathbf{b}$ in this case is

$$\begin{bmatrix} 14 & -3 \\ -3 & 21 \end{bmatrix} \begin{bmatrix} x_1 \\ x_2 \end{bmatrix} = \begin{bmatrix} 1 \\ 10 \end{bmatrix}$$

Solving this system yields the least squares solution

$$x_1 = \tfrac{17}{95}, \qquad x_2 = \tfrac{143}{285} \quad \blacktriangle$$

From (5), the orthogonal projection of $\mathbf{b}$ on the column space of A is

$$A\mathbf{x} = \begin{bmatrix} 1 & -1 \\ 3 & 2 \\ -2 & 4 \end{bmatrix} \begin{bmatrix} \frac{17}{95} \\ \frac{143}{285} \end{bmatrix} = \begin{bmatrix} -\frac{92}{285} \\ \frac{439}{285} \\ \frac{94}{57} \end{bmatrix}$$

Example 2   Find the orthogonal projection of the vector $\mathbf{u} = (-3, -3, 8, 9)$ on the subspace of R^4 spanned by the vectors

$$\mathbf{u}_1 = (3, 1, 0, 1), \qquad \mathbf{u}_2 = (1, 2, 1, 1), \qquad \mathbf{u}_3 = (-1, 0, 2, -1)$$

Solution.   One could solve this problem by first using the Gram–Schmidt process to convert $\{\mathbf{u}_1, \mathbf{u}_2, \mathbf{u}_3\}$ into an orthonormal basis, and then applying the method used in Example 6 of Section 6.3. However, the following method is more efficient.

The subspace W of R^4 spanned by $\mathbf{u}_1$, $\mathbf{u}_2$, and $\mathbf{u}_3$ is the column space of the matrix

$$A = \begin{bmatrix} 3 & 1 & -1 \\ 1 & 2 & 0 \\ 0 & 1 & 2 \\ 1 & 1 & -1 \end{bmatrix}$$

Thus, if $\mathbf{u}$ is expressed as a column vector, we can find the orthogonal projection of $\mathbf{u}$ on W by finding a least squares solution of the system $A\mathbf{x} = \mathbf{u}$ and then calculating $\operatorname{proj}_W \mathbf{u} = A\mathbf{x}$ from the least squares solution. The computations are as follows: The system $A\mathbf{x} = \mathbf{u}$ is

$$\begin{bmatrix} 3 & 1 & -1 \\ 1 & 2 & 0 \\ 0 & 1 & 2 \\ 1 & 1 & -1 \end{bmatrix} \begin{bmatrix} x_1 \\ x_2 \\ x_3 \end{bmatrix} = \begin{bmatrix} -3 \\ -3 \\ 8 \\ 9 \end{bmatrix}$$

so

$$A^T A = \begin{bmatrix} 3 & 1 & 0 & 1 \\ 1 & 2 & 1 & 1 \\ -1 & 0 & 2 & -1 \end{bmatrix} \begin{bmatrix} 3 & 1 & -1 \\ 1 & 2 & 0 \\ 0 & 1 & 2 \\ 1 & 1 & -1 \end{bmatrix} = \begin{bmatrix} 11 & 6 & -4 \\ 6 & 7 & 0 \\ -4 & 0 & 6 \end{bmatrix}$$

$$A^T \mathbf{u} = \begin{bmatrix} 3 & 1 & 0 & 1 \\ 1 & 2 & 1 & 1 \\ -1 & 0 & 2 & -1 \end{bmatrix} \begin{bmatrix} -3 \\ -3 \\ 8 \\ 9 \end{bmatrix} = \begin{bmatrix} -3 \\ 8 \\ 10 \end{bmatrix}$$

The normal system $A^T A \mathbf{x} = A^T \mathbf{u}$ in this case is

$$\begin{bmatrix} 11 & 6 & -4 \\ 6 & 7 & 0 \\ -4 & 0 & 6 \end{bmatrix} \begin{bmatrix} x_1 \\ x_2 \\ x_3 \end{bmatrix} = \begin{bmatrix} -3 \\ 8 \\ 10 \end{bmatrix}$$

Solving this system yields as the least squares solution of $A\mathbf{x} = \mathbf{u}$

$$\mathbf{x} = \begin{bmatrix} x_1 \\ x_2 \\ x_3 \end{bmatrix} = \begin{bmatrix} -1 \\ 2 \\ 1 \end{bmatrix}$$

(verify), so

$$\text{proj}_W \mathbf{u} = A\mathbf{x} = \begin{bmatrix} 3 & 1 & -1 \\ 1 & 2 & 0 \\ 0 & 1 & 2 \\ 1 & 1 & -1 \end{bmatrix} \begin{bmatrix} -1 \\ 2 \\ 1 \end{bmatrix} = \begin{bmatrix} -2 \\ 3 \\ 4 \\ 0 \end{bmatrix}$$

or in horizontal notation (consistent with the original phrasing of the problem), $\text{proj}_W \mathbf{u} = (-2, 3, 4, 0)$. ▲

ORTHOGONAL PROJECTION OPERATORS

In Section 4.2 we discussed some basic orthogonal projection operators on R^2 and R^3 (Tables 4 and 5). The concept of an orthogonal projection operator can be extended to general inner product spaces as follows.

> **Definition.** If W is a subspace of R^m, then the transformation $P: R^m \to W$ that maps each vector in $\mathbf{x}$ in R^m into its orthogonal projection $\text{proj}_W \mathbf{x}$ in W is called the ***orthogonal projection of R^m onto W***.

We leave it as an exercise to show that orthogonal projections are linear operators. It follows from Formula (5) that the standard matrix for the orthogonal projection of R^m on W is

$$[P] = A(A^T A)^{-1} A^T \tag{6}$$

where A is constructed using any basis for W as its column vectors.

Example 3 In Table 5 of Section 4.2 we showed that the standard matrix for the orthogonal projection of R^3 on the xy-plane is

$$[P] = \begin{bmatrix} 1 & 0 & 0 \\ 0 & 1 & 0 \\ 0 & 0 & 0 \end{bmatrix} \tag{7}$$

To see that this is consistent with Formula (6), take the unit vectors along the positive x and y axes as a basis for the xy-plane, so that

$$A = \begin{bmatrix} 1 & 0 \\ 0 & 1 \\ 0 & 0 \end{bmatrix}$$

We leave it for the reader to verify that $A^T A$ is the 2×2 identity matrix; thus, (6) simplifies to

$$[P] = AA^T = \begin{bmatrix} 1 & 0 \\ 0 & 1 \\ 0 & 0 \end{bmatrix} \begin{bmatrix} 1 & 0 & 0 \\ 0 & 1 & 0 \end{bmatrix} = \begin{bmatrix} 1 & 0 & 0 \\ 0 & 1 & 0 \\ 0 & 0 & 0 \end{bmatrix}$$

which agrees with (7). ▲

Example 4 Find the standard matrix for the orthogonal projection P of R^2 onto the line l that passes through the origin and makes an angle θ with the positive x-axis.

Solution. The line l is a one-dimensional subspace of R^2. As illustrated in Figure 3, we can take $\mathbf{v} = (\cos \theta, \sin \theta)$ as a basis for this subspace, so

$$A = \begin{bmatrix} \cos \theta \\ \sin \theta \end{bmatrix}$$

We leave it for the reader to show that $A^T A$ is the 1×1 identity matrix; thus, (6) simplifies to

$$[P] = AA^T = \begin{bmatrix} \cos \theta \\ \sin \theta \end{bmatrix} [\cos \theta \quad \sin \theta] = \begin{bmatrix} \cos^2 \theta & \sin \theta \cos \theta \\ \sin \theta \cos \theta & \sin^2 \theta \end{bmatrix}$$ ▲

 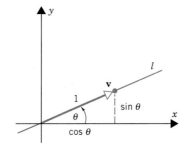

Figure 3

Theorem 6.4.3 enables us to add an additional result to Theorem 6.2.7.

Theorem 6.4.5. *If A is an $n \times n$ matrix, and if $T_A : R^n \rightarrow R^n$ is multiplication by A, then the following are equivalent.*

(a) A is invertible.
(b) $A\mathbf{x} = \mathbf{0}$ has only the trivial solution.
(c) The reduced row-echelon form of A is I_n.
(d) A is expressible as a product of elementary matrices.
(e) $A\mathbf{x} = \mathbf{b}$ is consistent for every $n \times 1$ matrix $\mathbf{b}$.
(f) $A\mathbf{x} = \mathbf{b}$ has exactly one solution for every $n \times 1$ matrix $\mathbf{b}$.
(g) $\det(A) \neq 0$.
(h) The range of T_A is R^n.

(i) T_A is one-to-one.

(j) *The column vectors of A are linearly independent.*

(k) *The row vectors of A are linearly independent.*

(l) *The column vectors of A span R^n.*

(m) *The row vectors of A span R^n.*

(n) *The column vectors of A form a basis for R^n.*

(o) *The row vectors of A form a basis for R^n.*

(p) *A has rank n.*

(q) *A has nullity 0.*

(r) *The orthogonal complement of the nullspace of A is R^n.*

(s) *The orthogonal complement of the row space of A is $\{0\}$.*

(t) $A^T A$ *is invertible.*

This theorem relates all of the major topics we have studied thus far.

EXERCISE SET 6.4

1. Find the normal system associated with the given linear system.

(a) $\begin{bmatrix} 1 & -1 \\ 2 & 3 \\ 4 & 5 \end{bmatrix} \begin{bmatrix} x_1 \\ x_2 \end{bmatrix} = \begin{bmatrix} 2 \\ -1 \\ 5 \end{bmatrix}$
(b) $\begin{bmatrix} 2 & -1 & 0 \\ 3 & 1 & 2 \\ -1 & 4 & 5 \\ 1 & 2 & 4 \end{bmatrix} \begin{bmatrix} x_1 \\ x_2 \\ x_3 \end{bmatrix} = \begin{bmatrix} -1 \\ 0 \\ 1 \\ 2 \end{bmatrix}$

2. In each part find $\det(A^T A)$, and apply Theorem 6.4.3 to determine whether A has linearly independent column vectors.

(a) $A = \begin{bmatrix} -1 & 3 & 2 \\ 2 & 1 & 3 \\ 0 & 1 & 1 \end{bmatrix}$
(b) $A = \begin{bmatrix} 2 & -1 & 3 \\ 0 & 1 & 1 \\ -1 & 0 & -2 \\ 4 & -5 & 3 \end{bmatrix}$

3. Find the least squares solution of the linear system $Ax = b$, and find the orthogonal projection of b on the column space of A.

(a) $A = \begin{bmatrix} 1 & 1 \\ -1 & 1 \\ -1 & 2 \end{bmatrix}$, $b = \begin{bmatrix} 7 \\ 0 \\ -7 \end{bmatrix}$
(b) $A = \begin{bmatrix} 2 & -2 \\ 1 & 1 \\ 3 & 1 \end{bmatrix}$, $b = \begin{bmatrix} 2 \\ -1 \\ 1 \end{bmatrix}$

(c) $A = \begin{bmatrix} 1 & 0 & -1 \\ 2 & 1 & -2 \\ 1 & 1 & 0 \\ 1 & 1 & -1 \end{bmatrix}$, $b = \begin{bmatrix} 6 \\ 0 \\ 9 \\ 3 \end{bmatrix}$
(d) $A = \begin{bmatrix} 2 & 0 & -1 \\ 1 & -2 & 2 \\ 2 & -1 & 0 \\ 0 & 1 & -1 \end{bmatrix}$, $b = \begin{bmatrix} 0 \\ 6 \\ 0 \\ 6 \end{bmatrix}$

4. Find the orthogonal projection of u on the subspace of R^3 spanned by the vectors v_1 and v_2.

(a) $u = (2, 1, 3)$; $v_1 = (1, 1, 0)$, $v_2 = (1, 2, 1)$

(b) $u = (1, -6, 1)$; $v_1 = (-1, 2, 1)$, $v_2 = (2, 2, 4)$

5. Find the orthogonal projection of **u** on the subspace of R^4 spanned by the vectors $\mathbf{v}_1$, $\mathbf{v}_2$, and $\mathbf{v}_3$.

(a) $\mathbf{u} = (6, 3, 9, 6)$; $\mathbf{v}_1 = (2, 1, 1, 1)$, $\mathbf{v}_2 = (1, 0, 1, 1)$, $\mathbf{v}_3 = (-2, -1, 0, -1)$

(b) $\mathbf{u} = (-2, 0, 2, 4)$; $\mathbf{v}_1 = (1, 1, 3, 0)$, $\mathbf{v}_2 = (-2, -1, -2, 1)$, $\mathbf{v}_3 = (-3, -1, 1, 3)$

6. Find the orthogonal projection of $\mathbf{u} = (5, 6, 7, 2)$ on the solution space of the homogeneous linear system

$$
\begin{aligned}
x_1 + x_2 + x_3 \quad\ \ &= 0 \\
2x_2 + x_3 + x_4 &= 0
\end{aligned}
$$

7. Use Formula (6) and the method of Example 3 to find the standard matrix for the orthogonal projection $P: R^2 \to R^2$ onto

(a) the x-axis   (b) the y-axis

[*Note.* Compare your results to Table 4 of Section 4.2.]

8. Use Formula (6) and the method of Example 3 to find the standard matrix for the orthogonal projection $P: R^3 \to R^3$ onto

(a) the xz-plane   (b) the yz-plane

[*Note.* Compare your results to Table 5 of Section 4.2.]

9. Let W be the plane with equation $5x - 3y + z = 0$.

(a) Find a basis for W.

(b) Use Formula (6) to find the standard matrix for the orthogonal projection onto W.

(c) Use the matrix obtained in (b) to find the orthogonal projection of a point $P_0(x_0, y_0, z_0)$ on W.

(d) Find the distance between the point $P_0(1, -2, 4)$ and the plane W, and check your result using Theorem 3.5.2.

10. Let W be the line with parametric equations

$$
x = 2t, \quad y = -t, \quad z = 4t \quad\quad (-\infty < t < \infty)
$$

(a) Find a basis for W.

(b) Use Formula (6) to find the standard matrix for the orthogonal projection onto W.

(c) Use the matrix obtained in (b) to find the orthogonal projection of a point $P_0(x_0, y_0, z_0)$ on W.

(d) Find the distance between the point $P_0(2, 1, -3)$ and the line W.

11. For the linear systems in Exercise 3, verify that the *error vector* $A\bar{\mathbf{x}} - \mathbf{b}$ resulting from the least squares solution $\bar{\mathbf{x}}$ is orthogonal to the column space of A.

12. Prove: If A has linearly independent column vectors, and if $A\mathbf{x} = \mathbf{b}$ is consistent, then the least squares solution of $A\mathbf{x} = \mathbf{b}$ and the exact solution of $A\mathbf{x} = \mathbf{b}$ are the same.

13. Prove: If A has linearly independent column vectors, and if $\mathbf{b}$ is orthogonal to the column space of A, then the least squares solution of $A\mathbf{x} = \mathbf{b}$ is $\mathbf{x} = \mathbf{0}$.

14. Let $P: R^m \to W$ be the orthogonal projection of R^m onto a subspace W.

(a) Prove that $[P]^2 = [P]$.

(b) What does the result in part (a) imply about the compositon $P \circ P$?

(c) Show that $[P]$ is symmetric.

(d) Verify that the matrices in Tables 4 and 5 of Section 4.2 have the properties in parts (a) and (c).

15. Let A be an $m \times n$ matrix with linearly independent row vectors. Find a standard matrix for the orthogonal projection of R^n onto the row space of A. [*Hint.* Start with Formula (6).]

6.5 ORTHOGONAL MATRICES; CHANGE OF BASIS

A basis that is suitable for one problem may not be suitable for another, so it is a common process in the study of vector spaces to change from one basis to another. Because a basis is the vector space generalization of a coordinate system, changing bases is akin to changing coordinate axes in R^2 and R^3. In this section we shall study various problems relating to change of basis. We shall also develop properties of square matrices with orthonormal column vectors. Such matrices arise in many contexts, including problems involving a change from one orthonormal basis to another.

ORTHOGONAL MATRICES

Matrices whose inverses can be obtained by transposition are sufficiently important that there is some terminology associated with them.

Definition. A square matrix A with the property

$$A^{-1} = A^T$$

is said to be an ***orthogonal matrix***.

It follows from this definition that *a square matrix A is orthogonal if and only if*

$$AA^T = A^T A = I \tag{1}$$

In fact, it follows from Theorem 1.6.3 that a square matrix A is orthogonal if *either* $AA^T = I$ or $A^T A = I$.

Example 1 The matrix

$$A = \begin{bmatrix} \frac{3}{7} & \frac{2}{7} & \frac{6}{7} \\ -\frac{6}{7} & \frac{3}{7} & \frac{2}{7} \\ \frac{2}{7} & \frac{6}{7} & -\frac{3}{7} \end{bmatrix}$$

is orthogonal, since

$$A^T A = \begin{bmatrix} \frac{3}{7} & -\frac{6}{7} & \frac{2}{7} \\ \frac{2}{7} & \frac{3}{7} & \frac{6}{7} \\ \frac{6}{7} & \frac{2}{7} & -\frac{3}{7} \end{bmatrix} \begin{bmatrix} \frac{3}{7} & \frac{2}{7} & \frac{6}{7} \\ -\frac{6}{7} & \frac{3}{7} & \frac{2}{7} \\ \frac{2}{7} & \frac{6}{7} & -\frac{3}{7} \end{bmatrix} = \begin{bmatrix} 1 & 0 & 0 \\ 0 & 1 & 0 \\ 0 & 0 & 1 \end{bmatrix}$$

Example 2 Recall from Table 6 of Section 4.2 that the standard matrix for the counterclockwise rotation of R^2 through an angle θ is

$$A = \begin{bmatrix} \cos \theta & -\sin \theta \\ \sin \theta & \cos \theta \end{bmatrix}$$

This matrix is orthogonal for all choices of θ, since

$$A^T A = \begin{bmatrix} \cos\theta & \sin\theta \\ -\sin\theta & \cos\theta \end{bmatrix} \begin{bmatrix} \cos\theta & -\sin\theta \\ \sin\theta & \cos\theta \end{bmatrix} = \begin{bmatrix} 1 & 0 \\ 0 & 1 \end{bmatrix}$$

In fact, it is a simple matter to check that all of the "reflection matrices" in Tables 2 and 3 and all of the "rotation matrices" in Tables 6 and 7 of Section 4.2 are orthogonal matrices.

Observe that for the orthogonal matrices in Examples 1 and 2, both the row vectors and the column vectors form orthonormal sets with respect to the Euclidean inner product (verify). This is not accidental; it is a consequence of the following theorem.

Theorem 6.5.1. *The following are equivalent for an $n \times n$ matrix A.*

(a) *A is orthogonal.*

(b) *The row vectors of A form an orthonormal set in R^n with the Euclidean inner product.*

(c) *The column vectors of A form an orthonormal set in R^n with the Euclidean inner product.*

Proof. We shall prove the equivalence of (a) and (b) and leave the equivalence of (a) and (c) as an exercise.

$(a) \Leftrightarrow (b)$: The entry in the ith row and jth column of the matrix product AA^T is the dot product of the ith row vector of A and the jth column vector of A^T. But, except for a difference in notation, the jth column vector of A^T is the jth row vector of A. Thus, if the row vectors of A are $\mathbf{r}_1, \mathbf{r}_2, \ldots, \mathbf{r}_n$, then the matrix product AA^T can be expressed as

$$AA^T = \begin{bmatrix} \mathbf{r}_1 \cdot \mathbf{r}_1 & \mathbf{r}_1 \cdot \mathbf{r}_2 & \cdots & \mathbf{r}_1 \cdot \mathbf{r}_n \\ \mathbf{r}_2 \cdot \mathbf{r}_1 & \mathbf{r}_2 \cdot \mathbf{r}_2 & \cdots & \mathbf{r}_2 \cdot \mathbf{r}_n \\ \vdots & \vdots & & \vdots \\ \mathbf{r}_n \cdot \mathbf{r}_1 & \mathbf{r}_n \cdot \mathbf{r}_2 & \cdots & \mathbf{r}_n \cdot \mathbf{r}_n \end{bmatrix}$$

Thus, $AA^T = I$ if and only if

$$\mathbf{r}_1 \cdot \mathbf{r}_1 = \mathbf{r}_2 \cdot \mathbf{r}_2 = \cdots = \mathbf{r}_n \cdot \mathbf{r}_n = 1$$

and

$$\mathbf{r}_i \cdot \mathbf{r}_j = 0 \quad \text{when } i \neq j$$

which are true if and only if $\{\mathbf{r}_1, \mathbf{r}_2, \ldots, \mathbf{r}_n\}$ is an orthonormal set in R^n.

REMARK. In light of Theorem 6.5.1, it seems more appropriate to call orthogonal matrices *orthonormal matrices*. However, we will not do so in deference to historical tradition.

SOME BASIC PROPERTIES OF ORTHOGONAL MATRICES

The following theorem lists some additional fundamental properties of orthogonal matrices. The proofs are all straightforward and are left for the reader.

Theorem 6.5.2

(*a*) *The inverse of an orthogonal matrix is orthogonal.*
(*b*) *A product of orthogonal matrices is orthogonal.*
(*c*) *If A is orthogonal, then* $det(A) = 1$ *or* $det(A) = -1$.

Example 3 The matrix

$$A = \begin{bmatrix} 1/\sqrt{2} & -1/\sqrt{2} \\ -1/\sqrt{2} & 1/\sqrt{2} \end{bmatrix}$$

is orthogonal since its row (and column) vectors form orthonormal sets in R^2. We leave it for the reader to check that $\det(A) = 1$. Interchanging the rows produces an orthogonal matrix for which $\det(A) = -1$. ▲

ORTHOGONAL MATRICES AS LINEAR OPERATORS

We observed in Example 2 that the standard matrices for the basic reflection and rotation operators on R^2 and R^3 are orthogonal. The next theorem will help explain why this is so.

Theorem 6.5.3. *If A is an* $n \times n$ *matrix, then the following are equivalent.*

(*a*) *A is orthogonal.*
(*b*) $\|A\mathbf{x}\| = \|\mathbf{x}\|$ *for all* $\mathbf{x}$ *in* R^n.
(*c*) $A\mathbf{x} \cdot A\mathbf{y} = \mathbf{x} \cdot \mathbf{y}$ *for all* $\mathbf{x}$ *and* $\mathbf{y}$ *in* R^n.

Proof. We shall prove the sequence of implications $(a) \Rightarrow (b) \Rightarrow (c) \Rightarrow (a)$.

$(a) \Rightarrow (b)$: Assume that A is orthogonal, so that $A^T A = I$. Then from Formula (8) of Section 4.1,

$$\|A\mathbf{x}\| = (A\mathbf{x} \cdot A\mathbf{x})^{1/2} = (\mathbf{x} \cdot A^T A\mathbf{x})^{1/2} = (\mathbf{x} \cdot \mathbf{x})^{1/2} = \|\mathbf{x}\|$$

$(b) \Rightarrow (c)$: Assume that $\|A\mathbf{x}\| = \|\mathbf{x}\|$ for all $\mathbf{x}$ in R^n. From Theorem 4.1.6 we have

$$A\mathbf{x} \cdot A\mathbf{y} = \tfrac{1}{4}\|A\mathbf{x} + A\mathbf{y}\|^2 - \tfrac{1}{4}\|A\mathbf{x} - A\mathbf{y}\|^2 = \tfrac{1}{4}\|A(\mathbf{x} + \mathbf{y})\|^2 - \tfrac{1}{4}\|A(\mathbf{x} - \mathbf{y})\|^2$$
$$= \tfrac{1}{4}\|\mathbf{x} + \mathbf{y}\|^2 - \tfrac{1}{4}\|\mathbf{x} - \mathbf{y}\|^2 = \mathbf{x} \cdot \mathbf{y}$$

$(c) \Rightarrow (a)$: Assume that $A\mathbf{x} \cdot A\mathbf{y} = \mathbf{x} \cdot \mathbf{y}$ for all $\mathbf{x}$ and $\mathbf{y}$ in R^n. Then from Formula (8) of Section 4.1 we have

$$\mathbf{x} \cdot \mathbf{y} = \mathbf{x} \cdot A^T A\mathbf{y}$$

which can be rewritten as

$$\mathbf{x} \cdot (A^T A \mathbf{y} - \mathbf{y}) = 0 \qquad \text{or} \qquad \mathbf{x} \cdot (A^T A - I) \mathbf{y} = 0$$

Since this holds for all $\mathbf{x}$ in R^n, it holds in particular if

$$\mathbf{x} = (A^T A - I) \mathbf{y}$$

so

$$(A^T A - I) \mathbf{y} \cdot (A^T A - I) \mathbf{y} = 0$$

from which we can conclude that

$$(A^T A - I) \mathbf{y} = \mathbf{0} \tag{2}$$

(why?). Thus, (2) is a homogeneous system of linear equations that is satisfied by every $\mathbf{y}$ in R^n. But this implies that the coefficient matrix must be zero (why?), so $A^T A = I$ and, consequently, A is orthogonal. ∎

If $T: R^n \to R^n$ is multiplication by an orthogonal matrix A, then T is called an ***orthogonal operator*** on R^n. It follows from parts (*a*) and (*b*) of the preceding theorem that the orthogonal operators on R^n are precisely those operators that leave the lengths of all vectors unchanged. Since reflections and rotations of R^2 and R^3 have this property, this explains our observation in Example 2 that the standard matrices for the basic reflections and rotations of R^2 and R^3 are orthogonal.

COORDINATE MATRICES

Recall from Theorem 5.4.1 that if $S = \{\mathbf{v}_1, \mathbf{v}_2, \ldots, \mathbf{v}_n\}$ is a basis for a vector space V, then each vector $\mathbf{v}$ in V can be expressed uniquely as a linear combination of the basis vectors, say

$$\mathbf{v} = k_1 \mathbf{v}_1 + k_2 \mathbf{v}_2 + \cdots + k_n \mathbf{v}_n$$

The scalars $k_1, k_2, \ldots, k_n$ are the coordinates of $\mathbf{v}$ relative to S, and the vector

$$(\mathbf{v})_S = (k_1, k_2, \ldots, k_n)$$

is the coordinate vector of $\mathbf{v}$ relative to S. In this section it will be convenient to list the coordinates as entries of an $n \times 1$ matrix. Thus, we define

$$[\mathbf{v}]_S = \begin{bmatrix} k_1 \\ k_2 \\ \vdots \\ k_n \end{bmatrix}$$

to be the ***coordinate matrix*** of $\mathbf{v}$ relative to S.

CHANGE OF BASIS

In applications it is common to work with more than one coordinate system, and in such cases it is usually necessary to know the relationships between the coordinates of a fixed point or vector in the various coordinate systems. Since a basis is the vector space generalization of a coordinate system, we are led to consider the following problem.

Change of Basis Problem. If we change the basis for a vector space from some old basis B to some new basis B', how is the old coordinate matrix $[\mathbf{v}]_B$ of a vector $\mathbf{v}$ related to the new coordinate matrix $[\mathbf{v}]_{B'}$?

For simplicity, we will solve this problem for two-dimensional spaces. The solution for n-dimensional spaces is similar and is left for the reader. Let

$$B = \{\mathbf{u}_1, \mathbf{u}_2\} \quad \text{and} \quad B' = \{\mathbf{u}_1', \mathbf{u}_2'\}$$

be the old and new bases, respectively. We will need the coordinate matrices for the new basis vectors relative to the old basis. Suppose they are

$$[\mathbf{u}_1']_B = \begin{bmatrix} a \\ b \end{bmatrix} \quad \text{and} \quad [\mathbf{u}_2']_B = \begin{bmatrix} c \\ d \end{bmatrix} \tag{3}$$

That is,

$$\begin{aligned} \mathbf{u}_1' &= a\mathbf{u}_1 + b\mathbf{u}_2 \\ \mathbf{u}_2' &= c\mathbf{u}_1 + d\mathbf{u}_2 \end{aligned} \tag{4}$$

Now let $\mathbf{v}$ be any vector in V and let

$$[\mathbf{v}]_{B'} = \begin{bmatrix} k_1 \\ k_2 \end{bmatrix} \tag{5}$$

be the new coordinate matrix, so that

$$\mathbf{v} = k_1\mathbf{u}_1' + k_2\mathbf{u}_2' \tag{6}$$

In order to find the old coordinates of $\mathbf{v}$ we must express $\mathbf{v}$ in terms of the old basis B. To do this, we substitute (4) into (6). This yields

$$\mathbf{v} = k_1(a\mathbf{u}_1 + b\mathbf{u}_2) + k_2(c\mathbf{u}_1 + d\mathbf{u}_2)$$

or

$$\mathbf{v} = (k_1a + k_2c)\mathbf{u}_1 + (k_1b + k_2d)\mathbf{u}_2$$

Thus, the old coordinate matrix for $\mathbf{v}$ is

$$[\mathbf{v}]_B = \begin{bmatrix} k_1a + k_2c \\ k_1b + k_2d \end{bmatrix}$$

which can be written as

$$[\mathbf{v}]_B = \begin{bmatrix} a & c \\ b & d \end{bmatrix}\begin{bmatrix} k_1 \\ k_2 \end{bmatrix}$$

or, from (5),

$$[\mathbf{v}]_B = \begin{bmatrix} a & c \\ b & d \end{bmatrix}[\mathbf{v}]_{B'}$$

This equation states that the old coordinate matrix $[\mathbf{v}]_B$ results when we multiply the new coordinate matrix $[\mathbf{v}]_{B'}$ on the left by the matrix

$$P = \begin{bmatrix} a & c \\ b & d \end{bmatrix}$$

The columns of this matrix are the coordinates of the new basis vectors relative to the old basis [see (3)]. Thus, we have the following solution of the change of basis problem.

Solution of the Change of Basis Problem. If we change the basis for a vector space V from some old basis $B = \{\mathbf{u}_1, \mathbf{u}_2, \ldots, \mathbf{u}_n\}$ to some new basis $B' = \{\mathbf{u}_1', \mathbf{u}_2', \ldots, \mathbf{u}_n'\}$, then the old coordinate matrix $[\mathbf{v}]_B$ of a vector $\mathbf{v}$ is related to the new coordinate matrix $[\mathbf{v}]_{B'}$ of the same vector $\mathbf{v}$ by the equation

$$[\mathbf{v}]_B = P[\mathbf{v}]_{B'} \tag{7}$$

where the columns of P are the coordinate matrices of the new basis vectors relative to the old basis; that is, the column vectors of P are

$$[\mathbf{u}_1']_B, [\mathbf{u}_2']_B, \ldots, [\mathbf{u}_n']_B$$

TRANSITION MATRICES

The matrix P is called the ***transition matrix*** from B' to B; it can be expressed in terms of its column vectors as

$$P = \left[[\mathbf{u}_1']_B \;\vdots\; [\mathbf{u}_2']_B \;\vdots\; \cdots \;\vdots\; [\mathbf{u}_n']_B \right] \tag{8}$$

Example 4   Consider the bases $B = \{\mathbf{u}_1, \mathbf{u}_2\}$ and $B' = \{\mathbf{u}_1', \mathbf{u}_2'\}$ for R^2, where

$$\mathbf{u}_1 = (1, 0); \quad \mathbf{u}_2 = (0, 1); \quad \mathbf{u}_1' = (1, 1); \quad \mathbf{u}_2' = (2, 1)$$

(a)   Find the transition matrix from B' to B.
(b)   Use (7) to find $[\mathbf{v}]_B$ if

$$[\mathbf{v}]_{B'} = \begin{bmatrix} -3 \\ 5 \end{bmatrix}$$

Solution (a).   First we must find the coordinate matrices for the new basis vectors $\mathbf{u}_1'$ and $\mathbf{u}_2'$ relative to the old basis B. By inspection

$$\mathbf{u}_1' = \mathbf{u}_1 + \mathbf{u}_2$$
$$\mathbf{u}_2' = 2\mathbf{u}_1 + \mathbf{u}_2$$

so that

$$[\mathbf{u}_1']_B = \begin{bmatrix} 1 \\ 1 \end{bmatrix} \qquad \text{and} \qquad [\mathbf{u}_2']_B = \begin{bmatrix} 2 \\ 1 \end{bmatrix}$$

Thus, the transition matrix from B' to B is

$$P = \begin{bmatrix} 1 & 2 \\ 1 & 1 \end{bmatrix}$$

Solution (*b*). Using (7) and the transition matrix in part (a),

$$[\mathbf{v}]_B = \begin{bmatrix} 1 & 2 \\ 1 & 1 \end{bmatrix} \begin{bmatrix} -3 \\ 5 \end{bmatrix} = \begin{bmatrix} 7 \\ 2 \end{bmatrix}$$

As a check, we should be able to recover the vector $\mathbf{v}$ either from $[\mathbf{v}]_B$ or $[\mathbf{v}]_{B'}$. We leave it for the reader to show that $-3\mathbf{u}_1' + 5\mathbf{u}_2' = 7\mathbf{u}_1 + 2\mathbf{u}_2 = \mathbf{v} = (7, 2)$. ▲

Example 5 Consider the vectors $\mathbf{u}_1 = (1, 0)$, $\mathbf{u}_2 = (0, 1)$, $\mathbf{u}_1' = (1, 1)$, $\mathbf{u}_2' = (2, 1)$. In Example 4 we found the transition matrix from the basis $B' = \{\mathbf{u}_1', \mathbf{u}_2'\}$ for R^2 to the basis $B = \{\mathbf{u}_1, \mathbf{u}_2\}$. However, we can just as well ask for the transition matrix from B to B'. To obtain this matrix, we simply change our point of view and regard B' as the old basis and B as the new basis. As usual, the columns of the transition matrix will be the coordinates of the new basis vectors relative to the old basis.

By equating corresponding components and solving the resulting linear system, the reader should be able to show that

$$\begin{aligned} \mathbf{u}_1 &= -\mathbf{u}_1' + \mathbf{u}_2' \\ \mathbf{u}_2 &= 2\mathbf{u}_1' - \mathbf{u}_2' \end{aligned}$$

so that

$$[\mathbf{u}_1]_{B'} = \begin{bmatrix} -1 \\ 1 \end{bmatrix} \quad \text{and} \quad [\mathbf{u}_2]_{B'} = \begin{bmatrix} 2 \\ -1 \end{bmatrix}$$

Thus, the transition matrix from B to B' is

$$Q = \begin{bmatrix} -1 & 2 \\ 1 & -1 \end{bmatrix} \quad ▲$$

If we multiply the transition matrix from B' to B obtained in Example 4 and the transition matrix from B to B' obtained in Example 5, we find

$$PQ = \begin{bmatrix} 1 & 2 \\ 1 & 1 \end{bmatrix} \begin{bmatrix} -1 & 2 \\ 1 & -1 \end{bmatrix} = \begin{bmatrix} 1 & 0 \\ 0 & 1 \end{bmatrix} = I$$

which shows that $Q = P^{-1}$. The following theorem shows that this is not accidental.

Theorem 6.5.4. *If P is the transition matrix from a basis B' to a basis B, then:*

(*a*) *P is invertible.*

(*b*) *P^{-1} is the transition matrix from B to B'.*

Proof. Let Q be the transition matrix from B to B'. We shall show that $PQ = I$ and thus conclude that $Q = P^{-1}$ to complete the proof.

Assume that $B = \{\mathbf{u}_1, \mathbf{u}_2, \ldots, \mathbf{u}_n\}$ and suppose that

$$PQ = \begin{bmatrix} c_{11} & c_{12} & \cdots & c_{1n} \\ c_{21} & c_{22} & \cdots & c_{2n} \\ \vdots & \vdots & & \vdots \\ c_{n1} & c_{n2} & \cdots & c_{nn} \end{bmatrix}$$

From (7)

$$[\mathbf{x}]_B = P[\mathbf{x}]_{B'}$$

and

$$[\mathbf{x}]_{B'} = Q[\mathbf{x}]_B$$

for all $\mathbf{x}$ in V. Multiplying the bottom equation through on the left by P and substituting the top equation gives

$$[\mathbf{x}]_B = PQ[\mathbf{x}]_B \tag{9}$$

for all $\mathbf{x}$ in V. Letting $\mathbf{x} = \mathbf{u}_1$ in (9) gives

$$\begin{bmatrix} 1 \\ 0 \\ \vdots \\ 0 \end{bmatrix} = \begin{bmatrix} c_{11} & c_{12} & \cdots & c_{1n} \\ c_{21} & c_{22} & \cdots & c_{2n} \\ \vdots & \vdots & & \vdots \\ c_{n1} & c_{n2} & \cdots & c_{nn} \end{bmatrix} \begin{bmatrix} 1 \\ 0 \\ \vdots \\ 0 \end{bmatrix}$$

or

$$\begin{bmatrix} 1 \\ 0 \\ \vdots \\ 0 \end{bmatrix} = \begin{bmatrix} c_{11} \\ c_{21} \\ \vdots \\ c_{n1} \end{bmatrix}$$

Similarly, successively substituting $\mathbf{x} = \mathbf{u}_2, \ldots, \mathbf{u}_n$ in (9) yields

$$\begin{bmatrix} c_{12} \\ c_{22} \\ \vdots \\ c_{n2} \end{bmatrix} = \begin{bmatrix} 0 \\ 1 \\ \vdots \\ 0 \end{bmatrix}, \ldots, \begin{bmatrix} c_{1n} \\ c_{2n} \\ \vdots \\ c_{nn} \end{bmatrix} = \begin{bmatrix} 0 \\ 0 \\ \vdots \\ 1 \end{bmatrix}$$

Therefore, $PQ = I$.

To summarize, if P is the transition matrix from a basis B' to a basis B, then for every vector $\mathbf{v}$ the following relationships hold:

$$[\mathbf{v}]_B = P[\mathbf{v}]_{B'} \tag{10}$$

$$[\mathbf{v}]_{B'} = P^{-1}[\mathbf{v}]_B \tag{11}$$

CHANGE OF
ORTHONORMAL
BASIS

The following theorem shows that in an inner product space, the transition matrix from one orthonormal basis to another is orthogonal.

Theorem 6.5.5. *If P is the transition matrix from one orthonormal basis to another orthonormal basis for an inner product space, then P is an orthogonal matrix; that is,*

$$P^{-1} = P^T$$

Proof. Assume that V is an n-dimensional inner product space and that P is the transition matrix from an orthonormal basis B' to an orthonormal basis B. To prove that P is orthogonal we shall use Theorem 6.5.3b and show that $\|P\mathbf{x}\| = \|\mathbf{x}\|$ for every vector $\mathbf{x}$ in R^n.

Recall from Theorem 6.3.2a that for any orthonormal basis for V, the norm of any vector $\mathbf{u}$ in V is the same as the norm of its coordinate vector in R^n with respect to the Euclidean inner product. Thus, for any vector $\mathbf{u}$ in V we have

$$\|\mathbf{u}\| = \|[\mathbf{u}]_{B'}\| = \|[\mathbf{u}]_B\|$$

or from (10)

$$\|\mathbf{u}\| = \|[\mathbf{u}]_{B'}\| = \|P[\mathbf{u}]_{B'}\| \tag{12}$$

where the first norm is with respect to the inner product on V and the second and third are with respect to the Euclidean inner product on R^n.

Now let $\mathbf{x}$ be any vector in R^n, and let $\mathbf{u}$ be the vector in V whose coordinate matrix with respect to the basis B' is $\mathbf{x}$; that is, $[\mathbf{u}]_{B'} = \mathbf{x}$. Thus, from (12)

$$\|\mathbf{u}\| = \|\mathbf{x}\| = \|P\mathbf{x}\|$$

which proves that P is orthogonal. ▪

ROTATION OF
COORDINATE
AXES

Example 6 (*Application to Rotation of Coordinate Axes.*) In many problems a rectangular xy-coordinate system is given and a new $x'y'$-coordinate system is obtained by rotating the xy-system counterclockwise about the origin through an angle θ. When this is done, each point Q in the plane has two sets of coordinates: coordinates (x, y) relative to the xy-system and coordinates (x', y') relative to the $x'y'$-system (Figure 1a).

By introducing unit vectors $\mathbf{u}_1$ and $\mathbf{u}_2$ along the positive x and y axes and unit vectors $\mathbf{u}_1'$ and $\mathbf{u}_2'$ along the positive x' and y' axes, we can regard this rotation as a change from an old basis $B = \{\mathbf{u}_1, \mathbf{u}_2\}$ to a new basis $B' = \{\mathbf{u}_1', \mathbf{u}_2'\}$ (Figure 1b). Thus, the new coordinates (x', y') and the old coordinates (x, y) of a point Q will be related by

$$\begin{bmatrix} x' \\ y' \end{bmatrix} = P^{-1} \begin{bmatrix} x \\ y \end{bmatrix} \tag{13}$$

where P is the transition from B' to B. To find P we must determine the coordinate matrices of the new basis vectors $\mathbf{u}_1'$ and $\mathbf{u}_2'$ relative to the old basis. As indicated in

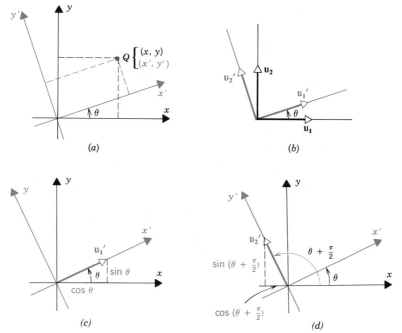

Figure 1

(a)

(b)

(c)

(d)

Figure 1c, the components of $\mathbf{u}_1'$ in the old basis are $\cos\theta$ and $\sin\theta$ so that

$$[\mathbf{u}_1']_B = \begin{bmatrix} \cos\theta \\ \sin\theta \end{bmatrix}$$

Similarly, from Figure 1d, we see that the components of $\mathbf{u}_2'$ in the old basis are $\cos(\theta + \pi/2) = -\sin\theta$ and $\sin(\theta + \pi/2) = \cos\theta$, so that

$$[\mathbf{u}_2']_B = \begin{bmatrix} -\sin\theta \\ \cos\theta \end{bmatrix}$$

Thus, the transition matrix from B' to B is

$$P = \begin{bmatrix} \cos\theta & -\sin\theta \\ \sin\theta & \cos\theta \end{bmatrix}$$

Observe that P is an orthogonal matrix, as expected, since B and B' are orthonormal bases. Thus,

$$P^{-1} = P^T = \begin{bmatrix} \cos\theta & \sin\theta \\ -\sin\theta & \cos\theta \end{bmatrix}$$

so (13) yields

$$\begin{bmatrix} x' \\ y' \end{bmatrix} = \begin{bmatrix} \cos\theta & \sin\theta \\ -\sin\theta & \cos\theta \end{bmatrix} \begin{bmatrix} x \\ y \end{bmatrix} \tag{14}$$

or equivalently,

$$x' = x \cos \theta + y \sin \theta$$
$$y' = -x \sin \theta + y \cos \theta$$

For example, if the axes are rotated $\theta = \pi/4$, then since

$$\sin \frac{\pi}{4} = \cos \frac{\pi}{4} = \frac{1}{\sqrt{2}}$$

Equation (14) becomes

$$\begin{bmatrix} x' \\ y' \end{bmatrix} = \begin{bmatrix} \dfrac{1}{\sqrt{2}} & \dfrac{1}{\sqrt{2}} \\ -\dfrac{1}{\sqrt{2}} & \dfrac{1}{\sqrt{2}} \end{bmatrix} \begin{bmatrix} x \\ y \end{bmatrix}$$

Thus, if the old coordinates of a point Q are $(x, y) = (2, -1)$, then

$$\begin{bmatrix} x' \\ y' \end{bmatrix} = \begin{bmatrix} \dfrac{1}{\sqrt{2}} & \dfrac{1}{\sqrt{2}} \\ -\dfrac{1}{\sqrt{2}} & \dfrac{1}{\sqrt{2}} \end{bmatrix} \begin{bmatrix} 2 \\ -1 \end{bmatrix} = \begin{bmatrix} \dfrac{1}{\sqrt{2}} \\ -\dfrac{3}{\sqrt{2}} \end{bmatrix}$$

so the new coordinates of Q are $(x', y') = (1/\sqrt{2}, -3/\sqrt{2})$.

REMARK. Observe that the coefficient matrix in (14) is the same as the standard matrix for the linear operator that rotates the vectors of R^2 through the angle $-\theta$ (Table 6 of Section 4.2). This is to be expected since rotating the coordinate axes through the angle θ with the vectors of R^2 kept fixed has the same effect as rotating the vectors through the angle $-\theta$ with the axes kept fixed.

Example 7 (*Application to Rotation of Axes in 3-space.*) Suppose that a rectangular xyz-coordinate system is rotated around its z-axis counterclockwise (looking down the positive z-axis) through an angle θ (Figure 2). If we introduce unit vectors $\mathbf{u}_1$, $\mathbf{u}_2$, and $\mathbf{u}_3$ along the positive x, y, and z axes and unit vectors $\mathbf{u}_1'$, $\mathbf{u}_2'$, and $\mathbf{u}_3'$ along the positive

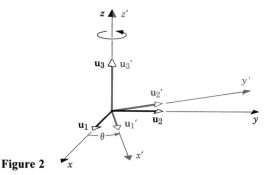

Figure 2

x', y', and z' axes, we can regard the rotation as a change from the old basis $B = \{\mathbf{u}_1, \mathbf{u}_2, \mathbf{u}_3\}$ to the new basis $B' = \{\mathbf{u}_1', \mathbf{u}_2', \mathbf{u}_3'\}$. In light of Example 6 it should be evident that

$$[\mathbf{u}_1']_B = \begin{bmatrix} \cos\theta \\ \sin\theta \\ 0 \end{bmatrix} \qquad \text{and} \qquad [\mathbf{u}_2']_B = \begin{bmatrix} -\sin\theta \\ \cos\theta \\ 0 \end{bmatrix}$$

Moreover, since $\mathbf{u}_3'$ extends 1 unit up the positive z'-axis,

$$[\mathbf{u}_3']_B = \begin{bmatrix} 0 \\ 0 \\ 1 \end{bmatrix}$$

Thus, the transition matrix from B' to B is

$$P = \begin{bmatrix} \cos\theta & -\sin\theta & 0 \\ \sin\theta & \cos\theta & 0 \\ 0 & 0 & 1 \end{bmatrix}$$

and the transition matrix from B to B' is

$$P^{-1} = \begin{bmatrix} \cos\theta & \sin\theta & 0 \\ -\sin\theta & \cos\theta & 0 \\ 0 & 0 & 1 \end{bmatrix}$$

(verify). Thus, the new coordinates (x', y', z') of a point Q can be computed from its old coordinates (x, y, z) by

$$\begin{bmatrix} x' \\ y' \\ z' \end{bmatrix} = \begin{bmatrix} \cos\theta & \sin\theta & 0 \\ -\sin\theta & \cos\theta & 0 \\ 0 & 0 & 1 \end{bmatrix} \begin{bmatrix} x \\ y \\ z \end{bmatrix}$$

EXERCISE SET 6.5

1. Show that

$$A = \begin{bmatrix} \frac{4}{5} & 0 & -\frac{3}{5} \\ -\frac{9}{25} & \frac{4}{5} & -\frac{12}{25} \\ \frac{12}{25} & \frac{3}{5} & \frac{16}{25} \end{bmatrix}$$

is an orthogonal matrix by
   (a) calculating $A^T A$
   (b) using part (b) of Theorem 6.5.1
   (c) using part (c) of Theorem 6.5.1

2. Find the inverse of the matrix in Exercise 1.

3. Determine which of the following matrices are orthogonal. For those that are orthogonal, find the inverse.

(a) $\begin{bmatrix} 1 & 0 \\ 0 & 1 \end{bmatrix}$ (b) $\begin{bmatrix} 1/\sqrt{2} & -1/\sqrt{2} \\ 1/\sqrt{2} & 1/\sqrt{2} \end{bmatrix}$ (c) $\begin{bmatrix} 0 & 1 & 1/\sqrt{2} \\ 1 & 0 & 0 \\ 0 & 0 & 1/\sqrt{2} \end{bmatrix}$

(d) $\begin{bmatrix} -1/\sqrt{2} & 1\sqrt{6} & 1/\sqrt{3} \\ 0 & -2/\sqrt{6} & 1/\sqrt{3} \\ 1/\sqrt{2} & 1/\sqrt{6} & 1/\sqrt{3} \end{bmatrix}$ (e) $\begin{bmatrix} \frac{1}{2} & \frac{1}{2} & \frac{1}{2} & \frac{1}{2} \\ \frac{1}{2} & -\frac{5}{6} & \frac{1}{6} & \frac{1}{6} \\ \frac{1}{2} & \frac{1}{6} & \frac{1}{6} & -\frac{5}{6} \\ \frac{1}{2} & \frac{1}{6} & -\frac{5}{6} & \frac{1}{6} \end{bmatrix}$ (f) $\begin{bmatrix} 1 & 0 & 0 & 0 \\ 0 & 1/\sqrt{3} & -1/2 & 0 \\ 0 & 1/\sqrt{3} & 0 & 1 \\ 0 & 1/\sqrt{3} & 1/2 & 0 \end{bmatrix}$

4. Verify that the rotation matrices and the reflection matrices in Tables 2 and 3 of Section 4.2 are orthogonal.

5. Find the coordinate matrix for **w** relative to the basis $S = \{\mathbf{u}_1, \mathbf{u}_2\}$ for R^2.
(a) $\mathbf{u}_1 = (1, 0)$, $\mathbf{u}_2 = (0, 1)$; $\mathbf{w} = (3, -7)$ (b) $\mathbf{u}_1 = (2, -4)$, $\mathbf{u}_2 = (3, 8)$; $\mathbf{w} = (1, 1)$
(c) $\mathbf{u}_1 = (1, 1)$, $\mathbf{u}_2 = (0, 2)$; $\mathbf{w} = (a, b)$

6. Find the coordinate matrix for **v** relative to $S = \{\mathbf{v}_1, \mathbf{v}_2, \mathbf{v}_3\}$.
(a) $\mathbf{v} = (2, -1, 3)$; $\mathbf{v}_1 = (1, 0, 0)$, $\mathbf{v}_2 = (2, 2, 0)$, $\mathbf{v}_3 = (3, 3, 3)$
(b) $\mathbf{v} = (5, -12, 3)$; $\mathbf{v}_1 = (1, 2, 3)$, $\mathbf{v}_2 = (-4, 5, 6)$, $\mathbf{v}_3 = (7, -8, 9)$

7. Find the coordinate matrix for **p** relative to $S = \{\mathbf{p}_1, \mathbf{p}_2, \mathbf{p}_3\}$.
(a) $\mathbf{p} = 4 - 3x + x^2$; $\mathbf{p}_1 = 1$, $\mathbf{p}_2 = x$, $\mathbf{p}_3 = x^2$
(b) $\mathbf{p} = 2 - x + x^2$; $\mathbf{p}_1 = 1 + x$, $\mathbf{p}_2 = 1 + x^2$, $\mathbf{p}_3 = x + x^2$

8. Find the coordinate matrix for A relative to $S = \{A_1, A_2, A_3, A_4\}$.

$$A = \begin{bmatrix} 2 & 0 \\ -1 & 3 \end{bmatrix}, \quad A_1 = \begin{bmatrix} -1 & 1 \\ 0 & 0 \end{bmatrix}, \quad A_2 = \begin{bmatrix} 1 & 1 \\ 0 & 0 \end{bmatrix}, \quad A_3 = \begin{bmatrix} 0 & 0 \\ 1 & 0 \end{bmatrix}, \quad A_4 = \begin{bmatrix} 0 & 0 \\ 0 & 1 \end{bmatrix}$$

9. Consider the coordinate matrices

$$[\mathbf{w}]_S = \begin{bmatrix} 6 \\ -1 \\ 4 \end{bmatrix}, \quad [\mathbf{q}]_S = \begin{bmatrix} 3 \\ 0 \\ 4 \end{bmatrix}, \quad [B]_S = \begin{bmatrix} -8 \\ 7 \\ 6 \\ 3 \end{bmatrix}$$

(a) Find **w** if S is the basis in Exercise 6(a).
(b) Find **q** if S is the basis in Exercise 7(a).
(c) Find B if S is the basis in Exercise 8.

10. Consider the bases $B = \{\mathbf{u}_1, \mathbf{u}_2\}$ and $B' = \{\mathbf{v}_1, \mathbf{v}_2\}$ for R^2, where

$$\mathbf{u}_1 = \begin{bmatrix} 1 \\ 0 \end{bmatrix}, \quad \mathbf{u}_2 = \begin{bmatrix} 0 \\ 1 \end{bmatrix}, \quad \mathbf{v}_1 = \begin{bmatrix} 2 \\ 1 \end{bmatrix}, \quad \text{and} \quad \mathbf{v}_2 = \begin{bmatrix} -3 \\ 4 \end{bmatrix}$$

(a) Find the transition matrix from B' to B.
(b) Find the transition matrix from B to B'.
(c) Compute the coordinate matrix $[\mathbf{w}]_B$, where

$$\mathbf{w} = \begin{bmatrix} 3 \\ -5 \end{bmatrix}$$

and use (11) to compute $[\mathbf{w}]_{B'}$.
(d) Check your work by computing $[\mathbf{w}]_{B'}$ directly.

11. Repeat the directions of Exercise 10 with

$$\mathbf{u}_1 = \begin{bmatrix} 2 \\ 2 \end{bmatrix}, \quad \mathbf{u}_2 = \begin{bmatrix} 4 \\ -1 \end{bmatrix}, \quad \mathbf{v}_1 = \begin{bmatrix} 1 \\ 3 \end{bmatrix}, \quad \mathbf{v}_2 = \begin{bmatrix} -1 \\ -1 \end{bmatrix}$$

12. Consider the bases $B = \{\mathbf{u}_1, \mathbf{u}_2, \mathbf{u}_3\}$ and $B' = \{\mathbf{v}_1, \mathbf{v}_2, \mathbf{v}_3\}$ for R^3, where

$$\mathbf{u}_1 = \begin{bmatrix} -3 \\ 0 \\ -3 \end{bmatrix}, \quad \mathbf{u}_2 = \begin{bmatrix} -3 \\ 2 \\ -1 \end{bmatrix}, \quad \mathbf{u}_3 = \begin{bmatrix} 1 \\ 6 \\ -1 \end{bmatrix}, \quad \mathbf{v}_1 = \begin{bmatrix} -6 \\ -6 \\ 0 \end{bmatrix}, \quad \mathbf{v}_2 = \begin{bmatrix} -2 \\ -6 \\ 4 \end{bmatrix}, \quad \mathbf{v}_3 = \begin{bmatrix} -2 \\ -3 \\ 7 \end{bmatrix}$$

(a) Find the transition matrix from B to B'.

(b) Compute the coordinate matrix $[\mathbf{w}]_B$, where

$$\mathbf{w} = \begin{bmatrix} -5 \\ 8 \\ -5 \end{bmatrix}$$

and use (11) to compute $[\mathbf{w}]_{B'}$.

(c) Check your work by computing $[\mathbf{w}]_{B'}$ directly.

13. Repeat the directions of Exercise 12 with the same vector $\mathbf{w}$, but with

$$\mathbf{u}_1 = \begin{bmatrix} 2 \\ 1 \\ 1 \end{bmatrix}, \quad \mathbf{u}_2 = \begin{bmatrix} 2 \\ -1 \\ 1 \end{bmatrix}, \quad \mathbf{u}_3 = \begin{bmatrix} 1 \\ 2 \\ 1 \end{bmatrix}, \quad \mathbf{v}_1 = \begin{bmatrix} 3 \\ 1 \\ -5 \end{bmatrix}, \quad \mathbf{v}_2 = \begin{bmatrix} 1 \\ 1 \\ -3 \end{bmatrix}, \quad \mathbf{v}_3 = \begin{bmatrix} -1 \\ 0 \\ 2 \end{bmatrix}$$

14. Consider the bases $B = \{\mathbf{p}_1, \mathbf{p}_2\}$ and $B' = \{\mathbf{q}_1, \mathbf{q}_2\}$ for P_1, where

$$\mathbf{p}_1 = 6 + 3x, \quad \mathbf{p}_2 = 10 + 2x, \quad \mathbf{q}_1 = 2, \quad \mathbf{q}_2 = 3 + 2x$$

(a) Find the transition matrix from B' to B.

(b) Find the transition matrix from B to B'.

(c) Compute the coordinate matrix $[\mathbf{p}]_B$, where $\mathbf{p} = -4 + x$, and use (11) to compute $[\mathbf{p}]_{B'}$.

(d) Check your work by computing $[\mathbf{p}]_{B'}$ directly.

15. Let V be the space spanned by $\mathbf{f}_1 = \sin x$ and $\mathbf{f}_2 = \cos x$.

(a) Show that $\mathbf{g}_1 = 2 \sin x + \cos x$ and $\mathbf{g}_2 = 3 \cos x$ form a basis for V.

(b) Find the transition matrix from $B' = \{\mathbf{g}_1, \mathbf{g}_2\}$ to $B = \{\mathbf{f}_1, \mathbf{f}_2\}$.

(c) Find the transition matrix from B to B'.

(d) Compute the coordinate matrix $[\mathbf{h}]_B$, where $\mathbf{h} = 2 \sin x - 5 \cos x$, and use (11) to obtain $[\mathbf{h}]_{B'}$.

(e) Check your work by computing $[\mathbf{h}]_{B'}$ directly.

16. Let a rectangular $x'y'$-coordinate system be obtained by rotating a rectangular xy-coordinate system counterclockwise through the angle $\theta = 3\pi/4$.

(a) Find the $x'y'$-coordinates of the point whose xy-coordinates are $(-2, 6)$.

(b) Find the xy-coordinates of the point whose $x'y'$-coordinates are $(5, 2)$.

17. Repeat Exercise 16 with $\theta = \pi/3$.

18. Let a rectangular $x'y'z'$-coordinate system be obtained by rotating a rectangular xyz-coordinate system counterclockwise about the z-axis (looking down the z-axis) through the angle $\theta = \pi/4$.

(a) Find the $x'y'z'$-coordinates of the point whose xyz-coordinates are $(-1, 2, 5)$.

(b) Find the xyz-coordinates of the point whose $x'y'z'$-coordinates are $(1, 6, -3)$.

19. Repeat Exercise 18 for a rotation of $\theta = \pi/3$ counterclockwise about the y-axis (looking along the positive y-axis toward the origin).

20. Repeat Exercise 18 for a rotation of $\theta = 3\pi/4$ counterclockwise about the x-axis (looking along the positive x-axis toward the origin).

21. (a) A rectangular $x'y'z'$-coordinate system is obtained by rotating an xyz-coordinate system counterclockwise about the y-axis through an angle θ (looking along the positive y-axis toward the origin). Find a matrix A such that

$$\begin{bmatrix} x' \\ y' \\ z' \end{bmatrix} = A \begin{bmatrix} x \\ y \\ z \end{bmatrix}$$

where (x, y, z) and (x', y', z') are the coordinates of the same point in the xyz- and $x'y'z'$-systems, respectively.
   (b) Repeat part (a) for a rotation about the x-axis.

22. A rectangular $x''y''z''$-coordinate system is obtained by first rotating a rectangular xyz-coordinate system $60°$ counterclockwise about the z-axis (looking down the positive z-axis) to obtain an $x'y'z'$-coordinate system, and then rotating the $x'y'z'$-coordinate system $45°$ counterclockwise about the y'-axis (looking along the positive y'-axis toward the origin). Find a matrix A such that

$$\begin{bmatrix} x'' \\ y'' \\ z'' \end{bmatrix} = A \begin{bmatrix} x \\ y \\ z \end{bmatrix}$$

where (x, y, z) and (x'', y'', z'') are the xyz- and $x''y''z''$-coordinates of the same point.

23. What conditions must a and b satisfy for the matrix

$$\begin{bmatrix} a+b & b-a \\ a-b & b+a \end{bmatrix}$$

to be orthogonal?

24. Prove that a 2×2 orthogonal matrix A has one of two possible forms:

$$A = \begin{bmatrix} \cos\theta & -\sin\theta \\ \sin\theta & \cos\theta \end{bmatrix} \quad \text{or} \quad A = \begin{bmatrix} \cos\theta & -\sin\theta \\ -\sin\theta & -\cos\theta \end{bmatrix}$$

where $0 \le \theta < 2\pi$. [**Hint.** Start with a general 2×2 matrix $A = (a_{ij})$, and use the fact that the column vectors form an orthonormal set in R^2.]

25. (a) Use the result in Exercise 24 to prove that multiplication by a 2×2 orthogonal matrix is either a rotation or a rotation followed by a reflection about the x-axis.
   (b) Show that multiplication by A is a rotation if $\det(A) = 1$ and a rotation followed by a reflection if $\det(A) = -1$.

26. Use the result in Exercise 25 to determine whether multiplication by A is a rotation or a rotation followed by a reflection. Find the angle of rotation in either case.

   (a) $A = \begin{bmatrix} -1/\sqrt{2} & 1/\sqrt{2} \\ -1/\sqrt{2} & -1/\sqrt{2} \end{bmatrix}$ (b) $A = \begin{bmatrix} \sqrt{3}/2 & -1/2 \\ 1/2 & \sqrt{3}/2 \end{bmatrix}$

27. The result in Exercise 25 has an analog for 3×3 orthogonal matrices: It can be proved that multiplication by a 3×3 orthogonal matrix A is a rotation about some axis if $\det(A) = 1$ and is a rotation about some axis followed by a reflection about some coordinate plane if

$\det(A) = -1$. Determine whether multiplication by A is a rotation or a rotation followed by a reflection.

(a) $A = \begin{bmatrix} \frac{3}{7} & \frac{2}{7} & \frac{6}{7} \\ -\frac{6}{7} & \frac{3}{7} & \frac{2}{7} \\ \frac{2}{7} & \frac{6}{7} & -\frac{3}{7} \end{bmatrix}$ (b) $A = \begin{bmatrix} \frac{2}{7} & \frac{3}{7} & \frac{6}{7} \\ \frac{3}{7} & -\frac{6}{7} & \frac{2}{7} \\ \frac{6}{7} & \frac{2}{7} & -\frac{3}{7} \end{bmatrix}$

28. Use the result in Exercise 27 and part (b) of Theorem 6.5.2 to show that a composition of rotations can always be accomplished by a single rotation about some appropriate axis.

29. Prove the equivalence of statements (a) and (c) in Theorem 6.5.1.

SUPPLEMENTARY EXERCISES

1. Let R^4 have the Euclidean inner product.
   (a) Find a vector in R^4 that is orthogonal to $\mathbf{u}_1 = (1, 0, 0, 0)$ and $\mathbf{u}_4 = (0, 0, 0, 1)$ and makes equal angles with $\mathbf{u}_2 = (0, 1, 0, 0)$ and $\mathbf{u}_3 = (0, 0, 1, 0)$.
   (b) Find a vector $\mathbf{x} = (x_1, x_2, x_3, x_4)$ of length 1 that is orthogonal to $\mathbf{u}_1$ and $\mathbf{u}_4$ above and such that the cosine of the angle between $\mathbf{x}$ and $\mathbf{u}_2$ is twice the cosine of the angle between $\mathbf{x}$ and $\mathbf{u}_3$.

2. Show that if $\mathbf{x}$ is a nonzero vector in R^n, then the $n \times n$ matrix

$$A = I_n - \frac{2}{\|\mathbf{x}\|^2} \mathbf{x}\mathbf{x}^T$$

   is both orthogonal and symmetric.

3. Let $A\mathbf{x} = \mathbf{0}$ be a system of m equations in n unknowns. Show that

$$\mathbf{x} = \begin{bmatrix} x_1 \\ x_2 \\ \vdots \\ x_n \end{bmatrix}$$

   is a solution of the system if and only if the vector $\mathbf{x} = (x_1, x_2, \ldots, x_n)$ is orthogonal to every row vector of A in the Euclidean inner product on R^n.

4. Use the Cauchy–Schwarz inequality to show that if $a_1, a_2, \ldots, a_n$ are positive real numbers, then

$$(a_1 + a_2 + \cdots + a_n)\left(\frac{1}{a_1} + \frac{1}{a_2} + \cdots + \frac{1}{a_n}\right) \geq n^2$$

5. Show that if $\mathbf{x}$ and $\mathbf{y}$ are vectors in an inner product space and c is any scalar, then

$$\|c\mathbf{x} + \mathbf{y}\|^2 = c^2\|\mathbf{x}\|^2 + 2c\langle \mathbf{x}, \mathbf{y}\rangle + \|\mathbf{y}\|^2$$

6. Let R^3 have the Euclidean inner product. Find two vectors of length 1 that are orthogonal to all three of the vectors $\mathbf{u}_1 = (1, 1, -1)$, $\mathbf{u}_2 = (-2, -1, 2)$, and $\mathbf{u}_3 = (-1, 0, 1)$.

7. Find a weighted Euclidean inner product on R^n such that the vectors

$$\mathbf{v}_1 = (1, 0, 0, \ldots, 0)$$
$$\mathbf{v}_2 = (0, \sqrt{2}, 0, \ldots, 0)$$
$$\mathbf{v}_3 = (0, 0, \sqrt{3}, \ldots, 0)$$
$$\vdots$$
$$\mathbf{v}_n = (0, 0, 0, \ldots, \sqrt{n})$$

form an orthonormal set.

8. Is there a weighted Euclidean inner product on R^2 for which the vectors $(1, 2)$ and $(3, -1)$ form an orthonormal set? Justify your answer.

9. Prove: If Q is an orthogonal matrix, then each entry of Q is the same as its cofactor if $\det(Q) = 1$ and is the negative of its cofactor if $\det(Q) = -1$.

10. If $\mathbf{u}$ and $\mathbf{v}$ are vectors in an inner product space V, then $\mathbf{u}$, $\mathbf{v}$, and $\mathbf{u} - \mathbf{v}$ can be regarded as sides of a "triangle" in V (Figure 1). Prove that the law of cosines holds for any such triangle; that is, $\|\mathbf{u} - \mathbf{v}\|^2 = \|\mathbf{u}\|^2 + \|\mathbf{v}\|^2 - 2\|\mathbf{u}\|\,\|\mathbf{v}\|\cos\theta$, where θ is the angle between $\mathbf{u}$ and $\mathbf{v}$.

Figure 1

11. (a) In R^3 the vectors $(k, 0, 0)$, $(0, k, 0)$, and $(0, 0, k)$ form the edges of a cube with diagonal (k, k, k) (Figure 4 of Section 3.3). Similarly, in R^n the vectors

$$(k, 0, 0, \ldots, 0), \quad (0, k, 0, \ldots, 0), \quad \ldots, \quad (0, 0, 0, \ldots, k)$$

can be regarded as edges of a "cube" with diagonal $(k, k, k, \ldots, k)$. Show that each of the above edges makes an angle of θ with the diagonal, where $\cos\theta = 1/\sqrt{n}$.

(b) **(For readers who have studied calculus.)** What happens to the angle θ in part (a) as the dimension of R^n approaches $+\infty$?

12. Let $\mathbf{u}$ and $\mathbf{v}$ be vectors in an inner product space.
(a) Prove that $\|\mathbf{u}\| = \|\mathbf{v}\|$ if and only if $\mathbf{u} + \mathbf{v}$ and $\mathbf{u} - \mathbf{v}$ are orthogonal.
(b) Give a geometric interpretation of this result in R^2 with the Euclidean inner product.

13. Let $\mathbf{u}$ be a vector in an inner product space V, and let $\{\mathbf{v}_1, \mathbf{v}_2, \ldots, \mathbf{v}_n\}$ be an orthonormal basis for V. Show that if α_i is the angle between $\mathbf{u}$ and $\mathbf{v}_i$, then

$$\cos^2\alpha_1 + \cos^2\alpha_2 + \cdots + \cos^2\alpha_n = 1$$

14. Prove: If $\langle \mathbf{u}, \mathbf{v} \rangle_1$ and $\langle \mathbf{u}, \mathbf{v} \rangle_2$ are two inner products on a vector space V, then the quantity $\langle \mathbf{u}, \mathbf{v} \rangle = \langle \mathbf{u}, \mathbf{v} \rangle_1 + \langle \mathbf{u}, \mathbf{v} \rangle_2$ is also an inner product.

15. Show that the inner product on R^n generated by any orthogonal matrix is the Euclidean inner product.

16. Find a, b, and c such that the matrix

$$A = \begin{bmatrix} a & 1/\sqrt{2} & -1/\sqrt{2} \\ b & 1/\sqrt{6} & 1/\sqrt{6} \\ c & 1/\sqrt{3} & 1/\sqrt{3} \end{bmatrix}$$

is orthogonal. Are the values of a, b, and c unique? Explain.

17. Prove part (c) of Theorem 6.2.5.

CHAPTER **7**

EIGENVALUES, EIGENVECTORS

7.1 EIGENVALUES AND EIGENVECTORS

*If A is an $n \times n$ matrix and **x** is a vector in R^n, then there is usually no general geometric relationship between the vector **x** and the vector A**x** (Figure 1a). However, there are often certain nonzero vectors **x** such that **x** and A**x** are scalar multiples of one another (Figure 1b). Such vectors arise naturally in the study of vibrations, electrical systems, genetics, chemical reactions, quantum mechanics, mechanical stress, economics, and geometry. In this section we shall show how to find these vectors, and in later sections we shall touch on some of their applications.*

Figure 1   (*a*)   (*b*)

REVIEW OF EIGENVECTORS AND EIGENVALUES

We begin with a review of some concepts developed in Sections 2.3 and 4.3.

Definition. If A is an $n \times n$ matrix, then a nonzero vector **x** in R^n is called an *eigenvector* of A if A**x** is a scalar multiple of **x**; that is,

$$A\mathbf{x} = \lambda\mathbf{x}$$

for some scalar λ. The scalar λ is called an *eigenvalue* of A, and **x** is said to be an eigenvector of A *corresponding* to λ.

In R^2 and R^3 multiplication by A maps each eigenvector $\mathbf{x}$ of A (if any) onto the same line through the origin as $\mathbf{x}$. Depending on the sign and the magnitude of the eigenvalue λ corresponding to $\mathbf{x}$, the linear operator $A\mathbf{x} = \lambda\mathbf{x}$ compresses or stretches $\mathbf{x}$ by a factor of λ, with a reversal of direction in the case where λ is negative (Figure 2).

Figure 2

| $0 \leq \lambda \leq 1$ | $\lambda \geq 1$ | $-1 \leq \lambda \leq 0$ | $\lambda \leq -1$ |

Example 1 The vector $\mathbf{x} = \begin{bmatrix} 1 \\ 2 \end{bmatrix}$ is an eigenvector of

$$A = \begin{bmatrix} 3 & 0 \\ 8 & -1 \end{bmatrix}$$

corresponding to the eigenvalue $\lambda = 3$, since

$$A\mathbf{x} = \begin{bmatrix} 3 & 0 \\ 8 & -1 \end{bmatrix}\begin{bmatrix} 1 \\ 2 \end{bmatrix} = \begin{bmatrix} 3 \\ 6 \end{bmatrix} = 3\mathbf{x} \quad \blacktriangle$$

To find the eigenvalues of an $n \times n$ matrix A we rewrite $A\mathbf{x} = \lambda\mathbf{x}$ as

$$A\mathbf{x} = \lambda I\mathbf{x}$$

or equivalently,

$$(\lambda I - A)\mathbf{x} = \mathbf{0} \tag{1}$$

For λ to be an eigenvalue, there must be a nonzero solution of this equation. However, by Theorem 6.2.7, Equation (1) has a nonzero solution if and only if

$$\det(\lambda I - A) = 0$$

This is called the ***characteristic equation*** of A; the scalars satisfying this equation are the eigenvalues of A. When expanded, the determinant $\det(\lambda I - A)$ is a polynomial in λ called the ***characteristic polynomial*** of A.

It can be shown (Exercise 15) that if A is an $n \times n$ matrix, then the characteristic polynomial of A has degree n and the coefficient of λ^n is 1; that is, the characteristic polynomial of an $n \times n$ matrix has the form

$$\det(\lambda I - A) = \lambda^n + c_1\lambda^{n-1} + \cdots + c_n$$

It follows from the Fundamental Theorem of Algebra that the characteristic equation

$$\lambda^n + c_1\lambda^{n-1} + \cdots + c_n = 0$$

has at most n distinct solutions, so an $n \times n$ matrix has at most n distinct eigenvalues.

The reader may wish to review Example 6 of Section 2.3, where we found the eigenvalues of a 2×2 matrix by solving the characteristic equation. The following example involves a 3×3 matrix.

Example 2 Find the eigenvalues of

$$A = \begin{bmatrix} 0 & 1 & 0 \\ 0 & 0 & 1 \\ 4 & -17 & 8 \end{bmatrix}$$

Solution. The characteristic polynomial of A is

$$\det(\lambda I - A) = \det \begin{bmatrix} \lambda & -1 & 0 \\ 0 & \lambda & -1 \\ -4 & 17 & \lambda - 8 \end{bmatrix} = \lambda^3 - 8\lambda^2 + 17\lambda - 4$$

The eigenvalues of A must therefore satisfy the cubic equation

$$\lambda^3 - 8\lambda^2 + 17\lambda - 4 = 0 \tag{2}$$

To solve this equation, we shall begin by searching for integer solutions. This task can be greatly simplified by exploiting the fact that all integer solutions (if there are any) to a polynomial equation with integer coefficients

$$\lambda^n + c_1 \lambda^{n-1} + \cdots + c_n = 0$$

must be divisors of the constant term, c_n. Thus, the only possible integer solutions of (2) are the divisors of -4, that is, ± 1, ± 2, ± 4. Successively substituting these values in (2) shows that $\lambda = 4$ is an integer solution. As a consequence, $\lambda - 4$ must be a factor of the left side of (2). Dividing $\lambda - 4$ into $\lambda^3 - 8\lambda^2 + 17\lambda - 4$ shows that (2) can be rewritten as

$$(\lambda - 4)(\lambda^2 - 4\lambda + 1) = 0$$

Thus, the remaining solutions of (2) satisfy the quadratic equation

$$\lambda^2 - 4\lambda + 1 = 0$$

which can be solved by the quadratic formula. Thus, the eigenvalues of A are

$$\lambda = 4, \quad \lambda = 2 + \sqrt{3}, \quad \text{and} \quad \lambda = 2 - \sqrt{3} \ \triangle$$

EIGENVALUES OF TRIANGULAR MATRICES

Example 3 Find the eigenvalues of the upper triangular matrix

$$A = \begin{bmatrix} a_{11} & a_{12} & a_{13} & a_{14} \\ 0 & a_{22} & a_{23} & a_{24} \\ 0 & 0 & a_{33} & a_{34} \\ 0 & 0 & 0 & a_{44} \end{bmatrix}$$

Solution. Recalling that the determinant of a triangular matrix is the product of the entries on the main diagonal (Theorem 2.2.2), we obtain

$$\det(\lambda I - A) = \det \begin{bmatrix} \lambda - a_{11} & -a_{12} & -a_{13} & -a_{14} \\ 0 & \lambda - a_{22} & -a_{23} & -a_{24} \\ 0 & 0 & \lambda - a_{33} & -a_{34} \\ 0 & 0 & 0 & \lambda - a_{44} \end{bmatrix}$$

$$= (\lambda - a_{11})(\lambda - a_{22})(\lambda - a_{33})(\lambda - a_{44})$$

Thus, the characteristic equation is

$$(\lambda - a_{11})(\lambda - a_{22})(\lambda - a_{33})(\lambda - a_{44}) = 0$$

and the eigenvalues are

$$\lambda = a_{11}, \qquad \lambda = a_{22}, \qquad \lambda = a_{33}, \qquad \lambda = a_{44}$$

which are precisely the diagonal entries of A. ▲

The following general theorem should be evident from the computations in the preceding example.

> **Theorem 7.1.1.** *If A is an $n \times n$ triangular matrix (upper triangular, lower triangular, or diagonal), then the eigenvalues of A are the entries on the main diagonal of A.*

Example 4 By inspection, the eigenvalues of the lower triangular matrix

$$A = \begin{bmatrix} \frac{1}{2} & 0 & 0 \\ -1 & \frac{2}{3} & 0 \\ 5 & -8 & -\frac{1}{4} \end{bmatrix}$$

are $\lambda = \frac{1}{2}$, $\lambda = \frac{2}{3}$, and $\lambda = -\frac{1}{4}$. ▲

REMARK. In practical problems, the matrix A is often so large that computing the characteristic equation is not practical. As a result, various approximation methods are used to obtain eigenvalues.

COMPLEX EIGENVALUES

It is possible for the characteristic equation of a matrix with real entries to have complex solutions. For example, the characteristic polynomial of the matrix

$$A = \begin{bmatrix} -2 & -1 \\ 5 & 2 \end{bmatrix}$$

is

$$\det(\lambda I - A) = \det \begin{bmatrix} \lambda + 2 & 1 \\ -5 & \lambda - 2 \end{bmatrix} = \lambda^2 + 1$$

so the characteristic equation is $\lambda^2 + 1 = 0$, the solutions of which are the imaginary numbers $\lambda = i$ and $\lambda = -i$. Thus, we are forced to consider complex eigenvalues, even for real matrices. This, in turn, leads us to consider the possibility of complex vector spaces, that is, vector spaces in which scalars are allowed to have complex values. Such vector spaces will be considered in Chapter 10. For now, we will allow complex eigenvalues, but our discussion of eigenvectors will be limited to matrices with real eigenvalues.

The following theorem summarizes our discussion thus far.

Theorem 7.1.2. *If A is an $n \times n$ matrix and λ is a real number, then the following are equivalent.*

(a) *λ is an eigenvalue of A.*
(b) *The system of equations $(\lambda I - A)\mathbf{x} = \mathbf{0}$ has nontrivial solutions.*
(c) *There is a nonzero vector $\mathbf{x}$ in R^n such that $A\mathbf{x} = \lambda\mathbf{x}$.*
(d) *λ is a solution of the characteristic equation $\det(\lambda I - A) = 0$.*

FINDING BASES FOR EIGENSPACES

Now that we know how to find eigenvalues, we turn to the problem of finding eigenvectors. The eigenvectors of A corresponding to an eigenvalue λ are the nonzero vectors $\mathbf{x}$ that satisfy $A\mathbf{x} = \lambda\mathbf{x}$. Equivalently, the eigenvectors corresponding to λ are the nonzero vectors in the solution space of $(\lambda I - A)\mathbf{x} = \mathbf{0}$. We call this solution space the ***eigenspace*** of A corresponding to λ.

Example 5 Find bases for the eigenspaces of

$$A = \begin{bmatrix} 0 & 0 & -2 \\ 1 & 2 & 1 \\ 1 & 0 & 3 \end{bmatrix}$$

Solution. The characteristic equation of A is $\lambda^3 - 5\lambda^2 + 8\lambda - 4 = 0$, or in factored form, $(\lambda - 1)(\lambda - 2)^2 = 0$ (verify); thus, the eigenvalues of A are $\lambda = 1$ and $\lambda = 2$, so there are two eigenspaces of A.

By definition,

$$\mathbf{x} = \begin{bmatrix} x_1 \\ x_2 \\ x_3 \end{bmatrix}$$

is an eigenvector of A corresponding to λ if and only if $\mathbf{x}$ is a nontrivial solution of $(\lambda I - A)\mathbf{x} = \mathbf{0}$, that is, of

$$\begin{bmatrix} \lambda & 0 & 2 \\ -1 & \lambda - 2 & -1 \\ -1 & 0 & \lambda - 3 \end{bmatrix} \begin{bmatrix} x_1 \\ x_2 \\ x_3 \end{bmatrix} = \begin{bmatrix} 0 \\ 0 \\ 0 \end{bmatrix} \tag{3}$$

If $\lambda = 2$, then (3) becomes

$$\begin{bmatrix} 2 & 0 & 2 \\ -1 & 0 & -1 \\ -1 & 0 & -1 \end{bmatrix} \begin{bmatrix} x_1 \\ x_2 \\ x_3 \end{bmatrix} = \begin{bmatrix} 0 \\ 0 \\ 0 \end{bmatrix}$$

Solving this system yields (verify)

$$x_1 = -s, \qquad x_2 = t, \qquad x_3 = s$$

Thus, the eigenvectors of A corresponding to $\lambda = 2$ are the nonzero vectors of the form

$$\mathbf{x} = \begin{bmatrix} -s \\ t \\ s \end{bmatrix} = \begin{bmatrix} -s \\ 0 \\ s \end{bmatrix} + \begin{bmatrix} 0 \\ t \\ 0 \end{bmatrix} = s \begin{bmatrix} -1 \\ 0 \\ 1 \end{bmatrix} + t \begin{bmatrix} 0 \\ 1 \\ 0 \end{bmatrix}$$

Since

$$\begin{bmatrix} -1 \\ 0 \\ 1 \end{bmatrix} \quad \text{and} \quad \begin{bmatrix} 0 \\ 1 \\ 0 \end{bmatrix}$$

are linearly independent, these vectors form a basis for the eigenspace corresponding to $\lambda = 2$.

If $\lambda = 1$, then (3) becomes

$$\begin{bmatrix} 1 & 0 & 2 \\ -1 & -1 & -1 \\ -1 & 0 & -2 \end{bmatrix} \begin{bmatrix} x_1 \\ x_2 \\ x_3 \end{bmatrix} = \begin{bmatrix} 0 \\ 0 \\ 0 \end{bmatrix}$$

Solving this system yields (verify)

$$x_1 = -2s, \qquad x_2 = s, \qquad x_3 = s$$

Thus, the eigenvectors corresponding to $\lambda = 1$ are the nonzero vectors of the form

$$\begin{bmatrix} -2s \\ s \\ s \end{bmatrix} = s \begin{bmatrix} -2 \\ 1 \\ 1 \end{bmatrix}$$

so that

$$\begin{bmatrix} -2 \\ 1 \\ 1 \end{bmatrix}$$

is a basis for the eigenspace corresponding to $\lambda = 1$. ▲

EIGENVALUES OF THE POWERS OF A MATRIX

Once the eigenvalues and eigenvectors of a matrix A are found, it is a simple matter to find the eigenvalues and eigenvectors of any positive integer power of A; for example, if λ is an eigenvalue of A and $\mathbf{x}$ is a corresponding eigenvector, then

$$A^2\mathbf{x} = A(A\mathbf{x}) = A(\lambda\mathbf{x}) = \lambda(A\mathbf{x}) = \lambda(\lambda\mathbf{x}) = \lambda^2\mathbf{x}$$

which shows that λ^2 is an eigenvalue of A^2 and $\mathbf{x}$ is a corresponding eigenvector. In general, we have the following result.

> **Theorem 7.1.3.** *If k is a positive integer, λ is an eigenvalue of a matrix A, and $\mathbf{x}$ is a corresponding eigenvector, then λ^k is an eigenvalue of A^k and $\mathbf{x}$ is a corresponding eigenvector.*

Example 6 In Example 5 we showed that the eigenvalues of

$$A = \begin{bmatrix} 0 & 0 & -2 \\ 1 & 2 & 1 \\ 1 & 0 & 3 \end{bmatrix}$$

are $\lambda = 2$ and $\lambda = 1$, so from Theorem 7.1.3 both $\lambda = 2^7 = 128$ and $\lambda = 1^7 = 1$ are eigenvalues of A^7. We also showed that

$$\begin{bmatrix} -1 \\ 0 \\ 1 \end{bmatrix} \quad \text{and} \quad \begin{bmatrix} 0 \\ 1 \\ 0 \end{bmatrix}$$

are eigenvectors of A corresponding to the eigenvalue $\lambda = 2$, so from Theorem 7.1.3 they are also eigenvectors of A^7 corresponding to $\lambda = 2^7 = 128$. Similarly, the eigenvector

$$\begin{bmatrix} -2 \\ 1 \\ 1 \end{bmatrix}$$

of A corresponding to the eigenvalue $\lambda = 1$ is also an eigenvector of A^7 corresponding to $\lambda = 1^7 = 1$. ▲

EIGENVALUES AND INVERTIBILITY

The next theorem establishes a relationship between the eigenvalues and the invertibility of a matrix.

> **Theorem 7.1.4.** *A square matrix A is invertible if and only if $\lambda = 0$ is not an eigenvalue of A.*

Proof. Assume that A is an $n \times n$ matrix and observe first that $\lambda = 0$ is a solution of the characteristic equation

$$\lambda^n + c_1 \lambda^{n-1} + \cdots + c_n = 0$$

if and only if the constant term c_n is zero. Thus, it suffices to prove that A is invertible if and only if $c_n \neq 0$. But

$$\det(\lambda I - A) = \lambda^n + c_1 \lambda^{n-1} + \cdots + c_n$$

or on setting $\lambda = 0$,

$$\det(-A) = c_n \quad \text{or} \quad (-1)^n \det(A) = c_n$$

It follows from the last equation that $\det(A) = 0$ if and only if $c_n = 0$, and this in turn implies that A is invertible if and only if $c_n \neq 0$. ▮

Example 7 The matrix A in Example 5 is invertible since it has eigenvalues $\lambda = 1$ and $\lambda = 2$, neither of which is zero. We leave it for the reader to check this conclusion showing that $\det(A) \neq 0$. ▲

SUMMARY   Theorem 7.1.4 enables us to add an additional result to Theorem 6.4.5.

Theorem 7.1.5. *If A is an n × n matrix, and if $T_A : R^n \to R^n$ is multiplication by A, then the following are equivalent.*

(a) A is invertible.

(b) $A\mathbf{x} = \mathbf{0}$ has only the trivial solution.

(c) The reduced row-echelon form of A is I_n.

(d) A is expressible as a product of elementary matrices.

(e) $A\mathbf{x} = \mathbf{b}$ is consistent for every n × 1 matrix $\mathbf{b}$.

(f) $A\mathbf{x} = \mathbf{b}$ has exactly one solution for every n × 1 matrix $\mathbf{b}$.

(g) $\det(A) \neq 0$.

(h) The range of T_A is R^n.

(i) T_A is one-to-one.

(j) The column vectors of A are linearly independent.

(k) The row vectors of A are linearly independent.

(l) The column vectors of A span R^n.

(m) The row vectors of A span R^n.

(n) The column vectors of A form a basis for R^n.

(o) The row vectors of A form a basis for R^n.

(p) A has rank n.

(q) A has nullity 0.

(r) The orthogonal complement of the nullspace of A is R^n.

(s) The orthogonal complement of the row space of A is $\{\mathbf{0}\}$.

(t) $A^T A$ is invertible.

(u) $\lambda = 0$ is not an eigenvalue of A.

This theorem relates all of the major topics we have studied thus far.

EXERCISE SET 7.1

1. Find the characteristic equations of the following matrices:

(a) $\begin{bmatrix} 3 & 0 \\ 8 & -1 \end{bmatrix}$ (b) $\begin{bmatrix} 10 & -9 \\ 4 & -2 \end{bmatrix}$ (c) $\begin{bmatrix} 0 & 3 \\ 4 & 0 \end{bmatrix}$ (d) $\begin{bmatrix} -2 & -7 \\ 1 & 2 \end{bmatrix}$ (e) $\begin{bmatrix} 0 & 0 \\ 0 & 0 \end{bmatrix}$ (f) $\begin{bmatrix} 1 & 0 \\ 0 & 1 \end{bmatrix}$

2. Find the eigenvalues of the matrices in Exercise 1.

3. Find bases for the eigenspaces of the matrices in Exercise 1.

4. Find the characteristic equations of the following matrices:

(a) $\begin{bmatrix} 4 & 0 & 1 \\ -2 & 1 & 0 \\ -2 & 0 & 1 \end{bmatrix}$ (b) $\begin{bmatrix} 3 & 0 & -5 \\ \frac{1}{5} & -1 & 0 \\ 1 & 1 & -2 \end{bmatrix}$ (c) $\begin{bmatrix} -2 & 0 & 1 \\ -6 & -2 & 0 \\ 19 & 5 & -4 \end{bmatrix}$

(d) $\begin{bmatrix} -1 & 0 & 1 \\ -1 & 3 & 0 \\ -4 & 13 & -1 \end{bmatrix}$ (e) $\begin{bmatrix} 5 & 0 & 1 \\ 1 & 1 & 0 \\ -7 & 1 & 0 \end{bmatrix}$ (f) $\begin{bmatrix} 5 & 6 & 2 \\ 0 & -1 & -8 \\ 1 & 0 & -2 \end{bmatrix}$

5. Find the eigenvalues of the matrices in Exercise 4.

6. Find bases for the eigenspaces of the matrices in Exercise 4.

7. Find the characteristic equations of the following matrices:

(a) $\begin{bmatrix} 0 & 0 & 2 & 0 \\ 1 & 0 & 1 & 0 \\ 0 & 1 & -2 & 0 \\ 0 & 0 & 0 & 1 \end{bmatrix}$ (b) $\begin{bmatrix} 10 & -9 & 0 & 0 \\ 4 & -2 & 0 & 0 \\ 0 & 0 & -2 & -7 \\ 0 & 0 & 1 & 2 \end{bmatrix}$

8. Find the eigenvalues of the matrices in Exercise 7.

9. Find bases for the eigenspaces of the matrices in Exercise 7.

10. By inspection, find the eigenvalues of the following matrices:

(a) $\begin{bmatrix} -1 & 6 \\ 0 & 5 \end{bmatrix}$ (b) $\begin{bmatrix} 3 & 0 & 0 \\ -2 & 7 & 0 \\ 4 & 8 & 1 \end{bmatrix}$ (c) $\begin{bmatrix} -\frac{1}{3} & 0 & 0 & 0 \\ 0 & -\frac{1}{3} & 0 & 0 \\ 0 & 0 & 1 & 0 \\ 0 & 0 & 0 & \frac{1}{2} \end{bmatrix}$

11. Find the eigenvalues of A^9 for

$$A = \begin{bmatrix} 1 & 3 & 7 & 11 \\ 0 & \frac{1}{2} & 3 & 8 \\ 0 & 0 & 0 & 4 \\ 0 & 0 & 0 & 2 \end{bmatrix}$$

12. Find the eigenvalues and bases for the eigenspaces of A^{25} for

$$A = \begin{bmatrix} -1 & -2 & -2 \\ 1 & 2 & 1 \\ -1 & -1 & 0 \end{bmatrix}$$

13. Let A be a 2 × 2 matrix, and call a line through the origin of R^2 *invariant* under A if $A\mathbf{x}$ lies on the line when $\mathbf{x}$ does. Find equations for all lines in R^2, if any, that are invariant under the given matrix.

(a) $A = \begin{bmatrix} 4 & -1 \\ 2 & 1 \end{bmatrix}$ (b) $A = \begin{bmatrix} 0 & 1 \\ -1 & 0 \end{bmatrix}$ (c) $A = \begin{bmatrix} 2 & 3 \\ 0 & 2 \end{bmatrix}$

14. Find $\det(A)$ given that A has $p(\lambda)$ as its characteristic polynomial.
(a) $p(\lambda) = \lambda^3 - 2\lambda^2 + \lambda + 5$ (b) $p(\lambda) = \lambda^4 - \lambda^3 + 7$
[***Hint.*** See the proof of Theorem 7.1.4.]

15. Let A be an $n \times n$ matrix.
(a) Prove that the characteristic polynomial of A has degree n.
(b) Prove that the coefficient of λ^n in the characteristic polynomial is 1.

16. Show that the characteristic equation of a 2×2 matrix A can be expressed as $\lambda^2 - \text{tr}(A)\lambda + \det(A) = 0$, where $\text{tr}(A)$ is the trace of A.

17. Use the result in Exercise 16 to show that if

$$A = \begin{bmatrix} a & b \\ c & d \end{bmatrix}$$

then the solutions of the characteristic equation of A are

$$\lambda = \tfrac{1}{2}\left[(a+d) \pm \sqrt{(a-d)^2 + 4bc}\,\right]$$

Use this result to show that A has
(a) two distinct real eigenvalues if $(a-d)^2 + 4bc > 0$ (b) one real eigenvalue if $(a-d)^2 + 4bc = 0$
(c) no real eigenvalues if $(a-d)^2 + 4bc < 0$

18. Let A be the matrix in Exercise 17. Show that if $(a-d)^2 + 4bc > 0$ and $b \neq 0$, then eigenvectors of A corresponding to the eigenvalues

$$\lambda_1 = \tfrac{1}{2}\left[(a+d) + \sqrt{(a-d)^2 + 4bc}\,\right] \qquad \text{and} \qquad \lambda_2 = \tfrac{1}{2}\left[(a+d) - \sqrt{(a-d)^2 + 4bc}\,\right]$$

are

$$\begin{bmatrix} -b \\ a - \lambda_1 \end{bmatrix} \qquad \text{and} \qquad \begin{bmatrix} -b \\ a - \lambda_2 \end{bmatrix}$$

respectively.

19. Prove: If a, b, c, and d are integers such that $a + b = c + d$, then

$$A = \begin{bmatrix} a & b \\ c & d \end{bmatrix}$$

has integer eigenvalues, namely, $\lambda_1 = a + b$ and $\lambda_2 = a - c$. [**Hint.** See Exercise 17.]

20. Prove: If λ is an eigenvalue of an invertible matrix A and $\mathbf{x}$ is a corresponding eigenvector, then $1/\lambda$ is an eigenvalue of A^{-1}, and $\mathbf{x}$ is a corresponding eigenvector.

21. Prove: If λ is an eigenvalue of A, $\mathbf{x}$ is a corresponding eigenvector, and s is a scalar, then $\lambda - s$ is an eigenvalue of $A - sI$, and $\mathbf{x}$ is a corresponding eigenvector.

22. Find the eigenvalues and bases for the eigenspaces of

$$A = \begin{bmatrix} -2 & 2 & 3 \\ -2 & 3 & 2 \\ -4 & 2 & 5 \end{bmatrix}$$

Then use Exercises 20 and 21 to find the eigenvalues and bases for the eigenspaces of
(a) A^{-1} (b) $A - 3I$ (c) $A + 2I$

23. (a) Prove that if A is a square matrix, then A and A^T have the same eigenvalues. [**Hint.** Look at the characteristic equation $\det(\lambda I - A) = 0$.]
(b) Show that A and A^T need not have the same eigenspaces. [**Hint.** Use the result in Exercise 18 to find a 2×2 matrix for which A and A^T have different eigenspaces.]

7.2 DIAGONALIZATION

In this section we shall be concerned with the problem of finding a basis for R^n that consists of eigenvectors of a given $n \times n$ matrix A. Such bases can be used to study geometric properties of A and to simplify various numerical computations involving A. These bases are also of physical significance in a wide variety of applications, some of which will be considered later in this text.

THE MATRIX DIAGONAL- IZATION PROBLEM

Our first objective in this section is to show that the following two problems, which on the surface seem quite different, are actually equivalent.

The Eigenvector Problem. Given an $n \times n$ matrix A, does there exist a basis for R^n consisting of eigenvectors of A?

The Diagonalization Problem (*Matrix Form*). Given an $n \times n$ matrix A, does there exist an invertible matrix P such that $P^{-1}AP$ is a diagonal matrix?

The latter problem suggests the following terminology.

Definition. A square matrix A is called *diagonalizable* if there is an invertible matrix P such that $P^{-1}AP$ is a diagonal matrix; the matrix P is said to *diagonalize* A.

The following theorem shows that the eigenvector problem and the diagonalization problem are equivalent.

Theorem 7.2.1. *If A is an $n \times n$ matrix, then the following are equivalent.*

(a) A is diagonalizable.

(b) A has n linearly independent eigenvectors.

Proof $(a) \Rightarrow (b)$: Since A is assumed diagonalizable, there is an invertible matrix

$$P = \begin{bmatrix} p_{11} & p_{12} & \cdots & p_{1n} \\ p_{21} & p_{22} & \cdots & p_{2n} \\ \vdots & \vdots & & \vdots \\ p_{n1} & p_{n2} & \cdots & p_{nn} \end{bmatrix}$$

such that $P^{-1}AP$ is diagonal, say $P^{-1}AP = D$, where

$$D = \begin{bmatrix} \lambda_1 & 0 & \cdots & 0 \\ 0 & \lambda_2 & \cdots & 0 \\ \vdots & \vdots & & \vdots \\ 0 & 0 & \cdots & \lambda_n \end{bmatrix}$$

It follows from the formula $P^{-1}AP = D$ that $AP = PD$; that is,

$$AP = \begin{bmatrix} p_{11} & p_{12} & \cdots & p_{1n} \\ p_{21} & p_{22} & \cdots & p_{2n} \\ \vdots & \vdots & & \vdots \\ p_{n1} & p_{n2} & \cdots & p_{nn} \end{bmatrix} \begin{bmatrix} \lambda_1 & 0 & \cdots & 0 \\ 0 & \lambda_2 & \cdots & 0 \\ \vdots & \vdots & & \vdots \\ 0 & 0 & \cdots & \lambda_n \end{bmatrix} = \begin{bmatrix} \lambda_1 p_{11} & \lambda_2 p_{12} & \cdots & \lambda_n p_{1n} \\ \lambda_1 p_{21} & \lambda_2 p_{22} & \cdots & \lambda_n p_{2n} \\ \vdots & \vdots & & \vdots \\ \lambda_1 p_{n1} & \lambda_2 p_{n2} & \cdots & \lambda_n p_{nn} \end{bmatrix}$$

(1)

If we now let $\mathbf{p}_1, \mathbf{p}_2, \ldots, \mathbf{p}_n$ denote the column vectors of P, then from (1) the successive columns of AP are $\lambda_1 \mathbf{p}_1, \lambda_2 \mathbf{p}_2, \ldots, \lambda_n \mathbf{p}_n$. However, from Formula (3) of Section 1.3, the successive columns of AP are $A\mathbf{p}_1, A\mathbf{p}_2, \ldots, A\mathbf{p}_n$. Thus, we must have

$$A\mathbf{p}_1 = \lambda_1 \mathbf{p}_1, \qquad A\mathbf{p}_2 = \lambda_2 \mathbf{p}_2, \quad \ldots, \quad A\mathbf{p}_n = \lambda_n \mathbf{p}_n \tag{2}$$

Since P is invertible, its column vectors are all nonzero; thus, it follows from (2) that $\lambda_1, \lambda_2, \ldots, \lambda_n$ are eigenvalues of A, and $\mathbf{p}_1, \mathbf{p}_2, \ldots, \mathbf{p}_n$ are corresponding eigenvectors. Since P is invertible, it follows from Theorem 7.1.5 that $\mathbf{p}_1, \mathbf{p}_2, \ldots, \mathbf{p}_n$ are linearly independent. Thus, A has n linearly independent eigenvectors.

$(b) \Rightarrow (a)$: Assume that A has n linearly independent eigenvectors, $\mathbf{p}_1, \mathbf{p}_2, \ldots, \mathbf{p}_n$, with corresponding eigenvalues $\lambda_1, \lambda_2, \ldots, \lambda_n$, and let

$$P = \begin{bmatrix} p_{11} & p_{12} & \cdots & p_{1n} \\ p_{21} & p_{22} & \cdots & p_{2n} \\ \vdots & \vdots & & \vdots \\ p_{n1} & p_{n2} & \cdots & p_{nn} \end{bmatrix}$$

be the matrix whose column vectors are $\mathbf{p}_1, \mathbf{p}_2, \ldots, \mathbf{p}_n$. By Formula (3) of Section 1.3, the column vectors of the product AP are

$$A\mathbf{p}_1, A\mathbf{p}_2, \ldots, A\mathbf{p}_n$$

But

$$A\mathbf{p}_1 = \lambda_1 \mathbf{p}_1, \quad A\mathbf{p}_2 = \lambda_2 \mathbf{p}_2, \quad \ldots, \quad A\mathbf{p}_n = \lambda_n \mathbf{p}_n$$

so that

$$AP = \begin{bmatrix} \lambda_1 p_{11} & \lambda_2 p_{12} & \cdots & \lambda_n p_{1n} \\ \lambda_1 p_{21} & \lambda_2 p_{22} & \cdots & \lambda_n p_{2n} \\ \vdots & \vdots & & \vdots \\ \lambda_1 p_{n1} & \lambda_2 p_{n2} & \cdots & \lambda_n p_{nn} \end{bmatrix}$$

$$= \begin{bmatrix} p_{11} & p_{12} & \cdots & p_{1n} \\ p_{21} & p_{22} & \cdots & p_{2n} \\ \vdots & \vdots & & \vdots \\ p_{n1} & p_{n2} & \cdots & p_{nn} \end{bmatrix} \begin{bmatrix} \lambda_1 & 0 & \cdots & 0 \\ 0 & \lambda_2 & \cdots & 0 \\ \vdots & \vdots & & \vdots \\ 0 & 0 & \cdots & \lambda_n \end{bmatrix} = PD$$

(3)

where D is the diagonal matrix having the eigenvalues $\lambda_1, \lambda_2, \ldots, \lambda_n$ on the main diagonal. Since the column vectors of P are linearly independent, P is invertible; thus, (3) can be rewritten as $P^{-1}AP = D$; that is, A is diagonalizable. ▮

PROCEDURE FOR DIAGONALIZING A MATRIX

The preceding theorem guarantees that an $n \times n$ matrix A with n linearly independent eigenvectors is diagonalizable, and the proof provides the following method for diagonalizing A.

Step 1. Find n linearly independent eigenvectors of A, say, $\mathbf{p}_1, \mathbf{p}_2, \ldots, \mathbf{p}_n$.

Step 2. Form the matrix P having $\mathbf{p}_1, \mathbf{p}_2, \ldots, \mathbf{p}_n$ as its column vectors.

Step 3. The matrix $P^{-1}AP$ will then be diagonal with $\lambda_1, \lambda_2, \ldots, \lambda_n$ as its successive diagonal entries, where λ_i is the eigenvalue corresponding to $\mathbf{p}_i$, for $i = 1, 2, \ldots, n$.

In order to carry out Step 1 of this procedure, one first needs a way of determining whether a given $n \times n$ matrix A has n linearly independent eigenvectors, and then one needs a method for finding them. One can address both problems at once by finding bases for the eigenspaces of A. Later in this section we will show that those basis vectors, as a combined set, are linearly independent, so that if there is a total of n such vectors, then A is diagonalizable, and the n basis vectors can be used as the column vectors of the diagonalizing matrix P. If there are fewer than n basis vectors, the A is not diagonalizable.

Example 1 Find a matrix P that diagonalizes

$$A = \begin{bmatrix} 0 & 0 & -2 \\ 1 & 2 & 1 \\ 1 & 0 & 3 \end{bmatrix}$$

Solution. From Example 5 of the preceding section we found the characteristic equation of A to be

$$(\lambda - 1)(\lambda - 2)^2 = 0$$

and we found the following bases for the eigenspaces:

$$\lambda = 2: \quad \mathbf{p}_1 = \begin{bmatrix} -1 \\ 0 \\ 1 \end{bmatrix}, \quad \mathbf{p}_2 = \begin{bmatrix} 0 \\ 1 \\ 0 \end{bmatrix}$$

$$\lambda = 1: \quad \mathbf{p}_3 = \begin{bmatrix} -2 \\ 1 \\ 1 \end{bmatrix}$$

There are three basis vectors in total, so the matrix A is diagonalizable and

$$P = \begin{bmatrix} -1 & 0 & -2 \\ 0 & 1 & 1 \\ 1 & 0 & 1 \end{bmatrix}$$

diagonalizes A. As a check, the reader should verify that

$$P^{-1}AP = \begin{bmatrix} 1 & 0 & 2 \\ 1 & 1 & 1 \\ -1 & 0 & -1 \end{bmatrix}\begin{bmatrix} 0 & 0 & -2 \\ 1 & 2 & 1 \\ 1 & 0 & 3 \end{bmatrix}\begin{bmatrix} -1 & 0 & -2 \\ 0 & 1 & 1 \\ 1 & 0 & 1 \end{bmatrix} = \begin{bmatrix} 2 & 0 & 0 \\ 0 & 2 & 0 \\ 0 & 0 & 1 \end{bmatrix} \qquad \blacktriangle$$

There is no preferred order for the columns of P. Since the ith diagonal entry of $P^{-1}AP$ is an eigenvalue for the ith column vector of P, changing the order of the columns of P just changes the order of the eigenvalues on the diagonal of $P^{-1}AP$. Thus, had we written

$$P = \begin{bmatrix} -1 & -2 & 0 \\ 0 & 1 & 1 \\ 1 & 1 & 0 \end{bmatrix}$$

in Example 1, we would have obtained

$$P^{-1}AP = \begin{bmatrix} 2 & 0 & 0 \\ 0 & 1 & 0 \\ 0 & 0 & 2 \end{bmatrix}$$

Example 2 Find a matrix P that diagonalizes

$$A = \begin{bmatrix} 1 & 0 & 0 \\ 1 & 2 & 0 \\ -3 & 5 & 2 \end{bmatrix}$$

Solution. The characteristic polynomial of A is

$$\det(\lambda I - A) = \begin{vmatrix} \lambda - 1 & 0 & 0 \\ -1 & \lambda - 2 & 0 \\ 3 & -5 & \lambda - 2 \end{vmatrix} = (\lambda - 1)(\lambda - 2)^2$$

so the characteristic equation is

$$(\lambda - 1)(\lambda - 2)^2 = 0$$

Thus, the eigenvalues of A are $\lambda = 1$ and $\lambda = 2$. We leave it for the reader to show that bases for the eigenspaces are

$$\lambda = 1: \qquad \mathbf{p}_1 = \begin{bmatrix} \frac{1}{8} \\ -\frac{1}{8} \\ 1 \end{bmatrix}$$

$$\lambda = 2: \qquad \mathbf{p}_2 = \begin{bmatrix} 0 \\ 0 \\ 1 \end{bmatrix}$$

Since A is a 3×3 matrix and there are only two basis vectors in total, A is not diagonalizable.

Alternative Solution. If one is interested only in determining whether a matrix is diagonalizable and is not concerned with actually finding a diagonalizing matrix P, then it is not necessary to compute bases for the eigenspaces; it suffices to find the dimensions of the eigenspaces. For this example, the eigenspace corresponding to $\lambda = 1$ is the solution space of the system

$$\begin{bmatrix} 0 & 0 & 0 \\ -1 & -1 & 0 \\ 3 & -5 & -1 \end{bmatrix} \begin{bmatrix} x_1 \\ x_2 \\ x_3 \end{bmatrix} = \begin{bmatrix} 0 \\ 0 \\ 0 \end{bmatrix}$$

The coefficient matrix has rank 2 (verify). Thus, the nullity of this matrix is 1 and by Theorem 5.6.4 the solution space is one-dimensional.

The eigenspace corresponding to $\lambda = 2$ is the solution space of the system

$$\begin{bmatrix} 1 & 0 & 0 \\ -1 & 0 & 0 \\ 3 & -5 & 0 \end{bmatrix} \begin{bmatrix} x_1 \\ x_2 \\ x_3 \end{bmatrix} = \begin{bmatrix} 0 \\ 0 \\ 0 \end{bmatrix}$$

This coefficient matrix also has rank 2 and nullity 1 (verify), so the eigenspace corresponding to $\lambda = 2$ is also one-dimensional. Since the eigenspaces produce a total of two basis vectors, the matrix A is not diagonalizable. ▲

There is an assumption in Example 1 that the column vectors of P, which are made up of basis vectors from the various eigenspaces of A, are linearly independent. The following theorem addresses this issue.

Theorem 7.2.2. *If* $\mathbf{v}_1, \mathbf{v}_2, \ldots, \mathbf{v}_k$ *are eigenvectors of A corresponding to distinct eigenvalues* $\lambda_1, \lambda_2, \ldots, \lambda_k$, *then* $\{\mathbf{v}_1, \mathbf{v}_2, \ldots, \mathbf{v}_k\}$ *is a linearly independent set.*

Proof. Let $\mathbf{v}_1, \mathbf{v}_2, \ldots, \mathbf{v}_k$ be eigenvectors of A corresponding to distinct eigenvalues $\lambda_1, \lambda_2, \ldots, \lambda_k$. We shall assume that $\mathbf{v}_1, \mathbf{v}_2, \ldots, \mathbf{v}_k$ are linearly dependent and obtain a contradiction. We can then conclude that $\mathbf{v}_1, \mathbf{v}_2, \ldots, \mathbf{v}_k$ are linearly independent.

Since an eigenvector is nonzero by definition, $\{\mathbf{v}_1\}$ is linearly independent. Let r be the largest integer such that $\{\mathbf{v}_1, \mathbf{v}_2, \ldots, \mathbf{v}_r\}$ is linearly independent. Since we are assuming that $\{\mathbf{v}_1, \mathbf{v}_2, \ldots, \mathbf{v}_k\}$ is linearly dependent, r satisfies $1 \leq r < k$. Moreover, by definition of r, $\{\mathbf{v}_1, \mathbf{v}_2, \ldots, \mathbf{v}_{r+1}\}$ is linearly dependent. Thus, there are scalars $c_1, c_2, \ldots, c_{r+1}$, not all zero, such that

$$c_1\mathbf{v}_1 + c_2\mathbf{v}_2 + \cdots + c_{r+1}\mathbf{v}_{r+1} = \mathbf{0} \tag{4}$$

Multiplying both sides of (4) by A and using

$$A\mathbf{v}_1 = \lambda_1\mathbf{v}_1, \quad A\mathbf{v}_2 = \lambda_2\mathbf{v}_2, \quad \ldots, \quad A\mathbf{v}_{r+1} = \lambda_{r+1}\mathbf{v}_{r+1}$$

we obtain

$$c_1\lambda_1\mathbf{v}_1 + c_2\lambda_2\mathbf{v}_2 + \cdots + c_{r+1}\lambda_{r+1}\mathbf{v}_{r+1} = \mathbf{0} \tag{5}$$

Multiplying both sides of (4) by λ_{r+1} and subtracting the resulting equation from (5) yields

$$c_1(\lambda_1 - \lambda_{r+1})\mathbf{v}_1 + c_2(\lambda_2 - \lambda_{r+1})\mathbf{v}_2 + \cdots + c_r(\lambda_r - \lambda_{r+1})\mathbf{v}_r = \mathbf{0}$$

Since $\{\mathbf{v}_1, \mathbf{v}_2, \ldots, \mathbf{v}_r\}$ is a linearly independent set, this equation implies that

$$c_1(\lambda_1 - \lambda_{r+1}) = c_2(\lambda_2 - \lambda_{r+1}) = \cdots = c_r(\lambda_r - \lambda_{r+1}) = 0$$

and since $\lambda_1, \lambda_2, \ldots, \lambda_{r+1}$ are distinct, it follows that

$$c_1 = c_2 = \cdots = c_r = 0 \tag{6}$$

Substituting these values in (4) yields

$$c_{r+1}\mathbf{v}_{r+1} = \mathbf{0}$$

Since the eigenvector $\mathbf{v}_{r+1}$ is nonzero, it follows that

$$c_{r+1} = 0 \tag{7}$$

Equations (6) and (7) contradict the fact that $c_1, c_2, \ldots, c_{r+1}$ are not all zero; this completes the proof. ▮

REMARK. Theorem 7.2.2 is a special case of a more general result: Suppose that λ_1, $\lambda_2, \ldots, \lambda_k$ are distinct eigenvalues and that we choose a linearly independent set in each of the corresponding eigenspaces. If we then merge all these vectors into a single set, the result will still be a linearly independent set. For example, if we choose three linearly independent vectors from one eigenspace and two linearly independent vectors from another eigenspace, then the five vectors together form a linearly independent set. We omit the proof.

As a consequence of Theorem 7.2.2, we obtain the following important result.

Theorem 7.2.3. *If an $n \times n$ matrix A has n distinct eigenvalues, then A is diagonalizable.*

Proof. If $\mathbf{v}_1, \mathbf{v}_2, \ldots, \mathbf{v}_n$ are eigenvectors corresponding to the distinct eigenvalues $\lambda_1, \lambda_2, \ldots, \lambda_n$, then by Theorem 7.2.2, $\mathbf{v}_1, \mathbf{v}_2, \ldots, \mathbf{v}_n$ are linearly independent. Thus, A is diagonalizable by Theorem 7.2.1. ▮

Example 3   We saw in Example 2 of the preceding section that

$$A = \begin{bmatrix} 0 & 1 & 0 \\ 0 & 0 & 1 \\ 4 & -17 & 8 \end{bmatrix}$$

has three distinct eigenvalues, $\lambda = 4$, $\lambda = 2 + \sqrt{3}$, $\lambda = 2 - \sqrt{3}$. Therefore, A is diagonalizable. Further,

$$P^{-1}AP = \begin{bmatrix} 4 & 0 & 0 \\ 0 & 2 + \sqrt{3} & 0 \\ 0 & 0 & 2 - \sqrt{3} \end{bmatrix}$$

for some invertible matrix P. If desired, the matrix P can be found using the method shown in Example 1 of this section. ▲

Example 4   From Theorem 7.1.1 the eigenvalues of a triangular matrix are the entries on its main diagonal. Thus, a triangular matrix with distinct entries on the main diagonal is diagonalizable. For example,

$$A = \begin{bmatrix} -1 & 2 & 4 & 0 \\ 0 & 3 & 1 & 7 \\ 0 & 0 & 5 & 8 \\ 0 & 0 & 0 & -2 \end{bmatrix}$$

is a diagonalizable matrix. ▲

GEOMETRIC AND ALGEBRAIC MULTIPLICITY

Theorem 7.2.3 does not completely settle the diagonalization problem, since it is possible for an $n \times n$ matrix A to be diagonalizable without having n distinct eigenvalues. We saw this in Example 1, where the given 3×3 matrix had only two distinct eigenvalues, yet was diagonalizable. What really matters for diagonalizability are the dimensions of the eigenspaces—those dimensions must add up to n in order for an $n \times n$ matrix to be diagonalizable. Examples 1 and 2 illustrate this; the matrices in those examples have the same characteristic equation and the same eigenvalues, but the matrix in Example 1 is diagonalizable because the dimensions of the eigenspaces add to 3 and the matrix in Example 2 is not diagonalizable because the dimensions only add to 2.

A full excursion into the study of diagonalizability is left for more advanced courses, but we shall touch on one theorem that is important to a fuller understanding of diagonalizability. It can be proved that if λ_0 is an eigenvalue of A, then the dimension of the eigenspace corresponding to λ_0 cannot exceed the number of times that $\lambda - \lambda_0$ appears as a factor in the characteristic polynomial of A. For example, in Examples 1 and 2 the characteristic polynomial is

$$(\lambda - 1)(\lambda - 2)^2$$

Thus, the eigenspace corresponding to $\lambda = 1$ is at most (hence exactly) one-dimensional and the eigenspace corresponding to $\lambda = 2$ is at most two-dimensional. In Example 1, the eigenspace corresponding to $\lambda = 2$ actually had dimension 2, resulting in diagonalizability, but in Example 2 that eigenspace had only dimension 1, resulting in the failure of diagonalizability.

There is some terminology that relates to these ideas. If λ_0 is an eigenvalue of an $n \times n$ matrix A, then the dimension of the eigenspace corresponding to λ_0 is called the *geometric multiplicity* of λ_0, and the number of times that $\lambda - \lambda_0$ appears as a factor in the characteristic polynomial of A is called the *algebraic multiplicity* of A. The following theorem, which we state without proof, summarizes the preceding discussion.

Theorem 7.2.4. *If A is a square matrix, then:*

(a) For every eigenvalue of A the geometric multiplicity is less than or equal to the algebraic multiplicity.

(b) A is diagonalizable if and only if the geometric multiplicity is equal to the algebraic multiplicity for every eigenvalue.

COMPUTING POWERS OF A MATRIX

There are numerous problems in applied mathematics that require the computation of high powers of a square matrix. We shall conclude this section by showing how diagonalization can be used to simplify such computations for diagonalizable matrices.

If A is an $n \times n$ matrix and P is an invertible matrix, then

$$(P^{-1}AP)^2 = P^{-1}APP^{-1}AP = P^{-1}AIAP = P^{-1}A^2P$$

More generally, for any positive integer k

$$(P^{-1}AP)^k = P^{-1}A^kP \tag{8}$$

It follows from this equation that if A is diagonalizable, and $P^{-1}AP = D$ is a diagonal matrix, then

$$P^{-1}A^kP = (P^{-1}AP)^k = D^k \tag{9}$$

Solving this equation for A^k yields

$$A^k = PD^kP^{-1} \tag{10}$$

This last equation expresses the kth power of A in terms of the kth power of the diagonal matrix D. But D^k is easy to compute; for example, if

$$D = \begin{bmatrix} d_1 & 0 & \cdots & 0 \\ 0 & d_2 & \cdots & 0 \\ \vdots & \vdots & & \vdots \\ 0 & 0 & \cdots & d_n \end{bmatrix}$$

then

$$D^k = \begin{bmatrix} d_1^k & 0 & \cdots & 0 \\ 0 & d_2^k & \cdots & 0 \\ \vdots & \vdots & & \vdots \\ 0 & 0 & \cdots & d_n^k \end{bmatrix}$$

Example 5 Use (10) to find A^{13}, where

$$A = \begin{bmatrix} 0 & 0 & -2 \\ 1 & 2 & 1 \\ 1 & 0 & 3 \end{bmatrix}$$

Solution. We showed in Example 1 that the matrix A is diagonalized by

$$P = \begin{bmatrix} -1 & 0 & -2 \\ 0 & 1 & 1 \\ 1 & 0 & 1 \end{bmatrix}$$

and that

$$D = P^{-1}AP = \begin{bmatrix} 2 & 0 & 0 \\ 0 & 2 & 0 \\ 0 & 0 & 1 \end{bmatrix}$$

Thus, from (10)

$$A^{13} = PD^{13}P^{-1} = \begin{bmatrix} -1 & 0 & -2 \\ 0 & 1 & 1 \\ 1 & 0 & 1 \end{bmatrix} \begin{bmatrix} 2^{13} & 0 & 0 \\ 0 & 2^{13} & 0 \\ 0 & 0 & 1^{13} \end{bmatrix} \begin{bmatrix} 1 & 0 & 2 \\ 1 & 1 & 1 \\ -1 & 0 & -1 \end{bmatrix}$$ (11)

$$= \begin{bmatrix} -8190 & 0 & -16382 \\ 8191 & 8192 & 8191 \\ 8191 & 0 & 16383 \end{bmatrix}$$

REMARK. With the method in the preceding example, most of the work is in diagonalizing A. Once that work is done, it can be used to compute any power of A. Thus, to compute A^{1000} we need only change the exponents from 13 to 1000 in (11).

EXERCISE SET 7.2

1. Let A be a 6×6 matrix with characteristic equation $\lambda^2(\lambda - 1)(\lambda - 2)^3 = 0$. What are the possible dimensions for eigenspaces of A?

2. Let

$$A = \begin{bmatrix} 4 & 0 & 1 \\ 2 & 3 & 2 \\ 1 & 0 & 4 \end{bmatrix}$$

(a) Find the eigenvalues of A.
(b) For each eigenvalue λ find the rank of the matrix $\lambda I - A$.
(c) Is A diagonalizable? Justify your conclusion.

In Exercises 3–7 use the method of Exercise 2 to determine whether the matrix is diagonalizable.

3. $\begin{bmatrix} 2 & 0 \\ 1 & 2 \end{bmatrix}$ **4.** $\begin{bmatrix} 2 & -3 \\ 1 & -1 \end{bmatrix}$ **5.** $\begin{bmatrix} 3 & 0 & 0 \\ 0 & 2 & 0 \\ 0 & 1 & 2 \end{bmatrix}$ **6.** $\begin{bmatrix} -1 & 0 & 1 \\ -1 & 3 & 0 \\ -4 & 13 & -1 \end{bmatrix}$ **7.** $\begin{bmatrix} 2 & -1 & 0 & 1 \\ 0 & 2 & 1 & -1 \\ 0 & 0 & 3 & 2 \\ 0 & 0 & 0 & 3 \end{bmatrix}$

In Exercises 8–11 find a matrix P that diagonalizes A, and determine $P^{-1}AP$.

8. $A = \begin{bmatrix} -14 & 12 \\ -20 & 17 \end{bmatrix}$ **9.** $A = \begin{bmatrix} 1 & 0 \\ 6 & -1 \end{bmatrix}$ **10.** $A = \begin{bmatrix} 1 & 0 & 0 \\ 0 & 1 & 1 \\ 0 & 1 & 1 \end{bmatrix}$ **11.** $A = \begin{bmatrix} 2 & 0 & -2 \\ 0 & 3 & 0 \\ 0 & 0 & 3 \end{bmatrix}$

In Exercises 12–17 determine whether A is diagonalizable. If so, find a matrix P that diagonalizes A, and determine $P^{-1}AP$.

12. $A = \begin{bmatrix} 19 & -9 & -6 \\ 25 & -11 & -9 \\ 17 & -9 & -4 \end{bmatrix}$ **13.** $A = \begin{bmatrix} -1 & 4 & -2 \\ -3 & 4 & 0 \\ -3 & 1 & 3 \end{bmatrix}$ **14.** $A = \begin{bmatrix} 5 & 0 & 0 \\ 1 & 5 & 0 \\ 0 & 1 & 5 \end{bmatrix}$

15. $A = \begin{bmatrix} 0 & 0 & 0 \\ 0 & 0 & 0 \\ 3 & 0 & 1 \end{bmatrix}$ **16.** $A = \begin{bmatrix} -2 & 0 & 0 & 0 \\ 0 & -2 & 0 & 0 \\ 0 & 0 & 3 & 0 \\ 0 & 0 & 1 & 3 \end{bmatrix}$ **17.** $A = \begin{bmatrix} -2 & 0 & 0 & 0 \\ 0 & -2 & 5 & -5 \\ 0 & 0 & 3 & 0 \\ 0 & 0 & 0 & 3 \end{bmatrix}$

18. Use the method of Example 5 to compute A^{10}, where

$$A = \begin{bmatrix} 1 & 0 \\ -1 & 2 \end{bmatrix}$$

19. Use the method of Example 5 to compute A^{11}, where

$$A = \begin{bmatrix} -1 & 7 & -1 \\ 0 & 1 & 0 \\ 0 & 15 & -2 \end{bmatrix}$$

20. In each part compute the stated power of

$$A = \begin{bmatrix} 1 & -2 & 8 \\ 0 & -1 & 0 \\ 0 & 0 & -1 \end{bmatrix}$$

(a) A^{1000} (b) A^{-1000} (c) A^{2301} (d) A^{-2301}

21. Find A^n if n is a positive integer and

$$A = \begin{bmatrix} 3 & -1 & 0 \\ -1 & 2 & -1 \\ 0 & -1 & 3 \end{bmatrix}$$

22. Let

$$A = \begin{bmatrix} a & b \\ c & d \end{bmatrix}$$

Show:
(a) A is diagonalizable if $(a - d)^2 + 4bc > 0$.
(b) A is not diagonalizable if $(a - d)^2 + 4bc < 0$.
    [***Hint.*** See Exercises 17 and 18 of Section 7.1.]

23. In the case where the matrix A in Exercise 22 is diagonalizable, find a matrix P that diagonalizes A.

24. Prove that if A is a diagonalizable matrix, then the rank of A is the number of nonzero eigenvalues of A.

25. Prove: If A is invertible and diagonalizable, then A^{-1} is diagonalizable, and a matrix P that diagonalizes A also diagonalizes A^{-1}.

7.3 ORTHOGONAL DIAGONALIZATION

In this section we shall be concerned with the problem of finding an orthonormal basis for R^n with the Euclidean inner product consisting of eigenvectors of a given $n \times n$ matrix A. Our earlier work on symmetric matrices and orthogonal matrices will play an important role here.

THE MATRIX ORTHOGONAL DIAGONAL-IZATION PROBLEM

Our first objective in this section is to show that the following two problems are equivalent.

> ***The Orthonormal Eigenvector Problem.*** Given an $n \times n$ matrix A, does there exist an orthonormal basis for R^n with the Euclidean inner product consisting of eigenvectors of A?

> ***The Orthogonal Diagonalization Problem*** (***Matrix Form***). Given an $n \times n$ matrix A, does there exist an orthogonal matrix P such that the matrix $P^{-1}AP = P^TAP$ is diagonal? If there is such a matrix, then A is said to be ***orthogonally diagonalizable*** and P is said to ***orthogonally diagonalize*** A.

For the latter problem we have two questions to consider:

- Which matrices are orthogonally diagonalizable?
- How do we find an orthogonal matrix to carry out the diagonalization?

With regard to the first question, we note that there is no hope of orthogonally diagonalizing a matrix A unless A is symmetric (i.e., $A = A^T$). To see why this is so, suppose that

$$P^TAP = D \tag{1}$$

where P is an orthogonal matrix and D is a diagonal matrix. Since P is orthogonal, $PP^T = P^TP = I$, so it follows that (1) can be written as

$$A = PDP^T \tag{2}$$

Since D is a diagonal matrix, we have $D = D^T$, so transposing both sides of (2) yields

$$A^T = (PDP^T)^T = (P^T)^TD^TP^T = PDP^T = A$$

so A must be symmetric.

CONDITIONS FOR ORTHOGONAL DIAGONALIZ-ABILITY

The following theorem shows that every symmetric matrix is, in fact, orthogonally diagonalizable. In this theorem, and for the remainder of this section, *orthogonal* will mean orthogonal with respect to the Euclidean inner product on R^n.

> **Theorem 7.3.1.** *If A is an n × n matrix, then the following are equivalent.*
>
> (*a*) *A is orthogonally diagonalizable.*
> (*b*) *A has an orthonormal set of n eigenvectors.*
> (*c*) *A is symmetric.*

Proof (*a*)⟹(*b*): Since A is orthogonally diagonalizable, there is an orthogonal matrix P such that $P^{-1}AP$ is diagonal. As shown in the proof of Theorem 7.2.1, the n column vectors of P are eigenvectors of A. Since P is orthogonal, these column vectors are orthonormal (see Theorem 6.5.1), so that A has n orthonormal eigenvectors.

(*b*)⟹(*a*): Assume that A has an orthonormal set of n eigenvectors $\{\mathbf{p}_1, \mathbf{p}_2, \ldots, \mathbf{p}_n\}$. As shown in the proof of Theorem 7.2.1, the matrix P with these eigenvectors as columns diagonalizes A. Since these eigenvectors are orthonormal, P is orthogonal and thus orthogonally diagonalizes A.

(*a*)⟹(*c*): In the proof that (*a*)⟹(*b*) we showed that an orthogonally diagonalizable $n \times n$ matrix A is orthogonally diagonalized by an $n \times n$ matrix P whose columns form an orthonormal set of eigenvectors of A. Let D be the diagonal matrix

$$D = P^{-1}AP$$

Thus,

$$A = PDP^{-1}$$

or, since P is orthogonal,

$$A = PDP^T$$

Therefore,

$$A^T = (PDP^T)^T = PD^TP^T = PDP^T = A$$

which shows that A is symmetric.

(*c*)⟹(*a*): The proof of this part is beyond the scope of this text and will be omitted. ∎

SOME PROPERTIES OF SYMMETRIC MATRICES

Our next goal is to devise a procedure for orthogonally diagonalizing a symmetric matrix, but before we can do so, we need a critical theorem about eigenvalues and eigenvectors of symmetric matrices.

> **Theorem 7.3.2.** *If A is a symmetric matrix, then:*
>
> (*a*) *The eigenvalues of A are all real numbers.*
> (*b*) *Eigenvectors from different eigenspaces are orthogonal.*

Proof (a). The proof of part (a), which requires results about complex vector spaces, is discussed in Section 10.6.

Proof (b). Let $\mathbf{v}_1$ and $\mathbf{v}_2$ be eigenvectors corresponding to distinct eigenvalues λ_1 and λ_2 of the matrix A. We want to show that $\mathbf{v}_1 \cdot \mathbf{v}_2 = 0$. The proof of this involves the trick of starting with the expression $A\mathbf{v}_1 \cdot \mathbf{v}_2$. It follows from Formula (8) of Section 4.1 and the symmetry of A that

$$A\mathbf{v}_1 \cdot \mathbf{v}_2 = \mathbf{v}_1 \cdot A^T\mathbf{v}_2 = \mathbf{v}_1 \cdot A\mathbf{v}_2 \tag{3}$$

But $\mathbf{v}_1$ is an eigenvector of A corresponding to λ_1 and $\mathbf{v}_2$ is an eigenvector of A corresponding to λ_2, so (3) yields the relationship

$$\lambda_1\mathbf{v}_1 \cdot \mathbf{v}_2 = \mathbf{v}_1 \cdot \lambda_2\mathbf{v}_2$$

which can be rewritten as

$$(\lambda_1 - \lambda_2)(\mathbf{v}_1 \cdot \mathbf{v}_2) = 0 \tag{4}$$

But $\lambda_1 - \lambda_2 \neq 0$, since λ_1 and λ_2 were assumed distinct. Thus, it follows from (4) that $\mathbf{v}_1 \cdot \mathbf{v}_2 = 0$. ▌

REMARK. We remind the reader that we have assumed to this point that all of our matrices have real entries. Indeed, we shall see in Chapter 10 that part (*a*) of Theorem 7.3.2 is false for matrices with complex entries.

DIAGONALIZ-ATION OF SYMMETRIC MATRICES

As a consequence of the preceding theorem we obtain the following procedure for orthogonally diagonalizing a symmetric matrix.

Step 1. Find a basis for each eigenspace of A.

Step 2. Apply the Gram–Schmidt process to each of these bases to obtain an orthonormal basis for each eigenspace.

Step 3. Form the matrix P whose columns are the basis vectors constructed in Step 2; this matrix orthogonally diagonalizes A.

The justification of this procedure should be clear: Theorem 7.3.2 ensures that eigenvectors from *different* eigenspaces are orthogonal, while the application of the Gram–Schmidt process ensures that the eigenvectors obtained within the *same* eigenspace are orthonormal. Thus, the *entire* set of eigenvectors obtained by this procedure is orthonormal.

Example 1 Find an orthogonal matrix P that diagonalizes

$$A = \begin{bmatrix} 4 & 2 & 2 \\ 2 & 4 & 2 \\ 2 & 2 & 4 \end{bmatrix}$$

Solution. The characteristic equation of A is

$$\det(\lambda I - A) = \det \begin{bmatrix} \lambda - 4 & -2 & -2 \\ -2 & \lambda - 4 & -2 \\ -2 & -2 & \lambda - 4 \end{bmatrix} = (\lambda - 2)^2(\lambda - 8) = 0$$

Thus, the eigenvalues of A are $\lambda = 2$ and $\lambda = 8$. By the method used in Example 5 of Section 7.1, it can be shown that

$$\mathbf{u}_1 = \begin{bmatrix} -1 \\ 1 \\ 0 \end{bmatrix} \quad \text{and} \quad \mathbf{u}_2 = \begin{bmatrix} -1 \\ 0 \\ 1 \end{bmatrix}$$

form a basis for the eigenspace corresponding to $\lambda = 2$. Applying the Gram–Schmidt process to $\{\mathbf{u}_1, \mathbf{u}_2\}$ yields the following orthonormal eigenvectors (verify):

$$\mathbf{v}_1 = \begin{bmatrix} -1/\sqrt{2} \\ 1/\sqrt{2} \\ 0 \end{bmatrix} \quad \text{and} \quad \mathbf{v}_2 = \begin{bmatrix} -1/\sqrt{6} \\ -1/\sqrt{6} \\ 2/\sqrt{6} \end{bmatrix}$$

The eigenspace corresponding to $\lambda = 8$ has

$$\mathbf{u}_3 = \begin{bmatrix} 1 \\ 1 \\ 1 \end{bmatrix}$$

as a basis. Applying the Gram–Schmidt process to $\{\mathbf{u}_3\}$ yields

$$\mathbf{v}_3 = \begin{bmatrix} 1/\sqrt{3} \\ 1/\sqrt{3} \\ 1/\sqrt{3} \end{bmatrix}$$

Finally, using $\mathbf{v}_1$, $\mathbf{v}_2$, and $\mathbf{v}_3$ as column vectors we obtain

$$P = \begin{bmatrix} -1/\sqrt{2} & -1/\sqrt{6} & 1/\sqrt{3} \\ 1/\sqrt{2} & -1/\sqrt{6} & 1/\sqrt{3} \\ 0 & 2/\sqrt{6} & 1/\sqrt{3} \end{bmatrix}$$

which orthogonally diagonalizes A. (As a check, the reader may wish to verify that $P^T A P$ is a diagonal matrix.) ▲

EXERCISE SET 7.3

1. Find the characteristic equation of the given symmetric matrix, and then by inspection determine the dimensions of the eigenspaces.

(a) $\begin{bmatrix} 1 & 2 \\ 2 & 4 \end{bmatrix}$ (b) $\begin{bmatrix} 1 & -4 & 2 \\ -4 & 1 & -2 \\ 2 & -2 & -2 \end{bmatrix}$ (c) $\begin{bmatrix} 1 & 1 & 1 \\ 1 & 1 & 1 \\ 1 & 1 & 1 \end{bmatrix}$

(d) $\begin{bmatrix} 4 & 2 & 2 \\ 2 & 4 & 2 \\ 2 & 2 & 4 \end{bmatrix}$ (e) $\begin{bmatrix} 4 & 4 & 0 & 0 \\ 4 & 4 & 0 & 0 \\ 0 & 0 & 0 & 0 \\ 0 & 0 & 0 & 0 \end{bmatrix}$ (f) $\begin{bmatrix} 2 & -1 & 0 & 0 \\ -1 & 2 & 0 & 0 \\ 0 & 0 & 2 & -1 \\ 0 & 0 & -1 & 2 \end{bmatrix}$

In Exercises 2–9 find a matrix P that orthogonally diagonalizes A, and determine $P^{-1}AP$.

2. $A = \begin{bmatrix} 3 & 1 \\ 1 & 3 \end{bmatrix}$ **3.** $A = \begin{bmatrix} 6 & 2\sqrt{3} \\ 2\sqrt{3} & 7 \end{bmatrix}$ **4.** $A = \begin{bmatrix} 6 & -2 \\ -2 & 3 \end{bmatrix}$ **5.** $A = \begin{bmatrix} -2 & 0 & -36 \\ 0 & -3 & 0 \\ -36 & 0 & -23 \end{bmatrix}$

6. $A = \begin{bmatrix} 1 & 1 & 0 \\ 1 & 1 & 0 \\ 0 & 0 & 0 \end{bmatrix}$ **7.** $A = \begin{bmatrix} 2 & -1 & -1 \\ -1 & 2 & -1 \\ -1 & -1 & 2 \end{bmatrix}$ **8.** $A = \begin{bmatrix} 3 & 1 & 0 & 0 \\ 1 & 3 & 0 & 0 \\ 0 & 0 & 0 & 0 \\ 0 & 0 & 0 & 0 \end{bmatrix}$ **9.** $A = \begin{bmatrix} -7 & 24 & 0 & 0 \\ 24 & 7 & 0 & 0 \\ 0 & 0 & -7 & 24 \\ 0 & 0 & 24 & 7 \end{bmatrix}$

10. Assuming that $b \neq 0$, find a matrix that orthogonally diagonalizes

$$\begin{bmatrix} a & b \\ b & a \end{bmatrix}$$

11. Prove that if A is any $m \times n$ matrix, then $A^T A$ has an orthonormal set of n eigenvectors.

12. (a) Show that if $\mathbf{v}$ is any $n \times 1$ matrix and I is the $n \times n$ identity matrix, then $I - \mathbf{v}\mathbf{v}^T$ is orthogonally diagonalizable.
(b) Find a matrix P that orthogonally diagonalizes $I - \mathbf{v}\mathbf{v}^T$ if

$$\mathbf{v} = \begin{bmatrix} 1 \\ 0 \\ 1 \end{bmatrix}$$

13. Use the result in Exercise 17 of Section 7.1 to prove Theorem 7.3.2a for 2×2 symmetric matrices.

SUPPLEMENTARY EXERCISES

1. (a) Show that if $0 < \theta < \pi$, then

$$A = \begin{bmatrix} \cos \theta & -\sin \theta \\ \sin \theta & \cos \theta \end{bmatrix}$$

has no eigenvalues and consequently no eigenvectors.
(b) Give a geometric explanation of the result in (a).

2. Find the eigenvalues of

$$A = \begin{bmatrix} 0 & 1 & 0 \\ 0 & 0 & 1 \\ k^3 & -3k^2 & 3k \end{bmatrix}$$

3. (a) Show that if D is a diagonal matrix with nonnegative entries on the main diagonal, then there is a matrix S such that $S^2 = D$.

(b) Show that if A is a diagonalizable matrix with nonnegative eigenvalues, then there is a matrix S such that $S^2 = A$.

(c) Find a matrix S such that $S^2 = A$, if

$$A = \begin{bmatrix} 1 & 3 & 1 \\ 0 & 4 & 5 \\ 0 & 0 & 9 \end{bmatrix}$$

4. (a) Prove: If A is a square matrix, then A and A^T have the same eigenvalues.

(b) Show that A and A^T need not have the same eigenvectors. [**Hint.** Use Exercise 18 of Section 7.1 to find a 2×2 matrix A in which A and A^T have different eigenvectors.]

5. Prove: If A is a square matrix and $p(\lambda) = \det(\lambda I - A)$ is the characteristic polynomial of A, then the coefficient of λ^{n-1} in $p(\lambda)$ is the negative of the trace of A.

6. Prove: If $b \neq 0$, then

$$A = \begin{bmatrix} a & b \\ 0 & a \end{bmatrix}$$

is not diagonalizable.

7. In advanced linear algebra, one proves the *Cayley–Hamilton Theorem*, which states that a square matrix A satisfies its characteristic equation; that is, if

$$c_0 + c_1\lambda + c_2\lambda^2 + \cdots + c_{n-1}\lambda^{n-1} + \lambda^n = 0$$

is the characteristic equation of A, then

$$c_0 I + c_1 A + c_2 A^2 + \cdots + c_{n-1}A^{n-1} + A^n = 0.$$

Verify this result for

(a) $A = \begin{bmatrix} 3 & 6 \\ 1 & 2 \end{bmatrix}$ (b) $A = \begin{bmatrix} 0 & 1 & 0 \\ 0 & 0 & 1 \\ 1 & -3 & 3 \end{bmatrix}$

Exercises 8–10 use the Cayley–Hamilton Theorem, stated in Exercise 7.

8. Use Exercise 16 of Section 7.1 to prove the Cayley–Hamilton Theorem for 2×2 matrices.

9. The Cayley–Hamilton Theorem provides an efficient method for calculating powers of a matrix. For example, if A is a 2×2 matrix with characteristic equation

$$c_0 + c_1\lambda + \lambda^2 = 0$$

then $c_0 I + c_1 A + A^2 = 0$, so

$$A^2 = -c_1 A - c_0 I$$

Multiplying through by A yields $A^3 = -c_1 A^2 - c_0 A$, which expresses A^3 in terms of A^2 and A, and multiplying through by A^2 yields $A^4 = -c_1 A^3 - c_0 A^2$, which expresses A^4 in terms of A^3 and A^2. Continuing in this way, we can calculate successive powers of A simply by expressing them in terms of lower powers. Use this procedure to calculate

$$A^2, \quad A^3, \quad A^4, \quad \text{and} \quad A^5$$

for

$$A = \begin{bmatrix} 3 & 6 \\ 1 & 2 \end{bmatrix}$$

10. Use the method of the preceding exercise to calculate A^3 and A^4 for

$$A = \begin{bmatrix} 0 & 1 & 0 \\ 0 & 0 & 1 \\ 1 & -3 & 3 \end{bmatrix}$$

11. Find the eigenvalues of the matrix

$$A = \begin{bmatrix} c_1 & c_2 & \cdots & c_n \\ c_1 & c_2 & \cdots & c_n \\ \vdots & \vdots & & \vdots \\ c_1 & c_2 & \cdots & c_n \end{bmatrix}$$

12. (a) It was shown in Exercise 15 of Section 7.1 that if A is an $n \times n$ matrix, then the coefficient of λ^n in the characteristic polynomial of A is 1. (A polynomial with this property is called *monic*.) Show that the matrix

$$\begin{bmatrix} 0 & 0 & 0 & \cdots & 0 & -c_0 \\ 1 & 0 & 0 & \cdots & 0 & -c_1 \\ 0 & 1 & 0 & \cdots & 0 & -c_2 \\ \vdots & \vdots & \vdots & & \vdots & \vdots \\ 0 & 0 & 0 & \cdots & 1 & -c_{n-1} \end{bmatrix}$$

has characteristic polynomial $p(\lambda) = c_0 + c_1 \lambda + \cdots + c_{n-1} \lambda^{n-1} + \lambda^n$. This shows that every monic polynomial is the characteristic polynomial of some matrix. The matrix in this example is called the *companion matrix* of $p(\lambda)$. [*Hint.* Evaluate all determinants in the problem by adding a multiple of the second row to the first to introduce a zero at the top of the first column, and then expanding by cofactors along the first column.]

(b) Find a matrix with characteristic polynomial $p(\lambda) = 1 - 2\lambda + \lambda^2 + 3\lambda^3 + \lambda^4$.

13. A square matrix A is called *nilpotent* if $A^n = 0$ for some positive integer n. What can you say about the eigenvalues of a nilpotent matrix?

14. Prove: If A is an $n \times n$ matrix and n is odd, then A has at least one real eigenvalue.

15. Find a 3×3 matrix A that has eigenvalues $\lambda = 0, 1$, and -1 with corresponding eigenvectors

$$\begin{bmatrix} 0 \\ 1 \\ -1 \end{bmatrix}, \quad \begin{bmatrix} 1 \\ -1 \\ 1 \end{bmatrix}, \quad \begin{bmatrix} 0 \\ 1 \\ 1 \end{bmatrix}$$

respectively.

16. Suppose that a 4×4 matrix A has eigenvalues $\lambda_1 = 1$, $\lambda_2 = -2$, $\lambda_3 = 3$, and $\lambda_4 = -3$.
   (a) Use Exercise 14 of Section 7.1 to find $\det(A)$.
   (b) Use Exercise 5 above to find $\mathrm{tr}(A)$.

17. Let A be a square matrix such that $A^3 = A$. What can you say about the eigenvalues of A?

CHAPTER **8**

LINEAR TRANSFORMATIONS

8.1 GENERAL LINEAR TRANSFORMATIONS

In Sections 4.2 and 4.3 we studied linear transformations from R^n to R^m. In this section we shall define and study linear transformations from an arbitrary vector space V to an arbitrary vector space W. The results we obtain here have important applications in physics, engineering, and various branches of mathematics.

DEFINITIONS AND TERMINOLOGY

Recall that a linear transformation from R^n to R^m was first defined as a function

$$T(x_1, x_2, \ldots, x_n) = (w_1, w_2, \ldots, w_m)$$

for which the equations relating $w_1, w_2, \ldots, w_m$ and $x_1, x_2, \ldots, x_n$ are linear. Subsequently, we showed that a transformation $T: R^n \to R^m$ is linear if and only if the following relationships hold for all vectors **u** and **v** in R^n and every scalar c (see Theorem 4.3.2):

$$T(\mathbf{u} + \mathbf{v}) = T(\mathbf{u}) + T(\mathbf{v})$$
$$T(c\mathbf{u}) = cT(\mathbf{u})$$

We shall use these properties as the starting point for general linear transformations.

Definition. If $T: V \to W$ is a function from a vector space V into a vector space W, then T is called a ***linear transformation*** from V to W if for all vectors **u** and **v** in V and all scalars c

(a) $T(\mathbf{u} + \mathbf{v}) = T(\mathbf{u}) + T(\mathbf{v})$
(b) $T(c\mathbf{u}) = cT(\mathbf{u})$

In the special case where $V = W$, the linear transformation $T: V \to V$ is called a ***linear operator*** on V.

Example 1 Because the preceding definition of a linear transformation was based on Theorem 4.3.2, linear transformations from R^n to R^m, as defined in Section 4.2, are linear transformations under this more general definition as well. We shall call linear transformations from R^n to R^m **matrix transformations**, since they can be carried out by matrix multiplication. ▲

Example 2 Let V and W be any two vector spaces. The mapping $T: V \rightarrow W$ such that $T(\mathbf{v}) = \mathbf{0}$ for every $\mathbf{v}$ in V is a linear transformation called the **zero transformation**. To see that T is linear, observe that

$$T(\mathbf{u} + \mathbf{v}) = \mathbf{0}, \quad T(\mathbf{u}) = \mathbf{0}, \quad T(\mathbf{v}) = \mathbf{0}, \quad \text{and} \quad T(k\mathbf{u}) = \mathbf{0}$$

Therefore,

$$T(\mathbf{u} + \mathbf{v}) = T(\mathbf{u}) + T(\mathbf{v}) \quad \text{and} \quad T(k\mathbf{u}) = kT(\mathbf{u}) \quad ▲$$

Example 3 Let V be any vector space. The mapping $I: V \rightarrow V$ defined by $I(\mathbf{v}) = \mathbf{v}$ is called the **identity operator** on V. The verification that I is linear is left as an exercise. ▲

Example 4 Let V be any vector space and k any fixed scalar. We leave it as an exercise to check that the function $T: V \rightarrow V$ defined by

$$T(\mathbf{v}) = k\mathbf{v}$$

is a linear operator on V. This linear operator is called a **dilation** of V with factor k if $k > 1$ and is called a **contraction** of V with factor k if $0 < k < 1$. Geometrically, the dilation "stretches" each vector in V by a factor of k, and the contraction of V "compresses" each vector by a factor of k (Figure 1). ▲

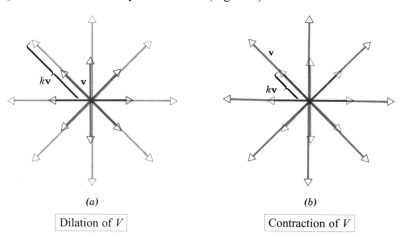

(a)

(b)

Figure 1

Dilation of V

Contraction of V

Example 5 In Section 6.4 we defined the orthogonal projection of R^m onto a subspace W. [See Formula (6) and the definition preceding it in that section.] Orthogonal projections can also be defined in general inner product spaces as follows: Suppose that W is a finite-dimensional subspace of an inner product space V; then the **orthogonal projection of V onto W** is the transformation defined by

$$T(\mathbf{v}) = \text{proj}_W \mathbf{v}$$

(Figure 2). It follows from Theorem 6.3.5 that if

$$S = \{\mathbf{w}_1, \mathbf{w}_2, \ldots, \mathbf{w}_r\}$$

is any orthonormal basis for W, then $T(\mathbf{v})$ is given by the formula

$$T(\mathbf{v}) = \text{proj}_W \mathbf{v} = \langle \mathbf{v}, \mathbf{w}_1 \rangle \mathbf{w}_1 + \langle \mathbf{v}, \mathbf{w}_2 \rangle \mathbf{w}_2 + \cdots + \langle \mathbf{v}, \mathbf{w}_r \rangle \mathbf{w}_r$$

Figure 2   The orthogonal projection of V onto W

The proof that T is a linear transformation follows from properties of the inner product. For example,

$$T(\mathbf{u} + \mathbf{v}) = \langle \mathbf{u} + \mathbf{v}, \mathbf{w}_1 \rangle \mathbf{w}_1 + \langle \mathbf{u} + \mathbf{v}, \mathbf{w}_2 \rangle \mathbf{w}_2 + \cdots + \langle \mathbf{u} + \mathbf{v}, \mathbf{w}_r \rangle \mathbf{w}_r$$
$$= \langle \mathbf{u}, \mathbf{w}_1 \rangle \mathbf{w}_1 + \langle \mathbf{u}, \mathbf{w}_2 \rangle \mathbf{w}_2 + \cdots + \langle \mathbf{u}, \mathbf{w}_r \rangle \mathbf{w}_r$$
$$+ \langle \mathbf{v}, \mathbf{w}_1 \rangle \mathbf{w}_1 + \langle \mathbf{v}, \mathbf{w}_2 \rangle \mathbf{w}_2 + \cdots + \langle \mathbf{v}, \mathbf{w}_r \rangle \mathbf{w}_r$$
$$= T(\mathbf{u}) + T(\mathbf{v})$$

Similarly, $T(k\mathbf{u}) = kT(\mathbf{u})$. ▲

Example 6   As a special case of the preceding example, let $V = R^3$ have the Euclidean inner product. The vectors $\mathbf{w}_1 = (1, 0, 0)$ and $\mathbf{w}_2 = (0, 1, 0)$ form an orthonormal basis for the xy-plane. Thus, if $\mathbf{v} = (x, y, z)$ is any vector in R^3, the orthogonal projection of R^3 onto the xy-plane is given by

$$T(\mathbf{v}) = \langle \mathbf{v}, \mathbf{w}_1 \rangle \mathbf{w}_1 + \langle \mathbf{v}, \mathbf{w}_2 \rangle \mathbf{w}_2$$
$$= x(1, 0, 0) + y(0, 1, 0)$$
$$= (x, y, 0)$$

(See Figure 3.) ▲

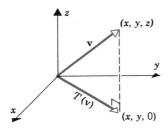

Figure 3   The orthogonal projection of R^3 onto the xy-plane

Example 7 Let $S = \{\mathbf{w}_1, \mathbf{w}_2, \ldots, \mathbf{w}_n\}$ be a basis for an n-dimensional vector space V, and let

$$(\mathbf{v})_S = (k_1, k_2, \ldots, k_n)$$

be the coordinate vector relative to S of a vector $\mathbf{v}$ in V; thus

$$\mathbf{v} = k_1\mathbf{w}_1 + k_2\mathbf{w}_2 + \cdots + k_n\mathbf{w}_n$$

Define $T: V \to R^n$ to be the function that maps $\mathbf{v}$ into its coordinate vector relative to S; that is,

$$T(\mathbf{v}) = (\mathbf{v})_S = (k_1, k_2, \ldots, k_n)$$

The function T is a linear transformation. To see that this is so, suppose that $\mathbf{u}$ and $\mathbf{v}$ are vectors in V and that

$$\mathbf{u} = c_1\mathbf{w}_1 + c_2\mathbf{w}_2 + \cdots + c_n\mathbf{w}_n \qquad \text{and} \qquad \mathbf{v} = d_1\mathbf{w}_1 + d_2\mathbf{w}_2 + \cdots + d_n\mathbf{w}_n$$

Thus,

$$(\mathbf{u})_S = (c_1, c_2, \ldots, c_n)$$
$$(\mathbf{v})_S = (d_1, d_2, \ldots, d_n)$$

But

$$\mathbf{u} + \mathbf{v} = (c_1 + d_1)\mathbf{w}_1 + (c_2 + d_2)\mathbf{w}_2 + \cdots + (c_n + d_n)\mathbf{w}_n$$
$$k\mathbf{u} = (kc_1)\mathbf{w}_1 + (kc_2)\mathbf{w}_2 + \cdots + (kc_n)\mathbf{w}_n$$

so that

$$(\mathbf{u} + \mathbf{v})_S = (c_1 + d_1, c_2 + d_2, \ldots, c_n + d_n)$$
$$(k\mathbf{u})_S = (kc_1, kc_2, \ldots, kc_n)$$

Therefore,

$$(\mathbf{u} + \mathbf{v})_S = (\mathbf{u})_S + (\mathbf{v})_S \qquad \text{and} \qquad (k\mathbf{u})_S = k(\mathbf{u})_S$$

Expressing these equations in terms of T, we obtain

$$T(\mathbf{u} + \mathbf{v}) = T(\mathbf{u}) + T(\mathbf{v}) \qquad \text{and} \qquad T(k\mathbf{u}) = kT(\mathbf{u})$$

which shows that T is a linear transformation. ▲

REMARK. The computations in the preceding example could just as well have been performed using coordinate matrices rather than coordinate vectors; that is,

$$[\mathbf{u} + \mathbf{v}]_S = [\mathbf{u}]_S + [\mathbf{v}]_S$$

and

$$[k\mathbf{u}]_S = k[\mathbf{u}]_S$$

Example 8 Let $\mathbf{p} = p(x) = c_0 + c_1 x + \cdots + c_n x^n$ be a polynomial in P_n, and define the function $T: P_n \to P_{n+1}$ by

$$T(\mathbf{p}) = T(p(x)) = xp(x) = c_0 x + c_1 x^2 + \cdots + c_n x^{n+1}$$

The function T is a linear transformation, since for any scalar k and any polynomials $\mathbf{p}_1$ and $\mathbf{p}_2$ in P_n we have

$$T(\mathbf{p}_1 + \mathbf{p}_2) = T(p_1(x) + p_2(x)) = x(p_1(x) + p_2(x))$$
$$= xp_1(x) + xp_2(x) = T(\mathbf{p}_1) + T(\mathbf{p}_2)$$

and

$$T(k\mathbf{p}) = T(kp(x)) = x(kp(x)) = k(xp(x)) = kT(\mathbf{p}) \quad \blacktriangle$$

Example 9   Let $\mathbf{p} = p(x) = c_0 + c_1 x + \cdots + c_n x^n$ be a polynomial in P_n, and let a and b be any scalars. We leave it as an exercise to show that the function T defined by

$$T(\mathbf{p}) = T(p(x)) = p(ax + b) = c_0 + c_1(ax + b) + \cdots + c_n(ax + b)^n$$

is a linear operator. For example, if $ax + b = 3x - 5$, then $T: P_2 \to P_2$ would be the linear operator given by the formula

$$T(c_0 + c_1 x + c_2 x^2) = c_0 + c_1(3x - 5) + c_2(3x - 5)^2 \quad \blacktriangle$$

Example 10   Let V be an inner product space and let $\mathbf{v}_0$ be any fixed vector in V. Let $T: V \to R$ be the transformation that maps a vector $\mathbf{v}$ into its inner product with $\mathbf{v}_0$; that is,

$$T(\mathbf{v}) = \langle \mathbf{v}, \mathbf{v}_0 \rangle$$

From the properties of an inner product

$$T(\mathbf{u} + \mathbf{v}) = \langle \mathbf{u} + \mathbf{v}, \mathbf{v}_0 \rangle = \langle \mathbf{u}, \mathbf{v}_0 \rangle + \langle \mathbf{v}, \mathbf{v}_0 \rangle = T(\mathbf{u}) + T(\mathbf{v})$$

and

$$T(k\mathbf{u}) = \langle k\mathbf{u}, \mathbf{v}_0 \rangle = k\langle \mathbf{u}, \mathbf{v}_0 \rangle = kT(\mathbf{u})$$

so that T is a linear transformation. $\blacktriangle$

Example 11 (*For readers who have studied calculus.*)   Let $V = C^1(-\infty, \infty)$ be the vector space of functions with continuous first derivatives on $(-\infty, \infty)$ and let $W = F(-\infty, \infty)$ be the vector space of all real-valued functions defined on $(-\infty, \infty)$. Let $D: V \to W$ be the transformation that maps a function $\mathbf{f} = f(x)$ into its derivative; that is,

$$D(\mathbf{f}) = f'(x)$$

From the properties of differentiation, we have

$$D(\mathbf{f} + \mathbf{g}) = D(\mathbf{f}) + D(\mathbf{g})$$

and

$$D(k\mathbf{f}) = kD(\mathbf{f})$$

Thus, D is a linear transformation. $\blacktriangle$

Example 12 (*For readers who have studied calculus.*)   Let $V = C(-\infty, \infty)$ be the vector space of continuous functions $(-\infty, \infty)$ and let $W = C^1(-\infty, \infty)$ be the vector space of functions with continuous first derivatives on $(-\infty, \infty)$. Let $J: V \rightarrow W$ be the transformation that maps $\mathbf{f} = f(x)$ into the integral

$$\int_0^x f(t)\, dt$$

For example, if $\mathbf{f} = x^2$, then

$$J(\mathbf{f}) = \int_0^x t^2\, dt = \left.\frac{t^3}{3}\right|_0^x = \frac{x^3}{3}$$

From the properties of integration, we have

$$J(\mathbf{f} + \mathbf{g}) = \int_0^x (f(t) + g(t))\, dt = \int_0^x f(t)\, dt + \int_0^x g(t)\, dt = J(\mathbf{f}) + J(\mathbf{g})$$

$$J(c\mathbf{f}) = \int_0^x cf(t)\, dt = c\int_0^x f(t)\, dt = cJ(\mathbf{f})$$

so J is a linear transformation.   ▲

Example 13   Let $T: M_{nn} \rightarrow R$ be the transformation that maps an $n \times n$ matrix into its determinant; that is,

$$T(A) = \det(A)$$

This transformation does not satisfy either of the properties required of a linear transformation. For example, we saw in Example 1 of Section 2.3 that

$$\det(A_1 + A_2) \neq \det(A_1) + \det(A_2)$$

in general. Moreover, $\det(cA) = c^n \det(A)$, so

$$\det(cA) \neq c\det(A)$$

in general. Thus, T is *not* a linear transformation.   ▲

PROPERTIES OF LINEAR TRANSFOR- MATIONS

If $T: V \rightarrow W$ is a linear transformation, then for any vectors $\mathbf{v}_1$ and $\mathbf{v}_2$ in V and any scalars c_1 and c_2, we have

$$T(c_1\mathbf{v}_1 + c_2\mathbf{v}_2) = T(c_1\mathbf{v}_1) + T(c_2\mathbf{v}_2) = c_1 T(\mathbf{v}_1) + c_2 T(\mathbf{v}_2)$$

and more generally, if $\mathbf{v}_1, \mathbf{v}_2, \ldots, \mathbf{v}_n$ are vectors in V and $c_1, c_2, \ldots, c_n$ are scalars, then

$$T(c_1\mathbf{v}_1 + c_2\mathbf{v}_2 + \cdots + c_n\mathbf{v}_n) = c_1 T(\mathbf{v}_1) + c_2 T(\mathbf{v}_2) + \cdots + c_n T(\mathbf{v}_n) \qquad (1)$$

Formula (1) is sometimes described by saying that *linear transformations preserve linear combinations.*

The following theorem lists three basic properties common to all linear transformations.

Theorem 8.1.1. *If $T: V \rightarrow W$ is a linear transformation, then:*

(a) $T(\mathbf{0}) = \mathbf{0}$

(b) $T(-\mathbf{v}) = -T(\mathbf{v})$ *for all* $\mathbf{v}$ *in* V

(c) $T(\mathbf{v} - \mathbf{w}) = T(\mathbf{v}) - T(\mathbf{w})$ *for all* $\mathbf{v}$ *and* $\mathbf{w}$ *in* V

Proof. Let $\mathbf{v}$ be any vector in V. Since $0\mathbf{v} = \mathbf{0}$, we have

$$T(\mathbf{0}) = T(0\mathbf{v}) = 0T(\mathbf{v}) = \mathbf{0}$$

which proves (a). Also,

$$T(-\mathbf{v}) = T((-1)\mathbf{v}) = (-1)T(\mathbf{v}) = -T(\mathbf{v})$$

which proves (b). Finally, $\mathbf{v} - \mathbf{w} = \mathbf{v} + (-1)\mathbf{w}$; thus,

$$T(\mathbf{v} - \mathbf{w}) = T(\mathbf{v} + (-1)\mathbf{w})$$
$$= T(\mathbf{v}) + (-1)T(\mathbf{w})$$
$$= T(\mathbf{v}) - T(\mathbf{w})$$

which proves (c). ▨

In words, part (a) of the preceding theorem states that a linear transformation maps $\mathbf{0}$ into $\mathbf{0}$. This property is useful for identifying transformations that are *not* linear. For example, if $\mathbf{x}_0$ is a fixed nonzero vector in R^2, then the transformation

$$T(\mathbf{x}) = \mathbf{x} + \mathbf{x}_0$$

has the geometric effect of translating each point $\mathbf{x}$ in a direction parallel to $\mathbf{x}_0$ through a distance of $\|\mathbf{x}_0\|$ (Figure 4). This is not a linear transformation, since $T(\mathbf{0}) = x_0$, so T does not map $\mathbf{0}$ to $\mathbf{0}$.

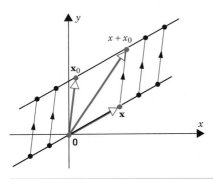

Figure 4 | $T(\mathbf{x}) = \mathbf{x} + \mathbf{x}_0$ translates each point $\mathbf{x}$ along a line parallel to $\mathbf{x}_0$ through a distance $\|\mathbf{x}_0\|$.

Theorem 4.3.3 shows that if T is a matrix transformation, then the standard matrix for T can be obtained from the images of the standard basis vectors. Stated another way, *a matrix transformation is completely determined by its images of the standard basis vectors*. This is a special case of a more general result: If $T: V \to W$ is a linear transformation, and if $\{\mathbf{v}_1, \mathbf{v}_2, \ldots, \mathbf{v}_n\}$ is any basis for V, then the image $T(\mathbf{v})$ of any vector $\mathbf{v}$ in V can be calculated from the images

$$T(\mathbf{v}_1), T(\mathbf{v}_2), \ldots, T(\mathbf{v}_n)$$

of the basis vectors. This can be done by first expressing $\mathbf{v}$ as a linear combination of the basis vectors, say

$$\mathbf{v} = c_1 \mathbf{v}_1 + c_2 \mathbf{v}_2 + \cdots + c_n \mathbf{v}_n$$

and then using Formula (1) to write

$$T(\mathbf{v}) = c_1 T(\mathbf{v}_1) + c_2 T(\mathbf{v}_2) + \cdots + c_n T(\mathbf{v}_n)$$

In words, *a linear transformation is completely determined by its images of any basis vectors*.

Example 14 Consider the basis $S = \{\mathbf{v}_1, \mathbf{v}_2, \mathbf{v}_3\}$ for R^3, where $\mathbf{v}_1 = (1, 1, 1)$, $\mathbf{v}_2 = (1, 1, 0)$, $\mathbf{v}_3 = (1, 0, 0)$; and let $T: R^3 \to R^2$ be the linear transformation such that

$$T(\mathbf{v}_1) = (1, 0), \qquad T(\mathbf{v}_2) = (2, -1), \qquad T(\mathbf{v}_3) = (4, 3)$$

Find a formula for $T(x_1, x_2, x_3)$; then use this formula to compute $T(2, -3, 5)$.

Solution. We first express $\mathbf{x} = (x_1, x_2, x_3)$ as a linear combination of $\mathbf{v}_1 = (1, 1, 1)$, $\mathbf{v}_2 = (1, 1, 0)$, and $\mathbf{v}_3 = (1, 0, 0)$. If we write

$$(x_1, x_2, x_3) = c_1(1, 1, 1) + c_2(1, 1, 0) + c_3(1, 0, 0)$$

then on equating corresponding components we obtain

$$\begin{aligned}
c_1 + c_2 + c_3 &= x_1 \\
c_1 + c_2 \phantom{{}+ c_3} &= x_2 \\
c_1 \phantom{{}+ c_2 + c_3} &= x_3
\end{aligned}$$

which yields $c_1 = x_3$, $c_2 = x_2 - x_3$, $c_3 = x_1 - x_2$, so that

$$\begin{aligned}
(x_1, x_2, x_3) &= x_3(1, 1, 1) + (x_2 - x_3)(1, 1, 0) + (x_1 - x_2)(1, 0, 0) \\
&= x_3 \mathbf{v}_1 + (x_2 - x_3) \mathbf{v}_2 + (x_1 - x_2) \mathbf{v}_3
\end{aligned}$$

Thus,

$$\begin{aligned}
T(x_1, x_2, x_3) &= x_3 T(\mathbf{v}_1) + (x_2 - x_3) T(\mathbf{v}_2) + (x_1 - x_2) T(\mathbf{v}_3) \\
&= x_3(1, 0) + (x_2 - x_3)(2, -1) + (x_1 - x_2)(4, 3) \\
&= (4x_1 - 2x_2 - x_3, \; 3x_1 - 4x_2 + x_3)
\end{aligned}$$

From this formula we obtain

$$T(2, -3, 5) = (9, 23) \quad \blacktriangle$$

COMPOSITIONS OF LINEAR TRANSFOR- MATIONS

In Section 4.2 we defined the composition of matrix transformations. The following definition extends that concept to general linear transformations.

Definition. If $T_1: U \rightarrow V$ and $T_2: V \rightarrow W$ are linear transformations, the ***composition of T_2 with T_1***, denoted by $T_2 \circ T_1$ (read "T_2 circle T_1"), is the function defined by the formula

$$(T_2 \circ T_1)(\mathbf{u}) = T_2(T_1(\mathbf{u})) \tag{2}$$

where $\mathbf{u}$ is a vector in U.

REMARK. Observe that this definition requires that the domain of T_2 (which is V) contain the range of T_1; this is essential for the formula $T_2(T_1(\mathbf{u}))$ to make sense (Figure 5). The reader should compare (2) to Formula (18) in Section 4.2.

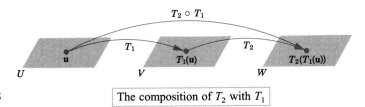

Figure 5

The composition of T_2 with T_1

The next result shows that the composition of two linear transformations is itself a linear transformation.

Theorem 8.1.2. *If $T_1: U \rightarrow V$ and $T_2: V \rightarrow W$ are linear transformations, then $(T_2 \circ T_1): U \rightarrow W$ is also a linear transformation.*

Proof. If $\mathbf{u}$ and $\mathbf{v}$ are vectors in U and c is a scalar, then it follows from (2) and the linearity of T_1 and T_2 that

$$(T_2 \circ T_1)(\mathbf{u} + \mathbf{v}) = T_2(T_1(\mathbf{u} + \mathbf{v})) = T_2(T_1(\mathbf{u}) + T_1(\mathbf{v}))$$
$$= T_2(T_1(\mathbf{u})) + T_2(T_1(\mathbf{v}))$$
$$= (T_2 \circ T_1)(\mathbf{u}) + (T_2 \circ T_1)(\mathbf{v})$$

and

$$(T_2 \circ T_1)(c\mathbf{u}) = T_2(T_1(c\mathbf{u})) = T_2(cT_1(\mathbf{u}))$$
$$= cT_2(T_1(\mathbf{u})) = c(T_2 \circ T_1)(\mathbf{u})$$

Thus, $T_2 \circ T_1$ satisfies the two requirements of a linear transformation.

Example 15   Let $T_1 : P_1 \rightarrow P_2$ and $T_2 : P_2 \rightarrow P_2$ be the linear transformations given by the formulas

$$T_1(p(x)) = xp(x) \quad \text{and} \quad T_2(p(x)) = p(2x + 4)$$

Then the composition $(T_2 \circ T_1) : P_1 \rightarrow P_2$ is given by the formula

$$(T_2 \circ T_1)(p(x)) = T_2(T_1(p(x))) = T_2(xp(x)) = (2x + 4)p(2x + 4)$$

In particular, if $p(x) = c_0 + c_1 x$, then

$$(T_2 \circ T_1)(p(x)) = (T_2 \circ T_1)(c_0 + c_1 x) = (2x + 4)(c_0 + c_1(2x + 4))$$
$$= c_0(2x + 4) + c_1(2x + 4)^2 \quad \blacktriangle$$

Example 16   If $T : V \rightarrow V$ is any linear operator, and if $I : V \rightarrow V$ is the identity operator (Example 3), then for all vectors $\mathbf{v}$ in V we have

$$(T \circ I)(\mathbf{v}) = T(I(\mathbf{v})) = T(\mathbf{v})$$
$$(I \circ T)(\mathbf{v}) = I(T(\mathbf{v})) = T(\mathbf{v})$$

It follows that $T \circ I$ and $I \circ T$ are the same as T; that is,

$$T \circ I = T$$
$$I \circ T = T \quad \blacktriangle \tag{3}$$

We conclude this section by noting that compositions can be defined for more than two linear transformations. For example, if

$$T_1 : U \rightarrow V, \quad T_2 : V \rightarrow W, \quad \text{and} \quad T_3 : W \rightarrow Y$$

are linear transformations, then the composition $T_3 \circ T_2 \circ T_1$ is defined by

$$(T_3 \circ T_2 \circ T_1)(\mathbf{u}) = T_3(T_2(T_1(\mathbf{u}))) \tag{4}$$

(Figure 6).

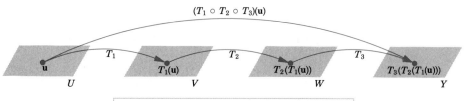

Figure 6   The composition of three linear transformations

EXERCISE SET 8.1

1. Use the definition of a linear operator given in this section to show that the function $T : R^2 \rightarrow R^2$ given by the formula $T(x_1, x_2) = (x_1 + 2x_2, 3x_1 - x_2)$ is a linear operator.

2. Use the definition of a linear transformation given in this section to show that the function $T: R^3 \to R^2$ given by the formula $T(x_1, x_2, x_3) = (2x_1 - x_2 + x_3, x_2 - 4x_3)$ is a linear transformation.

In Exercises 3–10 determine whether the function is a linear transformation. Justify your answer.

3. $T: V \to R$, where V is an inner product space, and $T(\mathbf{u}) = \|\mathbf{u}\|$.

4. $T: R^3 \to R^3$, where $\mathbf{v}_0$ is a fixed vector in R^3 and $T(\mathbf{u}) = \mathbf{u} \times \mathbf{v}_0$.

5. $T: M_{22} \to M_{23}$, where B is a fixed 2×3 matrix and $T(A) = AB$.

6. $T: M_{nn} \to R$, where $T(A) = \mathrm{tr}(A)$.

7. $F: M_{mn} \to M_{nm}$, where $F(A) = A^T$.

8. $T: M_{22} \to R$, where

   (a) $T\left(\begin{bmatrix} a & b \\ c & d \end{bmatrix}\right) = 3a - 4b + c - d$ (b) $T\left(\begin{bmatrix} a & b \\ c & d \end{bmatrix}\right) = a^2 + b^2$

9. $T: P_2 \to P_2$, where
   (a) $T(a_0 + a_1 x + a_2 x^2) = a_0 + a_1(x + 1) + a_2(x + 1)^2$
   (b) $T(a_0 + a_1 x + a_2 x^2) = (a_0 + 1) + (a_1 + 1)x + (a_2 + 1)x^2$

10. $T: F(-\infty, \infty) \to F(-\infty, \infty)$, where
    (a) $T(f(x)) = 1 + f(x)$ (b) $T(f(x)) = f(x + 1)$

11. Show that the function T in Example 9 is a linear operator.

12. Consider the basis $S = \{\mathbf{v}_1, \mathbf{v}_2\}$ for R^2, where $\mathbf{v}_1 = (1, 1)$ and $\mathbf{v}_2 = (1, 0)$, and let $T: R^2 \to R^2$ be the linear operator such that

    $$T(\mathbf{v}_1) = (1, -2) \quad \text{and} \quad T(\mathbf{v}_2) = (-4, 1)$$

    Find a formula for $T(x_1, x_2)$, and use that formula to find $T(5, -3)$.

13. Consider the basis $S = \{\mathbf{v}_1, \mathbf{v}_2\}$ for R^2, where $\mathbf{v}_1 = (-2, 1)$ and $\mathbf{v}_2 = (1, 3)$, and let $T: R^2 \to R^3$ be the linear transformation such that

    $$T(\mathbf{v}_1) = (-1, 2, 0) \quad \text{and} \quad T(\mathbf{v}_2) = (0, -3, 5)$$

    Find a formula for $T(x_1, x_2)$, and use that formula to find $T(2, -3)$.

14. Consider the basis $S = \{\mathbf{v}_1, \mathbf{v}_2, \mathbf{v}_3\}$ for R^3, where $\mathbf{v}_1 = (1, 1, 1)$, $\mathbf{v}_2 = (1, 1, 0)$, and $\mathbf{v}_3 = (1, 0, 0)$ and let $T: R^3 \to R^3$ be the linear operator such that

    $$T(\mathbf{v}_1) = (2, -1, 4), \quad T(\mathbf{v}_2) = (3, 0, 1), \quad T(\mathbf{v}_3) = (-1, 5, 1)$$

    Find a formula for $T(x_1, x_2, x_3)$, and use that formula to find $T(2, 4, -1)$.

15. Consider the basis $S = \{\mathbf{v}_1, \mathbf{v}_2, \mathbf{v}_3\}$ for R^3, where $\mathbf{v}_1 = (1, 2, 1)$, $\mathbf{v}_2 = (2, 9, 0)$, and $\mathbf{v}_3 = (3, 3, 4)$, and let $T: R^3 \to R^2$ be the linear transformation such that

    $$T(\mathbf{v}_1) = (1, 0), \quad T(\mathbf{v}_2) = (-1, 1), \quad T(\mathbf{v}_3) = (0, 1)$$

    Find a formula for $T(x_1, x_2, x_3)$, and use that formula to find $T(7, 13, 7)$.

16. Let $\mathbf{v}_1$, $\mathbf{v}_2$, and $\mathbf{v}_3$ be vectors in a vector space V and $T: V \to R^3$ a linear transformation for which

    $$T(\mathbf{v}_1) = (1, -1, 2), \quad T(\mathbf{v}_2) = (0, 3, 2), \quad T(\mathbf{v}_3) = (-3, 1, 2)$$

    Find $T(2\mathbf{v}_1 - 3\mathbf{v}_2 + 4\mathbf{v}_3)$.

17. Find the domain and codomain of $T_2 \circ T_1$, and find $(T_2 \circ T_1)(x, y)$.
    (a) $T_1(x, y) = (2x, 3y)$, $T_2(x, y) = (x - y, x + y)$
    (b) $T_1(x, y) = (x - 3y, 0)$, $T_2(x, y) = (4x - 5y, 3x - 6y)$
    (c) $T_1(x, y) = (2x, -3y, x + y)$, $T_2(x, y, z) = (x - y, y + z)$
    (d) $T_1(x, y) = (x - y, y + z, x - z)$, $T_2(x, y, z) = (0, x + y + z)$

18. Find the domain and codomain of $T_3 \circ T_2 \circ T_1$, and find $(T_3 \circ T_2 \circ T_1)(x, y)$.
    (a) $T_1(x, y) = (-2y, 3x, x - 2y)$, $T_2(x, y, z) = (y, z, x)$, $T_3(x, y, z) = (x + z, y - z)$
    (b) $T_1(x, y) = (x + y, y, -x)$, $T_2(x, y, z) = (0, x + y + z, 3y)$,
        $T_3(x, y, z) = (3x + 2y, 4z - x - 3y)$

19. Let $T_1 : M_{22} \rightarrow R$ and $T_2 : M_{22} \rightarrow M_{22}$ be the linear transformations given by $T_1(A) = \text{tr}(A)$ and $T_2(A) = A^T$.
    (a) Find $(T_1 \circ T_2)(A)$, where $A = \begin{bmatrix} a & b \\ c & d \end{bmatrix}$.
    (b) Can you find $(T_2 \circ T_1)(A)$? Explain.

20. Let $T_1 : P_n \rightarrow P_n$ and $T_2 : P_n \rightarrow P_n$ be the linear operators given by $T_1(p(x)) = p(x - 1)$ and $T_2(p(x)) = p(x + 1)$. Find $(T_1 \circ T_2)(p(x))$ and $(T_2 \circ T_1)(p(x))$.

21. Let $T_1 : V \rightarrow V$ be the dilation $T_1(\mathbf{v}) = 4\mathbf{v}$. Find a linear operator $T_2 : V \rightarrow V$ such that $T_1 \circ T_2 = I$ and $T_2 \circ T_1 = I$.

22. Suppose that the linear transformations $T_1 : P_2 \rightarrow P_2$ and $T_2 : P_2 \rightarrow P_3$ are given by the formulas $T_1(p(x)) = p(x + 1)$ and $T_2(p(x)) = xp(x)$. Find $(T_2 \circ T_1)(a_0 + a_1 x + a_2 x^2)$.

23. Let $q_0(x)$ be a fixed polynomial of degree m, and define a function T with domain P_n by the formula $T(p(x)) = p(q_0(x))$.
    (a) Show that T is a linear transformation.   (b) What is the codomain of T?

24. Use the definition of $T_3 \circ T_2 \circ T_1$ given by Formula (4) to prove that
    (a) $T_3 \circ T_2 \circ T_1$ is a linear transformation
    (b) $T_3 \circ T_2 \circ T_1 = (T_3 \circ T_2) \circ T_1$
    (c) $T_3 \circ T_2 \circ T_1 = T_3 \circ (T_2 \circ T_1)$

25. Let $T : R^3 \rightarrow R^3$ be the orthogonal projection of R^3 onto the xy-plane. Show that $T \circ T = T$.

26. (a) Let $T : V \rightarrow W$ be a linear transformation and k a scalar. Define the function $(kT) : V \rightarrow W$ by $(kT)(\mathbf{v}) = k(T(\mathbf{v}))$. Show that kT is a linear transformation.
    (b) Find $(3T)(x_1, x_2)$ if $T : R^2 \rightarrow R^2$ is given by the formula $T(x_1, x_2) = (2x_1 - x_2, x_2 + x_1)$.

27. (a) Let $T_1 : V \rightarrow W$ and $T_2 : V \rightarrow W$ be linear transformations. Define the functions $(T_1 + T_2) : V \rightarrow W$ and $(T_1 - T_2) : V \rightarrow W$ by

$$(T_1 + T_2)(\mathbf{v}) = T_1(\mathbf{v}) + T_2(\mathbf{v})$$
$$(T_1 - T_2)(\mathbf{v}) = T_1(\mathbf{v}) - T_2(\mathbf{v})$$

    Show that $T_1 + T_2$ and $T_1 - T_2$ are linear transformations.
    (b) Find $(T_1 + T_2)(x, y)$ and $(T_1 - T_2)(x, y)$ if $T_1 : R^2 \rightarrow R^2$ and $T_2 : R^2 \rightarrow R^2$ are given by the formulas $T_1(x, y) = (2y, 3x)$ and $T_2(x, y) = (y, x)$.

28. (a) Prove that if a_1, a_2, b_1, and b_2 are any scalars, then the formula

$$F(x, y) = (a_1 x + b_1 y, a_2 x + b_2 y)$$

    defines a linear operator on R^2.
    (b) Does the formula $F(x, y) = (a_1 x^2 + b_1 y^2, a_2 x^2 + b_2 y^2)$ define a linear operator on R^2? Explain.

29. (*For readers who have studied calculus.*) Let

$$D(\mathbf{f}) = f'(x) \quad \text{and} \quad J(\mathbf{f}) = \int_0^x f(t)\, dt$$

be the linear transformations in Examples 11 and 12. Find $(J \circ D)(\mathbf{f})$ for
(a) $\mathbf{f}(x) = x^2 + 3x + 2$ (b) $\mathbf{f}(x) = \sin x$ (c) $\mathbf{f}(x) = x$

30. Let $\{\mathbf{v}_1, \mathbf{v}_2, \ldots, \mathbf{v}_n\}$ be a basis for a vector space V, and let $T: V \rightarrow W$ be a linear transformation. Show that if $T(\mathbf{v}_1) = T(\mathbf{v}_2) = \cdots = T(\mathbf{v}_n) = \mathbf{0}$, then T is the zero transformation.

31. Let $\{\mathbf{v}_1, \mathbf{v}_2, \ldots, \mathbf{v}_n\}$ be a basis for a vector space V, and let $T: V \rightarrow V$ be a linear operator. Show that if $T(\mathbf{v}_1) = \mathbf{v}_1$, $T(\mathbf{v}_2) = \mathbf{v}_2$, $\ldots$, $T(\mathbf{v}_n) = \mathbf{v}_n$, then T is the identity transformation on V.

8.2 KERNEL AND RANGE

In this section we shall develop some basic properties of linear transformations that generalize properties of matrix transformations obtained earlier in the text.

KERNEL AND RANGE

Recall that if A is an $m \times n$ matrix, then the nullspace of A consists of all vectors $\mathbf{x}$ in R^n such that $A\mathbf{x} = \mathbf{0}$, and by Theorem 5.5.1 the column space of A consists of all vectors $\mathbf{b}$ in R^m for which there is at least one vector $\mathbf{x}$ in R^n such that $A\mathbf{x} = \mathbf{b}$. From the viewpoint of matrix transformations, the nullspace of A consists of all vectors in R^n that multiplication by A maps into $\mathbf{0}$, and the column space of A consists of all vectors in R^m that are images of at least one vector in R^n under multiplication by A. The following definition extends these ideas to general linear transformations.

> **Definition.** If $T: V \rightarrow W$ is a linear transformation, then the set of vectors in V that T maps into $\mathbf{0}$ is called the *kernel* of T; it is denoted by $\ker(T)$. The set of all vectors in W that are images under T of at least one vector in V is called the *range* of T; it is denoted by $R(T)$.

Example 1 If $T_A: R^n \rightarrow R^m$ is multiplication by the $m \times n$ matrix A, then from the discussion preceding the definition above, the kernel of T_A is the nullspace of A, and the range of T_A is the column space of A. ▲

Example 2 Let $T: V \rightarrow W$ be the zero transformation (Example 2 of Section 8.1). Since T maps *every* vector in V into $\mathbf{0}$, it follows that $\ker(T) = V$. Moreover, since $\mathbf{0}$ is the *only* image under T of vectors in V, we have $R(T) = \{\mathbf{0}\}$. ▲

Example 3 Let $I: V \rightarrow V$ be the identity operator (Example 3 of Section 8.1). Since $I(\mathbf{v}) = \mathbf{v}$ for all vectors in V, *every* vector in V is the image of some vector (namely, itself); thus, $R(I) = V$. Since the *only* vector that I maps into $\mathbf{0}$ is $\mathbf{0}$, it follows that $\ker(I) = \{\mathbf{0}\}$. ▲

Example 4 Let $T: R^3 \to R^3$ be the orthogonal projection on the xy-plane. The kernel of T is the set of points that T maps into $\mathbf{0} = (0, 0, 0)$; these are the points on the z-axis (Figure 1a). Since T maps every point in R^3 into the xy-plane, the range of T must be some subset of this plane. But every point $(x_0, y_0, 0)$ in the xy-plane is the image under T of some point; in fact, it is the image of all points on the vertical line that passes through $(x_0, y_0, 0)$ (Figure 1b). Thus $R(T)$ is the entire xy-plane. ▲

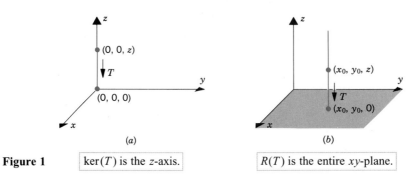

(a)   (b)

Figure 1 $\ker(T)$ is the z-axis. $R(T)$ is the entire xy-plane.

Example 5 Let $T: R^2 \to R^2$ be the linear operator that rotates each vector in the xy-plane through the angle θ (Figure 2). Since *every* vector in the xy-plane can be obtained by rotating some vector through the angle θ (why?), we have $R(T) = R^2$. Moreover, the only vector that rotates into $\mathbf{0}$ is $\mathbf{0}$, so $\ker(T) = \{\mathbf{0}\}$. ▲

Figure 2

Example 6 (*For readers who have studied calculus.*) Let $V = C^1(-\infty, \infty)$ be the vector space of functions with continuous first derivatives on $(-\infty, \infty)$, let $W = F(-\infty, \infty)$ be the vector space of all real-valued functions defined on $(-\infty, \infty)$, and let $D: V \to W$ be the differentiation transformation $D(f) = f'(x)$. The kernel of D is the set of functions in V with derivative zero. From calculus, this is the set of constant functions on $(-\infty, \infty)$. ▲

PROPERTIES OF KERNEL AND RANGE

In all of the preceding examples, $\ker(T)$ and $R(T)$ turned out to be *subspaces*. In Examples 2, 3, and 5 they were either the zero subspace or the entire vector space. In Example 4 the kernel was a line through the origin and the range was a plane through the origin, both of which are subspaces of R^3. All of this is not accidental; it is a consequence of the following general result.

Theorem 8.2.1. *If* $T:V \to W$ *is a linear transformation, then:*

(a) *The kernel of T is a subspace of V.*

(b) *The range of T is a subspace of W.*

Proof (a). To show that $\ker(T)$ is a subspace, we must show that it contains at least one vector and is closed under addition and scalar multiplication. By part (a) of Theorem 8.1.1, the vector **0** is in $\ker(T)$, so this set contains at least one vector. Let $\mathbf{v}_1$ and $\mathbf{v}_2$ be vectors in $\ker(T)$, and let k be any scalar. Then

$$T(\mathbf{v}_1 + \mathbf{v}_2) = T(\mathbf{v}_1) + T(\mathbf{v}_2) = \mathbf{0} + \mathbf{0} = \mathbf{0}$$

so that $\mathbf{v}_1 + \mathbf{v}_2$ is in $\ker(T)$. Also,

$$T(k\mathbf{v}_1) = kT(\mathbf{v}_1) = k\mathbf{0} = \mathbf{0}$$

so that $k\mathbf{v}_1$ is in $\ker(T)$.

Proof (b). Since $T(\mathbf{0}) = \mathbf{0}$, there is at least one vector in $R(T)$. Let $\mathbf{w}_1$ and $\mathbf{w}_2$ be vectors in the range of T, and let k be any scalar. To prove this part we must show that $\mathbf{w}_1 + \mathbf{w}_2$ and $k\mathbf{w}_1$ are in the range of T; that is, we must find vectors **a** and **b** in V such that $T(\mathbf{a}) = \mathbf{w}_1 + \mathbf{w}_2$ and $T(\mathbf{b}) = k\mathbf{w}_1$.

Since $\mathbf{w}_1$ and $\mathbf{w}_2$ are in the range of T, there are vectors $\mathbf{a}_1$ and $\mathbf{a}_2$ in V such that $T(\mathbf{a}_1) = \mathbf{w}_1$ and $T(\mathbf{a}_2) = \mathbf{w}_2$. Let $\mathbf{a} = \mathbf{a}_1 + \mathbf{a}_2$ and $\mathbf{b} = k\mathbf{a}_1$. Then

$$T(\mathbf{a}) = T(\mathbf{a}_1 + \mathbf{a}_2) = T(\mathbf{a}_1) + T(\mathbf{a}_2) = \mathbf{w}_1 + \mathbf{w}_2$$

and

$$T(\mathbf{b}) = T(k\mathbf{a}_1) = kT(\mathbf{a}_1) = k\mathbf{w}_1$$

which completes the proof. ▮

RANK AND NULLITY OF LINEAR TRANSFORMATIONS

In Section 5.6 we defined the rank of a matrix to be the dimension of its column (or row) space and the nullity to be the dimension of its nullspace. The following definition extends these definitions to general linear transformations.

Definition. If $T:V \to W$ is a linear transformation, then the dimension of the range of T is called the ***rank of T*** and is denoted by $\text{rank}(T)$; the dimension of the kernel is called the ***nullity of T*** and is denoted by $\text{nullity}(T)$.

If A is an $m \times n$ matrix and $T_A : R^n \to R^m$ is multiplication by A, then we know from Example 1 that the kernel of T_A is the nullspace of A and the range of T_A is the column space of A. Thus, we have the following relationship between the rank and nullity of a matrix and the rank and nullity of the corresponding matrix transformation.

Theorem 8.2.2. *If A is an $m \times n$ matrix and $T_A : R^n \to R^m$ is multiplication by A, then:*

(a) *$\text{nullity}(T_A) = \text{nullity}(A)$*

(b) *$\text{rank}(T_A) = \text{rank}(A)$*

Example 7 Let $T_A : R^6 \to R^4$ be multiplication by

$$
A = \begin{bmatrix}
-1 & 2 & 0 & 4 & 5 & -3 \\
3 & -7 & 2 & 0 & 1 & 4 \\
2 & -5 & 2 & 4 & 6 & 1 \\
4 & -9 & 2 & -4 & -4 & 7
\end{bmatrix}
$$

Find the rank and nullity of T_A.

Solution. In Example 1 of Section 5.6 we showed that $\text{rank}(A) = 2$ and $\text{nullity}(A) = 4$. Thus, from Theorem 8.2.2 we have $\text{rank}(T_A) = 2$ and $\text{nullity}(T_A) = 4$. ▲

Example 8 Let $T : R^3 \to R^3$ be the orthogonal projection on the xy-plane. From Example 4, the kernel of T is the z-axis, which is one-dimensional; and the range of T is the xy-plane, which is two-dimensional. Thus,

$$
\text{nullity}(T) = 1 \quad \text{and} \quad \text{rank}(T) = 2 \quad ▲
$$

DIMENSION THEOREM FOR LINEAR TRANSFORMATIONS

Recall from the Dimension Theorem for Matrices (Theorem 5.6.3) that if A is a matrix with n columns, then

$$
\text{rank}(A) + \text{nullity}(A) = n
$$

The following theorem, whose proof is deferred to the end of the section, extends this result to general linear transformations.

Theorem 8.2.3 (*Dimension Theorem for Linear Transformations*). *If $T : V \to W$ is a linear transformation from an n-dimensional vector space V to a vector space W, then*

$$
rank(T) + nullity(T) = n \tag{1}
$$

In words, this theorem states that *for linear transformations the rank plus the nullity is equal to the dimension of the domain.*

REMARK. If A is an $m \times n$ matrix and $T_A : R^n \to R^m$ is multiplication by A, then the domain of T_A has dimension n, so Theorem 8.2.3 agrees with Theorem 5.6.3 in this case.

Example 9 Let $T : R^2 \to R^2$ be the linear operator that rotates each vector in the xy-plane through an angle θ. We showed in Example 5 that $\ker(T) = \{\mathbf{0}\}$ and $R(T) = R^2$. Thus,

$$
\text{rank}(T) + \text{nullity}(T) = 0 + 2 = 2
$$

which is consistent with the fact that the domain of T is two-dimensional. ▲

ADDITIONAL PROOF

Proof of Theorem 8.2.3. We must show that

$$\dim(R(T)) + \dim(\ker(T)) = n$$

We shall give the proof for the case where $1 \leq \dim(\ker(T)) < n$. The cases $\dim(\ker(T)) = 0$ and $\dim(\ker(T)) = n$ are left as exercises. Assume $\dim(\ker(T)) = r$, and let $\mathbf{v}_1, \ldots, \mathbf{v}_r$ be a basis for the kernel. Since $\{\mathbf{v}_1, \ldots, \mathbf{v}_r\}$ is linearly independent, Theorem 5.4.6b states that there are $n - r$ vectors, $\mathbf{v}_{r+1}, \ldots, \mathbf{v}_n$, such that $\{\mathbf{v}_1, \ldots, \mathbf{v}_r, \mathbf{v}_{r+1}, \ldots, \mathbf{v}_n\}$ is a basis for V. To complete the proof, we shall show that the $n - r$ vectors in the set $S = \{T(\mathbf{v}_{r+1}), \ldots, T(\mathbf{v}_n)\}$ form a basis for the range of T. It will then follow that

$$\dim(R(T)) + \dim(\ker(T)) = (n - r) + r = n$$

First we show that S spans the range of T. If $\mathbf{b}$ is any vector in the range of T, then $\mathbf{b} = T(\mathbf{v})$ for some vector $\mathbf{v}$ in V. Since $\{\mathbf{v}_1, \ldots, \mathbf{v}_r, \mathbf{v}_{r+1}, \ldots, \mathbf{v}_n\}$ is a basis for V, the vector $\mathbf{v}$ can be written in the form

$$\mathbf{v} = c_1 \mathbf{v}_1 + \cdots + c_r \mathbf{v}_r + c_{r+1} \mathbf{v}_{r+1} + \cdots + c_n \mathbf{v}_n$$

Since $\mathbf{v}_1, \ldots, \mathbf{v}_r$ lie in the kernel of T, we have $T(\mathbf{v}_1) = \cdots = T(\mathbf{v}_r) = \mathbf{0}$, so that

$$\mathbf{b} = T(\mathbf{v}) = c_{r+1} T(\mathbf{v}_{r+1}) + \cdots + c_n T(\mathbf{v}_n)$$

Thus, S spans the range of T.

Finally, we show that S is a linearly independent set and consequently forms a basis for the range of T. Suppose that some linear combination of the vectors in S is zero; that is,

$$k_{r+1} T(\mathbf{v}_{r+1}) + \cdots + k_n T(\mathbf{v}_n) = \mathbf{0} \tag{2}$$

We must show $k_{r+1} = \cdots = k_n = 0$. Since T is linear, (2) can be rewritten as

$$T(k_{r+1} \mathbf{v}_{r+1} + \cdots + k_n \mathbf{v}_n) = \mathbf{0}$$

which says that $k_{r+1} \mathbf{v}_{r+1} + \cdots + k_n \mathbf{v}_n$ is in the kernel of T. This vector can therefore be written as a linear combination of the basis vectors $\{\mathbf{v}_1, \ldots, \mathbf{v}_r\}$, say

$$k_{r+1} \mathbf{v}_{r+1} + \cdots + k_n \mathbf{v}_n = k_1 \mathbf{v}_1 + \cdots + k_r \mathbf{v}_r$$

Thus,

$$k_1 \mathbf{v}_1 + \cdots + k_r \mathbf{v}_r - k_{r+1} \mathbf{v}_{r+1} - \cdots - k_n \mathbf{v}_n = \mathbf{0}$$

Since $\{\mathbf{v}_1, \ldots, \mathbf{v}_n\}$ is linearly independent, all of the k's are zero; in particular, $k_{r+1} = \cdots = k_n = 0$, which completes the proof.

EXERCISE SET 8.2

1. Let $T : R^2 \to R^2$ be the linear operator given by the formula

$$T(x, y) = (2x - y, -8x + 4y)$$

Which of the following vectors are in $R(T)$?
(a) $(1, -4)$ (b) $(5, 0)$ (c) $(-3, 12)$

2. Let $T: R^2 \to R^2$ be the linear operator in Exercise 1. Which of the following vectors are in $\ker(T)$?
(a) $(5, 10)$ (b) $(3, 2)$ (c) $(1, 1)$

3. Let $T: R^4 \to R^3$ be the linear transformation given by the formula

$$T(x_1, x_2, x_3, x_4) = (4x_1 + x_2 - 2x_3 - 3x_4, \ 2x_1 + x_2 + x_3 - 4x_4, \ 6x_1 - 9x_3 + 9x_4)$$

Which of the following are in $R(T)$?
(a) $(0, 0, 6)$ (b) $(1, 3, 0)$ (c) $(2, 4, 1)$

4. Let $T: R^4 \to R^3$ be the linear transformation in Exercise 3. Which of the following are in $\ker(T)$?
(a) $(3, -8, 2, 0)$ (b) $(0, 0, 0, 1)$ (c) $(0, -4, 1, 0)$

5. Let $T: P_2 \to P_3$ be the linear transformation defined by $T(p(x)) = xp(x)$. Which of the following are in $\ker(T)$?
(a) x^2 (b) 0 (c) $1 + x$

6. Let $T: P_2 \to P_3$ be the linear transformation in Exercise 5. Which of the following are in $R(T)$?
(a) $x + x^2$ (b) $1 + x$ (c) $3 - x^2$

7. Find a basis for the kernel of
(a) the linear operator in Exercise 1   (b) the linear transformation in Exercise 3
(c) the linear transformation in Exercise 5

8. Find a basis for the range of
(a) the linear operator in Exercise 1   (b) the linear transformation in Exercise 3
(c) the linear transformation in Exercise 5

9. Verify Formula (1) of the dimension theorem for
(a) the linear operator in Exercise 1   (b) the linear transformation in Exercise 3
(c) the linear transformation in Exercise 5.

In Exercises 10–13 let T be multiplication by the matrix A. Find
(a) a basis for the range of T (b) a basis for the kernel of T
(c) the rank and nullity of T (d) the rank and nullity of A

10. $A = \begin{bmatrix} 1 & -1 & 3 \\ 5 & 6 & -4 \\ 7 & 4 & 2 \end{bmatrix}$ 11. $A = \begin{bmatrix} 2 & 0 & -1 \\ 4 & 0 & -2 \\ 0 & 0 & 0 \end{bmatrix}$

12. $A = \begin{bmatrix} 4 & 1 & 5 & 2 \\ 1 & 2 & 3 & 0 \end{bmatrix}$ 13. $A = \begin{bmatrix} 1 & 4 & 5 & 0 & 9 \\ 3 & -2 & 1 & 0 & -1 \\ -1 & 0 & -1 & 0 & -1 \\ 2 & 3 & 5 & 1 & 8 \end{bmatrix}$

14. Describe the range and nullspace of
    (a) the orthogonal projection on the xz-plane
    (b) the orthogonal projection on the yz-plane
    (c) the orthogonal projection on the plane with the equation $y = x$

15. Let V be any vector space, and let $T: V \rightarrow V$ be defined by $T(\mathbf{v}) = 3\mathbf{v}$.
    (a) What is the kernel of T?   (b) What is the range of T?

16. In each part use the given information to find the nullity of T.
    (a) $T: R^5 \rightarrow R^7$ has rank 3.   (b) $T: P_4 \rightarrow P_3$ has rank 1.
    (c) The range of $T: R^6 \rightarrow R^3$ is R^3.   (d) $T: M_{22} \rightarrow M_{22}$ has rank 3.

17. Let A be a 7×6 matrix such that $A\mathbf{x} = \mathbf{0}$ has only the trivial solution, and let $T: R^6 \rightarrow R^7$ be multiplication by A. Find the rank and nullity of T.

18. Let A be a 5×7 matrix with rank 4.
    (a) What is the dimension of the solution space of $A\mathbf{x} = \mathbf{0}$?
    (b) Is $A\mathbf{x} = \mathbf{b}$ consistent for all vectors $\mathbf{b}$ in R^5? Explain.

19. Let $T: R^3 \rightarrow V$ be a linear transformation from R^3 to any vector space. Show that the kernel of T is a line through the origin, a plane through the origin, the origin only, or all of R^3.

20. Let $T: V \rightarrow R^3$ be a linear transformation from any vector space to R^3. Show that the range of T is a line through the origin, a plane through the origin, the origin only, or all of R^3.

21. Let $T: R^3 \rightarrow R^3$ be multiplication by

$$\begin{bmatrix} 1 & 3 & 4 \\ 3 & 4 & 7 \\ -2 & 2 & 0 \end{bmatrix}$$

    (a) Show that the kernel of T is a line through the origin, and find parametric equations for it.
    (b) Show that the range of T is a plane through the origin, and find an equation for it.

22. Prove: If $\{\mathbf{v}_1, \mathbf{v}_2, \ldots, \mathbf{v}_n\}$ is a basis for V and $\mathbf{w}_1, \mathbf{w}_2, \ldots, \mathbf{w}_n$ are vectors in W, not necessarily distinct, then there exists a linear transformation $T: V \rightarrow W$ such that $T(\mathbf{v}_1) = \mathbf{w}_1$, $T(\mathbf{v}_2) = \mathbf{w}_2, \ldots, T(\mathbf{v}_n) = \mathbf{w}_n$.

23. Prove the dimension theorem in the cases
    (a) $\dim(\ker(T)) = 0$ (b) $\dim(\ker(T)) = n$

24. Let $T: V \rightarrow V$ be a linear operator on a finite-dimensional vector space V. Prove that $R(T) = V$ if and only if $\ker(T) = \{\mathbf{0}\}$.

25. **(For readers who have studied calculus.)** Let $D: P_3 \rightarrow P_2$ be the differentiation transformation $D(\mathbf{p}) = p'(x)$. Describe the kernel of D.

26. **(For readers who have studied calculus.)** Let $J: P_1 \rightarrow R$ be the integration transformation $J(\mathbf{p}) = \int_{-1}^{1} p(x)\, dx$. Describe the kernel of J.

27. **(For readers who have studied calculus.)** Let $D: V \rightarrow W$ be the differentiation transformation $D(\mathbf{f}) = f'(x)$, where $V = C^2(-\infty, \infty)$ and $W = F(-\infty, \infty)$. Describe the kernel of $D \circ D$.

8.3 INVERSE LINEAR TRANSFORMATIONS

In section 4.3 we discussed properties of one-to-one linear transformations from R^n to R^m. In this section we shall extend those ideas to general linear transformations.

ONE-TO-ONE LINEAR TRANS-FORMATIONS

Recall from Section 4.3 that a linear transformation from R^n to R^m is called *one-to-one* if it maps distinct vectors in R^n into distinct vectors in R^m. The following definition generalizes that idea.

> **Definition.** A linear transformation $T : V \rightarrow W$ is said to be **one-to-one** if T maps distinct vectors in V into distinct vectors in W.

Example 1 Recall from Theorem 4.3.1 that if A is an $n \times n$ matrix and $T_A : R^n \rightarrow R^n$ is multiplication by A, then T_A is one-to-one if and only if A is an invertible matrix. ▲

Example 2 Let $T : P_n \rightarrow P_{n+1}$ be the linear transformation

$$T(\mathbf{p}) = T(p(x)) = xp(x)$$

discussed in Example 8 of Section 8.1. If

$$\mathbf{p} = p(x) = c_0 + c_1 x + \cdots + c_n x^n \qquad \text{and} \qquad \mathbf{q} = q(x) = d_0 + d_1 x + \cdots + d_n x^n$$

are distinct polynomials, then they differ in at least one coefficient. Thus,

$$T(\mathbf{p}) = c_0 x + c_1 x^2 + \cdots + c_n x^{n+1} \qquad \text{and} \qquad T(\mathbf{q}) = d_0 x + d_1 x^2 + \cdots + d_n x^{n+1}$$

also differ in at least one coefficient. Thus, T is one-to-one, since it maps distinct polynomials $\mathbf{p}$ and $\mathbf{q}$ into distinct polynomials $T(\mathbf{p})$ and $T(\mathbf{q})$. ▲

Example 3 (*For readers who have studied calculus.*) Let

$$D : C^1(-\infty, \infty) \rightarrow F(-\infty, \infty)$$

be the differentiation transformation discussed in Example 11 of Section 8.1. This linear transformation is *not* one-to-one because it maps functions that differ by a constant into the same function. For example,

$$D(x^2) = D(x^2 + 1) = 2x \quad ▲$$

The following theorem establishes a relationship between a one-to-one linear transformation and its kernel.

> **Theorem 8.3.1.** *If $T : V \rightarrow W$ is a linear transformation, then the following are equivalent.*
>
> (a) *T is one-to-one.*
> (b) *The kernel of T contains only the zero vector; that is, $\ker(T) = \{\mathbf{0}\}$.*
> (c) *nullity$(T) = 0$.*

Proof. We leave it as a simple exercise to show the equivalence of (b) and (c); we shall complete the proof by proving the equivalence of (a) and (b).

$(a) \Rightarrow (b)$: Assume that T is one-to-one, and let $\mathbf{v}$ be any vector in ker(T). Since $\mathbf{v}$ and $\mathbf{0}$ both lie in ker(T), we have $T(\mathbf{v}) = \mathbf{0}$ and $T(\mathbf{0}) = \mathbf{0}$, so $T(\mathbf{v}) = T(\mathbf{0})$. But this implies that $\mathbf{v} = \mathbf{0}$, since T is one-to-one; thus, ker(T) contains only the zero vector.

$(b) \Rightarrow (a)$: Assume that ker(T) = {$\mathbf{0}$} and that $\mathbf{v}$ and $\mathbf{w}$ are distinct vectors in V; that is,

$$\mathbf{v} - \mathbf{w} \neq \mathbf{0} \tag{1}$$

To prove that T is one-to-one we must show that $T(\mathbf{v})$ and $T(\mathbf{w})$ are distinct vectors. But if this were not so, then we would have

$$T(\mathbf{v}) = T(\mathbf{w})$$
$$T(\mathbf{v}) - T(\mathbf{w}) = \mathbf{0}$$
$$T(\mathbf{v} - \mathbf{w}) = \mathbf{0}$$

which implies that $\mathbf{v} - \mathbf{w}$ is in the kernel of T. Since ker(T) = {$\mathbf{0}$}, this implies that

$$\mathbf{v} - \mathbf{w} = \mathbf{0}$$

which contradicts (1). Thus, $T(\mathbf{v})$ and $T(\mathbf{w})$ must be distinct. ▨

Example 4 In each part determine whether the linear transformation is one-to-one by finding the kernel or the nullity and applying Theorem 8.3.1.

(a) $T: R^2 \rightarrow R^2$ rotates each vector through the angle θ.

(b) $T: R^3 \rightarrow R^3$ is the orthogonal projection on the xy-plane.

(c) $T: R^6 \rightarrow R^4$ is multiplication by the matrix

$$A = \begin{bmatrix} -1 & 2 & 0 & 4 & 5 & -3 \\ 3 & -7 & 2 & 0 & 1 & 4 \\ 2 & -5 & 2 & 4 & 6 & 1 \\ 4 & -9 & 2 & -4 & -4 & 7 \end{bmatrix}$$

Solution (a). From Example 5 of Section 8.2, ker(T) = {$\mathbf{0}$}, so T is one-to-one.

Solution (b). From Example 4 of Section 8.2, ker(T) contains nonzero vectors, so T is not one-to-one.

Solution (c). From Example 7 of Section 8.2, nullity(T) = 4, so T is not one-to-one. ▲

In the special case where T is a *linear operator* on a *finite-dimensional* vector space, a fourth equivalent statement can be added to those in Theorem 8.3.1.

> **Theorem 8.3.2.** *If V is a finite-dimensional vector space, and T: V→V is a linear operator, then the following are equivalent.*
>
> (*a*) *T is one-to-one.*
> (*b*) *ker(T) = {**0**}.*
> (*c*) *nullity(T) = 0.*
> (*d*) *The range of T is V; that is, R(T) = V.*

Proof. We already know that (*a*), (*b*), and (*c*) are equivalent, so we can complete the proof by proving the equivalence of (*c*) and (*d*).

(*c*)⇒(*d*): Suppose that $\dim(V) = n$ and $\text{nullity}(T) = 0$. It follows from the dimension theorem (Theorem 8.2.3) that

$$\text{rank}(T) = n - \text{nullity}(T) = n$$

By definition, $\text{rank}(T)$ is the dimension of the range of T, so the range of T has dimension n. It now follows from Theorem 5.4.7 that the range of T is V, since the two spaces have the same dimension.

(*d*)⇒(*c*): Suppose that $\dim(V) = n$ and $R(T) = V$. It follows from these relationships that $\dim(R(T)) = n$, or equivalently, $\text{rank}(T) = n$. Thus, it follows from the dimension theorem (Theorem 8.2.3) that

$$\text{nullity}(T) = n - \text{rank}(T) = n - n = 0 \quad \blacksquare$$

Example 5 Let $T_A: R^4 \rightarrow R^4$ be multiplication by

$$A = \begin{bmatrix} 1 & 3 & -2 & 4 \\ 2 & 6 & -4 & 8 \\ 3 & 9 & 1 & 5 \\ 1 & 1 & 4 & 8 \end{bmatrix}$$

Determine whether T_A is one-to-one.

Solution. As noted in Example 1, the given problem is equivalent to determining whether A is invertible. But $\det(A) = 0$, since the first two rows of A are proportional, and consequently A is not invertible. Thus, T_A is not one-to-one. ▲

INVERSE LINEAR TRANSFOR-MATIONS

In Section 4.3 we defined the *inverse* of a one-to-one matrix operator $T_A: R^n \rightarrow R^n$ to be the matrix operator $T_{A^{-1}}: R^n \rightarrow R^n$, and we showed that if **w** is the image of a vector **x** under T_A, then T_A^{-1} maps **w** back into **x**. We shall now extend these ideas to general linear transformations.

Recall that if $T: V \rightarrow W$ is a linear transformation, then the range of T, denoted by $R(T)$, is the subspace of W consisting of all images under T of vectors in V. If T is one-to-one, then each vector **v** in V has a unique image $\mathbf{w} = T(\mathbf{v})$ in $R(T)$. This uniqueness of the image vector allows us to define a new function, called the ***inverse of T***, and denoted by T^{-1}, which maps **w** back into **v** (Figure 1).

Figure 1 | The inverse of T maps $T(\mathbf{v})$ back into $\mathbf{v}$. |

It can be proved (Exercise 19) that $T^{-1}: R(T) \rightarrow V$ is a linear transformation. Moreover, it follows from the definition of T^{-1} that

$$T^{-1}(T(\mathbf{v})) = T^{-1}(\mathbf{w}) = \mathbf{v} \tag{2a}$$

$$T(T^{-1}(\mathbf{w})) = T(\mathbf{v}) = \mathbf{w} \tag{2b}$$

so that T and T^{-1}, when applied in succession in either order, cancel the effect of one another.

REMARK. It is important to note that if $T: V \rightarrow W$ is a one-to-one linear transformation, then the domain of T^{-1} is the *range* of T. The range may or may not be all of W. However, in the special case where $T: V \rightarrow V$ is a one-to-one linear operator it follows from Theorem 8.3.2 that $R(T) = V$; that is, the domain of T^{-1} is *all* of V.

Example 6 In Example 2 we showed that the linear transformation $T: P_n \rightarrow P_{n+1}$ given by

$$T(\mathbf{p}) = T(p(x)) = xp(x)$$

is one-to-one; thus, T has an inverse. Here, the range of T is *not* all of P_{n+1}; rather, $R(T)$ is the subspace of P_{n+1} consisting of polynomials with a zero constant term. This is evident from the formula for T:

$$T(c_0 + c_1 x + \cdots + c_n x^n) = c_0 x + c_1 x^2 + \cdots + c_n x^{n+1}$$

It follows that $T^{-1}: R(T) \rightarrow P_n$ is given by the formula

$$T^{-1}(c_0 x + c_1 x^2 + \cdots + c_n x^{n+1}) = c_0 + c_1 x + \cdots + c_n x^n$$

For example, in the case where $n = 4$,

$$T^{-1}(2x - x^2 + 5x^3 + 3x^4) = 2 - x + 5x^2 + 3x^3 \;\; \blacktriangle$$

Example 7 Let $T: R^3 \rightarrow R^3$ be the linear operator defined by the formula

$$T(x_1, x_2, x_3) = (3x_1 + x_2, \; -2x_1 - 4x_2 + 3x_3, \; 5x_1 + 4x_2 - 2x_3)$$

Determine whether T is one-to-one; if so, find $T^{-1}(x_1, x_2, x_3)$.

Solution. From Theorem 4.3.3 the standard matrix for T is

$$[T] = \begin{bmatrix} 3 & 1 & 0 \\ -2 & -4 & 3 \\ 5 & 4 & -2 \end{bmatrix}$$

(verify). This matrix is invertible and from Formula (1) of Section 4.3 the standard matrix for T^{-1} is

$$[T^{-1}] = [T]^{-1} = \begin{bmatrix} 4 & -2 & -3 \\ -11 & 6 & 9 \\ -12 & 7 & 10 \end{bmatrix}$$

It follows that

$$T^{-1}\left(\begin{bmatrix} x_1 \\ x_2 \\ x_3 \end{bmatrix}\right) = [T^{-1}]\begin{bmatrix} x_1 \\ x_2 \\ x_3 \end{bmatrix} = \begin{bmatrix} 4 & -2 & -3 \\ -11 & 6 & 9 \\ -12 & 7 & 10 \end{bmatrix}\begin{bmatrix} x_1 \\ x_2 \\ x_3 \end{bmatrix} = \begin{bmatrix} 4x_1 - 2x_2 - 3x_3 \\ -11x_1 + 6x_2 + 9x_3 \\ -12x_1 + 7x_2 + 10x_3 \end{bmatrix}$$

Expressing this result in horizontal notation yields

$$T^{-1}(x_1, x_2, x_3) = (4x_1 - 2x_2 - 3x_3, \ -11x_1 + 6x_2 + 9x_3, \ -12x_1 + 7x_2 + 10x_3) \ \blacktriangle$$

INVERSES OF COMPOSITIONS

The following theorem shows that a composition of one-to-one linear transformations is one-to-one, and it relates the inverse of the composition to the inverses of the individual linear transformations.

Theorem 8.3.3. *If $T_1 : U \rightarrow V$ and $T_2 : V \rightarrow W$ are one-to-one linear transformations, then:*

(a) $T_2 \circ T_1$ is one-to-one.
(b) $(T_2 \circ T_1)^{-1} = T_1^{-1} \circ T_2^{-1}$.

Proof (a). We want to show that $T_2 \circ T_1$ maps distinct vectors in U into distinct vectors in W. But if $\mathbf{u}$ and $\mathbf{v}$ are distinct vectors in U, then $T_1(\mathbf{u})$ and $T_1(\mathbf{v})$ are distinct vectors in V since T_1 is one-to-one. This and the fact that T_2 is one-to-one implies that

$$T_2(T_1(\mathbf{u})) \qquad \text{and} \qquad T_2(T_1(\mathbf{v}))$$

are also distinct vectors. But these expressions can also be written as

$$(T_2 \circ T_1)(\mathbf{u}) \qquad \text{and} \qquad (T_2 \circ T_1)(\mathbf{v})$$

so $T_2 \circ T_1$ maps $\mathbf{u}$ and $\mathbf{v}$ into distinct vectors in W.

Proof (b). We want to show that

$$(T_2 \circ T_1)^{-1}(\mathbf{w}) = (T_1^{-1} \circ T_2^{-1})(\mathbf{w})$$

for every vector $\mathbf{w}$ in the range of $T_2 \circ T_1$. For this purpose let

$$\mathbf{u} = (T_2 \circ T_1)^{-1}(\mathbf{w}) \tag{3}$$

so our goal is to show that

$$\mathbf{u} = (T_1^{-1} \circ T_2^{-1})(\mathbf{w})$$

But it follows from (3) that

$$(T_2 \circ T_1)(\mathbf{u}) = \mathbf{w}$$

or equivalently,

$$T_2(T_1(\mathbf{u})) = \mathbf{w}$$

Now, taking T_2^{-1} of each side of this equation and then T_1^{-1} of each side of the result yields (verify)

$$\mathbf{u} = T_1^{-1}(T_2^{-1}(\mathbf{w}))$$

or equivalently,

$$\mathbf{u} = (T_1^{-1} \circ T_2^{-1})(\mathbf{w}) \quad \blacksquare$$

In words, part (*b*) of Theorem 8.3.3 states that *the inverse of a composition is the composition of the inverses in the reverse order.* This result can be extended to compositions of three or more linear transformations; for example,

$$(T_3 \circ T_2 \circ T_1)^{-1} = T_1^{-1} \circ T_2^{-1} \circ T_3^{-1} \tag{4}$$

In the special case where T_A, T_B, and T_C are matrix operators on R^n, Formula (4) can be written as

$$(T_C \circ T_B \circ T_A)^{-1} = T_A^{-1} \circ T_B^{-1} \circ T_C^{-1}$$

or equivalently,

$$(T_{CBA})^{-1} = T_{A^{-1}B^{-1}C^{-1}} \tag{5}$$

In words, this formula states that *the standard matrix for the inverse of a composition is the product of the inverses of the standard matrices of the individual operators in the reverse order.*

Some problems that use Formulas (4) and (5) are given in the exercises.

EXERCISE SET 8.3

1. In each part find $\ker(T)$, and determine whether the linear transformation T is one-to-one.
   (a) $T: R^2 \to R^2$, where $T(x, y) = (y, x)$
   (b) $T: R^2 \to R^2$, where $T(x, y) = (0, 2x + 3y)$
   (c) $T: R^2 \to R^2$, where $T(x, y) = (x + y, x - y)$
   (d) $T: R^2 \to R^3$, where $T(x, y) = (x, y, x + y)$
   (e) $T: R^2 \to R^3$, where $T(x, y) = (x - y, y - x, 2x - 2y)$
   (f) $T: R^3 \to R^2$, where $T(x, y, z) = (x + y + z, x - y - z)$

2. In each part let $T: R^2 \to R^2$ be multiplication by A. Determine whether T has an inverse; if so, find

$$T^{-1}\left(\begin{bmatrix} x_1 \\ x_2 \end{bmatrix}\right)$$

(a) $A = \begin{bmatrix} 5 & 2 \\ 2 & 1 \end{bmatrix}$ (b) $A = \begin{bmatrix} 6 & -3 \\ 4 & -2 \end{bmatrix}$ (c) $A = \begin{bmatrix} 4 & 7 \\ -1 & 3 \end{bmatrix}$

3. In each part let $T: R^3 \to R^3$ be multiplication by A. Determine whether T has an inverse; if so, find

$$T^{-1}\left(\begin{bmatrix} x_1 \\ x_2 \\ x_3 \end{bmatrix}\right)$$

(a) $A = \begin{bmatrix} 1 & 5 & 2 \\ 1 & 2 & 1 \\ -1 & 1 & 0 \end{bmatrix}$ (b) $A = \begin{bmatrix} 1 & 4 & -1 \\ 1 & 2 & 1 \\ -1 & 1 & 0 \end{bmatrix}$ (c) $A = \begin{bmatrix} 1 & 0 & 1 \\ 0 & 1 & 1 \\ 1 & 1 & 0 \end{bmatrix}$ (d) $A = \begin{bmatrix} 1 & -1 & 1 \\ 0 & 2 & -1 \\ 2 & 3 & 0 \end{bmatrix}$

4. In each part determine whether multiplication by A is a one-to-one linear transformation.

(a) $A = \begin{bmatrix} 1 & -2 \\ 2 & -4 \\ -3 & 6 \end{bmatrix}$ (b) $A = \begin{bmatrix} 1 & 3 & 5 & 7 \\ 2 & -1 & 2 & 4 \\ -1 & 3 & 0 & 0 \end{bmatrix}$ (c) $A = \begin{bmatrix} 4 & -2 \\ 1 & 5 \\ 5 & 3 \end{bmatrix}$

5. Let $T: R^2 \to R^2$ be the orthogonal projection on the line $y = x$ (Figure 2).
   (a) Find the kernel of T.
   (b) Is T one-to-one? Justify your conclusion.

Figure 2

6. Let $T: R^2 \to R^2$ be the linear operator $T(x, y) = (-x, y)$ that reflects each point about the y-axis (Figure 3).
   (a) Find the kernel of T.
   (b) Is T one-to-one? Justify your conclusion.

Figure 3

7. In each part use the given information to determine whether T is one-to-one.
   (a) $T: R^m \to R^m$; nullity$(T) = 0$ (b) $T: R^n \to R^n$; rank$(T) = n - 1$
   (c) $T: R^m \to R^n$; $n < m$ (d) $T: R^n \to R^n$; $R(T) = R^n$

8. In each part determine whether the linear transformation T is one-to-one.
   (a) $T: P_2 \to P_3$, where $T(a_0 + a_1 x + a_2 x^2) = x(a_0 + a_1 x + a_2 x^2)$
   (b) $T: P_2 \to P_2$, where $T(p(x)) = p(x + 1)$

9. Let A be a square matrix such that $\det(A) = 0$. Is multiplication by A a one-to-one linear transformation? Justify your conclusion.

10. In each part determine whether the linear operator $T: R^n \to R^n$ is one-to-one; if so, find $T^{-1}(x_1, x_2, \ldots, x_n)$.
    (a) $T(x_1, x_2, \ldots, x_n) = (0, x_1, x_2, \ldots, x_{n-1})$ (b) $T(x_1, x_2, \ldots, x_n) = (x_n, x_{n-1}, \ldots, x_2, x_1)$
    (c) $T(x_1, x_2, \ldots, x_n) = (x_2, x_3, \ldots, x_n, x_1)$

11. Let $T: R^n \to R^n$ be the linear operator defined by the formula

    $$T(x_1, x_2, \ldots, x_n) = (a_1 x_1, a_2 x_2, \ldots, a_n x_n)$$

    (a) Under what conditions will T have an inverse?
    (b) Assuming that the conditions determined in part (a) are satisfied, find a formula for $T^{-1}(x_1, x_2, \ldots, x_n)$.

12. Let $T_1: R^2 \to R^2$ and $T_2: R^2 \to R^2$ be the linear operators given by the formulas

    $$T_1(x, y) = (x + y, x - y) \quad \text{and} \quad T_2(x, y) = (2x + y, x - 2y)$$

    (a) Show that T_1 and T_2 are one-to-one.
    (b) Find formulas for $T_1^{-1}(x, y)$, and $T_2^{-1}(x, y)$, and $(T_2 \circ T_1)^{-1}(x, y)$.
    (c) Verify that $(T_2 \circ T_1)^{-1} = T_1^{-1} \circ T_2^{-1}$.

13. Let $T_1: P_2 \to P_3$ and $T_2: P_3 \to P_3$ be the linear transformations given by the formulas

    $$T_1(p(x)) = xp(x) \quad \text{and} \quad T_2(p(x)) = p(x + 1)$$

    (a) Find formulas for $T_1^{-1}(p(x))$, $T_2^{-1}(p(x))$, and $(T_2 \circ T_1)^{-1}(p(x))$.
    (b) Verify that $(T_2 \circ T_1)^{-1} = T_1^{-1} \circ T_2^{-1}$.

14. Let $T_A: R^3 \to R^3$, $T_B: R^3 \to R^3$, and $T_C: R^3 \to R^3$ be the reflections about the xy-plane, the xz-plane, and the yz-plane, respectively. Verify Formula (5) for these linear operators.

15. Let $T: P_1 \to R^2$ be the function defined by the formula

    $$T(p(x)) = (p(0), p(1))$$

    (a) Find $T(1 - 2x)$.
    (b) Show that T is a linear transformation.
    (c) Show that T is one-to-one.
    (d) Find $T^{-1}(2, 3)$, and sketch its graph.

16. Prove: If V and W are finite-dimensional vector spaces such that $\dim W < \dim V$, then there is no one-to-one linear transformation $T: V \to W$.

17. In each part determine whether the linear operator $T: M_{22} \to M_{22}$ is one-to-one. If so, find

    $$T^{-1}\left(\begin{bmatrix} a & b \\ c & d \end{bmatrix} \right)$$

    (a) $T\left(\begin{bmatrix} a & b \\ c & d \end{bmatrix} \right) = \begin{bmatrix} a & 0 \\ 0 & d \end{bmatrix}$ (b) $T\left(\begin{bmatrix} a & b \\ c & d \end{bmatrix} \right) = \begin{bmatrix} a & c \\ b & d \end{bmatrix}$ (c) $T\left(\begin{bmatrix} a & b \\ c & d \end{bmatrix} \right) = \begin{bmatrix} d & -b \\ -c & a \end{bmatrix}$

18. Let $T: R^2 \to R^2$ be the linear operator given by the formula $T(x, y) = (x + ky, -y)$. Show that T is one-to-one for every real value of k and that $T^{-1} = T$.

19. Prove that if $T: V \rightarrow W$ is a one-to-one linear transformation, then $T^{-1}: R(T) \rightarrow V$ is a linear transformation.

20. (*For readers who have studied calculus.*) Let $J: P_1 \rightarrow R$ be the integration transformation $J(\mathbf{p}) = \int_{-1}^{1} p(x)\, dx$. Determine whether J is one-to-one. Justify your conclusion.

8.4 MATRICES OF GENERAL LINEAR TRANSFORMATIONS

In this section we shall show that if V and W are finite-dimensional vector spaces (not necessarily R^n and R^m), then with a little ingenuity any linear transformation $T: V \rightarrow W$ can be regarded as a matrix transformation. The basic idea is to work with coordinate matrices of the vectors rather than with the vectors themselves.

MATRICES OF LINEAR TRANSFOR-MATIONS

Suppose that V is an n-dimensional vector space and W an m-dimensional vector space. If we choose bases B and B' for V and W, respectively, then for each $\mathbf{x}$ in V, the coordinate matrix $[\mathbf{x}]_B$ will be a vector in R^n, and the coordinate matrix $[T(\mathbf{x})]_{B'}$ will be a vector in R^m (Figure 1).

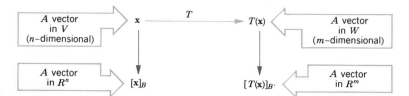

Figure 1

If, as illustrated in Figure 2, we complete the rectangle suggested by Figure 1, we obtain a mapping from R^n to R^m, which can be shown to be a linear transformation. If we let A be the standard matrix for this transformation, then

$$A[\mathbf{x}]_B = [T(\mathbf{x})]_{B'} \tag{1}$$

The matrix A in (1) is called the *matrix for T with respect to the bases B and B'*.

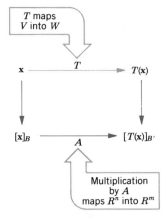

Figure 2

Later in this section, we shall give some of the uses of the matrix A in (1), but first, let us show how it can be computed. For this purpose, let us suppose that $B = \{\mathbf{u}_1, \mathbf{u}_2, \ldots, \mathbf{u}_n\}$ is a basis for the n-dimensional space V and $B' = \{\mathbf{v}_1, \mathbf{v}_2, \ldots, \mathbf{v}_m\}$ is a basis for the m-dimensional space W. We are looking for an $m \times n$ matrix

$$
A = \begin{bmatrix}
a_{11} & a_{12} & \cdots & a_{1n} \\
a_{21} & a_{22} & \cdots & a_{2n} \\
\vdots & \vdots & & \vdots \\
a_{m1} & a_{m2} & \cdots & a_{mn}
\end{bmatrix}
$$

such that (1) holds for all vectors $\mathbf{x}$ in V. In particular, we want this equation to hold for the basis vectors $\mathbf{u}_1, \mathbf{u}_2, \ldots, \mathbf{u}_n$; that is,

$$
A[\mathbf{u}_1]_B = [T(\mathbf{u}_1)]_{B'}, \quad A[\mathbf{u}_2]_B = [T(\mathbf{u}_2)]_{B'}, \quad \ldots, \quad A[\mathbf{u}_n]_B = [T(\mathbf{u}_n)]_{B'} \tag{2}
$$

But

$$
[\mathbf{u}_1]_B = \begin{bmatrix} 1 \\ 0 \\ 0 \\ \vdots \\ 0 \end{bmatrix}, \quad [\mathbf{u}_2]_B = \begin{bmatrix} 0 \\ 1 \\ 0 \\ \vdots \\ 0 \end{bmatrix}, \quad \ldots, \quad [\mathbf{u}_n]_B = \begin{bmatrix} 0 \\ 0 \\ 0 \\ \vdots \\ 1 \end{bmatrix}
$$

so

$$
A[\mathbf{u}_1]_B = \begin{bmatrix}
a_{11} & a_{12} & \cdots & a_{1n} \\
a_{21} & a_{22} & \cdots & a_{2n} \\
\vdots & \vdots & & \vdots \\
a_{m1} & a_{m2} & \cdots & a_{mn}
\end{bmatrix}
\begin{bmatrix} 1 \\ 0 \\ 0 \\ \vdots \\ 0 \end{bmatrix}
= \begin{bmatrix} a_{11} \\ a_{21} \\ \vdots \\ a_{m1} \end{bmatrix}
$$

$$
A[\mathbf{u}_2]_B = \begin{bmatrix}
a_{11} & a_{12} & \cdots & a_{1n} \\
a_{21} & a_{22} & \cdots & a_{2n} \\
\vdots & \vdots & & \vdots \\
a_{m1} & a_{m2} & \cdots & a_{mn}
\end{bmatrix}
\begin{bmatrix} 0 \\ 1 \\ 0 \\ \vdots \\ 0 \end{bmatrix}
= \begin{bmatrix} a_{12} \\ a_{22} \\ \vdots \\ a_{m2} \end{bmatrix}
$$

$$\vdots$$

$$
A[\mathbf{u}_n]_B = \begin{bmatrix}
a_{11} & a_{12} & \cdots & a_{1n} \\
a_{21} & a_{22} & \cdots & a_{2n} \\
\vdots & \vdots & & \vdots \\
a_{m1} & a_{m2} & \cdots & a_{mn}
\end{bmatrix}
\begin{bmatrix} 0 \\ 0 \\ 0 \\ \vdots \\ 1 \end{bmatrix}
= \begin{bmatrix} a_{1n} \\ a_{2n} \\ \vdots \\ a_{mn} \end{bmatrix}
$$

Substituting these results into (2) yields

$$\begin{bmatrix} a_{11} \\ a_{21} \\ \vdots \\ a_{m1} \end{bmatrix} = [\,T(\mathbf{u}_1)\,]_{B'}, \quad \begin{bmatrix} a_{12} \\ a_{22} \\ \vdots \\ a_{m2} \end{bmatrix} = [\,T(\mathbf{u}_2)\,]_{B'}, \quad \ldots, \quad \begin{bmatrix} a_{1n} \\ a_{2n} \\ \vdots \\ a_{mn} \end{bmatrix} = [\,T(\mathbf{u}_n)\,]_{B'}$$

which shows that the successive columns of A are the coordinate matrices of

$$T(\mathbf{u}_1), T(\mathbf{u}_2), \ldots, T(\mathbf{u}_n)$$

with respect to the basis B'. Thus, the matrix for T with respect to the bases B and B' is

$$A = \left[\, [\,T(\mathbf{u}_1)\,]_{B'} \,\vdots\, [\,T(\mathbf{u}_2)\,]_{B'} \,\vdots\, \cdots \,\vdots\, [\,T(\mathbf{u}_n)\,]_{B'} \,\right] \tag{3}$$

This matrix is commonly denoted by the symbol

$$[\,T\,]_{B',B}$$

so that the preceding formula can also be written as

$$[\,T\,]_{B',B} = \left[\, [\,T(\mathbf{u}_1)\,]_{B'} \,\vdots\, [\,T(\mathbf{u}_2)\,]_{B'} \,\vdots\, \cdots \,\vdots\, [\,T(\mathbf{u}_n)\,]_{B'} \,\right] \tag{4}$$

and from (1) this matrix has the property

$$[\,T\,]_{B',B}[\mathbf{x}]_B = [\,T(\mathbf{x})\,]_{B'} \tag{4a}$$

REMARK. Observe that in the notation $[\,T\,]_{B',B}$ the right subscript is a basis for the domain of T, and the left subscript is a basis for the image space of T (Figure 3).

Figure 3

Moreover, observe how the subscript B seems to "cancel out" in Formula (4a) (Figure 4).

Figure 4

**MATRICES OF
LINEAR
OPERATORS**

In the special case where $V = W$ (so that $T:V \rightarrow V$ is a linear operator) it is usual to take $B = B'$ when constructing a matrix for T. In this case the resulting matrix is called the *matrix for T with respect to the basis B* and is usually denoted by $[T]_B$ rather than $[T]_{B,B}$. If $B = \{\mathbf{u}_1, \mathbf{u}_2, \ldots, \mathbf{u}_n\}$, then in this case Formulas (4) and (4a) become

$$[T]_B = \left[[T(\mathbf{u}_1)]_B \ \vdots \ [T(\mathbf{u}_2)]_B \ \vdots \ \cdots \ \vdots \ [T(\mathbf{u}_n)]_B \right] \tag{5}$$

and

$$[T]_B[\mathbf{x}]_B = [T(\mathbf{x})]_B \tag{5a}$$

Phrased informally, (4a) and (5a) state that *the matrix for T times the coordinate matrix for* $\mathbf{x}$ *is the coordinate matrix for* $T(\mathbf{x})$.

Example 1 Let $T:P_1 \rightarrow P_2$ be the linear transformation defined by

$$T(p(x)) = xp(x)$$

Find the matrix for T with respect to the standard bases

$$B = \{\mathbf{u}_1, \mathbf{u}_2\} \qquad \text{and} \qquad B' = \{\mathbf{v}_1, \mathbf{v}_2, \mathbf{v}_3\}$$

where

$$\mathbf{u}_1 = 1, \quad \mathbf{u}_2 = x; \quad \mathbf{v}_1 = 1, \quad \mathbf{v}_2 = x, \quad \mathbf{v}_3 = x^2$$

Solution. From the given formula for T we obtain

$$T(\mathbf{u}_1) = T(1) = (x)(1) = x$$
$$T(\mathbf{u}_2) = T(x) = (x)(x) = x^2$$

By inspection, we can determine the coordinate matrices for $T(\mathbf{u}_1)$ and $T(\mathbf{u}_2)$ relative to B'; they are

$$[T(\mathbf{u}_1)]_{B'} = \begin{bmatrix} 0 \\ 1 \\ 0 \end{bmatrix}, \qquad [T(\mathbf{u}_2)]_{B'} = \begin{bmatrix} 0 \\ 0 \\ 1 \end{bmatrix}$$

Thus, the matrix for T with respect to B and B' is

$$[T]_{B',B} = \left[[T(\mathbf{u}_1)]_{B'} \ \vdots \ [T(\mathbf{u}_2)]_{B'} \right] = \begin{bmatrix} 0 & 0 \\ 1 & 0 \\ 0 & 1 \end{bmatrix} \quad \triangle$$

Example 2 Let $T:P_1 \rightarrow P_2$ be the linear transformation in Example 1. Show that the matrix

$$[T]_{B',B} = \begin{bmatrix} 0 & 0 \\ 1 & 0 \\ 0 & 1 \end{bmatrix}$$

(obtained in Example 1) satisfies (4a) for every vector $\mathbf{x} = a + bx$ in P_1.

Solution. Since $x = p(x) = a + bx$, we have

$$T(x) = xp(x) = ax + bx^2$$

For the bases B and B' in Example 1, it follows by inspection that

$$[x]_B = [ax + b]_B = \begin{bmatrix} a \\ b \end{bmatrix}$$

$$[T(x)]_{B'} = [ax + bx^2] = \begin{bmatrix} 0 \\ a \\ b \end{bmatrix}$$

Thus,

$$[T]_{B',B}[x]_B = \begin{bmatrix} 0 & 0 \\ 1 & 0 \\ 0 & 1 \end{bmatrix} \begin{bmatrix} a \\ b \end{bmatrix} = \begin{bmatrix} 0 \\ a \\ b \end{bmatrix} = [T(x)]_{B'}$$

so (4a) holds. ▲

Example 3 Let $T:R^2 \rightarrow R^3$ be the linear transformation defined by

$$T\left(\begin{bmatrix} x_1 \\ x_2 \end{bmatrix}\right) = \begin{bmatrix} x_2 \\ -5x_1 + 13x_2 \\ -7x_1 + 16x_2 \end{bmatrix}$$

Find the matrix for the transformation T with respect to the bases $B = \{u_1, u_2\}$ for R^2 and $B' = \{v_1, v_2, v_3\}$ for R^3, where

$$u_1 = \begin{bmatrix} 3 \\ 1 \end{bmatrix}, \quad u_2 = \begin{bmatrix} 5 \\ 2 \end{bmatrix}; \quad v_1 = \begin{bmatrix} 1 \\ 0 \\ -1 \end{bmatrix}, \quad v_2 = \begin{bmatrix} -1 \\ 2 \\ 2 \end{bmatrix}, \quad v_3 = \begin{bmatrix} 0 \\ 1 \\ 2 \end{bmatrix}$$

Solution. From the formula for T,

$$T(u_1) = \begin{bmatrix} 1 \\ -2 \\ -5 \end{bmatrix}, \quad T(u_2) = \begin{bmatrix} 2 \\ 1 \\ -3 \end{bmatrix}$$

Expressing these vectors as linear combinations of v_1, v_2, and v_3 we obtain (verify)

$$T(u_1) = v_1 - 2v_3, \quad T(u_2) = 3v_1 + v_2 - v_3$$

Thus,

$$[T(u_1)]_{B'} = \begin{bmatrix} 1 \\ 0 \\ -2 \end{bmatrix}, \quad [T(u_2)]_{B'} = \begin{bmatrix} 3 \\ 1 \\ -1 \end{bmatrix}$$

so

$$[T]_{B',B} = \left[[T(\mathbf{u}_1)]_{B'} \mid [T(\mathbf{u}_2)]_{B'}\right] = \begin{bmatrix} 1 & 3 \\ 0 & 1 \\ -2 & -1 \end{bmatrix} \quad \blacktriangle$$

Example 4 Let $T: R^2 \to R^2$ be the linear operator defined by

$$T\left(\begin{bmatrix} x_1 \\ x_2 \end{bmatrix}\right) = \begin{bmatrix} x_1 + x_2 \\ -2x_1 + 4x_2 \end{bmatrix}$$

and let $B = \{\mathbf{u}_1, \mathbf{u}_2\}$ be the basis, where

$$\mathbf{u}_1 = \begin{bmatrix} 1 \\ 1 \end{bmatrix}, \qquad \mathbf{u}_2 = \begin{bmatrix} 1 \\ 2 \end{bmatrix}$$

(a) Find $[T]_B$.
(b) Verify that (5a) holds for every vector $\mathbf{x}$ in R^2.

Solution (a). From the given formula for T,

$$T(\mathbf{u}_1) = \begin{bmatrix} 2 \\ 2 \end{bmatrix} = 2\mathbf{u}_1, \qquad T(\mathbf{u}_2) = \begin{bmatrix} 3 \\ 6 \end{bmatrix} = 3\mathbf{u}_2$$

Therefore,

$$[T(\mathbf{u}_1)]_B = \begin{bmatrix} 2 \\ 0 \end{bmatrix} \qquad \text{and} \qquad [T(\mathbf{u}_2)]_B = \begin{bmatrix} 0 \\ 3 \end{bmatrix}$$

Consequently,

$$[T]_B = \left[[T(\mathbf{u}_1)]_B \mid [T(\mathbf{u}_2)]_B\right] = \begin{bmatrix} 2 & 0 \\ 0 & 3 \end{bmatrix}$$

Solution (b). If

$$\mathbf{x} = \begin{bmatrix} x_1 \\ x_2 \end{bmatrix} \tag{6}$$

is any vector in R^2, then from the given formula for T

$$T(\mathbf{x}) = \begin{bmatrix} x_1 + x_2 \\ -2x_1 + 4x_2 \end{bmatrix} \tag{7}$$

To find $[\mathbf{x}]_B$ and $[T(\mathbf{x})]_B$, we must express (6) and (7) as linear combinations of $\mathbf{u}_1$ and $\mathbf{u}_2$. This yields the vector equations

$$\begin{bmatrix} x_1 \\ x_2 \end{bmatrix} = k_1 \begin{bmatrix} 1 \\ 1 \end{bmatrix} + k_2 \begin{bmatrix} 1 \\ 2 \end{bmatrix} \tag{8}$$

$$\begin{bmatrix} x_1 + x_2 \\ -2x_1 + 4x_2 \end{bmatrix} = c_1 \begin{bmatrix} 1 \\ 1 \end{bmatrix} + c_2 \begin{bmatrix} 1 \\ 2 \end{bmatrix} \tag{9}$$

Equating corresponding entries yields the linear systems

$$
\begin{aligned}
k_1 + k_2 &= x_1 \\
k_1 + 2k_2 &= x_2
\end{aligned}
\tag{10}
$$

and

$$
\begin{aligned}
c_1 + c_2 &= x_1 + x_2 \\
c_1 + 2c_2 &= -2x_1 + 4x_2
\end{aligned}
\tag{11}
$$

Solving (10) for k_1 and k_2 yields

$$
k_1 = 2x_1 - x_2, \qquad k_2 = -x_1 + x_2
$$

so that

$$
[\mathbf{x}]_B = \begin{bmatrix} 2x_1 - x_2 \\ -x_1 + x_2 \end{bmatrix}
$$

and solving (11) for c_1 and c_2 yields

$$
c_1 = 4x_1 - 2x_2, \qquad c_2 = -3x_1 + 3x_2
$$

so that

$$
[T(\mathbf{x})]_B = \begin{bmatrix} 4x_1 - 2x_2 \\ -3x_1 + 3x_2 \end{bmatrix}
$$

Thus,

$$
[T]_B[\mathbf{x}]_B = \begin{bmatrix} 2 & 0 \\ 0 & 3 \end{bmatrix} \begin{bmatrix} 2x_1 - x_2 \\ -x_1 + x_2 \end{bmatrix} = \begin{bmatrix} 4x_1 - 2x_2 \\ -3x_1 + 3x_2 \end{bmatrix} = [T(\mathbf{x})]_B
$$

so (5a) holds. ▲

MATRICES OF IDENTITY OPERATORS

Example 5 If $B = \{\mathbf{u}_1, \mathbf{u}_2, \ldots, \mathbf{u}_n\}$ is any basis for a finite-dimensional vector space V and $I: V \rightarrow V$ is the identity operator on V, then

$$
I(\mathbf{u}_1) = \mathbf{u}_1, \, I(\mathbf{u}_2) = \mathbf{u}_2, \ldots, \, I(\mathbf{u}_n) = \mathbf{u}_n
$$

Therefore,

$$
[I(\mathbf{u}_1)]_B = \begin{bmatrix} 1 \\ 0 \\ 0 \\ \vdots \\ 0 \end{bmatrix}, \quad [I(\mathbf{u}_2)]_B = \begin{bmatrix} 0 \\ 1 \\ 0 \\ \vdots \\ 0 \end{bmatrix}, \quad \ldots, \quad [I(\mathbf{u}_n)]_B = \begin{bmatrix} 0 \\ 0 \\ 0 \\ \vdots \\ 1 \end{bmatrix}
$$

Thus,

$$
[I]_B = \begin{bmatrix} 1 & 0 & \cdots & 0 \\ 0 & 1 & \cdots & 0 \\ 0 & 0 & \cdots & 0 \\ \vdots & \vdots & & \vdots \\ 0 & 0 & \cdots & 1 \end{bmatrix} = I
$$

Consequently, the matrix of the identity operator with respect to any basis is the $n \times n$ identity matrix. This result could have been anticipated from Formula (5a), since the formula yields

$$[I]_B[\mathbf{x}]_B = [I(\mathbf{x})]_B = [\mathbf{x}]_B$$

which is consistent with the fact that $[I]_B = I$. ▲

We leave it as an exercise to prove the following result.

Theorem 8.4.1. *If $T: R^n \to R^m$ is a linear transformation and if B and B' are the standard bases for R^n and R^m, respectively, then*

$$[T]_{B', B} = [T] \tag{12}$$

This theorem tells us that in the special case where T maps R^n into R^m, the matrix for T with respect to the standard bases is the standard matrix for T. In this special case Formula (4a) of this section reduces to

$$[T]\mathbf{x} = T(\mathbf{x})$$

WHY MATRICES OF LINEAR TRANSFORMATIONS ARE IMPORTANT

There are two primary reasons for studying matrices of general linear transformations, one theoretical and the other quite practical:

- Answers to theoretical questions about the structure of general linear transformations on finite-dimensional vector spaces can often be obtained by studying just the matrix transformations. Such matters are considered in detail in more advanced linear algebra courses, but will be touched on in later sections.
- These matrices make it possible to compute images of vectors using matrix multiplication. Such computations can be performed rapidly on computers.

To focus on the latter idea, let $T: V \to W$ be a linear transformation. As shown in Figure 5, the matrix $[T]_{B', B}$ can be used to calculate $T(\mathbf{x})$ in three steps by the following *indirect* procedure:

Figure 5

(1) Compute the coordinate matrix $[\mathbf{x}]_B$.
(2) Multiply $[\mathbf{x}]_B$ on the left by $[T]_{B', B}$ to produce $[T(\mathbf{x})]_{B'}$.
(3) Reconstruct $T(\mathbf{x})$ from its coordinate matrix $[T(\mathbf{x})]_{B'}$.

Example 6 Let $T: P_2 \to P_2$ be the linear operator defined by

$$T(p(x)) = p(3x - 5)$$

that is, $T(c_0 + c_1 x + c_2 x^2) = c_0 + c_1(3x - 5) + c_2(3x - 5)^2$.

(a) Find $[T]_B$ with respect to the basis $B = \{1, x, x^2\}$.
(b) Use the indirect procedure to compute $T(1 + 2x + 3x^2)$.
(c) Check the result in (b) by computing $T(1 + 2x + 3x^2)$ directly.

Solution (a). From the formula for T,

$$T(1) = 1, \qquad T(x) = 3x - 5, \qquad T(x^2) = (3x - 5)^2 = 9x^2 - 30x + 25$$

so that

$$[T(1)]_B = \begin{bmatrix} 1 \\ 0 \\ 0 \end{bmatrix}, \qquad [T(x)]_B = \begin{bmatrix} -5 \\ 3 \\ 0 \end{bmatrix}, \qquad [T(x^2)]_B = \begin{bmatrix} 25 \\ -30 \\ 9 \end{bmatrix}$$

Thus,

$$[T]_B = \begin{bmatrix} 1 & -5 & 25 \\ 0 & 3 & -30 \\ 0 & 0 & 9 \end{bmatrix}$$

Solution (b). The coordinate matrix relative to B for the vector $\mathbf{p} = 1 + 2x + 3x^2$ is

$$[\mathbf{p}]_B = \begin{bmatrix} 1 \\ 2 \\ 3 \end{bmatrix}$$

Thus, from (5a)

$$[T(1 + 2x + 3x^2)]_B = [T(\mathbf{p})]_B = [T]_B [\mathbf{p}]_B$$

$$= \begin{bmatrix} 1 & -5 & 25 \\ 0 & 3 & -30 \\ 0 & 0 & 9 \end{bmatrix} \begin{bmatrix} 1 \\ 2 \\ 3 \end{bmatrix} = \begin{bmatrix} 66 \\ -84 \\ 27 \end{bmatrix}$$

from which it follows that

$$T(1 + 2x + 3x^2) = 66 - 84x + 27x^2$$

Solution (c). By direct computation

$$\begin{aligned} T(1 + 2x + 3x^2) &= 1 + 2(3x - 5) + 3(3x - 5)^2 \\ &= 1 + 6x - 10 + 27x^2 - 90x + 75 \\ &= 66 - 84x + 27x^2 \end{aligned}$$

which agrees with the result in (b). ▲

MATRICES OF COMPOSITIONS AND INVERSE TRANSFOR-MATIONS

We shall now mention two theorems that are generalizations of Formula (21) of Section 4.2 and Formula (1) of Section 4.3. The proofs are omitted.

Theorem 8.4.2. *If $T_1 : U \to V$ and $T_2 : V \to W$ are linear transformations, and if B, B'', and B' are bases for U, V, and W, respectively, then*

$$[T_2 \circ T_1]_{B', B} = [T_2]_{B', B''}[T_1]_{B'', B} \qquad (13)$$

Theorem 8.4.3. *If $T : V \to V$ is a linear operator, and if B is a basis for V, then the following are equivalent.*

(a) T is one-to-one.
(b) $[T]_B$ is invertible.

Moreover, when these equivalent conditions hold

$$[T^{-1}]_B = [T]_B^{-1} \qquad (14)$$

REMARK. In (13), observe how the interior subscript B'' (the basis for the intermediate space V) seems to "cancel out," leaving only the bases for the domain and image space of the composition as subscripts (Figure 6).

Figure 6

This cancellation of interior subscripts suggests the following extension of Formula (13) to compositions of three linear transformations (Figure 7).

Figure 7

Basis B $\quad T_1 \quad$ Basis B'' $\quad T_2 \quad$ Basis B''' $\quad T_3 \quad$ Basis B'

$$[T_3 \circ T_2 \circ T_1]_{B', B} = [T_3]_{B', B'''}[T_2]_{B''', B''}[T_1]_{B'', B} \qquad (15)$$

The following example illustrates Theorem 8.4.2.

Example 7 Let $T_1 : P_1 \to P_2$ be the linear transformation defined by

$$T_1(p(x)) = xp(x)$$

and let $T_2 : P_2 \to P_2$ be the linear operator defined by

$$T_2(p(x)) = p(3x - 5)$$

Then the composition $(T_2 \circ T_1) : P_1 \to P_2$ is given by

$$(T_2 \circ T_1)(p(x)) = T_2(T_1(p(x))) = T_2(xp(x)) = (3x - 5)p(3x - 5)$$

Thus, if $p(x) = c_0 + c_1 x$, then

$$
\begin{aligned}
(T_2 \circ T_1)(c_0 + c_1 x) &= (3x - 5)(c_0 + c_1(3x - 5)) \\
&= c_0(3x - 5) + c_1(3x - 5)^2
\end{aligned}
\tag{16}
$$

In this example, P_1 plays the role of U in Theorem 8.4.2 and P_2 plays the roles of both V and W; thus we can take $B' = B''$ in (13) so that the formula simplifies to

$$[T_2 \circ T_1]_{B', B} = [T_2]_{B'}[T_1]_{B', B} \tag{17}$$

Let us choose $B = \{1, x\}$ to be the basis for P_1 and $B' = \{1, x, x^2\}$ to be the basis for P_2. We showed in Examples 1 and 6 that

$$
[T_1]_{B', B} = \begin{bmatrix} 0 & 0 \\ 1 & 0 \\ 0 & 1 \end{bmatrix}
\qquad \text{and} \qquad
[T_2]_{B'} = \begin{bmatrix} 1 & -5 & 25 \\ 0 & 3 & -30 \\ 0 & 0 & 9 \end{bmatrix}
$$

Thus, it follows from (17) that

$$
[T_2 \circ T_1]_{B', B} = \begin{bmatrix} 1 & -5 & 25 \\ 0 & 3 & -30 \\ 0 & 0 & 9 \end{bmatrix} \begin{bmatrix} 0 & 0 \\ 1 & 0 \\ 0 & 1 \end{bmatrix} = \begin{bmatrix} -5 & 25 \\ 3 & -30 \\ 0 & 9 \end{bmatrix}
\tag{18}
$$

As a check, we will calculate $[T_2 \circ T_1]_{B', B}$ directly from Formula (4). Since $B = \{1, x\}$, it follows from Formula (4) with $\mathbf{u}_1 = 1$ and $\mathbf{u}_2 = x$ that

$$
[T_2 \circ T_1]_{B', B} = \left[[(T_2 \circ T_1)(1)]_{B'} \ \vdots \ [(T_2 \circ T_1)(x)]_{B'} \right]
\tag{19}
$$

Using (16) yields

$$(T_2 \circ T_1)(1) = 3x - 5 \qquad \text{and} \qquad (T_2 \circ T_1)(x) = (3x - 5)^2 = 9x^2 - 30x + 25$$

Since $B' = \{1, x, x^2\}$, it follows from this that

$$
[(T_2 \circ T_1)(1)]_{B'} = \begin{bmatrix} -5 \\ 3 \\ 0 \end{bmatrix}
\qquad \text{and} \qquad
[(T_2 \circ T_1)(x)]_{B'} = \begin{bmatrix} 25 \\ -30 \\ 9 \end{bmatrix}
$$

Substituting in (19) yields

$$
[T_2 \circ T_1]_{B', B} = \begin{bmatrix} -5 & 25 \\ 3 & -30 \\ 0 & 9 \end{bmatrix}
$$

which agrees with (18). ▲

EXERCISE SET 8.4

1. Let $T: P_2 \rightarrow P_3$ be the linear transformation defined by $T(p(x)) = xp(x)$.

(a) Find the matrix for T with respect to the standard bases

$$B = \{\mathbf{u}_1, \mathbf{u}_2, \mathbf{u}_3\} \qquad \text{and} \qquad B' = \{\mathbf{v}_1, \mathbf{v}_2, \mathbf{v}_3, \mathbf{v}_4\}$$

where

$$\mathbf{u}_1 = 1, \qquad \mathbf{u}_2 = x, \qquad \mathbf{u}_3 = x^2$$
$$\mathbf{v}_1 = 1, \qquad \mathbf{v}_2 = x, \qquad \mathbf{v}_3 = x^2, \qquad \mathbf{v}_4 = x^3$$

(b) Verify that the matrix $[T]_{B',B}$ obtained in part (a) satisfies Formula (4a) for every vector $\mathbf{x} = c_0 + c_1 x + c_2 x^2$ in P_2.

2. Let $T: P_2 \rightarrow P_1$ be the linear transformation defined by

$$T(a_0 + a_1 x + a_2 x^2) = (a_0 + a_1) - (2a_1 + 3a_2)x$$

(a) Find the matrix for T with respect to the standard bases $B = \{1, x, x^2\}$ and $B' = \{1, x\}$ for P_2 and P_1.

(b) Verify that the matrix $[T]_{B',B}$ obtained in part (a) satisfies Formula (4a) for every vector $\mathbf{x} = c_0 + c_1 x + c_2 x^2$ in P_2.

3. Let $T: P_2 \rightarrow P_2$ be the linear operator defined by

$$T(a_0 + a_1 x + a_2 x^2) = a_0 + a_1(x - 1) + a_2(x - 1)^2$$

(a) Find the matrix for T with respect to the standard basis $B = \{1, x, x^2\}$ for P_2.

(b) Verify that the matrix $[T]_B$ obtained in part (a) satisfies Formula (5a) for every vector $\mathbf{x} = a_0 + a_1 x + a_2 x^2$ in P_2.

4. Let $T: R^2 \rightarrow R^2$ be the linear operator defined by

$$T\left(\begin{bmatrix} x_1 \\ x_2 \end{bmatrix}\right) = \begin{bmatrix} x_1 - x_2 \\ x_1 + x_2 \end{bmatrix}$$

and let $B = \{\mathbf{u}_1, \mathbf{u}_2\}$ be the basis for which

$$\mathbf{u}_1 = \begin{bmatrix} 1 \\ 1 \end{bmatrix} \qquad \text{and} \qquad \mathbf{u}_2 = \begin{bmatrix} -1 \\ 0 \end{bmatrix}$$

(a) Find $[T]_B$.

(b) Verify that Formula (5a) holds for every vector $\mathbf{x}$ in R^2.

5. Let $T: R^2 \rightarrow R^3$ be defined by

$$T\left(\begin{bmatrix} x_1 \\ x_2 \end{bmatrix}\right) = \begin{bmatrix} x_1 + 2x_2 \\ -x_1 \\ 0 \end{bmatrix}$$

(a) Find the matrix $[T]_{B',B}$ with respect to the bases $B = \{\mathbf{u}_1, \mathbf{u}_2\}$ and $B' = \{\mathbf{v}_1, \mathbf{v}_2, \mathbf{v}_3\}$, where

$$\mathbf{u}_1 = \begin{bmatrix} 1 \\ 3 \end{bmatrix}, \qquad \mathbf{u}_2 = \begin{bmatrix} -2 \\ 4 \end{bmatrix}, \qquad \mathbf{v}_1 = \begin{bmatrix} 1 \\ 1 \\ 1 \end{bmatrix}, \qquad \mathbf{v}_2 = \begin{bmatrix} 2 \\ 2 \\ 0 \end{bmatrix}, \qquad \mathbf{v}_3 = \begin{bmatrix} 3 \\ 0 \\ 0 \end{bmatrix}$$

(b) Verify that Formula (4a) holds for every vector

$$\mathbf{x} = \begin{bmatrix} x_1 \\ x_2 \end{bmatrix}$$

in R^2.

6. Let $T: R^3 \rightarrow R^3$ be defined by $T(x_1, x_2, x_3) = (x_1 - x_2, x_2 - x_1, x_1 - x_3)$.
   (a) Find the matrix for T with respect to the basis $B = \{\mathbf{v}_1, \mathbf{v}_2, \mathbf{v}_3\}$, where

$$\mathbf{v}_1 = (1, 0, 1), \qquad \mathbf{v}_2 = (0, 1, 1), \qquad \mathbf{v}_3 = (1, 1, 0)$$

   (b) Verify that Formula (5a) holds for every vector $\mathbf{x} = (x_1, x_2, x_3)$ in R^3.

7. Let $T: P_2 \rightarrow P_2$ be the linear operator defined by $T(p(x)) = p(2x + 1)$; that is,

$$T(c_0 + c_1 x + c_2 x^2) = c_0 + c_1(2x + 1) + c_2(2x + 1)^2$$

   (a) Find $[T]_B$ with respect to the basis $B = \{1, x, x^2\}$.
   (b) Use the indirect procedure illustrated in Figure 5 to compute $T(2 - 3x + 4x^2)$.
   (c) Check the result obtained in part (b) by computing $T(2 - 3x + 4x^2)$ directly.

8. Let $T: P_2 \rightarrow P_3$ be the linear transformation defined by $T(p(x)) = xp(x - 3)$; that is,

$$T(c_0 + c_1 x + c_2 x^2) = x(c_0 + c_1(x - 3) + c_2(x - 3)^2)$$

   (a) Find $[T]_{B', B}$ with respect to the bases $B = \{1, x, x^2\}$ and $B' = \{1, x, x^2, x^3\}$.
   (b) Use the indirect procedure illustrated in Figure 5 to compute $T(1 + x - x^2)$.
   (c) Check the result obtained in part (b) by computing $T(1 + x - x^2)$ directly.

9. Let $\mathbf{v}_1 = \begin{bmatrix} 1 \\ 3 \end{bmatrix}$ and $\mathbf{v}_2 = \begin{bmatrix} -1 \\ 4 \end{bmatrix}$, and let

$$A = \begin{bmatrix} 1 & 3 \\ -2 & 5 \end{bmatrix}$$

   be the matrix for $T: R^2 \rightarrow R^2$ with respect to the basis $B = \{\mathbf{v}_1, \mathbf{v}_2\}$.
   (a) Find $[T(\mathbf{v}_1)]_B$ and $[T(\mathbf{v}_2)]_B$.   (b) Find $T(\mathbf{v}_1)$ and $T(\mathbf{v}_2)$.

   (c) Find a formula for $T\left(\begin{bmatrix} x_1 \\ x_2 \end{bmatrix} \right)$.

   (d) Use the formula obtained in (c) to compute $T\left(\begin{bmatrix} 1 \\ 1 \end{bmatrix} \right)$.

10. Let $A = \begin{bmatrix} 3 & -2 & 1 & 0 \\ 1 & 6 & 2 & 1 \\ -3 & 0 & 7 & 1 \end{bmatrix}$ be the matrix of $T: R^4 \rightarrow R^3$ with respect to the bases

$B = \{\mathbf{v}_1, \mathbf{v}_2, \mathbf{v}_3, \mathbf{v}_4\}$ and $B' = \{\mathbf{w}_1, \mathbf{w}_2, \mathbf{w}_3\}$, where

$$\mathbf{v}_1 = \begin{bmatrix} 0 \\ 1 \\ 1 \\ 1 \end{bmatrix}, \qquad \mathbf{v}_2 = \begin{bmatrix} 2 \\ 1 \\ -1 \\ -1 \end{bmatrix}, \qquad \mathbf{v}_3 = \begin{bmatrix} 1 \\ 4 \\ -1 \\ 2 \end{bmatrix}, \qquad \mathbf{v}_4 = \begin{bmatrix} 6 \\ 9 \\ 4 \\ 2 \end{bmatrix}$$

$$\mathbf{w}_1 = \begin{bmatrix} 0 \\ 8 \\ 8 \end{bmatrix}, \qquad \mathbf{w}_2 = \begin{bmatrix} -7 \\ 8 \\ 1 \end{bmatrix}, \qquad \mathbf{w}_3 = \begin{bmatrix} -6 \\ 9 \\ 1 \end{bmatrix}$$

(a) Find $[T(\mathbf{v}_1)]_{B'}$, $[T(\mathbf{v}_2)]_{B'}$, $[T(\mathbf{v}_3)]_{B'}$, and $[T(\mathbf{v}_4)]_{B'}$.   (b) Find $T(\mathbf{v}_1)$, $T(\mathbf{v}_2)$, $T(\mathbf{v}_3)$, and $T(\mathbf{v}_4)$.

(c) Find a formula for $T\left(\begin{bmatrix} x_1 \\ x_2 \\ x_3 \\ x_4 \end{bmatrix}\right)$.   (d) Use the formula obtained in (c) to compute $T\left(\begin{bmatrix} 2 \\ 2 \\ 0 \\ 0 \end{bmatrix}\right)$.

11. Let $A = \begin{bmatrix} 1 & 3 & -1 \\ 2 & 0 & 5 \\ 6 & -2 & 4 \end{bmatrix}$ be the matrix of $T: P_2 \rightarrow P_2$ with respect to the basis

$B = \{\mathbf{v}_1, \mathbf{v}_2, \mathbf{v}_3\}$, where $\mathbf{v}_1 = 3x + 3x^2$, $\mathbf{v}_2 = -1 + 3x + 2x^2$, $\mathbf{v}_3 = 3 + 7x + 2x^2$.
(a) Find $[T(\mathbf{v}_1)]_B$, $[T(\mathbf{v}_2)]_B$, and $[T(\mathbf{v}_3)]_B$.   (b) Find $T(\mathbf{v}_1)$, $T(\mathbf{v}_2)$, and $T(\mathbf{v}_3)$.
(c) Find a formula for $T(a_0 + a_1 x + a_2 x^2)$.
(d) Use the formula obtained in (c) to compute $T(1 + x^2)$.

12. Let $T_1: P_1 \rightarrow P_2$ be the linear transformation defined by

$$T_1(p(x)) = xp(x)$$

and let $T_2: P_2 \rightarrow P_2$ be the linear operator defined by

$$T_2(p(x)) = p(2x + 1)$$

Let $B = \{1, x\}$ and $B' = \{1, x, x^2\}$ be the standard bases for P_1 and P_2.
(a) Find $[T_2 \circ T_1]_{B', B}$, $[T_2]_{B'}$, and $[T_1]_{B', B}$.
(b) State a formula relating the matrices in part (a).
(c) Verify that the matrices in part (a) satisfy the formula you stated in part (b).

13. Let $T_1: P_1 \rightarrow P_2$ be the linear transformation defined by

$$T_1(c_0 + c_1 x) = 2c_0 - 3c_1 x$$

and let $T_2: P_2 \rightarrow P_3$ be the linear transformation defined by

$$T_1(c_0 + c_1 x + c_2 x^2) = 3c_0 x + 3c_1 x^2 + 3c_2 x^3$$

Let $B = \{1, x\}$, $B'' = \{1, x, x^2\}$, $B' = \{1, x, x^2, x^3\}$
(a) Find $[T_2 \circ T_1]_{B', B}$, $[T_2]_{B', B''}$, and $[T_1]_{B'', B}$.
(b) State a formula relating the matrices in part (a).
(c) Verify that the matrices in part (a) satisfy the formula you stated in part (b).

14. Show that if $T: V \rightarrow W$ is the zero transformation, then the matrix of T with respect to any bases for V and W is a zero matrix.

15. Show that if $T: V \rightarrow V$ is a contraction or a dilation of V (Example 4 of Section 8.1), then the matrix of T with respect to any basis for V is a diagonal matrix.

16. Let $B = \{\mathbf{v}_1, \mathbf{v}_2, \mathbf{v}_3, \mathbf{v}_4\}$ be a basis for a vector space V. Find the matrix with respect to B of the linear operator $T: V \rightarrow V$ defined by $T(\mathbf{v}_1) = \mathbf{v}_2$, $T(\mathbf{v}_2) = \mathbf{v}_3$, $T(\mathbf{v}_3) = \mathbf{v}_4$, $T(\mathbf{v}_4) = \mathbf{v}_1$.

17. (*For readers who have studied calculus.*) Let $D: P_2 \rightarrow P_2$ be the differentiation operator $D(\mathbf{p}) = p'(x)$. In parts (a) and (b) find the matrix of D with respect to the basis $B = \{\mathbf{p}_1, \mathbf{p}_2, \mathbf{p}_3\}$.
(a) $\mathbf{p}_1 = 1$, $\mathbf{p}_2 = x$, $\mathbf{p}_3 = x^2$ (b) $\mathbf{p}_1 = 2$, $\mathbf{p}_2 = 2 - 3x$, $\mathbf{p}_3 = 2 - 3x + 8x^2$
(c) Use the matrix in part (a) to compute $D(6 - 6x + 24x^2)$.
(d) Repeat the directions of part (c) for the matrix in part (b).

18. (**For readers who have studied calculus.**) In each part, $B = \{\mathbf{f}_1, \mathbf{f}_2, \mathbf{f}_3\}$ is a basis for a subspace V of the vector space of real-valued functions defined on the real line. Find the matrix with respect to B of the differentiation operator $D: V \to V$.
   (a) $\mathbf{f}_1 = 1$, $\mathbf{f}_2 = \sin x$, $\mathbf{f}_3 = \cos x$ (b) $\mathbf{f}_1 = 1$, $\mathbf{f}_2 = e^x$, $\mathbf{f}_3 = e^{2x}$
   (c) $\mathbf{f}_1 = e^{2x}$, $\mathbf{f}_2 = xe^{2x}$, $\mathbf{f}_3 = x^2 e^{2x}$

19. Prove: If B and B' are the standard bases for R^n and R^m, respectively, then the matrix for a linear transformation $T: R^n \to R^m$ with respect to the bases B and B' is the standard matrix for T.

8.5   SIMILARITY

The matrix of a linear operator $T: V \to V$ depends on the basis selected for V. One of the fundamental problems of linear algebra is to choose a basis for V that makes the matrix for T as simple as possible—diagonal or triangular, for example. In this section we shall study this problem.

CHOOSING BASES TO PRODUCE SIMPLE MATRICES FOR LINEAR OPERATORS

Standard bases do not necessarily produce the simplest matrices for linear operators. For example, consider the linear operator $T: R^2 \to R^2$ defined by

$$T\left(\begin{bmatrix} x_1 \\ x_2 \end{bmatrix}\right) = \begin{bmatrix} x_1 + x_2 \\ -2x_1 + 4x_2 \end{bmatrix} \tag{1}$$

and the standard basis $B = \{\mathbf{e}_1, \mathbf{e}_2\}$ for R^2, where

$$\mathbf{e}_1 = \begin{bmatrix} 1 \\ 0 \end{bmatrix}, \qquad \mathbf{e}_2 = \begin{bmatrix} 0 \\ 1 \end{bmatrix}$$

By Theorem 8.4.1, the matrix for T with respect to this basis is the standard matrix for T; that is,

$$[T]_B = [T] = [T(\mathbf{e}_1) \mid T(\mathbf{e}_2)]$$

From (1),

$$T(\mathbf{e}_1) = \begin{bmatrix} 1 \\ -2 \end{bmatrix}, \qquad T(\mathbf{e}_2) = \begin{bmatrix} 1 \\ 4 \end{bmatrix}$$

so

$$[T]_B = \begin{bmatrix} 1 & 1 \\ -2 & 4 \end{bmatrix} \tag{2}$$

In comparison, we showed in Example 4 of Section 8.4 that if

$$\mathbf{u}_1 = \begin{bmatrix} 1 \\ 1 \end{bmatrix}, \qquad \mathbf{u}_2 = \begin{bmatrix} 1 \\ 2 \end{bmatrix} \tag{3}$$

then the matrix for T with respect to the basis $B' = \{\mathbf{u}_1, \mathbf{u}_2\}$ is the diagonal matrix

$$[T]_{B'} = \begin{bmatrix} 2 & 0 \\ 0 & 3 \end{bmatrix} \tag{4}$$

This matrix is "simpler" than (2) in the sense that diagonal matrices enjoy special properties that more general matrices do not.

One of the major themes in more advanced linear algebra courses is to determine the "simplest possible form" that can be obtained for the matrix of a linear operator by choosing the basis appropriately. Sometimes it is possible to obtain a diagonal matrix (as above, for example); other times one must settle for a triangular matrix or some other form. We will only be able to touch on this important topic in this text.

The problem of finding a basis that produces the simplest possible matrix for a linear operator $T: V \rightarrow V$ can be attacked by first finding a matrix for T relative to *any* basis, say a standard basis, where applicable, then changing the basis in a manner that simplifies the matrix. Before pursuing this idea, it will be helpful to review some concepts about changing bases.

Recall from Formula (8) in Section 6.5 that if $B = \{\mathbf{u}_1, \mathbf{u}_2, \ldots, \mathbf{u}_n\}$ and $B' = \{\mathbf{u}'_1, \mathbf{u}'_2, \ldots, \mathbf{u}'_n\}$ are bases for a vector space V, then the *transition matrix* from B' to B is given by the formula

$$P = \left[[\mathbf{u}'_1]_B \mid [\mathbf{u}'_2]_B \mid \cdots \mid [\mathbf{u}'_n]_B \right] \tag{5}$$

This matrix has the property that for every vector $\mathbf{v}$ in V

$$P[\mathbf{v}]_{B'} = [\mathbf{v}]_B \tag{6}$$

that is, multiplication by P maps the coordinate matrix for $\mathbf{v}$ relative to B' into the coordinate matrix for $\mathbf{v}$ relative to B [see Formula (7) in Section 6.5]. We showed in Theorem 6.5.4 that P is invertible and P^{-1} is the transition matrix from B to B'.

RELATIONSHIP BETWEEN TRANSITION MATRICES AND IDENTITY OPERATORS

The following theorem gives a useful alternative viewpoint about transition matrices; it shows that the transition matrix from a basis B' to a basis B can be regarded as the matrix of an identity operator.

> **Theorem 8.5.1.** *If B and B' are bases for a finite-dimensional vector space V, and if $I: V \rightarrow V$ is the identity operator, then $[I]_{B, B'}$ is the transition matrix from B' to B.*

Proof. Suppose that $B = \{\mathbf{u}_1, \mathbf{u}_2, \ldots, \mathbf{u}_n\}$ and $B' = \{\mathbf{u}'_1, \mathbf{u}'_2, \ldots, \mathbf{u}'_n\}$ are bases for V. Using the fact that $I(\mathbf{v}) = \mathbf{v}$ for all $\mathbf{v}$ in V, it follows from Formula (4) of Section 8.4 with B and B' *reversed* that

$$[I]_{B, B'} = \left[[I(\mathbf{u}'_1)]_B \mid [I(\mathbf{u}'_2)]_B \mid \cdots \mid [I(\mathbf{u}'_n)]_B \right]$$

$$= \left[[\mathbf{u}'_1]_B \mid [\mathbf{u}'_2]_B \mid \cdots \mid [\mathbf{u}'_n]_B \right]$$

Thus, from (5), we have $[I]_{B,B'} = P$, which shows that $[I]_{B,B'}$ is the transition matrix from B' to B. ▌

The result in this theorem is illustrated in Figure 1.

Basis $= B'$ Basis $= B$

Figure 1 $[I]_{B,B'}$ is the transition matrix from B' to B.

EFFECT OF CHANGING BASES ON MATRICES OF LINEAR OPERATORS

We are now ready to consider the main problem in this section.

> **Problem.** If B and B' are two bases for a finite-dimensional vector space V, and if $T: V \rightarrow V$ is a linear operator, what relationship, if any, exists between the matrices $[T]_B$ and $[T]_{B'}$?

The answer to this question can be obtained by considering the composition of the three linear operators on V pictured in Figure 2.

Figure 2   Basis $= B'$ Basis $= B$ Basis $= B$ Basis $= B'$

     In this figure **v** is first mapped into itself by the identity operator, then **v** is mapped into $T(\mathbf{v})$ by T, then $T(\mathbf{v})$ is mapped into itself by the identity operator. All four vector spaces involved in the composition are the same (namely, V); however, the bases for the spaces vary. Since the starting vector is **v** and the final vector is $T(\mathbf{v})$, the composition is the same as T; that is,

$$T = I \circ T \circ I \tag{7}$$

If, as illustrated in Figure 2, the first and last vector spaces are assigned the basis B' and the middle two spaces are assigned the basis B, then it follows from (7) and Formula (15) of Section 8.4 (with an appropriate adjustment in the names of the bases) that

$$[T]_{B',B'} = [I \circ T \circ I]_{B',B'} = [I]_{B',B}[T]_{B,B}[I]_{B,B'} \tag{8}$$

or in simpler notation,

$$[T]_{B'} = [I]_{B',B}[T]_B[I]_{B,B'} \tag{9}$$

But it follows from Theorem 8.5.1 that $[I]_{B,B'}$ is the transition matrix from B' to B and consequently $[I]_{B',B}$ is the transition matrix from B to B'. Thus, if we let $P =$

$[I]_{B,B'}$, then $P^{-1} = [I]_{B',B}$, so (9) can be written as

$$[T]_{B'} = P^{-1}[T]_B P$$

In summary, we have the following theorem.

Theorem 8.5.2. *Let* $T: V \rightarrow V$ *be a linear operator on a finite-dimensional vector space* V, *and let* B *and* B' *be bases for* V. *Then*

$$[T]_{B'} = P^{-1}[T]_B P \qquad (10)$$

where P *is the transition matrix from* B' *to* B.

Warning. When applying Theorem 8.5.2, it is easy to forget whether P is the transition matrix from B to B' (incorrect) or from B' to B (correct). It may help to write (10) in form (9), keeping in mind that the three "interior" subscripts are the same, and the two exterior subscripts are the same:

$$[T]_{B'} = [I]_{B',B}[T]_B[I]_{B,B'}$$

The interior subscripts are the same

The exterior subscripts are the same

Once you master this pattern, you need only remember that $P = [I]_{B,B'}$ is the transition matrix from B' to B and $P^{-1} = [I]_{B',B}$ is its inverse.

Example 1   Let $T: R^2 \rightarrow R^2$ be defined by

$$T\left(\begin{bmatrix} x_1 \\ x_2 \end{bmatrix}\right) = \begin{bmatrix} x_1 + x_2 \\ -2x_1 + 4x_2 \end{bmatrix}$$

Find the matrix of T with respect to the standard basis $B = \{\mathbf{e}_1, \mathbf{e}_2\}$ for R^2, then use Theorem 8.5.2 to find the matrix of T with respect to the basis $B' = \{\mathbf{u}_1', \mathbf{u}_2'\}$, where

$$\mathbf{u}_1' = \begin{bmatrix} 1 \\ 1 \end{bmatrix} \qquad \text{and} \qquad \mathbf{u}_2' = \begin{bmatrix} 1 \\ 2 \end{bmatrix}$$

Solution. We showed earlier in this section [see (2)] that

$$[T]_B = \begin{bmatrix} 1 & 1 \\ -2 & 4 \end{bmatrix}$$

To find $[T]_{B'}$ from (10), we will need to find the transition matrix

$$P = [I]_{B,B'} = \left[[\mathbf{u}_1']_B \; \vdots \; [\mathbf{u}_2']_B \right]$$

[see (5)]. By inspection

$$\mathbf{u}_1' = \mathbf{e}_1 + \mathbf{e}_2$$
$$\mathbf{u}_2' = \mathbf{e}_1 + 2\mathbf{e}_2$$

so that

$$[\mathbf{u}_1']_B = \begin{bmatrix} 1 \\ 1 \end{bmatrix} \quad \text{and} \quad [\mathbf{u}_2']_B = \begin{bmatrix} 1 \\ 2 \end{bmatrix}$$

Thus, the transition matrix from B' to B is

$$P = \begin{bmatrix} 1 & 1 \\ 1 & 2 \end{bmatrix}$$

The reader can check that

$$P^{-1} = \begin{bmatrix} 2 & -1 \\ -1 & 1 \end{bmatrix}$$

so that by Theorem 8.5.2 the matrix of T relative to the basis B' is

$$[T]_{B'} = P^{-1}[T]_B P = \begin{bmatrix} 2 & -1 \\ -1 & 1 \end{bmatrix}\begin{bmatrix} 1 & 1 \\ -2 & 4 \end{bmatrix}\begin{bmatrix} 1 & 1 \\ 1 & 2 \end{bmatrix} = \begin{bmatrix} 2 & 0 \\ 0 & 3 \end{bmatrix}$$

which agrees with (4). △

SIMILARITY

The relationship in Formula (10) is of such importance that there is some terminology associated with it.

> **Definition.** If A and B are square matrices, we say that **B is similar to A** if there is an invertible matrix P such that $B = P^{-1}AP$.

REMARK. Note that the equation $B = P^{-1}AP$ can be rewritten as

$$A = PBP^{-1} = (P^{-1})^{-1}BP^{-1}$$

Letting $Q = P^{-1}$ yields

$$A = Q^{-1}BQ$$

which says that A is similar to B. Therefore, B is similar to A if and only if A is similar to B; consequently, we shall usually simply say that **A and B are similar**.

SIMILARITY INVARIANTS

Similar matrices often have properties in common; for example, if A and B are similar matrices, then A and B have the same determinant. To see that this is so, suppose that

$$B = P^{-1}AP$$

Thus,

$$\det(B) = \det(P^{-1}AP) = \det(P^{-1})\det(A)\det(P)$$
$$= \frac{1}{\det(P)}\det(A)\det(P) = \det(A)$$

We make the following definition.

> **Definition.** A property of square matrices is said to be a ***similarity invariant*** or ***invariant under similarity*** if that property is shared by any two similar matrices.

In the terminology of this definition, the determinant of a square matrix is a similarity invariant. Table 1 lists some other important similarity invariants. The proofs of some of the results in Table 1 are given in the exercises.

It follows from Theorem 8.5.2 that *two matrices representing the same linear operator* $T : V \rightarrow V$ *with respect to different bases are similar.* Thus, if B is a basis for V, and the matrix $[T]_B$ has some property that is invariant under similarity, then for every basis B', the matrix $[T]_{B'}$ has that same property. For example, for any two bases B and B' we must have

$$\det([T]_B) = \det([T]_{B'})$$

It follows from this equation that the value of the determinant depends on T, but not on the particular basis that is used to obtain the matrix for T. Thus, the determinant can be regarded as a property of the linear operator T; indeed, if V is a finite-dimensional vector space, then we can *define* the ***determinant of the linear operator*** T to be

$$\det(T) = \det([T]_B) \tag{11}$$

where B is any basis for V.

TABLE 1 *Similarity Invariants*

Property	Description
Determinant	A and $P^{-1}AP$ have the same determinant.
Invertibility	A is invertible if and only if $P^{-1}AP$ is invertible.
Rank	A and $P^{-1}AP$ have the same rank.
Nullity	A and $P^{-1}AP$ have the same nullity.
Trace	A and $P^{-1}AP$ have the same trace.
Characteristic polynomial	A and $P^{-1}AP$ have the same characteristic polynomial.
Eigenvalues	A and $P^{-1}AP$ have the same eigenvalues.
Eigenspace dimension	If λ is an eigenvalue of A and $P^{-1}AP$, then the eigenspace of A corresponding to λ and the eigenspace of $P^{-1}AP$ corresponding to λ have the same dimension.

Example 2 Let $T: R^2 \to R^2$ be defined by

$$T\left(\begin{bmatrix} x_1 \\ x_2 \end{bmatrix}\right) = \begin{bmatrix} x_1 + x_2 \\ -2x_1 + 4x_2 \end{bmatrix}$$

Find $\det(T)$.

Solution. We can choose any basis B and calculate $\det([T]_B)$. If we take the standard basis, then from Example 1

$$[T]_B = \begin{bmatrix} 1 & 1 \\ -2 & 4 \end{bmatrix}$$

so

$$\det(T) = \begin{vmatrix} 1 & 1 \\ -2 & 4 \end{vmatrix} = 6$$

Had we chosen the basis $B' = \{\mathbf{u}_1, \mathbf{u}_2\}$ of Example 1, then we would have obtained

$$[T]_{B'} = \begin{bmatrix} 2 & 0 \\ 0 & 3 \end{bmatrix}$$

so

$$\det(T) = \begin{vmatrix} 2 & 0 \\ 0 & 3 \end{vmatrix} = 6$$

which agrees with the preceding computation. ▲

A GEOMETRIC EXAMPLE

Example 3 Let l be the line in the xy-plane that passes through the origin and makes an angle θ with the positive x-axis, where $0 \le \theta < \pi$. As illustrated in Figure 3, let $T: R^2 \to R^2$ be the linear operator that maps each vector into its reflection about the line l.

Figure 3

(a)   Find the standard matrix for T.
(b)   Find the reflection of the vector $\mathbf{x} = (1, 2)$ about the line l through the origin that makes an angle of $\theta = \pi/6$ with the positive x-axis.

Solution (a). We could proceed as in Example 5 of Section 4.3 and try to construct the standard matrix from the formula

$$[T]_B = [T] = [T(\mathbf{e}_1) \ \vdots \ T(\mathbf{e}_2)]$$

where $B = \{\mathbf{e}_1, \mathbf{e}_2\}$ is the standard basis for R^2. However, it is easier to use a different strategy: Instead of finding $[T]_B$ directly, we shall first find the matrix $[T]_{B'}$, where

$$B' = \{\mathbf{u}_1', \mathbf{u}_2'\}$$

is the basis consisting of a unit vector $\mathbf{u}_1'$ along l and a unit vector $\mathbf{u}_2'$ perpendicular to l (Figure 4).

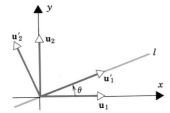

Figure 4

Once we have found $[T]_{B'}$, we shall perform a change of basis to find $[T]_B$. The computations are as follows:

$$T(\mathbf{u}_1') = \mathbf{u}_1' \quad \text{and} \quad T(\mathbf{u}_2') = -\mathbf{u}_2'$$

so

$$[T(\mathbf{u}_1')]_{B'} = \begin{bmatrix} 1 \\ 0 \end{bmatrix} \quad \text{and} \quad T[(\mathbf{u}_2')]_{B'} = \begin{bmatrix} 0 \\ -1 \end{bmatrix}$$

Thus,

$$[T]_{B'} = \begin{bmatrix} 1 & 0 \\ 0 & -1 \end{bmatrix}$$

From the computations in Example 6 of Section 6.5, the transition matrix from B' to B is

$$P = \begin{bmatrix} [\mathbf{u}_1']_B & \vdots & [\mathbf{u}_2']_B \end{bmatrix} = \begin{bmatrix} \cos\theta & -\sin\theta \\ \sin\theta & \cos\theta \end{bmatrix} \tag{12}$$

It follows from Formula (10) that

$$[T]_B = P[T]_{B'}P^{-1}$$

Thus, from (12) the standard matrix for T is

$$[T] = P[T]_{B'}P^{-1} = \begin{bmatrix} \cos\theta & -\sin\theta \\ \sin\theta & \cos\theta \end{bmatrix} \begin{bmatrix} 1 & 0 \\ 0 & -1 \end{bmatrix} \begin{bmatrix} \cos\theta & \sin\theta \\ -\sin\theta & \cos\theta \end{bmatrix}$$

$$= \begin{bmatrix} \cos^2\theta - \sin^2\theta & 2\sin\theta\cos\theta \\ 2\sin\theta\cos\theta & \sin^2\theta - \cos^2\theta \end{bmatrix}$$

$$= \begin{bmatrix} \cos 2\theta & \sin 2\theta \\ \sin 2\theta & -\cos 2\theta \end{bmatrix}$$

Solution (b). It follows from part (a) that the formula for T in matrix notation is

$$T\left(\begin{bmatrix} x \\ y \end{bmatrix}\right) = \begin{bmatrix} \cos 2\theta & \sin 2\theta \\ \sin 2\theta & -\cos 2\theta \end{bmatrix}\begin{bmatrix} x \\ y \end{bmatrix}$$

Substituting $\theta = \pi/6$ in this formula yields

$$T\left(\begin{bmatrix} x \\ y \end{bmatrix}\right) = \begin{bmatrix} \frac{1}{2} & \frac{\sqrt{3}}{2} \\ \frac{\sqrt{3}}{2} & -\frac{1}{2} \end{bmatrix}\begin{bmatrix} x \\ y \end{bmatrix}$$

so

$$T\left(\begin{bmatrix} 1 \\ 2 \end{bmatrix}\right) = \begin{bmatrix} \frac{1}{2} & \frac{\sqrt{3}}{2} \\ \frac{\sqrt{3}}{2} & -\frac{1}{2} \end{bmatrix}\begin{bmatrix} 1 \\ 2 \end{bmatrix} = \begin{bmatrix} \frac{1}{2} + \sqrt{3} \\ \frac{\sqrt{3}}{2} - 1 \end{bmatrix}$$

Thus, $T(1, 2) = (\frac{1}{2} + \sqrt{3}, \frac{\sqrt{3}}{2} - 1)$. ▲

EIGENVALUES OF A LINEAR OPERATOR

Eigenvectors and eigenvalues can be defined for linear operators as well as matrices. A scalar λ is called an *eigenvalue* of a linear operator $T: V \rightarrow V$ if there is a nonzero vector $\mathbf{x}$ in V such that $T\mathbf{x} = \lambda\mathbf{x}$. The vector $\mathbf{x}$ is called an *eigenvector* of T corresponding to λ. Equivalently, the eigenvectors of T corresponding to λ are the nonzero vectors in the kernel of $\lambda I - T$ (Exercise 15). This kernel is called the *eigenspace* of T corresponding to λ.

It can be shown that if V is a finite-dimensional vector space, and B is *any* basis for V, then

1. The eigenvalues of T are the same as the eigenvalues of $[T]_B$.
2. A vector $\mathbf{x}$ is an eigenvector of T corresponding to λ if and only if its coordinate matrix $[\mathbf{x}]_B$ is an eigenvector of $[T]_B$ corresponding to λ.

We omit the proofs.

Example 4 Find the eigenvalues and bases for the eigenspaces of the linear operator $T: P_2 \rightarrow P_2$ defined by

$$T(a + bx + cx^2) = -2c + (a + 2b + c)x + (a + 3c)x^2$$

Solution. The matrix of T with respect to the standard basis $B = \{1, x, x^2\}$ is

$$[T]_B = \begin{bmatrix} 0 & 0 & -2 \\ 1 & 2 & 1 \\ 1 & 0 & 3 \end{bmatrix}$$

(verify). The eigenvalues of T are $\lambda = 1$ and $\lambda = 2$ (Example 5 of Section 7.1). Also from that example, the eigenspace of $[T]_B$ corresponding to $\lambda = 2$ has the basis

$\{\mathbf{u}_1, \mathbf{u}_2\}$, where

$$\mathbf{u}_1 = \begin{bmatrix} -1 \\ 0 \\ 1 \end{bmatrix}, \qquad \mathbf{u}_2 = \begin{bmatrix} 0 \\ 1 \\ 0 \end{bmatrix}$$

and the eigenspace of $[\,T\,]_B$ corresponding to $\lambda = 1$ has the basis $\{\mathbf{u}_3\}$, where

$$\mathbf{u}_3 = \begin{bmatrix} -2 \\ 1 \\ 1 \end{bmatrix}$$

The matrices $\mathbf{u}_1$, $\mathbf{u}_2$, and $\mathbf{u}_3$ are the coordinate matrices relative to B of

$$\mathbf{p}_1 = -1 + x^2, \qquad \mathbf{p}_2 = x, \qquad \mathbf{p}_3 = -2 + x + x^2$$

Thus, the eigenspace of T corresponding to $\lambda = 2$ has the basis

$$\{\mathbf{p}_1, \mathbf{p}_2\} = \{-1 + x^2, x\}$$

and that corresponding to $\lambda = 1$ has the basis

$$\{\mathbf{p}_3\} = \{-2 + x + x^2\}$$

As a check, the reader should use the given formula for T to verify that $T(\mathbf{p}_1) = 2\mathbf{p}_1$, $T(\mathbf{p}_2) = 2\mathbf{p}_2$, and $T(\mathbf{p}_3) = \mathbf{p}_3$. ▲

Example 5 Let $T: R^3 \to R^3$ be the linear operator given by

$$T\left(\begin{bmatrix} x_1 \\ x_2 \\ x_3 \end{bmatrix}\right) = \begin{bmatrix} -2x_3 \\ x_1 + 2x_2 + x_3 \\ x_1 + 3x_3 \end{bmatrix}$$

Find a basis for R^3 relative to which the matrix for T is diagonal.

Solution. First we will find the standard matrix for T; then we will look for a change of basis that diagonalizes the standard matrix.

  If $B = \{\mathbf{e}_1, \mathbf{e}_2, \mathbf{e}_3\}$ denotes the standard basis for R^3, then

$$T(\mathbf{e}_1) = T\left(\begin{bmatrix} 1 \\ 0 \\ 0 \end{bmatrix}\right) = \begin{bmatrix} 0 \\ 1 \\ 1 \end{bmatrix}, \quad T(\mathbf{e}_2) = T\left(\begin{bmatrix} 0 \\ 1 \\ 0 \end{bmatrix}\right) = \begin{bmatrix} 0 \\ 2 \\ 0 \end{bmatrix}, \quad T(\mathbf{e}_3) = T\left(\begin{bmatrix} 0 \\ 0 \\ 1 \end{bmatrix}\right) = \begin{bmatrix} -2 \\ 1 \\ 3 \end{bmatrix}$$

so that the standard matrix for T is

$$[\,T\,] = \begin{bmatrix} 0 & 0 & -2 \\ 1 & 2 & 1 \\ 1 & 0 & 3 \end{bmatrix} \tag{13}$$

We now want to change from the standard basis B to a new basis $B' = \{\mathbf{u}_1', \mathbf{u}_2', \mathbf{u}_3'\}$ in order to obtain a diagonal matrix for T. If we let P be the transition matrix from the unknown basis B' to the standard basis B, then by Theorem 8.5.2 the matrices $[\,T\,]$ and

$[T]_{B'}$ will be related by

$$[T]_{B'} = P^{-1}[T]P \tag{14}$$

In Example 1 of Section 7.2 we found that the matrix in (13) is diagonalized by

$$P = \begin{bmatrix} -1 & 0 & -2 \\ 0 & 1 & 1 \\ 1 & 0 & 1 \end{bmatrix}$$

Since P represents the transition matrix from the basis $B' = \{u_1', u_2', u_3'\}$ to the standard basis $B = \{e_1, e_2, e_3\}$, the columns of P are $[u_1']_B$, $[u_2']_B$, and $[u_3']_B$, so that

$$[u_1']_B = \begin{bmatrix} -1 \\ 0 \\ 1 \end{bmatrix}, \qquad [u_2']_B = \begin{bmatrix} 0 \\ 1 \\ 0 \end{bmatrix}, \qquad [u_3']_B = \begin{bmatrix} -2 \\ 1 \\ 1 \end{bmatrix}.$$

Thus,

$$u_1' = (-1)e_1 + (0)e_2 + (1)e_3 = \begin{bmatrix} -1 \\ 0 \\ 1 \end{bmatrix}$$

$$u_2' = (0)e_1 + (1)e_2 + (0)e_3 = \begin{bmatrix} 0 \\ 1 \\ 0 \end{bmatrix}$$

$$u_3' = (-2)e_1 + (1)e_2 + (1)e_3 = \begin{bmatrix} -2 \\ 1 \\ 1 \end{bmatrix}$$

are basis vectors that produce a diagonal matrix for $[T]_{B'}$. As a check, let us compute $[T]_{B'}$ directly. From the given formula for T we have

$$T(u_1') = \begin{bmatrix} -2 \\ 0 \\ 2 \end{bmatrix} = 2u_1', \qquad T(u_2') = \begin{bmatrix} 0 \\ 2 \\ 0 \end{bmatrix} = 2u_2', \qquad T(u_3') = \begin{bmatrix} -2 \\ 1 \\ 1 \end{bmatrix} = u_3'$$

so that

$$[T(u_1')]_{B'} = \begin{bmatrix} 2 \\ 0 \\ 0 \end{bmatrix}, \qquad [T(u_2')]_{B'} = \begin{bmatrix} 0 \\ 2 \\ 0 \end{bmatrix}, \qquad [T(u_3')]_{B'} = \begin{bmatrix} 0 \\ 0 \\ 1 \end{bmatrix}$$

Thus,

$$[T]_{B'} = \left[[T(u_1')]_{B'} \mid [T(u_2')]_{B'} \mid [T(u_3')]_{B'} \right] = \begin{bmatrix} 2 & 0 & 0 \\ 0 & 2 & 0 \\ 0 & 0 & 1 \end{bmatrix}$$

This is consistent with (14) since

$$P^{-1}[T]P = \begin{bmatrix} 1 & 0 & 2 \\ 1 & 1 & 1 \\ -1 & 0 & -1 \end{bmatrix} \begin{bmatrix} 0 & 0 & -2 \\ 1 & 2 & 1 \\ 1 & 0 & 3 \end{bmatrix} \begin{bmatrix} -1 & 0 & -2 \\ 0 & 1 & 1 \\ 1 & 0 & 1 \end{bmatrix}$$

$$= \begin{bmatrix} 2 & 0 & 0 \\ 0 & 2 & 0 \\ 0 & 0 & 1 \end{bmatrix} \quad \blacktriangle$$

EXERCISE SET 8.5

In Exercises 1–7 find the matrix of T with respect to B, and use Theorem 8.5.2 to compute the matrix of T with respect to B'.

1. $T: R^2 \to R^2$ is defined by

$$T\left(\begin{bmatrix} x_1 \\ x_2 \end{bmatrix}\right) = \begin{bmatrix} x_1 - 2x_2 \\ -x_2 \end{bmatrix}$$

$B = \{\mathbf{u}_1, \mathbf{u}_2\}$ and $B' = \{\mathbf{v}_1, \mathbf{v}_2\}$, where

$$\mathbf{u}_1 = \begin{bmatrix} 1 \\ 0 \end{bmatrix}, \quad \mathbf{u}_2 = \begin{bmatrix} 0 \\ 1 \end{bmatrix}, \quad \mathbf{v}_1 = \begin{bmatrix} 2 \\ 1 \end{bmatrix}, \quad \mathbf{v}_2 = \begin{bmatrix} -3 \\ 4 \end{bmatrix}$$

2. $T: R^2 \to R^2$ is defined by

$$T\left(\begin{bmatrix} x_1 \\ x_2 \end{bmatrix}\right) = \begin{bmatrix} x_1 + 7x_2 \\ 3x_1 - 4x_2 \end{bmatrix}$$

$B = \{\mathbf{u}_1, \mathbf{u}_2\}$ and $B' = \{\mathbf{v}_1, \mathbf{v}_2\}$, where

$$\mathbf{u}_1 = \begin{bmatrix} 2 \\ 2 \end{bmatrix}, \quad \mathbf{u}_2 = \begin{bmatrix} 4 \\ -1 \end{bmatrix}, \quad \mathbf{v}_1 = \begin{bmatrix} 1 \\ 3 \end{bmatrix}, \quad \mathbf{v}_2 = \begin{bmatrix} -1 \\ -1 \end{bmatrix}$$

3. $T: R^2 \to R^2$ is the rotation about the origin through 45°; B and B' are the bases in Exercise 1.

4. $T: R^3 \to R^3$ is defined by

$$T\left(\begin{bmatrix} x_1 \\ x_2 \\ x_3 \end{bmatrix}\right) = \begin{bmatrix} x_1 + 2x_2 - x_3 \\ -x_2 \\ x_1 + 7x_3 \end{bmatrix}$$

B is the standard basis for R^3 and $B' = \{\mathbf{v}_1, \mathbf{v}_2, \mathbf{v}_3\}$, where

$$\mathbf{v}_1 = \begin{bmatrix} 1 \\ 0 \\ 0 \end{bmatrix}, \quad \mathbf{v}_2 = \begin{bmatrix} 1 \\ 1 \\ 0 \end{bmatrix}, \quad \mathbf{v}_3 = \begin{bmatrix} 1 \\ 1 \\ 1 \end{bmatrix}$$

5. $T: R^3 \to R^3$ is the orthogonal projection on the xy-plane; B and B' are as in Exercise 4.

6. $T: R^2 \to R^2$ is defined by $T(\mathbf{x}) = 5\mathbf{x}$; B and B' are the bases in Exercise 2.

7. $T: P_1 \to P_1$ is defined by $T(a_0 + a_1 x) = a_0 + a_1(x + 1)$; $B = \{\mathbf{p}_1, \mathbf{p}_2\}$ and $B' = \{\mathbf{q}_1, \mathbf{q}_2\}$, where $\mathbf{p}_1 = 6 + 3x$, $\mathbf{p}_2 = 10 + 2x$, $\mathbf{q}_1 = 2$, $\mathbf{q}_2 = 3 + 2x$.

8. Find $\det(T)$.
   (a) $T: R^2 \to R^2$, where $T(x_1, x_2) = (3x_1 - 4x_2, -x_1 + 7x_2)$
   (b) $T: R^3 \to R^3$, where $T(x_1, x_2, x_3) = (x_1 - x_2, x_2 - x_3, x_3 - x_1)$
   (c) $T: P_2 \to P_2$, where $T(p(x)) = p(x - 1)$

9. Prove that the following are similarity invariants:
   (a) rank   (b) nullity   (c) invertibility

10. Let $T: P_4 \to P_4$ be the linear operator given by the formula $T(p(x)) = p(2x + 1)$.
   (a) Find a matrix for T with respect to some convenient basis; then use the result in Exercise 9 to find the rank and nullity of T.
   (b) Use the result in part (a) to determine whether T is one-to-one.

11. In each part find a basis for R^2 relative to which the matrix for T is diagonal.

   (a) $T\left(\begin{bmatrix} x_1 \\ x_2 \end{bmatrix}\right) = \begin{bmatrix} x_1 - x_2 \\ 2x_1 + 4x_2 \end{bmatrix}$ (b) $T\left(\begin{bmatrix} x_1 \\ x_2 \end{bmatrix}\right) = \begin{bmatrix} 4x_1 - x_2 \\ -3x_1 + x_2 \end{bmatrix}$

12. In each part find a basis for R^3 relative to which the matrix for T is diagonal.

   (a) $T\left(\begin{bmatrix} x_1 \\ x_2 \\ x_3 \end{bmatrix}\right) = \begin{bmatrix} -2x_1 + x_2 - x_3 \\ x_1 - 2x_2 - x_3 \\ -x_1 - x_2 - 2x_3 \end{bmatrix}$ (b) $T\left(\begin{bmatrix} x_1 \\ x_2 \\ x_3 \end{bmatrix}\right) = \begin{bmatrix} -x_2 + x_3 \\ -x_1 + x_3 \\ x_1 + x_2 \end{bmatrix}$

   (c) $T\left(\begin{bmatrix} x_1 \\ x_2 \\ x_3 \end{bmatrix}\right) = \begin{bmatrix} 4x_1 + x_3 \\ 2x_1 + 3x_2 + 2x_3 \\ x_1 + 4x_3 \end{bmatrix}$

13. Let $T: P_2 \to P_2$ be defined by

$$T(a_0 + a_1 x + a_2 x^2) = (5a_0 + 6a_1 + 2a_2) - (a_1 + 8a_2)x + (a_0 - 2a_2)x^2$$

   (a) Find the eigenvalues of T.   (b) Find bases for the eigenspaces of T.

14. Let $T: M_{22} \to M_{22}$ be defined by

$$T\left(\begin{bmatrix} a & b \\ c & d \end{bmatrix}\right) = \begin{bmatrix} 2c & a + c \\ b - 2c & d \end{bmatrix}$$

   (a) Find the eigenvalues of T.   (b) Find bases for the eigenspaces of T.

15. Let λ be an eigenvalue of a linear operator $T: V \to V$. Prove that the eigenvectors of T corresponding to λ are the nonzero vectors in the kernel of $\lambda I - T$.

16. Prove that if A and B are similar matrices, then A^2 and B^2 are also similar. More generally, prove that A^k and B^k are similar, where k is any positive integer.

17. Let C and D be $m \times n$ matrices, and let $B = \{\mathbf{v}_1, \mathbf{v}_2, \ldots, \mathbf{v}_n\}$ be a basis for a vector space V. Show that if $C[\mathbf{x}]_B = D[\mathbf{x}]_B$ for all $\mathbf{x}$ in V, then $C = D$.

18. Let l be a line in the xy-plane that passes through the origin and makes an angle θ with the positive x-axis. As illustrated in Figure 5, let $T: R^2 \to R^2$ be the orthogonal projection of R^2 onto l. Use the method of Example 3 to show that

$$T\left(\begin{bmatrix} x \\ y \end{bmatrix}\right) = \begin{bmatrix} \cos^2 \theta & \sin \theta \cos \theta \\ \sin \theta \cos \theta & \sin^2 \theta \end{bmatrix}\begin{bmatrix} x \\ y \end{bmatrix}$$

[*Note.* See Example 5 of Section 4.3.]

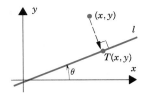

Figure 5

SUPPLEMENTARY EXERCISES

1. Let A be an $n \times n$ matrix, B a nonzero $n \times 1$ matrix, and $\mathbf{x}$ a vector in R^n expressed in matrix notation. Is $T(\mathbf{x}) = A\mathbf{x} + B$ a linear operator on R^n? Justify your answer.

2. Let

$$A = \begin{bmatrix} \cos\theta & -\sin\theta \\ \sin\theta & \cos\theta \end{bmatrix}$$

(a) Show that

$$A^2 = \begin{bmatrix} \cos 2\theta & -\sin 2\theta \\ \sin 2\theta & \cos 2\theta \end{bmatrix} \quad \text{and} \quad A^3 = \begin{bmatrix} \cos 3\theta & -\sin 3\theta \\ \sin 3\theta & \cos 3\theta \end{bmatrix}$$

(b) Guess the form of the matrix A^n for any positive integer n.
(c) By considering the geometric effect of $T: R^2 \rightarrow R^2$, where T is multiplication by A, obtain the result in (b) geometrically.

3. Let $\mathbf{v}_0$ be a fixed vector in an inner product space V, and let $T: V \rightarrow V$ be defined by $T(\mathbf{v}) = \langle \mathbf{v}, \mathbf{v}_0 \rangle \mathbf{v}_0$. Show that T is a linear operator on V.

4. Let $\mathbf{v}_1, \mathbf{v}_2, \ldots, \mathbf{v}_m$ be fixed vectors in R^n, and let $T: R^n \rightarrow R^m$ be the function defined by $T(\mathbf{x}) = (\mathbf{x} \cdot \mathbf{v}_1, \mathbf{x} \cdot \mathbf{v}_2, \ldots, \mathbf{x} \cdot \mathbf{v}_m)$, where $\mathbf{x} \cdot \mathbf{v}_i$ is the Euclidean inner product on R^n.
(a) Show that T is a linear transformation.
(b) Show that the matrix with row vectors $\mathbf{v}_1, \mathbf{v}_2, \ldots, \mathbf{v}_m$ is the standard matrix for T.

5. Let $\{\mathbf{e}_1, \mathbf{e}_2, \mathbf{e}_3, \mathbf{e}_4\}$ be the standard basis for R^4 and $T: R^4 \rightarrow R^3$ the linear transformation for which

$$T(\mathbf{e}_1) = (1, 2, 1), \qquad T(\mathbf{e}_2) = (0, 1, 0),$$
$$T(\mathbf{e}_3) = (1, 3, 0), \qquad T(\mathbf{e}_4) = (1, 1, 1)$$

(a) Find bases for the range and kernel of T.   (b) Find the rank and nullity of T.

6. Suppose that vectors in R^3 are denoted by 1×3 matrices, and define $T: R^3 \rightarrow R^3$ by

$$T([x_1 \quad x_2 \quad x_3]) = [x_1 \quad x_2 \quad x_3] \begin{bmatrix} -1 & 2 & 4 \\ 3 & 0 & 1 \\ 2 & 2 & 5 \end{bmatrix}$$

(a) Find a basis for the kernel of T.   (b) Find a basis for the range of T.

7. Let $B = \{v_1, v_2, v_3, v_4\}$ be a basis for a vector space V and $T: V \to V$ the linear operator for which

$$T(v_1) = v_1 + v_2 + v_3 + 3v_4$$
$$T(v_2) = v_1 - v_2 + 2v_3 + 2v_4$$
$$T(v_3) = 2v_1 - 4v_2 + 5v_3 + 3v_4$$
$$T(v_4) = -2v_1 + 6v_2 - 6v_3 - 2v_4$$

(a) Find the rank and nullity of T.   (b) Determine whether T is one-to-one.

8. Let V and W be vector spaces, T, T_1, and T_2 linear transformations from V to W, and k a scalar. Define new transformations, $T_1 + T_2$ and kT, by the formulas

$$(T_1 + T_2)(x) = T_1(x) + T_2(x)$$
$$(kT)(x) = k(T(x))$$

(a) Show that $(T_1 + T_2): V \to W$ and $kT: V \to W$ are linear transformations.
(b) Show that the set of all linear transformations from V to W with the operations in (a) forms a vector space.

9. Let A and B be similar matrices. Prove:
(a) A^T and B^T are similar.
(b) If A and B are invertible, then A^{-1} and B^{-1} are similar.

10. (*Fredholm Alternative Theorem.*) Let $T: V \to V$ be a linear operator on an n-dimensional vector space. Prove that exactly one of the following statements holds:
 (i) The equation $T(x) = b$ has a solution for all vectors b in V.
 (ii) Nullity of $T > 0$.

11. Let $T: M_{22} \to M_{22}$ be the linear operator defined by

$$T(X) = \begin{bmatrix} 1 & 1 \\ 0 & 0 \end{bmatrix} X + X \begin{bmatrix} 0 & 0 \\ 1 & 1 \end{bmatrix}$$

Find the rank and nullity of T.

12. Prove: If A and B are similar matrices, and if B and C are similar matrices, then A and C are similar matrices.

13. Let $T: M_{22} \to M_{22}$ be the linear operator defined by $T(M) = M^T$. Find the matrix for T with respect to the standard basis for M_{22}.

14. Let $B = \{u_1, u_2, u_3\}$ and $B' = \{v_1, v_2, v_3\}$ be bases for a vector space V, and let

$$P = \begin{bmatrix} 2 & -1 & 3 \\ 1 & 1 & 4 \\ 0 & 1 & 2 \end{bmatrix}$$

be the transition matrix from B' to B.
(a) Express v_1, v_2, v_3 as linear combinations of u_1, u_2, u_3.
(b) Express u_1, u_2, u_3 as linear combinations of v_1, v_2, v_3.

15. Let $B = \{u_1, u_2, u_3\}$ be a basis for a vector space V and $T: V \to V$ a linear operator such that

$$[T]_B = \begin{bmatrix} -3 & 4 & 7 \\ 1 & 0 & -2 \\ 0 & 1 & 0 \end{bmatrix}$$

Find $[T]_{B'}$, where $B' = \{v_1, v_2, v_3\}$ is the basis for V defined by

$$v_1 = u_1, \qquad v_2 = u_1 + u_2, \qquad v_3 = u_1 + u_2 + u_3$$

16. Show that the matrices

$$\begin{bmatrix} 1 & 1 \\ -1 & 4 \end{bmatrix} \quad \text{and} \quad \begin{bmatrix} 2 & 1 \\ 1 & 3 \end{bmatrix}$$

are similar, but that

$$\begin{bmatrix} 3 & 1 \\ -6 & -2 \end{bmatrix} \quad \text{and} \quad \begin{bmatrix} -1 & 2 \\ 1 & 0 \end{bmatrix}$$

are not.

17. Suppose that $T: V \to V$ is a linear operator and B is a basis for V such that for any vector x in V

$$[T(\mathbf{x})]_B = \begin{bmatrix} x_1 - x_2 + x_3 \\ x_2 \\ x_1 - x_3 \end{bmatrix} \quad \text{if} \quad [\mathbf{x}]_B = \begin{bmatrix} x_1 \\ x_2 \\ x_3 \end{bmatrix}$$

Find $[T]_B$.

18. Let $T: V \to V$ be a linear operator. Prove that T is one-to-one if and only if $\det(T) \neq 0$.

19. (*For readers who have studied calculus.*)
   (a) Show that the function $D: C^2(-\infty, \infty) \to F(-\infty, \infty)$ defined by $D(\mathbf{f}) = f''(x)$ is a linear transformation.
   (b) Find a basis for the kernel of D.
   (c) Show that the function satisfying the equation $D(\mathbf{f}) = f(x)$ form a two-dimensional subspace of $C^2(-\infty, \infty)$, and find a basis for this subspace.

20. Let $T: P_2 \to R^3$ be the function defined by the formula

$$T(p(x)) = \begin{bmatrix} p(-1) \\ p(0) \\ p(1) \end{bmatrix}$$

   (a) Find $T(x^2 + 5x + 6)$.
   (b) Show that T is a linear transformation.
   (c) Show that T is one-to-one.
   (d) Find

$$T^{-1}\left(\begin{bmatrix} 0 \\ 3 \\ 0 \end{bmatrix} \right)$$

   (e) Sketch the graph of the polynomial in part (d).

21. Let x_1, x_2, and x_3 be distinct real numbers such that $x_1 < x_2 < x_3$, and let $T: P_2 \to R^3$ be the function defined by the formula

$$T(p(x)) = \begin{bmatrix} p(x_1) \\ p(x_2) \\ p(x_3) \end{bmatrix}$$

   (a) Show that T is a linear transformation.

(b) Show that T is one-to-one.

(c) Verify that if a_1, a_2, and a_3 are any real numbers, then

$$T^{-1}\left(\begin{bmatrix} a_1 \\ a_2 \\ a_3 \end{bmatrix}\right) = a_1 P_1(x) + a_2 P_2(x) + a_3 P_3(x)$$

where

$$P_1(x) = \frac{(x - x_2)(x - x_3)}{(x_1 - x_2)(x_1 - x_3)}, \qquad P_2(x) = \frac{(x - x_1)(x - x_3)}{(x_2 - x_1)(x_2 - x_3)}, \qquad P_3(x) = \frac{(x - x_1)(x - x_2)}{(x_3 - x_1)(x_3 - x_2)}$$

(d) What relationship exists between the graph of the function

$$a_1 P_1(x) + a_2 P_2(x) + a_3 P_3(x)$$

and the points (x_1, a_1), (x_2, a_2), and (x_3, a_3)?

22. (*For readers who have studied calculus.*) Let $p(x)$ and $q(x)$ be continuous functions, and let V be the subspace of $C(-\infty, +\infty)$ consisting of all twice differentiable functions. Define $L: V \rightarrow V$ by

$$L(y(x)) = y''(x) + p(x)y'(x) + q(x)y(x)$$

(a) Show that L is a linear operator.

(b) Consider the special case where $p(x) = 0$ and $q(x) = 1$. Show that the function $\phi(x) = c_1 \sin x + c_2 \cos x$ is in the nullspace of L for all real values of c_1 and c_2.

23. (*For readers who have studied calculus.*) Let $D: P_n \rightarrow P_n$ be the differentiation operator $D(\mathbf{p}) = \mathbf{p}'$. Show that the matrix for D with respect to the basis $B = \{1, x, x^2, \ldots, x^n\}$ is

$$\begin{bmatrix} 0 & 1 & 0 & 0 & \cdots & 0 \\ 0 & 0 & 2 & 0 & \cdots & 0 \\ 0 & 0 & 0 & 3 & \cdots & 0 \\ \vdots & \vdots & \vdots & \vdots & & \vdots \\ 0 & 0 & 0 & 0 & \cdots & n \\ 0 & 0 & 0 & 0 & \cdots & 0 \end{bmatrix}$$

24. (*For readers who have studied calculus.*) It can be shown that for any real number c, the vectors

$$1, x - c, \frac{(x - c)^2}{2!}, \ldots, \frac{(x - c)^n}{n!}$$

form a basis for P_n. Find the matrix for the differentiation operator of Exercise 23 with respect to this basis.

25. (*For readers who have studied calculus.*) Let $J: P_n \rightarrow P_{n+1}$ be the integration transformation defined by

$$J(\mathbf{p}) = \int (a_0 + a_1 x + \cdots + a_n x^n)\, dx = a_0 x + \frac{a_1}{2} x^2 + \cdots + \frac{a_n}{n+1} x^{n+1}$$

where $\mathbf{p} = a_0 + a_1 x + \cdots + a_n x^n$. Find the matrix for T with respect to the standard bases for P_n and P_{n+1}.

CHAPTER **9**

ADDITIONAL TOPICS

9.1 APPLICATION TO DIFFERENTIAL EQUATIONS

*Many laws of physics, chemistry, biology, and economics are described in terms of **differential equations**, that is, equations involving functions and their derivatives. The purpose of this section is to illustrate one way in which linear algebra can be applied to solve certain systems of differential equations. The scope of this section is narrow, but it illustrates an important area of application of linear algebra.*

TERMINOLOGY   One of the simplest differential equations is

$$y' = ay \tag{1}$$

where $y = f(x)$ is an unknown function to be determined, $y' = dy/dx$ is its derivative, and a is a constant. Like most differential equations, (1) has infinitely many solutions; they are the functions of the form

$$y = ce^{ax} \tag{2}$$

where c is an arbitrary constant. Each function of this form is a solution of $y' = ay$ since

$$y' = cae^{ax} = ay$$

Conversely, every solution of $y' = ay$ must be a function of the form ce^{ax} (Exercise 7), so that (2) describes all solutions of $y' = ay$. We call (2) the **general solution** of $y' = ay$.

Sometimes the physical problem that generates a differential equation imposes some added condition that enables us to isolate one ***particular solution*** from the general solution. For example, if we require that the solution of $y' = ay$ satisfy the added condition

$$y(0) = 3 \tag{3}$$

that is, $y = 3$ when $x = 0$, then on substituting these values in the general solution $y = ce^{ax}$ we obtain a value for c, namely

$$3 = ce^0 = c$$

Thus,

$$y = 3e^{ax}$$

is the only solution of $y' = ay$ that satisfies the added condition. A condition such as (3), which specifies the value of the solution at a point, is called an ***initial condition***, and the problem of solving a differential equation subject to an initial condition is called an ***initial-value problem***.

LINEAR SYSTEMS OF FIRST-ORDER EQUATIONS

In this section we will be concerned with solving systems of differential equations having the form

$$
\begin{aligned}
y_1' &= a_{11}y_1 + a_{12}y_2 + \cdots + a_{1n}y_n \\
y_2' &= a_{21}y_1 + a_{22}y_2 + \cdots + a_{2n}y_n \\
&\;\;\vdots \qquad \vdots \qquad \vdots \qquad\qquad \vdots \\
y_n' &= a_{n1}y_1 + a_{n2}y_2 + \cdots + a_{nn}y_n
\end{aligned}
\tag{4}
$$

where $y_1 = f_1(x)$, $y_2 = f_2(x)$, $\ldots$, $y_n = f_n(x)$ are functions to be determined, and the a_{ij}'s are constants. In matrix notation (4) can be written as

$$
\begin{bmatrix} y_1' \\ y_2' \\ \vdots \\ y_n' \end{bmatrix}
=
\begin{bmatrix}
a_{11} & a_{12} & \cdots & a_{1n} \\
a_{21} & a_{22} & \cdots & a_{2n} \\
\vdots & \vdots & & \vdots \\
a_{n1} & a_{n2} & \cdots & a_{nn}
\end{bmatrix}
\begin{bmatrix} y_1 \\ y_2 \\ \vdots \\ y_n \end{bmatrix}
$$

or more briefly,

$$Y' = AY$$

Example 1

(a) Write the following system in matrix form:

$$
\begin{aligned}
y_1' &= 3y_1 \\
y_2' &= -2y_2 \\
y_3' &= 5y_3
\end{aligned}
$$

(b) Solve the system.

(c) Find a solution of the system that satisfies the initial conditions $y_1(0) = 1$, $y_2(0) = 4$, and $y_3(0) = -2$.

Solution (a).

$$\begin{bmatrix} y_1' \\ y_2' \\ y_3' \end{bmatrix} = \begin{bmatrix} 3 & 0 & 0 \\ 0 & -2 & 0 \\ 0 & 0 & 5 \end{bmatrix} \begin{bmatrix} y_1 \\ y_2 \\ y_3 \end{bmatrix} \tag{5}$$

or

$$Y' = \begin{bmatrix} 3 & 0 & 0 \\ 0 & -2 & 0 \\ 0 & 0 & 5 \end{bmatrix} Y$$

Solution (b). Because each equation involves only one unknown function, we can solve the equations individually. From (2), we obtain

$$y_1 = c_1 e^{3x}$$
$$y_2 = c_2 e^{-2x}$$
$$y_3 = c_3 e^{5x}$$

or in matrix notation,

$$Y = \begin{bmatrix} y_1 \\ y_2 \\ y_3 \end{bmatrix} = \begin{bmatrix} c_1 e^{3x} \\ c_2 e^{-2x} \\ c_3 e^{5x} \end{bmatrix}$$

Solution (c). From the given initial conditions, we obtain

$$1 = y_1(0) = c_1 e^0 = c_1$$
$$4 = y_2(0) = c_2 e^0 = c_2$$
$$-2 = y_3(0) = c_3 e^0 = c_3$$

so that the solution satisfying the initial conditions is

$$y_1 = e^{3x}, \qquad y_2 = 4e^{-2x}, \qquad y_3 = -2e^{5x}$$

or in matrix notation,

$$Y = \begin{bmatrix} y_1 \\ y_2 \\ y_3 \end{bmatrix} = \begin{bmatrix} e^{3x} \\ 4e^{-2x} \\ -2e^{5x} \end{bmatrix}$$

The system in the preceding example is easy to solve because each equation involves only one unknown function, and this is the case because the matrix of coefficients (5) for the system is diagonal. But how do we handle a system

$$Y' = AY$$

in which the matrix A is not diagonal? The idea is simple: Try to make a substitution for Y that will yield a new system with a diagonal coefficient matrix; solve this new simpler system, and then use this solution to determine the solution of the original system.

The kind of substitution we have in mind is

$$
\begin{aligned}
y_1 &= p_{11}u_1 + p_{12}u_2 + \cdots + p_{1n}u_n \\
y_2 &= p_{21}u_1 + p_{22}u_2 + \cdots + p_{2n}u_n \\
&\ \ \vdots \qquad \vdots \qquad \vdots \qquad\qquad \vdots \\
y_n &= p_{n1}u_1 + p_{n2}u_2 + \cdots + p_{nn}u_n
\end{aligned}
\tag{6}
$$

or in matrix notation,

$$
\begin{bmatrix} y_1 \\ y_2 \\ \vdots \\ y_n \end{bmatrix}
=
\begin{bmatrix}
p_{11} & p_{12} & \cdots & p_{1n} \\
p_{21} & p_{22} & \cdots & p_{2n} \\
\vdots & \vdots & & \vdots \\
p_{n1} & p_{n2} & \cdots & p_{nn}
\end{bmatrix}
\begin{bmatrix} u_1 \\ u_2 \\ \vdots \\ u_n \end{bmatrix}
$$

or more briefly,

$$
Y = PU
$$

In this substitution the p_{ij}'s are constants to be determined in such a way that the new system involving the unknown functions $u_1, u_2, \ldots, u_n$ has a diagonal coefficient matrix. We leave it as an exercise for the reader to differentiate each equation in (6) and deduce

$$
Y' = PU'
$$

If we make the substitutions $Y = PU$ and $Y' = PU'$ in the original system

$$
Y' = AY
$$

and if we assume P to be invertible, we obtain

$$
PU' = A(PU)
$$

or

$$
U' = (P^{-1}AP)U
$$

or

$$
U' = DU
$$

where $D = P^{-1}AP$. The choice for P is now clear; if we want the new coefficient matrix D to be diagonal, we must choose P to be a matrix that diagonalizes A.

PROCEDURE FOR SOLVING A SYSTEM OF FIRST-ORDER LINEAR DIFFERENTIAL EQUATIONS

This suggests the following procedure for solving a system

$$
Y' = AY
$$

with a diagonalizable coefficient matrix A.

Step 1. Find a matrix P that diagonalizes A.

Step 2. Make the substitutions $Y = PU$ and $Y' = PU'$ to obtain a new "diagonal system" $U' = DU$, where $D = P^{-1}AP$.

Step 3. Solve $U' = DU$.

Step 4. Determine Y from the equation $Y = PU$.

Example 2

(a) Solve the system

$$y_1' = y_1 + y_2$$
$$y_2' = 4y_1 - 2y_2$$

(b) Find the solution that satisfies the initial conditions $y_1(0) = 1$, $y_2(0) = 6$.

Solution (a). The coefficient matrix for the system is

$$A = \begin{bmatrix} 1 & 1 \\ 4 & -2 \end{bmatrix}$$

As discussed in Section 7.2, A will be diagonalized by any matrix P whose columns are linearly independent eigenvectors of A. Since

$$\det(\lambda I - A) = \begin{vmatrix} \lambda - 1 & -1 \\ -4 & \lambda + 2 \end{vmatrix} = \lambda^2 + \lambda - 6 = (\lambda + 3)(\lambda - 2)$$

the eigenvalues of A are $\lambda = 2$, $\lambda = -3$. By definition,

$$\mathbf{x} = \begin{bmatrix} x_1 \\ x_2 \end{bmatrix}$$

is an eigenvector of A corresponding to λ if and only if $\mathbf{x}$ is a nontrivial solution of $(\lambda I - A)\mathbf{x} = \mathbf{0}$, that is, of

$$\begin{bmatrix} \lambda - 1 & -1 \\ -4 & \lambda + 2 \end{bmatrix} \begin{bmatrix} x_1 \\ x_2 \end{bmatrix} = \begin{bmatrix} 0 \\ 0 \end{bmatrix}$$

If $\lambda = 2$, this system becomes

$$\begin{bmatrix} 1 & -1 \\ -4 & 4 \end{bmatrix} \begin{bmatrix} x_1 \\ x_2 \end{bmatrix} = \begin{bmatrix} 0 \\ 0 \end{bmatrix}$$

Solving this system yields

$$x_1 = t, \qquad x_2 = t$$

so that

$$\begin{bmatrix} x_1 \\ x_2 \end{bmatrix} = \begin{bmatrix} t \\ t \end{bmatrix} = t \begin{bmatrix} 1 \\ 1 \end{bmatrix}$$

Thus,

$$\mathbf{p}_1 = \begin{bmatrix} 1 \\ 1 \end{bmatrix}$$

is a basis for the eigenspace corresponding to $\lambda = 2$. Similarly, the reader can show that

$$\mathbf{p}_2 = \begin{bmatrix} -\frac{1}{4} \\ 1 \end{bmatrix}$$

is a basis for the eigenspace corresponding to $\lambda = -3$. Thus,

$$P = \begin{bmatrix} 1 & -\frac{1}{4} \\ 1 & 1 \end{bmatrix}$$

diagonalizes A, and

$$D = P^{-1}AP = \begin{bmatrix} 2 & 0 \\ 0 & -3 \end{bmatrix}$$

Therefore, the substitution

$$Y = PU \quad \text{and} \quad Y' = PU'$$

yields the new "diagonal system"

$$U' = DU = \begin{bmatrix} 2 & 0 \\ 0 & -3 \end{bmatrix} U \quad \text{or} \quad \begin{aligned} u_1' &= 2u_1 \\ u_2' &= -3u_2 \end{aligned}$$

From (2) the solution of this system is

$$\begin{aligned} u_1 &= c_1 e^{2x} \\ u_2 &= c_2 e^{-3x} \end{aligned} \quad \text{or} \quad U = \begin{bmatrix} c_1 e^{2x} \\ c_2 e^{-3x} \end{bmatrix}$$

so that the equation $Y = PU$ yields as the solution for Y

$$Y = \begin{bmatrix} y_1 \\ y_2 \end{bmatrix} = \begin{bmatrix} 1 & -\frac{1}{4} \\ 1 & 1 \end{bmatrix} \begin{bmatrix} c_1 e^{2x} \\ c_2 e^{-3x} \end{bmatrix} = \begin{bmatrix} c_1 e^{2x} - \frac{1}{4} c_2 e^{-3x} \\ c_1 e^{2x} + c_2 e^{-3x} \end{bmatrix}$$

or

$$\begin{aligned} y_1 &= c_1 e^{2x} - \frac{1}{4} c_2 e^{-3x} \\ y_2 &= c_1 e^{2x} + c_2 e^{-3x} \end{aligned} \tag{7}$$

Solution (b). If we substitute the given initial conditions in (7), we obtain

$$\begin{aligned} c_1 - \tfrac{1}{4} c_2 &= 1 \\ c_1 + c_2 &= 6 \end{aligned}$$

Solving this system we obtain

$$c_1 = 2, \quad c_2 = 4$$

so that from (7) the solution satisfying the initial conditions is

$$\begin{aligned} y_1 &= 2e^{2x} - e^{-3x} \\ y_2 &= 2e^{2x} + 4e^{-3x} \end{aligned} \;\blacktriangle$$

We have assumed in this section that the coefficient matrix of $Y' = AY$ is diagonalizable. If this is not the case, other methods must be used to solve the system. Such methods are discussed in more advanced texts.

EXERCISE SET 9.1

1. (a) Solve the system

$$y_1' = y_1 + 4y_2$$
$$y_2' = 2y_1 + 3y_2$$

(b) Find the solution that satisfies the initial conditions $y_1(0) = 0$, $y_2(0) = 0$.

2. (a) Solve the system

$$y_1' = y_1 + 3y_2$$
$$y_2' = 4y_1 + 5y_2$$

(b) Find the solution that satisfies the conditions $y_1(0) = 2$, $y_2'(0) = 1$.

3. (a) Solve the system

$$y_1' = 4y_1 + y_3$$
$$y_2' = -2y_1 + y_2$$
$$y_3' = -2y_1 + y_3$$

(b) Find the solution that satisfies the initial conditions $y_1(0) = -1$, $y_2(0) = 1$, $y_3(0) = 0$.

4. Solve the system

$$y_1' = 4y_1 + 2y_2 + 2y_3$$
$$y_2' = 2y_1 + 4y_2 + 2y_3$$
$$y_3' = 2y_1 + 2y_2 + 4y_3$$

5. Solve the differential equation $y'' - y' - 6y = 0$. [***Hint.*** Let $y_1 = y$, $y_2 = y'$ and then show that

$$y_1' = y_2$$
$$y_2' = y'' = y' + 6y = 6y_1 + y_2]$$

6. Solve the differential equation $y''' - 6y'' + 11y' - 6y = 0$. [***Hint.*** Let $y_1 = y$, $y_2 = y'$, $y_3 = y''$ and then show that

$$y_1' = y_2$$
$$y_2' = y_3$$
$$y_3' = 6y_1 - 11y_2 + 6y_3]$$

7. Prove: Every solution of $y' = ay$ has the form $y = ce^{ax}$. [***Hint.*** Let $y = f(x)$ be a solution and show that $f(x)e^{-ax}$ is constant.]

8. Prove: If A is diagonalizable and

$$Y = \begin{bmatrix} y_1 \\ y_2 \\ \vdots \\ y_n \end{bmatrix}$$

satisfies $Y' = AY$, then each y_i is a linear combination of $e^{\lambda_1 x}$, $e^{\lambda_2 x}$, ..., $e^{\lambda_n x}$, where λ_1, λ_2, ..., λ_n are eigenvalues of A.

9.2 GEOMETRY OF LINEAR OPERATORS ON R^2

In Section 4.2 we studied some of the geometric properties of linear operators on R^2 and R^3. In this section we shall study linear operators on R^2 in a little more depth. Some of the ideas that will be developed here have important applications to the growing field of computer graphics.

If $T: R^2 \rightarrow R^2$ is the matrix operator whose standard matrix is

$$A = \begin{bmatrix} a & b \\ c & d \end{bmatrix}$$

then

$$T\left(\begin{bmatrix} x \\ y \end{bmatrix}\right) = \begin{bmatrix} a & b \\ c & d \end{bmatrix}\begin{bmatrix} x \\ y \end{bmatrix} = \begin{bmatrix} ax + by \\ cx + dy \end{bmatrix} \tag{1}$$

There are two equally good geometric interpretations of this formula. We may view the entries in the matrices

$$\begin{bmatrix} x \\ y \end{bmatrix} \quad \text{and} \quad \begin{bmatrix} ax + by \\ cx + dy \end{bmatrix}$$

either as components of vectors or coordinates of points. With the first interpretation, T maps arrows to arrows, and with the second, points to points (Figure 1). The choice is a matter of taste.

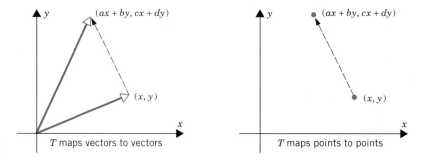

Figure 1

In this section we shall view linear operators on R^2 as mapping points to points. One useful device for visualizing the behavior of a linear operator is to observe its effect on the points of simple figures in the plane. For example, Table 1 shows the effect of some basic linear operators on a unit square that has been partially colored.

TABLE 1

Operator	Standard Matrix	Effect on the Unit Square
Reflection about the *y*-axis	$\begin{bmatrix} -1 & 0 \\ 0 & 1 \end{bmatrix}$	
Reflection about the *x*-axis	$\begin{bmatrix} 1 & 0 \\ 0 & -1 \end{bmatrix}$	
Reflection about the line $y = x$	$\begin{bmatrix} 0 & 1 \\ 1 & 0 \end{bmatrix}$	
Counterclockwise rotation through an angle θ	$\begin{bmatrix} \cos\theta & -\sin\theta \\ \sin\theta & \cos\theta \end{bmatrix}$	

In Section 4.2 we discussed reflections, projections, rotations, contractions, and dilations of R^2. We shall now consider some other basic linear operators on R^2.

EXPANSIONS AND COMPRESSIONS

If the *x*-coordinate of each point in the plane is multiplied by a positive constant k, then the effect is to expand or compress each plane figure in the *x*-direction. If $0 < k < 1$, the result is a compression, and if $k > 1$, an expansion (Figure 2). We call such an operator an *expansion* (or *compression*) *in the x-direction with factor k*. Similarly, if the y-coordinate of each point is multiplied by a positive constant k, we obtain an **expansion** (or **compression**) **in the y-direction with factor k**. It can be shown that expansions and compressions along the coordinate axes are linear transformations.

Figure 2   (Unit square)   (Compression) $k = \frac{1}{2}$ (Expansion) $k = 2$

If $T: R^2 \to R^2$ is an expansion or compression in the x-direction with factor k, then

$$T(\mathbf{e}_1) = T\left(\begin{bmatrix} 1 \\ 0 \end{bmatrix}\right) = \begin{bmatrix} k \\ 0 \end{bmatrix}, \qquad T(\mathbf{e}_2) = T\left(\begin{bmatrix} 0 \\ 1 \end{bmatrix}\right) = \begin{bmatrix} 0 \\ 1 \end{bmatrix}$$

so the standard matrix for T is

$$\begin{bmatrix} k & 0 \\ 0 & 1 \end{bmatrix}$$

Similarly, the standard matrix for an expansion or compression in the y-direction is

$$\begin{bmatrix} 1 & 0 \\ 0 & k \end{bmatrix}$$

Example 1   Suppose that the xy-plane is first expanded or compressed by a factor of k_1 in the x-direction and then is expanded or compressed by a factor of k_2 in the y-direction. Find a single matrix operator that performs both operations.

Solution. The standard matrices for the two operations are

$$\begin{bmatrix} k_1 & 0 \\ 0 & 1 \end{bmatrix} \qquad\qquad \begin{bmatrix} 1 & 0 \\ 0 & k_2 \end{bmatrix}$$

x-expansion (compression) y-expansion (compression)

Thus, the standard matrix for the composition of the x-operation followed by the y-operation is

$$A = \begin{bmatrix} 1 & 0 \\ 0 & k_2 \end{bmatrix}\begin{bmatrix} k_1 & 0 \\ 0 & 1 \end{bmatrix} = \begin{bmatrix} k_1 & 0 \\ 0 & k_2 \end{bmatrix} \tag{2}$$

In the special case where k_1 and k_2 are the same, say $k_1 = k_2 = k$, note that (2) simplifies to

$$A = \begin{bmatrix} k & 0 \\ 0 & k \end{bmatrix}$$

which is a dilation or a contraction (Table 8 of Section 4.2). ▲

SHEARS

A **shear in the x-direction with factor k** is a transformation that moves each point (x, y) parallel to the x-axis by an amount ky to the new position $(x + ky, y)$. Under such a transformation, points on the x-axis are unmoved since $y = 0$. However, as we progress away from the x-axis, the magnitude of y increases, so that points farther from the x-axis move a greater distance than those closer (Figure 3).

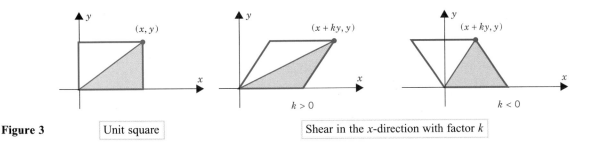

Figure 3

Unit square   Shear in the x-direction with factor k

A **shear in the y-direction with factor k** is a transformation that moves each point (x, y) parallel to the y-axis by an amount kx to the new position $(x, y + kx)$. Under such a transformation, points on the y-axis remain fixed, and points farther from the y-axis move a greater distance than do those closer.

It can be shown that shears are linear transformations. If $T: R^2 \rightarrow R^2$ is a shear of factor k in the x-direction, then

$$T(\mathbf{e}_1) = T\left(\begin{bmatrix} 1 \\ 0 \end{bmatrix}\right) = \begin{bmatrix} 1 \\ 0 \end{bmatrix}, \qquad T(\mathbf{e}_2) = T\left(\begin{bmatrix} 0 \\ 1 \end{bmatrix}\right) = \begin{bmatrix} k \\ 1 \end{bmatrix}$$

so the standard matrix for T is

$$\begin{bmatrix} 1 & k \\ 0 & 1 \end{bmatrix}$$

Similarly, the standard matrix for a shear in the y-direction of factor k is

$$\begin{bmatrix} 1 & 0 \\ k & 1 \end{bmatrix}$$

REMARK. Multiplication by the 2×2 identity matrix is the identity operator on R^2. This operator can be viewed as a rotation through $0°$, or as a shear along either axis with $k = 0$, or as a compression or expansion along either axis with factor $k = 1$.

Example 2

(a) Find a matrix transformation from R^2 to R^2 that first shears by a factor of 2 in the x-direction and then reflects about $y = x$.

(b) Find a matrix transformation from R^2 to R^2 that first reflects about $y = x$ and then shears by a factor of 2 in the x-direction.

Solution (a). The standard matrix for the shear is

$$A_1 = \begin{bmatrix} 1 & 2 \\ 0 & 1 \end{bmatrix}$$

and for the reflection is

$$A_2 = \begin{bmatrix} 0 & 1 \\ 1 & 0 \end{bmatrix}$$

Thus, the standard matrix for the shear followed by the reflection is

$$A_2 A_1 = \begin{bmatrix} 0 & 1 \\ 1 & 0 \end{bmatrix} \begin{bmatrix} 1 & 2 \\ 0 & 1 \end{bmatrix} = \begin{bmatrix} 0 & 1 \\ 1 & 2 \end{bmatrix}$$

Solution (b). The reflection followed by the shear is represented by

$$A_1 A_2 = \begin{bmatrix} 1 & 2 \\ 0 & 1 \end{bmatrix} \begin{bmatrix} 0 & 1 \\ 1 & 0 \end{bmatrix} = \begin{bmatrix} 2 & 1 \\ 1 & 0 \end{bmatrix} \quad \blacktriangle$$

In the last example, note that $A_1 A_2 \neq A_2 A_1$, so that the effect of shearing and then reflecting is different from reflecting and then shearing. This is illustrated geometrically in Figure 4, where we show the effect of the transformations on a unit square.

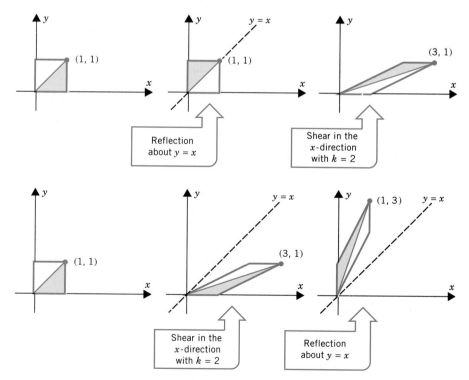

Figure 4

Example 3   Show that if $T:R^2 \to R^2$ is multiplication by an *elementary matrix*, then the transformation is one of the following:

(a)   a shear along a coordinate axis
(b)   a reflection about $y = x$
(c)   a compression along a coordinate axis
(d)   an expansion along a coordinate axis
(e)   a reflection about a coordinate axis
(f)   a compression or expansion along a coordinate axis followed by a reflection about a coordinate axis.

Solution.   Because a 2×2 elementary matrix results from performing a single elementary row operation on the 2×2 identity matrix, it must have one of the following forms (verify):

$$\begin{bmatrix} 1 & 0 \\ k & 1 \end{bmatrix}, \quad \begin{bmatrix} 1 & k \\ 0 & 1 \end{bmatrix}, \quad \begin{bmatrix} 0 & 1 \\ 1 & 0 \end{bmatrix}, \quad \begin{bmatrix} k & 0 \\ 0 & 1 \end{bmatrix}, \quad \begin{bmatrix} 1 & 0 \\ 0 & k \end{bmatrix}$$

The first two matrices represent shears along coordinate axes, and the third, a reflection about $y = x$. If $k > 0$, the last two matrices represent compressions or expansions along coordinate axes depending on whether $0 \le k \le 1$ or $k \ge 1$. If $k < 0$ and if we express k in the form $k = -k_1$ where $k_1 > 0$, then the last two matrices can be written as

$$\begin{bmatrix} k & 0 \\ 0 & 1 \end{bmatrix} = \begin{bmatrix} -k_1 & 0 \\ 0 & 1 \end{bmatrix} = \begin{bmatrix} -1 & 0 \\ 0 & 1 \end{bmatrix}\begin{bmatrix} k_1 & 0 \\ 0 & 1 \end{bmatrix} \tag{3}$$

$$\begin{bmatrix} 1 & 0 \\ 0 & k \end{bmatrix} = \begin{bmatrix} 1 & 0 \\ 0 & -k_1 \end{bmatrix} = \begin{bmatrix} 1 & 0 \\ 0 & -1 \end{bmatrix}\begin{bmatrix} 1 & 0 \\ 0 & k_1 \end{bmatrix} \tag{4}$$

Since $k_1 > 0$, the product in (3) represents a compression or expansion along the x-axis followed by a reflection about the y-axis, and (4) represents a compression or expansion along the y-axis followed by a reflection about the x-axis. In the case where $k = -1$, transformations (3) and (4) are simply reflections about the y-axis and x-axis, respectively.   ▲

Reflections, rotations, expansions, compressions, and shears are all one-to-one linear operators. This is evident geometrically, since all of those operators map distinct points into distinct points. This can also be checked algebraically by verifying that the standard matrices for those operators are invertible.

Example 4   It is intuitively clear that if we compress the xy-plane by a factor of $\frac{1}{2}$ in the y-direction, then we must expand the xy-plane by a factor of 2 in the y-direction to move each point back to its original position. This is indeed the case, since

$$A = \begin{bmatrix} 1 & 0 \\ 0 & \frac{1}{2} \end{bmatrix}$$

represents a compression of factor $\frac{1}{2}$ in the y-direction, and

$$A^{-1} = \begin{bmatrix} 1 & 0 \\ 0 & 2 \end{bmatrix}$$

is an expansion of factor 2 in the y-direction. ▲

GEOMETRIC PROPERTIES OF LINEAR OPERATORS ON R^2

We conclude this section with two theorems that provide some insight into the geometric properties of linear operators on R^2.

Theorem 9.2.1. *If $T: R^2 \to R^2$ is multiplication by an invertible matrix A, then the geometric effect of T is the same as an appropriate succession of shears, compressions, expansions, and reflections.*

Proof. Since A is invertible, it can be reduced to the identity by a finite sequence of elementary row operations. An elementary row operation can be performed by multiplying on the left by an elementary matrix, and so there exist elementary matrices E_1, E_2, ..., E_k such that

$$E_k \cdots E_2 E_1 A = I$$

Solving for A yields

$$A = E_1^{-1} E_2^{-1} \cdots E_k^{-1} I$$

or equivalently,

$$A = E_1^{-1} E_2^{-1} \cdots E_k^{-1} \tag{5}$$

This equation expresses A as a product of elementary matrices (since the inverse of an elementary matrix is also elementary by Theorem 1.5.2). The result now follows from Example 3. ▊

Example 5 Assuming that k_1 and k_2 are positive, express the diagonal matrix

$$A = \begin{bmatrix} k_1 & 0 \\ 0 & k_2 \end{bmatrix}$$

as a product of elementary matrices and describe the geometric effect of multiplication by A in terms of expansions and compressions.

Solution. From Example 1 we have

$$A = \begin{bmatrix} k_1 & 0 \\ 0 & k_2 \end{bmatrix} = \begin{bmatrix} 1 & 0 \\ 0 & k_2 \end{bmatrix}\begin{bmatrix} k_1 & 0 \\ 0 & 1 \end{bmatrix}$$

which shows that multiplication by A has the geometric effect of expanding or compressing by a factor of k_1 in the x-direction and then expanding or compressing by a factor of k_2 in the y-direction. ▲

Example 6 Express

$$A = \begin{bmatrix} 1 & 2 \\ 3 & 4 \end{bmatrix}$$

as a product of elementary matrices, and then describe the geometric effect of multiplication by A in terms of shears, compressions, expansions, and reflections.

Solution. A can be reduced to I as follows:

$$\begin{bmatrix} 1 & 2 \\ 3 & 4 \end{bmatrix} \longrightarrow \begin{bmatrix} 1 & 2 \\ 0 & -2 \end{bmatrix} \longrightarrow \begin{bmatrix} 1 & 2 \\ 0 & 1 \end{bmatrix} \longrightarrow \begin{bmatrix} 1 & 0 \\ 0 & 1 \end{bmatrix}$$

| Add -3 times the first row to the second. | Multiply the second row by $-\frac{1}{2}$. | Add -2 times the second row to the first. |

The three successive row operations can be performed by multiplying on the left successively by

$$E_1 = \begin{bmatrix} 1 & 0 \\ -3 & 1 \end{bmatrix}, \qquad E_2 = \begin{bmatrix} 1 & 0 \\ 0 & -\frac{1}{2} \end{bmatrix}, \qquad E_3 = \begin{bmatrix} 1 & -2 \\ 0 & 1 \end{bmatrix}$$

Inverting these matrices and using (5) yields

$$A = E_1^{-1} E_2^{-1} E_3^{-1} = \begin{bmatrix} 1 & 0 \\ 3 & 1 \end{bmatrix} \begin{bmatrix} 1 & 0 \\ 0 & -2 \end{bmatrix} \begin{bmatrix} 1 & 2 \\ 0 & 1 \end{bmatrix}$$

Reading from right to left and noting that

$$\begin{bmatrix} 1 & 0 \\ 0 & -2 \end{bmatrix} = \begin{bmatrix} 1 & 0 \\ 0 & -1 \end{bmatrix} \begin{bmatrix} 1 & 0 \\ 0 & 2 \end{bmatrix}$$

it follows that the effect of multiplying by A is equivalent to

    (1) shearing by a factor of 2 in the x-direction,
    (2) then expanding by a factor of 2 in the y-direction,
    (3) then reflecting about the x-axis,
    (4) then shearing by a factor of 3 in the y-direction. ▲

The proofs for parts of the following theorem are discussed in the exercises.

Theorem 9.2.2. *If* $T: R^2 \to R^2$ *is multiplication by an invertible matrix, then:*

(a) *The image of a straight line is a straight line.*

(b) *The image of a straight line through the origin is a straight line through the origin.*

(c) *The images of parallel straight lines are parallel straight lines.*

(d) *The image of the line segment joining points P and Q is the line segment joining the images of P and Q.*

(e) *The images of three points lie on a line if and only if the points themselves do.*

REMARK. It follows from parts (*c*), (*d*), and (*e*) that multiplication by an invertible 2×2 matrix A maps triangles into triangles and parallelograms into parallelograms.

Example 7   Sketch the image of the square with vertices $P_1(0, 0)$, $P_2(1, 0)$, $P_3(0, 1)$, and $P_4(1, 1)$ under multiplication by

$$A = \begin{bmatrix} -1 & 2 \\ 2 & -1 \end{bmatrix}$$

Solution.   Since

$$\begin{bmatrix} -1 & 2 \\ 2 & -1 \end{bmatrix}\begin{bmatrix} 0 \\ 0 \end{bmatrix} = \begin{bmatrix} 0 \\ 0 \end{bmatrix} \qquad \begin{bmatrix} -1 & 2 \\ 2 & -1 \end{bmatrix}\begin{bmatrix} 1 \\ 0 \end{bmatrix} = \begin{bmatrix} -1 \\ 2 \end{bmatrix}$$

$$\begin{bmatrix} -1 & 2 \\ 2 & -1 \end{bmatrix}\begin{bmatrix} 0 \\ 1 \end{bmatrix} = \begin{bmatrix} 2 \\ -1 \end{bmatrix} \qquad \begin{bmatrix} -1 & 2 \\ 2 & -1 \end{bmatrix}\begin{bmatrix} 1 \\ 1 \end{bmatrix} = \begin{bmatrix} 1 \\ 1 \end{bmatrix}$$

the image of the square is a parallelogram with vertices $(0, 0)$, $(-1, 2)$, $(2, -1)$, and $(1, 1)$ (Figure 5). ▲

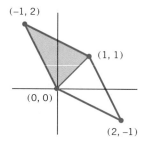

Figure 5

Example 8   According to Theorem 9.2.2, the invertible matrix

$$A = \begin{bmatrix} 3 & 1 \\ 2 & 1 \end{bmatrix}$$

maps the line $y = 2x + 1$ into another line. Find its equation.

Solution.   Let (x, y) be a point on the line $y = 2x + 1$ and let (x', y') be its image under multiplication by A. Then

$$\begin{bmatrix} x' \\ y' \end{bmatrix} = \begin{bmatrix} 3 & 1 \\ 2 & 1 \end{bmatrix}\begin{bmatrix} x \\ y \end{bmatrix} \qquad \text{and} \qquad \begin{bmatrix} x \\ y \end{bmatrix} = \begin{bmatrix} 3 & 1 \\ 2 & 1 \end{bmatrix}^{-1}\begin{bmatrix} x' \\ y' \end{bmatrix} = \begin{bmatrix} 1 & -1 \\ -2 & 3 \end{bmatrix}\begin{bmatrix} x' \\ y' \end{bmatrix}$$

so

$$x = \quad x' - y'$$
$$y = -2x' + 3y'$$

Substituting in $y = 2x + 1$ yields

$$-2x' + 3y' = 2(x' - y') + 1$$

or equivalently,

$$y' = \tfrac{4}{5}x' + \tfrac{1}{5}$$

Thus, (x', y') satisfies

$$y = \tfrac{4}{5}x + \tfrac{1}{5}$$

which is the equation we want. ▲

EXERCISE SET 9.2

1. Find the standard matrix for the plane linear transformation $T: R^2 \to R^2$ that maps a point (x, y) into (see Figure 6)
   (a) its reflection about the line $y = -x$
   (b) its reflection through the origin
   (c) its orthogonal projection on the x-axis
   (d) its orthogonal projection on the y-axis

(a)　　　　　　(b)　　　　　　(c)　　　　　　(d)　　　**Figure 6**

2. For each part of Exercise 1, use the matrix you have obtained to compute $T(2, 1)$. Check your answers geometrically by plotting the points $(2, 1)$ and $T(2, 1)$.

3. Find the standard matrix for the linear operator $T: R^3 \to R^3$ that maps a point (x, y, z) into
   (a) its reflection through the xy-plane　(b) its reflection through the xz-plane
   (c) its reflection through the yz-plane

4. For each part of Exercise 3, use the matrix you have obtained to compute $T(1, 1, 1)$. Check your answers geometrically by sketching the vectors $(1, 1, 1)$ and $T(1, 1, 1)$.

5. Find the standard matrix for the linear operator $T: R^3 \to R^3$ that
   (a) rotates each vector 90° counterclockwise about the z-axis (looking along the positive z-axis toward the origin)
   (b) rotates each vector 90° counterclockwise about the x-axis (looking along the positive x-axis toward the origin)
   (c) rotates each vector 90° counterclockwise about the y-axis (looking along the positive y-axis toward the origin)

6. Sketch the image of the rectangle with vertices $(0, 0)$, $(1, 0)$, $(1, 2)$, and $(0, 2)$ under
   (a) a reflection about the x-axis
   (b) a reflection about the y-axis
   (c) a compression of factor $k = \frac{1}{4}$ in the y-direction
   (d) an expansion of factor $k = 2$ in the x-direction
   (e) a shear of factor $k = 3$ in the x-direction
   (f) a shear of factor $k = 2$ in the y-direction

7. Sketch the image of the square with vertices $(0, 0)$, $(1, 0)$, $(0, 1)$, and $(1, 1)$ under multiplication by
$$A = \begin{bmatrix} -3 & 0 \\ 0 & 1 \end{bmatrix}$$

8. Find the matrix that rotates a point (x, y) about the origin through
   (a) $45°$ (b) $90°$ (c) $180°$ (d) $270°$ (e) $-30°$

9. Find the matrix that shears by
   (a) a factor of $k = 4$ in the y-direction   (b) a factor of $k = -2$ in the x-direction

10. Find the matrix that compresses or expands by
   (a) a factor of $\frac{1}{3}$ in the y-direction   (b) a factor of 6 in the x-direction

11. In each part, describe the geometric effect of multiplication by the given matrix.

(a) $\begin{bmatrix} 3 & 0 \\ 0 & 1 \end{bmatrix}$ (b) $\begin{bmatrix} 1 & 0 \\ 0 & -5 \end{bmatrix}$ (c) $\begin{bmatrix} 1 & 4 \\ 0 & 1 \end{bmatrix}$

12. Express the matrix as a product of elementary matrices, and then describe the effect of multiplication by the given matrix in terms of compressions, expansions, reflections, and shears.

(a) $\begin{bmatrix} 2 & 0 \\ 0 & 3 \end{bmatrix}$ (b) $\begin{bmatrix} 1 & 4 \\ 2 & 9 \end{bmatrix}$ (c) $\begin{bmatrix} 0 & -2 \\ 4 & 0 \end{bmatrix}$ (d) $\begin{bmatrix} 1 & -3 \\ 4 & 6 \end{bmatrix}$

13. In each part, find a single matrix that performs the indicated succession of operations:
   (a) compresses by a factor of $\frac{1}{2}$ in the x-direction, then expands by a factor of 5 in the y-direction
   (b) expands by a factor of 5 in the y-direction, then shears by a factor of 2 in the y-direction
   (c) reflects about $y = x$, then rotates through an angle of $180°$

14. In each part, find a single matrix that performs the indicated succession of operations:
   (a) reflects about the y-axis, then expands by a factor of 5 in the x-direction, and then reflects about $y = x$
   (b) rotates through $30°$, then shears by a factor of -2 in the y-direction, and then expands by a factor of 3 in the y-direction

15. By matrix inversion, show the following:
   (a) The inverse transformation for a reflection about $y = x$ is a reflection about $y = x$.
   (b) The inverse transformation for a compression along an axis is an expansion along that axis.
   (c) The inverse transformation for a reflection about a coordinate axis is a reflection about that axis.
   (d) The inverse transformation for a shear along a coordinate axis is a shear along that axis.

16. Find the equation of the image of the line $y = -4x + 3$ under multiplication by

$$A = \begin{bmatrix} 4 & -3 \\ 3 & -2 \end{bmatrix}$$

17. In parts (a) through (e) find the equation of the image of the line $y = 2x$ under
(a) a shear of factor 3 in the x-direction   (b) a compression of factor $\frac{1}{2}$ in the y-direction
(c) a reflection about $y = x$ (d) a reflection about the y-axis
(e) a rotation of 60°

18. Find the matrix for a shear in the x-direction that transforms the triangle with vertices (0, 0), (2, 1), and (3, 0) into a right triangle with the right angle at the origin.

19. (a) Show that multiplication by

$$A = \begin{bmatrix} 3 & 1 \\ 6 & 2 \end{bmatrix}$$

maps every point in the plane onto the line $y = 2x$.
(b) It follows from (a) that the noncollinear points (1, 0), (0, 1), (−1, 0) are mapped on a line. Does this violate part (e) of Theorem 9.2.2?

20. Prove part (a) of Theorem 9.2.2. [**Hint.** A line in the plane has an equation of the form $Ax + By + C = 0$, where A and B are not both zero. Use the method of Example 8 to show that the image of this line under multiplication by the invertible matrix

$$\begin{bmatrix} a & b \\ c & d \end{bmatrix}$$

has the equation $A'x + B'y + C = 0$, where

$$A' = (dA - cB)/(ad - bc) \qquad \text{and} \qquad B' = (-bA + aB)/(ad - bc)$$

Then show that A' and B' are not both zero to conclude that the image is a line.]

21. Use the hint in Exercise 20 to prove parts (b) and (c) of Theorem 9.2.2.

22. In each part find the standard matrix for the linear operator $T: R^3 \to R^3$ described by Figure 7.

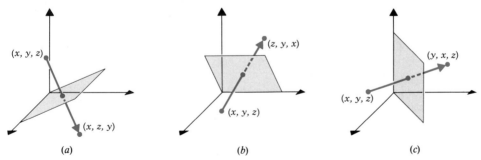

(a)   (b)   (c)   **Figure 7**

23. In R^3 the *shear in the xy-direction with factor k* is the linear transformation that moves each point (x, y, z) parallel to the xy-plane to the new position $(x + kz, y + kz, z)$. (See Figure 8.)
(a) Find the standard matrix for the shear in the xy-direction with factor k.
(b) How would you define the shear in the xz-direction with factor k and the shear in the yz-direction with factor k? Find the standard matrix for each of these linear transformations.

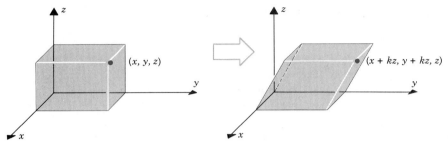

Figure 8

24. In each part find as many linearly independent eigenvectors as you can by inspection (by visualizing the geometric effect of the transformation on R^2). For each of your eigenvectors find the corresponding eigenvalue by inspection, then check your results by computing the eigenvalues and bases for the eigenspaces from the standard matrix for the transformation.
  (a) reflection about the x-axis   (b) reflection about the y-axis
  (c) reflection about $y = x$ (d) shear in the x-direction with factor k
  (e) shear in the y-direction with factor k (f) rotation through the angle θ

9.3 LEAST SQUARES FITTING TO DATA

In this section we shall use results about orthogonal projections in inner product spaces to obtain a technique for fitting a line or other polynomial curve to a set of experimentally determined points in the plane.

FITTING A CURVE TO EXPERIMENTAL DATA

A common problem in experimental work is to obtain a mathematical relationship $y = f(x)$ between two variables x and y by "fitting" a curve to points in the plane corresponding to various experimentally determined values of x and y, say

$$(x_1, y_1), (x_2, y_2), \ldots, (x_n, y_n)$$

On the basis of theoretical considerations or simply by the pattern of the points, one decides on the general form of the curve $y = f(x)$ to be fitted. Some possibilities are (Figure 1)

(a)

(b)

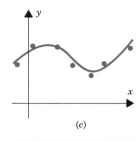

(c)

Figure 1 $\boxed{y = a + bx}$ $\boxed{y = a + bx + cx^2}$ $\boxed{y = a + bx + cx^2 + dx^3}$

(a) A straight line: $y = a + bx$

(b) A quadratic polynomial: $y = a + bx + cx^2$

(c) A cubic polynomial: $y = a + bx + cx^2 + dx^3$

Because the points are obtained experimentally, there is usually some measurement "error" in the data making it impossible to find a curve of the desired form that passes through all the points. Thus, the idea is to choose the curve (by determining its coefficients) that "best" fits the data. We begin with the simplest case: fitting a straight line to the data points.

LEAST SQUARES FIT OF A STRAIGHT LINE

Suppose we want to fit a straight line

$$y = a + bx$$

to the experimentally determined points

$$(x_1, y_1), (x_2, y_2), \ldots, (x_n, y_n)$$

If the data points are collinear, the line would pass through all n points, and so the unknown coefficients a and b would satisfy

$$y_1 = a + bx_1$$
$$y_2 = a + bx_2$$
$$\vdots$$
$$y_n = a + bx_n$$

We can write this system in matrix form as

$$\begin{bmatrix} 1 & x_1 \\ 1 & x_2 \\ \vdots & \vdots \\ 1 & x_n \end{bmatrix} \begin{bmatrix} a \\ b \end{bmatrix} = \begin{bmatrix} y_1 \\ y_2 \\ \vdots \\ y_n \end{bmatrix}$$

or, more compactly, as

$$M\mathbf{v} = \mathbf{y} \qquad (1)$$

where

$$\mathbf{y} = \begin{bmatrix} y_1 \\ y_2 \\ \vdots \\ y_n \end{bmatrix}, \qquad M = \begin{bmatrix} 1 & x_1 \\ 1 & x_2 \\ \vdots & \vdots \\ 1 & x_n \end{bmatrix}, \qquad \mathbf{v} = \begin{bmatrix} a \\ b \end{bmatrix} \qquad (2)$$

If the data points are not collinear, then it is impossible to find coefficients a and b that satisfy system (1) exactly; that is, the system is inconsistent. In this case we shall look for a least squares solution

$$\mathbf{v} = \mathbf{v}^* = \begin{bmatrix} a^* \\ b^* \end{bmatrix}$$

We call a line $y = a^* + b^*x$ whose coefficients come from a least squares solution a *least squares straight line fit* to the data. To explain this terminology, recall that a least squares solution of (1) minimizes

$$\|\mathbf{y} - M\mathbf{v}\| \tag{3}$$

If we express the square of (3) in terms of components, we obtain

$$\|\mathbf{y} - M\mathbf{v}\|^2 = (y_1 - a - bx_1)^2 + (y_2 - a - bx_2)^2 + \cdots + (y_n - a - bx_n)^2 \tag{4}$$

If we now let

$$d_1 = |y_1 - a - bx_1|, \, d_2 = |y_2 - a - bx_2|, \, \ldots, \, d_n = |y_n - a - bx_n|$$

then (4) can be written as

$$\|\mathbf{y} - M\mathbf{v}\|^2 = d_1^2 + d_2^2 + \cdots + d_n^2 \tag{5}$$

As illustrated in Figure 2, d_i can be interpreted as the vertical distance between the line $y = a + bx$ and the data point (x_i, y_i). This distance is a measure of the "error" at the point (x_i, y_i) resulting from the inexact fit of $y = a + bx$ to the data points. Since (3) and (5) are minimized by the same vector $\mathbf{v}^*$, the least squares straight line fit minimizes the sum of the squares of these errors, hence the name *least squares straight line fit*.

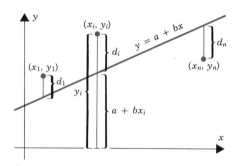

Figure 2 d_i measures the vertical error in the least squares straight line fit.

NORMAL EQUATIONS

Recall from Theorem 6.4.2 that the least squares solutions of (1) can be obtained by solving the associated normal system

$$M^T M \mathbf{v} = M^T \mathbf{y}$$

the equations of which are called the *normal equations*.

In the exercises it will be shown that column vectors of M are linearly independent if and only if the n data points do not lie on a vertical line in the xy-plane. In this case it follows from Theorem 6.4.4 that the least squares solution is unique and is given by

$$\mathbf{v}^* = (M^TM)^{-1}M^T\mathbf{y}$$

In summary, we have the following theorem.

Theorem 9.3.1. *Let $(x_1, y_1), (x_2, y_2), \ldots, (x_n, y_n)$ be a set of two or more data points, not all lying on a vertical line, and let*

$$M = \begin{bmatrix} 1 & x_1 \\ 1 & x_2 \\ \vdots & \vdots \\ 1 & x_n \end{bmatrix} \quad and \quad \mathbf{y} = \begin{bmatrix} y_1 \\ y_2 \\ \vdots \\ y_n \end{bmatrix}$$

Then there is a unique least squares straight line fit

$$y = a^* + b^*x$$

to the data points. Moreover,

$$\mathbf{v}^* = \begin{bmatrix} a^* \\ b^* \end{bmatrix}$$

is given by the formula

$$\mathbf{v}^* = (M^TM)^{-1}M^T\mathbf{y} \tag{6}$$

which expresses the fact that $\mathbf{v} = \mathbf{v}^*$ is the unique solution of the normal equations

$$M^TM\mathbf{v} = M^T\mathbf{y} \tag{7}$$

Example 1 Find the least squares straight line fit to the four points $(0, 1)$, $(1, 3)$, $(2, 4)$, and $(3, 4)$. (See Figure 3.)

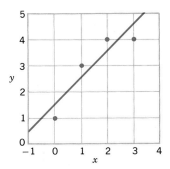

Figure 3

Solution. We have

$$M = \begin{bmatrix} 1 & 0 \\ 1 & 1 \\ 1 & 2 \\ 1 & 3 \end{bmatrix}$$

$$M^T M = \begin{bmatrix} 4 & 6 \\ 6 & 14 \end{bmatrix}$$

$$(M^T M)^{-1} = \frac{1}{10} \begin{bmatrix} 7 & -3 \\ -3 & 2 \end{bmatrix}$$

$$\mathbf{v}^* = (M^T M)^{-1} M^T \mathbf{y} = \frac{1}{10} \begin{bmatrix} 7 & -3 \\ -3 & 2 \end{bmatrix} \begin{bmatrix} 1 & 1 & 1 & 1 \\ 0 & 1 & 2 & 3 \end{bmatrix} \begin{bmatrix} 1 \\ 3 \\ 4 \\ 4 \end{bmatrix} = \begin{bmatrix} 1.5 \\ 1 \end{bmatrix}$$

And so the desired line is $y = 1.5 + x$. ▲

Example 2 Hooke's law in physics states that the length x of a uniform spring is a linear function of the force y applied to it. If we write $y = a + bx$, then the coefficient b is called the spring constant. Suppose a particular unstretched spring has a measured length of 6.1 inches (i.e., $x = 6.1$ when $y = 0$). Forces of 2 pounds, 4 pounds, and 6 pounds are then applied to the spring, and the corresponding lengths are found to be 7.6 inches, 8.7 inches, and 10.4 inches (see Figure 4). Find the spring constant of this spring.

Figure 4   Force y

x_i	6.1	7.6	8.7	10.4
y_i	0	2	4	6

Solution. We have

$$M = \begin{bmatrix} 1 & 6.1 \\ 1 & 7.6 \\ 1 & 8.7 \\ 1 & 10.4 \end{bmatrix}, \qquad \mathbf{y} = \begin{bmatrix} 0 \\ 2 \\ 4 \\ 6 \end{bmatrix},$$

and

$$\mathbf{v}^* = \begin{bmatrix} a^* \\ b^* \end{bmatrix} = (M^T M)^{-1} M^T \mathbf{y} \simeq \begin{bmatrix} -8.6 \\ 1.4 \end{bmatrix}$$

where the numerical values have been rounded to one decimal place. Thus, the estimated value of the spring constant is $b^* \simeq 1.4$ pounds/inch. ▲

LEAST SQUARES FIT OF A POLYNOMIAL

The technique described for fitting a straight line to data points generalizes easily to fitting a polynomial of any specified degree to data points. Let us attempt to fit a polynomial of fixed degree m

$$y = a_0 + a_1 x + \cdots + a_m x^m \tag{8}$$

to n points

$$(x_1, y_1), (x_2, y_2), \ldots, (x_n, y_n)$$

Substituting these n values of x and y into (8) yields the n equations

$$y_1 = a_0 + a_1 x_1 + \cdots + a_m x_1^m$$
$$y_2 = a_0 + a_1 x_2 + \cdots + a_m x_2^m$$
$$\vdots \qquad \vdots \qquad \vdots \qquad \qquad \vdots$$
$$y_n = a_0 + a_1 x_n + \cdots + a_m x_n^m$$

or in matrix form,

$$M\mathbf{v} = \mathbf{y} \tag{9}$$

where

$$\mathbf{y} = \begin{bmatrix} y_1 \\ y_2 \\ \vdots \\ y_n \end{bmatrix}, \qquad M = \begin{bmatrix} 1 & x_1 & x_1^2 & \cdots & x_1^m \\ 1 & x_2 & x_2^2 & \cdots & x_2^m \\ \vdots & \vdots & \vdots & & \vdots \\ 1 & x_n & x_n^2 & \cdots & x_n^m \end{bmatrix}, \qquad \mathbf{v} = \begin{bmatrix} a_0 \\ a_1 \\ \vdots \\ a_m \end{bmatrix}$$

As before, the solutions of the normal equations

$$M^T M \mathbf{v} = M^T \mathbf{y}$$

determine coefficients of polynomials that minimize

$$\|\mathbf{y} - M\mathbf{v}\|$$

Conditions that guarantee the invertibility of $M^T M$ are discussed in the exercises. If $M^T M$ is invertible, then the normal equations have a unique solution $\mathbf{v} = \mathbf{v}^*$, which is given by

$$\mathbf{v}^* = (M^T M)^{-1} M^T \mathbf{y}$$

Example 3   According to Newton's second law of motion, a body near the earth's surface falls vertically downward according to the equation

$$s = s_0 + v_0 t + \tfrac{1}{2} g t^2 \tag{10}$$

where

$$s = \text{vertical displacement downward relative to some fixed point}$$
$$s_0 = \text{initial displacement at time } t = 0$$
$$v_0 = \text{initial velocity at time } t = 0$$
$$g = \text{acceleration of gravity at the earth's surface}$$

Suppose that a laboratory experiment is performed to evaluate g using this equation. A weight is released with unknown initial displacement and velocity, and at certain times the distances fallen from some fixed reference point are measured. In particular, suppose that at times $t = .1, .2, .3, .4$, and $.5$ seconds it is found that the weight has fallen $s = -0.18, 0.31, 1.03, 2.48$, and 3.73 feet, respectively, from the reference point. Find an approximate value of g using these data.

Solution. The mathematical problem is to fit a quadratic curve

$$s = a_0 + a_1 t + a_2 t^2 \tag{11}$$

to the five data points:

$$(.1, -0.18), \quad (.2, 0.31), \quad (.3, 1.03), \quad (.4, 2.48), \quad (.5, 3.73)$$

The necessary calculations are

$$M = \begin{bmatrix} 1 & t_1 & t_1^2 \\ 1 & t_2 & t_2^2 \\ 1 & t_3 & t_3^2 \\ 1 & t_4 & t_4^2 \\ 1 & t_5 & t_5^2 \end{bmatrix} = \begin{bmatrix} 1 & .1 & .01 \\ 1 & .2 & .04 \\ 1 & .3 & .09 \\ 1 & .4 & .16 \\ 1 & .5 & .25 \end{bmatrix}$$

$$\mathbf{y} = \begin{bmatrix} s_1 \\ s_2 \\ s_3 \\ s_4 \\ s_5 \end{bmatrix} = \begin{bmatrix} -0.18 \\ 0.31 \\ 1.03 \\ 2.48 \\ 3.73 \end{bmatrix}$$

and

$$\mathbf{v}^* = \begin{bmatrix} a_0^* \\ a_1^* \\ a_2^* \end{bmatrix} = (M^T M)^{-1} M^T \mathbf{y} \approx \begin{bmatrix} -0.40 \\ 0.35 \\ 16.1 \end{bmatrix}$$

From (10) and (11) we have $a_2 = \frac{1}{2}g$, and so the estimated value of g is

$$g = 2a_2^* = 2(16.1) = 32.2 \text{ feet/second}^2$$

If desired, we can also estimate the initial displacement and initial velocity of the weight:

$$s_0 = a_0^* = -0.40 \text{ feet}$$
$$v_0 = a_1^* = \quad 0.35 \text{ feet/second}$$

In Figure 5 we have plotted the five data points and the approximating polynomial. ▲

Figure 5

EXERCISE SET 9.3

1. Find the least squares straight line fit to the three points $(0, 0)$, $(1, 2)$, and $(2, 7)$.

2. Find the least squares straight line fit to the four points $(0, 1)$, $(2, 0)$, $(3, 1)$, and $(3, 2)$.

3. Find the quadratic polynomial that best fits the four points $(2, 0)$, $(3, -10)$, $(5, -48)$, and $(6, -76)$.

4. Find the cubic polynomial that best fits the five points $(-1, -14)$, $(0, -5)$, $(1, -4)$, $(2, 1)$, and $(3, 22)$.

5. Show that the matrix M in Equation (2) has linearly independent columns if and only if at least two of the numbers $x_1, x_2, \ldots, x_n$ are distinct.

6. Show that the columns of the $n \times (m + 1)$ matrix M in Equation (9) are linearly independent if $n > m$ and at least $m + 1$ of the numbers $x_1, x_2, \ldots, x_n$ are distinct.

7. Let M be the matrix in Equation (9). Using Exercise 6, show that a sufficient condition for the matrix $M^T M$ to be invertible is that $n > m$ and at least $m + 1$ of the numbers $x_1, x_2, \ldots, x_n$ are distinct.

8. The owner of a rapidly expanding business finds that for the first five months of the year the sales (in thousands) are $4.0, $4.4, $5.2, $6.4, and $8.0. The owner plots these figures on a graph and conjectures that for the rest of the year the sales curve can be approximated by a quadratic polynomial. Find the least squares quadratic polynomial fit to the sales curve, and use it to project the sales for the twelfth month of the year.

9.4 APPROXIMATION PROBLEMS; FOURIER SERIES

In this section we shall use results about orthogonal projections in inner product spaces to solve problems that involve approximating a given function by simpler functions. Such problems arise in a variety of engineering and scientific applications.

**BEST APPROXI-
MATIONS**

All of the problems that we will study in this section will be special cases of the following general problem.

> **Approximation Problem.** Given a function f that is continuous on an interval $[a, b]$, find the "best possible approximation" to f using only functions from a specified subspace W of $C[a, b]$.

Here are some examples of such problems:

(a) Find the best possible approximation to e^x over $[0, 1]$ by a polynomial of the form $a_0 + a_1 x + a_2 x^2$.

(b) Find the best possible approximation to $\sin \pi x$ over $[-1, 1]$ by a function of the form $a_0 + a_1 e^x + a_2 e^{2x} + a_3 e^{3x}$.

(c) Find the best possible approximation to x over $[0, 2\pi]$ by a function of the form $a_0 + a_1 \sin x + a_2 \sin 2x + b_1 \cos x + b_2 \cos 2x$.

In the first example W is the subspace of $C[0, 1]$ spanned by 1, x, and x^2; in the second example W is the subspace of $C[-1, 1]$ spanned by 1, e^x, e^{2x}, and e^{3x}; and in the third example W is the subspace of $C[0, 2\pi]$ spanned by 1, $\sin x$, $\sin 2x$, $\cos x$, and $\cos 2x$.

MEASUREMENTS OF ERROR

To solve approximation problems of the preceding types, we must make the phrase "best approximation over $[a, b]$" mathematically precise; to do this we need a precise way of measuring the error that results when one continuous function is approximated by another over $[a, b]$. If we were concerned only with approximating $f(x)$ at a single point x_0, then the error at x_0 by an approximation $g(x)$ would be simply

$$\text{error} = |f(x_0) - g(x_0)|$$

sometimes called the **deviation** between f and g at x_0 (Figure 1). However, we are concerned with approximation over the entire interval $[a, b]$, not at a single point. Consequently, in one part of the interval an approximation g_1 to f may have smaller deviations from f than an approximation g_2 to f, and in another part of the interval it might be the other way around. How does one decide which is the better overall approximation? What we need is some way of measuring the overall error in an approximation $g(x)$. One possible measure of overall error is obtained by integrating the deviation $|f(x) - g(x)|$ over the entire interval $[a, b]$; that is,

$$\text{error} = \int_a^b |f(x) - g(x)| \, dx \tag{1}$$

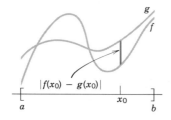

Figure 1 The deviation between f and g at x_0

Geometrically (1) is the area between the graphs of $f(x)$ and $g(x)$ over the interval $[a, b]$ (Figure 2); the greater the area, the greater the overall error.

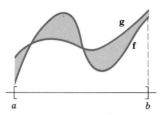

Figure 2

The area between the graphs of f and g over $[a, b]$ measures the error in approximating f by g over $[a, b]$.

While (1) is natural and appealing geometrically, most mathematicians and scientists generally favor the following alternative measure of error, called the ***mean square error***.

$$\text{mean square error} = \int_a^b [f(x) - g(x)]^2 \, dx$$

Mean square error emphasizes the effect of larger errors because of the squaring and has the added advantage that it allows us to bring to bear the theory of inner product spaces. To see how, suppose that **f** is a continuous function on $[a, b]$ that we want to approximate by a function **g** from a subspace W of $C[a, b]$, and suppose that $C[a, b]$ is given the inner product

$$\langle \mathbf{f}, \mathbf{g} \rangle = \int_a^b f(x)g(x) \, dx$$

It follows that

$$\|\mathbf{f} - \mathbf{g}\|^2 = \langle \mathbf{f} - \mathbf{g}, \mathbf{f} - \mathbf{g} \rangle = \int_a^b [f(x) - g(x)]^2 \, dx = \text{mean square error}$$

so that minimizing the mean square error is the same as minimizing $\|\mathbf{f} - \mathbf{g}\|^2$. Thus, the approximation problem posed informally at the beginning of this section can be restated more precisely as follows:

LEAST SQUARES APPROXIMATION

Least Squares Approximation Problem. Let **f** be a function that is continuous on an interval $[a, b]$, let $C[a, b]$ have the inner product

$$\langle \mathbf{f}, \mathbf{g} \rangle = \int_a^b f(x)g(x) \, dx$$

and let W be a finite-dimensional subspace of $C[a, b]$. Find a function **g** in W that minimizes

$$\|\mathbf{f} - \mathbf{g}\|^2 = \int_a^b [f(x) - g(x)]^2 \, dx$$

Since $\|\mathbf{f} - \mathbf{g}\|^2$ and $\|\mathbf{f} - \mathbf{g}\|$ are minimized by the same function $\mathbf{g}$, the preceding problem is equivalent to looking for a function $\mathbf{g}$ in W that is closest to $\mathbf{f}$. But we know from Theorem 6.4.1 that $\mathbf{g} = \text{proj}_W \mathbf{f}$ is such a function (Figure 3).

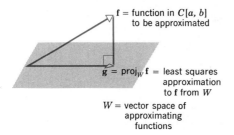

f = function in $C[a, b]$
to be approximated

$\mathbf{g} = \text{proj}_W \mathbf{f}$ = least squares
approximation
to f from W

W = vector space of
approximating
functions

Figure 3

Thus, we have the following result.

Solution of the Least Squares Approximation Problem. If $\mathbf{f}$ is a continuous function on $[a, b]$, and W is a finite-dimensional subspace of $C[a, b]$, then the function $\mathbf{g}$ in W that minimizes the mean square error

$$\int_a^b [f(x) - g(x)]^2 \, dx$$

is $\mathbf{g} = \text{proj}_W \mathbf{f}$, where the orthogonal projection is relative to the inner product

$$\langle \mathbf{f}, \mathbf{g} \rangle = \int_a^b f(x)g(x) \, dx$$

The function $\mathbf{g} = \text{proj}_W \mathbf{f}$ is called the ***least squares approximation*** to $\mathbf{f}$ from W.

FOURIER SERIES   A function of the form

$$t(x) = c_0 + c_1 \cos x + c_2 \cos 2x + \cdots + c_n \cos nx$$
$$+ \, d_1 \sin x + d_2 \sin 2x + \cdots + d_n \sin nx \qquad (2)$$

is called a ***trigonometric polynomial***; if c_n and d_n are not both zero, then $t(x)$ is said to have ***order n***. For example,

$$t(x) = 2 + \cos x - 3 \cos 2x + 7 \sin 4x$$

is a trigonometric polynomial with

$$c_0 = 2, \quad c_1 = 1, \quad c_2 = -3, \quad d_1 = 0, \quad d_2 = 0, \quad d_3 = 0, \quad d_4 = 7$$

The order of $t(x)$ is 4.

It is evident from (2) that the trigonometric polynomials of order n or less are the various possible linear combinations of

$$1, \quad \cos x, \quad \cos 2x, \quad \ldots, \quad \cos nx, \quad \sin x, \quad \sin 2x, \quad \ldots, \quad \sin nx \qquad (3)$$

It can be shown that these $2n + 1$ functions are linearly independent and that consequently for any interval $[a, b]$ they form a basis for a $(2n + 1)$-dimensional subspace of $C[a, b]$.

Let us now consider the problem of finding the least squares approximation of a continuous function $f(x)$ over the interval $[0, 2\pi]$ by a trigonometric polynomial of order n or less. As noted above, the least squares approximation to $\mathbf{f}$ from W is the orthogonal projection of $\mathbf{f}$ on W. To find this orthogonal projection, we must find an orthonormal basis $\mathbf{g}_0, \mathbf{g}_1, \ldots, \mathbf{g}_{2n}$ for W, after which we can compute the orthogonal projection on W from the formula

$$\operatorname{proj}_W \mathbf{f} = \langle \mathbf{f}, \mathbf{g}_0 \rangle \mathbf{g}_0 + \langle \mathbf{f}, \mathbf{g}_1 \rangle \mathbf{g}_1 + \cdots + \langle \mathbf{f}, \mathbf{g}_{2n} \rangle \mathbf{g}_{2n} \tag{4}$$

[see Theorem 6.3.5]. An orthonormal basis for W can be obtained by applying the Gram–Schmidt process to the basis (3), using the inner product

$$\langle \mathbf{f}, \mathbf{g} \rangle = \int_0^{2\pi} f(x)g(x)\, dx$$

This yields (Exercise 6) the orthonormal basis

$$\mathbf{g}_0 = \frac{1}{\sqrt{2\pi}}, \quad \mathbf{g}_1 = \frac{1}{\sqrt{\pi}} \cos x, \quad \ldots, \quad \mathbf{g}_n = \frac{1}{\sqrt{\pi}} \cos nx,$$

$$\mathbf{g}_{n+1} = \frac{1}{\sqrt{\pi}} \sin x, \quad \ldots, \quad \mathbf{g}_{2n} = \frac{1}{\sqrt{\pi}} \sin nx \tag{5}$$

If we introduce the notation

$$a_0 = \frac{2}{\sqrt{2\pi}} \langle \mathbf{f}, \mathbf{g}_0 \rangle, \quad a_1 = \frac{1}{\sqrt{\pi}} \langle \mathbf{f}, \mathbf{g}_1 \rangle, \quad \ldots, \quad a_n = \frac{1}{\sqrt{\pi}} \langle \mathbf{f}, \mathbf{g}_n \rangle$$

$$b_1 = \frac{1}{\sqrt{\pi}} \langle \mathbf{f}, \mathbf{g}_{n+1} \rangle, \quad \ldots, \quad b_n = \frac{1}{\sqrt{\pi}} \langle \mathbf{f}, \mathbf{g}_{2n} \rangle$$

then on substituting (5) in (4) we obtain

$$\operatorname{proj}_W \mathbf{f} = \frac{a_0}{2} + [a_1 \cos x + \cdots + a_n \cos nx] + [b_1 \sin x + \cdots + b_n \sin nx]$$

where

$$a_0 = \frac{2}{\sqrt{2\pi}} \langle \mathbf{f}, \mathbf{g}_0 \rangle = \frac{2}{\sqrt{2\pi}} \int_0^{2\pi} f(x) \frac{1}{\sqrt{2\pi}}\, dx = \frac{1}{\pi} \int_0^{2\pi} f(x)\, dx$$

$$a_1 = \frac{1}{\sqrt{\pi}} \langle \mathbf{f}, \mathbf{g}_1 \rangle = \frac{1}{\sqrt{\pi}} \int_0^{2\pi} f(x) \frac{1}{\sqrt{\pi}} \cos x\, dx = \frac{1}{\pi} \int_0^{2\pi} f(x) \cos x\, dx$$

$$\vdots$$

$$a_n = \frac{1}{\sqrt{\pi}} \langle \mathbf{f}, \mathbf{g}_n \rangle = \frac{1}{\sqrt{\pi}} \int_0^{2\pi} f(x) \frac{1}{\sqrt{\pi}} \cos nx\, dx = \frac{1}{\pi} \int_0^{2\pi} f(x) \cos nx\, dx$$

$$b_1 = \frac{1}{\sqrt{\pi}} \langle \mathbf{f}, \mathbf{g}_{n+1} \rangle = \frac{1}{\sqrt{\pi}} \int_0^{2\pi} f(x) \frac{1}{\sqrt{\pi}} \sin x \, dx = \frac{1}{\pi} \int_0^{2\pi} f(x) \sin x \, dx$$

$$\vdots$$

$$b_n = \frac{1}{\sqrt{\pi}} \langle \mathbf{f}, \mathbf{g}_{2n} \rangle = \frac{1}{\sqrt{\pi}} \int_0^{2\pi} f(x) \frac{1}{\sqrt{\pi}} \sin nx \, dx = \frac{1}{\pi} \int_0^{2\pi} f(x) \sin nx \, dx$$

In short,

$$a_k = \frac{1}{\pi} \int_0^{2\pi} f(x) \cos kx \, dx, \qquad b_k = \frac{1}{\pi} \int_0^{2\pi} f(x) \sin kx \, dx \qquad (6)$$

The numbers $a_0, a_1, \ldots, a_n, b_1, \ldots, b_n$ are called the **Fourier* coefficients** of **f**.

Example 1   Find the least squares approximation of $f(x) = x$ on $[0, 2\pi]$ by

(a)   a trigonometric polynomial of order 2 or less;
(b)   a trigonometric polynomial of order n or less.

Solution (a).

$$a_0 = \frac{1}{\pi} \int_0^{2\pi} f(x) \, dx = \frac{1}{\pi} \int_0^{2\pi} x \, dx = 2\pi \qquad (7a)$$

For $k = 1, 2, \ldots$ integration by parts yields (verify)

$$a_k = \frac{1}{\pi} \int_0^{2\pi} f(x) \cos kx \, dx = \frac{1}{\pi} \int_0^{2\pi} x \cos kx \, dx = 0 \qquad (7b)$$

$$b_k = \frac{1}{\pi} \int_0^{2\pi} f(x) \sin kx \, dx = \frac{1}{\pi} \int_0^{2\pi} x \sin kx \, dx = -\frac{2}{k} \qquad (7c)$$

Thus, the least squares approximation to x on $[0, 2\pi]$ by a trigonometric polynomial of order 2 or less is

$$x \approx \frac{a_0}{2} + a_1 \cos x + a_2 \cos 2x + b_1 \sin x + b_2 \sin 2x$$

or from (7a), (7b), and (7c)

$$x \approx \pi - 2 \sin x - \sin 2x$$

---

* *Jean Baptiste Joseph Fourier* (1768–1830) was a French mathematician and physicist who discovered the Fourier series and related ideas while working on problems of heat diffusion. This discovery is one of the most influential in the history of mathematics; it is the cornerstone of many fields of mathematical research and a basic tool in many branches of engineering. Fourier, a political activist during the French revolution, spent time in jail for his defense of many victims during the Terror. He later became a favorite of Napoleon and was named both a baron and a count.

Solution (*b*). The least squares approximation to x on $[0, 2\pi]$ by a trigonometric polynomial of order n or less is

$$x \approx \frac{a_0}{2} + [a_1 \cos x + \cdots + a_n \cos nx] + [b_1 \sin x + \cdots + b_n \sin nx]$$

or from (7a), (7b), and (7c)

$$x \approx \pi - 2\left(\sin x + \frac{\sin 2x}{2} + \frac{\sin 3x}{3} + \cdots + \frac{\sin nx}{n} \right)$$

The graphs of $y = x$ and some of these approximations are shown in Figure 4.

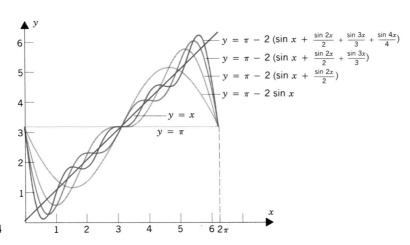

Figure 4

It is natural to expect that the mean square error will diminish as the number of terms in the least squares approximation

$$f(x) \approx \frac{a_0}{2} + \sum_{k=1}^{n} (a_k \cos kx + b_k \sin kx)$$

increases. It can be proved that for functions f in $C[0, \pi]$ the mean square error approaches zero as $n \to +\infty$; this is denoted by writing

$$f(x) = \frac{a_0}{2} + \sum_{k=1}^{\infty} (a_k \cos kx + b_k \sin kx)$$

The right side of this equation is called the **Fourier series** for f over the interval $[0, \pi]$. Such series are of major importance in engineering, science, and mathematics.

EXERCISE SET 9.4

1. Find the least squares approximation of $f(x) = 1 + x$ over the interval $[0, 2\pi]$ by
   (a) a trigonometric polynomial of order 2 or less
   (b) a trigonometric polynomial of order n or less

2. Find the least squares approximation of $f(x) = x^2$ over the interval $[0, 2\pi]$ by
   (a) a trigonometric polynomial of order 3 or less
   (b) a trigonometric polynomial of order n or less

3. (a) Find the least squares approximation of x over the interval $[0, 1]$ by a function of the form $a + be^x$.
   (b) Find the mean square error of the approximation.

4. (a) Find the least squares approximation of e^x over the interval $[0, 1]$ by a polynomial of the form $a_0 + a_1 x$.
   (b) Find the mean square error of the approximation.

5. (a) Find the least squares approximation of $\sin \pi x$ over the interval $[-1, 1]$ by a polynomial of the form $a_0 + a_1 x + a_2 x^2$.
   (b) Find the mean square error of the approximation.

6. Use the Gram–Schmidt process to obtain the orthonormal basis (5) from the basis (3).

7. Carry out the integrations in (7a), (7b), and (7c).

8. Find the Fourier series of $f(x) = \pi - x$ over the interval $[0, 2\pi]$.

9.5 QUADRATIC FORMS

Up to now, our emphasis in this text has been on linear equations, that is, equations of the form

$$a_1 x_1 + a_2 x_2 + \cdots + a_n x_n = b$$

The left side of this equation,

$$a_1 x_1 + a_2 x_2 + \cdots + a_n x_n$$

*is a function of n variables, called a **linear form**. In a linear form all variables occur to the first power, and there are no products of variables in the expression. In this section we shall study functions in which the terms are squares of variables or products of two variables. Such functions arise in a variety of applications, including geometry, vibrations of mechanical systems, statistics, and electrical engineering.*

QUADRATIC FORMS IN TWO VARIABLES

A **quadratic form in two variables**, x and y, is defined to be an expression that can be written as

$$ax^2 + 2bxy + cy^2 \tag{1}$$

Example 1 The following are quadratic forms in x and y:

$$2x^2 + 6xy - 7y^2 \quad (a = 2,\ b = 3,\ c = -7)$$
$$4x^2 - 5y^2 \quad (a = 4,\ b = 0,\ c = -5)$$
$$xy \quad (a = 0,\ b = \tfrac{1}{2},\ c = 0)$$

If we agree to omit the brackets on 1×1 matrices, then (1) can be written in matrix form as

$$ax^2 + 2bxy + cy^2 = [x \quad y]\begin{bmatrix} a & b \\ b & c \end{bmatrix}\begin{bmatrix} x \\ y \end{bmatrix} \tag{2}$$

(Verify by multiplying the matrices.) Note that the 2×2 matrix in (2) is symmetric, the diagonal entries are the coefficients of the squared terms, and the entries off the main diagonal are each half the coefficient of the product term xy.

Example 2

$$2x^2 + 6xy - 7y^2 = [x \quad y]\begin{bmatrix} 2 & 3 \\ 3 & -7 \end{bmatrix}\begin{bmatrix} x \\ y \end{bmatrix}$$

$$4x^2 - 5y^2 = [x \quad y]\begin{bmatrix} 4 & 0 \\ 0 & -5 \end{bmatrix}\begin{bmatrix} x \\ y \end{bmatrix}$$

$$xy = [x \quad y]\begin{bmatrix} 0 & \tfrac{1}{2} \\ \tfrac{1}{2} & 0 \end{bmatrix}\begin{bmatrix} x \\ y \end{bmatrix}$$

QUADRATIC FORMS IN n VARIABLES

Quadratic forms are not limited to two variables. A general quadratic form is defined as follows.

Definition. A *quadratic form* in the n variables $x_1, x_2, \ldots, x_n$ is an expression that can be written as

$$[x_1 \quad x_2 \quad \cdots \quad x_n]A\begin{bmatrix} x_1 \\ x_2 \\ \vdots \\ x_n \end{bmatrix} \tag{3}$$

where A is a symmetric $n \times n$ matrix.

If we let

$$\mathbf{x} = \begin{bmatrix} x_1 \\ x_2 \\ \vdots \\ x_n \end{bmatrix}$$

then (3) can be written more compactly as

$$\mathbf{x}^T A \mathbf{x} \tag{4}$$

Moreover, it can be shown that if the matrices in (4) are multiplied out, the resulting expression has the form

$$\mathbf{x}^T A \mathbf{x} = a_{11}x_1^2 + a_{22}x_2^2 + \cdots + a_{nn}x_n^2 + \sum_{i \neq j} a_{ij}x_i x_j$$

where

$$\sum_{i \neq j} a_{ij}x_i x_j$$

denotes the sum of all terms of the form $a_{ij}x_i x_j$, where x_i and x_j are different variables. The terms $a_{ij}x_i x_j$ are called the **cross-product terms** of the quadratic form.

Symmetric matrices are useful, but not essential, for representing quadratic forms in matrix notation. For example, for the quadratic form $2x^2 + 6xy - 7y^2$ in Example 2, we might split the coefficient of the cross-product term into $5 + 1$ or $4 + 2$ and write

$$2x^2 + 6xy - 7y^2 = [x \quad y] \begin{bmatrix} 2 & 5 \\ 1 & -7 \end{bmatrix} \begin{bmatrix} x \\ y \end{bmatrix}$$

or

$$2x^2 + 6xy - 7y^2 = [x \quad y] \begin{bmatrix} 2 & 4 \\ 2 & -7 \end{bmatrix} \begin{bmatrix} x \\ y \end{bmatrix}$$

However, symmetric matrices generally produce the simplest results, so we will always use them. Thus, when we denote a quadratic form by $\mathbf{x}^T A \mathbf{x}$ it will be understood that A is symmetric, even if this is not stated explicitly.

REMARK. If we use the fact that A is symmetric, that is, $A = A^T$, then (4) can be expressed in terms of the Euclidean inner product by writing

$$\mathbf{x}^T A \mathbf{x} = \mathbf{x}^T (A\mathbf{x}) = \langle A\mathbf{x}, \mathbf{x} \rangle = \langle \mathbf{x}, A\mathbf{x} \rangle \tag{5}$$

Example 3 The following is a quadratic form in x_1, x_2, and x_3:

$$x_1^2 + 7x_2^2 - 3x_3^2 + 4x_1x_2 - 2x_1x_3 + 6x_2x_3 = [x_1 \quad x_2 \quad x_3] \begin{bmatrix} 1 & 2 & -1 \\ 2 & 7 & 3 \\ -1 & 3 & -3 \end{bmatrix} \begin{bmatrix} x_1 \\ x_2 \\ x_3 \end{bmatrix}$$

Note that the coefficients of the squared terms appear on the main diagonal of the 3×3 matrix, and the coefficients of the cross-product terms are each split in half and appear in the off-diagonal positions as follows:

Coefficient of	Positions in Matrix A
x_1x_2	a_{12} and a_{21}
x_1x_3	a_{13} and a_{31}
x_2x_3	a_{23} and a_{32}

The study of quadratic forms is an extensive topic that we can only touch on in this section. The following are some of the important mathematical problems relating to quadratic forms.

- Find the maximum and minimum values of the quadratic form $\mathbf{x}^T A\mathbf{x}$ if $\mathbf{x}$ is constrained so that

$$\|\mathbf{x}\| = (x_1^2 + x_2^2 + \cdots + x_n^2)^{1/2} = 1$$

- What conditions must A satisfy in order for a quadratic form to satisfy the inequality $\mathbf{x}^T A\mathbf{x} > 0$ for all $\mathbf{x} \neq \mathbf{0}$?
- If $\mathbf{x}^T A\mathbf{x}$ is a quadratic form in two or three variables and c is a constant, what does the graph of the equation $\mathbf{x}^T A\mathbf{x} = c$ look like?
- If P is an orthogonal matrix, the change of variables $\mathbf{x} = P\mathbf{y}$ converts the quadratic form $\mathbf{x}^T A\mathbf{x}$ to $(P\mathbf{y})^T A(P\mathbf{y}) = \mathbf{y}^T (P^T AP)\mathbf{y}$. But $P^T AP$ is a symmetric matrix if A is (verify), so $\mathbf{y}^T (P^T AP)\mathbf{y}$ is a new quadratic form in the variables of $\mathbf{y}$. It is important to know if P can be chosen so that this new quadratic form has no cross-product terms.

In this section we shall study the first two problems, and in the following sections we shall study the last two. The following theorem provides a solution to the first problem. The proof is deferred to the end of this section.

Theorem 9.5.1. *Let A be a symmetric $n \times n$ matrix whose eigenvalues in decreasing size order are $\lambda_1 \geq \lambda_2 \geq \cdots \geq \lambda_n$. If $\mathbf{x}$ is constrained so that $\|\mathbf{x}\| = 1$ relative to the Euclidean inner product on R^n, then:*

(a) $\lambda_1 \geq \mathbf{x}^T A\mathbf{x} \geq \lambda_n$.

(b) $\mathbf{x}^T A\mathbf{x} = \lambda_n$ if $\mathbf{x}$ is an eigenvector of A corresponding to λ_n and $\mathbf{x}^T A\mathbf{x} = \lambda_1$ if $\mathbf{x}$ is an eigenvector of A corresponding to λ_1.

It follows from this theorem that subject to the constraint

$$\|\mathbf{x}\| = (x_1^2 + x_2^2 + \cdots + x_n^2)^{1/2} = 1$$

the quadratic form $\mathbf{x}^T A\mathbf{x}$ has a maximum value of λ_1 (the largest eigenvalue) and a minimum value of λ_n (the smallest eigenvalue).

Example 4 Find the maximum and minimum values of the quadratic form

$$x_1^2 + x_2^2 + 4x_1x_2$$

subject to the constraint $x_1^2 + x_2^2 = 1$, and determine values of x_1 and x_2 at which the maximum and minimum occur.

Solution. The quadratic form can be written as

$$x_1^2 + x_2^2 + 4x_1x_2 = \mathbf{x}^T A\mathbf{x} = [x_1 \quad x_2]\begin{bmatrix} 1 & 2 \\ 2 & 1 \end{bmatrix}\begin{bmatrix} x_1 \\ x_2 \end{bmatrix}$$

The characteristic equation of A is

$$\det(\lambda I - A) = \det \begin{bmatrix} \lambda - 1 & -2 \\ -2 & \lambda - 1 \end{bmatrix} = \lambda^2 - 2\lambda - 3 = (\lambda - 3)(\lambda + 1) = 0$$

Thus, the eigenvalues of A are $\lambda = 3$ and $\lambda = -1$, which are the maximum and minimum values, respectively, of the quadratic form subject to the constraint. To find values of x_1 and x_2 at which these extreme values occur, we must find eigenvectors corresponding to these eigenvalues and then normalize these eigenvectors to satisfy the condition $x_1^2 + x_2^2 = 1$.

We leave it for the reader to show that bases for the eigenspaces are

$$\lambda = 3 \colon \begin{bmatrix} 1 \\ 1 \end{bmatrix}, \qquad \lambda = -1 \colon \begin{bmatrix} 1 \\ -1 \end{bmatrix}$$

Normalizing these eigenvectors yields

$$\begin{bmatrix} 1/\sqrt{2} \\ 1/\sqrt{2} \end{bmatrix}, \qquad \begin{bmatrix} 1/\sqrt{2} \\ -1\sqrt{2} \end{bmatrix}$$

Thus, subject to the constraint $x_1^2 + x_2^2 = 1$, the maximum value of the quadratic form is $\lambda = 3$, which occurs if $x_1 = 1/\sqrt{2}$, $x_2 = 1/\sqrt{2}$; and the minimum value is $\lambda = -1$, which occurs if $x_1 = 1/\sqrt{2}$, $x_2 = -1/\sqrt{2}$. Moreover, alternative bases for the eigenspaces can be obtained by multiplying the basis vectors above by -1. Thus, the maximum value, $\lambda = 3$, also occurs if $x_1 = -1/\sqrt{2}$, $x_2 = -1/\sqrt{2}$, and the minimum value, $\lambda = -1$, also occurs if $x_1 = -1/\sqrt{2}$, $x_2 = 1/\sqrt{2}$. ▲

POSITIVE DEFINITE MATRICES AND QUADRATIC FORMS

> **Definition.** A quadratic form $\mathbf{x}^T A \mathbf{x}$ is called *positive definite* if $\mathbf{x}^T A \mathbf{x} > 0$ for all $\mathbf{x} \neq \mathbf{0}$, and a symmetric matrix A is called a *positive definite matrix* if $\mathbf{x}^T A \mathbf{x}$ is a positive definite quadratic form.

The following theorem is the main result about positive definite matrices.

> **Theorem 9.5.2.** *A symmetric matrix A is positive definite if and only if all the eigenvalues of A are positive.*

Proof. Assume that A is positive definite, and let λ be any eigenvalue of A. If $\mathbf{x}$ is an eigenvector of A corresponding to λ, then $\mathbf{x} \neq \mathbf{0}$ and $A\mathbf{x} = \lambda \mathbf{x}$, so

$$0 < \mathbf{x}^T A \mathbf{x} = \mathbf{x}^T \lambda \mathbf{x} = \lambda \mathbf{x}^T \mathbf{x} = \lambda \|\mathbf{x}\|^2 \tag{6}$$

where $\|\mathbf{x}\|$ is the Euclidean norm of $\mathbf{x}$. Since $\|\mathbf{x}\|^2 > 0$ it follows that $\lambda > 0$, which is what we wanted to show.

Conversely, assume that all eigenvalues of A are positive. We must show that $\mathbf{x}^T A \mathbf{x} > 0$ for all $\mathbf{x} \neq \mathbf{0}$. But if $\mathbf{x} \neq \mathbf{0}$, we can normalize $\mathbf{x}$ to obtain the vector $\mathbf{y} = \mathbf{x}/\|\mathbf{x}\|$ with the property $\|\mathbf{y}\| = 1$. It now follows from Theorem 9.5.1 that

$$\mathbf{y}^T A \mathbf{y} \geq \lambda_n > 0$$

where λ_n is the smallest eigenvalue of A. Thus,

$$\mathbf{y}^T A \mathbf{y} = \left(\frac{\mathbf{x}}{\|\mathbf{x}\|}\right)^T A \left(\frac{\mathbf{x}}{\|\mathbf{x}\|}\right) = \frac{1}{\|\mathbf{x}\|^2} \mathbf{x}^T A \mathbf{x} > 0$$

Multiplying through by $\|\mathbf{x}\|^2$ yields

$$\mathbf{x}^T A \mathbf{x} > 0$$

which is what we wanted to show.

Example 5 In Example 1 of Section 7.3 we showed that the symmetric matrix

$$A = \begin{bmatrix} 4 & 2 & 2 \\ 2 & 4 & 2 \\ 2 & 2 & 4 \end{bmatrix}$$

has eigenvalues $\lambda = 2$ and $\lambda = 8$. Since these are positive, the matrix A is positive definite, and for all $\mathbf{x} \neq \mathbf{0}$

$$\mathbf{x}^T A \mathbf{x} = 4x_1^2 + 4x_2^2 + 4x_3^2 + 4x_1x_2 + 4x_1x_3 + 4x_2x_3 > 0 \quad \blacktriangle$$

Our next objective is to give a criterion that can be used to determine whether a symmetric matrix is positive definite without finding its eigenvalues. To do this it will be helpful to introduce some terminology. If

$$A = \begin{bmatrix} a_{11} & a_{12} & \cdots & a_{1n} \\ a_{21} & a_{22} & \cdots & a_{2n} \\ \vdots & \vdots & & \vdots \\ a_{n1} & a_{n2} & \cdots & a_{nn} \end{bmatrix}$$

is a square matrix, then the ***principal submatrices*** of A are the submatrices formed from the first r rows and r columns of A for $r = 1, 2, \ldots, n$. These submatrices are

$$A_1 = [a_{11}], \quad A_2 = \begin{bmatrix} a_{11} & a_{12} \\ a_{21} & a_{22} \end{bmatrix}, \quad A_3 = \begin{bmatrix} a_{11} & a_{12} & a_{13} \\ a_{21} & a_{22} & a_{23} \\ a_{31} & a_{32} & a_{33} \end{bmatrix}, \ldots,$$

$$A_n = A = \begin{bmatrix} a_{11} & a_{12} & \cdots & a_{1n} \\ a_{21} & a_{22} & \cdots & a_{2n} \\ \vdots & \vdots & & \vdots \\ a_{n1} & a_{n2} & \cdots & a_{nn} \end{bmatrix}$$

Theorem 9.5.3. *A symmetric matrix A is positive definite if and only if the determinant of every principal submatrix is positive.*

We omit the proof.

Example 6 The matrix

$$A = \begin{bmatrix} 2 & -1 & -3 \\ -1 & 2 & 4 \\ -3 & 4 & 9 \end{bmatrix}$$

is positive definite since

$$|2| = 2, \quad \begin{vmatrix} 2 & -1 \\ -1 & 2 \end{vmatrix} = 3, \quad \begin{vmatrix} 2 & -1 & -3 \\ -1 & 2 & 4 \\ -3 & 4 & 9 \end{vmatrix} = 1$$

all of which are positive. Thus, we are guaranteed that all eigenvalues of A are positive and $\mathbf{x}^T A \mathbf{x} > 0$ for all $\mathbf{x} \neq \mathbf{0}$. ▲

REMARK. A symmetric matrix A and the quadratic form $\mathbf{x}^T A \mathbf{x}$ are called

positive semidefinite	if $\mathbf{x}^T A \mathbf{x} \geq 0$ for all $\mathbf{x}$
negative definite	if $\mathbf{x}^T A \mathbf{x} < 0$ for $\mathbf{x} \neq \mathbf{0}$
negative semidefinite	if $\mathbf{x}^T A \mathbf{x} \leq 0$ for all $\mathbf{x}$
indefinite	if $\mathbf{x}^T A \mathbf{x}$ has both positive and negative values

Theorems 9.5.2 and 9.5.3 can be modified in an obvious way to apply to matrices of the first three types. For example, a symmetric matrix A is positive semidefinite if and only if all of its eigenvalues are *nonnegative*. Also, A is positive semidefinite if and only if all its principal submatrices have *nonnegative* determinants.

OPTIONAL

Proof of Theorem 9.5.1a. Since A is symmetric, it follows from Theorem 7.3.1 that there is an orthonormal basis for R^n consisting of eigenvectors of A. Suppose that $S = \{\mathbf{v}_1, \mathbf{v}_2, \ldots, \mathbf{v}_n\}$ is such a basis, where $\mathbf{v}_i$ is the eigenvector corresponding to the eigenvalue λ_i. If $\langle\ ,\ \rangle$ denotes the Euclidean inner product, then it follows from Theorem 6.3.1 that for any $\mathbf{x}$ in R^n

$$\mathbf{x} = \langle \mathbf{x}, \mathbf{v}_1 \rangle \mathbf{v}_1 + \langle \mathbf{x}, \mathbf{v}_2 \rangle \mathbf{v}_2 + \cdots + \langle \mathbf{x}, \mathbf{v}_n \rangle \mathbf{v}_n$$

Thus,

$$\begin{aligned} A\mathbf{x} &= \langle \mathbf{x}, \mathbf{v}_1 \rangle A\mathbf{v}_1 + \langle \mathbf{x}, \mathbf{v}_2 \rangle A\mathbf{v}_2 + \cdots + \langle \mathbf{x}, \mathbf{v}_n \rangle A\mathbf{v}_n \\ &= \langle \mathbf{x}, \mathbf{v}_1 \rangle \lambda_1 \mathbf{v}_1 + \langle \mathbf{x}, \mathbf{v}_2 \rangle \lambda_2 \mathbf{v}_2 + \cdots + \langle \mathbf{x}, \mathbf{v}_n \rangle \lambda_n \mathbf{v}_n \\ &= \lambda_1 \langle \mathbf{x}, \mathbf{v}_1 \rangle \mathbf{v}_1 + \lambda_2 \langle \mathbf{x}, \mathbf{v}_2 \rangle \mathbf{v}_2 + \cdots + \lambda_n \langle \mathbf{x}, \mathbf{v}_n \rangle \mathbf{v}_n \end{aligned}$$

It follows that the coordinate vectors for $\mathbf{x}$ and $A\mathbf{x}$ relative to the basis S are

$$(\mathbf{x})_S = (\langle \mathbf{x}, \mathbf{v}_1 \rangle, \langle \mathbf{x}, \mathbf{v}_2 \rangle \ldots, \langle \mathbf{x}, \mathbf{v}_n \rangle)$$
$$(A\mathbf{x})_S = (\lambda_1 \langle \mathbf{x}, \mathbf{v}_1 \rangle, \lambda_2 \langle \mathbf{x}, \mathbf{v}_2 \rangle, \ldots, \lambda_n \langle \mathbf{x}, \mathbf{v}_n \rangle)$$

Thus, from Theorem 6.3.2a and the fact that $\|\mathbf{x}\| = 1$, we obtain

$$\|\mathbf{x}\|^2 = \langle \mathbf{x}, \mathbf{v}_1 \rangle^2 + \langle \mathbf{x}, \mathbf{v}_2 \rangle^2 + \cdots + \langle \mathbf{x}, \mathbf{v}_n \rangle^2 = 1$$

$$\langle \mathbf{x}, A\mathbf{x} \rangle = \lambda_1 \langle \mathbf{x}, \mathbf{v}_1 \rangle^2 + \lambda_2 \langle \mathbf{x}, \mathbf{v}_2 \rangle^2 + \cdots + \lambda_n \langle \mathbf{x}, \mathbf{v}_n \rangle^2$$

Using these two equations and Formula (5), we can prove that $\mathbf{x}^T A\mathbf{x} \le \lambda_1$ as follows:

$$\begin{aligned}
\mathbf{x}^T A\mathbf{x} = \langle \mathbf{x}, A\mathbf{x} \rangle &= \lambda_1 \langle \mathbf{x}, \mathbf{v}_1 \rangle^2 + \lambda_2 \langle \mathbf{x}, \mathbf{v}_2 \rangle^2 + \cdots + \lambda_n \langle \mathbf{x}, \mathbf{v}_n \rangle^2 \\
&\le \lambda_1 \langle \mathbf{x}, \mathbf{v}_1 \rangle^2 + \lambda_1 \langle \mathbf{x}, \mathbf{v}_2 \rangle^2 + \cdots + \lambda_1 \langle \mathbf{x}, \mathbf{v}_n \rangle^2 \\
&= \lambda_1 (\langle \mathbf{x}, \mathbf{v}_1 \rangle^2 + \langle \mathbf{x}, \mathbf{v}_2 \rangle^2 + \cdots + \langle \mathbf{x}, \mathbf{v}_n \rangle^2) \\
&= \lambda_1
\end{aligned}$$

The proof that $\lambda_n \le \mathbf{x}^T A\mathbf{x}$ is similar and is left as an exercise.

Proof of Theorem 9.5.1b. If $\mathbf{x}$ is an eigenvector of A corresponding to λ_1 and $\|\mathbf{x}\| = 1$, then

$$\mathbf{x}^T A\mathbf{x} = \langle \mathbf{x}, A\mathbf{x} \rangle = \langle \mathbf{x}, \lambda_1 \mathbf{x} \rangle = \lambda_1 \langle \mathbf{x}, \mathbf{x} \rangle = \lambda_1 \|\mathbf{x}\|^2 = \lambda_1$$

Similarly, $\mathbf{x}^T A\mathbf{x} = \lambda_n$ if $\|\mathbf{x}\| = 1$ and $\mathbf{x}$ is an eigenvector of A corresponding to λ_n. ∎

EXERCISE SET 9.5

1. Which of the following are quadratic forms?
   (a) $x^2 - \sqrt{2}xy$
   (b) $5x_1^2 - 2x_2^3 + 4x_1x_2$
   (c) $4x_1^2 - 3x_2^2 + x_3^2 - 5x_1x_3$
   (d) $x_1^2 - 7x_2^2 + x_3^2 + 4x_1x_2x_3$
   (e) $x_1x_2 - 3x_1x_3 + 2x_2x_3$
   (f) $x_1^2 - 6x_2^2 + x_1 - 5x_2$
   (g) $(x_1 - 3x_2)^2$
   (h) $(x_1 - x_3)^2 + 2(x_1 + 4x_2)^2$

2. Express the following quadratic forms in the matrix notation $\mathbf{x}^T A\mathbf{x}$, where A is a symmetric matrix.
   (a) $3x_1^2 + 7x_2^2$
   (b) $4x_1^2 - 9x_2^2 - 6x_1x_2$
   (c) $5x_1^2 + 5x_1x_2$
   (d) $-7x_1x_2$

3. Express the following quadratic forms in the matrix notation $\mathbf{x}^T A\mathbf{x}$, where A is a symmetric matrix.
   (a) $9x_1^2 - x_2^2 + 4x_3^2 + 6x_1x_2 - 8x_1x_3 + x_2x_3$
   (b) $x_1^2 + x_2^2 - 3x_3^2 - 5x_1x_2 + 9x_1x_3$
   (c) $x_1x_2 + x_1x_3 + x_2x_3$
   (d) $\sqrt{2}x_1^2 - \sqrt{3}x_3^2 + 2\sqrt{2}x_1x_2 - 8\sqrt{3}x_1x_3$
   (e) $x_1^2 + x_2^2 - x_3^2 - x_4^2 + 2x_1x_2 - 10x_1x_4 + 4x_3x_4$

4. In each part find a formula for the quadratic form that does not use matrices.

   (a) $\begin{bmatrix} x & y \end{bmatrix} \begin{bmatrix} 2 & -3 \\ -3 & 5 \end{bmatrix} \begin{bmatrix} x \\ y \end{bmatrix}$

   (b) $\begin{bmatrix} x_1 & x_2 \end{bmatrix} \begin{bmatrix} 7 & \frac{5}{2} \\ \frac{5}{2} & 0 \end{bmatrix} \begin{bmatrix} x_1 \\ x_2 \end{bmatrix}$

   (c) $\begin{bmatrix} x & y & z \end{bmatrix} \begin{bmatrix} 1 & 0 & 0 \\ 0 & -3 & 0 \\ 0 & 0 & 5 \end{bmatrix} \begin{bmatrix} x \\ y \\ z \end{bmatrix}$

   (d) $\begin{bmatrix} x_1 & x_2 & x_3 \end{bmatrix} \begin{bmatrix} -2 & \frac{7}{2} & \frac{1}{2} \\ \frac{7}{2} & 0 & 6 \\ \frac{1}{2} & 6 & 3 \end{bmatrix} \begin{bmatrix} x_1 \\ x_2 \\ x_3 \end{bmatrix}$

   (e) $\begin{bmatrix} x_1 & x_2 & x_3 & x_4 \end{bmatrix} \begin{bmatrix} 0 & 1 & 1 & 1 \\ 1 & 0 & 1 & 1 \\ 1 & 1 & 0 & 1 \\ 1 & 1 & 1 & 0 \end{bmatrix} \begin{bmatrix} x_1 \\ x_2 \\ x_3 \\ x_4 \end{bmatrix}$

5. In each part find the maximum and minimum values of the quadratic form subject to the constraint $x_1^2 + x_2^2 = 1$, and determine the values of x_1 and x_2 at which the maximum and minimum occur.
(a) $5x_1^2 - x_2^2$ (b) $7x_1^2 + 4x_2^2 + x_1x_2$ (c) $5x_1^2 + 2x_2^2 - x_1x_2$ (d) $2x_1^2 + x_2^2 + 3x_1x_2$

6. In each part find the maximum and minimum values of the quadratic form subject to the constraint $x_1^2 + x_2^2 + x_3^2 = 1$, and determine the values of x_1, x_2, and x_3 at which the maximum and minimum occur.
(a) $x_1^2 + x_2^2 + 2x_3^2 - 2x_1x_2 + 4x_1x_3 + 4x_2x_3$ (b) $2x_1^2 + x_2^2 + x_3^2 + 2x_1x_3 + 2x_1x_2$
(c) $3x_1^2 + 2x_2^2 + 3x_3^2 + 2x_1x_3$

7. Use Theorem 9.5.2 to determine which of the following matrices are positive definite.
(a) $\begin{bmatrix} 2 & 3 \\ 3 & 2 \end{bmatrix}$ (b) $\begin{bmatrix} 5 & -1 \\ -1 & 5 \end{bmatrix}$ (c) $\begin{bmatrix} 2 & -2 \\ -2 & -1 \end{bmatrix}$

8. Use Theorem 9.5.3 to determine which of the matrices in Exercise 7 are positive definite.

9. Use Theorem 9.5.2 to determine which of the following matrices are positive definite.
(a) $\begin{bmatrix} 3 & -1 & 0 \\ -1 & 2 & -1 \\ 0 & -1 & 3 \end{bmatrix}$ (b) $\begin{bmatrix} 0 & 1 & 1 \\ 1 & 0 & 1 \\ 1 & 1 & 0 \end{bmatrix}$ (c) $\begin{bmatrix} 1 & 2 & 1 \\ 2 & 1 & 1 \\ 1 & 1 & 3 \end{bmatrix}$

10. Use Theorem 9.5.3 to determine which of the matrices in Exercise 9 are positive definite.

11. In each part classify the quadratic form as positive definite, positive semidefinite, negative definite, negative semidefinite, or indefinite.
(a) $x_1^2 + x_2^2$ (b) $-x_1^2 - 3x_2^2$ (c) $(x_1 - x_2)^2$
(d) $-(x_1 - x_2)^2$ (e) $x_1^2 - x_2^2$ (f) x_1x_2

12. In each part classify the matrix as positive definite, positive semidefinite, negative definite, negative semidefinite, or indefinite.
(a) $\begin{bmatrix} 3 & 0 & 0 \\ 0 & -2 & 0 \\ 0 & 0 & 1 \end{bmatrix}$ (b) $\begin{bmatrix} -5 & 0 & 0 \\ 0 & 0 & 0 \\ 0 & 0 & 1 \end{bmatrix}$ (c) $\begin{bmatrix} 6 & 7 & 1 \\ 7 & 9 & 2 \\ 1 & 2 & 1 \end{bmatrix}$

(d) $\begin{bmatrix} -4 & 7 & 8 \\ 7 & -3 & 9 \\ 8 & 9 & -1 \end{bmatrix}$ (e) $\begin{bmatrix} 0 & 0 & 0 \\ 0 & 0 & 0 \\ 0 & 0 & 0 \end{bmatrix}$ (f) $\begin{bmatrix} 1 & 0 & 0 \\ 0 & 1 & 0 \\ 0 & 0 & 1 \end{bmatrix}$

13. Let $\mathbf{x}^T A \mathbf{x}$ be a quadratic form in $x_1, x_2, \ldots, x_n$ and define $T: R^n \to R$ by $T(\mathbf{x}) = \mathbf{x}^T A \mathbf{x}$.
(a) Show that $T(\mathbf{x} + \mathbf{y}) = T(\mathbf{x}) + 2\mathbf{x}^T A \mathbf{y} + T(\mathbf{y})$.   (b) Show that $T(k\mathbf{x}) = k^2 T(\mathbf{x})$.
(c) Is T a linear transformation? Explain.

14. In each part find all values of k for which the quadratic form is positive definite.
(a) $x_1^2 + kx_2^2 - 4x_1x_2$ (b) $5x_1^2 + x_2^2 + kx_3^2 + 4x_1x_2 - 2x_1x_3 - 2x_2x_3$
(c) $3x_1^2 + x_2^2 + 2x_3^2 + 2x_1x_3 + 2kx_2x_3$

15. Express the quadratic form $(c_1x_1 + c_2x_2 + \cdots + c_nx_n)^2$ in the matrix notation $\mathbf{x}^T A \mathbf{x}$, where A is symmetric.

16. Let $\mathbf{x} = (x_1, x_2, \ldots, x_n)$. In statistics the quantity

$$\bar{x} = \frac{1}{n}(x_1 + x_2 + \cdots + x_n)$$

is called the ***sample mean*** of $x_1, x_2, \ldots, x_n$, and

$$s_{\mathbf{x}}^2 = \frac{1}{n-1}[(x_1 - \bar{x})^2 + (x_2 - \bar{x})^2 + \cdots + (x_n - \bar{x})^2]$$

is called the ***sample variance***.
(a) Express the quadratic form $s_{\mathbf{x}}^2$ in the matrix notation $\mathbf{x}^T A\mathbf{x}$, where A is symmetric.
(b) Is $s_{\mathbf{x}}^2$ a positive definite quadratic form? Explain.

17. Complete the proof of Theorem 9.5.1 by showing that $\lambda_n \leq \mathbf{x}^T A\mathbf{x}$ if $\|\mathbf{x}\| = 1$ and $\lambda_n = \mathbf{x}^T A\mathbf{x}$ if $\mathbf{x}$ is an eigenvector of A corresponding to λ_n.

9.6 DIAGONALIZING QUADRATIC FORMS; CONIC SECTIONS

In this section we shall show how to remove the cross-product terms from a quadratic form by changing variables, and we shall use our results to study the graphs of the conic sections.

DIAGONALIZA-TION OF QUADRATIC FORMS

Let

$$\mathbf{x}^T A\mathbf{x} = [x_1 \quad x_2 \quad \cdots \quad x_n] \begin{bmatrix} a_{11} & a_{12} & \cdots & a_{1n} \\ a_{21} & a_{22} & \cdots & a_{2n} \\ \vdots & \vdots & & \vdots \\ a_{n1} & a_{n2} & \cdots & a_{nn} \end{bmatrix} \begin{bmatrix} x_1 \\ x_2 \\ \vdots \\ x_n \end{bmatrix} \tag{1}$$

be a quadratic form, where A is a symmetric matrix. We know from Theorem 7.3.1 that there is an orthogonal matrix P that diagonalizes A; that is,

$$P^T A P = D = \begin{bmatrix} \lambda_1 & 0 & \cdots & 0 \\ 0 & \lambda_2 & \cdots & 0 \\ \vdots & \vdots & & \vdots \\ 0 & 0 & \cdots & \lambda_n \end{bmatrix}$$

where $\lambda_1, \lambda_2, \lambda_n$ are the eigenvalues of A. If we let

$$\mathbf{y} = \begin{bmatrix} y_1 \\ y_2 \\ \vdots \\ y_n \end{bmatrix}$$

where $y_1, y_2, \ldots, y_n$ are new variables, and if we make the substitution $\mathbf{x} = P\mathbf{y}$ in (1), then we obtain

$$\mathbf{x}^T A \mathbf{x} = (P\mathbf{y})^T A P \mathbf{y} = \mathbf{y}^T P^T A P \mathbf{y} = \mathbf{y}^T D \mathbf{y}$$

But

$$\mathbf{y}^T D \mathbf{y} = [y_1 \;\; y_2 \;\; \cdots \;\; y_n] \begin{bmatrix} \lambda_1 & 0 & \cdots & 0 \\ 0 & \lambda_2 & \cdots & 0 \\ \vdots & \vdots & & \vdots \\ 0 & 0 & \cdots & \lambda_n \end{bmatrix} \begin{bmatrix} y_1 \\ y_2 \\ \vdots \\ y_n \end{bmatrix}$$

$$= \lambda_1 y_1^2 + \lambda_2 y_2^2 + \cdots + \lambda_n y_n^2$$

which is a quadratic form with no cross-product terms.

In summary, we have the following result.

Theorem 9.6.1. *Let $\mathbf{x}^T A \mathbf{x}$ be a quadratic form in the variables $x_1, x_2, \ldots, x_n$, where A is symmetric. If P orthogonally diagonalizes A, and if the new variables y_1, $y_2, \ldots, y_n$ are defined by the equation $\mathbf{x} = P\mathbf{y}$, then substituting this equation in $\mathbf{x}^T A \mathbf{x}$ yields*

$$\mathbf{x}^T A \mathbf{x} = \mathbf{y}^T D \mathbf{y} = \lambda_1 y_1^2 + \lambda_2 y_2^2 + \cdots + \lambda_n y_n^2$$

where $\lambda_1, \lambda_2, \ldots, \lambda_n$ are the eigenvalues of A and

$$D = P^T A P = \begin{bmatrix} \lambda_1 & 0 & \cdots & 0 \\ 0 & \lambda_2 & \cdots & 0 \\ \vdots & \vdots & & \vdots \\ 0 & 0 & \cdots & \lambda_n \end{bmatrix}$$

The matrix P in this theorem is said to **orthogonally diagonalize** the quadratic form or **reduce the quadratic form to a sum of squares**.

Example 1 Find a change of variables that will reduce the quadratic form $x_1^2 - x_3^2 - 4x_1 x_2 + 4x_2 x_3$ to a sum of squares, and express the quadratic form in terms of the new variables.

Solution. The quadratic form can be written as

$$[x_1 \;\; x_2 \;\; x_3] \begin{bmatrix} 1 & -2 & 0 \\ -2 & 0 & 2 \\ 0 & 2 & -1 \end{bmatrix} \begin{bmatrix} x_1 \\ x_2 \\ x_3 \end{bmatrix}$$

The characteristic equation of the 3×3 matrix is

$$\begin{vmatrix} \lambda - 1 & 2 & 0 \\ 2 & \lambda & -2 \\ 0 & -2 & \lambda + 1 \end{vmatrix} = \lambda^3 - 9\lambda = \lambda(\lambda + 3)(\lambda - 3) = 0$$

so the eigenvalues are $\lambda = 0$, $\lambda = -3$, $\lambda = 3$. We leave it for the reader to show that orthonormal bases for the three eigenspaces are

$$\lambda = 0: \begin{bmatrix} \frac{2}{3} \\ \frac{1}{3} \\ \frac{2}{3} \end{bmatrix}, \qquad \lambda = -3: \begin{bmatrix} -\frac{1}{3} \\ -\frac{2}{3} \\ \frac{2}{3} \end{bmatrix}, \qquad \lambda = 3: \begin{bmatrix} -\frac{2}{3} \\ \frac{2}{3} \\ \frac{1}{3} \end{bmatrix}$$

Thus, the substitution $\mathbf{x} = P\mathbf{y}$ that eliminates cross-product terms is

$$\begin{bmatrix} x_1 \\ x_2 \\ x_3 \end{bmatrix} = \begin{bmatrix} \frac{2}{3} & -\frac{1}{3} & -\frac{2}{3} \\ \frac{1}{3} & -\frac{2}{3} & \frac{2}{3} \\ \frac{2}{3} & \frac{2}{3} & \frac{1}{3} \end{bmatrix} \begin{bmatrix} y_1 \\ y_2 \\ y_3 \end{bmatrix}$$

or equivalently,

$$x_1 = \tfrac{2}{3}y_1 - \tfrac{1}{3}y_2 - \tfrac{2}{3}y_3$$
$$x_2 = \tfrac{1}{3}y_1 - \tfrac{2}{3}y_2 + \tfrac{2}{3}y_3$$
$$x_3 = \tfrac{2}{3}y_1 + \tfrac{2}{3}y_2 + \tfrac{1}{3}y_3$$

The new quadratic form is

$$\begin{bmatrix} y_1 & y_2 & y_3 \end{bmatrix} \begin{bmatrix} 0 & 0 & 0 \\ 0 & -3 & 0 \\ 0 & 0 & 3 \end{bmatrix} \begin{bmatrix} y_1 \\ y_2 \\ y_3 \end{bmatrix}$$

or equivalently,

$$-3y_2^2 + 3y_3^2 \quad \blacktriangle$$

REMARK. There are other methods for eliminating the cross-product terms from a quadratic form, which we shall not discuss here. Two such methods, ***Lagrange's reduction*** and ***Kronecker's reduction***, are discussed in more advanced books.

CONIC SECTIONS

We shall now apply our work on quadratic forms to the study of equations of the form

$$ax^2 + 2bxy + cy^2 + dx + ey + f = 0 \tag{2}$$

where $a, b, \ldots, f$ are real numbers, and at least one of the numbers a, b, c is not zero. An equation of this type is called a ***quadratic equation*** in x and y, and

$$ax^2 + 2bxy + cy^2$$

is called the ***associated quadratic form***.

Example 2 In the quadratic equation

$$3x^2 + 5xy - 7y^2 + 2x + 7 = 0$$

the constants in (2) are

$$a = 3, \quad b = \tfrac{5}{2}, \quad c = -7, \quad d = 2, \quad e = 0, \quad f = 7 \quad \blacktriangle$$

Example 3

Quadratic Equation	Associated Quadratic Form
$3x^2 + 5xy - 7y^2 + 2x + 7 = 0$	$3x^2 + 5xy - 7y^2$
$4x^2 - 5y^2 + 8y + 9 = 0$	$4x^2 - 5y^2$
$xy + y = 0$	xy

Graphs of quadratic equations in x and y are called **conics** or **conic sections**. The most important conics are ellipses, circles, hyperbolas, and parabolas; these are called the **nondegenerate** conics. The remaining conics are called **degenerate** and include single points and pairs of lines (see Exercise 15).

A nondegenerate conic is said to be in **standard position** relative to the coordinate axes if its equation can be expressed in one of the forms given in Figure 1.

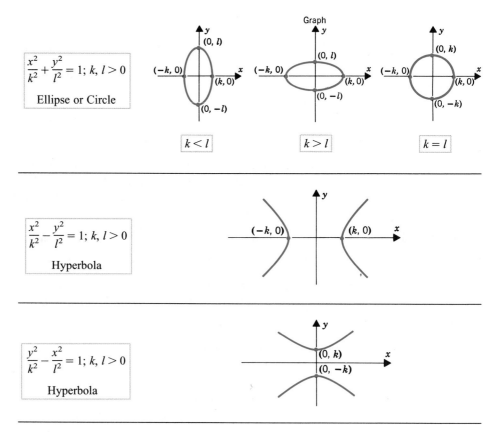

Figure 1 (continued on p. 487)

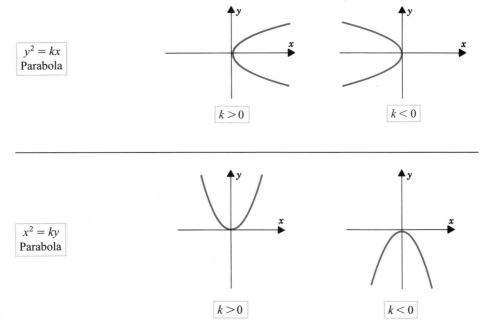

Figure 1

Example 4 The equation

$$\frac{x^2}{4} + \frac{y^2}{9} = 1 \text{ is of the form } \frac{x^2}{k^2} + \frac{y^2}{l^2} = 1 \text{ with } k = 2, l = 3$$

Thus, its graph is an ellipse in standard position intersecting the x-axis at $(-2, 0)$ and $(2, 0)$ and intersecting the y-axis at $(0, -3)$ and $(0, 3)$.

The equation $x^2 - 8y^2 = -16$ can be rewritten as $y^2/2 - x^2/16 = 1$, which is of the form $y^2/k^2 - x^2/l^2 = 1$ with $k = \sqrt{2}, l = 4$. Its graph is thus a hyperbola in standard position intersecting the y-axis at $(0, -\sqrt{2})$ and $(0, \sqrt{2})$.

The equation $5x^2 + 2y = 0$ can be rewritten as $x^2 = -\frac{2}{5}y$, which is of the form $x^2 = ky$ with $k = -\frac{2}{5}$. Since $k < 0$, its graph is a parabola in standard position opening downward. ▲

SIGNIFICANCE
OF THE CROSS-
PRODUCT TERM

Observe that no conic in standard position has an xy-term (i.e., a cross-product term) in its equation; the presence of an xy-term in the equation of a nondegenerate conic indicates that the conic is rotated out of standard position (Figure 2*a*). Also, no conic in standard position has both an x^2 and x term or both a y^2 and y term. If there is no cross-product term, the occurrence of either of these pairs in the equation of a nonde-generate conic indicates that the conic is translated out of standard position (Figure 2*b*).

One technique for identifying the graph of a nondegenerate conic that is not in standard position consists of rotating and translating the xy-coordinate axes to obtain an $x'y'$-coordinate system relative to which the conic is in standard position. Once this is done, the equation of the conic in the $x'y'$-system will have one of the forms given in Figure 1 and can then easily be identified.

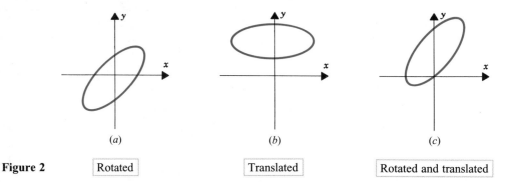

| (a) | (b) | (c) |

Figure 2   Rotated   Translated   Rotated and translated

Example 5   Since the quadratic equation

$$2x^2 + y^2 - 12x - 4y + 18 = 0$$

contains x^2, x, y^2, and y terms but no cross-product term, its graph is a conic that is translated out of standard position but not rotated. This conic can be brought into standard position by suitably translating the coordinate axes. To do this, first collect x and y terms. This yields

$$(2x^2 - 12x) + (y^2 - 4y) + 18 = 0$$

or

$$2(x^2 - 6x) + (y^2 - 4y) = -18$$

By completing the squares* on the two expressions in parentheses, we obtain

$$2(x^2 - 6x + 9) + (y^2 - 4y + 4) = -18 + 18 + 4$$

or

$$2(x - 3)^2 + (y - 2)^2 = 4 \tag{3}$$

If we translate the coordinate axes by means of the translation equations

$$x' = x - 3, \qquad y' = y - 2$$

then (3) becomes

$$2x'^2 + y'^2 = 4$$

---

* To complete the square on an expression of the form $x^2 + px$, add and subtract the constant $(p/2)^2$ to obtain

$$x^2 + px = x^2 + px + \left(\frac{p}{2}\right)^2 - \left(\frac{p}{2}\right)^2 = \left(x + \frac{p}{2}\right)^2 - \left(\frac{p}{2}\right)^2$$

or

$$\frac{x'^2}{2} + \frac{y'^2}{4} = 1$$

which is the equation of an ellipse in standard position in the $x'y'$-system. This ellipse is sketched in Figure 3. ▲

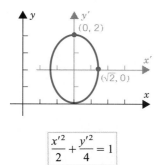

Figure 3

$$\frac{x'^2}{2} + \frac{y'^2}{4} = 1$$

ELIMINATING THE CROSS-PRODUCT TERM

We shall now show how to identify conics that are rotated out of standard position. If we omit the brackets on 1×1 matrices, then (2) can be written in the matrix form

$$[x \quad y] \begin{bmatrix} a & b \\ b & c \end{bmatrix} \begin{bmatrix} x \\ y \end{bmatrix} + [d \quad e] \begin{bmatrix} x \\ y \end{bmatrix} + f = 0$$

or

$$\mathbf{x}^T A \mathbf{x} + K \mathbf{x} + f = 0$$

where

$$\mathbf{x} = \begin{bmatrix} x \\ y \end{bmatrix}, \qquad A = \begin{bmatrix} a & b \\ b & c \end{bmatrix}, \qquad K = [d \quad e]$$

Now consider a conic C whose equation in xy-coordinates is

$$\mathbf{x}^T A \mathbf{x} + K \mathbf{x} + f = 0 \qquad (4)$$

We would like to rotate the xy-coordinate axes so that the equation of the conic in the new $x'y'$-coordinate system has no cross-product term. This can be done as follows.

Step 1. Find a matrix

$$P = \begin{bmatrix} p_{11} & p_{12} \\ p_{21} & p_{22} \end{bmatrix}$$

that orthogonally diagonalizes the quadratic form $\mathbf{x}^T A \mathbf{x}$.

Step 2. Interchange the columns of P, if necessary, to make $\det(P) = 1$. This assures that the orthogonal coordinate transformation

$$\mathbf{x} = P\mathbf{x}', \quad \text{that is,} \quad \begin{bmatrix} x \\ y \end{bmatrix} = P\begin{bmatrix} x' \\ y' \end{bmatrix} \tag{5}$$

is a rotation.

Step 3. To obtain the equation for C in the $x'y'$-system, substitute (5) into (4). This yields

$$(P\mathbf{x}')^T A(P\mathbf{x}') + K(P\mathbf{x}') + f = 0$$

or

$$(\mathbf{x}')^T (P^T A P)\mathbf{x}' + (KP)\mathbf{x}' + f = 0 \tag{6}$$

Since P orthogonally diagonalizes A,

$$P^T A P = \begin{bmatrix} \lambda_1 & 0 \\ 0 & \lambda_2 \end{bmatrix}$$

where λ_1 and λ_2 are eigenvalues of A. Thus, (6) can be rewritten as

$$[x' \quad y']\begin{bmatrix} \lambda_1 & 0 \\ 0 & \lambda_2 \end{bmatrix}\begin{bmatrix} x' \\ y' \end{bmatrix} + [d \quad e]\begin{bmatrix} p_{11} & p_{12} \\ p_{21} & p_{22} \end{bmatrix}\begin{bmatrix} x' \\ y' \end{bmatrix} + f = 0$$

or

$$\lambda_1 x'^2 + \lambda_2 y'^2 + d'x' + e'y' + f = 0$$

(where $d' = dp_{11} + ep_{21}$ and $e' = dp_{12} + ep_{22}$). This equation has no cross-product term.

The following theorem summarizes this discussion.

Theorem 9.6.2 (*Principal Axes Theorem for R^2*). *Let*

$$ax^2 + 2bxy + cy^2 + dx + ey + f = 0$$

be the equation of a conic C, and let

$$\mathbf{x}^T A\mathbf{x} = ax^2 + 2bxy + cy^2$$

be the associated quadratic form. Then the coordinate axes can be rotated so that the equation for C in the new $x'y'$-coordinate system has the form

$$\lambda_1 x'^2 + \lambda_2 y'^2 + d'x' + e'y' + f = 0$$

where λ_1 and λ_2 are the eigenvalues of A. The rotation can be accomplished by the substitution

$$\mathbf{x} = P\mathbf{x}'$$

where P orthogonally diagonalizes $\mathbf{x}^T A\mathbf{x}$ and $\det(P) = 1$.

Example 6 Describe the conic C whose equation is $5x^2 - 4xy + 8y^2 - 36 = 0$.

Solution. The matrix form of this equation is

$$\mathbf{x}^T A \mathbf{x} - 36 = 0 \tag{7}$$

where

$$A = \begin{bmatrix} 5 & -2 \\ -2 & 8 \end{bmatrix}$$

The characteristic equation of A is

$$\det(\lambda I - A) = \det \begin{bmatrix} \lambda - 5 & 2 \\ 2 & \lambda - 8 \end{bmatrix} = (\lambda - 9)(\lambda - 4) = 0$$

so the eigenvalues of A are $\lambda = 4$ and $\lambda = 9$. We leave it for the reader to show that orthonormal bases for the eigenspaces are

$$\lambda = 4: \quad \mathbf{v}_1 = \begin{bmatrix} 2/\sqrt{5} \\ 1/\sqrt{5} \end{bmatrix}, \qquad \lambda = 9: \quad \mathbf{v}_2 = \begin{bmatrix} -1/\sqrt{5} \\ 2/\sqrt{5} \end{bmatrix}$$

Thus,

$$P = \begin{bmatrix} 2/\sqrt{5} & -1/\sqrt{5} \\ 1/\sqrt{5} & 2/\sqrt{5} \end{bmatrix}$$

orthogonally diagonalizes $\mathbf{x}^T A \mathbf{x}$. Moreover, $\det(P) = 1$ so that the orthogonal coordinate transformation

$$\mathbf{x} = P\mathbf{x}' \tag{8}$$

is a rotation. Substituting (8) into (7) yields

$$(P\mathbf{x}')^T A (P\mathbf{x}') - 36 = 0$$

or

$$(\mathbf{x}')^T (P^T A P)\mathbf{x}' - 36 = 0$$

Since

$$P^T A P = \begin{bmatrix} 4 & 0 \\ 0 & 9 \end{bmatrix}$$

this equation can be written as

$$[x' \quad y'] \begin{bmatrix} 4 & 0 \\ 0 & 9 \end{bmatrix} \begin{bmatrix} x' \\ y' \end{bmatrix} - 36 = 0$$

or

$$4x'^2 + 9y'^2 - 36 = 0$$

or

$$\frac{x'^2}{9} + \frac{y'^2}{4} = 1$$

which is the equation of the ellipse sketched in Figure 4. In that figure, the vectors $\mathbf{v}_1$ and $\mathbf{v}_2$ are the column vectors of P. ▲

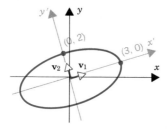

Figure 4

$$\frac{x'^2}{9} + \frac{y'^2}{4} = 1$$

Example 7 Describe the conic C whose equation is

$$5x^2 - 4xy + 8y^2 + \frac{20}{\sqrt{5}}x - \frac{80}{\sqrt{5}}y + 4 = 0$$

Solution. The matrix form of this equation is

$$\mathbf{x}^T A\mathbf{x} + K\mathbf{x} + 4 = 0 \tag{9}$$

where

$$A = \begin{bmatrix} 5 & -2 \\ -2 & 8 \end{bmatrix} \quad \text{and} \quad K = \begin{bmatrix} \dfrac{20}{\sqrt{5}} & -\dfrac{80}{\sqrt{5}} \end{bmatrix}$$

As shown in Example 6

$$P = \begin{bmatrix} \dfrac{2}{\sqrt{5}} & -\dfrac{1}{\sqrt{5}} \\ \dfrac{1}{\sqrt{5}} & \dfrac{2}{\sqrt{5}} \end{bmatrix}$$

orthogonally diagonalizes $\mathbf{x}^T A\mathbf{x}$. Substituting $\mathbf{x} = P\mathbf{x}'$ into (9) gives

$$(P\mathbf{x}')^T A(P\mathbf{x}') + K(P\mathbf{x}') + 4 = 0$$

or

$$(\mathbf{x}')^T (P^T A P)\mathbf{x}' + (KP)\mathbf{x}' + 4 = 0 \tag{10}$$

Since

$$P^T A P = \begin{bmatrix} 4 & 0 \\ 0 & 9 \end{bmatrix} \quad \text{and} \quad KP = \begin{bmatrix} \dfrac{20}{\sqrt{5}} & -\dfrac{80}{\sqrt{5}} \end{bmatrix} \begin{bmatrix} \dfrac{2}{\sqrt{5}} & -\dfrac{1}{\sqrt{5}} \\ \dfrac{1}{\sqrt{5}} & \dfrac{2}{\sqrt{5}} \end{bmatrix} = \begin{bmatrix} -8 & -36 \end{bmatrix}$$

(10) can be written as

$$4x'^2 + 9y'^2 - 8x' - 36y' + 4 = 0 \tag{11}$$

To bring the conic into standard position, the $x'y'$-axes must be translated. Proceeding as in Example 5, we rewrite (11) as

$$4(x'^2 - 2x') + 9(y'^2 - 4y') = -4$$

Completing the squares yields

$$4(x'^2 - 2x' + 1) + 9(y'^2 - 4y' + 4) = -4 + 4 + 36$$

or

$$4(x' - 1)^2 + 9(y' - 2)^2 = 36 \qquad (12)$$

If we translate the coordinate axes by means of the translation equations

$$x'' = x' - 1, \qquad y'' = y' - 2$$

then (12) becomes

$$4x''^2 + 9y''^2 = 36$$

or

$$\frac{x''^2}{9} + \frac{y''^2}{4} = 1$$

which is the equation of the ellipse sketched in Figure 5. In that figure, the vectors $\mathbf{v}_1$ and $\mathbf{v}_2$ are the column vectors of P. ◢

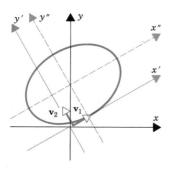

Figure 5

$$\boxed{\frac{x''^2}{9} + \frac{y''^2}{4} = 1}$$

EXERCISE SET 9.6

1. In each part find a change of variables that reduces the quadratic form to a sum or difference of squares, and express the quadratic form in terms of the new variables.
   (a) $2x_1^2 + 2x_2^2 - 2x_1x_2$ (b) $5x_1^2 + 2x_2^2 + 4x_1x_2$ (c) $2x_1x_2$ (d) $-3x_1^2 + 5x_2^2 + 2x_1x_2$

2. In each part find a change of variables that reduces the quadratic form to a sum or difference of squares, and express the quadratic form in terms of the new variables.
   (a) $3x_1^2 + 4x_2^2 + 5x_3^2 + 4x_1x_2 - 4x_2x_3$ (b) $2x_1^2 + 5x_2^2 + 5x_3^2 + 4x_1x_2 - 4x_1x_3 - 8x_2x_3$
   (c) $-5x_1^2 + x_2^2 - x_3^2 + 6x_1x_3 + 4x_1x_2$ (d) $2x_1x_3 + 6x_2x_3$

3. Find the quadratic forms associated with the following quadratic equations.
   (a) $2x^2 - 3xy + 4y^2 - 7x + 2y + 7 = 0$
   (b) $x^2 - xy + 5x + 8y - 3 = 0$
   (c) $5xy = 8$
   (d) $4x^2 - 2y^2 = 7$
   (e) $y^2 + 7x - 8y - 5 = 0$

4. Find the matrices of the quadratic forms in Exercise 3.

5. Express each of the quadratic equations in Exercise 3 in the matrix form
   $\mathbf{x}^T A \mathbf{x} + K \mathbf{x} + f = 0$.

6. Name the following conics.
   (a) $2x^2 + 5y^2 = 20$
   (b) $4x^2 + 9y^2 = 1$
   (c) $x^2 - y^2 - 8 = 0$
   (d) $4y^2 - 5x^2 = 20$
   (e) $x^2 + y^2 - 25 = 0$
   (f) $7y^2 - 2x = 0$
   (g) $-x^2 = 2y$
   (h) $3x - 11y^2 = 0$
   (i) $y - x^2 = 0$
   (j) $x^2 - 3 = -y^2$

7. In each part a translation will put the conic in standard position. Name the conic and give
   its equation in the translated coordinate system.
   (a) $9x^2 + 4y^2 - 36x - 24y + 36 = 0$
   (b) $x^2 - 16y^2 + 8x + 128y = 256$
   (c) $y^2 - 8x - 14y + 49 = 0$
   (d) $x^2 + y^2 + 6x - 10y + 18 = 0$
   (e) $2x^2 - 3y^2 + 6x + 20y = -41$
   (f) $x^2 + 10x + 7y = -32$

8. The following nondegenerate conics are rotated out of standard position. In each part rotate
   the coordinate axes to remove the xy-term. Name the conic and give its equation in the
   rotated coordinate system.
   (a) $2x^2 - 4xy - y^2 + 8 = 0$
   (b) $x^2 + 2xy + y^2 + 8x + y = 0$
   (c) $5x^2 + 4xy + 5y^2 = 9$
   (d) $11x^2 + 24xy + 4y^2 - 15 = 0$

In Exercises 9–14 translate and rotate the coordinate axes, if necessary, to put the conic in
standard position. Name the conic and give its equation in the final coordinate system.

9. $9x^2 - 4xy + 6y^2 - 10x - 20y = 5$

10. $3x^2 - 8xy - 12y^2 - 30x - 64y = 0$

11. $2x^2 - 4xy - y^2 - 4x - 8y = -14$

12. $21x^2 + 6xy + 13y^2 - 114x + 34y + 73 = 0$

13. $x^2 - 6xy - 7y^2 + 10x + 2y + 9 = 0$

14. $4x^2 - 20xy + 25y^2 - 15x - 6y = 0$

15. The graph of a quadratic equation in x and y can, in certain cases, be a point, a line, or a
    pair of lines. These are called *degenerate* conics. It is also possible that the equation is not
    satisfied by any real values of x and y. In such cases the equation has no graph; it is said to
    represent an *imaginary conic*. Each of the following represents a degenerate or imaginary
    conic. Where possible, sketch the graph.
    (a) $x^2 - y^2 = 0$
    (b) $x^2 + 3y^2 + 7 = 0$
    (c) $8x^2 + 7y^2 = 0$
    (d) $x^2 - 2xy + y^2 = 0$
    (e) $9x^2 + 12xy + 4y^2 - 52 = 0$
    (f) $x^2 + y^2 - 2x - 4y = -5$

9.7 QUADRIC SURFACES

In this section we shall apply the diagonalization techniques developed in the preceding section to quadratic equations in three variables, and we shall use our results to study quadric surfaces.

QUADRIC SURFACES

An equation of the form

$$ax^2 + by^2 + cz^2 + 2dxy + 2exz + 2fyz + gx + hy + iz + j = 0 \qquad (1)$$

where $a, b, \ldots, f$ are not all zero, is called a **quadratic equation in x, y, and z**; the expression

$$ax^2 + by^2 + cz^2 + 2dxy + 2exz + 2fyz$$

is called the **associated quadratic form**.

Equation (1) can be written in the matrix form

$$\begin{bmatrix} x & y & z \end{bmatrix} \begin{bmatrix} a & d & e \\ d & b & f \\ e & f & c \end{bmatrix} \begin{bmatrix} x \\ y \\ z \end{bmatrix} + \begin{bmatrix} g & h & i \end{bmatrix} \begin{bmatrix} x \\ y \\ z \end{bmatrix} + j = 0$$

or

$$\mathbf{x}^T A \mathbf{x} + K\mathbf{x} + j = 0$$

where

$$\mathbf{x} = \begin{bmatrix} x \\ y \\ z \end{bmatrix}, \qquad A = \begin{bmatrix} a & d & e \\ d & b & f \\ e & f & c \end{bmatrix}, \qquad K = \begin{bmatrix} g & h & i \end{bmatrix}$$

Example 1 The quadratic form associated with the quadratic equation

$$3x^2 + 2y^2 - z^2 + 4xy + 3xz - 8yz + 7x + 2y + 3z - 7 = 0$$

is

$$3x^2 + 2y^2 - z^2 + 4xy + 3xz - 8yz \quad \blacktriangle$$

Graphs of quadratic equations in $x, y,$ and z are called **quadrics** or **quadric surfaces**. The simplest equations for quadric surfaces occur when those surfaces are placed in certain **standard positions** relative to the coordinate axes. Figure 1 shows the six basic quadric surfaces and the equations for those surfaces when the surfaces are in the standard positions shown in the figure. If a quadric surface is cut by a plane, then the curve of intersection is called the **trace** of the plane on the surface. To help visualize the quadric surfaces in Figure 1, we have shown and described the traces made by planes parallel to the coordinate planes. The presence of one or more of the cross-product terms xy, xz, and yz in the equation of a quadric indicates that the quadric is rotated out of standard position; the presence of both x^2 and x terms, y^2 and y terms, or z^2 and z terms in a quadric with no cross-product term indicates the quadric is translated out of standard position.

Figure 1

Surface	Equation	Surface	Equation
Ellipsoid	$\dfrac{x^2}{l^2} + \dfrac{y^2}{m^2} + \dfrac{z^2}{n^2} = 1$ The traces in the coordinate planes are ellipses, as are the traces in planes parallel to the coordinate planes.	Elliptic cone	$z^2 = \dfrac{x^2}{l^2} + \dfrac{y^2}{m^2}$ The trace in the xy-plane is a point (the origin), and the traces in planes parallel to the xy-plane are ellipses. The traces in the yz- and xz-planes are pairs of lines intersecting at the origin. The traces in planes parallel to these are hyperbolas.
Hyperboloid of one sheet	$\dfrac{x^2}{l^2} + \dfrac{y^2}{m^2} - \dfrac{z^2}{n^2} = 1$ The trace in the xy-plane is an ellipse, as are the traces in planes parallel to the xy-plane. The traces in the yz-plane and xz-plane are hyperbolas, as are the traces in planes parallel to these.	Elliptic paraboloid	$z = \dfrac{x^2}{l^2} + \dfrac{y^2}{m^2}$ The trace in the xy-plane is a point (the origin), and the traces in planes parallel to and above the xy-plane are ellipses. The traces in the yz- and xz-planes are parabolas, as are the traces in planes parallel to these.
Hyperboloid of two sheets	$\dfrac{x^2}{l^2} + \dfrac{y^2}{m^2} - \dfrac{z^2}{n^2} = -1$ There is no trace in the xy-plane. In planes parallel to the xy-plane, which intersect the surface, the traces are ellipses. In the yz- and xz-planes, the traces are hyperbolas, as are the traces in planes parallel to these.	Hyperbolic paraboloid	$z = \dfrac{y^2}{m^2} - \dfrac{x^2}{l^2}$ The trace in the xy-plane is a pair of lines intersecting at the origin. The traces in planes parallel to the xy-plane are hyperbolas. The hyperbolas above the xy-plane open in the y-direction, and those below in the x-direction. The traces in the yz- and xz-planes are parabolas, as are the traces in planes parallel to these.

Example 2 Describe the quadric surface whose equation is

$$4x^2 + 36y^2 - 9z^2 - 16x - 216y + 304 = 0$$

Solution. Rearranging terms gives

$$4(x^2 - 4x) + 36(y^2 - 6y) - 9z^2 = -304$$

Completing the squares yields

$$4(x^2 - 4x + 4) + 36(y^2 - 6y + 9) - 9z^2 = -304 + 16 + 324$$

or

$$4(x - 2)^2 + 36(y - 3)^2 - 9z^2 = 36$$

or

$$\frac{(x - 2)^2}{9} + (y - 3)^2 - \frac{z^2}{4} = 1$$

Translating the axes by means of the translation equations

$$x' = x - 2, \qquad y' = y - 3, \qquad z' = z$$

yields

$$\frac{x'^2}{9} + y'^2 - \frac{z'^2}{4} = 1$$

which is the equation of a hyperboloid of one sheet. ▲

ELIMINATING CROSS-PRODUCT TERMS

The procedure for identifying quadrics that are rotated out of standard position is similar to the procedure for conics. Let Q be a quadric surface whose equation in *xyz*-coordinates is

$$\mathbf{x}^T A \mathbf{x} + K \mathbf{x} + j = 0 \tag{2}$$

We want to rotate the *xyz*-coordinate axes so that the equation of the quadric in the new *x'y'z'*-coordinate system has no cross-product terms. This can be done as follows:

Step 1. Find a matrix P that orthogonally diagonalizes $\mathbf{x}^T A \mathbf{x}$.

Step 2. Interchange two columns of P, if necessary, to make $\det(P) = 1$. This assures that the orthogonal coordinate transformation

$$\mathbf{x} = P\mathbf{x}', \quad \text{that is,} \quad \begin{bmatrix} x \\ y \\ z \end{bmatrix} = P \begin{bmatrix} x' \\ y' \\ z' \end{bmatrix}, \tag{3}$$

is a rotation.

Step 3. Substitute (3) into (2). This will produce an equation for the quadric in *x'y'z'*-coordinates with no cross-product terms. (The proof is similar to that for conics and is left as an exercise.)

The following theorem summarizes this discussion.

Theorem 9.7.1 (***Principal Axes Theorem for R^3***). *Let*

$$ax^2 + by^2 + cz^2 + 2dxy + 2exz + 2fyz + gx + hy + iz + j = 0$$

be the equation of a quadric Q, and let

$$\mathbf{x}^T A\mathbf{x} = ax^2 + by^2 + cz^2 + 2dxy + 2exz + 2fyz$$

be the associated quadratic form. The coordinate axes can be rotated so that the equation of Q in the x'y'z'-coordinate system has the form

$$\lambda_1 x'^2 + \lambda_2 y'^2 + \lambda_3 z'^2 + g'x' + h'y' + i'z' + j = 0$$

where λ_1, λ_2, and λ_3 are the eigenvalues of A. The rotation can be accomplished by the substitution

$$\mathbf{x} = P\mathbf{x}'$$

where P orthogonally diagonalizes $\mathbf{x}^T A\mathbf{x}$ and $\det(P) = 1$.

Example 3 Describe the quadric surface whose equation is

$$4x^2 + 4y^2 + 4z^2 + 4xy + 4xz + 4yz - 3 = 0$$

Solution. The matrix form of the above quadratic equation is

$$\mathbf{x}^T A\mathbf{x} - 3 = 0 \tag{4}$$

where

$$A = \begin{bmatrix} 4 & 2 & 2 \\ 2 & 4 & 2 \\ 2 & 2 & 4 \end{bmatrix}$$

As shown in Example 1 of Section 7.3, the eigenvalues of A are $\lambda = 2$ and $\lambda = 8$, and A is orthogonally diagonalized by the matrix

$$P = \begin{bmatrix} -1/\sqrt{2} & -1/\sqrt{6} & 1/\sqrt{3} \\ 1/\sqrt{2} & -1/\sqrt{6} & 1/\sqrt{3} \\ 0 & 2/\sqrt{6} & 1/\sqrt{3} \end{bmatrix}$$

where the first two column vectors in P are eigenvectors corresponding to $\lambda = 2$, and the third column vector is an eigenvector corresponding to $\lambda = 8$.

Since $\det(P) = 1$ (verify), the orthogonal coordinate transformation $\mathbf{x} = P\mathbf{x}'$ is a rotation. Substituting this expression in (4) yields

$$(P\mathbf{x}')^T A(P\mathbf{x}') - 3 = 0$$

or equivalently,

$$(\mathbf{x}')^T (P^T A P)\mathbf{x}' - 3 = 0 \tag{5}$$

But

$$P^T A P = \begin{bmatrix} 2 & 0 & 0 \\ 0 & 2 & 0 \\ 0 & 0 & 8 \end{bmatrix}$$

so (5) becomes

$$[x' \quad y' \quad z'] \begin{bmatrix} 2 & 0 & 0 \\ 0 & 2 & 0 \\ 0 & 0 & 8 \end{bmatrix} \begin{bmatrix} x' \\ y' \\ z' \end{bmatrix} - 3 = 0$$

or

$$2x'^2 + 2y'^2 + 8z'^2 = 3$$

This can be rewritten as

$$\frac{x'^2}{3/2} + \frac{y'^2}{3/2} + \frac{z'^2}{3/8} = 1$$

which is the equation of an ellipsoid. ▲

EXERCISE SET 9.7

1. Find the quadratic forms associated with the following quadratic equations.
   (a) $x^2 + 2y^2 - z^2 + 4xy - 5yz + 7x + 2z = 3$ (b) $3x^2 + 7z^2 + 2xy - 3xz + 4yz - 3x = 4$
   (c) $xy + xz + yz = 1$ (d) $x^2 + y^2 - z^2 = 7$
   (e) $3z^2 + 3xz - 14y + 9 = 0$ (f) $2z^2 + 2xz + y^2 + 2x - y + 3z = 0$

2. Find the matrices of the quadratic forms in Exercise 1.

3. Express each of the quadratic equations given in Exercise 1 in the matrix form
   $\mathbf{x}^T A \mathbf{x} + K\mathbf{x} + j = 0$.

4. Name the following quadrics.
   (a) $36x^2 + 9y^2 + 4z^2 - 36 = 0$ (b) $2x^2 + 6y^2 - 3z^2 = 18$ (c) $6x^2 - 3y^2 - 2z^2 - 6 = 0$
   (d) $9x^2 + 4y^2 - z^2 = 0$ (e) $16x^2 + y^2 = 16z$ (f) $7x^2 - 3y^2 + z = 0$
   (g) $x^2 + y^2 + z^2 = 25$

5. In each part determine the translation equations that will put the quadric in standard position.
   Name the quadric.
   (a) $9x^2 + 36y^2 + 4z^2 - 18x - 144y - 24z + 153 = 0$ (b) $6x^2 + 3y^2 - 2z^2 + 12x - 18y - 8z = -7$
   (c) $3x^2 - 3y^2 - z^2 + 42x + 144 = 0$ (d) $4x^2 + 9y^2 - z^2 - 54y - 50z = 544$
   (e) $x^2 + 16y^2 + 2x - 32y - 16z - 15 = 0$ (f) $7x^2 - 3y^2 + 126x + 72y + z + 135 = 0$
   (g) $x^2 + y^2 + z^2 - 2x + 4y - 6z = 11$

6. In each part find a rotation $\mathbf{x} = P\mathbf{x}'$ that removes the cross-product terms. Name the quadric
   and give its equation in the $x'y'z'$-system.
   (a) $2x^2 + 3y^2 + 23z^2 + 72xz + 150 = 0$ (b) $4x^2 + 4y^2 + 4z^2 + 4xy + 4xz + 4yz - 5 = 0$
   (c) $144x^2 + 100y^2 + 81z^2 - 216xz - 540x - 720z = 0$ (d) $2xy + z = 0$

In Exercises 7–10 translate and rotate the coordinate axes to put the quadric in standard position. Name the quadric and give its equation in the final coordinate system.

7. $2xy + 2xz + 2yz - 6x - 6y - 4z = -9$

8. $7x^2 + 7y^2 + 10z^2 - 2xy - 4xz + 4yz - 12x + 12y + 60z = 24$

9. $2xy - 6x + 10y + z - 31 = 0$

10. $2x^2 + 2y^2 + 5z^2 - 4xy - 2xz + 2yz + 10x - 26y - 2z = 0$

11. Prove Theorem 9.7.1.

9.8 COMPARISON OF PROCEDURES FOR SOLVING LINEAR SYSTEMS

In this section we shall discuss some practical aspects of solving systems of linear equations, inverting matrices, and finding eigenvalues. Although we have previously discussed methods for performing these computations, those methods are not directly applicable to the computer solution of the large-scale problems that arise in real-world applications.

COUNTING OPERATIONS

Since computers are limited in the number of decimal places they can carry, they round off or truncate most numerical quantities. For example, a computer designed to store eight decimal places might record $\frac{2}{3}$ either as .66666667 (rounded off) or .66666666 (truncated). In either case, an error is introduced that we shall call **roundoff error** or **rounding error**.

The main practical considerations in solving linear algebra problems on digital computers are minimizing the computer time (and thus cost) needed to obtain the solution, and minimizing inaccuracies due to roundoff errors. Thus, a good computer algorithm uses as few operations as possible and performs those operations in a way that minimizes the effect of roundoff errors.

In this text we have studied four methods for solving a linear system, $A\mathbf{x} = \mathbf{b}$, of n equations in n unknowns:

1. Gaussian elimination with back-substitution
2. Gauss–Jordan elimination
3. Computing A^{-1}, then $\mathbf{x} = A^{-1}\mathbf{b}$
4. Cramer's Rule

To determine how these methods compare as computational tools, we need to know how many arithmetic operations each requires. On a large modern computer, typical execution times in microseconds (1 microsecond $= 10^{-6}$ second) for the basic arithmetic operations are

$$\text{Multiplication} = 1.0 \text{ microsecond}$$
$$\text{Division} = 3.0 \text{ microseconds}$$
$$\text{Addition} = 0.5 \text{ microsecond}$$
$$\text{Subtraction} = 0.5 \text{ microsecond}$$

In our analysis we shall group divisions and multiplications together (average execution time = 2.0 microseconds), and we shall also group additions and subtractions together (average execution time = 0.5 microsecond). We shall refer to either multiplications or divisions as "multiplications" and to additions or subtractions as "additions."

In Table 1 we list the number of operations required to solve a linear system $A\mathbf{x} = \mathbf{b}$ of n equations in n unknowns by each of the four methods discussed in the text, as well as the number of operations required to invert A or to compute its determinant by row reduction.

TABLE 1 *Operation Counts for an Invertible $n \times n$ Matrix A*

Method	Number of Additions	Number of Multiplications
Solve $A\mathbf{x} = \mathbf{b}$ by Gauss–Jordan elimination	$\frac{1}{3}n^3 + \frac{1}{2}n^2 - \frac{5}{6}n$	$\frac{1}{3}n^3 + n^2 - \frac{1}{3}n$
Solve $A\mathbf{x} = \mathbf{b}$ by Gaussian elimination	$\frac{1}{3}n^3 + \frac{1}{2}n^2 - \frac{5}{6}n$	$\frac{1}{3}n^3 + n^2 - \frac{1}{3}n$
Find A^{-1} by reducing $[A \mid I]$ to $[I \mid A^{-1}]$	$n^3 - 2n^2 + n$	n^3
Solve $A\mathbf{x} = \mathbf{b}$ as $\mathbf{x} = A^{-1}\mathbf{b}$	$n^3 - n^2$	$n^3 + n^2$
Find $\det(A)$ by row reduction	$\frac{1}{3}n^3 - \frac{1}{2}n^2 + \frac{1}{6}n$	$\frac{1}{3}n^3 + \frac{2}{3}n - 1$
Solve $A\mathbf{x} = \mathbf{b}$ by Cramer's Rule	$\frac{1}{3}n^4 - \frac{1}{6}n^3 - \frac{1}{3}n^2 + \frac{1}{6}n$	$\frac{1}{3}n^4 + \frac{1}{3}n^3 + \frac{2}{3}n^2 + \frac{2}{3}n - 1$

Note that the text methods of Gauss–Jordan elimination and Gaussian elimination have the same operation counts. It is not hard to see why this is so. Both methods begin by reducing the augmented matrix to row-echelon form. This is called the *forward phase* or *forward pass*. Then the solution is completed by back-substitution in Gaussian elimination and by continued reduction to reduced row-echelon form in Gauss–Jordan elimination. This is called the *backward phase* or *backward pass*. It turns out that the number of operations required for the backward phase is the same whether one uses back-substitution or continued reduction to reduced row-echelon form. Thus, the text methods of Gaussian elimination and Gauss–Jordan elimination have the same operation counts.

REMARK. There is a common variation of Gauss–Jordan elimination that is less efficient than the one presented in this text. In our method the augmented matrix is first reduced to reduced row-echelon form by introducing zeros *below* the leading 1's; then the reduction is completed by introducing zeros above the leading 1's. An alternative procedure is to introduce zeros *above* and *below* a leading 1 as soon as it is obtained. This method requires

$$\frac{n^3}{2} - \frac{n}{2} \quad \text{additions} \qquad \text{and} \qquad \frac{n^3}{2} + \frac{n^2}{2} \quad \text{multiplications}$$

both of which are larger than our values for all $n \geq 3$.

To illustrate how the results in Table 1 are computed, we shall derive the operation counts for Gauss–Jordan elimination. For this discussion we need the following formulas for the sum of the first n positive integers and the sum of the squares of the first n positive integers:

$$1 + 2 + 3 + \cdots + n = \frac{n(n + 1)}{2} \tag{1}$$

$$1^2 + 2^2 + 3^2 + \cdots + n^2 = \frac{n(n + 1)(2n + 1)}{6} \tag{2}$$

Derivations of these formulas are discussed in the exercises. We also need formulas for the sum of the first $n - 1$ positive integers and the sum of the squares of the first $n - 1$ positive integers. These can be obtained by substituting $n - 1$ for n in (1) and (2).

$$1 + 2 + 3 + \cdots + (n - 1) = \frac{(n - 1)n}{2} \tag{3}$$

$$1^2 + 2^2 + 3^2 + \cdots + (n - 1)^2 = \frac{(n - 1)n(2n - 1)}{6} \tag{4}$$

**OPERATION
COUNT FOR
GAUSS–JORDAN
ELIMINATION**

Let $A\mathbf{x} = \mathbf{b}$ be a system of n linear equations in n unknowns, and assume that A is invertible, so that the system has a unique solution. Also assume, for simplicity, that no row interchanges are required to put the augmented matrix $[A \mid \mathbf{b}]$ in reduced row-echelon form. This assumption is justified by the fact that row interchanges are performed as bookkeeping operations on a computer and require much less time than arithmetic operations.

Since no row interchanges are required, the first step in the Gauss–Jordan elimination process is to introduce a leading 1 in the first row by multiplying the elements in that row by the reciprocal of the leftmost entry in the row. We shall represent this step schematically as follows:

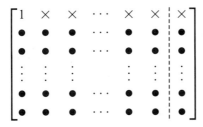

$$\begin{bmatrix} 1 & \times & \times & \cdots & \times & \times & \vdots & \times \\ \bullet & \bullet & \bullet & \cdots & \bullet & \bullet & \vdots & \bullet \\ \bullet & \bullet & \bullet & \cdots & \bullet & \bullet & \vdots & \bullet \\ \vdots & \vdots & \vdots & & \vdots & \vdots & \vdots & \vdots \\ \bullet & \bullet & \bullet & \cdots & \bullet & \bullet & \vdots & \bullet \\ \bullet & \bullet & \bullet & \cdots & \bullet & \bullet & \vdots & \bullet \end{bmatrix}$$

$\times$ denotes a quantity to be computed
$\bullet$ denotes a quantity that is not computed
The matrix size is $n \times (n + 1)$

Note that the leading 1 is simply recorded and requires no computation; only the remaining n entries in the first row must be computed.

The following is a schematic description of the steps and the number of operations required to reduce $[A \mid \mathbf{b}]$ to row-echelon form.

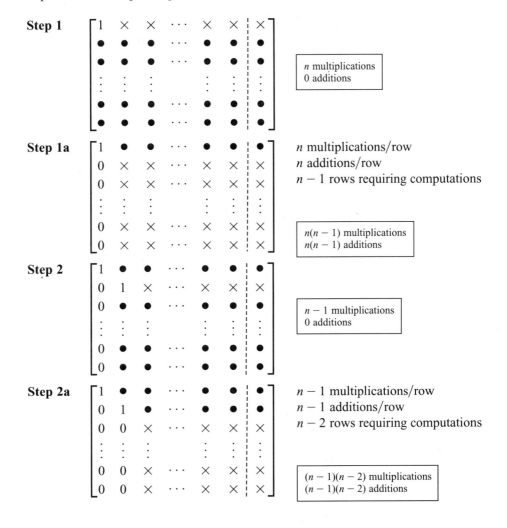

Step 1

n multiplications
0 additions

Step 1a

n multiplications/row
n additions/row
$n - 1$ rows requiring computations

$n(n - 1)$ multiplications
$n(n - 1)$ additions

Step 2

$n - 1$ multiplications
0 additions

Step 2a

$n - 1$ multiplications/row
$n - 1$ additions/row
$n - 2$ rows requiring computations

$(n - 1)(n - 2)$ multiplications
$(n - 1)(n - 2)$ additions

Step 3
$$\begin{bmatrix} 1 & \bullet & \bullet & \cdots & \bullet & \bullet & \vdots & \bullet \\ 0 & 1 & \bullet & \cdots & \bullet & \bullet & \vdots & \bullet \\ 0 & 0 & 1 & \cdots & \times & \times & \vdots & \times \\ \vdots & \vdots & \vdots & & \vdots & \vdots & \vdots & \vdots \\ 0 & 0 & \bullet & \cdots & \bullet & \bullet & \vdots & \bullet \\ 0 & 0 & \bullet & \cdots & \bullet & \bullet & \vdots & \bullet \end{bmatrix}$$

| $n - 2$ multiplications |
| 0 additions |

Step 3a
$$\begin{bmatrix} 1 & \bullet & \bullet & \cdots & \bullet & \bullet & \vdots & \bullet \\ 0 & 1 & \bullet & \cdots & \bullet & \bullet & \vdots & \bullet \\ 0 & 0 & 1 & \cdots & \bullet & \bullet & \vdots & \bullet \\ \vdots & \vdots & \vdots & & \vdots & \vdots & \vdots & \vdots \\ 0 & 0 & 0 & \cdots & \times & \times & \vdots & \times \\ 0 & 0 & 0 & \cdots & \times & \times & \vdots & \times \end{bmatrix}$$

$n - 2$ multiplications/row
$n - 2$ additions/row
$n - 3$ rows requiring computations

| $(n - 2)(n - 3)$ multiplications |
| $(n - 2)(n - 3)$ additions |

Step $(n - 1)$
$$\begin{bmatrix} 1 & \bullet & \bullet & \cdots & \bullet & \bullet & \vdots & \bullet \\ 0 & 1 & \bullet & \cdots & \bullet & \bullet & \vdots & \bullet \\ 0 & 0 & 1 & \cdots & \bullet & \bullet & \vdots & \bullet \\ \vdots & \vdots & \vdots & & \vdots & \vdots & \vdots & \vdots \\ 0 & 0 & 0 & \cdots & 1 & \times & \vdots & \times \\ 0 & 0 & 0 & \cdots & \bullet & \bullet & \vdots & \bullet \end{bmatrix}$$

| 2 multiplications |
| 0 additions |

Step $(n - 1)$a
$$\begin{bmatrix} 1 & \bullet & \bullet & \cdots & \bullet & \bullet & \vdots & \bullet \\ 0 & 1 & \bullet & \cdots & \bullet & \bullet & \vdots & \bullet \\ 0 & 0 & 1 & \cdots & \bullet & \bullet & \vdots & \bullet \\ \vdots & \vdots & \vdots & & \vdots & \vdots & \vdots & \vdots \\ 0 & 0 & 0 & \cdots & 1 & \bullet & \vdots & \bullet \\ 0 & 0 & 0 & \cdots & 0 & \times & \vdots & \times \end{bmatrix}$$

2 multiplications/row
2 additions/row
1 row requiring computations

| 2 multiplications |
| 2 additions |

Step n
$$\begin{bmatrix} 1 & \bullet & \bullet & \cdots & \bullet & \bullet & \vdots & \bullet \\ 0 & 1 & \bullet & \cdots & \bullet & \bullet & \vdots & \bullet \\ 0 & 0 & 1 & \cdots & \bullet & \bullet & \vdots & \bullet \\ \vdots & \vdots & \vdots & & \vdots & \vdots & \vdots & \vdots \\ 0 & 0 & 0 & \cdots & 1 & \bullet & \vdots & \bullet \\ 0 & 0 & 0 & \cdots & 0 & 1 & \vdots & \times \end{bmatrix}$$

| 1 multiplication |
| 0 additions |

Thus, the number of operations required to complete successive steps is as follows:

Steps 1 and 1a

| Multiplications: | $n + n(n - 1) = n^2$ |
| Additions: | $n(n - 1) = n^2 - n$ |

Steps 2 and 2a

| Multiplications: | $(n - 1) + (n - 1)(n - 2) = (n - 1)^2$ |
| Additions: | $(n - 1)(n - 2) = (n - 1)^2 - (n - 1)$ |

Steps 3 and 3a

Multiplications: $(n-2) + (n-2)(n-3) = (n-2)^2$

Additions: $(n-2)(n-3) = (n-2)^2 - (n-2)$

$$\vdots$$

Steps $(n-1)$ and $(n-1)$a

Multiplications: $4\ (=2^2)$

Additions: $2\ (=2^2-2)$

Step n

Multiplications: $1\ (=1^2)$

Additions: $0\ (=1^2-1)$

Therefore, the total number of operations required to reduce $[A \mid \mathbf{b}]$ to row-echelon form is

Multiplications: $n^2 + (n-1)^2 + (n-2)^2 + \cdots + 1^2$

Additions: $[n^2 + (n-1)^2 + (n-2)^2 + \cdots + 1^2]$
$$- [n + (n-1) + (n-2) + \cdots + 1]$$

or, on applying Formulas (1) and (2),

Multiplications: $\dfrac{n(n+1)(2n+1)}{6} = \dfrac{n^3}{3} + \dfrac{n^2}{2} + \dfrac{n}{6}$ (5)

Additions: $\dfrac{n(n+1)(2n+1)}{6} - \dfrac{n(n+1)}{2} = \dfrac{n^3}{3} - \dfrac{n}{3}$ (6)

This completes the operation count for the forward phase. For the backward phase we must put the row-echelon form of $[A \mid \mathbf{b}]$ into reduced row-echelon form by introducing zeros above the leading 1's. The operations are as follows:

Step 1
$$\begin{bmatrix} 1 & \bullet & \bullet & \cdots & \bullet & 0 & \mid & \times \\ 0 & 1 & \bullet & \cdots & \bullet & 0 & \mid & \times \\ 0 & 0 & 1 & \cdots & \bullet & 0 & \mid & \times \\ \vdots & \vdots & \vdots & & \vdots & \vdots & \mid & \vdots \\ 0 & 0 & 0 & \cdots & 1 & 0 & \mid & \times \\ 0 & 0 & 0 & \cdots & 0 & 1 & \mid & \bullet \end{bmatrix}$$

$n - 1$ multiplications
$n - 1$ additions

Step 2
$$\begin{bmatrix} 1 & \bullet & \bullet & \cdots & 0 & 0 & \mid & \times \\ 0 & 1 & \bullet & \cdots & 0 & 0 & \mid & \times \\ 0 & 0 & 1 & \cdots & 0 & 0 & \mid & \times \\ \vdots & \vdots & \vdots & & \vdots & \vdots & \mid & \vdots \\ 0 & 0 & 0 & \cdots & 1 & 0 & \mid & \bullet \\ 0 & 0 & 0 & \cdots & 0 & 1 & \mid & \bullet \end{bmatrix}$$

$n - 2$ multiplications
$n - 2$ additions

$\vdots$

Step
($n-2$)

$$\begin{bmatrix} 1 & \bullet & 0 & \cdots & 0 & 0 & \vdots & \times \\ 0 & 1 & 0 & \cdots & 0 & 0 & \vdots & \times \\ 0 & 0 & 1 & \cdots & 0 & 0 & \vdots & \bullet \\ \vdots & \vdots & \vdots & & \vdots & \vdots & \vdots & \vdots \\ 0 & 0 & 0 & \cdots & 1 & 0 & \vdots & \bullet \\ 0 & 0 & 0 & \cdots & 0 & 1 & \vdots & \bullet \end{bmatrix}$$

2 multiplications
2 additions

Step
($n-1$)

$$\begin{bmatrix} 1 & 0 & 0 & \cdots & 0 & 0 & \vdots & \times \\ 0 & 1 & 0 & \cdots & 0 & 0 & \vdots & \bullet \\ 0 & 0 & 1 & \cdots & 0 & 0 & \vdots & \bullet \\ \vdots & \vdots & \vdots & & \vdots & \vdots & \vdots & \vdots \\ 0 & 0 & 0 & \cdots & 1 & 0 & \vdots & \bullet \\ 0 & 0 & 0 & \cdots & 0 & 1 & \vdots & \bullet \end{bmatrix}$$

1 multiplication
1 addition

Thus, the number of operations required for the backward phase is

Multiplications: $(n-1)+(n-2)+\cdots+2+1$

Additions: $(n-1)+(n-2)+\cdots+2+1$

or, on applying Formula (3),

$$\text{Multiplications:} \quad \frac{(n-1)n}{2}=\frac{n^2}{2}-\frac{n}{2} \tag{7}$$

$$\text{Additions:} \quad \frac{(n-1)n}{2}=\frac{n^2}{2}-\frac{n}{2} \tag{8}$$

Thus, from (5), (6), (7), and (8) the total operation count for Gauss–Jordan elimination is

$$\text{Multiplications:} \quad \left(\frac{n^3}{3}+\frac{n^2}{2}+\frac{n}{6}\right)+\left(\frac{n^2}{2}-\frac{n}{2}\right)=\frac{n^3}{3}+n^2-\frac{n}{3} \tag{9}$$

$$\text{Additions:} \quad \left(\frac{n^3}{3}-\frac{n}{3}\right)+\left(\frac{n^2}{2}-\frac{n}{2}\right)=\frac{n^3}{3}+\frac{n^2}{2}-\frac{5n}{6} \tag{10}$$

COMPARISON OF METHODS FOR SOLVING LINEAR SYSTEMS

In practical applications it is not uncommon to encounter linear systems with thousands of equations in thousands of unknowns. Thus, we shall be interested in Table 1 for large values of n. It is a fact about polynomials that for large values of the variable, a polynomial can be approximated well by its term of highest degree; that is, if $a_k \neq 0$, then

$$a_0+a_1x+\cdots+a_kx^k \approx a_kx^k \qquad \text{for large } x$$

(Exercise 12). Thus, for large values of n the operation counts in Table 1 can be approximated as shown in Table 2.

It follows from Table 2 that for large n the best methods for solving $A\mathbf{x}=\mathbf{b}$ are Gaussian elimination and Gauss–Jordan elimination. The method of multiplying by A^{-1} is much worse than these (it requires three times as many operations), and the poorest of the four methods is Cramer's Rule.

TABLE 2 *Approximate Operation Counts for an Invertible n × n Matrix with Large n*

Method	Number of Additions	Number of Multiplications
Solve $A\mathbf{x} = \mathbf{b}$ by Gauss–Jordan elimination	$\approx \dfrac{n^3}{3}$	$\approx \dfrac{n^3}{3}$
Solve $A\mathbf{x} = \mathbf{b}$ by Gaussian elimination	$\approx \dfrac{n^3}{3}$	$\approx \dfrac{n^3}{3}$
Find A^{-1} by reducing $[A \mid I]$ to $[I \mid A^{-1}]$	$\approx n^3$	$\approx n^3$
Solve $A\mathbf{x} = \mathbf{b}$ as $\mathbf{x} = A^{-1}\mathbf{b}$	$\approx n^3$	$\approx n^3$
Find $\det(A)$ by row reduction	$\approx \dfrac{n^3}{3}$	$\approx \dfrac{n^3}{3}$
Solve $A\mathbf{x} = \mathbf{b}$ by Cramer's Rule	$\approx \dfrac{n^4}{3}$	$\approx \dfrac{n^4}{3}$

REMARK. We observed in the remark following Table 1 that if Gauss–Jordan elimination is performed by introducing zeros above and below leading 1's as soon as they are obtained, then the operation count is

$$\frac{n^3}{2} - \frac{n}{2} \quad \text{additions} \qquad \text{and} \qquad \frac{n^3}{2} + \frac{n^2}{2} \quad \text{multiplications}$$

Thus, for large n this procedure requires $\approx n^3/2$ multiplications, which is 50% greater than the $n^3/3$ multiplications required by the text method. Similarly for additions.

It is reasonable to ask if it is possible to devise other methods for solving linear systems that might require significantly fewer than the $\approx n^3/3$ additions and multiplications needed in Gaussian elimination and Gauss–Jordan elimination. The answer is a qualified "yes." In recent years methods have been devised that require $\approx Cn^q$ multiplications, where q is slightly larger than 2.5. However, these methods have little practical value because the programming is complicated, the constant C is very large, and the number of additions required is excessive. In short, there is currently no practical method for solving general linear systems that significantly improves on the operation counts for Gaussian elimination and the text method of Gauss–Jordan elimination.

EXERCISE SET 9.8

1. Find the number of additions and multiplications required to compute AB if A is an $m \times n$ matrix and B is an $n \times p$ matrix.

2. Use the result in Exercise 1 to find the number of additions and multiplications required to compute A^k by direct multiplication if A is an $n \times n$ matrix.

3. Assuming A to be an $n \times n$ matrix, use the formulas in Table 1 to determine the number of operations required for the procedures in Table 3.

TABLE 3

	$n = 5$		$n = 10$		$n = 100$		$n = 1000$	
	+	×	+	×	+	×	+	×
Solve $A\mathbf{x} = \mathbf{b}$ by Gauss–Jordan elimination								
Solve $A\mathbf{x} = \mathbf{b}$ by Gaussian elimination								
Find A^{-1} by reducing $[A \mid I]$ to $[I \mid A^{-1}]$								
Solve $A\mathbf{x} = \mathbf{b}$ as $\mathbf{x} = A^{-1}\mathbf{b}$								
Find $\det(A)$ by row reduction								
Solve $A\mathbf{x} = \mathbf{b}$ by Cramer's Rule								

4. Assuming a computer execution time of 2.0 microseconds for multiplications and 0.5 microsecond for additions, use the results in Exercise 3 to fill in the execution times in seconds for the procedures in Table 4.

TABLE 4

	$n = 5$	$n = 10$	$n = 100$	$n = 1000$
	Execution Time (sec)	Execution Time (sec)	Execution Time (sec)	Execution Time (sec)
Solve $A\mathbf{x} = \mathbf{b}$ by Gauss–Jordan elimination				
Solve $A\mathbf{x} = \mathbf{b}$ by Gaussian elimination				
Find A^{-1} by reducing $[A \mid I]$ to $[I \mid A^{-1}]$				
Solve $A\mathbf{x} = \mathbf{b}$ as $\mathbf{x} = A^{-1}\mathbf{b}$				
Find $\det(A)$ by row reduction				
Solve $A\mathbf{x} = \mathbf{b}$ by Cramer's Rule				

5. Derive the formula

$$1 + 2 + 3 + \cdots + n = \frac{n(n + 1)}{2}$$

[***Hint.*** Let $S_n = 1 + 2 + 3 + \cdots + n$. Write the terms of S_n in reverse order and add the two expressions for S_n.]

6. Use the result in Exercise 5 to show that

$$1 + 2 + 3 + \cdots + (n - 1) = \frac{(n - 1)n}{2}$$

7. Derive the formula

$$1^2 + 2^2 + 3^2 + \cdots + n^2 = \frac{n(n + 1)(2n + 1)}{6}$$

using the following steps.
(a) Show that $(k + 1)^3 - k^3 = 3k^2 + 3k + 1$.
(b) Show that

$$[2^3 - 1^3] + [3^3 - 2^3] + [4^3 - 3^3] + \cdots + [(n + 1)^3 - n^3] = (n + 1)^3 - 1$$

(c) Apply (a) to each term on the left side of (b) to show that

$$(n + 1)^3 - 1 = 3[1^2 + 2^2 + 3^2 + \cdots + n^2] + 3[1 + 2 + 3 + \cdots + n] + n$$

(d) Solve the equation in (c) for $1^2 + 2^2 + 3^2 + \cdots + n^2$, use the result of Exercise 5, and then simplify.

8. Use the result in Exercise 7 to show that

$$1^2 + 2^2 + 3^2 + \cdots + (n - 1)^2 = \frac{(n - 1)n(2n - 1)}{6}$$

9. Let R be a row-echelon form of an invertible $n \times n$ matrix. Show that solving the linear system $Rx = b$ by back-substitution requires

$$\frac{n^2}{2} - \frac{n}{2} \quad \text{multiplications}$$

$$\frac{n^2}{2} - \frac{n}{2} \quad \text{additions}$$

10. Show that to reduce an invertible $n \times n$ matrix to I_n by the text method requires

$$\frac{n^3}{3} - \frac{n}{3} \quad \text{multiplications}$$

$$\frac{n^3}{3} - \frac{n^2}{2} + \frac{n}{6} \quad \text{additions}$$

[*Note.* Assume that no row interchanges are required.]

11. Consider the variation of Gauss–Jordan elimination in which zeros are introduced above and below a leading 1 as soon as it is obtained, and let A be an invertible $n \times n$ matrix. Show that to solve a linear system $Ax = b$ using this version of Gauss–Jordan elimination requires

$$\frac{n^3}{2} + \frac{n^2}{2} \quad \text{multiplications}$$

$$\frac{n^3}{2} - \frac{n}{2} \quad \text{additions}$$

[*Note.* Assume that no row interchanges are required.]

12. (*For readers who have studied calculus.*) Show that if $p(x) = a_0 + a_1 x + \cdots + a_k x^k$, where $a_k \neq 0$, then

$$\lim_{x \to +\infty} \frac{p(x)}{a_k x^k} = 1$$

[*Note.* This result justifies the approximation $a_0 + a_1 x + \cdots + a_k x^k \approx a_k x^k$ for large x.]

9.9 *LU*-DECOMPOSITIONS

With Gaussian elimination and Gauss–Jordan elimination, a linear system is solved by operating systematically on the augmented matrix. In this section we shall discuss a different approach, one based on factoring the coefficient matrix into a product of lower and upper triangular matrices. This method is well suited for digital computers and is the basis for many practical computer programs. *

SOLVING LINEAR SYSTEMS BY FACTORIZATION

We shall proceed in two stages. First we shall show how a linear system $A\mathbf{x} = \mathbf{b}$ can be readily solved once A is factored into a product of lower and upper triangular matrices. Then we shall show how to construct such factorizations.

If an $n \times n$ matrix A can be factored into a product of $n \times n$ matrices as

$$A = LU$$

where L is lower triangular and U is upper triangular, then the linear system $A\mathbf{x} = \mathbf{b}$ can be solved as follows:

Step 1.   Rewrite the system $A\mathbf{x} = \mathbf{b}$ as

$$LU\mathbf{x} = \mathbf{b} \tag{1}$$

Step 2.   Define a new $n \times 1$ matrix $\mathbf{y}$ by

$$U\mathbf{x} = \mathbf{y} \tag{2}$$

Step 3.   Use (2) to rewrite (1) as $L\mathbf{y} = \mathbf{b}$ and solve this system for $\mathbf{y}$.

Step 4.   Substitute $\mathbf{y}$ in (2) and solve for $\mathbf{x}$.

Although this procedure replaces the problem of solving the single system $A\mathbf{x} = \mathbf{b}$ by the problem of solving the two systems, $L\mathbf{y} = \mathbf{b}$ and $U\mathbf{x} = \mathbf{y}$, the latter systems are easy to solve because the coefficient matrices are triangular. The following example illustrates this procedure.

---

* In 1979 an important library of machine-independent linear algebra programs called LINPAK, was developed at Argonne National Laboratories. Many of the programs in that library are based on the methods discussed in this section.

Example 1   Later in this section we will derive the factorization

$$\begin{bmatrix} 2 & 6 & 2 \\ -3 & -8 & 0 \\ 4 & 9 & 2 \end{bmatrix} = \begin{bmatrix} 2 & 0 & 0 \\ -3 & 1 & 0 \\ 4 & -3 & 7 \end{bmatrix}\begin{bmatrix} 1 & 3 & 1 \\ 0 & 1 & 3 \\ 0 & 0 & 1 \end{bmatrix}$$

Use this result and the method described above to solve the system

$$\begin{bmatrix} 2 & 6 & 2 \\ -3 & -8 & 0 \\ 4 & 9 & 2 \end{bmatrix}\begin{bmatrix} x_1 \\ x_2 \\ x_3 \end{bmatrix} = \begin{bmatrix} 2 \\ 2 \\ 3 \end{bmatrix} \tag{3}$$

Solution. Rewrite (3) as

$$\begin{bmatrix} 2 & 0 & 0 \\ -3 & 1 & 0 \\ 4 & -3 & 7 \end{bmatrix}\begin{bmatrix} 1 & 3 & 1 \\ 0 & 1 & 3 \\ 0 & 0 & 1 \end{bmatrix}\begin{bmatrix} x_1 \\ x_2 \\ x_3 \end{bmatrix} = \begin{bmatrix} 2 \\ 2 \\ 3 \end{bmatrix} \tag{4}$$

As specified in Step 2 above, define y_1, y_2, and y_3 by the equation

$$\begin{bmatrix} 1 & 3 & 1 \\ 0 & 1 & 3 \\ 0 & 0 & 1 \end{bmatrix}\begin{bmatrix} x_1 \\ x_2 \\ x_3 \end{bmatrix} = \begin{bmatrix} y_1 \\ y_2 \\ y_3 \end{bmatrix} \tag{5}$$

so (3) can be rewritten as

$$\begin{bmatrix} 2 & 0 & 0 \\ -3 & 1 & 0 \\ 4 & -3 & 7 \end{bmatrix}\begin{bmatrix} y_1 \\ y_2 \\ y_3 \end{bmatrix} = \begin{bmatrix} 2 \\ 2 \\ 3 \end{bmatrix}$$

or equivalently,

$$\begin{aligned} 2y_1 &= 2 \\ -3y_1 + y_2 &= 2 \\ 4y_1 - 3y_2 + 7y_3 &= 3 \end{aligned}$$

The procedure for solving this system is similar to back-substitution except that the equations are solved from the top down instead of from the bottom up. This procedure, called *forward-substitution*, yields

$$y_1 = 1, \qquad y_2 = 5, \qquad y_3 = 2$$

(Verify.) Substituting these values in (5) yields the linear system

$$\begin{bmatrix} 1 & 3 & 1 \\ 0 & 1 & 3 \\ 0 & 0 & 1 \end{bmatrix}\begin{bmatrix} x_1 \\ x_2 \\ x_3 \end{bmatrix} = \begin{bmatrix} 1 \\ 5 \\ 2 \end{bmatrix}$$

or equivalently,

$$\begin{aligned} x_1 + 3x_2 + x_3 &= 1 \\ x_2 + 3x_3 &= 5 \\ x_3 &= 2 \end{aligned}$$

Solving this system by back-substitution yields the solution

$$x_1 = 2, \qquad x_2 = -1, \qquad x_3 = 2$$

(Verify.) ▲

LU-DECOMPOSI-
TIONS

Now that we have seen how a linear system of n equations in n unknowns can be solved by factoring the coefficient matrix, we shall turn to the problem of constructing such factorizations. To motivate the method, suppose that an $n \times n$ matrix A has been reduced to a row-echelon form U by a sequence of elementary row operations. By Theorem 1.5.1 each of these operations can be accomplished by multiplying on the left by an appropriate elementary matrix. Thus, we can find elementary matrices $E_1, E_2, \ldots, E_k$ such that

$$E_k \cdots E_2 E_1 A = U \tag{6}$$

By Theorem 1.5.2, $E_1, E_2, \ldots, E_k$ are invertible, so we can multiply both sides of Equation (6) on the left successively by

$$E_k^{-1}, \ldots, E_2^{-1}, E_1^{-1}$$

to obtain

$$A = E_1^{-1} E_2^{-1} \cdots E_k^{-1} U \tag{7}$$

In Exercise 15 we will help the reader to show that the matrix L defined by

$$L = E_1^{-1} E_2^{-1} \cdots E_k^{-1} \tag{8}$$

is lower triangular provided that *no row interchanges are used in reducing A to U.* Assuming this to be so, substituting (8) into (7) yields

$$A = LU$$

which is a factorization of A into a product of a lower triangular matrix and an upper triangular matrix.

The following theorem summarizes the above result.

Theorem 9.9.1. *If A is a square matrix that can be reduced to a row-echelon form U without using row interchanges, then A can be factored as $A = LU$, where L is a lower triangular matrix.*

Definition. A factorization of a square matrix A as $A = LU$, where L is lower triangular and U is upper triangular, is called an ***LU-decomposition*** or ***triangular decomposition*** of A.

Example 2 Find an LU-decomposition of

$$A = \begin{bmatrix} 2 & 6 & 2 \\ -3 & -8 & 0 \\ 4 & 9 & 2 \end{bmatrix}$$

Solution. To obtain an LU-decomposition, $A = LU$, we shall reduce A to a row-echelon form U, then calculate L from (8). The steps are as follows:

	Reduction to Row-Echelon Form	Elementary Matrix Corresponding to the Row Operation	Inverse of the Elementary Matrix
	$\begin{bmatrix} 2 & 6 & 2 \\ -3 & -8 & 0 \\ 4 & 9 & 2 \end{bmatrix}$		
Step 1		$E_1 = \begin{bmatrix} \frac{1}{2} & 0 & 0 \\ 0 & 1 & 0 \\ 0 & 0 & 1 \end{bmatrix}$	$E_1^{-1} = \begin{bmatrix} 2 & 0 & 0 \\ 0 & 1 & 0 \\ 0 & 0 & 1 \end{bmatrix}$
	$\begin{bmatrix} 1 & 3 & 1 \\ -3 & -8 & 0 \\ 4 & 9 & 2 \end{bmatrix}$		
Step 2		$E_2 = \begin{bmatrix} 1 & 0 & 0 \\ 3 & 1 & 0 \\ 0 & 0 & 1 \end{bmatrix}$	$E_2^{-1} = \begin{bmatrix} 1 & 0 & 0 \\ -3 & 1 & 0 \\ 0 & 0 & 1 \end{bmatrix}$
	$\begin{bmatrix} 1 & 3 & 1 \\ 0 & 1 & 3 \\ 4 & 9 & 2 \end{bmatrix}$		
Step 3		$E_3 = \begin{bmatrix} 1 & 0 & 0 \\ 0 & 1 & 0 \\ -4 & 0 & 1 \end{bmatrix}$	$E_3^{-1} = \begin{bmatrix} 1 & 0 & 0 \\ 0 & 1 & 0 \\ 4 & 0 & 1 \end{bmatrix}$
	$\begin{bmatrix} 1 & 3 & 1 \\ 0 & 1 & 3 \\ 0 & -3 & -2 \end{bmatrix}$		
Step 4		$E_4 = \begin{bmatrix} 1 & 0 & 0 \\ 0 & 1 & 0 \\ 0 & 3 & 1 \end{bmatrix}$	$E_4^{-1} = \begin{bmatrix} 1 & 0 & 0 \\ 0 & 1 & 0 \\ 0 & -3 & 1 \end{bmatrix}$
	$\begin{bmatrix} 1 & 3 & 1 \\ 0 & 1 & 3 \\ 0 & 0 & 7 \end{bmatrix}$		
Step 5		$E_5 = \begin{bmatrix} 1 & 0 & 0 \\ 0 & 1 & 0 \\ 0 & 0 & \frac{1}{7} \end{bmatrix}$	$E_5^{-1} = \begin{bmatrix} 1 & 0 & 0 \\ 0 & 1 & 0 \\ 0 & 0 & 7 \end{bmatrix}$
	$\begin{bmatrix} 1 & 3 & 1 \\ 0 & 1 & 3 \\ 0 & 0 & 1 \end{bmatrix}$		

Thus,

$$U = \begin{bmatrix} 1 & 3 & 1 \\ 0 & 1 & 3 \\ 0 & 0 & 1 \end{bmatrix}$$

and, from (8),

$$L = \begin{bmatrix} 2 & 0 & 0 \\ 0 & 1 & 0 \\ 0 & 0 & 1 \end{bmatrix} \begin{bmatrix} 1 & 0 & 0 \\ -3 & 1 & 0 \\ 0 & 0 & 1 \end{bmatrix} \begin{bmatrix} 1 & 0 & 0 \\ 0 & 1 & 0 \\ 4 & 0 & 1 \end{bmatrix} \begin{bmatrix} 1 & 0 & 0 \\ 0 & 1 & 0 \\ 0 & -3 & 1 \end{bmatrix} \begin{bmatrix} 1 & 0 & 0 \\ 0 & 1 & 0 \\ 0 & 0 & 7 \end{bmatrix}$$

$$= \begin{bmatrix} 2 & 0 & 0 \\ -3 & 1 & 0 \\ 4 & -3 & 7 \end{bmatrix}$$

so

$$\begin{bmatrix} 2 & 6 & 2 \\ -3 & -8 & 0 \\ 4 & 9 & 2 \end{bmatrix} = \begin{bmatrix} 2 & 0 & 0 \\ -3 & 1 & 0 \\ 4 & -3 & 7 \end{bmatrix} \begin{bmatrix} 1 & 3 & 1 \\ 0 & 1 & 3 \\ 0 & 0 & 1 \end{bmatrix}$$

is an *LU*-decomposition of *A*. ▲

PROCEDURE FOR FINDING *LU*-DECOMPOSITIONS

As this example shows, most of the work in constructing an *LU*-decomposition is expended in the calculation of *L*. However, *all* this work can be eliminated by some careful bookkeeping of the operations used to reduce *A* to *U*. Because we are assuming that no row interchanges are required to reduce *A* to *U*, there are only two types of operations involved: multiplying a row by a nonzero constant, and adding a multiple of one row to another. The first operation is used to introduce the leading 1's and the second to introduce zeros below the leading 1's.

In Example 2, the multipliers needed to introduce the leading 1's in successive rows were as follows:

$\frac{1}{2}$ for the first row

1 for the second row

$\frac{1}{7}$ for the third row

Note that the successive diagonal entries in *L* were precisely the reciprocals of these multipliers (Figure 1).

$$L = \begin{bmatrix} ② & 0 & 0 \\ -3 & ① & 0 \\ 4 & -3 & ⑦ \end{bmatrix}$$

Figure 1

Next, observe that to introduce zeros below the leading 1 in the first row we used the following operations:

add 3 times the first row to the second

add -4 times the first row to the third

and to introduce the zero below the leading 1 in the second row, we used the operation

add 3 times the second row to the third

Now note that in each position below the main diagonal of L the entry is the *negative* of the multiplier in the operation that introduced the zero in that position in U (Figure 2).

$$L = \begin{bmatrix} 2 & 0 & 0 \\ -3 & 1 & 0 \\ 4 & -3 & 7 \end{bmatrix}$$

Figure 2

In summary, we have the following procedure for constructing an *LU*-decomposition of a square matrix A provided that A can be reduced to row-echelon form without row interchanges.

Step 1. Reduce A to a row-echelon form U without using row interchanges, keeping track of the multipliers used to introduce the leading 1's and the multipliers used to introduce the zeros below the leading 1's.

Step 2. In each position along the main diagonal of L, place the reciprocal of the multiplier that introduced the leading 1 in that position in U.

Step 3. In each position below the main diagonal of L, place the negative of the multiplier used to introduce the zero in that position in U.

Step 4. Form the decomposition $A = LU$.

Example 3 Find an *LU*-decomposition of

$$A = \begin{bmatrix} 6 & -2 & 0 \\ 9 & -1 & 1 \\ 3 & 7 & 5 \end{bmatrix}$$

Solution. We begin by reducing A to row-echelon form, keeping track of all multipliers.

$$\begin{bmatrix} 6 & -2 & 0 \\ 9 & -1 & 1 \\ 3 & 7 & 5 \end{bmatrix}$$

$$\begin{bmatrix} \textcircled{1} & -\frac{1}{3} & 0 \\ 9 & -1 & 1 \\ 3 & 7 & 5 \end{bmatrix} \longleftarrow \text{multiplier} = \frac{1}{6}$$

$$
\begin{bmatrix} 1 & -\frac{1}{3} & 0 \\ \textcircled{0} & 2 & 1 \\ \textcircled{0} & 8 & 5 \end{bmatrix}
\begin{array}{l} \\ \longleftarrow \text{multiplier} = -9 \\ \longleftarrow \text{multiplier} = -3 \end{array}
$$

$$
\begin{bmatrix} 1 & -\frac{1}{3} & 0 \\ 0 & \textcircled{1} & \frac{1}{2} \\ 0 & 8 & 5 \end{bmatrix}
\begin{array}{l} \\ \longleftarrow \text{multiplier} = \frac{1}{2} \\ \\ \end{array}
$$

$$
\begin{bmatrix} 1 & -\frac{1}{3} & 0 \\ 0 & 1 & \frac{1}{2} \\ 0 & \textcircled{0} & 1 \end{bmatrix}
\begin{array}{l} \\ \\ \longleftarrow \text{multiplier} = -8 \end{array}
$$

$$
\begin{bmatrix} 1 & -\frac{1}{3} & 0 \\ 0 & 1 & \frac{1}{2} \\ 0 & 0 & \textcircled{1} \end{bmatrix}
\begin{array}{l} \\ \\ \longleftarrow \text{multiplier} = 1 \end{array}
$$

> No actual operation is performed here, since there is already a leading 1 in the third row.

Constructing L from the multipliers yields the LU-decomposition.

$$
A = LU = \begin{bmatrix} 6 & 0 & 0 \\ 9 & 2 & 0 \\ 3 & 8 & 1 \end{bmatrix} \begin{bmatrix} 1 & -\frac{1}{3} & 0 \\ 0 & 1 & \frac{1}{2} \\ 0 & 0 & 1 \end{bmatrix} \quad \blacktriangle
$$

We conclude this section by briefly discussing two fundamental questions about LU-decompositions:

1. Does every square matrix have an LU-decomposition?
2. Can a square matrix have more than one LU-decomposition?

We already know that if a square matrix A can be reduced to row-echelon form without using row interchanges, then A has an LU-decomposition. In general, if row interchanges are required to reduce A to row-echelon form, then there is no LU-decomposition of A. However, in such cases it is possible to factor A in the form

$$
A = PLU
$$

where L is lower triangular, U is upper triangular, and P is a matrix obtained by interchanging the rows of I_n appropriately (see Exercise 17).

In the absence of additional restrictions, LU-decompositions are not unique. For example, if

$$
A = LU = \begin{bmatrix} l_{11} & 0 & 0 \\ l_{21} & l_{22} & 0 \\ l_{31} & l_{32} & l_{33} \end{bmatrix} \begin{bmatrix} 1 & u_{12} & u_{13} \\ 0 & 1 & u_{23} \\ 0 & 0 & 1 \end{bmatrix}
$$

and L has nonzero diagonal entries, then we can shift the diagonal entries from the left factor to the right factor by writing

$$A = \begin{bmatrix} 1 & 0 & 0 \\ \dfrac{l_{21}}{l_{11}} & 1 & 0 \\ \dfrac{l_{31}}{l_{11}} & \dfrac{l_{32}}{l_{22}} & 1 \end{bmatrix} \begin{bmatrix} l_{11} & 0 & 0 \\ 0 & l_{22} & 0 \\ 0 & 0 & l_{33} \end{bmatrix} \begin{bmatrix} 1 & u_{12} & u_{13} \\ 0 & 1 & u_{23} \\ 0 & 0 & 1 \end{bmatrix}$$

$$= \begin{bmatrix} 1 & 0 & 0 \\ \dfrac{l_{21}}{l_{11}} & 1 & 0 \\ \dfrac{l_{31}}{l_{11}} & \dfrac{l_{32}}{l_{22}} & 1 \end{bmatrix} \begin{bmatrix} l_{11} & l_{11}u_{12} & l_{11}u_{13} \\ 0 & l_{22} & l_{22}u_{23} \\ 0 & 0 & l_{33} \end{bmatrix}$$

which is another triangular decomposition of A.

EXERCISE SET 9.9

1. Use the method of Example 1 and the LU-decomposition

$$\begin{bmatrix} 3 & -6 \\ -2 & 5 \end{bmatrix} = \begin{bmatrix} 3 & 0 \\ -2 & 1 \end{bmatrix} \begin{bmatrix} 1 & -2 \\ 0 & 1 \end{bmatrix}$$

to solve the system

$$3x_1 - 6x_2 = 0$$
$$-2x_1 + 5x_2 = 1$$

2. Use the method of Example 1 and the LU-decomposition

$$\begin{bmatrix} 3 & -6 & -3 \\ 2 & 0 & 6 \\ -4 & 7 & 4 \end{bmatrix} = \begin{bmatrix} 3 & 0 & 0 \\ 2 & 4 & 0 \\ -4 & -1 & 2 \end{bmatrix} \begin{bmatrix} 1 & -2 & -1 \\ 0 & 1 & 2 \\ 0 & 0 & 1 \end{bmatrix}$$

to solve the system

$$3x_1 - 6x_2 - 3x_3 = -3$$
$$2x_1 \qquad + 6x_3 = -22$$
$$-4x_1 + 7x_2 + 4x_3 = 3$$

In Exercises 3–10 find an LU-decomposition of the coefficient matrix; then use the method of Example 1 to solve the system.

3. $\begin{bmatrix} 2 & 8 \\ -1 & -1 \end{bmatrix} \begin{bmatrix} x_1 \\ x_2 \end{bmatrix} = \begin{bmatrix} -2 \\ -2 \end{bmatrix}$

4. $\begin{bmatrix} -5 & -10 \\ 6 & 5 \end{bmatrix} \begin{bmatrix} x_1 \\ x_2 \end{bmatrix} = \begin{bmatrix} -10 \\ 19 \end{bmatrix}$

5. $\begin{bmatrix} 2 & -2 & -2 \\ 0 & -2 & 2 \\ -1 & 5 & 2 \end{bmatrix} \begin{bmatrix} x_1 \\ x_2 \\ x_3 \end{bmatrix} = \begin{bmatrix} -4 \\ -2 \\ 6 \end{bmatrix}$

6. $\begin{bmatrix} -3 & 12 & -6 \\ 1 & -2 & 2 \\ 0 & 1 & 1 \end{bmatrix} \begin{bmatrix} x_1 \\ x_2 \\ x_3 \end{bmatrix} = \begin{bmatrix} -33 \\ 7 \\ -1 \end{bmatrix}$

7. $\begin{bmatrix} 5 & 5 & 10 \\ -8 & -7 & -9 \\ 0 & 4 & 26 \end{bmatrix} \begin{bmatrix} x_1 \\ x_2 \\ x_3 \end{bmatrix} = \begin{bmatrix} 0 \\ 1 \\ 4 \end{bmatrix}$

8. $\begin{bmatrix} -1 & -3 & -4 \\ 3 & 10 & -10 \\ -2 & -4 & 11 \end{bmatrix} \begin{bmatrix} x_1 \\ x_2 \\ x_3 \end{bmatrix} = \begin{bmatrix} -6 \\ -3 \\ 9 \end{bmatrix}$

9. $\begin{bmatrix} -1 & 0 & 1 & 0 \\ 2 & 3 & -2 & 6 \\ 0 & -1 & 2 & 0 \\ 0 & 0 & 1 & 5 \end{bmatrix} \begin{bmatrix} x_1 \\ x_2 \\ x_3 \\ x_4 \end{bmatrix} = \begin{bmatrix} 5 \\ -1 \\ 3 \\ 7 \end{bmatrix}$

10. $\begin{bmatrix} 2 & -4 & 0 & 0 \\ 1 & 2 & 12 & 0 \\ 0 & -1 & -4 & -5 \\ 0 & 0 & 2 & 11 \end{bmatrix} \begin{bmatrix} x_1 \\ x_2 \\ x_3 \\ x_4 \end{bmatrix} = \begin{bmatrix} 8 \\ 0 \\ 1 \\ 0 \end{bmatrix}$

11. Let

$$A = \begin{bmatrix} 2 & 1 & -1 \\ -2 & -1 & 2 \\ 2 & 1 & 0 \end{bmatrix}$$

(a) Find an LU-decomposition of A.

(b) Express A in the form $A = L_1DU_1$, where L_1 is lower triangular with 1's along the main diagonal, U_1 is upper triangular, and D is a diagonal matrix.

(c) Express A in the form $A = L_2U_2$, where L_2 is lower triangular with 1's along the main diagonal and U_2 is upper triangular.

12. Show that the matrix

$$\begin{bmatrix} 0 & 1 \\ 1 & 0 \end{bmatrix}$$

has no LU-decomposition.

13. Let

$$A = \begin{bmatrix} a & b \\ c & d \end{bmatrix}$$

(a) Prove: If $A \neq 0$ then A has a unique LU-decomposition with 1's along the main diagonal of L.

(b) Find the LU-decomposition described in (a).

14. Let $A\mathbf{x} = \mathbf{b}$ be a linear system of n equations in n unknowns, and assume that A is an invertible matrix that can be reduced to row-echelon form without row interchanges. How many additions and multiplications are required to solve the system by the method of Example 1? [***Note.*** Count subtractions as additions and divisions as multiplications.]

15. (a) Prove: If L_1 and L_2 are $n \times n$ lower triangular matrices, then so is L_1L_2.

(b) The result in (a) is a special case of a general result that states that a product of finitely many lower triangular matrices is lower triangular. Use this fact to prove that the matrix L in (8) is lower triangular. [***Hint.*** See Exercise 27 of Section 2.4.]

16. Use the result in Exercise 15(b) to prove that a product of finitely many upper triangular matrices is upper triangular. [***Hint.*** Take transposes.]

17. Prove: If A is any $n \times n$ matrix, then A can be factored as $A = PLU$, where L is lower triangular, U is upper triangular, and P can be obtained by interchanging the rows of I_n appropriately. [**Hint.** Let U be a row-echelon form of A, and let all row interchanges required in the reduction of A to U be performed first.]

18. Factor

$$A = \begin{bmatrix} 3 & -1 & 0 \\ 3 & -1 & 1 \\ 0 & 2 & 1 \end{bmatrix}$$

as $A = PLU$, where P is obtained from I_3 by interchanging rows appropriately, L is lower triangular, and U is upper triangular.

CHAPTER **10**

COMPLEX VECTOR SPACES

10.1 COMPLEX NUMBERS

Up to now we have considered only vector spaces for which the scalars are real numbers. However, for many important applications of vectors it is desirable to allow the scalars to be complex numbers. A vector space that allows complex scalars is called a **complex vector space**, *and one that allows real scalars only is called a* **real vector space**. *One advantage of allowing complex scalars is that all matrices with scalar entries have eigenvalues, which is not true if only real scalars are allowed. For example, the matrix*

$$A = \begin{bmatrix} -2 & -1 \\ 5 & 2 \end{bmatrix}$$

has characteristic polynomial

$$\det(\lambda I - A) = \det \begin{bmatrix} \lambda + 2 & 1 \\ -5 & \lambda - 2 \end{bmatrix} = \lambda^2 + 1$$

so the characteristic equation, $\lambda^2 + 1 = 0$, has no real solutions and therefore no real eigenvalues.

In the first three sections of this chapter we will review some of the basic properties of complex numbers, and in subsequent sections we will discuss complex vector spaces.

COMPLEX
NUMBERS

Since $x^2 \geq 0$ for every real number x, the equation

$$x^2 = -1$$

has no real solutions. To deal with this problem, mathematicians of the eighteenth century introduced the "imaginary" number,

$$i = \sqrt{-1}$$

which they assumed had the property

$$i^2 = (\sqrt{-1})^2 = -1$$

but which otherwise could be treated as a real number. Expressions of the form

$$a + bi$$

where a and b are real numbers were called "complex numbers," and these were manipulated according to the standard rules of arithmetic with the added property that $i^2 = -1$.

By the beginning of the nineteenth century it was recognized that a complex number,

$$a + bi$$

could be regarded as an alternative symbol for the ordered pair

$$(a, b)$$

of real numbers, and that operations of addition, subtraction, multiplication, and division could be defined on these ordered pairs so that the familiar laws of arithmetic hold and $i^2 = -1$. This is the approach we will follow.

> **Definition.** A *complex number* is an ordered pair of real numbers, denoted either by (a, b) or $a + bi$.

Example 1 Some examples of complex numbers in both notations are as follows:

Ordered Pair	Equivalent Notation
$(3, 4)$	$3 + 4i$
$(-1, 2)$	$-1 + 2i$
$(0, 1)$	$0 + i$
$(2, 0)$	$2 + 0i$
$(4, -2)$	$4 + (-2)i$

For simplicity, the last three complex numbers would usually be abbreviated as

$$0 + i = i, \qquad 2 + 0i = 2, \qquad 4 + (-2)i = 4 - 2i \quad \triangle$$

Geometrically, a complex number can be viewed either as a point or a vector in the xy-plane (Figure 1).

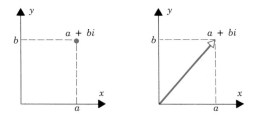

Figure 1 | A complex number can be viewed as a point or a vector. |

Example 2 Some complex numbers are shown as points in Figure 2*a* and as vectors in Figure 2*b*. ▲

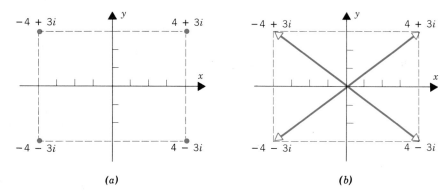

Figure 2   *(a)*   *(b)*

THE COMPLEX PLANE

Sometimes it is convenient to use a single letter, such as *z*, to denote a complex number. Thus, we might write

$$z = a + bi$$

The real number *a* is called the ***real part of z*** and the real number *b* the ***imaginary part of z***. These numbers are denoted by Re(z) and Im(z), respectively. Thus,

$$\text{Re}(4 - 3i) = 4 \qquad \text{and} \qquad \text{Im}(4 - 3i) = -3$$

When complex numbers are represented geometrically in an *xy*-coordinate system, the *x*-axis is called the ***real axis***, the *y*-axis the ***imaginary axis***, and the plane is called the ***complex plane*** (Figure 3).

Figure 3

OPERATIONS ON COMPLEX NUMBERS

Just as two vectors in R^2 are defined to be equal if they have the same components, so we define two complex numbers to be equal if their real parts are equal and their imaginary parts are equal:

> **Definition.** Two complex numbers, $a + bi$ and $c + di$, are defined to be *equal*, written
> $$a + bi = c + di$$
> if $a = c$ and $b = d$.

If $b = 0$, then the complex number $a + bi$ reduces to $a + 0i$, which we write simply as a. Thus, for any real number a,

$$a = a + 0i$$

so that the real numbers can be regarded as complex numbers with an imaginary part of zero. Geometrically, the real numbers correspond to points on the real axis. If we have $a = 0$, then $a + bi$ reduces to $0 + bi$, which we usually write as bi. These complex numbers, which correspond to points on the imaginary axis, are called *pure imaginary numbers*.

Just as vectors in R^2 are added by adding corresponding components, so complex numbers are added by adding their real parts and adding their imaginary parts:

$$(a + bi) + (c + di) = (a + c) + (b + d)i \tag{1}$$

The operations of subtraction and multiplication by a *real* number are also similar to the corresponding vector operations in R^2:

$$(a + bi) - (c + di) = (a - c) + (b - d)i \tag{2}$$

$$k(a + bi) = (ka) + (kb)i, \qquad k \text{ real} \tag{3}$$

Because the operations of addition, subtraction, and multiplication of a complex number by a real number parallel the corresponding operations for vectors in R^2, the familiar geometric interpretations of these operations hold for complex numbers (Figure 4).

It follows from (3) that $(-1)z + z = 0$ (verify), so we denote $(-1)z$ as $-z$ and call it the *negative of z*.

Example 3   If $z_1 = 4 - 5i$ and $z_2 = -1 + 6i$, find $z_1 + z_2$, $z_1 - z_2$, $3z_1$, and $-z_2$.

Solution.
$$z_1 + z_2 = (4 - 5i) + (-1 + 6i) = (4 - 1) + (-5 + 6)i = 3 + i$$
$$z_1 - z_2 = (4 - 5i) - (-1 + 6i) = (4 + 1) + (-5 - 6)i = 5 - 11i$$
$$3z_1 = 3(4 - 5i) = 12 - 15i$$
$$-z_2 = (-1)z_2 = (-1)(-1 + 6i) = 1 - 6i \quad \blacktriangle$$

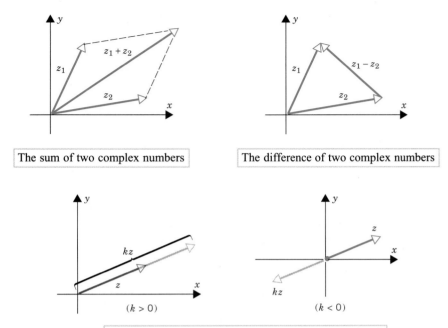

Figure 4

The product of a complex number z and a real number k

So far, there has been a parallel between complex numbers and vectors in R^2. However, we now define multiplication of complex numbers, an operation with no vector analog in R^2. To motivate the definition, we expand the product

$$(a + bi)(c + di)$$

following the usual rules of algebra, but treating i^2 as -1. This yields

$$(a + bi)(c + di) = ac + bdi^2 + adi + bci$$
$$= (ac - bd) + (ad + bc)i$$

which suggests the following *definition*:

$$(a + bi)(c + di) = (ac - bd) + (ad + bc)i \tag{4}$$

Example 4

$$(3 + 2i)(4 + 5i) = (3 \cdot 4 - 2 \cdot 5) + (3 \cdot 5 + 2 \cdot 4)i$$
$$= 2 + 23i$$
$$(4 - i)(2 - 3i) = [4 \cdot 2 - (-1)(-3)] + [(4)(-3) + (-1)(2)]i$$
$$= 5 - 14i$$
$$i^2 = (0 + i)(0 + i) = (0 \cdot 0 - 1 \cdot 1) + (0 \cdot 1 + 1 \cdot 0)i = -1 \quad \blacktriangle$$

We leave it as an exercise to verify the following rules of complex arithmetic:

$$z_1 + z_2 = z_2 + z_1$$
$$z_1 z_2 = z_2 z_1$$
$$z_1 + (z_2 + z_3) = (z_1 + z_2) + z_3$$
$$z_1(z_2 z_3) = (z_1 z_2)z_3$$
$$z_1(z_2 + z_3) = z_1 z_2 + z_1 z_3$$
$$0 + z = z$$
$$z + (-z) = 0$$
$$1 \cdot z = z$$

These rules make it possible to multiply complex numbers without using Formula (4) directly. Following the procedure used to motivate this formula, we can simply multiply each term of $a + bi$ by each term of $c + di$, set $i^2 = -1$, and simplify.

Example 5

$$(3 + 2i)(4 + i) = 12 + 3i + 8i + 2i^2 = 12 + 11i - 2 = 10 + 11i$$
$$(5 - \tfrac{1}{2}i)(2 + 3i) = 10 + 15i - i - \tfrac{3}{2}i^2 = 10 + 14i + \tfrac{3}{2} = \tfrac{23}{2} + 14i$$
$$i(1 + i)(1 - 2i) = i(1 - 2i + i - 2i^2) = i(3 - i) = 3i - i^2 = 1 + 3i \quad \blacktriangle$$

REMARK. Unlike the real numbers, there is no size ordering for the complex numbers. Thus, the order symbols $<$, $\leq$, $>$, and $\geq$ are not used with complex numbers.

Now that we have defined addition, subtraction, and multiplication of complex numbers, it is possible to add, subtract, and multiply matrices with complex entries and multiply a matrix by a complex number. Without going into detail, we note that the matrix operations and terminology discussed in Chapter 1 carry over without change to matrices with complex entries.

Example 6 If

$$A = \begin{bmatrix} 1 & -i \\ 1 + i & 4 - i \end{bmatrix} \quad \text{and} \quad B = \begin{bmatrix} i & 1 - i \\ 2 - 3i & 4 \end{bmatrix}$$

then

$$A + B = \begin{bmatrix} 1 + i & 1 - 2i \\ 3 - 2i & 8 - i \end{bmatrix}, \quad A - B = \begin{bmatrix} 1 - i & -1 \\ -1 + 4i & -i \end{bmatrix}$$

$$iA = \begin{bmatrix} i & -i^2 \\ i + i^2 & 4i - i^2 \end{bmatrix} = \begin{bmatrix} i & 1 \\ -1 + i & 1 + 4i \end{bmatrix}$$

$$AB = \begin{bmatrix} 1 & -i \\ 1 + i & 4 - i \end{bmatrix}\begin{bmatrix} i & 1 - i \\ 2 - 3i & 4 \end{bmatrix}$$

$$= \begin{bmatrix} 1 \cdot i + (-i) \cdot (2 - 3i) & 1 \cdot (1 - i) + (-i) \cdot 4 \\ (1 + i) \cdot i + (4 - i) \cdot (2 - 3i) & (1 + i) \cdot (1 - i) + (4 - i) \cdot 4 \end{bmatrix}$$

$$= \begin{bmatrix} -3 - i & 1 - 5i \\ 4 - 13i & 18 - 4i \end{bmatrix} \quad \blacktriangle$$

EXERCISE SET 10.1

1. In each part plot the point and sketch the vector that corresponds to the given complex number.
   (a) $2 + 3i$ (b) -4 (c) $-3 - 2i$ (d) $-5i$

2. Express each complex number in Exercise 1 as an ordered pair of real numbers.

3. In each part use the given information to find the real numbers x and y.
   (a) $x - iy = -2 + 3i$ (b) $(x + y) + (x - y)i = 3 + i$

4. Given that $z_1 = 1 - 2i$ and $z_2 = 4 + 5i$, find
   (a) $z_1 + z_2$ (b) $z_1 - z_2$ (c) $4z_1$ (d) $-z_2$ (e) $3z_1 + 4z_2$ (f) $\frac{1}{2}z_1 - \frac{3}{2}z_2$

5. In each part solve for z.
   (a) $z + (1 - i) = 3 + 2i$ (b) $-5z = 5 + 10i$ (c) $(i - z) + (2z - 3i) = -2 + 7i$

6. In each part sketch the vectors z_1, z_2, $z_1 + z_2$, and $z_1 - z_2$.
   (a) $z_1 = 3 + i$, $z_2 = 1 + 4i$ (b) $z_1 = -2 + 2i$, $z_2 = 4 + 5i$

7. In each part sketch the vectors z and kz.
   (a) $z = 1 + i$, $k = 2$ (b) $z = -3 - 4i$, $k = -2$ (c) $z = 4 + 6i$, $k = \frac{1}{2}$

8. In each part find real numbers k_1 and k_2 that satisfy the equation.
   (a) $k_1 i + k_2(1 + i) = 3 - 2i$ (b) $k_1(2 + 3i) + k_2(1 - 4i) = 7 + 5i$

9. In each part find $z_1 z_2$, z_1^2, and z_2^2.
   (a) $z_1 = 3i$, $z_2 = 1 - i$ (b) $z_1 = 4 + 6i$, $z_2 = 2 - 3i$ (c) $z_1 = \frac{1}{3}(2 + 4i)$, $z_2 = \frac{1}{2}(1 - 5i)$

10. Given that $z_1 = 2 - 5i$ and $z_2 = -1 - i$, find
   (a) $z_1 - z_1 z_2$ (b) $(z_1 + 3z_2)^2$ (c) $[z_1 + (1 + z_2)]^2$ (d) $iz_2 - z_1^2$

In Exercises 11–18 perform the calculations and express the result in the form $a + bi$.

11. $(1 + 2i)(4 - 6i)^2$

12. $(2 - i)(3 + i)(4 - 2i)$

13. $(1 - 3i)^3$

14. $i(1 + 7i) - 3i(4 + 2i)$

15. $[(2 + i)(\frac{1}{2} + \frac{3}{4}i)]^2$

16. $(\sqrt{2} + i) - i\sqrt{2}(1 + \sqrt{2}i)$

17. $(1 + i + i^2 + i^3)^{100}$

18. $(3 - 2i)^2 - (3 + 2i)^2$

19. Let
$$A = \begin{bmatrix} 1 & i \\ -i & 3 \end{bmatrix}, \qquad B = \begin{bmatrix} 2 & 2 + i \\ 3 - i & 4 \end{bmatrix}$$

Find
   (a) $A + 3iB$ (b) BA (c) AB (d) $B^2 - A^2$

20. Let
$$A = \begin{bmatrix} 3 + 2i & 0 \\ -i & 2 \\ 1 + i & 1 - i \end{bmatrix}, \qquad B = \begin{bmatrix} -i & 2 \\ 0 & i \end{bmatrix}, \qquad C = \begin{bmatrix} -1 - i & 0 & -i \\ 3 & 2i & -5 \end{bmatrix}$$

Find
   (a) $A(BC)$ (b) $(BC)A$ (c) $(CA)B^2$ (d) $(1 + i)(AB) + (3 - 4i)A$

21. Show that
   (a) $\text{Im}(iz) = \text{Re}(z)$ (b) $\text{Re}(iz) = -\text{Im}(z)$

22. In each part solve the equation by the quadratic formula and check your results by substituting the solutions into the given equation.

(a) $z^2 + 2z + 2 = 0$ (b) $z^2 - z + 1 = 0$

23. (a) Show that if n is a positive integer, then the only possible values for i^n are $1, -1, i$, and $-i$.

(b) Find i^{2509}. [***Hint.*** The value of i^n can be determined from the remainder when n is divided by 4.]

24. Prove: If $z_1 z_2 = 0$, then $z_1 = 0$ or $z_2 = 0$.

25. Use the result of Exercise 24 to prove: If $zz_1 = zz_2$ and $z \neq 0$, then $z_1 = z_2$.

26. Prove that for all complex numbers z_1, z_2, and z_3

(a) $z_1 + z_2 = z_2 + z_1$ (b) $z_1 + (z_2 + z_3) = (z_1 + z_2) + z_3$

27. Prove that for all complex numbers z_1, z_2, and z_3

(a) $z_1 z_2 = z_2 z_1$ (b) $z_1(z_2 z_3) = (z_1 z_2)z_3$

28. Prove that $z_1(z_2 + z_3) = z_1 z_2 + z_1 z_3$ for all complex numbers z_1, z_2, and z_3.

29. In quantum mechanics the ***Dirac* matrices*** are

$$\beta = \begin{bmatrix} 1 & 0 & 0 & 0 \\ 0 & 1 & 0 & 0 \\ 0 & 0 & -1 & 0 \\ 0 & 0 & 0 & -1 \end{bmatrix}, \quad \alpha_x = \begin{bmatrix} 0 & 0 & 0 & 1 \\ 0 & 0 & 1 & 0 \\ 0 & 1 & 0 & 0 \\ 1 & 0 & 0 & 0 \end{bmatrix},$$

$$\alpha_y = \begin{bmatrix} 0 & 0 & 0 & -i \\ 0 & 0 & i & 0 \\ 0 & -i & 0 & 0 \\ i & 0 & 0 & 0 \end{bmatrix}, \quad \alpha_z = \begin{bmatrix} 0 & 0 & 1 & 0 \\ 0 & 0 & 0 & -1 \\ 1 & 0 & 0 & 0 \\ 0 & -1 & 0 & 0 \end{bmatrix}$$

(a) Prove that $\beta^2 = \alpha_x^2 = \alpha_y^2 = \alpha_z^2 = I$.

(b) Two matrices A and B are called ***anticommutative*** if $AB = -BA$. Prove that any two Dirac matrices are anticommutative.

10.2 MODULUS; COMPLEX CONJUGATE; DIVISION

Our main objective in this section is to define the division of complex numbers.

COMPLEX CONJUGATES

We will begin with some preliminary ideas.

If $z = a + bi$ is any complex number, then the ***conjugate of z***, denoted by $\bar{z}$ (read "z bar"), is defined by

$$\bar{z} = a - bi$$

---

* *Paul Adrien Maurice Dirac* (1902–1984) was a British theoretical physicist who devised a new form of quantum mechanics and a theory that predicted electron spin and the existence of a fundamental atomic particle called a positron. In 1933 he received the Nobel prize for physics and in 1939, the medal of the Royal Society.

In words, $\bar{z}$ is obtained by reversing the sign of the imaginary part of z. Geometrically, $\bar{z}$ is the reflection of z about the real axis (Figure 1).

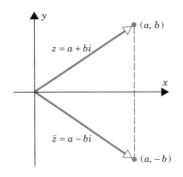

Figure 1

The conjugate of a complex number

Example 1

$$z = 3 + 2i \qquad \bar{z} = 3 - 2i$$
$$z = -4 - 2i \qquad \bar{z} = -4 + 2i$$
$$z = i \qquad \bar{z} = -i$$
$$z = 4 \qquad z = 4$$

REMARK. The last line in Example 1 illustrates the fact that a real number is the same as its conjugate. More precisely, it can be shown (Exercise 22) that $z = \bar{z}$ if and only if z is a real number.

If a complex number z is viewed as a vector in R^2, then the norm or length of the vector is called the modulus (or *absolute value*) of z. More precisely:

MODULUS

Definition. The **modulus** of a complex number $z = a + bi$, denoted by $|z|$, is defined by

$$|z| = \sqrt{a^2 + b^2} \tag{1}$$

If $b = 0$, then $z = a$ is a real number, and
$$|z| = \sqrt{a^2 + 0^2} = \sqrt{a^2} = |a|$$

so the modulus of a real number is simply its absolute value. Thus, the modulus of z is also called the **absolute value** of z.

Example 2 Find $|z|$ if $z = 3 - 4i$.

Solution. From (1) with $a = 3$ and $b = -4$, $|z| = \sqrt{(3)^2 + (-4)^2} = \sqrt{25} = 5$.

The following theorem establishes a basic relationship between $\bar{z}$ and $|z|$.

Theorem 10.2.1. *For any complex number z,*

$$z\bar{z} = |z|^2$$

Proof. If $z = a + bi$, then

$$z\bar{z} = (a + bi)(a - bi) = a^2 - abi + bai - b^2 i^2$$
$$= a^2 + b^2 = |z|^2 \quad \blacksquare$$

DIVISION OF COMPLEX NUMBERS

We now turn to the division of complex numbers. Our objective is to define division as the inverse of multiplication. Thus, if $z_2 \neq 0$, then our definition of $z = z_1/z_2$ should be such that

$$z_1 = z_2 z \tag{2}$$

Our procedure will be to prove that (2) has a unique solution for z if $z_2 \neq 0$, and then define z_1/z_2 to be this value of z. As with real numbers, division by zero is not allowed.

Theorem 10.2.2. *If $z_2 \neq 0$, then Equation (2) has a unique solution, which is*

$$z = \frac{1}{|z_2|^2} z_1 \bar{z}_2 \tag{3}$$

Proof. Let $z = x + iy$, $z_1 = x_1 + iy_1$, and $z_2 = x_2 + iy_2$. Then (2) can be written as

$$x_1 + iy_1 = (x_2 + iy_2)(x + iy)$$

or

$$x_1 + iy_1 = (x_2 x - y_2 y) + i(y_2 x + x_2 y)$$

or, on equating real and imaginary parts,

$$x_2 x - y_2 y = x_1$$
$$y_2 x + x_2 y = y_1$$

or

$$\begin{bmatrix} x_2 & -y_2 \\ y_2 & x_2 \end{bmatrix} \begin{bmatrix} x \\ y \end{bmatrix} = \begin{bmatrix} x_1 \\ y_1 \end{bmatrix} \tag{4}$$

Since $z_2 = x_2 + iy_2 \neq 0$, it follows that x_2 and y_2 are not both zero, so

$$\begin{vmatrix} x_2 & -y_2 \\ y_2 & x_2 \end{vmatrix} = x_2^2 + y_2^2 \neq 0$$

Thus, by Cramer's Rule (Theorem 2.4.3), system (4) has the unique solution

$$x = \frac{\begin{vmatrix} x_1 & -y_2 \\ y_1 & x_2 \end{vmatrix}}{\begin{vmatrix} x_2 & -y_2 \\ y_2 & x_2 \end{vmatrix}} = \frac{x_1 x_2 + y_1 y_2}{x_2^2 + y_2^2} = \frac{x_1 x_2 + y_1 y_2}{|z_2|^2}$$

$$y = \frac{\begin{vmatrix} x_2 & x_1 \\ y_2 & y_1 \end{vmatrix}}{\begin{vmatrix} x_2 & -y_2 \\ y_2 & x_2 \end{vmatrix}} = \frac{y_1 x_2 - x_1 y_2}{x_2^2 + y_2^2} = \frac{y_1 x_2 - x_1 y_2}{|z_2|^2}$$

Thus,

$$z = x + iy = \frac{1}{|z_2|^2} \left[(x_1 x_2 + y_1 y_2) + i(y_1 x_2 - x_1 y_2) \right]$$

$$= \frac{1}{|z_2|^2} (x_1 + iy_1)(x_2 - iy_2)$$

$$= \frac{1}{|z_2|^2} z_1 \bar{z}_2 \quad \blacksquare$$

Thus, for $z_2 \neq 0$ we define

$$\frac{z_1}{z_2} = \frac{1}{|z_2|^2} z_1 \bar{z}_2 \tag{5}$$

REMARK. To remember this formula, multiply numerator and denominator of z_1/z_2 by $\bar{z}_2$:

$$\frac{z_1}{z_2} = \frac{z_1 \bar{z}_2}{z_2 \bar{z}_2} = \frac{z_1 \bar{z}_2}{|z_2|^2} = \frac{1}{|z_2|^2} z_1 \bar{z}_2$$

Example 3   Express

$$\frac{3 + 4i}{1 - 2i}$$

in the form $a + bi$.

Solution. From (5) with $z_1 = 3 + 4i$ and $z_2 = 1 - 2i$,

$$\frac{3 + 4i}{1 - 2i} = \frac{1}{|1 - 2i|^2} (3 + 4i)\overline{(1 - 2i)} = \frac{1}{5}(3 + 4i)(1 + 2i)$$

$$= \frac{1}{5}(-5 + 10i) = -1 + 2i$$

Alternative Solution. As in the remark above, multiply numerator and denominator by the conjugate of the denominator:

$$\frac{3+4i}{1-2i} = \frac{3+4i}{1-2i} \cdot \frac{1+2i}{1+2i} = \frac{-5+10i}{5} = -1+2i \quad \blacktriangle$$

Systems of linear equations with complex coefficients arise in various applications. Without going into detail we note that all the results about linear systems studied in Chapters 1 and 2 carry over without change to systems with complex coefficients.

Example 4 Use Cramer's Rule to solve

$$ix + 2y = 1 - 2i$$
$$4x - iy = -1 + 3i$$

Solution.

$$x = \frac{\begin{vmatrix} 1-2i & 2 \\ -1+3i & -i \end{vmatrix}}{\begin{vmatrix} i & 2 \\ 4 & -i \end{vmatrix}} = \frac{(-i)(1-2i) - 2(-1+3i)}{i(-i) - 2(4)} = \frac{-7i}{-7} = i$$

$$y = \frac{\begin{vmatrix} i & 1-2i \\ 4 & -1+3i \end{vmatrix}}{\begin{vmatrix} i & 2 \\ 4 & -i \end{vmatrix}} = \frac{(i)(-1+3i) - 4(1-2i)}{i(-i) - 2(4)} = \frac{-7+7i}{-7} = 1-i$$

Thus, the solution is $x = i$, $y = 1 - i$. $\quad \blacktriangle$

PROPERTIES OF THE COMPLEX CONJUGATE

We conclude this section by listing some properties of the complex conjugate that will be useful in later sections.

Theorem 10.2.3. *For any complex numbers* z, z_1, *and* z_2

(a) $\overline{z_1 + z_2} = \bar{z}_1 + \bar{z}_2$

(b) $\overline{z_1 - z_2} = \bar{z}_1 - \bar{z}_2$

(c) $\overline{z_1 z_2} = \bar{z}_1 \bar{z}_2$

(d) $\overline{(z_1/z_2)} = \bar{z}_1/\bar{z}_2$

(e) $\bar{\bar{z}} = z$

We prove (a) and leave the rest as exercises.

Proof (a). Let $z_1 = a_1 + b_1 i$ and $z_2 = a_2 + b_2 i$; then

$$\overline{z_1 + z_2} = \overline{(a_1 + a_2) + (b_1 + b_2)i}$$
$$= (a_1 + a_2) - (b_1 + b_2)i$$
$$= (a_1 - b_1 i) + (a_2 - b_2 i)$$
$$= \bar{z}_1 + \bar{z}_2 \quad \blacksquare$$

REMARK. It is possible to extend part (a) of Theorem 10.2.3 to n terms and part (c) to n factors. More precisely,

$$\overline{z_1 + z_2 + \cdots + z_n} = \bar{z}_1 + \bar{z}_2 + \cdots + \bar{z}_n$$
$$\overline{z_1 z_2 \cdots z_n} = \bar{z}_1 \bar{z}_2 \cdots \bar{z}_n$$

EXERCISE SET 10.2

1. In each part find $\bar{z}$.
   (a) $z = 2 + 7i$ (b) $z = -3 - 5i$ (c) $z = 5i$ (d) $z = -i$ (e) $z = -9$ (f) $z = 0$

2. In each part find $|z|$.
   (a) $z = i$ (b) $z = -7i$ (c) $z = -3 - 4i$ (d) $z = 1 + i$ (e) $z = -8$ (f) $z = 0$

3. Verify that $z\bar{z} = |z|^2$ for
   (a) $z = 2 - 4i$ (b) $z = -3 + 5i$ (c) $z = \sqrt{2} - \sqrt{2}i$

4. Given that $z_1 = 1 - 5i$ and $z_2 = 3 + 4i$, find
   (a) z_1/z_2 (b) $\bar{z}_1/z_2$ (c) $z_1/\bar{z}_2$ (d) $\overline{(z_1/z_2)}$ (e) $z_1/|z_2|$ (f) $|z_1/z_2|$

5. In each part find $1/z$.

   (a) $z = i$ (b) $z = 1 - 5i$ (c) $z = \dfrac{-i}{7}$

6. Given that $z_1 = 1 + i$ and $z_2 = 1 - 2i$, find

   (a) $z_1 - \left(\dfrac{z_1}{z_2}\right)$ (b) $\dfrac{z_1 - 1}{z_2}$ (c) $z_1^2 - \left(\dfrac{iz_1}{z_2}\right)$ (d) $\dfrac{z_1}{iz_2}$

In Exercises 7–14 perform the calculations and express the result in the form $a + bi$.

7. $\dfrac{i}{1 + i}$

8. $\dfrac{2}{(1 - i)(3 + i)}$

9. $\dfrac{1}{(3 + 4i)^2}$

10. $\dfrac{2 + i}{i(-3 + 4i)}$

11. $\dfrac{\sqrt{3} + i}{(1 - i)(\sqrt{3} - i)}$

12. $\dfrac{1}{i(3 - 2i)(1 + i)}$

13. $\dfrac{i}{(1 - i)(1 - 2i)(1 + 2i)}$

14. $\dfrac{1 - 2i}{3 + 4i} - \dfrac{2 + i}{5i}$

15. In each part solve for z.
    (a) $iz = 2 - i$ (b) $(4 - 3i)\bar{z} = i$

16. Use Theorem 10.2.3 to prove the following identities:

(a) $\overline{\overline{z} + 5i} = z - 5i$ (b) $\overline{iz} = -i\overline{z}$ (c) $\dfrac{\overline{i + z}}{i - z} = -1$

17. In each part sketch the set of points in the complex plane that satisfies the equation.

(a) $|z| = 2$ (b) $|z - (1 + i)| = 1$ (c) $|z - i| = |z + i|$ (d) $\text{Im}(\overline{z} + i) = 3$

18. In each part sketch the set of points in the complex plane that satisfies the given condition(s).

(a) $|z + i| \leq 1$ (b) $1 < |z| < 2$ (c) $|2z - 4i| < 1$ (d) $|z| \leq |z + i|$

19. Given that $z = x + iy$, find

(a) $\text{Re}(i\overline{z})$ (b) $\text{Im}(i\overline{z})$ (c) $\text{Re}(i\overline{z})$ (d) $\text{Im}(i\overline{z})$

20. (a) Show that if n is a positive integer, then the only possible values for $(1/i)^n$ are 1, -1, i, and $-i$.

(b) Find $(1/i)^{2509}$. [**Hint.** See Exercise 23(b) of Section 10.1.]

21. Prove:

(a) $\dfrac{1}{2}(z + \overline{z}) = \text{Re}(z)$ (b) $\dfrac{1}{2i}(z - \overline{z}) = \text{Im}(z)$

22. Prove: $z = \overline{z}$ if and only if z is a real number.

23. Given that $z_1 = x_1 + iy_1$ and $z_2 = x_2 + iy_2$, find

(a) $\text{Re}\left(\dfrac{z_1}{z_2}\right)$ (b) $\text{Im}\left(\dfrac{z_1}{z_2}\right)$

24. Prove: If $(\overline{z})^2 = z^2$, then z is either real or pure imaginary.

25. Prove that $|z| = |\overline{z}|$.

26. Prove:

(a) $\overline{z_1 - z_2} = \overline{z}_1 - \overline{z}_2$ (b) $\overline{z_1 z_2} = \overline{z}_1 \overline{z}_2$ (c) $\overline{(z_1/z_2)} = \overline{z}_1/\overline{z}_2$ (d) $\overline{\overline{z}} = z$

27. (a) Prove that $\overline{z^2} = (\overline{z})^2$.

(b) Prove that if n is a positive integer, then $\overline{z^n} = (\overline{z})^n$.

(c) Is the result in (b) true if n is a negative integer? Explain.

In Exercises 28–31 solve the system of linear equations by Cramer's Rule.

28. $ix_1 - ix_2 = -2$
$2x_1 + x_2 = i$

29. $x_1 + x_2 = 2$
$x_1 - x_2 = 2i$

30. $x_1 + x_2 + x_3 = 3$
$x_1 + x_2 - x_3 = 2 + 2i$
$x_1 - x_2 + x_3 = -1$

31. $ix_1 + 3x_2 + (1 + i)x_3 = -i$
$x_1 + ix_2 + \qquad 3x_3 = -2i$
$x_1 + \quad x_2 + \qquad x_3 = 0$

In Exercises 32 and 33 solve the system of linear equations by Gauss–Jordan elimination.

32. $\begin{bmatrix} -1 & -1 - i \\ -1 + i & -2 \end{bmatrix}\begin{bmatrix} x_1 \\ x_2 \end{bmatrix} = \begin{bmatrix} 0 \\ 0 \end{bmatrix}$

33. $\begin{bmatrix} 2 & -1 - i \\ -1 + i & 1 \end{bmatrix}\begin{bmatrix} x_1 \\ x_2 \end{bmatrix} = \begin{bmatrix} 0 \\ 0 \end{bmatrix}$

34. Solve the following system of linear equations by Gauss–Jordan elimination.

$$x_1 + ix_2 - ix_3 = 0$$
$$-x_1 + (1 - i)x_2 + 2ix_3 = 0$$
$$2x_1 + (-1 + 2i)x_2 - 3ix_3 = 0$$

35. In each part use the formula in Theorem 1.4.5 to compute the inverse of the matrix, and check your result by showing that $AA^{-1} = A^{-1}A = I$.

(a) $A = \begin{bmatrix} i & -2 \\ 1 & i \end{bmatrix}$ (b) $A = \begin{bmatrix} 2 & i \\ 1 & 0 \end{bmatrix}$

36. Let $p(x) = a_0 + a_1 x + a_2 x^2 + \cdots + a_n x^n$ be a polynomial for which the coefficients $a_0, a_1, a_2, \ldots, a_n$ are real. Prove that if z is a solution of the equation $p(x) = 0$, then so is $\bar{z}$.

37. Prove: For any complex number z, $|\mathrm{Re}\,(z)| \le |z|$ and $|\mathrm{Im}\,(z)| \le |z|$.

38. Prove that

$$\frac{|\mathrm{Re}\,(z)| + |\mathrm{Im}\,(z)|}{\sqrt{2}} \le |z|$$

[**Hint.** Let $z = x + iy$ and use the fact that $(|x| - |y|)^2 \ge 0$.]

39. In each part use the method of Example 4 in Section 1.5 to find A^{-1}, and check your result by showing that $AA^{-1} = A^{-1}A = I$.

(a) $A = \begin{bmatrix} 1 & 1+i & 0 \\ 0 & 1 & i \\ -i & 1-2i & 2 \end{bmatrix}$ (b) $A = \begin{bmatrix} i & 0 & -i \\ 0 & 1 & -1-4i \\ 2-i & i & 3 \end{bmatrix}$

10.3 POLAR FORM; DEMOIVRE'S THEOREM

In this section we shall discuss a way to represent complex numbers using trigonometric properties. Our work will lead to a fundamental formula for powers of complex numbers and a method for finding nth roots of complex numbers.

POLAR FORM OF A COMPLEX NUMBER

If $z = x + iy$ is a nonzero complex number, $r = |z|$, and θ measures the angle from the positive real axis to the vector z, then, as suggested by Figure 1,

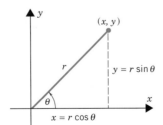

Figure 1

$$x = r \cos \theta$$
$$y = r \sin \theta \tag{1}$$

so that $z = x + iy$ can be written as

$$z = r \cos \theta + ir \sin \theta$$

or

$$z = r(\cos \theta + i \sin \theta) \tag{2}$$

This is called a **polar form of** z.

ARGUMENT OF
A COMPLEX
NUMBER

The angle θ is called an **argument of** z and is denoted by

$$\theta = \arg z$$

The argument of z is not uniquely determined because we can add or subtract any multiple of 2π from θ to produce another value of the argument. However, there is only one value of the argument in radians that satisfies

$$-\pi < \theta \le \pi$$

This is called the **principal argument of** z and is denoted by

$$\theta = \text{Arg } z$$

Example 1 Express the following complex numbers in polar form using their principal arguments:

(a) $z = 1 + \sqrt{3}i$ (b) $z = -1 - i$

Solution (a). The value of r is

$$r = |z| = \sqrt{1^2 + (\sqrt{3})^2} = \sqrt{4} = 2$$

and since $x = 1$ and $y = \sqrt{3}$, it follows from (1) that

$$1 = 2 \cos \theta$$
$$\sqrt{3} = 2 \sin \theta$$

so $\cos \theta = 1/2$ and $\sin \theta = \sqrt{3}/2$. The only value of θ that satisfies these relations and meets the requirement $-\pi < \theta \le \pi$ is $\theta = \pi/3 \ (= 60°)$ (see Figure 2a). Thus, a polar form of z is

$$z = 2\left(\cos \frac{\pi}{3} + i \sin \frac{\pi}{3} \right)$$

Solution (b). The value of r is

$$r = |z| = \sqrt{(-1)^2 + (-1)^2} = \sqrt{2}$$

Figure 2   (a)   (b)

and since $x = -1$, $y = -1$, it follows from (1) that

$$-1 = \sqrt{2}\cos\theta$$
$$-1 = \sqrt{2}\sin\theta$$

so $\cos\theta = -1/\sqrt{2}$ and $\sin\theta = -1/\sqrt{2}$. The only value of θ that satisfies these relations and meets the requirement $-\pi < \theta \le \pi$ is $\theta = -3\pi/4 \, (= -135°)$ (Figure 2b). Thus, a polar form of z is

$$z = \sqrt{2}\left(\cos\frac{-3\pi}{4} + i\sin\frac{-3\pi}{4}\right)$$

GEOMETRIC INTERPRETATION OF MULTIPLICATION AND DIVISION

We now show how polar forms can be used to give geometric interpretations of multiplication and division of complex numbers. Let

$$z_1 = r_1(\cos\theta_1 + i\sin\theta_1) \qquad \text{and} \qquad z_2 = r_2(\cos\theta_2 + i\sin\theta_2)$$

Multiplying, we obtain

$$z_1 z_2 = r_1 r_2[(\cos\theta_1\cos\theta_2 - \sin\theta_1\sin\theta_2) + i(\sin\theta_1\cos\theta_2 + \cos\theta_1\sin\theta_2)]$$

Recalling the trigonometric identities

$$\cos(\theta_1 + \theta_2) = \cos\theta_1\cos\theta_2 - \sin\theta_1\sin\theta_2$$
$$\sin(\theta_1 + \theta_2) = \sin\theta_1\cos\theta_2 + \cos\theta_1\sin\theta_2$$

We obtain

$$z_1 z_2 = r_1 r_2[\cos(\theta_1 + \theta_2) + i\sin(\theta_1 + \theta_2)] \tag{3}$$

which is a polar form of the complex number with modulus $r_1 r_2$ and argument $\theta_1 + \theta_2$. Thus, we have shown that

$$|z_1 z_2| = |z_1||z_2| \tag{4}$$

and

$$\arg(z_1 z_2) = \arg z_1 + \arg z_2$$

(Why?)

  In words, *the product of two complex numbers is obtained by multiplying their moduli and adding their arguments* (Figure 3).

Figure 3 The product of two complex numbers

We leave it as an exercise to show that if $z_2 \neq 0$, then

$$\frac{z_1}{z_2} = \frac{r_1}{r_2}[\cos(\theta_1 - \theta_2) + i\,\sin(\theta_1 - \theta_2)] \tag{5}$$

from which it follows that

$$\left|\frac{z_1}{z_2}\right| = \frac{|z_1|}{|z_2|} \qquad \text{if } z_2 \neq 0$$

and

$$\arg\left(\frac{z_1}{z_2}\right) = \arg z_1 - \arg z_2$$

In words, *the quotient of two complex numbers is obtained by dividing their moduli and subtracting their arguments* (*in the appropriate order*).

Example 2 Let

$$z_1 = 1 + \sqrt{3}i \qquad \text{and} \qquad z_2 = \sqrt{3} + i$$

Polar forms of these complex numbers are

$$z_1 = 2\left(\cos\frac{\pi}{3} + i\,\sin\frac{\pi}{3}\right)$$

$$z_2 = 2\left(\cos\frac{\pi}{6} + i\,\sin\frac{\pi}{6}\right)$$

(verify) so that from (3)

$$z_1 z_2 = 4\left[\cos\left(\frac{\pi}{3} + \frac{\pi}{6}\right) + i\,\sin\left(\frac{\pi}{3} + \frac{\pi}{6}\right)\right]$$

$$= 4\left[\cos\frac{\pi}{2} + i\,\sin\frac{\pi}{2}\right] = 4[0 + i] = 4i$$

and from (5)

$$\frac{z_1}{z_2} = 1 \cdot \left[\cos\left(\frac{\pi}{3} - \frac{\pi}{6}\right) + i \sin\left(\frac{\pi}{3} - \frac{\pi}{6}\right) \right]$$

$$= \cos\frac{\pi}{6} + i \sin\frac{\pi}{6} = \frac{\sqrt{3}}{2} + \frac{1}{2}i$$

As a check, we calculate $z_1 z_2$ and z_1/z_2 directly without using polar forms for z_1 and z_2:

$$z_1 z_2 = (1 + \sqrt{3}i)(\sqrt{3} + i) = (\sqrt{3} - \sqrt{3}) + (3 + 1)i = 4i$$

$$\frac{z_1}{z_2} = \frac{1 + \sqrt{3}i}{\sqrt{3} + i} \cdot \frac{\sqrt{3} - i}{\sqrt{3} - i} = \frac{(\sqrt{3} + \sqrt{3}) + (-i + 3i)}{4} = \frac{\sqrt{3}}{2} + \frac{1}{2}i$$

which agrees with our previous results. ◣

The complex number i has a modulus of 1 and an argument of $\pi/2$ ($= 90°$), so the product iz has the same modulus as z, but its argument is $90°$ greater than that of z. In short, *multiplying z by i rotates z counterclockwise by $90°$* (Figure 4).

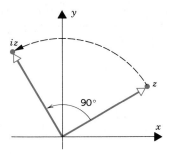

Figure 4 | Multiplying by i rotates z counterclockwise by $90°$.

DEMOIVRE'S FORMULA

If n is a positive integer and $z = r(\cos\theta + i\sin\theta)$, then from Formula (3),

$$z^n = \underbrace{z \cdot z \cdot z \cdots z}_{n\text{-factors}} = r^n[\cos(\underbrace{\theta + \theta + \cdots + \theta}_{n\text{-terms}}) + i\sin(\underbrace{\theta + \theta + \cdots + \theta}_{n\text{-terms}})]$$

or

$$z^n = r^n(\cos n\theta + i\sin n\theta) \qquad (6)$$

Moreover, if $z \neq 0$ we define $z^{-n} = 1/z^n$.

In the special case where $r = 1$, we have $z = \cos\theta + i\sin\theta$, so (6) becomes

$$(\cos\theta + i\sin\theta)^n = \cos n\theta + i\sin n\theta \qquad (7)$$

which is called **DeMoivre's* Formula**. Although we derived (7) assuming n to be a positive integer, it will be shown in the exercises that this formula is valid for all integers n.

FINDING nTH ROOTS

We now show how DeMoivre's Formula can be used to obtain roots of complex numbers. If n is a positive integer, and z is any complex number, then we define an **nth root of z** to be any complex number w that satisfies the equation

$$w^n = z \tag{8}$$

We denote an nth root of z by $z^{1/n}$. If $z \neq 0$, then we can derive formulas for the nth roots of z as follows. Let

$$w = \rho(\cos \alpha + i \sin \alpha) \qquad \text{and} \qquad z = r(\cos \theta + i \sin \theta)$$

If we assume that w satisfies (8), then it follows from (7) that

$$\rho^n(\cos n\alpha + i \sin n\alpha) = r(\cos \theta + i \sin \theta) \tag{9}$$

Comparing the moduli of the two sides, we see that $\rho^n = r$ or

$$\rho = \sqrt[n]{r}$$

where $\sqrt[n]{r}$ denotes the real positive nth root of r. Moreover, in order to have the equalities $\cos n\alpha = \cos \theta$ and $\sin n\alpha = \sin \theta$ in (9), the angles $n\alpha$ and θ must either be equal or differ by a multiple of 2π. That is,

$$n\alpha = \theta + 2k\pi, \qquad k = 0, \pm 1, \pm 2, \ldots$$

or

$$\alpha = \frac{\theta}{n} + \frac{2k\pi}{n}, \qquad k = 0, \pm 1, \pm 2, \ldots$$

Thus, the values of $w = \rho(\cos \alpha + i \sin \alpha)$ that satisfy (8) are given by

$$w = \sqrt[n]{r}\left[\cos\left(\frac{\theta}{n} + \frac{2k\pi}{n}\right) + i \sin\left(\frac{\theta}{n} + \frac{2k\pi}{n}\right)\right], \qquad k = 0, \pm 1, \pm 2, \ldots$$

Although there are infinitely many values of k, it can be shown (Exercise 16) that $k = 0, 1, 2, \ldots, n - 1$ produce distinct values of w satisfying (8), but all other choices of k yield duplicates of these. Therefore, there are exactly n different nth roots of $z = r(\cos \theta + i \sin \theta)$, and these are given by

$$z^{1/n} = \sqrt[n]{r}\left[\cos\left(\frac{\theta}{n} + \frac{2k\pi}{n}\right) + i \sin\left(\frac{\theta}{n} + \frac{2k\pi}{n}\right)\right], \qquad k = 0, 1, 2, \ldots, n - 1 \tag{10}$$

---

* *Abraham DeMoivre* (1667–1754) was a French mathematician who made important contributions to probability, statistics, and trigonometry. He developed the concept of statistically independent events, wrote a major influential treatise on probability, and helped transform trigonometry from a branch of geometry into a branch of analysis through his use of complex numbers. In spite of his important work, he barely managed to eke out a living as a tutor and a consultant on gambling and insurance.

Example 3 Find all cube roots of -8.

Solution. Since -8 lies on the negative real axis, we can use $\theta = \pi$ as an argument. Moreover, $r = |z| = |-8| = 8$, so a polar form of -8 is

$$-8 = 8(\cos \pi + i \sin \pi)$$

From (10) with $n = 3$ it follows that

$$(-8)^{1/3} = \sqrt[3]{8}\left[\cos\left(\frac{\pi}{3} + \frac{2k\pi}{3}\right) + i \sin\left(\frac{\pi}{3} + \frac{2k\pi}{3}\right)\right], \qquad k = 0, 1, 2$$

Thus, the cube roots of -8 are

$$2\left(\cos\frac{\pi}{3} + i \sin\frac{\pi}{3}\right) = 2\left(\frac{1}{2} + \frac{\sqrt{3}}{2}i\right) = 1 + \sqrt{3}i$$

$$2(\cos \pi + i \sin \pi) = 2(-1) = -2$$

$$2\cos\left(\frac{5\pi}{3} + i \sin\frac{5\pi}{3}\right) = 2\left(\frac{1}{2} - \frac{\sqrt{3}}{2}i\right) = 1 - \sqrt{3}i \quad \triangle$$

  As shown in Figure 5, the three cube roots of -8 obtained in Example 3 are equally spaced $\pi/3$ radians ($= 120°$) apart around the circle of radius 2 centered at the origin. This is not accidental. In general, it follows from Formula (10) that the nth roots of z lie on the circle of radius $\sqrt[n]{r}$ ($= \sqrt[n]{|z|}$), and are equally spaced $2\pi/n$ radians apart. (Can you see why?) Thus, once one nth root of z is found, the remaining $n - 1$ roots can be generated by rotating this root successively through increments of $2\pi/n$ radians.

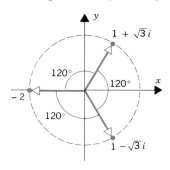

Figure 5   The cube roots of -8

Example 4 Find all fourth roots of 1.

Solution. We could apply Formula (10). Instead, we observe that $w = 1$ is one fourth root of 1, so that the remaining three roots can be generated by rotating this root through increments of $2\pi/4 = \pi/2$ radians ($= 90°$). From Figure 6 we see that the fourth roots of 1 are

$$1, \quad i, \quad -1, \quad -i \quad \triangle$$

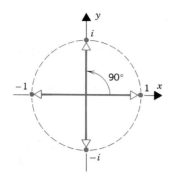

Figure 6   The fourth roots of 1

COMPLEX
EXPONENTS

We conclude this section with some comments on notation.

In more detailed studies of complex numbers, complex exponents are defined, and it is proved that

$$\cos\theta + i\sin\theta = e^{i\theta} \tag{11}$$

where e is an irrational real number given approximately by $e \approx 2.71828\ldots$. (For readers who have studied calculus, a proof of this result is given in Exercise 18.)

It follows from (11) that the polar form

$$z = r(\cos\theta + i\sin\theta)$$

can be written more briefly as

$$z = re^{i\theta} \tag{12}$$

Example 5   In Example 1 it was shown that

$$1 + \sqrt{3}i = 2\left(\cos\frac{\pi}{3} + i\sin\frac{\pi}{3}\right)$$

From (12) this can also be written as

$$1 + \sqrt{3}i = 2e^{i\pi/3} \quad \text{▲}$$

It can be proved that complex exponents follow the same laws as real exponents, so that if

$$z_1 = r_1 e^{i\theta_1} \qquad \text{and} \qquad z_2 = r_2 e^{i\theta_2}$$

are nonzero complex numbers, then

$$z_1 z_2 = r_1 r_2 e^{i\theta_1 + i\theta_2} = r_1 r_2 e^{i(\theta_1 + \theta_2)}$$

$$\frac{z_1}{z_2} = \frac{r_1}{r_2} e^{i\theta_1 - i\theta_2} = \frac{r_1}{r_2} e^{i(\theta_1 - \theta_2)}$$

But these are just Formulas (3) and (5) in a different notation.

We conclude this section with a useful formula for $\bar{z}$ in polar notation. If

$$z = re^{i\theta} = r(\cos\theta + i\sin\theta)$$

then

$$\bar{z} = r(\cos\theta - i\sin\theta) \tag{13}$$

Recalling the trigonometric identities

$$\sin(-\theta) = -\sin\theta \qquad \text{and} \qquad \cos(-\theta) = \cos\theta$$

we can rewrite (13) as

$$\bar{z} = r[\cos(-\theta) + i\sin(-\theta)] = re^{i(-\theta)}$$

or equivalently,

$$\bar{z} = re^{-i\theta} \tag{14}$$

In the special case where $r = 1$, the polar form of z is $z = e^{i\theta}$, and (14) yields the formula

$$\overline{e^{i\theta}} = e^{-i\theta} \tag{15}$$

EXERCISE SET 10.3

1. In each part find the principal argument of z.
   (a) $z = 1$ (b) $z = i$ (c) $z = -i$ (d) $z = 1 + i$ (e) $z = -1 + \sqrt{3}i$ (f) $z = 1 - i$

2. In each part find the value of $\theta = \arg(1 - \sqrt{3}i)$ that satisfies the given condition.

   (a) $0 < \theta \le 2\pi$ (b) $-\pi < \theta \le \pi$ (c) $-\dfrac{\pi}{6} \le \theta < \dfrac{11\pi}{6}$

3. In each part express the complex number in polar form using its principal argument.
   (a) $2i$ (b) -4 (c) $5 + 5i$ (d) $-6 + 6\sqrt{3}i$ (e) $-3 - 3i$ (f) $2\sqrt{3} - 2i$

4. Given that $z_1 = 2(\cos\pi/4 + i\sin\pi/4)$ and $z_2 = 3(\cos\pi/6 + i\sin\pi/6)$, find a polar form of

   (a) $z_1 z_2$ (b) $\dfrac{z_1}{z_2}$ (c) $\dfrac{z_2}{z_1}$ (d) $\dfrac{z_1^5}{z_2^2}$

5. Express $z_1 = i$, $z_2 = 1 - \sqrt{3}i$, and $z_3 = \sqrt{3} + i$ in polar form, and use your results to find $z_1 z_2 / z_3$. Check your results by performing the calculations without using polar forms.

6. Use Formula (6) to find

   (a) $(1 + i)^{12}$ (b) $\left(\dfrac{1}{\sqrt{2}} - \dfrac{1}{\sqrt{2}}i\right)^{-6}$ (c) $(\sqrt{3} + i)^7$ (d) $(1 - i\sqrt{3})^{-10}$

7. In each part find all the roots and sketch them as vectors in the complex plane.
   (a) $(-i)^{1/2}$ (b) $(1 + \sqrt{3}i)^{1/2}$ (c) $(-27)^{1/3}$ (d) $(i)^{1/3}$ (e) $(-1)^{1/4}$ (f) $(-8 + 8\sqrt{3}i)^{1/4}$

8. Use the method of Example 4 to find all cube roots of 1.

9. Use the method of Example 4 to find all sixth roots of 1.

10. Find all square roots of $1 + i$ and express your results in polar form.

11. In each part find all solutions of the equation.
   (a) $z^4 - 16 = 0$ (b) $z^{4/3} = -4$

12. Find four solutions of the equation $z^4 + 8 = 0$ and use your results to factor $z^4 + 8$ into two quadratic factors with real coefficients.

13. It was shown in the text that multiplying z by i rotates z counterclockwise by 90°. What is the geometric effect of dividing z by i?

14. In each part use (6) to calculate the given power.
   (a) $(1 + i)^8$ (b) $(-2\sqrt{3} + 2i)^{-9}$

15. In each part find $\text{Re}(z)$ and $\text{Im}(z)$.
   (a) $z = 3e^{i\pi}$ (b) $z = 3e^{-i\pi}$ (c) $\bar{z} = \sqrt{2}e^{\pi i/2}$ (d) $\bar{z} = -3e^{-2\pi i}$

16. (a) Show that the values of $z^{1/n}$ in Formula (10) are all different.
   (b) Show that integer values of k other than $k = 0, 1, 2, \ldots, n - 1$ produce values of $z^{1/n}$ that are duplicates of those in Formula (10).

17. Show that Formula (7) is valid if $n = 0$ or n is a negative integer.

18. (*For readers who have studied calculus.*) To prove Formula (11) recall the Maclaurin series for e^x:

$$e^x = 1 + x + \frac{x^2}{2!} + \cdots + \frac{x^n}{n!} + \cdots$$

   (a) By substituting $x = i\theta$ in this series and simplifying, obtain the formula

$$e^{i\theta} = \left(1 - \frac{\theta^2}{2!} + \frac{\theta^4}{4!} - \frac{\theta^6}{6!} + \cdots\right) + i\left(\theta - \frac{\theta^3}{3!} + \frac{\theta^5}{5!} - \frac{\theta^7}{7!} + \cdots\right)$$

   (b) Use the result in (a) to obtain (11).

19. Derive Formula (5).

10.4 COMPLEX VECTOR SPACES

In this section we shall develop the basic properties of complex vector spaces and discuss some of the ways in which they differ from real vector spaces. However, before going further the reader should review the definition of a complex vector space given at the beginning of Section 5.1.

BASIC PROPERTIES OF COMPLEX VECTOR SPACES

In complex vector spaces, linear combinations are defined exactly as in real vector spaces except the scalars are complex. More precisely, a vector $\mathbf{w}$ is called a *linear combination* of the vectors $\mathbf{v}_1, \mathbf{v}_2, \ldots, \mathbf{v}_r$ if $\mathbf{w}$ can be expressed in the form

$$\mathbf{w} = k_1\mathbf{v}_1 + k_2\mathbf{v}_2 + \cdots + k_r\mathbf{v}_r$$

where $k_1, k_2, \ldots, k_r$ are complex numbers.

The notions of *linear independence, spanning, basis, dimension*, and *subspace* carry over without change to complex vector spaces, and the theorems developed in Chapter 5 continue to hold.

Among the real vector spaces the most important one is R^n, the space of n-tuples of real numbers, with addition and scalar multiplication performed coordinatewise. Among the complex vector spaces the most important one is C^n, the space of n-tuples of complex numbers, with addition and scalar multiplication performed coordinatewise. A vector **u** in C^n can be written either in vector notation

$$\mathbf{u} = (u_1, u_2, \ldots, u_n)$$

or in matrix notation

$$\mathbf{u} = \begin{bmatrix} u_1 \\ u_2 \\ \vdots \\ u_n \end{bmatrix}$$

where

$$u_1 = a_1 + b_1 i, \quad u_2 = a_2 + b_2 i, \quad \ldots, \quad u_n = a_n + b_n i$$

Example 1 If

$$\mathbf{u} = (i, 1 + i, -2) \quad \text{and} \quad \mathbf{v} = (2 + i, 1 - i, 3 + 2i)$$

then

$$\mathbf{u} + \mathbf{v} = (i, 1 + i, -2) + (2 + i, 1 - i, 3 + 2i) = (2 + 2i, 2, 1 + 2i)$$

and

$$i\mathbf{u} = i(i, 1 + i, -2) = (i^2, i + i^2, -2i) = (-1, -1 + i, -2i) \; \blacktriangle$$

In C^n as in R^n, the vectors

$$\mathbf{e}_1 = (1, 0, 0, \ldots, 0), \quad \mathbf{e}_2 = (0, 1, 0, \ldots, 0), \quad \ldots, \quad \mathbf{e}_n = (0, 0, 0, \ldots, 1)$$

form a basis. It is called the *standard basis* for C^n. Since there are n vectors in this basis, C^n is an n-dimensional vector space.

REMARK. Do not confuse the complex number $i = \sqrt{-1}$ with the vector $\mathbf{i} = (1, 0, 0)$ from the standard basis for R^3 (see Example 3, Section 3.4). The complex number i will always be set in lightface type and the vector $\mathbf{i}$ in boldface.

Example 2 In Example 3 of Section 5.1 we defined the vector space M_{mn} of $m \times n$ matrices with real entries. The complex analog of this space is the vector space of matrices with complex entries and the operations of matrix addition and scalar multiplication. We refer to this space as complex M_{mn}. $\blacktriangle$

Example 3 If $f_1(x)$ and $f_2(x)$ are real-valued functions of the real variable x, then the expression

$$f(x) = f_1(x) + if_2(x)$$

is called a **complex-valued function of the real variable x**. Some examples are

$$f(x) = 2x + ix^3 \quad \text{and} \quad g(x) = 2 \sin x + i \cos x \tag{1}$$

Let V be the set of all complex-valued functions that are defined on the entire line. If $\mathbf{f} = f_1(x) + if_2(x)$ and $\mathbf{g} = g_1(x) + ig_2(x)$ are two such functions and k is any complex number, then we define the **sum** function $\mathbf{f} + \mathbf{g}$ and the **scalar multiple** $k\mathbf{f}$ by

$$(\mathbf{f} + \mathbf{g})(x) = [f_1(x) + g_1(x)] + i[f_2(x) + g_2(x)]$$
$$(k\mathbf{f})(x) = kf_1(x) + ikf_2(x)$$

In words, to form $\mathbf{f} + \mathbf{g}$ add the real parts of $\mathbf{f}$ and $\mathbf{g}$ and add the imaginary parts. To form $k\mathbf{f}$ multiply the real and imaginary parts of $\mathbf{f}$ by k. For example, if $\mathbf{f} = f(x)$ and $\mathbf{g} = g(x)$ are the functions in (1), then

$$(\mathbf{f} + \mathbf{g})(x) = (2x + 2 \sin x) + i(x^3 + \cos x)$$
$$(i\mathbf{f})(x) = 2xi + i^2 x^3 = -x^3 + 2xi$$

It can be shown that V together with the stated operations is a complex vector space. It is the complex analog of the vector space $F(-\infty, \infty)$ of real-valued functions discussed in Example 4 of Section 5.1. ▲

Example 4 (*For readers who have studied calculus.*) If $f(x) = f_1(x) + if_2(x)$ is a complex-valued function of the real variable x, then f is said to be **continuous** if $f_1(x)$ and $f_2(x)$ are continuous. We leave it as an exercise to show that the set of all continuous complex-valued functions of a real variable x is a subspace of the vector space of all complex-valued functions of x. This space is the complex analog of the vector space $C(-\infty, \infty)$ discussed in Example 7 of Section 5.2 and is called complex $C(-\infty, \infty)$. A closely related example is complex $C[a, b]$, the vector space of all complex-valued functions that are continuous on the closed interval $[a, b]$. ▲

Recall that in R^n the Euclidean inner product of two vectors

$$\mathbf{u} = (u_1, u_2, \ldots, u_n) \quad \text{and} \quad \mathbf{v} = (v_1, v_2, \ldots, v_n)$$

was defined as

$$\mathbf{u} \cdot \mathbf{v} = u_1 v_1 + u_2 v_2 + \cdots + u_n v_n \tag{2}$$

and the Euclidean norm (or length) of $\mathbf{u}$ as

$$\|\mathbf{u}\| = (\mathbf{u} \cdot \mathbf{u})^{1/2} = \sqrt{u_1^2 + u_2^2 + \cdots + u_n^2} \tag{3}$$

Unfortunately, these definitions are not appropriate for vectors in C^n. For example, if (3) were applied to the vector $\mathbf{u} = (i, 1)$ in C^2, we would obtain

$$\|\mathbf{u}\| = \sqrt{i^2 + 1} = \sqrt{0} = 0$$

so $\mathbf{u}$ would be a *nonzero* vector with zero length—a situation that is clearly unsatisfactory.

COMPLEX
EUCLIDEAN
INNER PRODUCTS

To extend the notions of norm, distance, and angle to C^n properly, we must modify the inner product slightly.

Definition. If $\mathbf{u} = (u_1, u_2, \ldots, u_n)$ and $\mathbf{v} = (v_1, v_2, \ldots, v_n)$ are vectors in C^n, then their **complex Euclidean inner product** $\mathbf{u} \cdot \mathbf{v}$ is defined by

$$\mathbf{u} \cdot \mathbf{v} = u_1 \bar{v}_1 + u_2 \bar{v}_2 + \cdots + u_n \bar{v}_n$$

where $\bar{v}_1, \bar{v}_2, \ldots, \bar{v}_n$ are the conjugates of $v_1, v_2, \ldots, v_n$.

REMARK. Observe that the Euclidean inner product of vectors in C^n is a complex number, whereas the Euclidean inner product of vectors in R^n is a real number.

Example 5 The complex Euclidean inner product of the vectors

$$\mathbf{u} = (-i, 2, 1 + 3i) \qquad \text{and} \qquad \mathbf{v} = (1 - i, 0, 1 + 3i)$$

is

$$\begin{aligned}
\mathbf{u} \cdot \mathbf{v} &= (-i)\overline{(1 - i)} + (2)(\bar{0}) + (1 + 3i)\overline{(1 + 3i)} \\
&= (-i)(1 + i) + (2)(0) + (1 + 3i)(1 - 3i) \\
&= -i - i^2 + 1 - 9i^2 = 11 - i
\end{aligned}$$

Theorem 4.1.2 listed the four main properties of the Euclidean inner product on R^n. The following theorem is the corresponding result for the complex Euclidean inner product on C^n.

Theorem 10.4.1. *If* $\mathbf{u}$, $\mathbf{v}$, *and* $\mathbf{w}$ *are vectors in* C^n, *and* k *is any complex number, then:*

(a) $\mathbf{u} \cdot \mathbf{v} = \overline{\mathbf{v} \cdot \mathbf{u}}$
(b) $(\mathbf{u} + \mathbf{v}) \cdot \mathbf{w} = \mathbf{u} \cdot \mathbf{w} + \mathbf{v} \cdot \mathbf{w}$
(c) $(k\mathbf{u}) \cdot \mathbf{v} = k(\mathbf{u} \cdot \mathbf{v})$
(d) $\mathbf{v} \cdot \mathbf{v} \geq 0$. *Further,* $\mathbf{v} \cdot \mathbf{v} = 0$ *if and only if* $\mathbf{v} = \mathbf{0}$.

Note the difference between part (*a*) of this theorem and part (*a*) of Theorem 4.1.2. We will prove parts (*a*) and (*d*) and leave the rest as exercises.

Proof (a). Let $\mathbf{u} = (u_1, u_2, \ldots, u_n)$ and $\mathbf{v} = (v_1, v_2, \ldots, v_n)$. Then

$$\mathbf{u} \cdot \mathbf{v} = u_1 \bar{v}_1 + u_2 \bar{v}_2 + \cdots + u_n \bar{v}_n$$

and

$$\mathbf{v} \cdot \mathbf{u} = v_1 \bar{u}_1 + v_2 \bar{u}_2 + \cdots + v_n \bar{u}_n$$

so

$$\overline{\mathbf{v} \cdot \mathbf{u}} = \overline{v_1 \overline{u}_1 + v_2 \overline{u}_2 + \cdots + v_n \overline{u}_n}$$

$$= \overline{v}_1 \overline{\overline{u}}_1 + \overline{v}_2 \overline{\overline{u}}_2 + \cdots + \overline{v}_n \overline{\overline{u}}_n \qquad [\text{Theorem } 10.2.3, \text{parts}\,(a)\,\text{and}\,(c)]$$

$$= \overline{v}_1 u_1 + \overline{v}_2 u_2 + \cdots + \overline{v}_n u_n \qquad [\text{Theorem } 10.2.3, \text{part}\,(e)]$$

$$= u_1 \overline{v}_1 + u_2 \overline{v}_2 + \cdots + u_n \overline{v}_n$$

$$= \mathbf{u} \cdot \mathbf{v}$$

Proof (d)

$$\mathbf{v} \cdot \mathbf{v} = v_1 \overline{v}_1 + v_2 \overline{v}_2 + \cdots + v_n \overline{v}_n = |v_1|^2 + |v_2|^2 + \cdots + |v_n|^2 \geq 0$$

Moreover, equality holds if and only if $|v_1| = |v_2| = \cdots = |v_n| = 0$. But this is true if and only if $v_1 = v_2 = \cdots = v_n = 0$, that is, if and only if $\mathbf{v} = \mathbf{0}$. ∎

REMARK. We leave it as an exercise to prove that

$$\mathbf{u} \cdot (k\mathbf{v}) = \overline{k}(\mathbf{u} \cdot \mathbf{v})$$

for vectors in C^n. Compare this to the corresponding formula

$$\mathbf{u} \cdot (k\mathbf{v}) = k(\mathbf{u} \cdot \mathbf{v})$$

for vectors in R^n.

NORM AND DISTANCE IN C^n

By analogy with (3), we define the *Euclidean norm* (or *Euclidean length*) of a vector $\mathbf{u} = (u_1, u_2, \ldots, u_n)$ in C^n by

$$\|\mathbf{u}\| = (\mathbf{u} \cdot \mathbf{u})^{1/2} = \sqrt{|u_1|^2 + |u_2|^2 + \cdots + |u_n|^2}$$

and we define the *Euclidean distance* between the points $\mathbf{u} = (u_1, u_2, \ldots, u_n)$ and $\mathbf{v} = (v_1, v_2, \ldots, v_n)$ by

$$d(\mathbf{u}, \mathbf{v}) = \|\mathbf{u} - \mathbf{v}\| = \sqrt{|u_1 - v_1|^2 + |u_2 - v_2|^2 + \cdots + |u_n - v_n|^2}$$

Example 6 If $\mathbf{u} = (i, 1 + i, 3)$ and $\mathbf{v} = (1 - i, 2, 4i)$, then

$$\|\mathbf{u}\| = \sqrt{|i|^2 + |1 + i|^2 + |3|^2} = \sqrt{1 + 2 + 9} = \sqrt{12} = 2\sqrt{3}$$

and

$$d(\mathbf{u}, \mathbf{v}) = \sqrt{|i - (1 - i)|^2 + |(1 + i) - 2|^2 + |3 - 4i|^2}$$

$$= \sqrt{|-1 + 2i|^2 + |-1 + i|^2 + |3 - 4i|^2} = \sqrt{5 + 2 + 25} = \sqrt{32} = 4\sqrt{2} \ \blacktriangle$$

The complex vector space C^n with norm and inner product defined above is called *complex Euclidean n-space*.

EXERCISE SET 10.4

1. Let $\mathbf{u} = (2i, 0, -1, 3)$, $\mathbf{v} = (-i, i, 1 + i, -1)$, and $\mathbf{w} = (1 + i, -i, -1 + 2i, 0)$. Find
   (a) $\mathbf{u} - \mathbf{v}$ (b) $i\mathbf{v} + 2\mathbf{w}$ (c) $-\mathbf{w} + \mathbf{v}$ (d) $3(\mathbf{u} - (1 + i)\mathbf{v})$ (e) $-i\mathbf{v} + 2i\mathbf{w}$ (f) $2\mathbf{v} - (\mathbf{u} + \mathbf{w})$

2. Let $\mathbf{u}$, $\mathbf{v}$, and $\mathbf{w}$ be the vectors in Exercise 1. Find the vector $\mathbf{x}$ that satisfies
   $\mathbf{u} - \mathbf{v} + i\mathbf{x} = 2i\mathbf{x} + \mathbf{w}$.

3. Let $\mathbf{u}_1 = (1 - i, i, 0)$, $\mathbf{u}_2 = (2i, 1 + i, 1)$, and $\mathbf{u}_3 = (0, 2i, 2 - i)$. Find scalars c_1, c_2, and c_3 such that $c_1\mathbf{u}_1 + c_2\mathbf{u}_2 + c_3\mathbf{u}_3 = (-3 + i, 3 + 2i, 3 - 4i)$.

4. Show that there do not exist scalars c_1, c_2, and c_3 such that

   $$c_1(i, 2 - i, 2 + i) + c_2(1 + i, -2i, 2) + c_3(3, i, 6 + i) = (i, i, i)$$

5. Find the Euclidean norm of $\mathbf{v}$ if
   (a) $\mathbf{v} = (1, i)$ (b) $\mathbf{v} = (1 + i, 3i, 1)$ (c) $\mathbf{v} = (2i, 0, 2i + 1, -1)$ (d) $\mathbf{v} = (-i, i, i, 3, 3 + 4i)$

6. Let $\mathbf{u} = (3i, 0, -i)$, $\mathbf{v} = (0, 3 + 4i, -2i)$, and $\mathbf{w} = (1 + i, 2i, 0)$. Find
   (a) $\|\mathbf{u} + \mathbf{v}\|$ (b) $\|\mathbf{u}\| + \|\mathbf{v}\|$ (c) $\|-i\mathbf{u}\| + i\|\mathbf{u}\|$ (d) $\|3\mathbf{u} - 5\mathbf{v} + \mathbf{w}\|$ (e) $\dfrac{1}{\|\mathbf{w}\|}\mathbf{w}$ (f) $\left\|\dfrac{1}{\|\mathbf{w}\|}\mathbf{w}\right\|$

7. Show that if $\mathbf{v}$ is a nonzero vector in C^n, then $(1/\|\mathbf{v}\|)\mathbf{v}$ has Euclidean norm 1.

8. Find all scalars k such that $\|k\mathbf{v}\| = 1$, where $\mathbf{v} = (3i, 4i)$.

9. Find the Euclidean inner product $\mathbf{u} \cdot \mathbf{v}$ if
   (a) $\mathbf{u} = (-i, 3i)$, $\mathbf{v} = (3i, 2i)$
   (b) $\mathbf{u} = (3 - 4i, 2 + i, -6i)$, $\mathbf{v} = (1 + i, 2 - i, 4)$
   (c) $\mathbf{u} = (1 - i, 1 + i, 2i, 3)$, $\mathbf{v} = (4 + 6i, -5i, -1 + i, i)$

In Exercises 10 and 11 a set of objects is given together with operations of addition and scalar multiplication. Determine which sets are complex vector spaces under the given operations. For those that are not, list all axioms that fail to hold.

10. The set of all triples of complex numbers (z_1, z_2, z_3) with the operations

    $$(z_1, z_2, z_3) + (z_1', z_2', z_3') = (z_1 + z_1', z_2 + z_2', z_3 + z_3')$$

    and

    $$k(z_1, z_2, z_3) = (\bar{k}z_1, \bar{k}z_2, \bar{k}z_3)$$

11. The set of all complex 2×2 matrices of the form

    $$\begin{bmatrix} z & 0 \\ 0 & \bar{z} \end{bmatrix}$$

    with the standard matrix operations of addition and scalar multiplication.

12. Is R^n a subspace of C^n? Explain.

13. Use Theorem 5.2.1 to determine which of the following are subspaces of C^3:
    (a) all vectors of the form $(z, 0, 0)$
    (b) all vectors of the form (z, i, i)
    (c) all vectors of the form (z_1, z_2, z_3), where $z_3 = \bar{z}_1 + \bar{z}_2$
    (d) all vectors of the form (z_1, z_2, z_3), where $z_3 = z_1 + z_2 + i$

14. Use Theorem 5.2.1 to determine which of the following are subspaces of complex M_{22}:
    (a) All complex matrices of the form

    $$\begin{bmatrix} z_1 & z_2 \\ z_3 & z_4 \end{bmatrix}$$

    where z_1 and z_2 are real.
    (b) All complex matrices of the form

    $$\begin{bmatrix} z_1 & z_2 \\ z_3 & z_4 \end{bmatrix}$$

    where $z_1 + z_4 = 0$.
    (c) All 2×2 complex matrices A such that $(\overline{A})^T = A$, where $\overline{A}$ is the matrix whose entries are the conjugates of the corresponding entries in A.

15. Use Theorem 5.2.1 to determine which of the following are subspaces of the vector space of complex-valued functions of the real variable x:
    (a) all f such that $f(1) = 0$
    (b) all f such that $f(0) = i$
    (c) all f such that $f(-x) = \overline{f(x)}$
    (d) all f of the form $k_1 + k_2 e^{ix}$, where k_1 and k_2 are complex numbers

16. Which of the following are linear combinations of $\mathbf{u} = (i, -i, 3i)$ and $\mathbf{v} = (2i, 4i, 0)$?
    (a) $(3i, 3i, 3i)$ (b) $(4i, 2i, 6i)$ (c) $(i, 5i, 6i)$ (d) $(0, 0, 0)$

17. Express the following as linear combinations of $\mathbf{u} = (1, 0, -i)$, $\mathbf{v} = (1 + i, 1, 1 - 2i)$, and $\mathbf{w} = (0, i, 2)$.
    (a) $(1, 1, 1)$ (b) $(i, 0, -i)$ (c) $(0, 0, 0)$ (d) $(2 - i, 1, 1 + i)$

18. In each part determine whether the given vectors span C^3.
    (a) $\mathbf{v}_1 = (i, i, i)$, $\mathbf{v}_2 = (2i, 2i, 0)$, $\mathbf{v}_3 = (3i, 0, 0)$
    (b) $\mathbf{v}_1 = (1 + i, 2 - i, 3 + i)$, $\mathbf{v}_2 = (2 + 3i, 0, 1 - i)$
    (c) $\mathbf{v}_1 = (1, 0, -i)$, $\mathbf{v}_2 = (1 + i, 1, 1 - 2i)$, $\mathbf{v}_3 = (0, i, 2)$
    (d) $\mathbf{v}_1 = (1, i, 0)$, $\mathbf{v}_2 = (0, -i, 1)$, $\mathbf{v}_3 = (1, 0, 1)$

19. Determine which of the following lie in the space spanned by

    $$\mathbf{f} = e^{ix} \quad \text{and} \quad \mathbf{g} = e^{-ix}$$

    (a) $\cos x$ (b) $\sin x$ (c) $\cos x + 3i \sin x$

20. Explain why the following are linearly dependent sets of vectors. (Solve this problem by inspection.)
    (a) $\mathbf{u}_1 = (1 - i, i)$ and $\mathbf{u}_2 = (1 + i, -1)$ in C^2
    (b) $\mathbf{u}_1 = (1, -i)$, $\mathbf{u}_2 = (2 + i, -1)$, $\mathbf{u}_3 = (4, 0)$ in C^2

    (c) $A = \begin{bmatrix} i & 3i \\ 2i & 0 \end{bmatrix}$ and $B = \begin{bmatrix} 1 & 3 \\ 2 & 0 \end{bmatrix}$ in complex M_{22}

21. Which of the following sets of vectors in C^3 are linearly independent?
    (a) $\mathbf{u}_1 = (1 - i, 1, 0)$, $\mathbf{u}_2 = (2, 1 + i, 0)$, $\mathbf{u}_3 = (1 + i, i, 0)$
    (b) $\mathbf{u}_1 = (1, 0, -i)$, $\mathbf{u}_2 = (1 + i, 1, 1 - 2i)$, $\mathbf{u}_3 = (0, i, 2)$
    (c) $\mathbf{u}_1 = (i, 0, 2 - i)$, $\mathbf{u}_2 = (0, 1, i)$, $\mathbf{u}_3 = (-i, -1 - 4i, 3)$

22. Let V be the vector space of all complex-valued functions of the real variable x. Show that the following vectors are linearly dependent.

$$\mathbf{f} = 3 + 3i \cos 2x, \qquad \mathbf{g} = \sin^2 x + i \cos^2 x, \qquad \mathbf{h} = \cos^2 x - i \sin^2 x$$

23. Explain why the following sets of vectors are not bases for the indicated vector spaces. (Solve this problem by inspection.)
(a) $\mathbf{u}_1 = (i, 2i)$, $\mathbf{u}_2 = (0, 3i)$, $\mathbf{u}_3 = (1, 7i)$ for C^2
(b) $\mathbf{u}_1 = (-1 + i, 0, 2 - i)$, $\mathbf{u}_2 = (1, -i, 1 + i)$ for C^3

24. Which of the following sets of vectors are bases for C^2?
(a) $(2i, -i)$, $(4i, 0)$ (b) $(1 + i, 1)$, $(1 + i, i)$
(c) $(0, 0)$, $(1 + i, 1 - i)$ (d) $(2 - 3i, i)$, $(3 + 2i, -1)$

25. Which of the following sets of vectors are bases for C^3?
(a) $(i, 0, 0)$, $(i, i, 0)$, (i, i, i) (b) $(1, 0, -i)$, $(1 + i, 1, 1 - 2i)$, $(0, i, 2)$
(c) $(i, 0, 2 - i)$, $(0, 1, i)$, $(-i, -1 - 4i, 3)$ (d) $(1, 0, i)$, $(2 - i, 1, 2 + i)$, $(0, 3i, 3i)$

In Exercises 26–29 determine the dimension of and a basis for the solution space of the system.

26.
$$\begin{aligned} x_1 + (1 + i)x_2 &= 0 \\ (1 - i)x_1 + \qquad 2x_2 &= 0 \end{aligned}$$

27.
$$\begin{aligned} 2x_1 - (1 + i)x_2 &= 0 \\ (-1 + i)x_1 + \qquad x_2 &= 0 \end{aligned}$$

28.
$$\begin{aligned} x_1 + (2 - i)x_2 \qquad\qquad &= 0 \\ x_2 + 3ix_3 &= 0 \\ ix_1 + (2 + 2i)x_2 + 3ix_3 &= 0 \end{aligned}$$

29.
$$\begin{aligned} x_1 + ix_2 - 2ix_3 + x_4 &= 0 \\ ix_1 + 3x_2 + 4x_3 - 2ix_4 &= 0 \end{aligned}$$

30. Prove: If $\mathbf{u}$ and $\mathbf{v}$ are vectors in complex Euclidean n-space, then

$$\mathbf{u} \cdot (k\mathbf{v}) = \bar{k}(\mathbf{u} \cdot \mathbf{v})$$

31. (a) Prove part (b) of Theorem 10.4.1.   (b) Prove part (c) of Theorem 10.4.1.

32. (*For readers who have studied calculus.*) Prove that complex $C(-\infty, \infty)$ is a subspace of the vector space of complex-valued functions of a real variable.

33. Establish the identity

$$\mathbf{u} \cdot \mathbf{v} = \frac{1}{4}\|\mathbf{u} + \mathbf{v}\|^2 - \frac{1}{4}\|\mathbf{u} - \mathbf{v}\|^2 + \frac{i}{4}\|\mathbf{u} + i\mathbf{v}\|^2 - \frac{i}{4}\|\mathbf{u} - i\mathbf{v}\|^2$$

for vectors in complex Euclidean n-space.

10.5 COMPLEX INNER PRODUCT SPACES

In Section 6.1 we defined the notion of an inner product on a real vector space by using the basic properties of the Euclidean inner product on R^n as axioms. In this section we shall define inner products on complex vector spaces by using the properties of the Euclidean inner product on C^n as axioms.

UNITARY SPACES   Motivated by Theorem 10.4.1 we make the following definition.

> **Definition.** An ***inner product on a complex vector space*** V is a function that associates a complex number $\langle \mathbf{u}, \mathbf{v} \rangle$ with each pair of vectors $\mathbf{u}$ and $\mathbf{v}$ in V in such a way that the following axioms are satisfied for all vectors $\mathbf{u}$, $\mathbf{v}$, and $\mathbf{w}$ in V and all scalars k.
>
> (1) $\langle \mathbf{u}, \mathbf{v} \rangle = \overline{\langle \mathbf{v}, \mathbf{u} \rangle}$
> (2) $\langle \mathbf{u} + \mathbf{v}, \mathbf{w} \rangle = \langle \mathbf{u}, \mathbf{w} \rangle + \langle \mathbf{v}, \mathbf{w} \rangle$
> (3) $\langle k\mathbf{u}, \mathbf{v} \rangle = k\langle \mathbf{u}, \mathbf{v} \rangle$
> (4) $\langle \mathbf{v}, \mathbf{v} \rangle \geq 0$ and $\langle \mathbf{v}, \mathbf{v} \rangle = 0$ if and only if $\mathbf{v} = \mathbf{0}$

A complex vector space with an inner product is called a ***complex inner product space*** or a ***unitary space***.

The following additional properties follow immediately from the four inner product axioms:

(i) $\langle \mathbf{0}, \mathbf{v} \rangle = \langle \mathbf{v}, \mathbf{0} \rangle = 0$
(ii) $\langle \mathbf{u}, \mathbf{v} + \mathbf{w} \rangle = \langle \mathbf{u}, \mathbf{v} \rangle + \langle \mathbf{u}, \mathbf{w} \rangle$
(iii) $\langle \mathbf{u}, k\mathbf{v} \rangle = \bar{k}\langle \mathbf{u}, \mathbf{v} \rangle$

Since only (iii) differs from the corresponding results for real inner products, we will prove it and leave the other proofs as exercises.

$$
\begin{aligned}
\langle \mathbf{u}, k\mathbf{v} \rangle &= \overline{\langle k\mathbf{v}, \mathbf{u} \rangle} && [\text{Axiom 1}] \\
&= \overline{k\langle \mathbf{v}, \mathbf{u} \rangle} && [\text{Axiom 3}] \\
&= \bar{k}\overline{\langle \mathbf{v}, \mathbf{u} \rangle} && [\text{Property of conjugates}] \\
&= \bar{k}\langle \mathbf{u}, \mathbf{v} \rangle && [\text{Axiom 1}]
\end{aligned}
$$

Example 1 Let $\mathbf{u} = (u_1, u_2, \ldots, u_n)$ and $\mathbf{v} = (v_1, v_2, \ldots, v_n)$ be vectors in C^n. The Euclidean inner product $\langle \mathbf{u}, \mathbf{v} \rangle = \mathbf{u} \cdot \mathbf{v} = u_1\bar{v}_1 + u_2\bar{v}_2 + \cdots + u_n\bar{v}_n$ satisfies all the inner product axioms by Theorem 10.4.1.

Example 2 If

$$
U = \begin{bmatrix} u_1 & u_2 \\ u_3 & u_4 \end{bmatrix} \quad \text{and} \quad V = \begin{bmatrix} v_1 & v_2 \\ v_3 & v_4 \end{bmatrix}
$$

are any 2×2 matrices with complex entries, then the following formula defines a complex inner product on complex M_{22} (verify):

$$
\langle U, V \rangle = u_1\bar{v}_1 + u_2\bar{v}_2 + u_3\bar{v}_3 + u_4\bar{v}_4
$$

For example, if

$$
U = \begin{bmatrix} 0 & i \\ 1 & 1+i \end{bmatrix} \quad \text{and} \quad V = \begin{bmatrix} 1 & -i \\ 0 & 2i \end{bmatrix}
$$

then

$$\langle U, V \rangle = (0)(\bar{1}) + i(\overline{-i}) + (1)(\bar{0}) + (1+i)(\overline{2i})$$
$$= (0)(1) + i(i) + (1)(0) + (1+i)(-2i)$$
$$= 0 + i^2 + 0 - 2i - 2i^2$$
$$= 1 - 2i \quad \blacktriangle$$

Example 3 (***For readers who have studied calculus.***)   If $f(x) = f_1(x) + if_2(x)$ is a complex-valued function of the real variable x, and if $f_1(x)$ and $f_2(x)$ are continuous on $[a, b]$, then we define

$$\int_a^b f(x)\, dx = \int_a^b [f_1(x) + if_2(x)]\, dx = \int_a^b f_1(x)\, dx + i \int_a^b f_2(x)\, dx$$

In words, *the integral of $f(x)$ is the integral of the real part of f plus i times the integral of the imaginary part of f.*

We leave it as an exercise to show that if the functions $\mathbf{f} = f_1(x) + if_2(x)$ and $\mathbf{g} = g_1(x) + ig_2(x)$ are vectors in complex $C[a, b]$, then the following formula defines an inner product on complex $C[a, b]$:

$$\langle \mathbf{f}, \mathbf{g} \rangle = \int_a^b [f_1(x) + if_2(x)]\overline{[g_1(x) + ig_2(x)]}\, dx$$

$$= \int_a^b [f_1(x) + if_2(x)][g_1(x) - ig_2(x)]\, dx$$

$$= \int_a^b [f_1(x)g_1(x) + f_2(x)g_2(x)]\, dx + i \int_a^b [f_2(x)g_1(x) - f_1(x)g_2(x)]\, dx \quad \blacktriangle$$

In complex inner product spaces, as in real inner product spaces, the ***norm*** (or ***length***) of a vector $\mathbf{u}$ is defined by

$$\|\mathbf{u}\| = \langle \mathbf{u}, \mathbf{u} \rangle^{1/2}$$

and the ***distance*** between two vectors $\mathbf{u}$ and $\mathbf{v}$ is defined by

$$d(\mathbf{u}, \mathbf{v}) = \|\mathbf{u} - \mathbf{v}\|$$

It can be shown that with these definitions Theorems 6.2.2 and 6.2.3 remain true in complex inner product spaces (Exercise 35).

Example 4   If $\mathbf{u} = (u_1, u_2, \ldots, u_n)$ and $\mathbf{v} = (v_1, v_2, \ldots, v_n)$ are vectors in C^n with the Euclidean inner product, then

$$\|\mathbf{u}\| = \langle \mathbf{u}, \mathbf{u} \rangle^{1/2} = \sqrt{|u_1|^2 + |u_2|^2 + \cdots + |u_n|^2}$$

and

$$d(\mathbf{u}, \mathbf{v}) = \|\mathbf{u} - \mathbf{v}\| = \langle \mathbf{u} - \mathbf{v}, \mathbf{u} - \mathbf{v} \rangle^{1/2}$$
$$= \sqrt{|u_1 - v_1|^2 + |u_2 - v_2|^2 + \cdots + |u_n - v_n|^2}$$

Observe that these are just the formulas for the Euclidean norm and distance discussed in Section 10.4. ▲

Example 5 (***For readers who have studied calculus.***) If complex $C[0, 2\pi]$ has the inner product of Example 3, and if $\mathbf{f} = e^{imx}$, where m is any integer, then with the help of Formula (15) of Section 10.3 we obtain

$$\|\mathbf{f}\| = \langle \mathbf{f}, \mathbf{f} \rangle^{1/2} = \left[\int_0^{2\pi} e^{imx}\overline{e^{imx}} \, dx \right]^{1/2}$$

$$= \left[\int_0^{2\pi} e^{imx} e^{-imx} \, dx \right]^{1/2} = \left[\int_0^{2\pi} dx \right]^{1/2} = \sqrt{2\pi} \quad \blacktriangle$$

ORTHOGONAL SETS

The definitions of such terms as ***orthogonal vectors***, ***orthogonal set***, ***orthonormal set***, and ***orthonormal basis*** carry over to unitary spaces without change. Moreover, Theorem 6.2.4, and the theorems in Section 6.3, and Theorem 6.5.4 remain valid in complex inner product spaces, and the Gram–Schmidt process can be used to convert an arbitrary basis for a complex inner product space into an orthonormal basis.

Example 6 The vectors

$$\mathbf{u} = (i, 1) \qquad \text{and} \qquad \mathbf{v} = (1, i)$$

in C^2 are orthogonal with respect to the Euclidean inner product, since

$$\mathbf{u} \cdot \mathbf{v} = (i)(\overline{1}) + (1)(\overline{i}) = (i)(1) + (1)(-i) = 0 \quad \blacktriangle$$

Example 7 Consider the vector space C^3 with the Euclidean inner product. Apply the Gram–Schmidt process to transform the basis vectors $\mathbf{u}_1 = (i, i, i)$, $\mathbf{u}_2 = (0, i, i)$, $\mathbf{u}_3 = (0, 0, i)$ into an orthonormal basis.

Solution.

Step 1. $\mathbf{v}_1 = \mathbf{u}_1 = (i, i, i)$

Step 2. $\mathbf{v}_2 = \mathbf{u}_2 - \text{proj}_{W_1} \mathbf{u}_2 = \mathbf{u}_2 - \dfrac{\langle \mathbf{u}_2, \mathbf{v}_1 \rangle}{\|\mathbf{v}_1\|^2} \mathbf{v}_1$

$$= (0, i, i) - \frac{2}{3}(i, i, i) = \left(-\frac{2}{3}i, \frac{1}{3}i, \frac{1}{3}i \right)$$

Step 3. $\mathbf{v}_3 = \mathbf{u}_3 - \text{proj}_{W_2} \mathbf{u}_3 = \mathbf{u}_3 - \dfrac{\langle \mathbf{u}_3, \mathbf{v}_1 \rangle}{\|\mathbf{v}_1\|^2} \mathbf{v}_1 - \dfrac{\langle \mathbf{u}_3, \mathbf{v}_2 \rangle}{\|\mathbf{v}_2\|^2} \mathbf{v}_2$

$$= (0, 0, i) - \frac{1}{3}(i, i, i) - \frac{1/3}{2/3}\left(-\frac{2}{3}i, \frac{1}{3}i, \frac{1}{3}i \right)$$

$$= \left(0, -\frac{1}{2}i, \frac{1}{2}i \right)$$

Thus,

$$\mathbf{v}_1 = (i, i, i), \qquad \mathbf{v}_2 = \left(-\frac{2}{3}i, \frac{1}{3}i, \frac{1}{3}i\right), \qquad \mathbf{v}_3 = \left(0, -\frac{1}{2}i, \frac{1}{2}i\right)$$

form an orthogonal basis for C^3. The norms of these vectors are

$$\|\mathbf{v}_1\| = \sqrt{3}, \qquad \|\mathbf{v}_2\| = \frac{\sqrt{6}}{3}, \qquad \|\mathbf{v}_3\| = \frac{1}{\sqrt{2}}$$

so an orthonormal basis for C^3 is

$$\frac{\mathbf{v}_1}{\|\mathbf{v}_1\|} = \left(\frac{i}{\sqrt{3}}, \frac{i}{\sqrt{3}}, \frac{i}{\sqrt{3}}\right), \qquad \frac{\mathbf{v}_2}{\|\mathbf{v}_2\|} = \left(-\frac{2i}{\sqrt{6}}, \frac{i}{\sqrt{6}}, \frac{i}{\sqrt{6}}\right),$$

$$\frac{\mathbf{v}_3}{\|\mathbf{v}_3\|} = \left(0, -\frac{i}{\sqrt{2}}, \frac{i}{\sqrt{2}}\right) \quad \blacktriangle$$

Example 8 (***For readers who have studied calculus.***)   Let complex $C[0, 2\pi]$ have the inner product of Example 3, and let W be the set of vectors in $C[0, 2\pi]$ of the form

$$e^{imx} = \cos mx + i \sin mx$$

where m is an integer. The set W is orthogonal because if

$$\mathbf{f} = e^{ikx} \qquad \text{and} \qquad \mathbf{g} = e^{ilx}$$

are distinct vectors in W, then

$$\langle \mathbf{f}, \mathbf{g} \rangle = \int_0^{2\pi} e^{ikx}\overline{e^{ilx}}\, dx = \int_0^{2\pi} e^{ikx}e^{-ilx}\, dx = \int_0^{2\pi} e^{i(k-l)x}\, dx$$

$$= \int_0^{2\pi} \cos(k-l)x\, dx + i\int_0^{2\pi} \sin(k-l)x\, dx$$

$$= \left[\frac{1}{k-l}\sin(k-l)x\right]_0^{2\pi} - i\left[\frac{1}{k-l}\cos(k-l)x\right]_0^{2\pi}$$

$$= (0) - i(0) = 0$$

If we normalize each vector in the orthogonal set W, we obtain an orthonormal set. But in Example 5 we showed that each vector in W has norm $\sqrt{2\pi}$, so the vectors

$$\frac{1}{\sqrt{2\pi}}e^{imx} \qquad m = 0, \pm 1, \pm 2, \dots$$

form an orthonormal set in complex $C[0, 2\pi]$. $\blacktriangle$

EXERCISE SET 10.5

1. Let $\mathbf{u} = (u_1, u_2)$ and $\mathbf{v} = (v_1, v_2)$. Show that $\langle \mathbf{u}, \mathbf{v} \rangle = 3u_1\bar{v}_1 + 2u_2\bar{v}_2$ defines an inner product on C^2.

2. Compute $\langle \mathbf{u}, \mathbf{v} \rangle$ using the inner product in Exercise 1.
   (a) $\mathbf{u} = (2i, -i)$, $\mathbf{v} = (-i, 3i)$ (b) $\mathbf{u} = (0, 0)$, $\mathbf{v} = (1 - i, 7 - 5i)$
   (c) $\mathbf{u} = (1 + i, 1 - i)$, $\mathbf{v} = (1 - i, 1 + i)$ (d) $\mathbf{u} = (3i, -1 + 2i)$, $\mathbf{v} = (3i, -1 + 2i)$

3. Let $\mathbf{u} = (u_1, u_2)$ and $\mathbf{v} = (v_1, v_2)$. Show that

$$\langle \mathbf{u}, \mathbf{v} \rangle = u_1\bar{v}_1 + (1 + i)u_1\bar{v}_2 + (1 - i)u_2\bar{v}_1 + 3u_2\bar{v}_2$$

   defines an inner product on C^2.

4. Compute $\langle \mathbf{u}, \mathbf{v} \rangle$ using the inner product in Exercise 3.
   (a) $\mathbf{u} = (2i, -i)$, $\mathbf{v} = (-i, 3i)$ (b) $\mathbf{u} = (0, 0)$, $\mathbf{v} = (1 - i, 7 - 5i)$
   (c) $\mathbf{u} = (1 + i, 1 - i)$, $\mathbf{v} = (1 - i, 1 + i)$ (d) $\mathbf{u} = (3i, -1 + 2i)$, $\mathbf{v} = (3i, -1 + 2i)$

5. Let $\mathbf{u} = (u_1, u_2)$ and $\mathbf{v} = (v_1, v_2)$. Determine which of the following are inner products on C^2. For those that are not, list the axioms that do not hold.
   (a) $\langle \mathbf{u}, \mathbf{v} \rangle = u_1\bar{v}_1$ (b) $\langle \mathbf{u}, \mathbf{v} \rangle = u_1\bar{v}_1 - u_2\bar{v}_2$ (c) $\langle \mathbf{u}, \mathbf{v} \rangle = |u_1|^2|v_1|^2 + |u_2|^2|v_2|^2$
   (d) $\langle \mathbf{u}, \mathbf{v} \rangle = 2u_1\bar{v}_1 + iu_1\bar{v}_2 + iu_2\bar{v}_1 + 2u_2\bar{v}_2$ (e) $\langle \mathbf{u}, \mathbf{v} \rangle = 2u_1\bar{v}_1 + iu_1\bar{v}_2 - iu_2\bar{v}_1 + 2u_2\bar{v}_2$

6. Use the inner product of Example 2 to find $\langle U, V \rangle$ if

$$U = \begin{bmatrix} -i & 1 + i \\ 1 - i & i \end{bmatrix} \quad \text{and} \quad V = \begin{bmatrix} 3 & -2 - 3i \\ 4i & 1 \end{bmatrix}$$

7. Let $\mathbf{u} = (u_1, u_2, u_3)$ and $\mathbf{v} = (v_1, v_2, v_3)$. Does $\langle \mathbf{u}, \mathbf{v} \rangle = u_1\bar{v}_1 + u_2\bar{v}_2 + u_3\bar{v}_3 - iu_3\bar{v}_1$ define an inner product on C^3? If not, list all axioms that fail to hold.

8. Let V be the vector space of complex-valued functions of the real variable x, and let $\mathbf{f} = f_1(x) + if_2(x)$ and $\mathbf{g} = g_1(x) + ig_2(x)$ be vectors in V. Does

$$\langle \mathbf{f}, \mathbf{g} \rangle = (f_1(0) + if_2(0))\overline{(g_1(0) + ig_2(0))}$$

   define an inner product on V? If not, list all axioms that fail to hold.

9. Let C^2 have the inner product of Exercise 1. Find $\|\mathbf{w}\|$ if
   (a) $\mathbf{w} = (-i, 3i)$ (b) $\mathbf{w} = (1 - i, 1 + i)$ (c) $\mathbf{w} = (0, 2 - i)$ (d) $\mathbf{w} = (0, 0)$

10. For each vector in Exercise 9, use the Euclidean inner product to find $\|\mathbf{w}\|$.

11. Use the inner product of Exercise 3 to find $\|\mathbf{w}\|$ if
    (a) $\mathbf{w} = (1, -i)$ (b) $\mathbf{w} = (1 - i, 1 + i)$ (c) $\mathbf{w} = (3 - 4i, 0)$ (d) $\mathbf{w} = (0, 0)$

12. Use the inner product of Example 2 to find $\|A\|$ if

   (a) $A = \begin{bmatrix} -i & 7i \\ 6i & 2i \end{bmatrix}$ (b) $A = \begin{bmatrix} -1 & 1 + i \\ 1 - i & 3 \end{bmatrix}$

13. Let C^2 have the inner product of Exercise 1. Find $d(\mathbf{x}, \mathbf{y})$ if
    (a) $\mathbf{x} = (1, 1)$, $\mathbf{y} = (i, -i)$ (b) $\mathbf{x} = (1 - i, 3 + 2i)$, $\mathbf{y} = (1 + i, 3)$

14. Repeat the directions of Exercise 13 using the Euclidean inner product on C^2.

15. Repeat the directions of Exercise 13 using the inner product of Exercise 3.

16. Let complex M_{22} have the inner product of Example 2. Find $d(A, B)$ if

   (a) $A = \begin{bmatrix} i & 5i \\ 8i & 3i \end{bmatrix}$ and $B = \begin{bmatrix} -5i & 0 \\ 7i & -3i \end{bmatrix}$

   (b) $A = \begin{bmatrix} -1 & 1 - i \\ 1 + i & 2 \end{bmatrix}$ and $B = \begin{bmatrix} 2i & 2 - 3i \\ i & 1 \end{bmatrix}$

17. Let C^3 have the Euclidean inner product. For which complex values of k are **u** and **v** orthogonal?

(a) $\mathbf{u} = (2i, i, 3i)$, $\mathbf{v} = (i, 6i, k)$ (b) $\mathbf{u} = (k, k, 1+i)$, $\mathbf{v} = (1, -1, 1-i)$

18. Let M_{22} have the inner product of Example 2. Determine which of the following are orthogonal to

$$A = \begin{bmatrix} 2i & i \\ -i & 3i \end{bmatrix}$$

(a) $\begin{bmatrix} -3 & 1-i \\ 1-i & 2 \end{bmatrix}$ (b) $\begin{bmatrix} 1 & 1 \\ 0 & -1 \end{bmatrix}$ (c) $\begin{bmatrix} 0 & 0 \\ 0 & 0 \end{bmatrix}$ (d) $\begin{bmatrix} 0 & 1 \\ 3-i & 0 \end{bmatrix}$

19. Let C^3 have the Euclidean inner product. Show that for all values of θ the vector $\mathbf{x} = e^{i\theta}\left(\dfrac{i}{\sqrt{3}}, \dfrac{1}{\sqrt{3}}, \dfrac{1}{\sqrt{3}}\right)$ has norm 1 and is orthogonal to both $(1, i, 0)$ and $(0, i, -i)$.

20. Let C^2 have the Euclidean inner product. Which of the following form orthonormal sets?

(a) $(i, 0)$, $(0, 1-i)$ (b) $\left(\dfrac{i}{\sqrt{2}}, -\dfrac{i}{\sqrt{2}}\right)$, $\left(\dfrac{i}{\sqrt{2}}, \dfrac{i}{\sqrt{2}}\right)$ (c) $\left(\dfrac{i}{\sqrt{2}}, \dfrac{i}{\sqrt{2}}\right)$, $\left(-\dfrac{i}{\sqrt{2}}, -\dfrac{i}{\sqrt{2}}\right)$ (d) $(i, 0)$, $(0, 0)$

21. Let C^3 have the Euclidean inner product. Which of the following form orthonormal sets?

(a) $\left(\dfrac{i}{\sqrt{2}}, 0, \dfrac{i}{\sqrt{2}}\right)$, $\left(\dfrac{i}{\sqrt{3}}, \dfrac{i}{\sqrt{3}}, -\dfrac{i}{\sqrt{3}}\right)$, $\left(-\dfrac{i}{\sqrt{2}}, 0, \dfrac{i}{\sqrt{2}}\right)$ (b) $\left(\dfrac{2}{3}i, -\dfrac{2}{3}i, \dfrac{1}{3}i\right)$, $\left(\dfrac{2}{3}i, \dfrac{1}{3}i, -\dfrac{2}{3}i\right)$, $\left(\dfrac{1}{3}i, \dfrac{2}{3}i, \dfrac{2}{3}i\right)$

(c) $\left(\dfrac{i}{\sqrt{6}}, \dfrac{i}{\sqrt{6}}, -\dfrac{2i}{\sqrt{6}}\right)$, $\left(\dfrac{i}{\sqrt{2}}, -\dfrac{i}{\sqrt{2}}, 0\right)$

22. Let

$$\mathbf{x} = \left(\dfrac{i}{\sqrt{5}}, -\dfrac{i}{\sqrt{5}}\right) \quad \text{and} \quad \mathbf{y} = \left(\dfrac{2i}{\sqrt{30}}, \dfrac{3i}{\sqrt{30}}\right)$$

Show that $\{\mathbf{x}, \mathbf{y}\}$ is an orthonormal set if C^2 has the inner product

$$\langle \mathbf{u}, \mathbf{v} \rangle = 3u_1\bar{v}_1 + 2u_2\bar{v}_2$$

but is not orthonormal if C^2 has the Euclidean inner product.

23. Show that

$$\mathbf{u}_1 = (i, 0, 0, i), \quad \mathbf{u}_2 = (-i, 0, 2i, i), \quad \mathbf{u}_3 = (2i, 3i, 2i, -2i), \quad \mathbf{u}_4 = (-i, 2i, -i, i)$$

is an orthogonal set in C^4 with the Euclidean inner product. By normalizing each of these vectors, obtain an orthonormal set.

24. Let C^2 have the Euclidean inner product. Use the Gram–Schmidt process to transform the basis $\{\mathbf{u}_1, \mathbf{u}_2\}$ into an orthonormal basis.

(a) $\mathbf{u}_1 = (i, -3i)$, $\mathbf{u}_2 = (2i, 2i)$ (b) $\mathbf{u}_1 = (i, 0)$, $\mathbf{u}_2 = (3i, -5i)$

25. Let C^3 have the Euclidean inner product. Use the Gram–Schmidt process to transform the basis $\{\mathbf{u}_1, \mathbf{u}_2, \mathbf{u}_3\}$ into an orthonormal basis.

(a) $\mathbf{u}_1 = (i, i, i)$, $\mathbf{u}_2 = (-i, i, 0)$, $\mathbf{u}_3 = (i, 2i, i)$ (b) $\mathbf{u}_1 = (i, 0, 0)$, $\mathbf{u}_2 = (3i, 7i, -2i)$, $\mathbf{u}_3 = (0, 4i, i)$

26. Let C^4 have the Euclidean inner product. Use the Gram–Schmidt process to transform the basis $\{\mathbf{u}_1, \mathbf{u}_2, \mathbf{u}_3, \mathbf{u}_4\}$ into an orthonormal basis.

$$\mathbf{u}_1 = (0, 2i, i, 0), \quad \mathbf{u}_2 = (i, -i, 0, 0), \quad \mathbf{u}_3 = (i, 2i, 0, -i), \quad \mathbf{u}_4 = (i, 0, i, i)$$

27. Let C^3 have the Euclidean inner product. Find an orthonormal basis for the subspace spanned by $(0, i, 1 - i)$ and $(-i, 0, 1 + i)$.

28. Let C^4 have the Euclidean inner product. Express $\mathbf{w} = (-i, 2i, 6i, 0)$ in the form $\mathbf{w} = \mathbf{w}_1 + \mathbf{w}_2$, where the vector $\mathbf{w}_1$ is in the space W spanned by $\mathbf{u}_1 = (-i, 0, i, 2i)$ and $\mathbf{u}_2 = (0, i, 0, i)$, and $\mathbf{w}_2$ is orthogonal to W.

29. (a) Prove: If k is a complex number and $\langle \mathbf{u}, \mathbf{v} \rangle$ is an inner product on a complex vector space, then $\langle \mathbf{u} - k\mathbf{v}, \mathbf{u} - k\mathbf{v} \rangle = \langle \mathbf{u}, \mathbf{u} \rangle - \bar{k}\langle \mathbf{u}, \mathbf{v} \rangle - k\overline{\langle \mathbf{u}, \mathbf{v} \rangle} + k\bar{k}\langle \mathbf{v}, \mathbf{v} \rangle$.

    (b) Use the result in (a) to prove that $0 \le \langle \mathbf{u}, \mathbf{u} \rangle - \bar{k}\langle \mathbf{u}, \mathbf{v} \rangle - k\overline{\langle \mathbf{u}, \mathbf{v} \rangle} + k\bar{k}\langle \mathbf{v}, \mathbf{v} \rangle$.

30. Prove that if $\mathbf{u}$ and $\mathbf{v}$ are vectors in a complex inner product space, then

    $$|\langle \mathbf{u}, \mathbf{v} \rangle|^2 \le \langle \mathbf{u}, \mathbf{u} \rangle \langle \mathbf{v}, \mathbf{v} \rangle$$

    This result, called the **Cauchy–Schwarz inequality for complex inner product spaces**, differs from its real analog (Theorem 6.2.1) in that an absolute value sign must be included in the left side. [**Hint.** Let $k = \langle \mathbf{u}, \mathbf{v} \rangle / \langle \mathbf{v}, \mathbf{v} \rangle$ in the inequality of Exercise 29(b).]

31. Prove: If $\mathbf{u} = (u_1, u_2, \ldots, u_n)$ and $\mathbf{v} = (v_1, v_2, \ldots, v_n)$ are vectors in C^n, then

    $$|u_1\bar{v}_1 + u_2\bar{v}_2 + \cdots + u_n\bar{v}_n| \le (|u_1|^2 + |u_2|^2 + \cdots + |u_n|^2)^{1/2}(|v_1|^2 + |v_2|^2 + \cdots + |v_n|^2)^{1/2}$$

    This is the complex version of **Cauchy's inequality** discussed in Example 1 of Section 6.2. [**Hint.** Use Exercise 30.]

32. Prove that equality holds in the Cauchy–Schwarz inequality for complex vector spaces if and only if $\mathbf{u}$ and $\mathbf{v}$ are linearly dependent.

33. Prove that if $\langle \mathbf{u}, \mathbf{v} \rangle$ is an inner product on a complex vector space, then

    $$\langle \mathbf{0}, \mathbf{v} \rangle = \langle \mathbf{v}, \mathbf{0} \rangle = 0$$

34. Prove that if $\langle \mathbf{u}, \mathbf{v} \rangle$ is an inner product on a complex vector space, then

    $$\langle \mathbf{u}, \mathbf{v} + \mathbf{w} \rangle = \langle \mathbf{u}, \mathbf{v} \rangle + \langle \mathbf{u}, \mathbf{w} \rangle$$

35. Theorems 6.2.2 and 6.2.3 remain true in complex inner product spaces. In each part prove that this is so.
    (a) Theorem 6.2.2a   (b) Theorem 6.2.2b   (c) Theorem 6.2.2c   (d) Theorem 6.2.2d
    (e) Theorem 6.2.3a   (f) Theorem 6.2.3b   (g) Theorem 6.2.3c   (h) Theorem 6.2.3d

36. In Example 7 it was shown that the vectors

    $$\mathbf{v}_1 = \left(\frac{i}{\sqrt{3}}, \frac{i}{\sqrt{3}}, \frac{i}{\sqrt{3}} \right), \qquad \mathbf{v}_2 = \left(-\frac{2i}{\sqrt{6}}, \frac{i}{\sqrt{6}}, \frac{i}{\sqrt{6}} \right), \qquad \mathbf{v}_3 = \left(0, -\frac{i}{\sqrt{2}}, \frac{i}{\sqrt{2}} \right)$$

    form an orthonormal basis for C^3. Use Theorem 6.3.1 to express $\mathbf{u} = (1 - i, 1 + i, 1)$ as a linear combination of these vectors.

37. Prove that if $\mathbf{u}$ and $\mathbf{v}$ are vectors in a complex inner product space, then

    $$\langle \mathbf{u}, \mathbf{v} \rangle = \frac{1}{4}\|\mathbf{u} + \mathbf{v}\|^2 - \frac{1}{4}\|\mathbf{u} - \mathbf{v}\|^2 + \frac{i}{4}\|\mathbf{u} + i\mathbf{v}\|^2 - \frac{i}{4}\|\mathbf{u} - i\mathbf{v}\|^2$$

38. Prove: If $\{\mathbf{v}_1, \mathbf{v}_2, \ldots, \mathbf{v}_n\}$ is an orthonormal basis for a complex inner product space V, and if $\mathbf{u}$ and $\mathbf{w}$ are any vectors in V, then

    $$\langle \mathbf{u}, \mathbf{w} \rangle = \langle \mathbf{u}, \mathbf{v}_1 \rangle \overline{\langle \mathbf{w}, \mathbf{v}_1 \rangle} + \langle \mathbf{u}, \mathbf{v}_2 \rangle \overline{\langle \mathbf{w}, \mathbf{v}_2 \rangle} + \cdots + \langle \mathbf{u}, \mathbf{v}_n \rangle \overline{\langle \mathbf{w}, \mathbf{v}_n \rangle}$$

    [**Hint.** Use Theorem 6.3.1 to express $\mathbf{u}$ and $\mathbf{w}$ as linear combinations of the basis vectors.]

39. (*For readers who have studied calculus.*) Prove that if $\mathbf{f} = f_1(x) + if_2(x)$ and $\mathbf{g} = g_1(x) + ig_2(x)$ are vectors in complex $C[a, b]$, then the formula

$$\langle \mathbf{f}, \mathbf{g} \rangle = \int_a^b [f_1(x) + if_2(x)]\overline{[g_1(x) + ig_2(x)]}\, dx$$

defines a complex inner product on $C[a, b]$.

40. (*For readers who have studied calculus.*)
(a) Let $\mathbf{f} = f_1(x) + if_2(x)$ and $\mathbf{g} = g_1(x) + ig_2(x)$ be vectors in $C[0, 1]$ and let this space have the inner product

$$\langle \mathbf{f}, \mathbf{g} \rangle = \int_0^1 [f_1(x) + if_2(x)]\overline{[g_1(x) + ig_2(x)]}\, dx$$

Show that the vectors

$$e^{2\pi imx}, \qquad m = 0, \pm 1, \pm 2, \ldots$$

form an orthogonal set.
(b) Obtain an orthonormal set by normalizing the vectors in part (a).

10.6 UNITARY, NORMAL, AND HERMITIAN MATRICES

*For matrices with real entries, the orthogonal matrices ($A^{-1} = A^T$) and the symmetric matrices ($A = A^T$) played an important role in the orthogonal diagonalization problem (Section 7.3). For matrices with complex entries, the orthogonal and symmetric matrices are of relatively little importance; they are superseded by two new classes of matrices, the **unitary** and **Hermitian** matrices, which we shall discuss in this section.*

UNITARY MATRICES

If A is a matrix with complex entries, then the **conjugate transpose** of A, denoted by A^*, is defined by

$$A^* = \overline{A}^T$$

where $\overline{A}$ is the matrix whose entries are the complex conjugates of the corresponding entries in A and $\overline{A}^T$ is the transpose of $\overline{A}$.

Example 1 If

$$A = \begin{bmatrix} 1 + i & -i & 0 \\ 2 & 3 - 2i & i \end{bmatrix}$$

then

$$\overline{A} = \begin{bmatrix} 1 - i & i & 0 \\ 2 & 3 + 2i & -i \end{bmatrix}$$

so

$$A^* = \overline{A}^T = \begin{bmatrix} 1-i & 2 \\ i & 3+2i \\ 0 & -i \end{bmatrix} \; \blacktriangle$$

The basic properties of the conjugate transpose operation are similar to those of the transpose:

Theorem 10.6.1. *If A and B are matrices with complex entries and k is any complex number, then:*

(a) $(A^*)^* = A$
(b) $(A + B)^* = A^* + B^*$
(c) $(kA)^* = \overline{k}A^*$
(d) $(AB)^* = B^*A^*$

The proofs are left as exercises.

Recall that a matrix with real entries is called *orthogonal* if $A^{-1} = A^T$. The complex analogs of the orthogonal matrices are called *unitary* matrices. They are defined as follows:

Definition. A square matrix A with complex entries is called **unitary** if

$$A^{-1} = A^*$$

The following theorem parallels Theorem 6.5.1.

Theorem 10.6.2. *If A is an $n \times n$ matrix with complex entries, then the following are equivalent.*

(a) *A is unitary.*
(b) *The row vectors of A form an orthonormal set in C^n with the Euclidean inner product.*
(c) *The column vectors of A form an orthonormal set in C^n with the Euclidean inner product.*

Example 2   The matrix

$$A = \begin{bmatrix} \dfrac{1+i}{2} & \dfrac{1+i}{2} \\ \dfrac{1-i}{2} & \dfrac{-1+i}{2} \end{bmatrix} \tag{1}$$

has row vectors

$$\mathbf{r}_1 = \left(\frac{1+i}{2}, \frac{1+i}{2}\right), \qquad \mathbf{r}_2 = \left(\frac{1-i}{2}, \frac{-1+i}{2}\right)$$

Relative to the Euclidean inner product on C^n, we have

$$\|\mathbf{r}_1\| = \sqrt{\left|\frac{1+i}{2}\right|^2 + \left|\frac{1+i}{2}\right|^2} = \sqrt{\frac{1}{2} + \frac{1}{2}} = 1$$

$$\|\mathbf{r}_2\| = \sqrt{\left|\frac{1-i}{2}\right|^2 + \left|\frac{-1+i}{2}\right|^2} = \sqrt{\frac{1}{2} + \frac{1}{2}} = 1$$

and

$$\mathbf{r}_1 \cdot \mathbf{r}_2 = \left(\frac{1+i}{2}\right)\overline{\left(\frac{1-i}{2}\right)} + \left(\frac{1+i}{2}\right)\overline{\left(\frac{-1+i}{2}\right)}$$

$$= \left(\frac{1+i}{2}\right)\left(\frac{1+i}{2}\right) + \left(\frac{1+i}{2}\right)\left(\frac{-1-i}{2}\right)$$

$$= \frac{i}{2} - \frac{i}{2} = 0$$

so the row vectors form an orthonormal set in C^2. Thus, A is unitary and

$$A^{-1} = A^* = \begin{bmatrix} \dfrac{1-i}{2} & \dfrac{1+i}{2} \\ \dfrac{1-i}{2} & \dfrac{-1-i}{2} \end{bmatrix} \tag{2}$$

The reader should verify that matrix (2) is the inverse of matrix (1) by showing that $AA^* = A^*A = I$. ▲

UNITARY DIAGONAL-IZATION

Recall that a square matrix A with real entries is called *orthogonally diagonalizable* if there is an orthogonal matrix P such that $P^{-1}AP\ (=P^TAP)$ is diagonal. For complex matrices we have an analogous concept.

> **Definition.** A square matrix A with complex entries is called **unitarily diagonalizable** if there is a unitary P such that $P^{-1}AP\ (=P^*AP)$ is diagonal; the matrix P is said to **unitarily diagonalize** A.

We have two questions to consider:

- Which matrices are unitarily diagonalizable?
- How do we find a unitary matrix P to carry out the diagonalization?

Before pursuing these questions, we note that our earlier definitions of the terms *eigenvector*, *eigenvalue*, *eigenspace*, *characteristic equation*, and *characteristic polynomial* carry over without change to complex vector spaces.

HERMITIAN MATRICES

In Section 7.3 we saw that the symmetric matrices played a fundamental role in the problem of orthogonally diagonalizing a matrix with real entries. The most natural complex analogs of the real symmetric matrices are the *Hermitian** matrices, which are defined as follows:

> **Definition.** A square matrix A with complex entries is called ***Hermitian*** if
> $$A = A*$$

Example 3 If

$$A = \begin{bmatrix} 1 & i & 1+i \\ -i & -5 & 2-i \\ 1-i & 2+i & 3 \end{bmatrix}$$

then

$$\overline{A} = \begin{bmatrix} 1 & -i & 1-i \\ i & -5 & 2+i \\ 1+i & 2-i & 3 \end{bmatrix}$$

so

$$A* = \overline{A}^T = \begin{bmatrix} 1 & i & 1+i \\ -i & -5 & 2-i \\ 1-i & 2+i & 3 \end{bmatrix} = A$$

which means that A is Hermitian. ▲

It is easy to recognize Hermitian matrices by inspection: The entries on the main diagonal are real numbers (Exercise 17), and the "mirror image" of each entry across the main diagonal is its complex conjugate (Figure 1).

Figure 1

NORMAL MATRICES

Hermitian matrices enjoy many but not all of the properties of real symmetric matrices. For example, just as the real symmetric matrices are orthogonally diagonalizable, so we shall see that the Hermitian matrices are unitarily diagonalizable. However, whereas the real symmetric matrices are the only matrices with real entries that are orthogonally

---

* *Charles Hermite* (1822–1901) was a French mathematician who made fundamental contributions to algebra, matrix theory, and various branches of analysis. He is noted for using integrals to solve a general fifth-degree polynomial equation. He also proved that the number e (the base for natural logarithms) is not the root of any polynomial equation with rational coefficients.

diagonalizable (Theorem 7.3.1), the Hermitian matrices do not constitute the entire class of unitarily diagonalizable matrices; that is, there are unitarily diagonalizable matrices that are not Hermitian. To explain why this is so we shall need the following definition:

Definition. A square matrix A with complex entries is called **normal** if

$$AA^* = A^*A$$

Example 4 Every Hermitian matrix A is normal since $AA^* = AA = A^*A$, and every unitary matrix A is normal since $AA^* = I = A^*A$. ▲

The following two theorems are the complex analogs of Theorems 7.3.1 and 7.3.2. The proofs will be omitted.

Theorem 10.6.3. *If A is a square matrix with complex entries, then the following are equivalent:*

(a) A is unitarily diagonalizable.
(b) A has an orthonormal set of n eigenvectors.
(c) A is normal.

Theorem 10.6.4. *If A is a normal matrix, then eigenvectors from different eigenspaces of A are orthogonal.*

Theorem 10.6.3 tells us that a square matrix A with complex entries is unitarily diagonalizable if and only if it is normal. Theorem 10.6.4 will be the key to constructing a matrix that unitarily diagonalizes a normal matrix.

DIAGONAL-IZATION PROCEDURE

We saw in Section 7.3 that a symmetric matrix A is orthogonally diagonalized by any orthogonal matrix whose column vectors are eigenvectors of A. Similarly, a normal matrix A is diagonalized by any unitary matrix whose column vectors are eigenvectors of A. The procedure for diagonalizing a normal matrix is as follows:

Step 1. Find a basis for each eigenspace of A.

Step 2. Apply the Gram–Schmidt process to each of these bases to obtain an orthonormal basis for each eigenspace.

Step 3. Form the matrix P whose columns are the basis vectors constructed in Step 2. This matrix unitarily diagonalizes A.

The justification of this procedure should be clear. Theorem 10.6.4 ensures that eigenvectors from *different* eigenspaces are orthogonal, and the application of the Gram–Schmidt process ensures that the eigenvectors within the *same* eigenspace are orthonormal. Thus, the *entire* set of eigenvectors obtained by this procedure is orthonormal. Theorem 10.6.3 ensures that this orthonormal set of eigenvectors is a basis.

Example 5   The matrix

$$A = \begin{bmatrix} 2 & 1+i \\ 1-i & 3 \end{bmatrix}$$

is unitarily diagonalizable because it is Hermitian and therefore normal. Find a matrix P that unitarily diagonalizes A.

Solution. The characteristic polynomial of A is

$$\det(\lambda I - A) = \det \begin{bmatrix} \lambda - 2 & -1-i \\ -1+i & \lambda - 3 \end{bmatrix} = (\lambda - 2)(\lambda - 3) - 2 = \lambda^2 - 5\lambda + 4$$

so the characteristic equation is

$$\lambda^2 - 5\lambda + 4 = (\lambda - 1)(\lambda - 4) = 0$$

and the eigenvalues are $\lambda = 1$ and $\lambda = 4$.

By definition,

$$\mathbf{x} = \begin{bmatrix} x_1 \\ x_2 \end{bmatrix}$$

will be an eigenvector of A corresponding to λ if and only if $\mathbf{x}$ is a nontrivial solution of

$$\begin{bmatrix} \lambda - 2 & -1-i \\ -1+i & \lambda - 3 \end{bmatrix} \begin{bmatrix} x_1 \\ x_2 \end{bmatrix} = \begin{bmatrix} 0 \\ 0 \end{bmatrix} \tag{3}$$

To find the eigenvectors corresponding to $\lambda = 1$, we substitute this value in (3):

$$\begin{bmatrix} -1 & -1-i \\ -1+i & -2 \end{bmatrix} \begin{bmatrix} x_1 \\ x_2 \end{bmatrix} = \begin{bmatrix} 0 \\ 0 \end{bmatrix}$$

Solving this system by Gauss–Jordan elimination yields (verify)

$$x_1 = (-1-i)s, \qquad x_2 = s$$

Thus, the eigenvectors of A corresponding to $\lambda = 1$ are the nonzero vectors in C^2 of the form

$$\mathbf{x} = \begin{bmatrix} (-1-i)s \\ s \end{bmatrix} = s \begin{bmatrix} -1-i \\ 1 \end{bmatrix}$$

Thus, this eigenspace is one-dimensional with basis

$$\mathbf{u} = \begin{bmatrix} -1-i \\ 1 \end{bmatrix} \tag{4}$$

In this case the Gram–Schmidt process involves only one step: normalizing this vector. Since

$$\|\mathbf{u}\| = \sqrt{|-1-i|^2 + |1|^2} = \sqrt{2+1} = \sqrt{3}$$

the vector

$$\mathbf{p}_1 = \frac{\mathbf{u}}{\|\mathbf{u}\|} = \begin{bmatrix} \dfrac{-1-i}{\sqrt{3}} \\ \dfrac{1}{\sqrt{3}} \end{bmatrix}$$

is an orthonormal basis for the eigenspace corresponding to $\lambda = 1$.

To find the eigenvectors corresponding to $\lambda = 4$, we substitute this value in (3):

$$\begin{bmatrix} 2 & -1-i \\ -1+i & 1 \end{bmatrix} \begin{bmatrix} x_1 \\ x_2 \end{bmatrix} = \begin{bmatrix} 0 \\ 0 \end{bmatrix}$$

Solving this system by Gauss–Jordan elimination yields (verify)

$$x_1 = \left(\frac{1+i}{2}\right)s, \qquad x_2 = s$$

so the eigenvectors of A corresponding to $\lambda = 4$ are the nonzero vectors in C^2 of the form

$$\mathbf{x} = \begin{bmatrix} \left(\dfrac{1+i}{2}\right)s \\ s \end{bmatrix} = s \begin{bmatrix} \dfrac{1+i}{2} \\ 1 \end{bmatrix}$$

Thus, the eigenspace is one-dimensional with basis

$$\mathbf{u} = \begin{bmatrix} \dfrac{1+i}{2} \\ 1 \end{bmatrix}$$

Applying the Gram–Schmidt process (i.e., normalizing this vector) yields

$$\mathbf{p}_2 = \frac{\mathbf{u}}{\|\mathbf{u}\|} = \begin{bmatrix} \dfrac{1+i}{\sqrt{6}} \\ \dfrac{2}{\sqrt{6}} \end{bmatrix}$$

Thus,

$$P = [\mathbf{p}_1 \;\vdots\; \mathbf{p}_2] = \begin{bmatrix} \dfrac{-1-i}{\sqrt{3}} & \dfrac{1+i}{\sqrt{6}} \\ \dfrac{1}{\sqrt{3}} & \dfrac{2}{\sqrt{6}} \end{bmatrix}$$

diagonalizes A and

$$P^{-1}AP = \begin{bmatrix} 1 & 0 \\ 0 & 4 \end{bmatrix} \quad \blacktriangle$$

EIGENVALUES OF HERMITIAN AND SYMMETRIC MATRICES

In Theorem 7.3.2 it was stated that the eigenvalues of a symmetric matrix with real entries are real numbers. This important result is a corollary of the following more general theorem.

Theorem 10.6.5. *The eigenvalues of a Hermitian matrix are real numbers.*

Proof. If λ is an eigenvalue and **v** a corresponding eigenvector of an $n \times n$ Hermitian matrix A, then

$$A\mathbf{v} = \lambda\mathbf{v}$$

Multiplying each side of this equation on the left by the conjugate transpose of **v** yields

$$\mathbf{v}^*A\mathbf{v} = \mathbf{v}^*(\lambda\mathbf{v}) = \lambda\mathbf{v}^*\mathbf{v} \tag{5}$$

We will show that the 1×1 matrices $\mathbf{v}^*A\mathbf{v}$ and $\mathbf{v}^*\mathbf{v}$ both have real entries, so it will follow from (5) that λ must be a real number.

Both $\mathbf{v}^*A\mathbf{v}$ and $\mathbf{v}^*\mathbf{v}$ are Hermitian, since

$$(\mathbf{v}^*A\mathbf{v})^* = \mathbf{v}^*A^*(\mathbf{v}^*)^* = \mathbf{v}^*A\mathbf{v}$$

and

$$(\mathbf{v}^*\mathbf{v})^* = \mathbf{v}^*(\mathbf{v}^*)^* = \mathbf{v}^*\mathbf{v}$$

Since Hermitian matrices have real entries on the main diagonal, and since $\mathbf{v}^*A\mathbf{v}$ and $\mathbf{v}^*\mathbf{v}$ are 1×1, it follows that these matrices have real entries, which completes the proof. ∎

Theorem 10.6.6. *The eigenvalues of a symmetric matrix with real entries are real numbers.*

Proof. Let A be a symmetric matrix with real entries. Because the entries in A are real, it follows that

$$\overline{A} = A$$

But this implies that A is Hermitian, since

$$A^* = (\overline{A})^T = A^T = A$$

Thus, A has real eigenvalues by Theorem 10.6.5. ∎

EXERCISE SET 10.6

1. In each part find A^*.

(a) $A = \begin{bmatrix} 2i & 1-i \\ 4 & 3+i \\ 5+i & 0 \end{bmatrix}$ (b) $A = \begin{bmatrix} 2i & 1-i & -1+i \\ 4 & 5-7i & -i \\ i & 3 & 1 \end{bmatrix}$

(c) $A = [7i \quad 0 \quad -3i]$ (d) $A = \begin{bmatrix} a_{11} & a_{12} & a_{13} \\ a_{21} & a_{22} & a_{23} \end{bmatrix}$

2. Which of the following are Hermitian matrices?

(a) $\begin{bmatrix} 0 & i \\ i & 2 \end{bmatrix}$ (b) $\begin{bmatrix} 1 & 1+i \\ 1-i & -3 \end{bmatrix}$ (c) $\begin{bmatrix} i & i \\ -i & i \end{bmatrix}$ (d) $\begin{bmatrix} -2 & 1-i & -1+i \\ 1+i & 0 & 3 \\ -1-i & 3 & 5 \end{bmatrix}$ (e) $\begin{bmatrix} 1 & 0 & 0 \\ 0 & 1 & 0 \\ 0 & 0 & 1 \end{bmatrix}$

3. Find k, l, and m to make A a Hermitian matrix.

$$A = \begin{bmatrix} -1 & k & -i \\ 3-5i & 0 & m \\ l & 2+4i & 2 \end{bmatrix}$$

4. Use Theorem 10.6.2 to determine which of the following are unitary matrices.

(a) $\begin{bmatrix} i & 0 \\ 0 & i \end{bmatrix}$ (b) $\begin{bmatrix} \dfrac{i}{\sqrt{2}} & \dfrac{1}{\sqrt{2}} \\ -\dfrac{i}{\sqrt{2}} & \dfrac{1}{\sqrt{2}} \end{bmatrix}$ (c) $\begin{bmatrix} 1+i & 1+i \\ 1-i & -1+i \end{bmatrix}$ (d) $\begin{bmatrix} -\dfrac{i}{\sqrt{2}} & \dfrac{i}{\sqrt{6}} & \dfrac{i}{\sqrt{3}} \\ 0 & -\dfrac{i}{\sqrt{6}} & \dfrac{i}{\sqrt{3}} \\ \dfrac{i}{\sqrt{2}} & \dfrac{i}{\sqrt{6}} & \dfrac{i}{\sqrt{3}} \end{bmatrix}$

5. In each part verify that the matrix is unitary and find its inverse.

(a) $\begin{bmatrix} \dfrac{3}{5} & \dfrac{4}{5}i \\ -\dfrac{4}{5} & \dfrac{3}{5}i \end{bmatrix}$ (b) $\begin{bmatrix} \dfrac{1}{\sqrt{2}} & \dfrac{1}{\sqrt{2}} \\ -\dfrac{1+i}{2} & \dfrac{1+i}{2} \end{bmatrix}$ (c) $\begin{bmatrix} \dfrac{1}{2\sqrt{2}}(\sqrt{3}+i) & \dfrac{1}{2\sqrt{2}}(1-i\sqrt{3}) \\ \dfrac{1}{2\sqrt{2}}(1+i\sqrt{3}) & \dfrac{1}{2\sqrt{2}}(i-\sqrt{3}) \end{bmatrix}$ (d) $\begin{bmatrix} \dfrac{1+i}{2} & -\dfrac{1}{2} & \dfrac{1}{2} \\ \dfrac{i}{\sqrt{3}} & \dfrac{1}{\sqrt{3}} & -\dfrac{i}{\sqrt{3}} \\ \dfrac{3+i}{2\sqrt{15}} & \dfrac{4+3i}{2\sqrt{15}} & \dfrac{5i}{2\sqrt{15}} \end{bmatrix}$

6. Show that the matrix

$$\frac{1}{\sqrt{2}} \begin{bmatrix} e^{i\theta} & e^{-i\theta} \\ ie^{i\theta} & -ie^{-i\theta} \end{bmatrix}$$

is unitary for every real value of θ.

In Exercises 7–12 find a unitary matrix P that diagonalizes A, and determine $P^{-1}AP$.

7. $A = \begin{bmatrix} 4 & 1-i \\ 1+i & 5 \end{bmatrix}$ 8. $A = \begin{bmatrix} 3 & -i \\ i & 3 \end{bmatrix}$ 9. $A = \begin{bmatrix} 6 & 2+2i \\ 2-2i & 4 \end{bmatrix}$

10. $\begin{bmatrix} 0 & 3+i \\ 3-i & -3 \end{bmatrix}$ 11. $A = \begin{bmatrix} 5 & 0 & 0 \\ 0 & -1 & -1+i \\ 0 & -1-i & 0 \end{bmatrix}$ 12. $A = \begin{bmatrix} 2 & \dfrac{i}{\sqrt{2}} & -\dfrac{i}{\sqrt{2}} \\ -\dfrac{i}{\sqrt{2}} & 2 & 0 \\ \dfrac{i}{\sqrt{2}} & 0 & 2 \end{bmatrix}$

13. Show that the eigenvalues of the symmetric matrix

$$A = \begin{bmatrix} 1 & 4i \\ 4i & 3 \end{bmatrix}$$

are not real. Does this violate Theorem 10.6.6?

14. Find a 2×2 matrix that is both Hermitian and unitary and whose entries are not all real numbers.

15. Prove: If A is an $n \times n$ matrix with complex entries, then $\det(\overline{A}) = \overline{\det(A)}$. [**Hint.** First show that the signed elementary products from $\overline{A}$ are the conjugates of the signed elementary products from A.]

16. (a) Use the result of Exercise 15 to prove that if A is an $n \times n$ matrix with complex entries, then $\det(A^*) = \overline{\det(A)}$.
   (b) Prove: If A is Hermitian, then $\det(A)$ is real.
   (c) Prove: If A is unitary, then $|\det(A)| = 1$.

17. Prove that the entries on the main diagonal of a Hermitian matrix are real numbers.

18. Let

$$A = \begin{bmatrix} a_{11} & a_{12} & a_{13} \\ a_{21} & a_{22} & a_{23} \\ a_{31} & a_{32} & a_{33} \end{bmatrix} \quad \text{and} \quad B = \begin{bmatrix} b_{11} & b_{12} & b_{13} \\ b_{21} & b_{22} & b_{23} \\ b_{31} & b_{32} & b_{33} \end{bmatrix}$$

be matrices with complex entries. Show that
   (a) $(A^*)^* = A$ (b) $(A + B)^* = A^* + B^*$ (c) $(kA)^* = \overline{k}A^*$ (d) $(AB)^* = B^*A^*$

19. Prove: If A is invertible, then so is A^*, in which case $(A^*)^{-1} = (A^{-1})^*$.

20. Show that if A is a unitary matrix, then A^* is also unitary.

21. Prove that an $n \times n$ matrix with complex entries is unitary if and only if its rows form an orthonormal set in C^n with the Euclidean inner product.

22. Use Exercises 20 and 21 to show that an $n \times n$ matrix is unitary if and only if its columns form an orthonormal set in C^n with the Euclidean inner product.

23. Prove: If $A = A^*$, then for every vector x in C^n, the entry in the 1×1 matrix $\mathbf{x}^*A\mathbf{x}$ is real.

24. Let λ and μ be distinct eigenvalues of a Hermitian matrix A.
   (a) Prove that if $\mathbf{x}$ is an eigenvector corresponding to λ and $\mathbf{y}$ an eigenvector corresponding to μ, then $\mathbf{x}^*A\mathbf{y} = \lambda\mathbf{x}^*\mathbf{y}$ and $\mathbf{x}^*A\mathbf{y} = \mu\mathbf{x}^*\mathbf{y}$.
   (b) Prove Theorem 10.6.4. [**Hint.** Subtract the equations in part (a).]

SUPPLEMENTARY EXERCISES

1. Let $\mathbf{u} = (u_1, u_2, \ldots, u_n)$ and $\mathbf{v} = (v_1, v_2, \ldots, v_n)$ be vectors in C^n, and let $\overline{\mathbf{u}} = (\overline{u}_1, \overline{u}_2, \ldots, \overline{u}_n)$ and $\overline{\mathbf{v}} = (\overline{v}_1, \overline{v}_2, \ldots, \overline{v}_n)$.
   (a) Prove: $\overline{\mathbf{u} \cdot \mathbf{v}} = \overline{\mathbf{u}} \cdot \overline{\mathbf{v}}$.
   (b) Prove: $\mathbf{u}$ and $\mathbf{v}$ are orthogonal if and only if $\overline{\mathbf{u}}$ and $\overline{\mathbf{v}}$ are orthogonal.

2. Show that if the matrix

$$\begin{bmatrix} a & b \\ -\overline{b} & \overline{a} \end{bmatrix}$$

   is nonzero, then it is invertible.

3. Find a basis for the solution space of the system

$$\begin{bmatrix} -1 & -i & 1 \\ -i & 1 & i \\ 1 & i & -1 \end{bmatrix} \begin{bmatrix} x_1 \\ x_2 \\ x_3 \end{bmatrix} = \begin{bmatrix} 0 \\ 0 \\ 0 \end{bmatrix}$$

4. Prove: If a and b are complex numbers such that $|a|^2 + |b|^2 = 1$, and if θ is a real number, then

$$A = \begin{bmatrix} a & b \\ -e^{i\theta}\overline{b} & e^{i\theta}\overline{a} \end{bmatrix}$$

   is a unitary matrix.

5. Find the eigenvalues of the matrix

$$\begin{bmatrix} 0 & 0 & 1 \\ 1 & 0 & \omega + 1 + \dfrac{1}{\omega} \\ 0 & 1 & -\omega - 1 - \dfrac{1}{\omega} \end{bmatrix}$$

   where $\omega = e^{2\pi i/3}$.

6. (a) Prove that if z is a complex number other than 1, then

$$1 + z + z^2 + \cdots + z^n = \frac{1 - z^{n+1}}{1 - z}$$

   [**Hint.** Let S be the sum on the left side of the equation and consider the quantity $S - zS$.]
   (b) Use the result in (a) to prove that if $z^n = 1$ and $z \neq 1$, then $1 + z + z^2 + \cdots + z^{n-1} = 0$.
   (c) Use the result in (a) to obtain Lagrange's trigonometric identity

$$1 + \cos\theta + \cos 2\theta + \cdots + \cos n\theta = \frac{1}{2} + \frac{\sin[(n + \frac{1}{2})\theta]}{2\sin(\theta/2)}$$

   for $0 < \theta < 2\pi$. [**Hint.** Let $z = \cos\theta + i\sin\theta$.]

7. Let $\omega = e^{2\pi i/3}$. Show that the vectors $\mathbf{v}_1 = (1/\sqrt{3})(1, 1, 1)$, $\mathbf{v}_2 = (1/\sqrt{3})(1, \omega, \omega^2)$, and $\mathbf{v}_3 = (1/\sqrt{3})(1, \omega^2, \omega^4)$ form an orthonormal set in C^3. [***Hint.*** Use part (b) of Exercise 6.]

8. Show that if U is an $n \times n$ unitary matrix and $|z_1| = |z_2| = \cdots = |z_n| = 1$, then the product

$$U \begin{bmatrix} z_1 & 0 & 0 & \cdots & 0 \\ 0 & z_2 & 0 & \cdots & 0 \\ \vdots & \vdots & \vdots & & \vdots \\ 0 & 0 & 0 & \cdots & z_n \end{bmatrix}$$

is also unitary.

9. Suppose that $A^* = -A$.
   (a) Show that iA is Hermitian.
   (b) Show that A is unitarily diagonalizable and has pure imaginary eigenvalues.

10. (a) Show that the set of complex numbers with the operations

$$(a + bi) + (c + di) = (a + c) + (b + d)i \text{ and } k(a + bi) = ka + kbi$$

where k is a real number, is a *real* vector space.
    (b) What is the dimension of this space?

ANSWERS
TO EXERCISES

EXERCISE SET 1.1 (page 6)

1. (a), (c), (f)   2. (a), (b), (c)

3. (a) $x = \frac{3}{7} + \frac{5}{7}t$ (b) $x_1 = \frac{5}{3}s - \frac{4}{3}t + \frac{7}{3}$ $x_1 = \frac{1}{4}r - \frac{5}{8}s + \frac{3}{4}t - \frac{1}{8}$ $v = \frac{8}{3}q - \frac{2}{3}r + \frac{1}{3}s - \frac{4}{3}t$
   $y = t$ $x_2 = s$ $x_2 = r$ $w = q$
   $x_3 = t$ $x_3 = s$ $x = r$
   $x_4 = t$ $y = s$
   $z = t$

4. (a) $\begin{bmatrix} 3 & -2 & -1 \\ 4 & 5 & 3 \\ 7 & 3 & 2 \end{bmatrix}$ (b) $\begin{bmatrix} 2 & 0 & 2 & 1 \\ 3 & -1 & 4 & 7 \\ 6 & 1 & -1 & 0 \end{bmatrix}$

   (c) $\begin{bmatrix} 1 & 2 & 0 & -1 & 1 & 1 \\ 0 & 3 & 1 & 0 & -1 & 2 \\ 0 & 0 & 1 & 7 & 0 & 1 \end{bmatrix}$ (d) $\begin{bmatrix} 1 & 0 & 0 & 1 \\ 0 & 1 & 0 & 2 \\ 0 & 0 & 1 & 3 \end{bmatrix}$

5. (a) $2x_1 \qquad\quad = 0$ (b) $3x_1 \qquad\; - 2x_3 = \quad 5$ (c) $7x_1 + 2x_2 + x_3 - 3x_4 = \quad 5$
   $3x_1 - 4x_2 = 0$ $7x_1 + x_2 + 4x_3 = -3$ $x_1 + 2x_2 + 4x_3 \qquad\;\; = \quad 1$
   $x_2 = 1$ $-2x_2 + x_3 = \quad 7$
   (d) $x_1 \qquad\qquad\qquad\qquad = \quad 7$
   $x_2 \qquad\qquad\quad = -2$
   $x_3 \qquad\; = \quad 3$
   $x_4 = \quad 4$

6. (a) $x - 2y = 5$ (b) Let $x = t$; then $t - 2y = 5$. Solving for y yields $y = \frac{1}{2}t - \frac{5}{2}$.

8. $k = 6$: infinitely many solutions
   $k \neq 6$: no solutions
   No value of k yields one solution.

9. (a) The lines have no common point of intersection.   (b) The lines intersect in exactly one point.
   (c) The three lines coincide.

EXERCISE SET 1.2 (page 20)

1. (a), (b), (c), (d), (h), (i), (j)   2. (a), (b), (d)   3. (a) Both   (b) Neither   (c) Both
   (d) Row-echelon   (e) Neither   (f) Both

4. (a) $x_1 = -3, x_2 = 0, x_3 = 7$
   (b) $x_1 = 7t + 8, x_2 = -3t + 2, x_3 = -t - 5, x_4 = t$
   (c) $x_1 = 6s - 3t - 2, x_2 = s, x_3 = -4t + 7, x_4 = -5t + 8, x_5 = t$
   (d) Inconsistent

5. (a) $x_1 = -37, x_2 = -8, x_3 = 5$
   (b) $x_1 = 13t - 10, x_2 = 13t - 5, x_3 = -t + 2, x_4 = t$
   (c) $x_1 = -7s + 2t - 11, x_2 = s, x_3 = -3t - 4, x_4 = -3t + 9, x_5 = t$
   (d) Inconsistent

6. (a) $x_1 = 3, x_2 = 1, x_3 = 2$ (b) $x_1 = -\frac{1}{7} - \frac{3}{7}t, x_2 = \frac{1}{7} - \frac{4}{7}, x_3 = t$
   (c) $x = t - 1, y = 2s, z = s, w = t$ (d) Inconsistent

8. (a) Inconsistent   (b) $x_1 = -4, x_2 = 2, x_3 = 7$ (c) $x_1 = 3 + 2t, x_2 = t$
   (d) $x = \frac{8}{5} - \frac{3}{5}t - \frac{2}{5}s, y = \frac{1}{10} + \frac{2}{5}t - \frac{1}{10}s, z = t, w = s$

10. (a) $x_1 = 2 - 12t, x_2 = 5 - 27t, x_3 = t$ (b) Inconsistent
    (c) $u = -2s - 3t - 6, v = s, w = -t - 2, x = t + 3, y = t$

12. (a), (c), (d)   13. (a) $x_1 = 0, x_2 = 0, x_3 = 0$
    (b) $x_1 = -s, x_2 = -t - s, x_3 = 4s, x_4 = t$
    (c) $w = t, x = -t, y = t, z = 0$

14. (a) Only the trivial solution   (b) $u = 7s - 5t, v = -6s + 4t, w = 2s, x = 2t$
    (c) Only the trivial solution

15. $I_1 = -1, I_2 = 0, I_3 = 1, I_4 = 2$ 16. (a) $x = \frac{2}{3}a - \frac{1}{3}b, y = -\frac{1}{3}a + \frac{2}{3}b$
    (b) $x_1 = a - \frac{1}{3}c, x_2 = a - \frac{1}{2}b, x_3 = -a + \frac{1}{2}b + \frac{1}{3}c$

17. $a = -4$, none; $a \neq \pm 4$, exactly one; $a = 4$, infinitely many

19. $\begin{bmatrix} 1 & 3 \\ 0 & 1 \end{bmatrix}$ and $\begin{bmatrix} 1 & 0 \\ 0 & 1 \end{bmatrix}$ are possible answers.

20. $\alpha = \pi/2, \beta = \pi, \gamma = 0$ 21. $x = \pm 1, y = \pm\sqrt{3}, z = \pm\sqrt{2}$

23. $\lambda = 4, \lambda = 2$ 25. $a = 1, b = -6, c = 2, d = 10$ 26. $x^2 + y^2 - 2x - 4y - 29 = 0$

30. (a) $x_1 = x_2 = -\frac{1}{2}s, x_3 = s$ 34. One possible answer: $x + y + z = 0$
    (b) $x_1 = -\frac{1}{2}s, x_2 = 0, x_3 = s$ $x + y + z = 1$

EXERCISE SET 1.3 (page 34)

1. (a) Undefined   (b) 4×2 (c) Undefined   (d) Undefined   2. $a = 5, b = -3, c = 4, d = 1$
   (e) 5×5 (f) 5×2 (g) Undefined   (h) 5×2

3. (a) $\begin{bmatrix} 7 & 6 & 5 \\ -2 & 1 & 3 \\ 7 & 3 & 7 \end{bmatrix}$ (b) $\begin{bmatrix} -5 & 4 & -1 \\ 0 & -1 & -1 \\ -1 & 1 & 1 \end{bmatrix}$ (c) $\begin{bmatrix} 15 & 0 \\ -5 & 10 \\ 5 & 5 \end{bmatrix}$ (d) $\begin{bmatrix} -7 & -28 & -14 \\ -21 & -7 & -35 \end{bmatrix}$

   (e) Undefined   (f) $\begin{bmatrix} 22 & -6 & 8 \\ -2 & 4 & 6 \\ 10 & 0 & 4 \end{bmatrix}$ (g) $\begin{bmatrix} -39 & -21 & -24 \\ 9 & -6 & -15 \\ -33 & -12 & -30 \end{bmatrix}$ (h) $\begin{bmatrix} 0 & 0 \\ 0 & 0 \\ 0 & 0 \end{bmatrix}$

(i) 5   (j) −25   (k) 168   (l) Undefined

4. (a) $\begin{bmatrix} 7 & 2 & 4 \\ 3 & 5 & 7 \end{bmatrix}$ (b) $\begin{bmatrix} -5 & 0 & -1 \\ 4 & -1 & 1 \\ -1 & -1 & 1 \end{bmatrix}$ (c) $\begin{bmatrix} -5 & 0 & -1 \\ 4 & -1 & 1 \\ -1 & -1 & 1 \end{bmatrix}$ (d) Undefined

(e) $\begin{bmatrix} -\frac{1}{4} & \frac{3}{2} \\ \frac{9}{4} & 0 \\ \frac{3}{4} & \frac{9}{4} \end{bmatrix}$ (f) $\begin{bmatrix} 0 & -1 \\ 1 & 0 \end{bmatrix}$ (g) $\begin{bmatrix} 9 & 1 & -1 \\ -13 & 2 & -4 \\ 0 & 1 & -6 \end{bmatrix}$ (h) $\begin{bmatrix} 9 & -13 & 0 \\ 1 & 2 & 1 \\ -1 & -4 & -6 \end{bmatrix}$

5. (a) $\begin{bmatrix} 12 & -3 \\ -4 & 5 \\ 4 & 1 \end{bmatrix}$ (b) Undefined   (c) $\begin{bmatrix} 42 & 108 & 75 \\ 12 & -3 & 21 \\ 36 & 78 & 63 \end{bmatrix}$ (d) $\begin{bmatrix} 3 & 45 & 9 \\ 11 & -11 & 17 \\ 7 & 17 & 13 \end{bmatrix}$

(e) $\begin{bmatrix} 3 & 45 & 9 \\ 11 & -11 & 17 \\ 7 & 17 & 13 \end{bmatrix}$ (f) $\begin{bmatrix} 21 & 17 \\ 17 & 35 \end{bmatrix}$ (g) $\begin{bmatrix} 0 & -2 & 11 \\ 12 & 1 & 8 \end{bmatrix}$ (h) $\begin{bmatrix} 12 & 6 & 9 \\ 48 & -20 & 14 \\ 24 & 8 & 16 \end{bmatrix}$

(i) 61   (j) 35   (k) 28

6. (a) $\begin{bmatrix} -6 & -3 \\ 36 & 0 \\ 4 & 7 \end{bmatrix}$ (b) Undefined   (c) $\begin{bmatrix} 2 & -10 & 11 \\ 13 & 2 & 5 \\ 4 & -3 & 13 \end{bmatrix}$ (d) $\begin{bmatrix} 10 & -6 \\ -14 & 2 \\ -1 & -8 \end{bmatrix}$

(e) $\begin{bmatrix} 40 & 72 \\ 26 & 42 \end{bmatrix}$ (f) $\begin{bmatrix} 0 & 0 & 0 \\ 0 & 0 & 0 \\ 0 & 0 & 0 \end{bmatrix}$

7. (a) [67 41 41]   (b) [63 67 57]   (c) $\begin{bmatrix} 41 \\ 21 \\ 67 \end{bmatrix}$ (d) $\begin{bmatrix} 6 \\ 6 \\ 63 \end{bmatrix}$ (e) [24 56 97]   (f) $\begin{bmatrix} 76 \\ 98 \\ 97 \end{bmatrix}$

8. (a) $\begin{bmatrix} 67 \\ 64 \\ 63 \end{bmatrix} = 6\begin{bmatrix} 3 \\ 6 \\ 0 \end{bmatrix} + 0\begin{bmatrix} -2 \\ 5 \\ 4 \end{bmatrix} + 7\begin{bmatrix} 7 \\ 4 \\ 9 \end{bmatrix}$ (b) $\begin{bmatrix} 6 \\ 6 \\ 63 \end{bmatrix} = 3\begin{bmatrix} 6 \\ 0 \\ 7 \end{bmatrix} + 6\begin{bmatrix} -2 \\ 1 \\ 7 \end{bmatrix} + 0\begin{bmatrix} 4 \\ 3 \\ 5 \end{bmatrix}$

$\begin{bmatrix} 41 \\ 21 \\ 67 \end{bmatrix} = -2\begin{bmatrix} 3 \\ 6 \\ 0 \end{bmatrix} + 1\begin{bmatrix} -2 \\ 5 \\ 4 \end{bmatrix} + 7\begin{bmatrix} 7 \\ 4 \\ 9 \end{bmatrix}$ $\begin{bmatrix} -6 \\ 17 \\ 41 \end{bmatrix} = -2\begin{bmatrix} 6 \\ 0 \\ 7 \end{bmatrix} + 5\begin{bmatrix} -2 \\ 1 \\ 7 \end{bmatrix} + 4\begin{bmatrix} 4 \\ 3 \\ 5 \end{bmatrix}$

$\begin{bmatrix} 41 \\ 59 \\ 57 \end{bmatrix} = 4\begin{bmatrix} 3 \\ 6 \\ 0 \end{bmatrix} + 3\begin{bmatrix} -2 \\ 5 \\ 4 \end{bmatrix} + 5\begin{bmatrix} 7 \\ 4 \\ 9 \end{bmatrix}$ $\begin{bmatrix} 70 \\ 31 \\ 122 \end{bmatrix} = 7\begin{bmatrix} 6 \\ 0 \\ 7 \end{bmatrix} + 4\begin{bmatrix} -2 \\ 1 \\ 7 \end{bmatrix} + 9\begin{bmatrix} 4 \\ 3 \\ 5 \end{bmatrix}$

10. (a) [67 41 41] = 3[6 −2 4] − 2 [0 1 3] + 7[7 7 5]
    [64 21 59] = 6[6 −2 4] + 5 [0 1 3] + 4[7 7 5]
    [63 67 57] = 0[6 −2 4] + 4[0 1 3] + 9[7 7 5]
  (b) [6 −6 70] = 6[3 −2 7] − 2[6 5 4] + 4[0 4 9]
    [6 17 31] = 0[3 −2 7] + 1[6 5 4] + 3[0 4 9]
    [63 41 122] = 7[3 −2 7] + 7[6 5 4] + 5[0 4 9]

13. (a) $A = \begin{bmatrix} 2 & -3 & 5 \\ 9 & -1 & 1 \\ 1 & 5 & 4 \end{bmatrix}$, $\mathbf{x} = \begin{bmatrix} x_1 \\ x_2 \\ x_3 \end{bmatrix}$, $\mathbf{b} = \begin{bmatrix} 7 \\ -1 \\ 0 \end{bmatrix}$ (b) $A = \begin{bmatrix} 4 & 0 & -3 & 1 \\ 5 & 1 & 0 & -8 \\ 2 & -5 & 9 & -1 \\ 0 & 3 & -1 & 7 \end{bmatrix}$, $\mathbf{x} = \begin{bmatrix} x_1 \\ x_2 \\ x_3 \\ x_4 \end{bmatrix}$, $\mathbf{b} = \begin{bmatrix} 1 \\ 3 \\ 0 \\ 2 \end{bmatrix}$

14. (a)
$$\begin{aligned} 3x_1 - x_2 + 2x_3 &= 2 \\ 4x_1 + 3x_2 + 7x_3 &= -1 \\ -2x_1 + x_2 + 5x_3 &= 4 \end{aligned}$$
(b)
$$\begin{aligned} 3w - 2x \quad\quad + z &= 0 \\ 5w \quad\quad + 2y - 2z &= 0 \\ 3w + x + 4y + 7z &= 0 \\ -2w + 5x + y + 6z &= 0 \end{aligned}$$

15. $\begin{bmatrix} -1 & 23 & -10 \\ 37 & -13 & 8 \\ 29 & 23 & 41 \end{bmatrix}$

16. (a) $\begin{bmatrix} -3 & -15 & -11 \\ 21 & -15 & 44 \end{bmatrix}$ (b) $\begin{bmatrix} 4 & -7 & -19 & -43 \\ 2 & 2 & 18 & 17 \\ 0 & 5 & 25 & 35 \\ 2 & 3 & 23 & 24 \end{bmatrix}$

(c) $\begin{bmatrix} 3 & 3 \\ -1 & 4 \\ 1 & 5 \\ 4 & -4 \\ 0 & 14 \end{bmatrix}$

17. (a) A_{11} is a 2×3 matrix and B_{11} is a 2×2 matrix. $A_{11}B_{11}$ does not exist.

(b) $\begin{bmatrix} -1 & 23 & -10 \\ 37 & -13 & 8 \\ 29 & 23 & 41 \end{bmatrix}$

21. (a) $\begin{bmatrix} a_{11} & 0 & 0 & 0 & 0 & 0 \\ 0 & a_{22} & 0 & 0 & 0 & 0 \\ 0 & 0 & a_{33} & 0 & 0 & 0 \\ 0 & 0 & 0 & a_{44} & 0 & 0 \\ 0 & 0 & 0 & 0 & a_{55} & 0 \\ 0 & 0 & 0 & 0 & 0 & a_{66} \end{bmatrix}$ (b) $\begin{bmatrix} a_{11} & a_{12} & a_{13} & a_{14} & a_{15} & a_{16} \\ 0 & a_{22} & a_{23} & a_{24} & a_{25} & a_{26} \\ 0 & 0 & a_{33} & a_{34} & a_{35} & a_{36} \\ 0 & 0 & 0 & a_{44} & a_{45} & a_{46} \\ 0 & 0 & 0 & 0 & a_{55} & a_{56} \\ 0 & 0 & 0 & 0 & 0 & a_{66} \end{bmatrix}$

(c) $\begin{bmatrix} a_{11} & 0 & 0 & 0 & 0 & 0 \\ a_{21} & a_{22} & 0 & 0 & 0 & 0 \\ a_{31} & a_{32} & a_{33} & 0 & 0 & 0 \\ a_{41} & a_{42} & a_{43} & a_{44} & 0 & 0 \\ a_{51} & a_{52} & a_{53} & a_{54} & a_{55} & 0 \\ a_{61} & a_{62} & a_{63} & a_{64} & a_{65} & a_{66} \end{bmatrix}$ (d) $\begin{bmatrix} a_{11} & a_{12} & 0 & 0 & 0 & 0 \\ a_{21} & a_{22} & a_{23} & 0 & 0 & 0 \\ 0 & a_{32} & a_{33} & a_{34} & 0 & 0 \\ 0 & 0 & a_{43} & a_{44} & a_{45} & 0 \\ 0 & 0 & 0 & a_{54} & a_{55} & a_{56} \\ 0 & 0 & 0 & 0 & a_{65} & a_{66} \end{bmatrix}$

22. (a) $\begin{bmatrix} 2 & 3 & 4 & 5 \\ 3 & 4 & 5 & 6 \\ 4 & 5 & 6 & 7 \\ 5 & 6 & 7 & 8 \end{bmatrix}$ **(b)** $\begin{bmatrix} 1 & 1 & 1 & 1 \\ 1 & 2 & 4 & 8 \\ 1 & 3 & 9 & 27 \\ 1 & 4 & 16 & 64 \end{bmatrix}$ **(c)** $\begin{bmatrix} -1 & -1 & 1 & 1 \\ -1 & -1 & -1 & 1 \\ 1 & -1 & -1 & -1 \\ 1 & 1 & -1 & -1 \end{bmatrix}$

EXERCISE SET 1.4 (page 48)

4. $A^{-1} = \begin{bmatrix} 2 & -1 \\ -5 & 3 \end{bmatrix}$, $B^{-1} = \begin{bmatrix} \frac{1}{5} & \frac{3}{20} \\ -\frac{1}{5} & \frac{1}{10} \end{bmatrix}$, $C^{-1} = \begin{bmatrix} \frac{1}{2} & 0 \\ 0 & \frac{1}{3} \end{bmatrix}$ **6.** No

7. (a) $A = \begin{bmatrix} \frac{5}{13} & \frac{1}{13} \\ -\frac{3}{13} & \frac{2}{13} \end{bmatrix}$ **(b)** $A = \begin{bmatrix} \frac{2}{7} & 1 \\ \frac{1}{7} & \frac{3}{7} \end{bmatrix}$ **(c)** $A = \begin{bmatrix} -\frac{2}{5} & 1 \\ -\frac{1}{5} & \frac{3}{5} \end{bmatrix}$ **(d)** $A = \begin{bmatrix} -\frac{9}{13} & \frac{1}{13} \\ \frac{2}{13} & -\frac{6}{13} \end{bmatrix}$

8. $A^3 = \begin{bmatrix} 8 & 0 \\ 28 & 1 \end{bmatrix}$, $A^{-3} = \begin{bmatrix} \frac{1}{8} & 0 \\ -\frac{7}{2} & 1 \end{bmatrix}$, $A^2 - 2A + I = \begin{bmatrix} 1 & 0 \\ 4 & 0 \end{bmatrix}$

9. (a) $p(A) = \begin{bmatrix} 1 & 1 \\ 2 & -1 \end{bmatrix}$ **(b)** $p(A) = \begin{bmatrix} 20 & 7 \\ 14 & 6 \end{bmatrix}$ **(c)** $p(A) = \begin{bmatrix} 39 & 13 \\ 26 & 13 \end{bmatrix}$

11. $\begin{bmatrix} \cos\theta & -\sin\theta \\ \sin\theta & \cos\theta \end{bmatrix}$ **12. (c)** $(A + B)^2 = A^2 + AB + BA + B^2$ **13.** $A^{-1} = \begin{bmatrix} \dfrac{1}{a_{11}} & 0 & \cdots & 0 \\ 0 & \dfrac{1}{a_{22}} & \cdots & 0 \\ \vdots & \vdots & & \vdots \\ 0 & 0 & \cdots & \dfrac{1}{a_{nn}} \end{bmatrix}$

18. $0A$ and $A0$ may not have the same size.   **19.** $\begin{bmatrix} \pm 1 & 0 & 0 \\ 0 & \pm 1 & 0 \\ 0 & 0 & \pm 1 \end{bmatrix}$

20. (a) One example is $\begin{bmatrix} 1 & 2 & 3 \\ 2 & 1 & 4 \\ 3 & 4 & 5 \end{bmatrix}$.

(b) One example is $\begin{bmatrix} 0 & -1 & -1 \\ 1 & 0 & -1 \\ 1 & 1 & 0 \end{bmatrix}$.

EXERCISE SET 1.5 (page 56)

1. (c), (d), (f)   **2. (a)** Add three times the first row to the second row.
   (b) Multiply the third row by $\frac{1}{3}$.
   (c) Interchange the first row and the fourth row.
   (d) Add $\frac{1}{7}$ times third row to the first row.

3. (a) $\begin{bmatrix} 0 & 0 & 1 \\ 0 & 1 & 0 \\ 1 & 0 & 0 \end{bmatrix}$ (b) $\begin{bmatrix} 0 & 0 & 1 \\ 0 & 1 & 0 \\ 1 & 0 & 0 \end{bmatrix}$ (c) $\begin{bmatrix} 1 & 0 & 0 \\ 0 & 1 & 0 \\ -2 & 0 & 1 \end{bmatrix}$ (d) $\begin{bmatrix} 1 & 0 & 0 \\ 0 & 1 & 0 \\ 2 & 0 & 1 \end{bmatrix}$

4. No, since C cannot be obtained by performing a single row operation on B.

5. (a) $\begin{bmatrix} -7 & 4 \\ 2 & -1 \end{bmatrix}$ (b) $\begin{bmatrix} -\frac{5}{39} & \frac{2}{13} \\ \frac{4}{39} & \frac{1}{13} \end{bmatrix}$ (c) Not invertible

6. (a) $\begin{bmatrix} \frac{3}{2} & -\frac{11}{10} & -\frac{6}{5} \\ -1 & 1 & 1 \\ -\frac{1}{2} & \frac{7}{10} & \frac{2}{5} \end{bmatrix}$ (b) Not invertible (c) $\begin{bmatrix} \frac{1}{2} & -\frac{1}{2} & \frac{1}{2} \\ -\frac{1}{2} & \frac{1}{2} & \frac{1}{2} \\ \frac{1}{2} & \frac{1}{2} & -\frac{1}{2} \end{bmatrix}$

(d) $\begin{bmatrix} \frac{7}{2} & 0 & -3 \\ -1 & 1 & 0 \\ 0 & -1 & 1 \end{bmatrix}$ (e) $\begin{bmatrix} \frac{1}{2} & -\frac{1}{2} & \frac{1}{2} \\ 0 & 0 & 1 \\ \frac{1}{2} & \frac{1}{2} & -\frac{1}{2} \end{bmatrix}$

7. (a) $\begin{bmatrix} 1 & 3 & 1 \\ 0 & 1 & -1 \\ -2 & 2 & 0 \end{bmatrix}$ (b) $\begin{bmatrix} \frac{\sqrt{2}}{26} & \frac{-3\sqrt{2}}{26} & 0 \\ \frac{4\sqrt{2}}{26} & \frac{\sqrt{2}}{26} & 0 \\ 0 & 0 & 1 \end{bmatrix}$ (c) $\begin{bmatrix} 1 & 0 & 0 & 0 \\ -\frac{1}{3} & \frac{1}{3} & 0 & 0 \\ 0 & -\frac{1}{5} & \frac{1}{5} & 0 \\ 0 & 0 & -\frac{1}{7} & \frac{1}{7} \end{bmatrix}$ (d) Not invertible

8. (a) $\begin{bmatrix} \frac{1}{k_1} & 0 & 0 & 0 \\ 0 & \frac{1}{k_2} & 0 & 0 \\ 0 & 0 & \frac{1}{k_3} & 0 \\ 0 & 0 & 0 & \frac{1}{k_4} \end{bmatrix}$ (b) $\begin{bmatrix} 0 & 0 & 0 & \frac{1}{k_4} \\ 0 & 0 & \frac{1}{k_3} & 0 \\ 0 & \frac{1}{k_2} & 0 & 0 \\ \frac{1}{k_1} & 0 & 0 & 0 \end{bmatrix}$ (c) $\begin{bmatrix} \frac{1}{k} & 0 & 0 & 0 \\ -\frac{1}{k^2} & \frac{1}{k} & 0 & 0 \\ \frac{1}{k^3} & -\frac{1}{k^2} & \frac{1}{k} & 0 \\ -\frac{1}{k^4} & \frac{1}{k^3} & -\frac{1}{k^2} & \frac{1}{k} \end{bmatrix}$

9. (a) $E_1 = \begin{bmatrix} 1 & 0 \\ 5 & 1 \end{bmatrix}$, $E_2 = \begin{bmatrix} 1 & 0 \\ 0 & \frac{1}{2} \end{bmatrix}$ (b) $A^{-1} = E_2 E_1$ (c) $A = E_1^{-1} E_2^{-1}$

11. $\begin{bmatrix} 1 & 0 & 0 \\ 0 & 1 & 0 \\ 0 & -2 & 1 \end{bmatrix} \begin{bmatrix} 1 & 0 & 0 \\ 0 & 1 & 0 \\ 1 & 0 & 1 \end{bmatrix} \begin{bmatrix} 0 & 1 & 0 \\ 1 & 0 & 0 \\ 0 & 0 & 1 \end{bmatrix} \begin{bmatrix} 1 & 3 & 3 & 8 \\ 0 & 1 & 7 & 8 \\ 0 & 0 & 0 & 0 \end{bmatrix}$

16. (b) Add -1 times the first row to the second row.
Add -1 times the first row to the third row.
Add -1 times the second row to the first row.
Add the second row to the third row.

EXERCISE SET 1.6 (page 64)

1. $x_1 = 3$, $x_2 = -1$ 2. $x_1 = -3$, $x_2 = -3$ 3. $x_1 = -1$, $x_2 = 4$, $x_3 = -7$ 4. $x_1 = 1$, $x_2 = -11$, $x_3 = 16$

5. $x_1 = 1$, $x_2 = 5$, $x_3 = -1$ 6. $w = -6$, $x = 1$, $y = 10$, $z = -7$ 7. $x_1 = 2b_1 - 5b_2$, $x_2 = -b_1 + 3b_2$

8. $x_1 = -\frac{15}{2}b_1 + \frac{1}{2}b_2 + \frac{5}{2}b_3$, $x_2 = \frac{1}{2}b_1 + \frac{1}{2}b_2 - \frac{1}{2}b_3$, $x_3 = \frac{5}{2}b_1 - \frac{1}{2}b_2 - \frac{1}{2}b_3$

9. (a) $x_1 = \frac{16}{3}, x_2 = -\frac{4}{3}, x_3 = -\frac{11}{3}$ (b) $x_1 = -\frac{5}{3}, x_2 = \frac{5}{3}, x_3 = \frac{10}{3}$ (c) $x_1 = 3, x_2 = 0, x_3 = -4$

11. (a) $x_1 = \frac{22}{17}, x_2 = \frac{1}{17}$ 12. (a) $x_1 = -18, x_2 = -1, x_3 = -14$
    (b) $x_1 = \frac{21}{17}, x_2 = \frac{11}{17}$ (b) $x_1 = -\frac{421}{2}, x_2 = -\frac{25}{2}, x_3 = -\frac{327}{2}$

13. (a) $x_1 = \frac{7}{15}, x_2 = \frac{4}{15}$ 14. (a) $x_1 = 18, x_2 = -9, x_3 = 2$ 15. (a) $x_1 = -12 - 3t, x_2 = -5 - t, x_3 = t$
    (b) $x_1 = \frac{34}{15}, x_2 = \frac{28}{15}$ (b) $x_1 = -23, x_2 = 11, x_3 = -2$ (b) $x_1 = 7 - 3t, x_2 = 3 - t, x_3 = t$
    (c) $x_1 = \frac{19}{15}, x_2 = \frac{13}{15}$ (c) $x_1 = 5, x_2 = -2, x_3 = 0$
    (d) $x_1 = -\frac{1}{5}, x_2 = \frac{3}{5}$

16. $b_1 = 2b_2$ 17. $b_1 = b_2 + b_3$ 18. No restrictions   19. $b_1 = b_3 + b_4, b_2 = 2b_3 + b_4$

21. $X = \begin{bmatrix} 11 & 12 & -3 & 27 & 26 \\ -6 & -8 & 1 & -18 & -17 \\ -15 & -21 & 9 & -38 & -35 \end{bmatrix}$ 22. (a) Only the trivial solution $x_1 = x_2 = x_3 = x_4 = 0$; invertible
    (b) Infinitely many solutions; not invertible

EXERCISE SET 1.7 (page 72)

1. (a) $\begin{bmatrix} \frac{1}{2} & 0 \\ 0 & -\frac{1}{5} \end{bmatrix}$ (b) Not invertible (c) $\begin{bmatrix} -1 & 0 & 0 \\ 0 & \frac{1}{2} & 0 \\ 0 & 0 & 3 \end{bmatrix}$

2. (a) $\begin{bmatrix} 6 & 3 \\ 4 & -1 \\ 4 & 10 \end{bmatrix}$ (b) $\begin{bmatrix} -24 & -10 & 12 \\ 3 & -10 & 0 \\ 60 & 20 & -16 \end{bmatrix}$

3. (a) $A^2 = \begin{bmatrix} 1 & 0 \\ 0 & 4 \end{bmatrix}$, $A^{-2} = \begin{bmatrix} 1 & 0 \\ 0 & \frac{1}{4} \end{bmatrix}$, $A^{-k} = \begin{bmatrix} 1 & 0 \\ 0 & 1/(-2)^k \end{bmatrix}$

   (b) $A^2 = \begin{bmatrix} \frac{1}{4} & 0 & 0 \\ 0 & \frac{1}{9} & 0 \\ 0 & 0 & \frac{1}{16} \end{bmatrix}$, $A^{-2} = \begin{bmatrix} 4 & 0 & 0 \\ 0 & 9 & 0 \\ 0 & 0 & 16 \end{bmatrix}$, $A^{-k} = \begin{bmatrix} 2^k & 0 & 0 \\ 0 & 3^k & 0 \\ 0 & 0 & 4^k \end{bmatrix}$

4. (b), (c)   5. (a)   6. $a = 11, b = -9, c = -13$ 7. $a = 2, b = -1$

8. (a) Does not commute (b) Commutes   10. (a) $\begin{bmatrix} 1 & 0 & 0 \\ 0 & -1 & 0 \\ 0 & 0 & -1 \end{bmatrix}$ (b) $\begin{bmatrix} \frac{1}{3} & 0 & 0 \\ 0 & \frac{1}{2} & 0 \\ 0 & 0 & 1 \end{bmatrix}$

11. (a) $\begin{bmatrix} a_{11} & a_{12} & a_{13} \\ a_{21} & a_{22} & a_{23} \\ a_{31} & a_{32} & a_{33} \end{bmatrix}\begin{bmatrix} 3 & 0 & 0 \\ 0 & 5 & 0 \\ 0 & 0 & 7 \end{bmatrix}$ (b) No   19. $\frac{n}{2}(1 + n)$

20. (a) Symmetric (b) Not symmetric (c) Symmetric (d) Not symmetric

24. (a) $x_1 = \frac{7}{4}, x_2 = 1, x_3 = -\frac{1}{2}$ (b) $x_1 = -\frac{3}{2}, x_2 = -\frac{5}{2}, x_3 = -3$

CHAPTER 1 SUPPLEMENTARY EXERCISES (page 74)

1. $x' = \frac{3}{5}x + \frac{4}{5}y, y' = -\frac{4}{5}x + \frac{3}{5}y$ 2. $x' = x \cos\theta + y \sin\theta, y' = -x \sin\theta + y \cos\theta$

3. One possible answer is   4. 3 pennies, 4 nickels, 6 dimes
   $x_1 - 2x_2 - x_3 - x_4 = 0$
   $x_1 + 5x_2 \quad + 2x_4 = 0$

5. $x = 4$, $y = 2$, $z = 3$ **6.** Infinitely many if $a = 2$ or $a = -\frac{3}{2}$; none otherwise

7. (a) $a \neq 0$, $b \neq 2$ (b) $a \neq 0$, $b = 2$ **8.** $x = \frac{5}{9}$, $y = 9$, $z = \frac{1}{3}$ **9.** $K = \begin{bmatrix} 0 & 2 \\ 1 & 1 \end{bmatrix}$
(c) $a = 0$, $b = 2$ (d) $a = 0$, $b \neq 2$

10. $a = 2$, $b = -1$, $c = 1$ **11.** (a) $X = \begin{bmatrix} -1 & 3 & -1 \\ 6 & 0 & 1 \end{bmatrix}$ (b) $X = \begin{bmatrix} 1 & -2 \\ 3 & 1 \end{bmatrix}$ (c) $X = \begin{bmatrix} -\frac{113}{37} & -\frac{160}{37} \\ -\frac{20}{37} & -\frac{46}{37} \end{bmatrix}$

12. (a) $Z = \begin{bmatrix} -1 & -7 & 11 \\ 14 & 10 & -26 \end{bmatrix} X$ (b) $z_1 = -x_1 - 7x_2 + 11x_3$
$\qquad\qquad\qquad\qquad\qquad\qquad z_2 = 14x_1 + 10x_2 - 26x_3$

13. *mpn* multiplications and $mp(n - 1)$ additions   **15.** $a = 1$, $b = -2$, $c = 3$

16. $a = 1$, $b = -4$, $c = -5$ **26.** $A = -\frac{7}{5}$, $B = \frac{4}{5}$, $C = \frac{3}{5}$

EXERCISE SET 2.1 (page 85)

1. (a) 5 (b) 9 (c) 6 (d) 10 (e) 0 (f) 2

2. (a) Odd (b) Odd (c) Even (d) Even (e) Even (f) Even

3. 22   **4.** 0   **5.** 52   **6.** $-3\sqrt{6}$ **7.** $a^2 - 5a + 21$ **8.** 0   **9.** -65 **10.** -4 **11.** -123

12. $-c^4 + c^3 - 16c^2 + 8c - 2$ **13.** (a) $\lambda = 1$, $\lambda = -3$ (b) $\lambda = -2$, $\lambda = 3$, $\lambda = 4$

16. 275   **17.** (a) $= -120$ (b) $= -120$ **18.** $x = \dfrac{3 \pm \sqrt{33}}{4}$

EXERCISE SET 2.2 (page 91)

1. (a) -30 (b) -2 (c) 0 (d) 0   **3.** (a) -5 (b) -1 (c) 1   **4.** 30   **5.** 5   **6.** -17

7. 33   **8.** 39   **9.** 6   **10.** $-\frac{1}{6}$ **11.** -2 **12.** (a) -6 (b) 72 (c) -6 (d) 18

EXERCISE SET 2.3 (page 99)

1. (a) $\det(2A) = -40 = 2^2 \det(A)$ (b) $\det(-2A) = -208 = (-2)^3 \det(A)$

2. $\det AB = -170 = (\det A)(\det B)$ **4.** (a) Invertible (b) Not invertible (c) Not invertible (d) Not invertible

5. (a) -189 (b) $-\frac{1}{7}$ (c) $-\frac{8}{7}$ (d) $-\frac{1}{56}$ (e) 7   **6.** If $x = 0$, the first and third rows are proportional.
$\qquad\qquad\qquad\qquad\qquad\qquad\qquad\qquad\qquad\qquad\qquad\qquad$ If $x = 2$, the first and second rows are proportional.

12. (a) $k = \dfrac{5 \pm \sqrt{17}}{2}$ (b) $k = -1$ **14.** (a) $\begin{bmatrix} \lambda - 1 & -2 \\ -2 & \lambda - 1 \end{bmatrix} \begin{bmatrix} x_1 \\ x_2 \end{bmatrix} = \begin{bmatrix} 0 \\ 0 \end{bmatrix}$

$\qquad\qquad\qquad\qquad\qquad\qquad\qquad\qquad\qquad\qquad$ (b) $\begin{bmatrix} \lambda - 2 & -3 \\ -4 & \lambda - 3 \end{bmatrix} \begin{bmatrix} x_1 \\ x_2 \end{bmatrix} = \begin{bmatrix} 0 \\ 0 \end{bmatrix}$

$\qquad\qquad\qquad\qquad\qquad\qquad\qquad\qquad\qquad\qquad$ (c) $\begin{bmatrix} \lambda - 3 & -1 \\ 5 & \lambda + 3 \end{bmatrix} \begin{bmatrix} x_1 \\ x_2 \end{bmatrix} = \begin{bmatrix} 0 \\ 0 \end{bmatrix}$

15. (a) $\lambda^2 - 2\lambda - 3 = 0$ (b) $\lambda = -1$, $\lambda = 3$ (c) $\begin{bmatrix} -t \\ t \end{bmatrix}$, $\begin{bmatrix} t \\ t \end{bmatrix}$

(a) $\lambda^2 - 5\lambda - 6 = 0$ (b) $\lambda = -1, \lambda = 6$ (c) $\begin{bmatrix} -t \\ t \end{bmatrix}, \begin{bmatrix} \frac{3}{4}t \\ t \end{bmatrix}$

(a) $\lambda^2 - 4 = 0$ (b) $\lambda = -2, \lambda = 2$ (c) $\begin{bmatrix} \frac{t}{5} \\ t \end{bmatrix}, \begin{bmatrix} -t \\ t \end{bmatrix}$

EXERCISE SET 2.4 (page 111)

1. (a) $M_{11} = 29, M_{12} = 21, M_{13} = 27, M_{21} = -11, M_{22} = 13, M_{23} = -5, M_{31} = -19, M_{32} = -19, M_{33} = 19$
 (b) $C_{11} = 29, C_{12} = -21, C_{13} = 27, C_{21} = 11, C_{22} = 13, C_{23} = 5, C_{31} = -19, C_{32} = 19, C_{33} = 19$

2. (a) $M_{13} = 0, C_{13} = 0$ (b) $M_{23} = -96, C_{23} = 96$ 3. 152
 (c) $M_{22} = -48, C_{22} = -48$ (d) $M_{21} = 72, C_{21} = -72$

4. (a) $\text{adj}(A) = \begin{bmatrix} 29 & 11 & -19 \\ -21 & 13 & 19 \\ 27 & 5 & 19 \end{bmatrix}$ (b) $A^{-1} = \begin{bmatrix} \frac{29}{152} & \frac{11}{152} & -\frac{19}{152} \\ -\frac{21}{152} & \frac{13}{152} & \frac{19}{152} \\ \frac{27}{152} & \frac{5}{152} & \frac{19}{152} \end{bmatrix}$

5. -40 6. -66 7. 0   8. $k^3 - 8k^2 - 10k + 95$ 9. -240 10. 0

11. $A^{-1} = \begin{bmatrix} 3 & -5 & -5 \\ -3 & 4 & 5 \\ 2 & -2 & -3 \end{bmatrix}$ 12. $A^{-1} = \begin{bmatrix} 2 & 0 & \frac{3}{2} \\ \frac{2}{3} & \frac{1}{3} & \frac{2}{3} \\ -1 & 0 & -1 \end{bmatrix}$ 13. $A^{-1} = \begin{bmatrix} \frac{1}{2} & \frac{3}{2} & 1 \\ 0 & 1 & \frac{3}{2} \\ 0 & 0 & \frac{1}{2} \end{bmatrix}$

14. $A^{-1} = \begin{bmatrix} \frac{1}{2} & 0 & 0 \\ -4 & 1 & 0 \\ \frac{29}{12} & -\frac{1}{2} & \frac{1}{6} \end{bmatrix}$ 15. $A^{-1} = \begin{bmatrix} -4 & 3 & 0 & -1 \\ 2 & -1 & 0 & 0 \\ -7 & 0 & -1 & 8 \\ 6 & 0 & 1 & -7 \end{bmatrix}$ 16. $x_1 = 1, x_2 = 2$

17. $x = \frac{3}{11}, y = \frac{2}{11}, z = -\frac{1}{11}$ 18. $x = -\frac{144}{55}, y = -\frac{61}{55}, z = \frac{46}{11}$

19. $x_1 = -\frac{30}{11}, x_2 = -\frac{38}{11}, x_3 = -\frac{40}{11}$ 20. $x_1 = 5, x_2 = 8, x_3 = 3, x_4 = -1$

21. Cramer's Rule does not apply.   22. $A^{-1} = \begin{bmatrix} \cos\theta & -\sin\theta & 0 \\ \sin\theta & \cos\theta & 0 \\ 0 & 0 & 1 \end{bmatrix}$ 23. $y = 0$

24. $x = 1, y = 0, z = 2, w = 0$

CHAPTER 2 SUPPLEMENTARY EXERCISES (page 113)

1. $x' = \frac{3}{5}x + \frac{4}{5}y, y' = -\frac{4}{5}x + \frac{3}{5}y$ 2. $x' = x\cos\theta + y\sin\theta, y' = -x\sin\theta + y\cos\theta$ 4. 2

5. $\cos\beta = \dfrac{c^2 + a^2 - b^2}{2ac}, \cos\gamma = \dfrac{a^2 + b^2 - c^2}{2ab}$ 10. (b) $\frac{19}{2}$ 12. $\det(B) = -1^{n(n-1)/2}\det(A)$

13. (a) The ith and jth columns will be interchanged.
 (b) The ith column will be divided by c.
 (c) $-c$ times the jth column will be added to the ith column.

15. (a) $\lambda^3 + (-a_{11} - a_{22} - a_{33})\lambda^2 + (a_{11}a_{22} + a_{11}a_{33} + a_{22}a_{33} - a_{12}a_{21} - a_{13}a_{31} - a_{23}a_{32})\lambda +$
 $(a_{11}a_{23}a_{32} + a_{12}a_{21}a_{33} + a_{13}a_{22}a_{31} - a_{11}a_{22}a_{33} - a_{12}a_{23}a_{31} - a_{13}a_{21}a_{32})$

18. (a) $\lambda = -5, \lambda = 2, \lambda = 4;$ $\begin{bmatrix} -2t \\ t \\ t \end{bmatrix}, \begin{bmatrix} 5t \\ t \\ t \end{bmatrix}, \begin{bmatrix} 7t \\ 19t \\ t \end{bmatrix}$ (b) $\lambda = 1;$ $\begin{bmatrix} \frac{1}{2}t \\ -\frac{1}{2}t \\ t \end{bmatrix}$

EXERCISE SET 3.1 (page 125)

3. (a) $\overrightarrow{P_1P_2} = (-1, -1)$ (b) $\overrightarrow{P_1P_2} = (-7, -2)$ (c) $\overrightarrow{P_1P_2} = (2, 1)$ (d) $\overrightarrow{P_1P_2} = (a, b)$
   (e) $\overrightarrow{P_1P_2} = (-5, 12, -6)$ (f) $\overrightarrow{P_1P_2} = (1, -1, -2)$ (g) $\overrightarrow{P_1P_2} = (-a, -b, -c)$ (h) $\overrightarrow{P_1P_2} = (a, b, c)$

4. (a) $Q(5, 10, -8)$ is one possible answer.   5. (a) $P(-1, 2, -4)$ is one possible answer.
   (b) $Q(-7, -4, -2)$ is one possible answer.   (b) $P(7, -2, -6)$ is one possible answer.

6. (a) $(-2, 1, -4)$ (b) $(-10, 6, 4)$ (c) $(-7, 1, 10)$ 7. $\mathbf{x} = (-\frac{8}{3}, \frac{1}{2}, \frac{8}{3})$
   (d) $(80, -20, -80)$ (e) $(132, -24, -72)$ (f) $(-77, 8, 94)$

8. $c_1 = 2, c_2 = -1, c_3 = 2$ 10. $c_1 = c_2 = c_3 = 0$ 11. (a) $(\frac{9}{2}, -\frac{1}{2}, -\frac{1}{2})$ 12. (a) $x' = 5, y' = 8$
   (b) $(\frac{23}{4}, -\frac{9}{4}, \frac{1}{4})$ (b) $x = -1, y = 3$

14. $\mathbf{u} = \left(\frac{\sqrt{3}}{2}, \frac{1}{2}\right), \mathbf{v} = \left(-\frac{1}{2}, -\frac{\sqrt{3}}{2}\right), \mathbf{u} + \mathbf{v} = \left(\frac{\sqrt{3}-1}{2}, \frac{1-\sqrt{3}}{2}\right), \mathbf{u} - \mathbf{v} = \left(\frac{\sqrt{3}+1}{2}, \frac{\sqrt{3}+1}{2}\right)$

EXERCISE SET 3.2 (page 130)

1. (a) 5   (b) $\sqrt{13}$ (c) 5   (d) $2\sqrt{3}$ (e) $3\sqrt{6}$ (f) 6
2. (a) $\sqrt{13}$ (b) $2\sqrt{26}$ (c) $\sqrt{209}$ (d) $3\sqrt{2}$

3. (a) $\sqrt{83}$ (b) $\sqrt{17} + \sqrt{26}$ (c) $4\sqrt{17}$ (d) $\sqrt{466}$ (e) $\left(\frac{3}{\sqrt{61}}, \frac{6}{\sqrt{61}}, -\frac{4}{\sqrt{61}}\right)$ (f) 1

4. $k = \pm \dfrac{4}{\sqrt{30}}$ 8. A sphere of radius 1 centered at (x_0, y_0, z_0)

EXERCISE SET 3.3 (page 139)

1. (a) -11 (b) -24 (c) 0   (d) 0   2. (a) $-\dfrac{11}{\sqrt{13}\sqrt{74}}$ (b) $-\dfrac{3}{\sqrt{10}}$ (c) 0   (d) 0

3. (a) Orthogonal   (b) Obtuse   (c) Acute   (d) Obtuse

4. (a) $(0, 0)$ (b) $(\frac{8}{13}, -\frac{12}{13})$ (c) $(-\frac{16}{13}, 0, -\frac{80}{13})$ (d) $(\frac{16}{89}, \frac{12}{89}, \frac{32}{89})$

5. (a) $(6, 2)$ (b) $(-\frac{21}{13}, -\frac{14}{13})$ (c) $(\frac{55}{13}, 1, -\frac{11}{13})$ (d) $(\frac{73}{89}, -\frac{12}{89}, -\frac{32}{89})$

6. (a) $\frac{2}{5}$ (b) $\dfrac{4\sqrt{5}}{5}$ (c) $\dfrac{18}{\sqrt{22}}$ (d) $\dfrac{43}{\sqrt{54}}$ 9. (a) 102   (b) $125\sqrt{2}$ (c) 170   (d) 170

11. $\cos\theta_1 = \dfrac{\sqrt{10}}{10}, \cos\theta_2 = \dfrac{3\sqrt{10}}{10}, \cos\theta_3 = 0$ 12. The right angle is at B.

13. No. The equation will hold if $\mathbf{b}$ and $\mathbf{c}$ are perpendicular to $\mathbf{a}$, and $\mathbf{b} \neq \mathbf{c}$.

15. (a) 1   (b) $\dfrac{1}{\sqrt{17}}$ (c) $\dfrac{6}{\sqrt{10}}$ 19. (b) $\cos\beta = \dfrac{b}{\|\mathbf{v}\|}, \cos\gamma = \dfrac{c}{\|\mathbf{v}\|}$ 20. $\theta_1 \approx 71°, \theta_2 \approx 61°, \theta_3 \approx 36°$

EXERCISE SET 3.4 (page 152)

1. (a) $(32, -6, -4)$ (b) $(-14, -20, -82)$ (c) $(27, 40, -42)$
   (d) $(0, 176, -264)$ (e) $(-44, 55, -22)$ (f) $(-8, -3, -8)$

2. (a) $(18, 36, -18)$ (b) $(-3, 9, -3)$ **3.** (a) $\sqrt{59}$ (b) $\sqrt{101}$ (c) 0 **4.** (a) $\dfrac{\sqrt{374}}{2}$ (b) $\sqrt{285}$

8. (a) -10 (b) -110 **9.** (a) -3 (b) 3 (c) 3 (d) -3 (e) -3 (f) 0 **10.** (a) 16 (b) 45

11. (a) No (b) Yes (c) No   **12.** $\pm\left(0, \dfrac{2}{\sqrt{5}}, \dfrac{1}{\sqrt{5}}\right)$

13. $\left(\dfrac{6}{\sqrt{61}}, -\dfrac{3}{\sqrt{61}}, \dfrac{4}{\sqrt{61}}\right), \left(-\dfrac{6}{\sqrt{61}}, \dfrac{3}{\sqrt{61}}, -\dfrac{4}{\sqrt{61}}\right)$ **15.** $2(\mathbf{v} \times \mathbf{u})$ **16.** $\dfrac{12\sqrt{13}}{49}$

17. (a) $\dfrac{\sqrt{26}}{2}$ (b) $\dfrac{\sqrt{26}}{3}$ **19.** (a) $\dfrac{2\sqrt{141}}{\sqrt{29}}$ (b) $\dfrac{\sqrt{137}}{3}$ **21.** (a) $\sqrt{122}$ (b) $\theta \approx 40°19''$

23. (a) $\mathbf{m} = (0, 1, 0)$ and $\mathbf{n} = (1, 0, 0)$ (b) $(-1, 0, 0)$ (c) $(0, 0, -1)$ **28.** $(-8, 0, -8)$ **31.** (a) $\frac{2}{3}$ (b) $\frac{1}{2}$

EXERCISE SET 3.5 (page 163)

1. (a) $-2(x + 1) + (y - 3) - (z + 2) = 0$
  (b) $(x - 1) + 9(y - 1) + 8(z - 4) = 0$
  (c) $2z = 0$
  (d) $x + 2y + 3z = 0$

2. (a) $-2x + y - z - 7 = 0$
  (b) $x + 9y + 8z - 42 = 0$
  (c) $2z = 0$
  (d) $x + 2y + 3y = 0$

3. (a) $(0, 0, 5)$ is a point in the plane and $\mathbf{n} = (-3, 7, 2)$ is a normal vector so that
  $-3(x - 0) + 7(y - 0) + 2(z - 5) = 0$ is a point-normal form; other points and normals yield other correct
  answers.

4. (a) $2y - z + 1 = 0$ (b) $x + 9y - 5z - 26 = 0$ **5.** (a) Not parallel (b) Parallel (c) Parallel

6. (a) Parallel (b) Not parallel   **7.** (a) Not perpendicular (b) Perpendicular

8. (a) Perpendicular (b) Not perpendicular

9. (a) $x = 3 + 2t, y = -1 + t, z = 2 + 3t$
  (b) $x = -2 + 6t, y = 3 - 6t, z = -3 - 2t$
  (c) $x = 2, y = 2 + t, z = 6$
  (d) $x = t, y = -2t, z = 3t$

10. (a) $x = 5 + t, y = -2 + 2t, z = 4 - 4t$
  (b) $x = 2t, y = -t, z = -3t$

11. (a) $x = -12 - 7t, y = -41 - 23t, z = t$
  (b) $x = \frac{5}{2}t, y = 0, z = t$

12. (a) $(-2, 4, 1) \cdot (x + 1, y - 2, z - 4) = 0$
  (b) $(-1, 4, 3) \cdot (x - 2, y, z + 5) = 0$
  (c) $(-1, 0, 0) \cdot (x - 5, y + 2, z - 1) = 0$
  (d) $(a, b, c) \cdot (x, y, z) = 0$

13. (a) Parallel (b) Not parallel

14. (a) Perpendicular (b) Not perpendicular

15. (a) $(x, y, z) = (-1, 2, 3) + t(7, -1, 5)$ $(-\infty < t < +\infty)$
  (b) $(x, y, z) = (2, 0, -1) + t(1, 1, 1)$ $(-\infty < t < +\infty)$
  (c) $(x, y, z) = (2, -4, 1) + t(0, 0, -2)$ $(-\infty < t < +\infty)$
  (d) $(x, y, z) = (0, 0, 0) + t(a, b, c)$ $(-\infty < t < +\infty)$

17. $2x + 3y - 5z + 36 = 0$ **18.** (a) $z = 0$ (b) $y = 0$ (c) $x = 0$

19. (a) $z - z_0 = 0$ (b) $x - x_0 = 0$ (c) $y - y_0 = 0$ **20.** $7x + 4y - 2z = 0$ **21.** $5x - 2y + z - 34 = 0$

22. $(-\frac{173}{3}, -\frac{43}{3}, \frac{49}{3})$ **23.** $y + 2z - 9 = 0$ **24.** $x - y - 4z - 2 = 0$ **26.** $x = \frac{11}{5}t - 2, y = -\frac{2}{5}t + 5, z = t$

27. $x + 5y + 3z - 18 = 0$ **28.** $(x - 2) + (y + 1) - 3(z - 4) = 0$ **29.** $4x + 13y - z - 17 = 0$

30. $3x + 10y + 4z - 53 = 0$ **31.** $3x - y - z - 2 = 0$ **32.** $5x - 3y + 2z - 5 = 0$

33. $2x + 4y + 8z + 13 = 0$ **36.** $x - 4y + 4z + 9 = 0$ **37.** (a) $x = \frac{11}{23} + \frac{7}{23}t$, $y = -\frac{41}{23} - \frac{1}{23}t$, $z = t$

    (b) $x = -\frac{2}{3}t$, $y = 0$, $z = t$

39. (a) $\frac{5}{3}$ (b) $\dfrac{1}{\sqrt{29}}$ (c) $\dfrac{4}{\sqrt{3}}$ **40.** (a) $\dfrac{1}{2\sqrt{26}}$ (b) 0 (c) $\dfrac{2}{\sqrt{6}}$

42. (a) $\dfrac{x-3}{2} = y + 1 = \dfrac{z-2}{3}$ (b) $\dfrac{x+2}{6} = -\dfrac{y-3}{6} = -\dfrac{z+3}{2}$

43. (a) $x - 2y - 17 = 0$ and $x + 4z - 27 = 0$ is one possible answer.
    (b) $x - 2y = 0$ and $-7y + 2z = 0$ is one possible answer.

44. (a) $\theta \approx 35°$ (b) $\theta \approx 79°$ **45.** $\theta \approx 75°$

EXERCISE SET 4.1 (page 178)

1. (a) $(-1, 9, -11, 1)$ (b) $(22, 53, -19, 14)$ (c) $(-13, 13, -36, -2)$
    (d) $(-90, -114, 60, -36)$ (e) $(-9, -5, -5, -3)$ (f) $(27, 29, -27, 9)$

2. $(\frac{6}{5}, \frac{2}{5}, \frac{2}{5}, \frac{2}{5})$ **3.** $c_1 = 1$, $c_2 = 1$, $c_3 = -1$, $c_4 = 1$ **5.** (a) $\sqrt{29}$ (b) 3 (c) 13 (d) $\sqrt{31}$

6. (a) $\sqrt{133}$ (b) $\sqrt{30} + \sqrt{77}$ (c) $4\sqrt{30}$ (d) $\sqrt{1811}$ (e) $\left(\dfrac{1}{\sqrt{2}}, \dfrac{1}{3\sqrt{2}}, \dfrac{2}{3\sqrt{2}}, \dfrac{2}{3\sqrt{2}}\right)$ (f) 1

8. $k = \pm\frac{5}{7}$ **9.** (a) 7 (b) 14 (c) 7 (d) 11 **10.** (a) $\left(\dfrac{1}{\sqrt{10}}, \dfrac{3}{\sqrt{10}}\right)$, $\left(-\dfrac{1}{\sqrt{10}}, -\dfrac{3}{\sqrt{10}}\right)$

11. (a) $\sqrt{10}$ (b) $2\sqrt{14}$ (c) $\sqrt{59}$ (d) 10

14. (a) Yes   (b) No   (c) Yes   (d) No   (e) No   (f) Yes   **15.** (a) $k = -3$ (b) $k = -2$, $k = -3$

16. $\pm\frac{1}{37}(-34, 44, -6, 11)$ **19.** $x_1 = 1$, $x_2 = -1$, $x_3 = 2$ **20.** -6

33. (a) Euclidean measure of "box" in R^n: $a_1 a_2 \cdots a_n$
    (b) Length of diagonal: $\sqrt{a_1^2 + a_2^2 + \cdots + a_n^2}$

EXERCISE SET 4.2 (page 196)

1. (a) Linear; $R^3 \to R^2$ (b) Nonlinear; $R^2 \to R^3$
    (c) Linear; $R^3 \to R^3$ (d) Nonlinear; $R^4 \to R^2$

2. (a) $\begin{bmatrix} 2 & -3 & 0 & 1 \\ 3 & 5 & 0 & -1 \end{bmatrix}$ (b) $\begin{bmatrix} 7 & 2 & -8 \\ 0 & -1 & 5 \\ 4 & 7 & -1 \end{bmatrix}$ (c) $\begin{bmatrix} -1 & 1 \\ 3 & -2 \\ 5 & -7 \end{bmatrix}$ (d) $\begin{bmatrix} 1 & 0 & 0 & 0 \\ 1 & 1 & 0 & 0 \\ 1 & 1 & 1 & 0 \\ 1 & 1 & 1 & 1 \end{bmatrix}$

3. $\begin{bmatrix} 3 & 5 & -1 \\ 4 & -1 & 1 \\ 3 & 2 & -1 \end{bmatrix}$; $T(-1, 2, 4) = (3, -2, -3)$

4. (a) $\begin{bmatrix} 2 & -1 \\ 1 & 1 \end{bmatrix}$ (b) $\begin{bmatrix} 1 & 0 \\ 0 & 1 \end{bmatrix}$ (c) $\begin{bmatrix} 1 & 2 & 1 \\ 1 & 5 & 0 \\ 0 & 0 & 1 \end{bmatrix}$ (d) $\begin{bmatrix} 4 & 0 & 0 \\ 0 & 7 & 0 \\ 0 & 0 & -8 \end{bmatrix}$

5. (a) $\begin{bmatrix} 0 & 1 \\ -1 & 0 \\ 1 & 3 \\ 1 & -1 \end{bmatrix}$ **(b)** $\begin{bmatrix} 7 & 2 & -1 & 1 \\ 0 & 1 & 1 & 0 \\ -1 & 0 & 0 & 0 \end{bmatrix}$ **(c)** $\begin{bmatrix} 0 & 0 & 0 \\ 0 & 0 & 0 \\ 0 & 0 & 0 \\ 0 & 0 & 0 \\ 0 & 0 & 0 \end{bmatrix}$ **(d)** $\begin{bmatrix} 0 & 0 & 0 & 1 \\ 1 & 0 & 0 & 0 \\ 0 & 0 & 1 & 0 \\ 0 & 1 & 0 & 0 \\ 1 & 0 & -1 & 0 \end{bmatrix}$

6. (a) $\begin{bmatrix} -1 \\ 1 \end{bmatrix}$ **(b)** $\begin{bmatrix} 3 \\ 13 \end{bmatrix}$ **(c)** $\begin{bmatrix} -2x_1 + x_2 + 4x_3 \\ 3x_1 + 5x_2 + 7x_3 \\ 6x_1 \quad - x_3 \end{bmatrix}$ **(d)** $\begin{bmatrix} -x_1 + x_2 \\ 2x_1 + 4x_2 \\ 7x_1 + 8x_2 \end{bmatrix}$

7. (a) $T(-1, 4) = (5, 4)$ **(b)** $T(2, 1, -3) = (0, -2, 0)$

8. (a) $(-1, -2)$ **(b)** $(1, 2)$ **(c)** $(2, -1)$ **9. (a)** $(2, -5, -3)$ **(b)** $(2, 5, 3)$ **(c)** $(-2, -5, 3)$

10. (a) $(2, 0)$ **(b)** $(0, -5)$ **11. (a)** $(-2, 1, 0)$ **(b)** $(-2, 0, 3)$ **(c)** $(0, 1, 3)$

12. (a) $\left(\dfrac{3\sqrt{3} + 4}{2}, \dfrac{3 - 4\sqrt{3}}{2} \right)$ **(b)** $\left(\dfrac{3 - 4\sqrt{3}}{2}, \dfrac{-3\sqrt{3} - 4}{2} \right)$ **(c)** $\left(\dfrac{7\sqrt{2}}{2}, \dfrac{-\sqrt{2}}{2} \right)$ **(d)** $(4, 3)$

13. (a) $\left(-2, \dfrac{\sqrt{3} - 2}{2}, \dfrac{1 + 2\sqrt{3}}{2} \right)$ **(b)** $(0, 1, 2\sqrt{2})$ **(c)** $(-1, -2, 2)$

14. (a) $\begin{bmatrix} 1 & 0 & 0 \\ 0 & 1/2 & \sqrt{3}/2 \\ 0 & -\sqrt{3}/2 & 1/2 \end{bmatrix}$ **(b)** $\begin{bmatrix} 1/2 & 0 & -\sqrt{3}/2 \\ 0 & 1 & 0 \\ \sqrt{3}/2 & 0 & 1/2 \end{bmatrix}$ **(c)** $\begin{bmatrix} 1/2 & \sqrt{3}/2 & 0 \\ -\sqrt{3}/2 & 1/2 & 0 \\ 0 & 0 & 1 \end{bmatrix}$

15. (a) $\left(-2, \dfrac{\sqrt{3} + 2}{2}, \dfrac{-1 + 2\sqrt{3}}{2} \right)$ **(b)** $(-2\sqrt{2}, 1, 0)$ **(c)** $(1, 2, 2)$

16. (a) $\begin{bmatrix} 1 & 0 \\ 0 & -1 \end{bmatrix}$ **(b)** $\begin{bmatrix} 0 & 0 \\ 0 & \frac{1}{2} \end{bmatrix}$ **(c)** $\begin{bmatrix} 3 & 0 \\ 0 & -3 \end{bmatrix}$

17. (a) $\begin{bmatrix} 0 & 0 \\ 1/2 & -\sqrt{3}/2 \end{bmatrix}$ **(b)** $\begin{bmatrix} -\sqrt{2} & \sqrt{2} \\ \sqrt{2} & \sqrt{2} \end{bmatrix}$ **(c)** $\begin{bmatrix} -1 & 0 \\ 0 & -1 \end{bmatrix}$

18. (a) $\begin{bmatrix} -1 & 0 & 0 \\ 0 & 0 & 0 \\ 0 & 0 & 1 \end{bmatrix}$ **(b)** $\begin{bmatrix} 1 & 0 & 1 \\ 0 & \sqrt{2} & 0 \\ -1 & 0 & 1 \end{bmatrix}$ **(c)** $\begin{bmatrix} -1 & 0 & 0 \\ 0 & 1 & 0 \\ 0 & 0 & 0 \end{bmatrix}$

19. (a) $\begin{bmatrix} \sqrt{3}/8 & -\sqrt{3}/16 & 1/16 \\ 1/8 & 3/16 & -\sqrt{3}/16 \\ 0 & 1/8 & \sqrt{3}/8 \end{bmatrix}$ **(b)** $\begin{bmatrix} 0 & 0 & 0 \\ 0 & -1 & 0 \\ 0 & 0 & -1 \end{bmatrix}$ **(c)** $\begin{bmatrix} 0 & 1 & 0 \\ 0 & 0 & -1 \\ -1 & 0 & 0 \end{bmatrix}$

20. (a) Yes **(b)** Yes **(c)** Yes **(d)** No **21. (a)** Yes **(b)** No

22. (a) $\begin{bmatrix} 1 & 0 & 0 \\ 0 & 0 & 0 \\ 0 & 0 & 0 \end{bmatrix} \begin{bmatrix} 0 & 0 & 0 \\ 0 & 1 & 0 \\ 0 & 0 & 0 \end{bmatrix} \begin{bmatrix} 0 & 0 & 0 \\ 0 & 0 & 0 \\ 0 & 0 & 1 \end{bmatrix}$

24. $\begin{bmatrix} \frac{1}{3}(1 - \cos\theta) + \cos\theta & \frac{1}{3}(1 - \cos\theta) - \dfrac{1}{\sqrt{3}}\sin\theta & \frac{1}{3}(1 - \cos\theta) - \dfrac{1}{\sqrt{3}}\sin\theta \\ \frac{1}{3}(1 - \cos\theta) - \dfrac{1}{\sqrt{3}}\sin\theta & \frac{1}{3}(1 - \cos\theta) + \cos\theta & \frac{1}{3}(1 - \cos\theta) - \dfrac{1}{\sqrt{3}}\sin\theta \\ \frac{1}{3}(1 - \cos\theta) - \dfrac{1}{\sqrt{3}}\sin\theta & \frac{1}{3}(1 - \cos\theta) - \dfrac{1}{\sqrt{3}}\sin\theta & \frac{1}{3}(1 - \cos\theta) + \cos\theta \end{bmatrix}$ **26.** $135°$ **28. (c)** $90°$

EXERCISE SET 4.3 (page 211)

1. (a) Not one-to-one (b) One-to-one (c) One-to-one (d) One-to-one
   (e) One-to-one (f) One-to-one (g) One-to-one

2. (a) $\begin{bmatrix} 8 & 4 \\ 2 & 1 \end{bmatrix}$; not one-to-one (b) $\begin{bmatrix} 2 & -3 \\ 5 & 1 \end{bmatrix}$; one-to-one

   (c) $\begin{bmatrix} -1 & 3 & 2 \\ 2 & 0 & 4 \\ 1 & 3 & 6 \end{bmatrix}$; not one-to-one (d) $\begin{bmatrix} 1 & 2 & 3 \\ 2 & 5 & 3 \\ 1 & 0 & 5 \end{bmatrix}$; one-to-one

3. For example, the vector $(1, 3)$ is not in the range.

4. For example, the vector $(1, 6, 2)$ is not in the range.

5. (a) One-to-one; $\begin{bmatrix} \frac{1}{3} & -\frac{2}{3} \\ \frac{1}{3} & \frac{1}{3} \end{bmatrix}$; $T^{-1}(w_1, w_2) = (\frac{1}{3}x_1 - \frac{2}{3}x_2, \frac{1}{3}x_1 + \frac{1}{3}x_2)$ (b) Not one-to-one

   (c) One-to-one; $\begin{bmatrix} 0 & -1 \\ -1 & 0 \end{bmatrix}$; $T^{-1}(w_1, w_2) = (-x_2, -x_1)$ (d) Not one-to-one

6. (a) One-to-one; $\begin{bmatrix} 1 & -2 & 4 \\ -1 & 2 & -3 \\ -1 & 3 & -5 \end{bmatrix}$; $T^{-1}(w_1, w_2, w_3) = (x_1 - 2x_2 + 4x_3, -x_1 + 2x_2 - 3x_3, -x_1 + 3x_2 - 5x_3)$

   (b) One-to-one; $\begin{bmatrix} \frac{1}{2} & \frac{1}{2} & -\frac{1}{2} \\ -\frac{5}{14} & \frac{5}{14} & \frac{3}{14} \\ -\frac{1}{7} & \frac{1}{7} & \frac{1}{7} \end{bmatrix}$; $T^{-1}(w_1, w_2, w_3) = \left(\frac{x_1 + x_2 - x_3}{2}, \frac{-5x_1 + 5x_2 + 3x_3}{14}, \frac{-x_1 + x_2 + x_3}{7} \right)$

   (c) One-to-one; $\begin{bmatrix} -\frac{3}{2} & -\frac{3}{2} & \frac{11}{2} \\ \frac{1}{2} & \frac{1}{2} & -\frac{3}{2} \\ -\frac{1}{2} & \frac{1}{2} & -\frac{1}{2} \end{bmatrix}$; $T^{-1}(w_1, w_2, w_3) = \left(\frac{-3x_1 - 3x_2 + 11x_3}{2}, \frac{x_1 + x_2 - 3x_3}{2}, \frac{-x_1 + x_2 - x_3}{2} \right)$

   (d) Not one-to-one

7. (a) Reflection about the x-axis (b) Rotation through the angle $-\pi/4$
   (c) Contraction by a factor of $\frac{1}{3}$ (d) Reflection about the yz-plane
   (e) Dilation by a factor of 5

8. (a) Linear (b) Nonlinear (c) Linear (d) Linear

9. (a) Linear (b) Nonlinear (c) Linear (d) Nonlinear

10. (a) Linear (b) Nonlinear   11. (a) Linear (b) Nonlinear

12. (a) For a reflection about the y-axis, $T(e_1) = \begin{bmatrix} -1 \\ 0 \end{bmatrix}$ and $T(e_2) = \begin{bmatrix} 0 \\ 1 \end{bmatrix}$. Thus, $T = \begin{bmatrix} -1 & 0 \\ 0 & 1 \end{bmatrix}$.

   (b) For a reflection about the xz-plane, $T(e_1) = \begin{bmatrix} 1 \\ 0 \\ 0 \end{bmatrix}$, $T(e_2) = \begin{bmatrix} 0 \\ -1 \\ 0 \end{bmatrix}$, and $T(e_3) = \begin{bmatrix} 0 \\ 0 \\ 1 \end{bmatrix}$. Thus, $T = \begin{bmatrix} 1 & 0 & 0 \\ 0 & -1 & 0 \\ 0 & 0 & 1 \end{bmatrix}$.

   (c) For an orthogonal projection on the x-axis, $T(e_1) = \begin{bmatrix} 1 \\ 0 \end{bmatrix}$ and $T(e_2) = \begin{bmatrix} 0 \\ 0 \end{bmatrix}$. Thus, $T = \begin{bmatrix} 1 & 0 \\ 0 & 0 \end{bmatrix}$.

(d) For an orthogonal projection on the *yz*-plane, $T(\mathbf{e}_1) = \begin{bmatrix} 0 \\ 0 \\ 0 \end{bmatrix}$, $T(\mathbf{e}_2) = \begin{bmatrix} 0 \\ 1 \\ 0 \end{bmatrix}$, and $T(\mathbf{e}_3) = \begin{bmatrix} 0 \\ 0 \\ 1 \end{bmatrix}$.

Thus, $T = \begin{bmatrix} 0 & 0 & 0 \\ 0 & 1 & 0 \\ 0 & 0 & 1 \end{bmatrix}$.

(e) For a rotation through a positive angle θ, $T(\mathbf{e}_1) = \begin{bmatrix} \cos\theta \\ \sin\theta \end{bmatrix}$ and $T(\mathbf{e}_2) = \begin{bmatrix} -\sin\theta \\ \cos\theta \end{bmatrix}$. Thus, $T = \begin{bmatrix} \cos\theta & -\sin\theta \\ \sin\theta & \cos\theta \end{bmatrix}$.

(f) For a dilation by a factor $k \geq 1$, $T(\mathbf{e}_1) = \begin{bmatrix} k \\ 0 \\ 0 \end{bmatrix}$, $T(\mathbf{e}_2) = \begin{bmatrix} 0 \\ k \\ 0 \end{bmatrix}$, $T(\mathbf{e}_3) = \begin{bmatrix} 0 \\ 0 \\ k \end{bmatrix}$. Thus, $T = \begin{bmatrix} k & 0 & 0 \\ 0 & k & 0 \\ 0 & 0 & k \end{bmatrix}$.

13. (a) $T(\mathbf{e}_1) = \begin{bmatrix} -1 \\ 0 \end{bmatrix}$ and $T(\mathbf{e}_2) = \begin{bmatrix} 0 \\ 0 \end{bmatrix}$. Thus, $T = \begin{bmatrix} -1 & 0 \\ 0 & 0 \end{bmatrix}$.

(b) $T(\mathbf{e}_1) = \begin{bmatrix} 0 \\ -1 \end{bmatrix}$ and $T(\mathbf{e}_2) = \begin{bmatrix} 1 \\ 0 \end{bmatrix}$. Thus, $T = \begin{bmatrix} 0 & 1 \\ -1 & 0 \end{bmatrix}$.

(c) $T(\mathbf{e}_1) = \begin{bmatrix} 0 \\ 3 \end{bmatrix}$ and $T(\mathbf{e}_2) = \begin{bmatrix} 0 \\ 0 \end{bmatrix}$. Thus, $T = \begin{bmatrix} 0 & 0 \\ 3 & 0 \end{bmatrix}$.

14. (a) $T(\mathbf{e}_1) = \begin{bmatrix} \frac{1}{5} \\ 0 \\ 0 \end{bmatrix}$, $T(\mathbf{e}_2) = \begin{bmatrix} 0 \\ -\frac{1}{5} \\ 0 \end{bmatrix}$, and $T(\mathbf{e}_3) = \begin{bmatrix} 0 \\ 0 \\ \frac{1}{5} \end{bmatrix}$. Thus, $T = \begin{bmatrix} \frac{1}{5} & 0 & 0 \\ 0 & -\frac{1}{5} & 0 \\ 0 & 0 & \frac{1}{5} \end{bmatrix}$.

(b) $T(\mathbf{e}_1) = \begin{bmatrix} 1 \\ 0 \\ 0 \end{bmatrix}$, $T(\mathbf{e}_2) = \begin{bmatrix} 0 \\ 0 \\ 0 \end{bmatrix}$, and $T(\mathbf{e}_3) = \begin{bmatrix} 0 \\ 0 \\ 0 \end{bmatrix}$. Thus, $T = \begin{bmatrix} 1 & 0 & 0 \\ 0 & 0 & 0 \\ 0 & 0 & 0 \end{bmatrix}$.

(c) $T(\mathbf{e}_1) = \begin{bmatrix} -1 \\ 0 \\ 0 \end{bmatrix}$, $T(\mathbf{e}_2) = \begin{bmatrix} 0 \\ -1 \\ 0 \end{bmatrix}$, and $T(\mathbf{e}_3) = \begin{bmatrix} 0 \\ 0 \\ -1 \end{bmatrix}$. Thus, $T = \begin{bmatrix} -1 & 0 & 0 \\ 0 & -1 & 0 \\ 0 & 0 & -1 \end{bmatrix}$.

15. (a) $T_A(\mathbf{e}_1) = \begin{bmatrix} -1 \\ 2 \\ 4 \end{bmatrix}$, $T_A(\mathbf{e}_2) = \begin{bmatrix} 3 \\ 1 \\ 5 \end{bmatrix}$, and $T_A(\mathbf{e}_3) = \begin{bmatrix} 0 \\ 2 \\ -3 \end{bmatrix}$.

(b) $T_A(\mathbf{e}_1 + \mathbf{e}_2 + \mathbf{e}_3) = T_A(\mathbf{e}_1) + T_A(\mathbf{e}_2) + T_A(\mathbf{e}_3) = \begin{bmatrix} 2 \\ 5 \\ 6 \end{bmatrix}$ (c) $T_A(7\mathbf{e}_3) = 7T_A(\mathbf{e}_3) = \begin{bmatrix} 0 \\ 14 \\ -21 \end{bmatrix}$

16. (a) Linear transformation from $R^2 \rightarrow R^3$ (b) Linear transformation from $R^3 \rightarrow R^2$

17. (a) $\left(\frac{1}{2}, \frac{1}{2} \right)$ (b) $\left(\frac{3}{4}, \frac{\sqrt{3}}{4} \right)$ (c) $\left(\frac{1 - 5\sqrt{3}}{4}, \frac{15 - \sqrt{3}}{4} \right)$

18. (a) $\lambda = 1$; $\begin{bmatrix} t \\ 0 \end{bmatrix}$ (b) $\lambda = 1$; $\begin{bmatrix} t \\ t \end{bmatrix}$ (c) $\lambda = 1$; $\begin{bmatrix} t \\ 0 \end{bmatrix}$ (d) $\lambda = \frac{1}{2}$; all vectors in R^2 are eigenvectors

$\lambda = -1$; $\begin{bmatrix} 0 \\ t \end{bmatrix}$ $\lambda = -1$; $\begin{bmatrix} t \\ -t \end{bmatrix}$ $\lambda = 0$; $\begin{bmatrix} 0 \\ t \end{bmatrix}$

19. (a) $\lambda = 1$; $\begin{bmatrix} 0 \\ t \\ 0 \end{bmatrix}$ and $\begin{bmatrix} 0 \\ 0 \\ t \end{bmatrix}$ (b) $\lambda = 1$; $\begin{bmatrix} t \\ 0 \\ 0 \end{bmatrix}$ and $\begin{bmatrix} 0 \\ 0 \\ t \end{bmatrix}$

$\lambda = -1$; $\begin{bmatrix} t \\ 0 \\ 0 \end{bmatrix}$ $\lambda = 0$; $\begin{bmatrix} 0 \\ t \\ 0 \end{bmatrix}$

(c) $\lambda = 2$; all vectors in R^3 are eigenvectors   (d) $\lambda = 1$; $\begin{bmatrix} 0 \\ 0 \\ t \end{bmatrix}$

20. (a) Yes (b) No

23. (a) $\begin{bmatrix} \cos 2\theta & \sin 2\theta \\ \sin 2\theta & -\cos 2\theta \end{bmatrix}$ (b) $\left(\dfrac{1 + 5\sqrt{3}}{2}, \dfrac{\sqrt{3} - 5}{2} \right)$

EXERCISE SET 5.1 (page 220)

1. Not a vector space. Axiom 8 fails.   2. Not a vector space. Axiom 10 fails.

3. Not a vector space. Axioms 9 and 10 fail.   4. The set is a vector space under the given operations.

5. The set is a vector space under the given operations.   6. Not a vector space. Axioms 5 and 6 fail.

7. The set is a vector space under the given operations.   8. Not a vector space. Axioms 7 and 8 fail.

9. Not a vector space. Axioms 1, 4, 5, and 6 fail.   10. The set is a vector space under the given operations.

11. The set is a vector space under the given operations.   12. The set is a vector space under the given operations.

13. The set is a vector space under the given operations.

EXERCISE SET 5.2 (page 230)

1. (a), (c)   2. (b)   3. (a), (b), (d)   4. (b), (d), (e)   5. (a), (b)

6. (a) Line; $x = -\frac{1}{2}t$, $y = -\frac{3}{2}t$, $z = t$
   (b) Line; $x = 2t$, $y = t$, $z = 0$
   (c) Origin (d) Origin
   (e) Line; $x = -3t$, $y = -2t$, $z = t$
   (f) Plane; $x - 3y + z = 0$

7. (a), (b), (d)

8. (a) $(-9, -7, -15) = -2\mathbf{u} + \mathbf{v} - 2\mathbf{w}$
   (b) $(6, 11, 6) = 4\mathbf{u} - 5\mathbf{v} + \mathbf{w}$
   (c) $(0, 0, 0) = 0\mathbf{u} + 0\mathbf{v} + 0\mathbf{w}$
   (d) $(7, 8, 9) = 0\mathbf{u} - 2\mathbf{v} + 3\mathbf{w}$

9. (a) $-9 - 7x - 15x^2 = -2\mathbf{p}_1 + \mathbf{p}_2 - 2\mathbf{p}_3$
   (b) $6 + 11x + 6x^2 = 4\mathbf{p}_1 - 5\mathbf{p}_2 + \mathbf{p}_3$
   (c) $0 = 0\mathbf{p}_1 + 0\mathbf{p}_2 + 0\mathbf{p}_3$
   (d) $7 + 8x + 9x^2 = 0\mathbf{p}_1 - 2\mathbf{p}_2 + 3\mathbf{p}_3$

11. (a) The vectors span.   (b) The vectors do not span.   12. (a), (c), (e)   13. No
    (c) The vectors do not span.   (d) The vectors span.

14. (a), (b), (d)   15. $y = z$ 16. $x = 3t$, $y = -2t$, $z = 5t$, where $-\infty < t < +\infty$

EXERCISE SET 5.3 (page 240)

1. (a) $\mathbf{u}_2$ is a scalar multiple of $\mathbf{u}_1$.   (b) The vectors are linearly dependent by Theorem 5.1.3.
   (c) $\mathbf{p}_2$ is a scalar multiple of $\mathbf{p}_1$.   (d) B is a scalar multiple of A.

2. (d)   3. None   4. (d)

5. (a) They do not lie in a plane.   6. (a) They do not lie on the same line.
   (b) They do lie in a plane.   (b) They do not lie on the same line.
                                       (c) They do lie on the same line.

7. (b) $\mathbf{v}_1 = \frac{2}{7}\mathbf{v}_2 - \frac{3}{7}\mathbf{v}_3$, $\mathbf{v}_2 = \frac{7}{2}\mathbf{v}_1 + \frac{3}{2}\mathbf{v}_3$, $\mathbf{v}_3 = -\frac{7}{3}\mathbf{v}_1 + \frac{2}{3}\mathbf{v}_2$ 8. $\lambda = -\frac{1}{2}$, $\lambda = 1$

17. If and only if the vector is not zero

18. (a) They are linearly independent since $\mathbf{v}_1$, $\mathbf{v}_2$, and $\mathbf{v}_3$ do not lie in the same plane when they are placed with their initial points at the origin.
    (b) They are not linearly independent since $\mathbf{v}_1$, $\mathbf{v}_2$, and $\mathbf{v}_3$ lie in the same plane when they are placed with their initial points at the origin.

19. (a), (d), (e), (f)

EXERCISE SET 5.4 (page 255)

1. (a) A basis for R^2 has two linearly independent vectors.   (b) A basis for R^3 has three linearly independent vectors.
   (c) A basis for P_2 has three linearly independent vectors.   (d) A basis for M_{22} has four linearly independent vectors.

2. (a), (b)   3. (a), (b)   4. (c), (d)   6. (b) Any two of the vectors $\mathbf{v}_1$, $\mathbf{v}_2$, $\mathbf{v}_3$

7. (a) $(\mathbf{w})_S = (3, -7)$ (b) $(\mathbf{w})_S = (\frac{5}{28}, \frac{3}{14})$ (c) $(\mathbf{w})_S = \left(a, \dfrac{b-a}{2}\right)$

8. (a) $(\mathbf{v})_S = (3, -2, 1)$ (b) $(\mathbf{v})_S = (-2, 0, 1)$ 9. (a) $(\mathbf{p})_S = (4, -3, 1)$ (b) $(\mathbf{p})_S = (0, 2, -1)$

10. $(A)_S = (-1, 1, -1, 3)$ 11. Basis: $(1, 0, 1)$; dimension $= 1$

12. Basis: $(-\frac{1}{4}, -\frac{1}{4}, 1, 0)$, $(0, -1, 0, 1)$; dimension $= 2$

13. Basis: $(4, 1, 0, 0)$, $(-3, 0, 1, 0)$, $(1, 0, 0, 1)$; dimension $= 3$

14. Basis: $(3, 1, 0)$, $(-1, 0, 1)$; dimension $= 2$ 15. No basis; dimension $= 0$

16. Basis: $(4, -5, 1)$; dimension $= 1$ 17. (a) $(\frac{2}{3}, 1, 0)$, $(-\frac{5}{3}, 0, 1)$ (b) $(1, 1, 0)$, $(0, 0, 1)$
                                               (c) $(2, -1, 4)$ (d) $(1, 1, 0)$, $(0, 1, 1)$

18. (a) 3-dimensional   (b) 2-dimensional   (c) 1-dimensional   19. 3-dimensional

20. (a) $\{\mathbf{v}_1, \mathbf{v}_2, \mathbf{e}_1\}$ or $\{\mathbf{v}_1, \mathbf{v}_2, \mathbf{e}_2\}$ (b) $\{\mathbf{v}_1, \mathbf{v}_2, \mathbf{e}_1\}$ or $\{\mathbf{v}_1, \mathbf{v}_2, \mathbf{e}_2\}$ or $\{\mathbf{v}_1, \mathbf{v}_2, \mathbf{e}_3\}$

21. $\{\mathbf{v}_1, \mathbf{v}_2, \mathbf{e}_2, \mathbf{e}_3\}$ or $\{\mathbf{v}_1, \mathbf{v}_2, \mathbf{e}_2, \mathbf{e}_4\}$ or $\{\mathbf{v}_1, \mathbf{v}_2, \mathbf{e}_3, \mathbf{e}_4\}$

26. (a) One possible answer is $\{-1 + x - 2x^2, 3 + 3x + 6x^2, 9\}$.
    (b) One possible answer is $\{1 + x, x^2, -2 + 2x^2\}$.
    (c) One possible answer is $\{1 + x - 3x^2\}$.

27. (a) $(0, \sqrt{2})$ (b) $(1, 0)$ (c) $(-1, \sqrt{2})$ (d) $(a - b, \sqrt{2}b)$

28. (a) $(2, 0)$ (b) $\left(\dfrac{2}{\sqrt{3}}, -\dfrac{1}{\sqrt{3}}\right)$ (c) $(0, 1)$ (d) $\left(\dfrac{2}{\sqrt{3}}a, b - \dfrac{a}{\sqrt{3}}\right)$

EXERCISE SET 5.5 (page 270)

1. $\mathbf{r}_1 = (2, -1, 0, 1)$, $\mathbf{r}_2 = (3, 5, 7, -1)$, $\mathbf{r}_3 = (1, 4, 2, 7)$; $\mathbf{c}_1 = \begin{bmatrix} 2 \\ 3 \\ 1 \end{bmatrix}$, $\mathbf{c}_2 = \begin{bmatrix} -1 \\ 5 \\ 4 \end{bmatrix}$, $\mathbf{c}_3 = \begin{bmatrix} 0 \\ 7 \\ 2 \end{bmatrix}$, $\mathbf{c}_4 = \begin{bmatrix} 1 \\ -1 \\ 7 \end{bmatrix}$

2. (a) $1\begin{bmatrix} 2 \\ -1 \end{bmatrix} + 2\begin{bmatrix} 3 \\ 4 \end{bmatrix} = \begin{bmatrix} 8 \\ 7 \end{bmatrix}$ (b) $-2\begin{bmatrix} 4 \\ 3 \\ 0 \end{bmatrix} + 3\begin{bmatrix} 0 \\ 6 \\ -1 \end{bmatrix} + 5\begin{bmatrix} -1 \\ 2 \\ 4 \end{bmatrix} = \begin{bmatrix} -13 \\ 22 \\ 17 \end{bmatrix}$

(c) $-1\begin{bmatrix} -3 \\ 5 \\ 2 \\ 1 \end{bmatrix} + 2\begin{bmatrix} 6 \\ -4 \\ 3 \\ 8 \end{bmatrix} + 5\begin{bmatrix} 2 \\ 0 \\ -1 \\ 3 \end{bmatrix} = \begin{bmatrix} 25 \\ -13 \\ -1 \\ 30 \end{bmatrix}$ (d) $3\begin{bmatrix} 2 \\ 6 \end{bmatrix} + 0\begin{bmatrix} 1 \\ 3 \end{bmatrix} + (-5)\begin{bmatrix} 5 \\ -8 \end{bmatrix} = \begin{bmatrix} -19 \\ 58 \end{bmatrix}$

3. (a) $\begin{bmatrix} -2 \\ 10 \end{bmatrix} = \begin{bmatrix} 1 \\ 4 \end{bmatrix} - \begin{bmatrix} 3 \\ -6 \end{bmatrix}$ (b) **b** is not in the column space of A.

(c) $\begin{bmatrix} 1 \\ 9 \\ 1 \end{bmatrix} - 3\begin{bmatrix} -1 \\ 3 \\ 1 \end{bmatrix} + \begin{bmatrix} 1 \\ 1 \\ 1 \end{bmatrix} = \begin{bmatrix} 5 \\ 1 \\ -1 \end{bmatrix}$ (d) $\begin{bmatrix} 2 \\ 0 \\ 0 \end{bmatrix} = \begin{bmatrix} 1 \\ 1 \\ -1 \end{bmatrix} + (t - 1)\begin{bmatrix} -1 \\ 1 \\ -1 \end{bmatrix} + t\begin{bmatrix} 1 \\ -1 \\ 1 \end{bmatrix}$

(e) $\begin{bmatrix} 4 \\ 3 \\ 5 \\ 7 \end{bmatrix} = -26\begin{bmatrix} 1 \\ 0 \\ 1 \\ 0 \end{bmatrix} + 13\begin{bmatrix} 2 \\ 1 \\ 2 \\ 1 \end{bmatrix} - 7\begin{bmatrix} 0 \\ 2 \\ 1 \\ 2 \end{bmatrix} + 4\begin{bmatrix} 1 \\ 1 \\ 3 \\ 2 \end{bmatrix}$

4. (a) $r\begin{bmatrix} -3 \\ 1 \\ 1 \\ 0 \end{bmatrix} + s\begin{bmatrix} 4 \\ -1 \\ 0 \\ 1 \end{bmatrix}$ (b) $\begin{bmatrix} -1 \\ 2 \\ 4 \\ -3 \end{bmatrix} + r\begin{bmatrix} -3 \\ 1 \\ 1 \\ 0 \end{bmatrix} + s\begin{bmatrix} 4 \\ -1 \\ 0 \\ 1 \end{bmatrix}$

5. (a) $\begin{bmatrix} 1 \\ 0 \end{bmatrix} + t\begin{bmatrix} 3 \\ 1 \end{bmatrix}$; $t\begin{bmatrix} 3 \\ 1 \end{bmatrix}$ (b) $\begin{bmatrix} -2 \\ 7 \\ 0 \end{bmatrix} + t\begin{bmatrix} -1 \\ -1 \\ 1 \end{bmatrix}$; $t\begin{bmatrix} -1 \\ -1 \\ 1 \end{bmatrix}$

(c) $\begin{bmatrix} -1 \\ 0 \\ 0 \\ 0 \end{bmatrix} + r\begin{bmatrix} 2 \\ 1 \\ 0 \\ 0 \end{bmatrix} + s\begin{bmatrix} -1 \\ 0 \\ 1 \\ 0 \end{bmatrix} + t\begin{bmatrix} -2 \\ 0 \\ 0 \\ 1 \end{bmatrix}$; $r\begin{bmatrix} 2 \\ 1 \\ 0 \\ 0 \end{bmatrix} + s\begin{bmatrix} -1 \\ 0 \\ 1 \\ 0 \end{bmatrix} + t\begin{bmatrix} -2 \\ 0 \\ 0 \\ 1 \end{bmatrix}$

(d) $\begin{bmatrix} \frac{6}{5} \\ \frac{7}{5} \\ 0 \\ 0 \end{bmatrix} + s\begin{bmatrix} \frac{7}{5} \\ \frac{4}{5} \\ 1 \\ 0 \end{bmatrix} + t\begin{bmatrix} \frac{1}{5} \\ -\frac{3}{5} \\ 0 \\ 1 \end{bmatrix}$; $s\begin{bmatrix} \frac{7}{5} \\ \frac{4}{5} \\ 1 \\ 0 \end{bmatrix} + t\begin{bmatrix} \frac{1}{5} \\ -\frac{3}{5} \\ 0 \\ 1 \end{bmatrix}$

6. (a) $\begin{bmatrix} 16 \\ 19 \\ 1 \end{bmatrix}$ (b) $\begin{bmatrix} 1 \\ 0 \\ 2 \end{bmatrix}, \begin{bmatrix} 0 \\ 1 \\ 0 \end{bmatrix}$ (c) $\begin{bmatrix} -1 \\ -1 \\ 1 \\ 0 \end{bmatrix}, \begin{bmatrix} 2 \\ -4 \\ 0 \\ 7 \end{bmatrix}$ (d) $\begin{bmatrix} -1 \\ -1 \\ 1 \\ 0 \\ 0 \end{bmatrix}, \begin{bmatrix} -2 \\ -1 \\ 0 \\ 1 \\ 0 \end{bmatrix}, \begin{bmatrix} -1 \\ -2 \\ 0 \\ 0 \\ 1 \end{bmatrix}$ (e) $\begin{bmatrix} -2 \\ 0 \\ 0 \\ 1 \\ 0 \end{bmatrix}, \begin{bmatrix} -16 \\ 2 \\ 5 \\ 0 \\ 12 \end{bmatrix}$

7. (a) $\mathbf{r}_1 = [1 \quad 0 \quad 2], \mathbf{r}_2 = [0 \quad 0 \quad 1], \mathbf{c} = \begin{bmatrix} 1 \\ 0 \\ 0 \end{bmatrix}, \mathbf{c}_2 = \begin{bmatrix} 2 \\ 1 \\ 0 \end{bmatrix}$

(b) $\mathbf{r}_1 = [1 \quad -3 \quad 0 \quad 0], \mathbf{r}_2 = [0 \quad 1 \quad 0 \quad 0], \mathbf{c}_1 = \begin{bmatrix} 1 \\ 0 \\ 0 \\ 0 \end{bmatrix}, \mathbf{c}_2 = \begin{bmatrix} -3 \\ 1 \\ 0 \\ 0 \end{bmatrix}$

(c) $\mathbf{r}_1 = [1 \quad 2 \quad 4 \quad 5], \mathbf{r}_2 = [0 \quad 1 \quad -3 \quad 0], \mathbf{r}_3 = [0 \quad 0 \quad 1 \quad -3], \mathbf{r}_4 = [0 \quad 0 \quad 0 \quad 1],$

$\mathbf{c}_1 = \begin{bmatrix} 1 \\ 0 \\ 0 \\ 0 \\ 0 \end{bmatrix}, \mathbf{c}_2 = \begin{bmatrix} 2 \\ 1 \\ 0 \\ 0 \\ 0 \end{bmatrix}, \mathbf{c}_3 = \begin{bmatrix} 4 \\ -3 \\ 1 \\ 0 \\ 0 \end{bmatrix}, \mathbf{c}_4 = \begin{bmatrix} 5 \\ 0 \\ -3 \\ 1 \\ 0 \end{bmatrix}$

(d) $\mathbf{r}_1 = [1 \quad 2 \quad -1 \quad 5], \mathbf{r}_2 = [0 \quad 1 \quad 4 \quad 3], \mathbf{r}_3 = [0 \quad 0 \quad 1 \quad -7], \mathbf{r}_4 = [0 \quad 0 \quad 0 \quad 1],$

$\mathbf{c}_1 = \begin{bmatrix} 1 \\ 0 \\ 0 \\ 0 \end{bmatrix}, \mathbf{c}_2 = \begin{bmatrix} 2 \\ 1 \\ 0 \\ 0 \end{bmatrix}, \mathbf{c}_3 = \begin{bmatrix} -1 \\ 4 \\ 1 \\ 0 \end{bmatrix}, \mathbf{c}_4 = \begin{bmatrix} 5 \\ 3 \\ -7 \\ 1 \end{bmatrix}$

8. (a) $(1, -1, 3), (0, 1, -19)$ (b) $(1, 0, -\frac{1}{2})$ (c) $(1, 4, 5, 2), (0, 1, 1, \frac{4}{7})$

(d) $(1, 4, 5, 6, 9), (0, 1, 1, 1, 2)$ (e) $(1, -3, 2, 2, 1), (0, 1, 2, 0, -1), (0, 0, 1, 0, -\frac{5}{12})$

9. (a) $\begin{bmatrix} 1 \\ 5 \\ 7 \end{bmatrix}, \begin{bmatrix} -1 \\ -4 \\ -6 \end{bmatrix}$ (b) $\begin{bmatrix} 2 \\ 4 \\ 0 \end{bmatrix}$ (c) $\begin{bmatrix} 1 \\ 2 \\ -1 \end{bmatrix}, \begin{bmatrix} 4 \\ 1 \\ 3 \end{bmatrix}$ (d) $\begin{bmatrix} 1 \\ 3 \\ -1 \\ 2 \end{bmatrix}, \begin{bmatrix} 4 \\ -2 \\ 0 \\ 3 \end{bmatrix}$ (e) $\begin{bmatrix} 1 \\ 0 \\ 2 \\ 3 \\ -2 \end{bmatrix}, \begin{bmatrix} -3 \\ 3 \\ -3 \\ -6 \\ 9 \end{bmatrix}, \begin{bmatrix} 2 \\ 6 \\ -2 \\ 0 \\ 2 \end{bmatrix}$

10. (a) $(1, -1, 3), (5, -4, -4)$ (b) $(2, 0, -1)$ (c) $(1, 4, 5, 2), (2, 1, 3, 0)$

(d) $(1, 4, 5, 6, 9), (3, -2, 1, 4, -1)$ (e) $(1, -3, 2, 2, 1), (0, 3, 6, 0, -3), (2, -3, -2, 4, 4)$

11. (a) $(1, 1, -4, -3), (0, 1, -5, -2), (0, 0, 1, -\frac{1}{2})$

(b) $(1, -1, 2, 0), (0, 1, 0, 0), (0, 0, 1, -\frac{1}{6})$

(c) $(1, 1, 0, 0), (0, 1, 1, 1), (0, 0, 1, 1), (0, 0, 0, 1)$

12. (a) $\{\mathbf{v}_1, \mathbf{v}_2\}; \mathbf{v}_3 = 2\mathbf{v}_1 + \mathbf{v}_2, \mathbf{v}_4 = -2\mathbf{v}_1 + \mathbf{v}_2$

(b) $\{\mathbf{v}_1, \mathbf{v}_3\}; \mathbf{v}_2 = 2\mathbf{v}_1, \mathbf{v}_4 = \mathbf{v}_1 + \mathbf{v}_3$

(c) $\{\mathbf{v}_1, \mathbf{v}_2, \mathbf{v}_4\}; \mathbf{v}_3 = 2\mathbf{v}_1 - \mathbf{v}_2, \mathbf{v}_5 = -\mathbf{v}_1 + 3\mathbf{v}_2 + 2\mathbf{v}_4$

14. (b) $\begin{bmatrix} 0 & 0 & 0 \\ 0 & 1 & 0 \\ 0 & 0 & 1 \end{bmatrix}$

EXERCISE SET 5.6 (page 282)

1. $\text{Rank}(A) = \text{rank}(A^T) = 2$

2. (a) Nullity = 1, rank = 2; $n = 3$ (b) Nullity = 2, rank = 1; $n = 3$
   (c) Nullity = 2, rank = 2; $n = 4$ (d) Nullity = 3, rank = 2; $n = 5$
   (e) Nullity = 2, rank = 3; $n = 5$

3. (a) 2; 1 (b) 1; 2 (c) 2; 2 (d) 2; 3 (e) 3; 2

4. (a) 3; 3; 0; 0 (b) 2; 2; 1; 1 (c) 1; 1; 2; 2 (d) 2; 2; 7; 3
   (e) 2; 2; 3; 7 (f) 0; 0; 4; 4 (g) 2; 2; 0; 4

5. (a) Rank = 4, nullity = 0 (b) Rank = 3, nullity = 2 (c) Rank = 3, nullity = 0

6. Rank = $\min(m, n)$, nullity = $n - \min(m, n)$

7. (a) Yes, 0   8. (a) Nullity = 0, number of parameters = 0
   (b) No   (b) Nullity = 1, number of parameters = 1
   (c) Yes, 2   (c) Nullity = 2, number of parameters = 2
   (d) Yes, 7   (d) Nullity = 7, number of parameters = 7
   (e) No   (e) Nullity = 7, number of parameters = 7
   (f) Yes, 4   (f) Nullity = 4, number of parameters = 4
   (g) Yes, 0   (g) Nullity = 0, number of parameters = 0

9. $b_1 = r$, $b_2 = s$, $b_3 = 4s - 3r$, $b_4 = 2r - s$, $b_5 = 8s - 7r$

12. (a) $\text{Rank}(A) = 1$ if $t = 1$; $\text{rank}(A) = 2$ if $t = -2$; $\text{rank}(A) = 3$ if $t \neq 1, -2$
    (b) $\text{Rank}(A) = 2$ if $t = 1, \frac{3}{2}$; $\text{rank}(A) = 3$ if $t \neq 1, \frac{3}{2}$

13. Rank is 2 if $r = 2$ and $s = 1$.

CHAPTER 5 SUPPLEMENTARY EXERCISES (page 284)

1. (a) All of R^3 2. A line through the origin: $s = -2$
   (b) Plane: $2x - 3y + z = 0$ A plane through the origin: $s = 1$
   (c) Line: $x = 2t$, $y = t$, $z = 0$ The origin only: $s \neq 1, -2$
   (d) The origin: $(0, 0, 0)$ All of R^3: no value of s

3. (a) $a(4, 1, 1) + b(0, -1, 2)$
   (b) $(a + c)(3, -1, 2) + b(1, 4, 1)$
   (c) $a(2, 3, 0) + b(-1, 0, 4) + c(4, -1, 1)$

5. (a) $\mathbf{v} = (-1 + r)\mathbf{v}_1 + (\frac{2}{3} - r)\mathbf{v}_2 + r\mathbf{v}_3$; r arbitrary   6. A must be invertible.   7. No

8. (a) Rank = 2, nullity = 1   9. (a) Rank = 2, nullity = 1
   (b) Rank = 2, nullity = 2   (b) Rank = 3, nullity = 2
   (c) Rank = 2, nullity = $n - 2$ (c) Rank = $n + 1$, nullity = n

11. $\{1, x^2, x^3, x^4, x^5, x^6, \ldots, x^n\}$ 12. (a) 2 (b) 1 (c) 2 (d) 3   13. 0, 1, or 2

EXERCISE SET 6.1 (page 296)

1. (a) 2 (b) 11 (c) -13 (d) -8 (e) 0   2. (a) -2 (b) 62 (c) -74 (d) 8 (e) 0

3. (a) 3 (b) 56   4. (a) -29 (b) -15 5. (b) 29   6. (b) -42

7. (a) $\begin{bmatrix} \sqrt{3} & 0 \\ 0 & \sqrt{5} \end{bmatrix}$ (b) $\begin{bmatrix} 2 & 0 \\ 0 & \sqrt{6} \end{bmatrix}$

9. (a) No. Axiom 4 fails.   (b) No. Axioms 2 and 3 fail.   (c) Yes   (d) No. Axiom 4 fails.

10. (a) $\sqrt{10}$ (b) $\sqrt{21}$ (c) $5\sqrt{5}$ 11. (a) $3\sqrt{2}$ (b) $3\sqrt{5}$ (c) $3\sqrt{13}$ 12. (a) $\sqrt{17}$ (b) 5

13. (a) $\sqrt{74}$ (b) 0   14. $3\sqrt{2}$ 15. (a) $\sqrt{105}$ (b) $\sqrt{47}$ 16. (a) 8   (b) -113 (c) -40
(d) 3   (e) $2\sqrt{53}$ (f) $\sqrt{881}$

17. (a) $\sqrt{2}, \frac{1}{3}\sqrt{6}, \frac{1}{5}\sqrt{10}$ (b) $\frac{2}{3}\sqrt{6}$

18. (a) $\dfrac{x^2}{4} + \dfrac{y^2}{16} = 1$ (b) $\dfrac{x^2}{\frac{1}{2}} + \dfrac{y^2}{1} = 1$

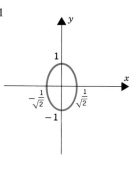

19. $\langle \mathbf{u}, \mathbf{v} \rangle = \frac{1}{9}u_1 v_1 + u_2 v_2$ 22. Axiom 4 fails.   27. (a) $-\frac{28}{15}$ (b) 0   28. (a) 0 (b) 1

EXERCISE SET 6.2 (page 308)

1. (a) Yes (b) No (c) Yes (d) No (e) No (f) Yes   2. No

3. (a) $-\dfrac{1}{\sqrt{2}}$ (b) $-\dfrac{3}{\sqrt{73}}$ (c) 0   4. (a) 0 (b) 0   6. (a) $\dfrac{19}{10\sqrt{7}}$ (b) 0

(d) $-\dfrac{20}{9\sqrt{10}}$ (e) $-\dfrac{1}{\sqrt{2}}$ (f) $\dfrac{2}{\sqrt{55}}$

7. (a) Orthogonal (b) Orthogonal   8. (a) $k = -3$ (b) $k = -2, k = -3$
(c) Orthogonal (d) Not orthogonal

9. $\pm\frac{1}{57}(-34, 44, -6, 11)$ 12. $y = -\frac{1}{2}x$ 13. (a) $x = t, y = -2t, z = -3t$ (b) $2x - 5y + 4z = 0$

14. (a) $(1, 2, -1, 2), (0, 1, -3, 2)$; $\begin{bmatrix} -5 \\ 3 \\ 1 \\ 0 \end{bmatrix}, \begin{bmatrix} 2 \\ -2 \\ 0 \\ 1 \end{bmatrix}$ 15. (a) $\begin{bmatrix} 1 \\ 3 \\ 1 \end{bmatrix}, \begin{bmatrix} 0 \\ 1 \\ 1 \end{bmatrix}; \begin{bmatrix} 2 \\ -1 \\ 1 \end{bmatrix}$

16. (a) $\begin{bmatrix} 16 \\ 19 \\ 1 \end{bmatrix}$ (b) $\begin{bmatrix} 0 \\ 1 \\ 0 \end{bmatrix}, \begin{bmatrix} \frac{1}{2} \\ 0 \\ 1 \end{bmatrix}$ (c) $\begin{bmatrix} -1 \\ -1 \\ 1 \\ 0 \end{bmatrix}, \begin{bmatrix} \frac{2}{7} \\ -\frac{4}{7} \\ 0 \\ 1 \end{bmatrix}$ (d) $\begin{bmatrix} -1 \\ -1 \\ 1 \\ 0 \\ 0 \end{bmatrix}, \begin{bmatrix} -2 \\ -1 \\ 0 \\ 1 \\ 0 \end{bmatrix}, \begin{bmatrix} -1 \\ -2 \\ 0 \\ 0 \\ 1 \end{bmatrix}$

33. $\langle \mathbf{u}, \mathbf{v} \rangle = \frac{1}{2}u_1 v_1 + \frac{1}{6}u_2 v_2$

EXERCISE SET 6.3 (page 325)

1. (a), (b), (d)   **2.** (b)   **3.** (b), (d)   **4.** (b), (d)   **5.** (a)   **6.** (a)

7. (a) $\left(-\dfrac{1}{\sqrt{5}}, \dfrac{2}{\sqrt{5}}\right), \left(\dfrac{2}{\sqrt{5}}, \dfrac{1}{\sqrt{5}}\right)$ (b) $\left(\dfrac{1}{\sqrt{2}}, 0, -\dfrac{1}{\sqrt{2}}\right), \left(\dfrac{1}{\sqrt{2}}, 0, \dfrac{1}{\sqrt{2}}\right), (0, 1, 0)$

   (c) $\left(\dfrac{1}{\sqrt{3}}, \dfrac{1}{\sqrt{3}}, \dfrac{1}{\sqrt{3}}\right), \left(-\dfrac{1}{\sqrt{2}}, \dfrac{1}{\sqrt{2}}, 0\right), \left(\dfrac{1}{\sqrt{6}}, \dfrac{1}{\sqrt{6}}, -\dfrac{2}{\sqrt{6}}\right)$

9. (a) $-\frac{7}{5}\mathbf{v}_1 + \frac{1}{5}\mathbf{v}_2 + 2\mathbf{v}_3$ (b) $-\frac{37}{5}\mathbf{v}_1 - \frac{9}{5}\mathbf{v}_2 + 4\mathbf{v}_3$ (c) $-\frac{3}{7}\mathbf{v}_1 - \frac{1}{7}\mathbf{v}_2 + \frac{5}{7}\mathbf{v}_3$

10. (a) $\frac{1}{7}\mathbf{v}_1 + \frac{5}{21}\mathbf{v}_2 + \frac{1}{3}\mathbf{v}_3 + \mathbf{v}_4$ (b) $\dfrac{15\sqrt{2}}{7}\mathbf{v}_1 + \dfrac{5\sqrt{2}}{21}\mathbf{v}_2 - \dfrac{2\sqrt{2}}{3}\mathbf{v}_3$ (c) $-\frac{3}{7}\mathbf{v}_1 + \frac{11}{63}\mathbf{v}_2 - \frac{1}{18}\mathbf{v}_3 + \frac{1}{2}\mathbf{v}_4$

11. (a) $(\mathbf{w})_S = (-2\sqrt{2}, 5\sqrt{2})$, $[\mathbf{w}]_S = \begin{bmatrix} -2\sqrt{2} \\ 5\sqrt{2} \end{bmatrix}$ (b) $(\mathbf{w})_S = (0, -2, 1)$, $[\mathbf{w}]_S = \begin{bmatrix} 0 \\ -2 \\ 1 \end{bmatrix}$

12. (a) $\mathbf{u} = (\frac{7}{5}, -\frac{1}{5})$, $\mathbf{v} = (\frac{13}{5}, \frac{16}{5})$ (b) $\|\mathbf{u}\| = \sqrt{2}$, $d(\mathbf{u}, \mathbf{v}) = \sqrt{13}$, $\langle \mathbf{u}, \mathbf{v} \rangle = 3$

13. (a) $\mathbf{u} = (1, \frac{14}{5}, -\frac{2}{5})$, $\mathbf{v} = (0, -\frac{17}{5}, \frac{6}{5})$, $\mathbf{w} = (-4, -\frac{11}{5}, \frac{23}{5})$ (b) $\|\mathbf{v}\| = \sqrt{13}$, $d(\mathbf{u}, \mathbf{v}) = 5\sqrt{3}$, $\langle \mathbf{w}, \mathbf{v} \rangle = 13$

14. (a) $\|\mathbf{u}\| = \sqrt{15}$, $\|\mathbf{v} - \mathbf{w}\| = 5$, $\|\mathbf{v} + \mathbf{w}\| = \sqrt{105}$, $\langle \mathbf{v}, \mathbf{w} \rangle = 20$
   (b) $\|\mathbf{u}\| = \sqrt{2}$, $\|\mathbf{v} - \mathbf{w}\| = \sqrt{34}$, $\|\mathbf{v} + \mathbf{w}\| = \sqrt{118}$, $\langle \mathbf{v}, \mathbf{w} \rangle = 21$

15. (b) $\mathbf{u} = -\frac{4}{5}\mathbf{v}_1 - \frac{11}{10}\mathbf{v}_2 + 0\mathbf{v}_3 + \frac{1}{2}\mathbf{v}_4$

16. (a) $\left(\dfrac{1}{\sqrt{10}}, -\dfrac{3}{\sqrt{10}}\right), \left(\dfrac{3}{\sqrt{10}}, \dfrac{1}{\sqrt{10}}\right)$ (b) $(1, 0), (0, -1)$

17. (a) $\left(\dfrac{1}{\sqrt{3}}, \dfrac{1}{\sqrt{3}}, \dfrac{1}{\sqrt{3}}\right), \left(-\dfrac{1}{\sqrt{2}}, \dfrac{1}{\sqrt{2}}, 0\right), \left(\dfrac{1}{\sqrt{6}}, \dfrac{1}{\sqrt{6}}, -\dfrac{2}{\sqrt{6}}\right)$

   (b) $(1, 0, 0), \left(0, \dfrac{7}{\sqrt{53}}, -\dfrac{2}{\sqrt{53}}\right), \left(0, \dfrac{2}{\sqrt{53}}, \dfrac{7}{53}\right)$

18. $\left(0, \dfrac{2}{\sqrt{5}}, \dfrac{1}{\sqrt{5}}, 0\right), \left(\dfrac{5}{\sqrt{30}}, -\dfrac{1}{\sqrt{30}}, \dfrac{2}{\sqrt{30}}, 0\right), \left(\dfrac{1}{\sqrt{10}}, \dfrac{1}{\sqrt{10}}, -\dfrac{2}{\sqrt{10}}, -\dfrac{2}{\sqrt{10}}\right), \left(\dfrac{1}{\sqrt{15}}, \dfrac{1}{\sqrt{15}}, -\dfrac{2}{\sqrt{15}}, \dfrac{3}{\sqrt{15}}\right)$

19. $\left(0, \dfrac{1}{\sqrt{5}}, \dfrac{2}{\sqrt{5}}\right), \left(-\dfrac{\sqrt{5}}{\sqrt{6}}, -\dfrac{2}{\sqrt{30}}, \dfrac{1}{\sqrt{30}}\right)$ **20.** $\left(\dfrac{1}{\sqrt{6}}, \dfrac{1}{\sqrt{6}}, \dfrac{1}{\sqrt{6}}\right), \left(\dfrac{1}{\sqrt{6}}, \dfrac{1}{\sqrt{6}}, -\dfrac{1}{\sqrt{6}}\right), \left(\dfrac{2}{\sqrt{6}}, -\dfrac{1}{\sqrt{6}}, 0\right)$

21. $\mathbf{w}_1 = (-\frac{4}{5}, 2, \frac{3}{5})$, $\mathbf{w}_2 = (\frac{9}{5}, 0, \frac{12}{5})$ **22.** $\mathbf{w}_1 = (\frac{39}{42}, \frac{93}{42}, \frac{120}{42})$, $\mathbf{w}_2 = (\frac{3}{42}, -\frac{9}{42}, \frac{6}{42})$

23. $\mathbf{w}_1 = (-\frac{5}{4}, -\frac{1}{4}, \frac{5}{4}, \frac{9}{4})$, $\mathbf{w}_2 = (\frac{1}{4}, \frac{9}{4}, \frac{19}{4}, -\frac{9}{4})$

24. (a) $\begin{bmatrix} \dfrac{1}{\sqrt{5}} & -\dfrac{2}{\sqrt{5}} \\ \dfrac{2}{\sqrt{5}} & \dfrac{1}{\sqrt{5}} \end{bmatrix} \begin{bmatrix} \sqrt{5} & \sqrt{5} \\ 0 & \sqrt{5} \end{bmatrix}$ (b) $\begin{bmatrix} \dfrac{1}{\sqrt{2}} & -\dfrac{1}{\sqrt{3}} \\ 0 & \dfrac{1}{\sqrt{3}} \\ \dfrac{1}{\sqrt{2}} & \dfrac{1}{\sqrt{3}} \end{bmatrix} \begin{bmatrix} \sqrt{2} & 3\sqrt{2} \\ 0 & \sqrt{3} \end{bmatrix}$ (c) $\begin{bmatrix} \dfrac{1}{3} & \dfrac{8}{\sqrt{234}} \\ -\dfrac{2}{3} & \dfrac{11}{\sqrt{234}} \\ \dfrac{2}{3} & \dfrac{7}{\sqrt{234}} \end{bmatrix} \begin{bmatrix} 3 & \dfrac{1}{3} \\ 0 & \dfrac{26}{\sqrt{234}} \end{bmatrix}$

(d) $\begin{bmatrix} \dfrac{1}{\sqrt{2}} & -\dfrac{1}{\sqrt{3}} & \dfrac{1}{\sqrt{6}} \\ 0 & \dfrac{1}{\sqrt{3}} & \dfrac{2}{\sqrt{6}} \\ \dfrac{1}{\sqrt{2}} & \dfrac{1}{\sqrt{3}} & -\dfrac{1}{\sqrt{6}} \end{bmatrix} \begin{bmatrix} \sqrt{2} & \sqrt{2} & \sqrt{2} \\ 0 & \sqrt{3} & -\dfrac{1}{\sqrt{3}} \\ 0 & 0 & \dfrac{4}{\sqrt{6}} \end{bmatrix}$ (e) $\begin{bmatrix} \dfrac{1}{\sqrt{2}} & \dfrac{\sqrt{2}}{2\sqrt{19}} & -\dfrac{3}{\sqrt{19}} \\ \dfrac{1}{\sqrt{2}} & -\dfrac{\sqrt{2}}{2\sqrt{19}} & \dfrac{3}{\sqrt{19}} \\ 0 & \dfrac{3\sqrt{2}}{\sqrt{19}} & \dfrac{1}{\sqrt{19}} \end{bmatrix} \begin{bmatrix} \sqrt{2} & \dfrac{3}{\sqrt{2}} & \sqrt{2} \\ 0 & \dfrac{\sqrt{19}}{\sqrt{2}} & \dfrac{3\sqrt{2}}{\sqrt{19}} \\ 0 & 0 & \dfrac{1}{\sqrt{19}} \end{bmatrix}$

(f) Columns not linearly independent

29. $\mathbf{v}_1 = \dfrac{1}{\sqrt{2}}, \mathbf{v}_2 = \sqrt{\dfrac{3}{2}}x, \mathbf{v}_3 = \dfrac{\sqrt{5}}{2\sqrt{2}}(3x^2 - 1)$

30. (a) $1 + x + 4x^2 = \frac{7}{3}\sqrt{2}\,\mathbf{v}_1 + \frac{1}{3}\sqrt{6}\,\mathbf{v}_2 + \frac{8}{15}\sqrt{10}\,\mathbf{v}_3$

(b) $2 - 7x^2 = -\dfrac{\sqrt{2}}{3}\mathbf{v}_1 - \frac{28}{15}\sqrt{\frac{5}{2}}\,\mathbf{v}_3$ (c) $4 + 3x = 4\sqrt{2}\,\mathbf{v}_1 + \sqrt{6}\,\mathbf{v}_2$

31. $\mathbf{v}_1 = 1, \mathbf{v}_2 = \sqrt{3}(2x - 1), \mathbf{v}_3 = \sqrt{5}(6x^2 - 6x + 1)$

EXERCISE SET 6.4 (page 336)

1. (a) $\begin{bmatrix} 21 & 25 \\ 25 & 35 \end{bmatrix}\begin{bmatrix} x_1 \\ x_2 \end{bmatrix} = \begin{bmatrix} 20 \\ 20 \end{bmatrix}$ (b) $\begin{bmatrix} 15 & -1 & 5 \\ -1 & 22 & 30 \\ 5 & 30 & 45 \end{bmatrix}\begin{bmatrix} x_1 \\ x_2 \\ x_3 \end{bmatrix} = \begin{bmatrix} -1 \\ 9 \\ 13 \end{bmatrix}$

2. (a) 0; column vectors are not linearly independent
(b) 0; column vectors are not linearly independent

3. (a) $x_1 = 3, x_2 = \frac{9}{2}; \begin{bmatrix} \frac{15}{2} \\ \frac{3}{2} \\ 6 \end{bmatrix}$ (b) $x_1 = \frac{3}{7}, x_2 = -\frac{2}{3}; \begin{bmatrix} \frac{46}{21} \\ -\frac{5}{21} \\ \frac{13}{21} \end{bmatrix}$

(c) $x_1 = 12, x_2 = -3, x_3 = 9; \begin{bmatrix} 3 \\ 3 \\ 9 \\ 0 \end{bmatrix}$ (d) $x_1 = 14, x_2 = 30, x_3 = 26; \begin{bmatrix} 2 \\ 6 \\ -2 \\ 4 \end{bmatrix}$

4. (a) $(\frac{16}{15}, \frac{29}{15}, \frac{13}{15})$ (b) $(-5, -4, -9)$ 5. (a) $(7, 2, 9, 5)$ (b) $(-\frac{12}{5}, -\frac{4}{5}, \frac{12}{5}, \frac{16}{5})$

6. $(0, -1, 1, 1)$

7. (a) $\begin{bmatrix} 1 & 0 \\ 0 & 0 \end{bmatrix}$ (b) $\begin{bmatrix} 0 & 0 \\ 0 & 1 \end{bmatrix}$ 8. (a) $\begin{bmatrix} 1 & 0 & 0 \\ 0 & 0 & 0 \\ 0 & 0 & 1 \end{bmatrix}$ (b) $\begin{bmatrix} 0 & 0 & 0 \\ 0 & 1 & 0 \\ 0 & 0 & 1 \end{bmatrix}$

9. (a) $\mathbf{v}_1 = (1, 0, -5), \mathbf{v}_2 = (0, 1, 3)$

(b) $\begin{bmatrix} \frac{10}{35} & \frac{15}{35} & -\frac{5}{35} \\ \frac{15}{35} & \frac{26}{35} & \frac{3}{35} \\ -\frac{5}{35} & \frac{3}{35} & \frac{34}{35} \end{bmatrix}$ (c) $\begin{bmatrix} \frac{10}{35}x_0 + \frac{15}{35}y_0 - \frac{5}{35}z_0 \\ \frac{15}{35}x_0 + \frac{26}{35}y_0 + \frac{3}{35}z_0 \\ -\frac{5}{35}x_0 + \frac{3}{35}y_0 + \frac{34}{35}z_0 \end{bmatrix}$ (d) $\dfrac{15}{\sqrt{35}}$

10. (a) $\mathbf{v}_1 = (2, -1, 4)$ (b) $\begin{bmatrix} \frac{4}{21} & -\frac{2}{21} & \frac{8}{21} \\ -\frac{2}{21} & \frac{1}{21} & -\frac{4}{21} \\ \frac{8}{21} & -\frac{4}{21} & \frac{16}{21} \end{bmatrix}$ (c) $\begin{bmatrix} \frac{4}{21}x_0 - \frac{2}{21}y_0 + \frac{8}{21}z_0 \\ -\frac{2}{21}x_0 + \frac{1}{21}y_0 - \frac{4}{21}z_0 \\ \frac{8}{21}x_0 - \frac{4}{21}y_0 + \frac{16}{21}z_0 \end{bmatrix}$

15. $P = A^T(AA^T)^{-1}A$

EXERCISE SET 6.5 (page 349)

2. $\begin{bmatrix} \frac{4}{5} & -\frac{9}{25} & \frac{12}{25} \\ 0 & \frac{4}{5} & \frac{3}{5} \\ -\frac{3}{5} & -\frac{12}{25} & \frac{16}{25} \end{bmatrix}$

3. (a) $\begin{bmatrix} 1 & 0 \\ 0 & 1 \end{bmatrix}$ (b) $\begin{bmatrix} \frac{1}{\sqrt{2}} & \frac{1}{\sqrt{2}} \\ -\frac{1}{\sqrt{2}} & \frac{1}{\sqrt{2}} \end{bmatrix}$ (d) $\begin{bmatrix} -\frac{1}{\sqrt{2}} & 0 & \frac{1}{\sqrt{2}} \\ \frac{1}{\sqrt{6}} & -\frac{2}{\sqrt{6}} & \frac{1}{\sqrt{6}} \\ \frac{1}{\sqrt{3}} & \frac{1}{\sqrt{3}} & \frac{1}{\sqrt{3}} \end{bmatrix}$ (e) $\begin{bmatrix} \frac{1}{2} & \frac{1}{2} & \frac{1}{2} & \frac{1}{2} \\ \frac{1}{2} & -\frac{5}{6} & \frac{1}{6} & \frac{1}{6} \\ \frac{1}{2} & \frac{1}{6} & \frac{1}{6} & -\frac{5}{6} \\ \frac{1}{2} & \frac{1}{6} & -\frac{5}{6} & \frac{1}{6} \end{bmatrix}$

5. (a) $(\mathbf{w})_S = (3, -7)$, $[\mathbf{w}]_S = \begin{bmatrix} 3 \\ -7 \end{bmatrix}$ (b) $(\mathbf{w})_S = (\frac{5}{28}, \frac{3}{14})$, $[\mathbf{w}]_S = \begin{bmatrix} \frac{5}{28} \\ \frac{3}{14} \end{bmatrix}$

   (c) $(\mathbf{w})_S = \left(a, \dfrac{b-a}{2}\right)$, $[\mathbf{w}]_S = \begin{bmatrix} a \\ \frac{b-a}{2} \end{bmatrix}$

6. (a) $(\mathbf{v})_S = (3, -2, 1)$, $[\mathbf{v}]_S = \begin{bmatrix} 3 \\ -2 \\ 1 \end{bmatrix}$ (b) $(\mathbf{v})_S = (-2, 0, 1)$, $[\mathbf{v}]_S = \begin{bmatrix} -2 \\ 0 \\ 1 \end{bmatrix}$

7. (a) $(\mathbf{p})_S = (4, -3, 1)$, $[\mathbf{p}]_S = \begin{bmatrix} 4 \\ -3 \\ 1 \end{bmatrix}$ (b) $(\mathbf{p})_S = (0, 2, -1)$, $[\mathbf{p}]_S = \begin{bmatrix} 0 \\ 2 \\ -1 \end{bmatrix}$

8. $(A)_S = (-1, 1, -1, 3)$, $[A]_S = \begin{bmatrix} -1 \\ 1 \\ -1 \\ 3 \end{bmatrix}$ 9. (a) $\mathbf{w} = (16, 10, 12)$ (b) $\mathbf{q} = 3 + 4x^2$ (c) $B = \begin{bmatrix} 15 & -1 \\ 6 & 3 \end{bmatrix}$

10. (a) $\begin{bmatrix} 2 & -3 \\ 1 & 4 \end{bmatrix}$ (b) $\begin{bmatrix} \frac{4}{11} & \frac{3}{11} \\ -\frac{1}{11} & \frac{2}{11} \end{bmatrix}$ (c) $[\mathbf{w}]_B = \begin{bmatrix} 3 \\ -5 \end{bmatrix}$, $[\mathbf{w}]_{B'} = \begin{bmatrix} -\frac{3}{11} \\ -\frac{13}{11} \end{bmatrix}$

11. (a) $\begin{bmatrix} \frac{13}{10} & -\frac{1}{2} \\ -\frac{2}{5} & 0 \end{bmatrix}$ (b) $\begin{bmatrix} 0 & -\frac{5}{2} \\ -2 & -\frac{13}{2} \end{bmatrix}$ (c) $[\mathbf{w}]_B = \begin{bmatrix} -\frac{17}{10} \\ \frac{8}{5} \end{bmatrix}$, $[\mathbf{w}]_{B'} = \begin{bmatrix} -4 \\ -7 \end{bmatrix}$

12. (a) $\begin{bmatrix} \frac{3}{4} & \frac{3}{4} & \frac{1}{12} \\ -\frac{3}{4} & -\frac{17}{12} & -\frac{17}{12} \\ 0 & \frac{2}{3} & \frac{2}{3} \end{bmatrix}$ (b) $\begin{bmatrix} \frac{19}{12} \\ -\frac{43}{12} \\ \frac{4}{3} \end{bmatrix}$ 13. (a) $\begin{bmatrix} 3 & 2 & \frac{5}{2} \\ -2 & -3 & -\frac{1}{2} \\ 5 & 1 & 6 \end{bmatrix}$ (b) $\begin{bmatrix} -\frac{7}{2} \\ \frac{23}{2} \\ 6 \end{bmatrix}$

14. (a) $\begin{bmatrix} -\frac{2}{9} & \frac{7}{9} \\ \frac{1}{3} & -\frac{1}{6} \end{bmatrix}$ (b) $\begin{bmatrix} \frac{3}{4} & \frac{7}{2} \\ \frac{3}{2} & 1 \end{bmatrix}$ (c) $[\mathbf{p}]_B = \begin{bmatrix} 1 \\ -1 \end{bmatrix}$ (d) $[\mathbf{p}]_{B'} = \begin{bmatrix} -\frac{11}{4} \\ \frac{1}{2} \end{bmatrix}$

15. (b) $\begin{bmatrix} 2 & 0 \\ 1 & 3 \end{bmatrix}$ (c) $\begin{bmatrix} \frac{1}{2} & 0 \\ -\frac{1}{6} & \frac{1}{3} \end{bmatrix}$ (d) $[\mathbf{h}]_B = \begin{bmatrix} 2 \\ -5 \end{bmatrix}$, $[\mathbf{h}]_{B'} = \begin{bmatrix} 1 \\ -2 \end{bmatrix}$

16. (a) $(4\sqrt{2}, -2\sqrt{2})$ (b) $(-\frac{7}{2}\sqrt{2}, \frac{3}{2}\sqrt{2})$ 17. (a) $(-1 + 3\sqrt{3}, 3 + \sqrt{3})$ (b) $(\frac{5}{2} - \sqrt{3}, \frac{5}{2}\sqrt{3} + 1)$

18. (a) $(\frac{1}{2}\sqrt{2}, \frac{3}{2}\sqrt{2}, 5)$ (b) $(-\frac{5}{2}\sqrt{2}, \frac{7}{2}\sqrt{2}, -3)$

19. (a) $(-\frac{1}{2} - \frac{5}{2}\sqrt{3}, 2, \frac{5}{2} - \frac{1}{2}\sqrt{3})$ (b) $(\frac{1}{2} - \frac{3}{2}\sqrt{3}, 6, -\frac{3}{2} - \frac{1}{2}\sqrt{3})$

20. (a) $(-1, \frac{3}{2}\sqrt{2}, -\frac{7}{2}\sqrt{2})$ (b) $(1, -\frac{3}{2}\sqrt{2}, \frac{9}{2}\sqrt{3})$

21. (a) $A = \begin{bmatrix} \cos\theta & 0 & -\sin\theta \\ 0 & 1 & 0 \\ \sin\theta & 0 & \cos\theta \end{bmatrix}$ (b) $A = \begin{bmatrix} 1 & 0 & 0 \\ 0 & \cos\theta & \sin\theta \\ 0 & -\sin\theta & \cos\theta \end{bmatrix}$ 22. $\begin{bmatrix} \frac{\sqrt{2}}{4} & \frac{\sqrt{6}}{4} & -\frac{\sqrt{2}}{2} \\ -\frac{\sqrt{3}}{2} & \frac{1}{2} & 0 \\ \frac{\sqrt{2}}{4} & \frac{\sqrt{6}}{4} & \frac{\sqrt{2}}{2} \end{bmatrix}$

23. $a^2 + b^2 = \frac{1}{2}$ 26. (a) Rotation   (b) Rotation

27. (a) Rotation followed by a reflection   (b) Rotation

CHAPTER 6 SUPPLEMENTARY EXERCISES (page 353)

1. (a) $(0, a, a, 0)$ with $a \neq 0$ (b) $\pm\left(0, \frac{2}{\sqrt{5}}, \frac{1}{\sqrt{5}}, 0\right)$ 6. $\pm\left(\frac{1}{\sqrt{2}}, 0, \frac{1}{\sqrt{2}}\right)$

7. $w_k = \frac{1}{k}$, $k = 1, 2, \ldots, n$ 8. No   11. (b) θ approaches $\frac{\pi}{2}$

12. (b) The diagonals of a parallelogram are perpendicular if and only if its sides have the same length.

16. $a = 0$, $b = -\frac{2}{\sqrt{6}}$, $c = \frac{1}{\sqrt{3}}$; not unique

EXERCISE SET 7.1 (page 362)

1. (a) $\lambda^2 - 2\lambda - 3 = 0$ (b) $\lambda^2 - 8\lambda + 16 = 0$ (c) $\lambda^2 - 12 = 0$
   (d) $\lambda^2 + 3 = 0$ (e) $\lambda^2 = 0$ (f) $\lambda^2 - 2\lambda + 1 = 0$

2. (a) $\lambda = 3$, $\lambda = -1$ (b) $\lambda = 4$ (c) $\lambda = \sqrt{12}$, $\lambda = -\sqrt{12}$
   (d) No real eigenvalues   (e) $\lambda = 0$ (f) $\lambda = 1$

3. (a) Basis for eigenspace corresponding to $\lambda = 3$: $\begin{bmatrix} \frac{1}{2} \\ 1 \end{bmatrix}$; basis for eigenspace corresponding to $\lambda = -1$: $\begin{bmatrix} 0 \\ 1 \end{bmatrix}$

   (b) Basis for eigenspace corresponding to $\lambda = 4$: $\begin{bmatrix} \frac{3}{2} \\ 1 \end{bmatrix}$

   (c) Basis for eigenspace corresponding to $\lambda = \sqrt{12}$: $\begin{bmatrix} \frac{3}{\sqrt{12}} \\ 1 \end{bmatrix}$;

   basis for eigenspace corresponding to $\lambda = -\sqrt{12}$: $\begin{bmatrix} -\frac{3}{\sqrt{12}} \\ 1 \end{bmatrix}$

(d) There are no eigenspaces. (e) Basis for eigenspace corresponding to $\lambda = 0$: $\begin{bmatrix} 1 \\ 0 \end{bmatrix}$, $\begin{bmatrix} 0 \\ 1 \end{bmatrix}$

(f) Basis for eigenspace corresponding to $\lambda = 1$: $\begin{bmatrix} 1 \\ 0 \end{bmatrix}\begin{bmatrix} 0 \\ 1 \end{bmatrix}$

4. (a) $\lambda^3 - 6\lambda^2 + 11\lambda - 6 = 0$ (b) $\lambda^3 - 2\lambda = 0$
   (c) $\lambda^3 + 8\lambda^2 + \lambda + 8 = 0$ (d) $\lambda^3 - \lambda^2 - \lambda - 2 = 0$
   (e) $\lambda^3 - 6\lambda^2 + 12\lambda - 8 = 0$ (f) $\lambda^3 - 2\lambda^2 - 15\lambda + 36 = 0$

5. (a) $\lambda = 1, \lambda = 2, \lambda = 3$ (b) $\lambda = 0, \lambda = \sqrt{2}, \lambda = -\sqrt{2}$ (c) $\lambda = -8$
   (d) $\lambda = 2$ (e) $\lambda = 2$ (f) $\lambda = -4, \lambda = 3$

6. (a) $\lambda = 1$: basis $\begin{bmatrix} 0 \\ 1 \\ 0 \end{bmatrix}$; $\lambda = 2$: basis $\begin{bmatrix} -\frac{1}{2} \\ 1 \\ 1 \end{bmatrix}$; $\lambda = 3$: basis $\begin{bmatrix} -1 \\ 1 \\ 1 \end{bmatrix}$

   (b) $\lambda = 0$: basis $\begin{bmatrix} \frac{5}{3} \\ \frac{1}{3} \\ 1 \end{bmatrix}$; $\lambda = \sqrt{2}$: basis $\begin{bmatrix} \frac{1}{7}(15 + 5\sqrt{2}) \\ \frac{1}{7}(-1 + 2\sqrt{2}) \\ 1 \end{bmatrix}$; $\lambda = -\sqrt{2}$: basis $\begin{bmatrix} \frac{1}{7}(15 - 5\sqrt{2}) \\ \frac{1}{7}(-1 - 2\sqrt{2}) \\ 1 \end{bmatrix}$

   (c) $\lambda = -8$: basis $\begin{bmatrix} -\frac{1}{6} \\ -\frac{1}{6} \\ 1 \end{bmatrix}$ (d) $\lambda = 2$: basis $\begin{bmatrix} \frac{1}{3} \\ \frac{1}{3} \\ 1 \end{bmatrix}$ (e) $\lambda = 2$: basis $\begin{bmatrix} -\frac{1}{3} \\ -\frac{1}{3} \\ 1 \end{bmatrix}$

   (f) $\lambda = -4$: basis $\begin{bmatrix} -2 \\ \frac{8}{3} \\ 1 \end{bmatrix}$; $\lambda = 3$: basis $\begin{bmatrix} 5 \\ -2 \\ 1 \end{bmatrix}$

7. (a) $(\lambda - 1)^2(\lambda + 2)(\lambda + 1) = 0$ (b) $(\lambda - 4)^2(\lambda^2 + 3) = 0$

8. (a) $\lambda = 1, \lambda = -2, \lambda = -1$ (b) $\lambda = 4$

9. (a) $\lambda = 1$: basis $\begin{bmatrix} 0 \\ 0 \\ 0 \\ 1 \end{bmatrix}$ and $\begin{bmatrix} 2 \\ 3 \\ 1 \\ 0 \end{bmatrix}$; $\lambda = -2$; basis $\begin{bmatrix} -1 \\ 0 \\ 1 \\ 0 \end{bmatrix}$; $\lambda = -1$: basis $\begin{bmatrix} -2 \\ 1 \\ 1 \\ 0 \end{bmatrix}$ (b) $\lambda = 4$: basis $\begin{bmatrix} \frac{3}{2} \\ 1 \\ 0 \\ 0 \end{bmatrix}$

10. (a) $\lambda = -1, \lambda = 5$ (b) $\lambda = 3, \lambda = 7, \lambda = 1$ (c) $\lambda = -\frac{1}{3}, \lambda = 1, \lambda = \frac{1}{2}$

11. $\lambda = 1, \lambda = \frac{1}{512}, \lambda = 512, \lambda = 0$

12. For A^{25}, $\lambda = 1, -1$; basis for $\lambda = 1$: $\begin{bmatrix} -1 \\ 1 \\ 0 \end{bmatrix}$, $\begin{bmatrix} -1 \\ 0 \\ 1 \end{bmatrix}$; basis for $\lambda = -1$: $\begin{bmatrix} 2 \\ -1 \\ 1 \end{bmatrix}$

13. (a) $y = x$ and $y = 2x$ (b) No lines (c) $y = 0$ 14. (a) -5 (b) 7

22. (a) $\lambda_1 = 1$: $\begin{bmatrix} 1 \\ 0 \\ 1 \end{bmatrix}$; $\lambda_2 = \frac{1}{2}$: $\begin{bmatrix} 1 \\ \frac{1}{2} \\ 0 \end{bmatrix}$; $\lambda_3 = \frac{1}{3}$: $\begin{bmatrix} 1 \\ 1 \\ 1 \end{bmatrix}$

   (b) $\lambda_1 = -2$: $\begin{bmatrix} 1 \\ 0 \\ 1 \end{bmatrix}$; $\lambda_2 = -1$: $\begin{bmatrix} 1 \\ \frac{1}{2} \\ 0 \end{bmatrix}$; $\lambda_3 = 0$: $\begin{bmatrix} 1 \\ 1 \\ 1 \end{bmatrix}$

(c) $\lambda_1 = 3$: $\begin{bmatrix} 1 \\ 0 \\ 1 \end{bmatrix}$; $\lambda_2 = 4$: $\begin{bmatrix} 1 \\ \frac{1}{2} \\ 0 \end{bmatrix}$; $\lambda_3 = 5$: $\begin{bmatrix} 1 \\ 1 \\ 1 \end{bmatrix}$

EXERCISE SET 7.2 (page 373)

1. 1, 2, or 3

2. (a) $\lambda = 3$, $\lambda = 5$
   (b) For $\lambda = 3$, the rank of $3I - A$ is 1 and the nullity is 2. For $\lambda = 5$, the rank of $5I - A$ is 2 and the nullity is 1.
   (c) A is diagonalizable since the eigenspaces produce a total of three basis vectors.

3. Not diagonalizable   4. Not diagonalizable   5. Not diagonalizable

6. Not diagonalizable   7. Not diagonalizable

8. $P = \begin{bmatrix} \frac{4}{5} & \frac{3}{4} \\ 1 & 1 \end{bmatrix}$; $P^{-1}AP = \begin{bmatrix} 1 & 0 \\ 0 & 2 \end{bmatrix}$ 9. $P = \begin{bmatrix} \frac{1}{3} & 0 \\ 1 & 1 \end{bmatrix}$; $P^{-1}AP = \begin{bmatrix} 1 & 0 \\ 0 & -1 \end{bmatrix}$

10. $P = \begin{bmatrix} 0 & 1 & 0 \\ 1 & 0 & 1 \\ -1 & 0 & 1 \end{bmatrix}$; $P^{-1}AP = \begin{bmatrix} 0 & 0 & 0 \\ 0 & 1 & 0 \\ 0 & 0 & 2 \end{bmatrix}$ 11. $P = \begin{bmatrix} -2 & 0 & 1 \\ 0 & 1 & 0 \\ 1 & 0 & 0 \end{bmatrix}$; $P^{-1}AP = \begin{bmatrix} 3 & 0 & 0 \\ 0 & 3 & 0 \\ 0 & 0 & 2 \end{bmatrix}$

12. Not diagonalizable   13. $P = \begin{bmatrix} 1 & 2 & 1 \\ 1 & 3 & 3 \\ 1 & 3 & 4 \end{bmatrix}$; $P^{-1}AP = \begin{bmatrix} 1 & 0 & 0 \\ 0 & 2 & 0 \\ 0 & 0 & 3 \end{bmatrix}$ 14. Not diagonalizable

15. $P = \begin{bmatrix} -\frac{1}{3} & 0 & 0 \\ 0 & 1 & 0 \\ 1 & 0 & 1 \end{bmatrix}$; $P^{-1}AP = \begin{bmatrix} 0 & 0 & 0 \\ 0 & 0 & 0 \\ 0 & 0 & 1 \end{bmatrix}$ 16. Not diagonalizable   17. $P = \begin{bmatrix} 1 & 1 & 0 & 0 \\ 0 & 1 & 1 & 0 \\ 0 & 0 & 1 & 1 \\ 0 & 0 & 0 & 1 \end{bmatrix}$;

$P^{-1}AP = \begin{bmatrix} -2 & 0 & 0 & 0 \\ 0 & -2 & 0 & 0 \\ 0 & 0 & 3 & 0 \\ 0 & 0 & 0 & 3 \end{bmatrix}$ 18. $\begin{bmatrix} 1 & 0 \\ -1023 & 1024 \end{bmatrix}$ 19. $\begin{bmatrix} -1 & 10237 & -2047 \\ 0 & 1 & 0 \\ 0 & 10245 & -2048 \end{bmatrix}$

20. (a) $\begin{bmatrix} 1 & 0 & 0 \\ 0 & 1 & 0 \\ 0 & 0 & 1 \end{bmatrix}$ (b) $\begin{bmatrix} 1 & 0 & 0 \\ 0 & 1 & 0 \\ 0 & 0 & 1 \end{bmatrix}$ (c) $\begin{bmatrix} 1 & -2 & 8 \\ 0 & -1 & 0 \\ 0 & 0 & -1 \end{bmatrix}$ (d) $\begin{bmatrix} 1 & -2 & 8 \\ 0 & -1 & 0 \\ 0 & 0 & -1 \end{bmatrix}$

21. $A^n = PD^nP^{-1} = \begin{bmatrix} 1 & 1 & 1 \\ 2 & 0 & -1 \\ 1 & -1 & 1 \end{bmatrix} \begin{bmatrix} 1^n & 0 & 0 \\ 0 & 3^n & 0 \\ 0 & 0 & 4^n \end{bmatrix} \begin{bmatrix} \frac{1}{6} & \frac{1}{3} & \frac{1}{6} \\ \frac{1}{2} & 0 & -\frac{1}{2} \\ \frac{1}{3} & -\frac{1}{3} & \frac{1}{3} \end{bmatrix}$

EXERCISE SET 7.3 (page 378)

1. (a) $\lambda^2 - 5\lambda = 0$; $\lambda = 0$: one-dimensional; $\lambda = 5$: one-dimensional
   (b) $\lambda^3 - 27\lambda - 54 = 0$; $\lambda = 6$: one-dimensional; $\lambda = -3$: two-dimensional
   (c) $\lambda^3 - 3\lambda^2 = 0$; $\lambda = 3$: one-dimensional; $\lambda = 0$: two-dimensional
   (d) $\lambda^3 - 12\lambda^2 + 36\lambda - 32 = 0$; $\lambda = 2$: two-dimensional; $\lambda = 8$: one-dimensional
   (e) $\lambda^4 - 8\lambda^3 = 0$; $\lambda = 0$: three-dimensional; $\lambda = 8$: one-dimensional

(f) $\lambda^4 - 8\lambda^3 + 22\lambda^2 - 24\lambda + 9 = 0$; $\lambda = 1$: two-dimensional; $\lambda = 3$: two-dimensional

2. $\begin{bmatrix} \dfrac{1}{\sqrt{2}} & -\dfrac{1}{\sqrt{2}} \\ \dfrac{1}{\sqrt{2}} & \dfrac{1}{\sqrt{2}} \end{bmatrix}$; $P^{-1}AP = \begin{bmatrix} 4 & 0 \\ 0 & 2 \end{bmatrix}$

3. $P = \begin{bmatrix} -\dfrac{2}{\sqrt{7}} & \dfrac{\sqrt{3}}{\sqrt{7}} \\ \dfrac{\sqrt{3}}{\sqrt{7}} & \dfrac{2}{\sqrt{7}} \end{bmatrix}$; $P^{-1}AP = \begin{bmatrix} 3 & 0 \\ 0 & 10 \end{bmatrix}$

4. $P = \begin{bmatrix} -\dfrac{2}{\sqrt{5}} & \dfrac{1}{\sqrt{5}} \\ \dfrac{1}{\sqrt{5}} & \dfrac{2}{\sqrt{5}} \end{bmatrix}$; $P^{-1}AP = \begin{bmatrix} 7 & 0 \\ 0 & 2 \end{bmatrix}$

5. $P = \begin{bmatrix} -\dfrac{4}{5} & 0 & \dfrac{3}{5} \\ 0 & 1 & 0 \\ \dfrac{3}{5} & 0 & \dfrac{4}{5} \end{bmatrix}$; $P^{-1}AP = \begin{bmatrix} 25 & 0 & 0 \\ 0 & -3 & 0 \\ 0 & 0 & -50 \end{bmatrix}$

6. $\begin{bmatrix} \dfrac{1}{\sqrt{2}} & \dfrac{1}{\sqrt{2}} & 0 \\ \dfrac{1}{\sqrt{2}} & -\dfrac{1}{\sqrt{2}} & 0 \\ 0 & 0 & 1 \end{bmatrix}$

7. $\begin{bmatrix} \dfrac{1}{\sqrt{3}} & \dfrac{1}{\sqrt{6}} & \dfrac{1}{\sqrt{2}} \\ \dfrac{1}{\sqrt{3}} & -\dfrac{2}{\sqrt{6}} & 0 \\ \dfrac{1}{\sqrt{3}} & \dfrac{1}{\sqrt{6}} & -\dfrac{1}{\sqrt{2}} \end{bmatrix}$

8. $\begin{bmatrix} 0 & 0 & \dfrac{1}{\sqrt{2}} & \dfrac{1}{\sqrt{2}} \\ 0 & 0 & \dfrac{1}{\sqrt{2}} & -\dfrac{1}{\sqrt{2}} \\ 1 & 0 & 0 & 0 \\ 0 & 1 & 0 & 0 \end{bmatrix}$

9. $P = \begin{bmatrix} -\dfrac{4}{5} & \dfrac{3}{5} & 0 & 0 \\ \dfrac{3}{5} & \dfrac{4}{5} & 0 & 0 \\ 0 & 0 & -\dfrac{4}{5} & \dfrac{3}{5} \\ 0 & 0 & \dfrac{3}{5} & \dfrac{4}{5} \end{bmatrix}$; $P^{-1}AP = \begin{bmatrix} -25 & 0 & 0 & 0 \\ 0 & 25 & 0 & 0 \\ 0 & 0 & -25 & 0 \\ 0 & 0 & 0 & 25 \end{bmatrix}$

10. $\begin{bmatrix} \dfrac{1}{\sqrt{2}} & -\dfrac{1}{\sqrt{2}} \\ \dfrac{1}{\sqrt{2}} & \dfrac{1}{\sqrt{2}} \end{bmatrix}$

12. (b) $\begin{bmatrix} \dfrac{1}{\sqrt{2}} & 0 & \dfrac{1}{\sqrt{2}} \\ 0 & 1 & 0 \\ -\dfrac{1}{\sqrt{2}} & 0 & \dfrac{1}{\sqrt{2}} \end{bmatrix}$

CHAPTER 7 SUPPLEMENTARY EXERCISES (page 379)

1. (b) The transformation rotates vectors through the angle θ; therefore, if $0 < \theta < \pi$, then no nonzero vector is transformed into a vector in the same or opposite direction.

2. $\lambda = k$ with multiplicity 3.

3. (c) $\begin{bmatrix} 1 & 1 & 0 \\ 0 & 2 & 1 \\ 0 & 0 & 3 \end{bmatrix}$

9. $A^2 = \begin{bmatrix} 15 & 30 \\ 5 & 10 \end{bmatrix}$, $A^3 = \begin{bmatrix} 75 & 150 \\ 25 & 50 \end{bmatrix}$, $A^4 = \begin{bmatrix} 375 & 750 \\ 125 & 250 \end{bmatrix}$, $A^5 = \begin{bmatrix} 1875 & 3750 \\ 625 & 1250 \end{bmatrix}$

10. $A^3 = \begin{bmatrix} 1 & -3 & 3 \\ 3 & -8 & 6 \\ 6 & -15 & 10 \end{bmatrix}$, $A^4 = \begin{bmatrix} 3 & -8 & 6 \\ 6 & -15 & 10 \\ 10 & -24 & 15 \end{bmatrix}$

12. (b) $\begin{bmatrix} 0 & 0 & 0 & -1 \\ 1 & 0 & 0 & 2 \\ 0 & 1 & 0 & -1 \\ 0 & 0 & 1 & -3 \end{bmatrix}$

15. $\begin{bmatrix} 1 & 0 & 0 \\ -1 & -\frac{1}{2} & -\frac{1}{2} \\ 1 & -\frac{1}{2} & -\frac{1}{2} \end{bmatrix}$

16. (a) 18 (b) -1

EXERCISE SET 8.1 (page 392)

3. Nonlinear   4. Linear   5. Linear   6. Linear

7. Linear   8. (a) Linear (b) Nonlinear

9. (a) Linear (b) Nonlinear   10. (a) Nonlinear (b) Linear

12. $T(x_1, x_2) = (-4x_1 + 5x_2, x_1 - 3x_2)$; $T(5, -3) = (-35, 14)$

13. $T(x_1, x_2) = \frac{1}{7}(3x_1 - x_2, -9x_1 - 4x_2, 5x_1 + 10x_2)$; $T(2, -3) = (\frac{9}{7}, -\frac{6}{7}, -\frac{20}{7})$

14. $T(x_1, x_2, x_3) = (-x_1 + 4x_2 - x_3, 5x_1 - 5x_2 - x_3, x_1 + 3x_3)$; $T(2, 4, -1) = (15, -9, -1)$

15. $T(x_1, x_2, x_3) = (-41x_1 + 9x_2 + 24x_3, 14x_1 - 3x_2 - 8x_3)$; $T(7, 13, 7) = (-2, 3)$

16. $T(2\mathbf{v}_1 - 3\mathbf{v}_2 + 4\mathbf{v}_3) = (-10, -7, 6)$

17. (a) Domain: R^2; image space: R^2; $(T_2 \circ T_1)(x, y) = (2x - 3y, 2x + 3y)$
    (b) Domain: R^2; image space: the line $y = \frac{3}{4}x$; $(T_2 \circ T_1)(x, y) = (4x - 12y, 3x - 9y)$
    (c) Domain: R^2; image space: R^2; $(T_2 \circ T_1)(x, y) = (2x + 3y, x - 2y)$
    (d) Domain: R^2; image space: the line $x = 0$; $(T_2 \circ T_1)(x, y) = (0, 2x)$

18. (a) Domain: R^2; image space: R^2; $(T_3 \circ T_2 \circ T_1)(x, y) = (3x - 2y, x)$
    (b) Domain: R^2; image space: the line $y = \frac{3}{2}x$; $(T_3 \circ T_2 \circ T_1)(x, y) = (4y, 6y)$

19. (a) $a + d$ (b) $(T_2 \circ T_1)(A)$ does not exist since $T_1(A)$ is not a 2×2 matrix.

20. $(T_1 \circ T_2)(p(x)) = p(x)$; $(T_2 \circ T_1)(p(x)) = p(x)$

21. $T_2(\mathbf{v}) = \frac{1}{4}\mathbf{v}$ 22. $(T_2 \circ T_1)(a_0 + a_1x + a_2x^2) = (a_0 + a_1 + a_2)x + (a_1 + 2a_2)x^2 + a_2x^3$

26. (b) $(3T)(x_1, x_2) = (6x_1 - 3x_2, 3x_2 + 3x_1)$

27. (b) $(T_1 + T_2)(x, y) = (3y, 4x)$; $(T_2 - T_1)(x, y) = (y, 2x)$

28. (b) Nonlinear   29. (a) 4 (b) .8415 (c) 1

EXERCISE SET 8.2 (page 399)

1. (a), (c)   2. (a)   3. (a), (b), (c)   4. (a)   5. (b)   6. (c)   7. (a) $\begin{bmatrix} \frac{1}{2} \\ 1 \end{bmatrix}$ (b) $\begin{bmatrix} \frac{3}{2} \\ -4 \\ 1 \\ 0 \end{bmatrix}$

(c) No basis exists.

8. (a) $\begin{bmatrix} 1 \\ -4 \end{bmatrix}$ (b) $\begin{bmatrix} 4 \\ 2 \\ 6 \end{bmatrix}, \begin{bmatrix} 1 \\ 1 \\ 0 \end{bmatrix}, \begin{bmatrix} -3 \\ -4 \\ 9 \end{bmatrix}$ (c) $\begin{bmatrix} x \\ x^2 \\ x^3 \end{bmatrix}$

10. (a) $\begin{bmatrix} 1 \\ 5 \\ 7 \end{bmatrix}, \begin{bmatrix} 0 \\ 1 \\ 1 \end{bmatrix}$ (b) $\begin{bmatrix} -\frac{14}{11} \\ \frac{19}{11} \\ 1 \end{bmatrix}$ (c) Rank$(T) = 2$, nullity$(T) = 1$

11. (a) $\begin{bmatrix} 1 \\ 2 \\ 0 \end{bmatrix}$ (b) $\begin{bmatrix} \frac{1}{2} \\ 0 \\ 1 \end{bmatrix}, \begin{bmatrix} 0 \\ 1 \\ 0 \end{bmatrix}$ (c) Rank$(T) = 1$, nullity$(T) = 2$

12. (a) $\begin{bmatrix} 1 \\ \frac{1}{4} \end{bmatrix}, \begin{bmatrix} 0 \\ 1 \end{bmatrix}$ (b) $\begin{bmatrix} -1 \\ -1 \\ 1 \\ 0 \end{bmatrix}, \begin{bmatrix} -\frac{4}{7} \\ \frac{2}{7} \\ 0 \\ 1 \end{bmatrix}$ (c) Rank$(T) = 2$, nullity$(T) = 2$

13. (a) $\begin{bmatrix} 1 \\ 3 \\ -1 \\ 2 \end{bmatrix}, \begin{bmatrix} 0 \\ 1 \\ -\frac{2}{7} \\ \frac{5}{14} \end{bmatrix}, \begin{bmatrix} 0 \\ 0 \\ 0 \\ 1 \end{bmatrix}$ (b) $\begin{bmatrix} -1 \\ -1 \\ 1 \\ 0 \\ 0 \end{bmatrix}, \begin{bmatrix} -1 \\ -2 \\ 0 \\ 0 \\ 1 \end{bmatrix}$ (c) Rank$(T) = 3$, nullity$(T) = 2$

14. (a) Range: xz-plane; nullspace: y-axis   15. ker$(T) = \{0\}$; $R(T) = V$
    (b) Range: yz-plane; nullspace: x-axis
    (c) Range: plane $y = x$; nullspace: the line $x = -t$, $y = t$, $z = 0$

16. (a) Nullity$(T) = 2$ (b) Nullity$(T) = 4$ 17. Nullity$(T) = 0$, rank$(T) = 6$
    (c) Nullity$(T) = 3$ (d) Nullity$(T) = 1$

18. (a) Dimension $=$ nullity$(T) = 3$
    (b) No. In order for $A\mathbf{x} = \mathbf{b}$ to be consistent for all $\mathbf{b}$ in R^5, we must have $R(T) = R^5$. But $R(T) \neq R^5$, since rank$(T) = \dim R(T) = 4$.

21. (a) $x = -t$, $y = -t$, $z = t$, $-\infty < t < +\infty$ (b) $14x - 8y - 5z = 0$

25. ker(D) consists of all constant polynomials.   26. ker(J) consists of all polynomials of the form kx.

27. ker$(D \circ D)$ consists of all functions of the form $ax + b$.

EXERCISE SET 8.3 (page 407)

1. (a) ker$(T) = \{0\}$; T is one-to-one. (b) ker$(T) = \left\{ k\begin{bmatrix} -\frac{3}{2} \\ 1 \end{bmatrix} \right\}$; T is not one-to-one.

    (c) ker$(T) = \{0\}$; T is one-to-one. (d) ker$(T) = \{0\}$; T is one-to-one.

    (e) ker$(T) = \left\{ k\begin{bmatrix} 1 \\ 1 \end{bmatrix} \right\}$; T is not one-to-one. (f) ker$(T) = \left\{ k\begin{bmatrix} 0 \\ 1 \\ -1 \end{bmatrix} \right\}$; T is not one-to-one.

2. (a) $T^{-1}\begin{bmatrix} x_1 \\ x_2 \end{bmatrix} = \begin{bmatrix} x_1 - 2x_2 \\ -2x_1 + 5x_2 \end{bmatrix}$ (b) T has no inverse. (c) $T^{-1}\begin{bmatrix} x_1 \\ x_2 \end{bmatrix} = \begin{bmatrix} \frac{3}{19}x_1 - \frac{7}{19}x_2 \\ \frac{1}{19}x_1 + \frac{4}{19}x_2 \end{bmatrix}$

3. (a) T has no inverse. (b) $T^{-1}\begin{bmatrix} x_1 \\ x_2 \\ x_3 \end{bmatrix} = \begin{bmatrix} \frac{1}{8}x_1 + \frac{1}{8}x_2 - \frac{3}{4}x_3 \\ \frac{1}{8}x_1 + \frac{1}{8}x_2 + \frac{1}{4}x_3 \\ -\frac{3}{8}x_1 + \frac{5}{8}x_2 + \frac{1}{4}x_3 \end{bmatrix}$ (c) $T^{-1}\begin{bmatrix} x_1 \\ x_2 \\ x_3 \end{bmatrix} = \begin{bmatrix} \frac{1}{2}x_1 - \frac{1}{2}x_2 + \frac{1}{2}x_3 \\ -\frac{1}{2}x_1 + \frac{1}{2}x_2 + \frac{1}{2}x_3 \\ \frac{1}{2}x_1 + \frac{1}{2}x_2 - \frac{1}{2}x_3 \end{bmatrix}$

    (d) $T^{-1}\begin{bmatrix} x_1 \\ x_2 \\ x_3 \end{bmatrix} = \begin{bmatrix} 3x_1 + 3x_2 - x_3 \\ -2x_1 - 2x_2 + x_3 \\ -4x_1 - 5x_2 + 2x_3 \end{bmatrix}$ 4. (a) Not one-to-one (b) Not one-to-one (c) One-to-one

5. (a) ker$(T) = \left\{ k\begin{bmatrix} -1 \\ 1 \end{bmatrix} \right\}$ (b) T is not one-to-one since ker$(T) \neq \{0\}$.

6. (a) ker$(T) = \{0\}$ (b) T is one-to-one from Theorem 8.3.2.

7. (a) T is one-to-one. (b) T is not one-to-one. (c) T is not one-to-one. (d) T is one-to-one.

8. (a) T is one-to-one. (b) T is one-to-one.   9. No. A is not invertible.

10. (a) T is not one-to-one. (b) $T^{-1}(x_1, x_2, x_3, \ldots, x_n) = (x_n, x_{n-1}, x_{n-2}, \ldots, x_1)$
    (c) $T^{-1}(x_1, x_2, x_3, \ldots, x_n) = (x_n, x_1, x_2, \ldots, x_{n-1})$

11. (a) $a_i \neq 0$ for $i = 1, 2, 3, \ldots, n$ (b) $T^{-1}(x_1, x_2, x_3, \ldots, x_n) = \left(\dfrac{1}{a_1}x_1, \dfrac{1}{a_2}x_2, \dfrac{1}{a_3}x_3, \ldots, \dfrac{1}{a_n}x_n \right)$

12. (b) $T_1^{-1}(x, y) = (\frac{1}{2}x + \frac{1}{2}y, \frac{1}{2}x - \frac{1}{2}y)$;
    $T_2^{-1}(x, y) = (\frac{2}{5}x + \frac{1}{5}y, \frac{1}{5}x - \frac{2}{5}y)$;
    $(T_2 \circ T_1)^{-1}(x, y) = (\frac{3}{10}x - \frac{1}{10}y, \frac{1}{10}x + \frac{3}{10}y)$

13. (b) $T_1^{-1}(p(x)) = \dfrac{p(x)}{x}$; $T_2^{-1}(p(x)) = p(x - 1)$;
    $$(T_2 \circ T_1)^{-1}(p(x)) = \frac{1}{x}p(x - 1)$$

15. (a) $(1, -1)$ (d) $T^{-1}(2, 3) = 2 + x$

17. (a) T is not one-to-one. (b) T is one-to-one. $T^{-1}\begin{bmatrix} a & b \\ c & d \end{bmatrix} = \begin{bmatrix} a & c \\ b & d \end{bmatrix}$

    (c) T is one-to-one. $T^{-1}\begin{bmatrix} a & b \\ c & d \end{bmatrix} = \begin{bmatrix} d & -b \\ -c & a \end{bmatrix}$ 20. J is not one-to-one since $J(x) = J(x^3)$.

EXERCISE SET 8.4 (page 421)

1. (a) $\begin{bmatrix} 0 & 0 & 0 \\ 1 & 0 & 0 \\ 0 & 1 & 0 \\ 0 & 0 & 1 \end{bmatrix}$ 2. (a) $\begin{bmatrix} 1 & 1 & 0 \\ 0 & -2 & -3 \end{bmatrix}$ 3. (a) $\begin{bmatrix} 1 & -1 & 1 \\ 0 & 1 & -2 \\ 0 & 0 & 1 \end{bmatrix}$ 4. (a) $\begin{bmatrix} 2 & -1 \\ 2 & 0 \end{bmatrix}$

5. (a) $\begin{bmatrix} 0 & 0 \\ -\frac{1}{2} & 1 \\ \frac{8}{3} & \frac{4}{3} \end{bmatrix}$ 6. (a) $\begin{bmatrix} 1 & -\frac{3}{2} & \frac{1}{2} \\ -1 & \frac{1}{2} & \frac{1}{2} \\ 0 & \frac{1}{2} & -\frac{1}{2} \end{bmatrix}$ 7. (a) $\begin{bmatrix} 1 & 1 & 1 \\ 0 & 2 & 4 \\ 0 & 0 & 4 \end{bmatrix}$ (b) $3 + 10x + 16x^2$

8. (a) $\begin{bmatrix} 0 & 0 & 0 \\ 1 & -3 & 9 \\ 0 & 1 & -6 \\ 0 & 0 & 1 \end{bmatrix}$ (b) $-11x + 7x^2 - x^3$

9. (a) $[T(\mathbf{v}_1)]_B = \begin{bmatrix} 1 \\ -2 \end{bmatrix}, [T(\mathbf{v}_2)]_B = \begin{bmatrix} 3 \\ 5 \end{bmatrix}$ (b) $T(\mathbf{v}_1) = \begin{bmatrix} 3 \\ -5 \end{bmatrix}, T(\mathbf{v}_2) = \begin{bmatrix} -2 \\ 29 \end{bmatrix}$

    (c) $T\left(\begin{bmatrix} x_1 \\ x_2 \end{bmatrix}\right) = \begin{bmatrix} \frac{18}{7} & \frac{1}{7} \\ -\frac{107}{7} & \frac{24}{7} \end{bmatrix}\begin{bmatrix} x_1 \\ x_2 \end{bmatrix}$ (d) $\begin{bmatrix} \frac{19}{7} \\ -\frac{83}{7} \end{bmatrix}$

10. (a) $[T(\mathbf{v}_1)]_{B'} = \begin{bmatrix} 3 \\ 1 \\ -3 \end{bmatrix}, [T(\mathbf{v}_2)]_{B'} = \begin{bmatrix} -2 \\ 6 \\ 0 \end{bmatrix}, [T(\mathbf{v}_3)]_{B'} = \begin{bmatrix} 1 \\ 2 \\ 7 \end{bmatrix}, [T(\mathbf{v}_4)]_{B'} = \begin{bmatrix} 0 \\ 1 \\ 1 \end{bmatrix}$

    (b) $T(\mathbf{v}_1) = \begin{bmatrix} 11 \\ 5 \\ 22 \end{bmatrix}, T(\mathbf{v}_2) = \begin{bmatrix} -42 \\ 32 \\ -10 \end{bmatrix}, T(\mathbf{v}_3) = \begin{bmatrix} -56 \\ 87 \\ 17 \end{bmatrix}, T(\mathbf{v}_4) = \begin{bmatrix} -13 \\ 17 \\ 2 \end{bmatrix}$

    (c) $T\left(\begin{bmatrix} x_1 \\ x_2 \\ x_3 \\ x_4 \end{bmatrix}\right) = \begin{bmatrix} -\frac{253}{10} & \frac{49}{5} & \frac{241}{10} & -\frac{229}{10} \\ \frac{115}{2} & -39 & -\frac{65}{2} & \frac{153}{2} \\ 66 & -60 & -9 & 91 \end{bmatrix}\begin{bmatrix} x_1 \\ x_2 \\ x_3 \\ x_4 \end{bmatrix}$ (d) $\begin{bmatrix} -31 \\ 37 \\ 12 \end{bmatrix}$

11. (a) $[T(\mathbf{v}_1)]_B = \begin{bmatrix} 1 \\ 2 \\ 6 \end{bmatrix}, [T(\mathbf{v}_2)]_B = \begin{bmatrix} 3 \\ 0 \\ -2 \end{bmatrix}, [T(\mathbf{v}_3)]_B = \begin{bmatrix} -1 \\ 5 \\ 4 \end{bmatrix}$

(b) $T(\mathbf{v}_1) = 16 + 51x + 19x^2, \; T(\mathbf{v}_2) = -6 - 5x + 5x^2, \; T(\mathbf{v}_3) = 7 + 40x + 15x^2$

(c) $T(a_0 + a_1x + a_2x^2) = \dfrac{239a_0 - 161a_1 + 289a_2}{24} + \dfrac{201a_0 - 111a_1 + 247a_2}{8}x + \dfrac{61a_0 - 31a_1 + 107a_2}{12}x^2$

(d) $T(1 + x^2) = 22 + 56x + 14x^2$

12. (a) $[T_2 \circ T_1]_{B',B} = \begin{bmatrix} 1 & 1 \\ 2 & 4 \\ 0 & 4 \end{bmatrix}, [T_2]_{B'} = \begin{bmatrix} 1 & 1 & 1 \\ 0 & 2 & 4 \\ 0 & 0 & 4 \end{bmatrix}, [T_1]_{B',B} = \begin{bmatrix} 0 & 0 \\ 1 & 0 \\ 0 & 1 \end{bmatrix}$ (b) $[T_2 \circ T_1]_{B',B} = [T_2]_{B'}[T_1]_{B',B}$

13. (a) $[T_2 \circ T_1]_{B',B} = \begin{bmatrix} 0 & 0 \\ 6 & 0 \\ 0 & 0 \\ 0 & -9 \end{bmatrix}, [T_2]_{B',B''} = \begin{bmatrix} 0 & 0 & 0 \\ 3 & 0 & 0 \\ 0 & 3 & 0 \\ 0 & 0 & 3 \end{bmatrix}, [T_1]_{B'',B} = \begin{bmatrix} 2 & 0 \\ 0 & 0 \\ 0 & -3 \end{bmatrix}$

(b) $[T_2 \circ T_1]_{B',B} = [T_2]_{B',B''}[T_1]_{B'',B}$ 16. $\begin{bmatrix} 0 & 0 & 0 & 1 \\ 1 & 0 & 0 & 0 \\ 0 & 1 & 0 & 0 \\ 0 & 0 & 1 & 0 \end{bmatrix}$

17. (a) $\begin{bmatrix} 0 & 1 & 0 \\ 0 & 0 & 2 \\ 0 & 0 & 0 \end{bmatrix}$ (b) $\begin{bmatrix} 0 & -\frac{3}{2} & \frac{23}{6} \\ 0 & 0 & -\frac{16}{3} \\ 0 & 0 & 0 \end{bmatrix}$ (c) $-6 + 48x$

18. (a) $\begin{bmatrix} 0 & 0 & 0 \\ 0 & 0 & -1 \\ 0 & 1 & 0 \end{bmatrix}$ (b) $\begin{bmatrix} 0 & 0 & 0 \\ 0 & 1 & 0 \\ 0 & 0 & 2 \end{bmatrix}$ (c) $\begin{bmatrix} 2 & 1 & 0 \\ 0 & 2 & 2 \\ 0 & 0 & 2 \end{bmatrix}$

EXERCISE SET 8.5 (page 435)

1. $[T]_B = \begin{bmatrix} 1 & -2 \\ 0 & -1 \end{bmatrix}, [T]_{B'} = \begin{bmatrix} -\frac{3}{11} & -\frac{56}{11} \\ -\frac{2}{11} & \frac{3}{11} \end{bmatrix}$ 2. $[T]_B = \begin{bmatrix} \frac{4}{5} & \frac{61}{10} \\ \frac{18}{5} & -\frac{19}{5} \end{bmatrix}, [T]_{B'} = \begin{bmatrix} -\frac{31}{2} & \frac{9}{2} \\ -\frac{75}{2} & \frac{25}{2} \end{bmatrix}$

3. $[T]_B = \begin{bmatrix} \dfrac{1}{\sqrt{2}} & -\dfrac{1}{\sqrt{2}} \\ \dfrac{1}{\sqrt{2}} & \dfrac{1}{\sqrt{2}} \end{bmatrix}, [T]_{B'} = \begin{bmatrix} \dfrac{13}{11\sqrt{2}} & -\dfrac{25}{11\sqrt{2}} \\ \dfrac{5}{11\sqrt{2}} & \dfrac{9}{11\sqrt{2}} \end{bmatrix}$

4. $[T]_B = \begin{bmatrix} 1 & 2 & -1 \\ 0 & -1 & 0 \\ 1 & 0 & 7 \end{bmatrix}, [T]_{B'} = \begin{bmatrix} 1 & 4 & 3 \\ -1 & -2 & -9 \\ 1 & 1 & 8 \end{bmatrix}$ 5. $[T]_B = \begin{bmatrix} 1 & 0 & 0 \\ 0 & 1 & 0 \\ 0 & 0 & 0 \end{bmatrix}, [T]_{B'} = \begin{bmatrix} 1 & 0 & 0 \\ 0 & 1 & 1 \\ 0 & 0 & 0 \end{bmatrix}$

6. $[T]_B = \begin{bmatrix} 5 & 0 \\ 0 & 5 \end{bmatrix}, [T]_{B'} = \begin{bmatrix} 5 & 0 \\ 0 & 5 \end{bmatrix}$ 7. $[T]_B = \begin{bmatrix} \frac{2}{3} & -\frac{2}{9} \\ \frac{1}{2} & \frac{4}{3} \end{bmatrix}, [T]_{B'} = \begin{bmatrix} 1 & 1 \\ 0 & 1 \end{bmatrix}$

8. (a) $\det(T) = 17$
   (b) $\det(T) = 0$
   (c) $\det(T) = 1$

10. (a) $[T]_B = \begin{bmatrix} 1 & 1 & 1 & 1 & 1 \\ 0 & 2 & 4 & 6 & 8 \\ 0 & 0 & 4 & 12 & 24 \\ 0 & 0 & 0 & 8 & 32 \\ 0 & 0 & 0 & 0 & 16 \end{bmatrix}$, where B is the standard basis for P_4; $\text{rank}(T) = 5$ and $\text{nullity}(T) = 0$.

(b) T is one-to-one.

11. (a) $\mathbf{u}_1' = \begin{bmatrix} -1 \\ 1 \end{bmatrix}, \mathbf{u}_2' = \begin{bmatrix} 1 \\ -2 \end{bmatrix}$ (b) $\mathbf{u}_1' = \begin{bmatrix} 1 \\ \dfrac{3 - \sqrt{21}}{2} \end{bmatrix}, \mathbf{u}_2' = \begin{bmatrix} 1 \\ \dfrac{3 + \sqrt{21}}{2} \end{bmatrix}$

12. (a) $\mathbf{u}_1' = \begin{bmatrix} -1 \\ 1 \\ 0 \end{bmatrix}, \mathbf{u}_2' = \begin{bmatrix} 1 \\ 0 \\ 1 \end{bmatrix}, \mathbf{u}_3' = \begin{bmatrix} -1 \\ -1 \\ 1 \end{bmatrix}$ (b) $\mathbf{u}_1' = \begin{bmatrix} -1 \\ 1 \\ 0 \end{bmatrix}, \mathbf{u}_2' = \begin{bmatrix} 1 \\ 0 \\ 1 \end{bmatrix}, \mathbf{u}_3' = \begin{bmatrix} -1 \\ -1 \\ 1 \end{bmatrix}$

(c) $\mathbf{u}_1' = \begin{bmatrix} 1 \\ 2 \\ 1 \end{bmatrix}, \mathbf{u}_2' = \begin{bmatrix} 0 \\ 1 \\ 0 \end{bmatrix}, \mathbf{u}_3' = \begin{bmatrix} -1 \\ 0 \\ 1 \end{bmatrix}$

13. (a) $\lambda = -4, \lambda = 3$ (b) Basis for eigenspace corresponding to $\lambda = -4$: $-2 + \frac{8}{3}x + x^2$; basis for eigenspace corresponding to $\lambda = 3$: $5 - 2x + x^2$

14. (a) $\lambda = 1, \lambda = -2, \lambda = -1$

(b) Basis for eigenspace corresponding to $\lambda = 1$: $\begin{bmatrix} 0 & 0 \\ 0 & 1 \end{bmatrix}$ and $\begin{bmatrix} 2 & 3 \\ 1 & 0 \end{bmatrix}$;

basis for eigenspace corresponding to $\lambda = -2$: $\begin{bmatrix} -1 & 0 \\ 1 & 0 \end{bmatrix}$;

basis for eigenspace corresponding to $\lambda = -1$: $\begin{bmatrix} -2 & 1 \\ 1 & 0 \end{bmatrix}$

18. (b) $\left(\dfrac{3 + 5\sqrt{3}}{4}, \dfrac{\sqrt{3} + 5}{4} \right)$

CHAPTER 8 SUPPLEMENTARY EXERCISES (page 437)

1. No. $T(\mathbf{x}_1 + \mathbf{x}_2) = A(\mathbf{x}_1 + \mathbf{x}_2) + B \neq (A\mathbf{x}_1 + B) + (A\mathbf{x}_2 + B) = T(\mathbf{x}_1) + T(\mathbf{x}_2)$, and if $c \neq 1$, then $T(c\mathbf{x}) = cA\mathbf{x} + B \neq c(A\mathbf{x} + B) = cT(\mathbf{x})$.

2. (b) $A^n = \begin{bmatrix} \cos n\theta & -\sin n\theta \\ \sin n\theta & \cos n\theta \end{bmatrix}$

5. (a) $T(\mathbf{e}_3)$ and any two of $T(\mathbf{e}_1)$, $T(\mathbf{e}_2)$, and $T(\mathbf{e}_4)$ form bases for the range; $(-1, 1, 0, 1)$ is a basis for the kernel.
   (b) Rank = 3, nullity = 1

6. (a) $(-\frac{1}{3}, -\frac{13}{6}, 1)$, (b) $(1, 0, 0)$ and $(-2, 1, 0)$

7. (a) $\text{Rank}(T) = 2$ and $\text{nullity}(T) = 2$ (b) T is not one-to-one.

11. Rank = 3, nullity = 1   13. $\begin{bmatrix} 1 & 0 & 0 & 0 \\ 0 & 0 & 1 & 0 \\ 0 & 1 & 0 & 0 \\ 0 & 0 & 0 & 1 \end{bmatrix}$

14. (a) $v_1 = 2u_1 + u_2$, $v_2 = -u_1 + u_2 + u_3$, $v_3 = 3u_1 + 4u_2 + 2u_3$
     (b) $u_1 = -2v_1 - 2v_2 + v_3$, $u_2 = 5v_1 + 4v_2 - 2v_3$, $u_3 = -7v_1 - 5v_2 + 3v_3$

15. $[T]_{B'} = \begin{bmatrix} -4 & 0 & 9 \\ 1 & 0 & -2 \\ 0 & 1 & 1 \end{bmatrix}$ 17. $[T]_B = \begin{bmatrix} 1 & -1 & 1 \\ 0 & 1 & 0 \\ 1 & 0 & -1 \end{bmatrix}$

19. (b) x (c) e^x

20. (a) $\begin{bmatrix} 2 \\ 6 \\ 12 \end{bmatrix}$ (d) $-3x^2 + 3$ (e)

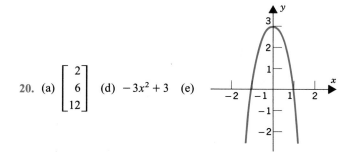

24. $\begin{bmatrix} 0 & 1 & 0 & 0 & \cdots & 0 \\ 0 & 0 & 1 & 0 & \cdots & 0 \\ 0 & 0 & 0 & 1 & \cdots & 0 \\ \vdots & \vdots & \vdots & \vdots & & \vdots \\ 0 & 0 & 0 & 0 & \cdots & 1 \\ 0 & 0 & 0 & 0 & \cdots & 0 \end{bmatrix}$ 25. $\begin{bmatrix} 0 & 0 & 0 & \cdots & 0 \\ 1 & 0 & 0 & \cdots & 0 \\ 0 & \frac{1}{2} & 0 & \cdots & 0 \\ 0 & 0 & \frac{1}{3} & \cdots & 0 \\ \vdots & \vdots & \vdots & & \vdots \\ 0 & 0 & 0 & \cdots & \dfrac{1}{n+1} \end{bmatrix}$

EXERCISE SET 9.1 (page 447)

1. (a) $y_1 = c_1 e^{5x} - 2c_2 e^{-x}$ (b) $y_1 = 0$ 2. (a) $y_1 = c_1 e^{7x} - 3c_2 e^{-x}$ (b) $y_1 = -\frac{1}{40}e^{7x} + \frac{81}{40}e^{-x}$
      $\quad\; y_2 = c_1 e^{5x} + c_2 e^{-x}$ $\quad\;\; y_2 = 0$ $\quad\;\; y_2 = 2c_1 e^{7x} + 2c_2 e^{-x}$ $\quad\;\; y_2 = -\frac{1}{20}e^{7x} - \frac{27}{20}e^{-x}$

3. (a) $y_1 = -c_2 e^{2x} + c_3 e^{3x}$ (b) $y_1 = e^{2x} - 2e^{3x}$ 4. $y_1 = (c_1 + c_2)e^{2x} + c_3 e^{8x}$
      $\quad\; y_2 = c_1 e^x + 2c_2 e^{2x} - c_3 e^{3x}$ $\quad\;\; y_2 = e^x - 2e^{2x} + 2e^{3x}$ $\quad\; y_2 = -c_2 e^{2x} + c_3 e^{8x}$
      $\quad\; y_3 = 2c_2 e^{2x} - c_3 e^{3x}$ $\quad\;\; y_3 = -2e^{2x} + 2e^{3x}$ $\quad\; y_3 = -c_1 e^{2x} + c_3 e^{8x}$

5. $y = c_1 e^{3x} + c_2 e^{-2x}$ 6. $y = c_1 e^x + c_2 e^{2x} + c_3 e^{3x}$

EXERCISE SET 9.2 (page 457)

1. (a) $\begin{bmatrix} 0 & -1 \\ -1 & 0 \end{bmatrix}$ (b) $\begin{bmatrix} -1 & 0 \\ 0 & -1 \end{bmatrix}$ (c) $\begin{bmatrix} 1 & 0 \\ 0 & 0 \end{bmatrix}$ (d) $\begin{bmatrix} 0 & 0 \\ 0 & 1 \end{bmatrix}$

2. (a) $(-1, -2)$ (b) $(-2, -1)$ (c) $(2, 0)$ (d) $(0, 1)$

3. (a) $\begin{bmatrix} 1 & 0 & 0 \\ 0 & 1 & 0 \\ 0 & 0 & -1 \end{bmatrix}$ (b) $\begin{bmatrix} 1 & 0 & 0 \\ 0 & -1 & 0 \\ 0 & 0 & 1 \end{bmatrix}$ (c) $\begin{bmatrix} -1 & 0 & 0 \\ 0 & 1 & 0 \\ 0 & 0 & 1 \end{bmatrix}$

4. (a) $(1, 1, -1)$ (b) $(1, -1, 1)$ (c) $(-1, 1, 1)$

5. (a) $\begin{bmatrix} 0 & -1 & 0 \\ 1 & 0 & 0 \\ 0 & 0 & 1 \end{bmatrix}$ (b) $\begin{bmatrix} 1 & 0 & 0 \\ 0 & 0 & -1 \\ 0 & 1 & 0 \end{bmatrix}$ (c) $\begin{bmatrix} 0 & 0 & 1 \\ 0 & 1 & 0 \\ -1 & 0 & 0 \end{bmatrix}$

6. (a) Rectangle with vertices at $(0, 0)$, $(1, 0)$, $(1, -2)$, $(0, -2)$
   (b) Rectangle with vertices at $(0, 0)$, $(-1, 0)$, $(-1, 2)$, $(0, 2)$
   (c) Rectangle with vertices at $(0, 0)$, $(1, 0)$, $(1, \frac{1}{2})$, $(0, \frac{1}{2})$
   (d) Square with vertices at $(0, 0)$, $(2, 0)$, $(2, 2)$, $(0, 2)$
   (e) Parallelogram with vertices at $(0, 0)$, $(1, 0)$, $(7, 2)$, $(6, 2)$
   (f) Parallelogram with vertices at $(0, 0)$, $(1, -2)$, $(1, 0)$, $(0, 2)$

7. Rectangle with vertices at $(0, 0)$, $(-3, 0)$, $(0, 1)$, $(-3, 1)$

8. (a) $\begin{bmatrix} \dfrac{1}{\sqrt{2}} & -\dfrac{1}{\sqrt{2}} \\ \dfrac{1}{\sqrt{2}} & \dfrac{1}{\sqrt{2}} \end{bmatrix}$ (b) $\begin{bmatrix} 0 & -1 \\ 1 & 0 \end{bmatrix}$ (c) $\begin{bmatrix} -1 & 0 \\ 0 & -1 \end{bmatrix}$ (d) $\begin{bmatrix} 0 & 1 \\ -1 & 0 \end{bmatrix}$ (e) $\begin{bmatrix} \dfrac{\sqrt{3}}{2} & \dfrac{1}{2} \\ -\dfrac{1}{2} & \dfrac{\sqrt{3}}{2} \end{bmatrix}$

9. (a) $\begin{bmatrix} 1 & 0 \\ 4 & 1 \end{bmatrix}$ (b) $\begin{bmatrix} 1 & -2 \\ 0 & 1 \end{bmatrix}$ 10. (a) $\begin{bmatrix} 1 & 0 \\ 0 & \frac{1}{3} \end{bmatrix}$ (b) $\begin{bmatrix} 6 & 0 \\ 0 & 1 \end{bmatrix}$

11. (a) Expansion by a factor of 3 in the x-direction   (b) Expansion by a factor of -5 in the y-direction
    (c) Shear by a factor of 4 in the x-direction

12. (a) $\begin{bmatrix} 2 & 0 \\ 0 & 1 \end{bmatrix}\begin{bmatrix} 3 & 0 \\ 0 & 1 \end{bmatrix}$; expansion in the y-direction by a factor of 3, then expansion in the x-direction by a factor of 2

   (b) $\begin{bmatrix} 1 & 0 \\ 2 & 1 \end{bmatrix}\begin{bmatrix} 1 & 4 \\ 0 & 1 \end{bmatrix}$; shear in the x-direction by a factor of 4, then shear in the y-direction by a factor of 2

   (c) $\begin{bmatrix} 0 & 1 \\ 1 & 0 \end{bmatrix}\begin{bmatrix} 4 & 0 \\ 0 & 1 \end{bmatrix}\begin{bmatrix} 1 & 0 \\ 0 & -2 \end{bmatrix}$; expansion in the y-direction by a factor of -2, then expansion in the x-direction by a factor of 4, then reflection about $y = x$

   (d) $\begin{bmatrix} 1 & 0 \\ 4 & 1 \end{bmatrix}\begin{bmatrix} 1 & 0 \\ 1 & 18 \end{bmatrix}\begin{bmatrix} 1 & -3 \\ 0 & 1 \end{bmatrix}$; shear in the x-direction by a factor of -3, then expansion in the y-direction by a factor of 18, then shear in the y-direction by a factor of 4

13. (a) $\begin{bmatrix} \frac{1}{2} & 0 \\ 0 & 5 \end{bmatrix}$ (b) $\begin{bmatrix} 1 & 0 \\ 2 & 5 \end{bmatrix}$ (c) $\begin{bmatrix} 0 & -1 \\ -1 & 0 \end{bmatrix}$ 14. (a) $\begin{bmatrix} 0 & 1 \\ -5 & 0 \end{bmatrix}$ (b) $\dfrac{1}{2}\begin{bmatrix} \sqrt{3} & -1 \\ -6\sqrt{3}+3 & 6+3\sqrt{3} \end{bmatrix}$

16. $16y - 11x - 3 = 0$ 17. (a) $y = \frac{2}{7}x$ (b) $y = x$ (c) $y = \frac{1}{2}x$ (d) $y = -2x$ 18. $\begin{bmatrix} 1 & -2 \\ 0 & 1 \end{bmatrix}$

19. (b) No. A is not invertible.   22. (a) $\begin{bmatrix} 1 & 0 & 0 \\ 0 & 0 & 1 \\ 0 & 1 & 0 \end{bmatrix}$ (b) $\begin{bmatrix} 0 & 0 & 1 \\ 0 & 1 & 0 \\ 1 & 0 & 0 \end{bmatrix}$ (c) $\begin{bmatrix} 0 & 1 & 0 \\ 1 & 0 & 0 \\ 0 & 0 & 1 \end{bmatrix}$

23. (a) $\begin{bmatrix} 1 & 0 & k \\ 0 & 1 & k \\ 0 & 0 & 1 \end{bmatrix}$ (b) xz-direction: $\begin{bmatrix} 1 & k & 0 \\ 0 & 1 & 0 \\ 0 & k & 1 \end{bmatrix}$; yz-direction: $\begin{bmatrix} 1 & 0 & 0 \\ k & 1 & 0 \\ k & 0 & 1 \end{bmatrix}$

24. (a) $\lambda_1 = 1$: $\begin{bmatrix} 1 \\ 0 \end{bmatrix}$; $\lambda_2 = -1$: $\begin{bmatrix} 0 \\ 1 \end{bmatrix}$ (b) $\lambda_1 = 1$: $\begin{bmatrix} 0 \\ 1 \end{bmatrix}$; $\lambda_2 = -1$: $\begin{bmatrix} 1 \\ 0 \end{bmatrix}$ (c) $\lambda_1 = 1$: $\begin{bmatrix} 1 \\ 1 \end{bmatrix}$; $\lambda_2 = -1$: $\begin{bmatrix} -1 \\ 1 \end{bmatrix}$

(d) $\lambda = 1$: $\begin{bmatrix} 1 \\ 0 \end{bmatrix}$ (e) $\lambda = 1$: $\begin{bmatrix} 0 \\ 1 \end{bmatrix}$ (f) (θ an odd integer multiple of π) $\lambda = -1$: $\begin{bmatrix} 1 \\ 0 \end{bmatrix}$

(θ an even integer with multiple of π) $\lambda = 1$: $\begin{bmatrix} 1 \\ 0 \end{bmatrix}, \begin{bmatrix} 0 \\ 1 \end{bmatrix}$

(θ not an integer multiple of π) no real eigenvalues

EXERCISE SET 9.3 (page 467)

1. $y = \frac{1}{2} + \frac{7}{2}x$ 2. $y = \frac{2}{3} + \frac{1}{6}x$ 3. $y = 2 + 5x - 3x^2$ 4. $y = -5 + 3x - 4x^2 + 2x^3$

8. $y = 4 - .2x + .2x^2$; if $x = 12$, then $y = 30.4$ (30.4 thousand)

EXERCISE SET 9.4 (page 474)

1. (a) $(1 + \pi) - 2 \sin x - \sin 2x$ (b) $(1 + \pi) - 2 \left[\sin x + \frac{\sin 2x}{2} + \frac{\sin 3x}{3} + \cdots + \frac{\sin nx}{n} \right]$

2. (a) $\frac{4}{3}\pi^2 + 4 \cos x + \cos 2x + \frac{4}{9} \cos 3x - 4\pi \sin x - 2\pi \sin 2x - \frac{4\pi}{3} \sin 3x$

(b) $\frac{4}{3}\pi^2 + 4 \sum_{k=1}^{n} \frac{\cos kx}{k^2} - 4\pi \sum_{k=1}^{n} \frac{\sin kx}{k}$ 3. (a) $-\frac{1}{2} + \frac{1}{e-1}e^x$ (b) $\frac{1}{12} - \frac{3-e}{2e-2}$

4. (a) $(4e - 10) + (18 - 6e)x$ (b) $\frac{(3-e)(7e-19)}{2}$ 5. (a) $\frac{3}{\pi}x$ (b) $1 - \frac{6}{\pi^2}$ 8. $\sum_{k=1}^{\infty} \frac{2}{k} \sin(kx)$

EXERCISE SET 9.5 (page 481)

1. (a), (c), (e), (g), (h)

2. (a) $A = \begin{bmatrix} 3 & 0 \\ 0 & 7 \end{bmatrix}$ (b) $A = \begin{bmatrix} 4 & -3 \\ -3 & -9 \end{bmatrix}$ (c) $A = \begin{bmatrix} 5 & \frac{5}{2} \\ \frac{5}{2} & 0 \end{bmatrix}$ (d) $A = \begin{bmatrix} 0 & -\frac{7}{2} \\ -\frac{7}{2} & 0 \end{bmatrix}$

3. (a) $A = \begin{bmatrix} 9 & 3 & -4 \\ 3 & -1 & \frac{1}{2} \\ -4 & \frac{1}{2} & 4 \end{bmatrix}$ (b) $\begin{bmatrix} 1 & -\frac{5}{2} & \frac{9}{2} \\ -\frac{5}{2} & 1 & 0 \\ \frac{9}{2} & 0 & -3 \end{bmatrix}$ (c) $A = \begin{bmatrix} 0 & \frac{1}{2} & \frac{1}{2} \\ \frac{1}{2} & 0 & \frac{1}{2} \\ \frac{1}{2} & \frac{1}{2} & 0 \end{bmatrix}$

(d) $A = \begin{bmatrix} \sqrt{2} & \sqrt{2} & -4\sqrt{3} \\ \sqrt{2} & 0 & 0 \\ -4\sqrt{3} & 0 & -\sqrt{3} \end{bmatrix}$ (e) $A = \begin{bmatrix} 1 & 1 & 0 & -5 \\ 1 & 1 & 0 & 0 \\ 0 & 0 & -1 & 2 \\ -5 & 0 & 2 & -1 \end{bmatrix}$

4. (a) $2x^2 + 5y^2 - 6xy$ (b) $7x_1^2 + 5x_1x_2$ (c) $x^2 - 3y^2 + 5z^2$

(d) $-2x_1^2 + 3x_3^2 + 7x_1x_2 + x_1x_3 + 12x_2x_3$ (e) $2x_1x_2 + 2x_1x_3 + 2x_1x_4 + 2x_2x_3 + 2x_2x_4 + 2x_3x_4$

5. (a) max value $= 5$ at $\pm(1, 0)$; min value $= -1$ at $\pm(0, 1)$

(b) max value $= \frac{11 + \sqrt{10}}{2}$ at $\pm \left(\frac{-1}{\sqrt{20 + 6\sqrt{10}}}, \frac{1}{\sqrt{20 + 6\sqrt{10}}} \right)$;

min value $= \frac{11 - \sqrt{10}}{2}$ at $\pm \left(\frac{-1}{\sqrt{20 + 6\sqrt{10}}}, \frac{1}{\sqrt{20 + 6\sqrt{10}}} \right)$

(c) max value $= \dfrac{7 + \sqrt{10}}{2}$ at $\pm \left(\dfrac{3 + \sqrt{10}}{\sqrt{20 + 6\sqrt{10}}}, \dfrac{-1}{\sqrt{20 + 6\sqrt{10}}} \right);$

min value $= \dfrac{7 - \sqrt{10}}{2}$ at $\pm \left(\dfrac{1}{\sqrt{20 + 6\sqrt{10}}}, \dfrac{3 + \sqrt{10}}{\sqrt{20 + 6\sqrt{10}}} \right)$

(d) max value $= \dfrac{7}{2}$ at $\pm \left(\dfrac{1}{\sqrt{2}}, \dfrac{1}{\sqrt{2}} \right)$; min value $= \dfrac{1}{2}$ at $\pm \left(\dfrac{1}{\sqrt{2}}, -\dfrac{1}{\sqrt{2}} \right)$

6. (a) max value $= 4$ at $\pm \left(\dfrac{1}{\sqrt{6}}, \dfrac{1}{\sqrt{6}}, \dfrac{2}{\sqrt{6}} \right)$; min value $= -2$ at $\pm \left(-\dfrac{1}{\sqrt{3}}, -\dfrac{1}{\sqrt{3}}, \dfrac{1}{\sqrt{3}} \right)$

(b) max value $= 3$ at $\left(\dfrac{2}{\sqrt{6}}, \dfrac{1}{\sqrt{6}}, \dfrac{1}{\sqrt{6}} \right)$; min value $= 0$ at $\left(\dfrac{1}{\sqrt{3}}, -\dfrac{1}{\sqrt{3}}, -\dfrac{1}{\sqrt{3}} \right)$

(c) max value $= 4$ at $\pm \left(\dfrac{1}{\sqrt{2}}, 0, \dfrac{1}{\sqrt{2}} \right)$; min value $= 2$ at $\pm \left(\dfrac{1}{\sqrt{2}}, 0, -\dfrac{1}{\sqrt{2}} \right)$ and $\pm (0, 1, 0)$

7. (b)   9. (a)

11. (a) Positive definite   (b) Negative definite   (c) Positive semidefinite
(d) Negative semidefinite   (e) Indefinite   (f) Indefinite

12. (a) Indefinite   (b) Indefinite   (c) Positive definite   (d) Indefinite
(e) Positive and negative semidefinite   (f) Positive definite

13. (c) No. $T(k\mathbf{x}) \neq kT(\mathbf{x})$, unless $k = 1$.

14. (a) $k > 4$ (b) $k > 2$ (c) $-\tfrac{1}{3}\sqrt{15} < k < \tfrac{1}{3}\sqrt{15}$

15. $A = \begin{bmatrix} c_1^2 & c_1 c_2 & c_1 c_3 & \cdots & c_1 c_n \\ c_1 c_2 & c_2^2 & c_2 c_3 & \cdots & c_2 c_n \\ \vdots & \vdots & \vdots & & \vdots \\ c_1 c_n & c_2 c_n & c_3 c_n & \cdots & c_n^2 \end{bmatrix}$

16. (a) $A = \begin{bmatrix} \dfrac{1}{n} & \dfrac{-1}{n(n-1)} & \dfrac{-1}{n(n-1)} & \cdots & \dfrac{-1}{n(n-1)} \\ \dfrac{-1}{n(n-1)} & \dfrac{1}{n} & \dfrac{-1}{n(n-1)} & \cdots & \dfrac{-1}{n(n-1)} \\ \vdots & \vdots & \vdots & & \vdots \\ \dfrac{-1}{n(n-1)} & \dfrac{-1}{n(n-1)} & \dfrac{-1}{n(n-1)} & \cdots & \dfrac{1}{n} \end{bmatrix}$ (b) Positive semidefinite

EXERCISE SET 9.6 (page 493)

1. (a) $\begin{bmatrix} x_1 \\ x_2 \end{bmatrix} = \begin{bmatrix} \dfrac{1}{\sqrt{2}} & \dfrac{1}{\sqrt{2}} \\ \dfrac{1}{\sqrt{2}} & -\dfrac{1}{\sqrt{2}} \end{bmatrix} \begin{bmatrix} y_1 \\ y_2 \end{bmatrix}$; $y_1^2 + 3y_2^2$ (b) $\begin{bmatrix} x_1 \\ x_2 \end{bmatrix} = \begin{bmatrix} \dfrac{1}{\sqrt{5}} & \dfrac{2}{\sqrt{5}} \\ -\dfrac{2}{\sqrt{5}} & \dfrac{1}{\sqrt{5}} \end{bmatrix} \begin{bmatrix} y_1 \\ y_2 \end{bmatrix}$; $y_1^2 + 6y_2^2$

(c) $\begin{bmatrix} x_1 \\ x_2 \end{bmatrix} = \begin{bmatrix} \dfrac{1}{\sqrt{2}} & \dfrac{1}{\sqrt{2}} \\ \dfrac{1}{\sqrt{2}} & -\dfrac{1}{\sqrt{2}} \end{bmatrix} \begin{bmatrix} y_1 \\ y_2 \end{bmatrix}; \; y_1^2 - y_2^2$

(d) $\begin{bmatrix} x_1 \\ x_2 \end{bmatrix} = \begin{bmatrix} \dfrac{\sqrt{17} - 4}{\sqrt{34 - 8\sqrt{17}}} & \dfrac{\sqrt{17} - 4}{\sqrt{34 + 8\sqrt{17}}} \\ \dfrac{1}{\sqrt{34 - 8\sqrt{17}}} & \dfrac{-1}{\sqrt{34 + 8\sqrt{17}}} \end{bmatrix} \begin{bmatrix} y_1 \\ y_2 \end{bmatrix}; \; (1 + \sqrt{17})y_1^2 + (1 - \sqrt{17})y_2^2$

2. (a) $\begin{bmatrix} x_1 \\ x_2 \\ x_3 \end{bmatrix} = \begin{bmatrix} \frac{2}{3} & \frac{1}{3} & \frac{2}{3} \\ -\frac{2}{3} & \frac{2}{3} & \frac{1}{3} \\ -\frac{1}{3} & -\frac{2}{3} & \frac{2}{3} \end{bmatrix} \begin{bmatrix} y_1 \\ y_2 \\ y_3 \end{bmatrix}; \; y_1^2 + 7y_2^2 + 4y_3^2$

(b) $\begin{bmatrix} x_1 \\ x_2 \\ x_3 \end{bmatrix} = \begin{bmatrix} \frac{1}{3} & \frac{2}{3} & \frac{2}{3} \\ \frac{2}{3} & \frac{1}{3} & -\frac{2}{3} \\ -\frac{2}{3} & \frac{2}{3} & -\frac{1}{3} \end{bmatrix} \begin{bmatrix} y_1 \\ y_2 \\ y_3 \end{bmatrix}; \; 7y_1^2 + 4y_2^2 + y_3^2$

(c) $\begin{bmatrix} x_1 \\ x_2 \\ x_3 \end{bmatrix} = \begin{bmatrix} \dfrac{1}{\sqrt{14}} & \dfrac{1}{\sqrt{6}} & -\dfrac{4}{\sqrt{21}} \\ -\dfrac{2}{\sqrt{14}} & \dfrac{2}{\sqrt{6}} & \dfrac{1}{\sqrt{21}} \\ \dfrac{3}{\sqrt{14}} & \dfrac{1}{\sqrt{6}} & \dfrac{2}{\sqrt{21}} \end{bmatrix} \begin{bmatrix} y_1 \\ y_2 \\ y_3 \end{bmatrix}; \; 2y_2^2 - 7y_3^2$

(d) $\begin{bmatrix} x_1 \\ x_2 \\ x_3 \end{bmatrix} = \begin{bmatrix} \dfrac{3}{\sqrt{10}} & \dfrac{1}{\sqrt{20}} & \dfrac{1}{\sqrt{20}} \\ -\dfrac{1}{\sqrt{10}} & \dfrac{3}{\sqrt{20}} & \dfrac{3}{\sqrt{20}} \\ 0 & \dfrac{1}{\sqrt{2}} & -\dfrac{1}{\sqrt{2}} \end{bmatrix} \begin{bmatrix} y_1 \\ y_2 \\ y_3 \end{bmatrix}; \; \sqrt{10}\,y_2^2 - \sqrt{10}\,y_3^2$

3. (a) $2x^2 - 3xy + 4y^2$ (b) $x^2 - xy$ (c) $5xy$ (d) $4x^2 - 2y^2$ (e) y^2

4. (a) $\begin{bmatrix} 2 & -\frac{3}{2} \\ -\frac{3}{2} & 4 \end{bmatrix}$ (b) $\begin{bmatrix} 1 & -\frac{1}{2} \\ -\frac{1}{2} & 0 \end{bmatrix}$ (c) $\begin{bmatrix} 0 & \frac{5}{2} \\ \frac{5}{2} & 0 \end{bmatrix}$ (d) $\begin{bmatrix} 4 & 0 \\ 0 & -2 \end{bmatrix}$ (e) $\begin{bmatrix} 0 & 0 \\ 0 & 1 \end{bmatrix}$

5. (a) $[x \;\; y]\begin{bmatrix} 2 & -\frac{3}{2} \\ -\frac{3}{2} & 4 \end{bmatrix}\begin{bmatrix} x \\ y \end{bmatrix} + [-7 \;\; 2]\begin{bmatrix} x \\ y \end{bmatrix} + 7 = 0$ (b) $[x \;\; y]\begin{bmatrix} 1 & -\frac{1}{2} \\ -\frac{1}{2} & 0 \end{bmatrix}\begin{bmatrix} x \\ y \end{bmatrix} + [5 \;\; 8]\begin{bmatrix} x \\ y \end{bmatrix} - 3 = 0$

(c) $[x \quad y]\begin{bmatrix} 0 & \frac{5}{2} \\ \frac{5}{2} & 0 \end{bmatrix}\begin{bmatrix} x \\ y \end{bmatrix} - 8 = 0$ (d) $[x \quad y]\begin{bmatrix} 4 & 0 \\ 0 & -2 \end{bmatrix}\begin{bmatrix} x \\ y \end{bmatrix} - 7 = 0$

(e) $[x \quad y]\begin{bmatrix} 0 & 0 \\ 0 & 1 \end{bmatrix}\begin{bmatrix} x \\ y \end{bmatrix} + [7 \quad -8]\begin{bmatrix} x \\ y \end{bmatrix} - 5 = 0$

6. (a) Ellipse   (b) Ellipse   (c) Hyperbola   (d) Hyperbola   (e) Circle
   (f) Parabola   (g) Parabola   (h) Parabola   (i) Parabola   (j) Circle

7. (a) $9x'^2 + 4y'^2 = 36$, ellipse   (b) $x'^2 - 16y'^2 = 16$, hyperbola
   (c) $y'^2 = 8x'$, parabola   (d) $x'^2 + y'^2 = 16$, circle
   (e) $18y'^2 - 12x'^2 = 419$, hyperbola   (f) $y' = -\frac{1}{7}x'^2$, parabola

8. (a) Hyperbola; possible equations are   (b) Parabola; possible equations are
    $3x'^2 - 2y'^2 + 8 = 0, \; -2x'^2 + 3y'^2 + 8 = 0$ $2\sqrt{2}x'^2 + 9x' - 7y' = 0, 2\sqrt{2}y'^2 + 7x' + 9y' = 0$
                                        $2\sqrt{2}y'^2 - 7x' - 9y' = 0, 2\sqrt{2}x'^2 - 9x' + 7y' = 0$

   (c) Ellipse; possible equations are   (d) Hyperbola; possible equations are
    $7x'^2 + 3y'^2 = 9, \; 3x'^2 + 7y'^2 = 9$ $4x'^2 - y'^2 = 3, \; 4y'^2 - x'^2 = 3$

9. $2x''^2 + y''^2 = 6$, ellipse   10. $13y''^2 - 4x''^2 = 81$, hyperbola   11. $2x''^2 - 3y''^2 = 24$, hyperbola

12. $6x''^2 + 11y''^2 = 66$, ellipse   13. $4y''^2 - x''^2 = 0$, hyperbola   14. $\sqrt{29}x'^2 - 3y' = 0$, parabola

15. (a) Two intersecting lines, $y = x$ and $y = -x$ (b) No graph
   (c) The graph is the single point $(0, 0)$.   (d) The graph is the line $y = x$.

   (e) The graph consists of two parallel lines $\dfrac{3}{\sqrt{13}}x + \dfrac{2}{\sqrt{13}}y = \pm 2$.   (f) The graph is the single point $(1, 2)$.

EXERCISE SET 9.7 (page 499)

1. (a) $x^2 + 2y^2 - z^2 + 4xy - 5yz$ (b) $3x^2 + 7z^2 + 2xy - 3xz + 4yz$
   (c) $xy + xz + yz$ (d) $x^2 + y^2 - z^2$
   (e) $3z^2 + 3xz$ (f) $2z^2 + 2xz + y^2$

2. (a) $\begin{bmatrix} 1 & 2 & 0 \\ 2 & 2 & -\frac{5}{2} \\ 0 & -\frac{5}{2} & -1 \end{bmatrix}$ (b) $\begin{bmatrix} 3 & 1 & -\frac{3}{2} \\ 1 & 0 & 2 \\ -\frac{3}{2} & 2 & 7 \end{bmatrix}$ (c) $\begin{bmatrix} 0 & \frac{1}{2} & \frac{1}{2} \\ \frac{1}{2} & 0 & \frac{1}{2} \\ \frac{1}{2} & \frac{1}{2} & 0 \end{bmatrix}$ (d) $\begin{bmatrix} 1 & 0 & 0 \\ 0 & 1 & 0 \\ 0 & 0 & -1 \end{bmatrix}$

   (e) $\begin{bmatrix} 0 & 0 & \frac{3}{2} \\ 0 & 0 & 0 \\ \frac{3}{2} & 0 & 3 \end{bmatrix}$ (f) $\begin{bmatrix} 0 & 0 & 1 \\ 0 & 1 & 0 \\ 1 & 0 & 2 \end{bmatrix}$

3. (a) $[x \quad y \quad z]\begin{bmatrix} 1 & 2 & 0 \\ 2 & 2 & -\frac{5}{2} \\ 0 & -\frac{5}{2} & -1 \end{bmatrix}\begin{bmatrix} x \\ y \\ z \end{bmatrix} + [7 \quad 0 \quad 2]\begin{bmatrix} x \\ y \\ z \end{bmatrix} - 3 = 0$

   (b) $[x \quad y \quad z]\begin{bmatrix} 3 & 1 & -\frac{3}{2} \\ 1 & 0 & 2 \\ -\frac{3}{2} & 2 & 7 \end{bmatrix}\begin{bmatrix} x \\ y \\ z \end{bmatrix} + [-3 \quad 0 \quad 0]\begin{bmatrix} x \\ y \\ z \end{bmatrix} - 4 = 0$

(c) $[x \quad y \quad z] \begin{bmatrix} 0 & \frac{1}{2} & \frac{1}{2} \\ \frac{1}{2} & 0 & \frac{1}{2} \\ \frac{1}{2} & \frac{1}{2} & 0 \end{bmatrix} \begin{bmatrix} x \\ y \\ z \end{bmatrix} - 1 = 0$ (d) $[x \quad y \quad z] \begin{bmatrix} 1 & 0 & 0 \\ 0 & 1 & 0 \\ 0 & 0 & -1 \end{bmatrix} \begin{bmatrix} x \\ y \\ z \end{bmatrix} - 7 = 0$

(e) $[x \quad y \quad z] \begin{bmatrix} 0 & 0 & \frac{3}{2} \\ 0 & 0 & 0 \\ \frac{3}{2} & 0 & 3 \end{bmatrix} \begin{bmatrix} x \\ y \\ z \end{bmatrix} + [0 \quad -14 \quad 0] \begin{bmatrix} x \\ y \\ z \end{bmatrix} + 9 = 0$

(f) $[x \quad y \quad z] \begin{bmatrix} 0 & 0 & 1 \\ 0 & 1 & 0 \\ 1 & 0 & 2 \end{bmatrix} \begin{bmatrix} x \\ y \\ z \end{bmatrix} + [2 \quad -1 \quad 3] \begin{bmatrix} x \\ y \\ z \end{bmatrix} = 0$

4. (a) Ellipsoid
   (b) Hyperboloid of one sheet
   (c) Hyperboloid of two sheets
   (d) Elliptic cone
   (e) Elliptic paraboloid
   (f) Hyperbolic paraboloid
   (g) Sphere

5. (a) $9x'^2 + 36y'^2 + 4z'^2 = 36$, ellipsoid
   (b) $6x'^2 + 3y'^2 - 2z'^2 = 18$, hyperboloid of one sheet
   (c) $3x'^2 - 3y'^2 - z'^2 = 3$, hyperboloid of two sheets
   (d) $4x'^2 + 9y'^2 - z'^2 = 0$, elliptic cone
   (e) $x'^2 + 16y'^2 - 16z' = 0$, elliptic paraboloid
   (f) $7x'^2 - 3y'^2 + z' = 0$, hyperbolic paraboloid
   (g) $x'^2 + y'^2 + z'^2 = 25$, sphere

6. (a) $25x'^2 - 3y'^2 - 50z'^2 - 150 = 0$, hyperboloid of two sheets
   (b) $2x'^2 + 2y'^2 + 8z'^2 - 5 = 0$, ellipsoid
   (c) $9x'^2 + 4y'^2 - 36z' = 0$, elliptic paraboloid
   (d) $x'^2 - y'^2 + z' = 0$, hyperbolic paraboloid

7. $x''^2 + y''^2 - 2z''^2 = -1$, hyperboloid of two sheets   8. $x''^2 + y''^2 + 2z''^2 = 4$, ellipsoid

9. $x''^2 - y''^2 + z'' = 0$, hyperbolic paraboloid   10. $6x''^2 + 3y''^2 - 8\sqrt{2}z'' = 0$, elliptic paraboloid

EXERCISE SET 9.8 (page 507)

1. Multiplications: mpn; additions: $mp(n - 1)$ 2. Multiplications: $(k - 1)n^3$; additions: $(k - 1)(n^3 - n^2)$

3.

	$n = 5$	$n = 10$	$n = 100$	$n = 1000$
Solve $Ax = b$ by Gauss–Jordan elimination	$+$: 50 $\times$: 65	$+$: 375 $\times$: 430	$+$: 383,250 $\times$: 343,300	$+$: 333,832,500 $\times$: 334,333,000
Solve $Ax = b$ by Gaussian elimination	$+$: 50 $\times$: 65	$+$: 375 $\times$: 430	$+$: 383,250 $\times$: 343,300	$+$: 333,832,500 $\times$: 334,333,000
Find A^{-1} by reducing $[A \mid I]$ to $[I \mid A^{-1}]$	$+$: 80 $\times$: 125	$+$: 810 $\times$: 1000	$+$: 980,100 $\times$: 1,000,000	$+$: 998,001,000 $\times$: 1,000,000,000
Solve $Ax = b$ as $x = A^{-1}b$	$+$: 100 $\times$: 150	$+$: 900 $\times$: 1100	$+$: 990,000 $\times$: 1,010,000	$+$: 999,000,000 $\times$: 1,001,000,000
Find $\det(A)$ by row reduction	$+$: 30 $\times$: 44	$+$: 285 $\times$: 339	$+$: 328,350 $\times$: 333,399	$+$: 332,833,500 $\times$: 333,333,999
Solve $Ax = b$ by Cramer's Rule	$+$: 180 $\times$: 264	$+$: 3135 $\times$: 3729	$+$: 33,163,350 $\times$: 33,673,399	$+$: 33,316,633 $\times 10^4$ $\times$: 33,366,733 $\times 10^4$

4.

	$n = 5$	$n = 10$	$n = 100$	$n = 1000$
	Execution Time (sec)	Execution Time (sec)	Execution Time (sec)	Execution Time (sec)
Solve $A\mathbf{x} = \mathbf{b}$ by Gauss–Jordan elimination	1.55×10^{-4}	1.05×10^{-3}	.878	836
Solve $A\mathbf{x} = \mathbf{b}$ by Gaussian elimination	1.55×10^{-4}	1.05×10^{-3}	.878	836
Find A^{-1} by reducing $[A\,|\,I]$ to $[I\,|\,A^{-1}]$	2.84×10^{-4}	2.41×10^{-3}	2.49	2499
Solve $A\mathbf{x} = \mathbf{b}$ as $\mathbf{x} = A^{-1}\mathbf{b}$	3.50×10^{-4}	2.65×10^{-3}	2.52	2502
Find $\det(A)$ by row reduction	1.03×10^{-4}	8.21×10^{-4}	.831	833
Solve $A\mathbf{x} = \mathbf{b}$ by Cramer's Rule	6.18×10^{-4}	90.3×10^{-4}	83.9	834×10^3

EXERCISE SET 9.9 (page 517)

1. $x_1 = 2, x_2 = 1$

2. $x_1 = -2, x_2 = 1, x_3 = -3$

3. $x_1 = 3, x_2 = -1$

4. $x_1 = 4, x_2 = -1$

5. $x_1 = -1, x_2 = 1, x_3 = 0$

6. $x_1 = 1, x_2 = -2, x_3 = 1$

7. $x_1 = -1, x_2 = 1, x_3 = 0$

8. $x_1 = -1, x_2 = 1, x_3 = 1$

9. $x_1 = -3, x_2 = 1, x_3 = 2, x_4 = 1$

10. $x_1 = 2, x_2 = -1, x_3 = 0, x_4 = 0$

11. (a) $A = L\mathbf{u} = \begin{bmatrix} 2 & 0 & 0 \\ -2 & 1 & 0 \\ 2 & 1 & 1 \end{bmatrix}\begin{bmatrix} 1 & \frac{1}{2} & -\frac{1}{2} \\ 0 & 0 & 1 \\ 0 & 0 & 0 \end{bmatrix}$

13. (b) $\begin{bmatrix} a & b \\ b & d \end{bmatrix} = \begin{bmatrix} 1 & 0 \\ \dfrac{c}{a} & 1 \end{bmatrix}\begin{bmatrix} a & b \\ 0 & \dfrac{ad - bc}{a} \end{bmatrix}$

(b) $A = L_1DU = \begin{bmatrix} 1 & 0 & 0 \\ -1 & 1 & 0 \\ 1 & 1 & 1 \end{bmatrix}\begin{bmatrix} 2 & 0 & 0 \\ 0 & 1 & 0 \\ 0 & 0 & 1 \end{bmatrix}\begin{bmatrix} 1 & \frac{1}{2} & -\frac{1}{2} \\ 0 & 0 & 1 \\ 0 & 0 & 0 \end{bmatrix}$

(c) $A = L_2U_2 = \begin{bmatrix} 1 & 0 & 0 \\ -1 & 1 & 0 \\ 1 & 1 & 1 \end{bmatrix}\begin{bmatrix} 2 & 1 & -1 \\ 0 & 0 & 1 \\ 0 & 0 & 0 \end{bmatrix}$

14. Additions: $\dfrac{n^3}{3} + \dfrac{n^2}{2} - \dfrac{5n}{6}$; multiplications: $\dfrac{n^3}{3} + n^2 - \dfrac{n}{3}$

18. $A = PLU = \begin{bmatrix} 1 & 0 & 0 \\ 0 & 0 & 1 \\ 0 & 1 & 0 \end{bmatrix}\begin{bmatrix} 3 & 0 & 0 \\ 0 & 2 & 0 \\ 3 & 0 & 1 \end{bmatrix}\begin{bmatrix} 1 & -\frac{1}{3} & 0 \\ 0 & 1 & \frac{1}{2} \\ 0 & 0 & 1 \end{bmatrix}$

EXERCISE SET 10.1 (page 527)

1. (a–d)

2. (a) $(2, 3)$ (b) $(-4, 0)$ (c) $(-3, -2)$ (d) $(0, -5)$ **3.** (a) $x = -2, y = -3$ (b) $x = 2, y = 1$

4. (a) $5 + 3i$ (b) $-3 - 7i$ (c) $4 - 8i$ (d) $-4 - 5i$ (e) $19 + 14i$ (f) $-\frac{11}{2} - \frac{17}{2}i$

5. (a) $2 + 3i$ (b) $-1 - 2i$ (c) $-2 + 9i$

6. (a)

(b)

7. (a)

(b)

(c)

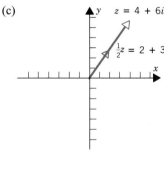

8. (a) $k_1 = -5, k_2 = 3$ **9.** (a) $z_1 z_2 = 3 + 3i, z_1^2 = -9, z_2^2 = -2i$
   (b) $k_1 = 3, k_2 = 1$ (b) $z_1 z_2 = 26, z_1^2 = -20 + 48i, z_2^2 = -5 - 12i$
                                  (c) $z_1 z_2 = \frac{11}{3} - i, z_1^2 = \frac{4}{9}(-3 + 4i), z_2^2 = -6 - \frac{5}{2}i$

10. (a) $9 - 8i$ (b) $-63 + 16i$ (c) $-32 - 24i$ (d) $22 + 19i$ **11.** $76 - 88i$ **12.** $26 - 18i$

13. $-26 + 18i$ **14.** $-1 - 11i$ **15.** $-\frac{63}{16} + i$ **16.** $(2 + \sqrt{2}) + i(1 - \sqrt{2})$ **17.** 0 **18.** $-24i$

19. (a) $\begin{bmatrix} 1 + 6i & -3 + 7i \\ 3 + 8i & 3 + 12i \end{bmatrix}$ (b) $\begin{bmatrix} 3 - 2i & 6 + 5i \\ 3 - 5i & 13 + 3i \end{bmatrix}$ (c) $\begin{bmatrix} 3 + 3i & 2 + 5i \\ 9 - 5i & 13 - 2i \end{bmatrix}$ (d) $\begin{bmatrix} 9 + i & 12 + 2i \\ 18 - 2i & 13 + i \end{bmatrix}$

20. (a) $\begin{bmatrix} 13 + 13i & -8 + 12i & -33 - 22i \\ 1 + i & 0 & i \\ 7 + 9i & -6 + 6i & -16 - 16i \end{bmatrix}$ (b) $\begin{bmatrix} 6 + 2i & -11 + 19i \\ -1 + 6i & -9 - 5i \end{bmatrix}$ (c) $\begin{bmatrix} 6i & 1 + i \\ -6 - i & 5 - 9i \end{bmatrix}$

(d) $\begin{bmatrix} 22 - 7i & 2 + 10i \\ -5 - 4i & 6 - 8i \\ 9 - i & -1 - i \end{bmatrix}$ **22.** (a) $z = -1 \pm i$ (b) $z = \frac{1}{2} \pm \frac{\sqrt{3}}{2}i$ **23.** (b) i

EXERCISE SET 10.2 (page 533)

1. (a) $2 - 7i$ (b) $-3 + 5i$ (c) $-5i$ **2.** (a) 1   (b) 7   (c) 5
(d) i (e) -9 (f) 0   (d) $\sqrt{2}$ (e) 8   (f) 0

4. (a) $-\frac{17}{25} - \frac{19}{25}i$ (b) $\frac{23}{25} + \frac{11}{25}i$ (c) $\frac{23}{25} - \frac{11}{25}i$ (d) $-\frac{17}{25} + \frac{19}{25}i$ (e) $\frac{1}{5} - i$ (f) $\frac{\sqrt{26}}{5}$

5. (a) $-i$ (b) $\frac{1}{26} + \frac{5}{26}i$ (c) $7i$ **6.** (a) $\frac{6}{5} + \frac{2}{5}i$ (b) $-\frac{2}{5} + \frac{1}{5}i$ (c) $\frac{3}{5} + \frac{11}{5}i$ (d) $\frac{3}{5} + \frac{1}{5}i$ **7.** $\frac{1}{2} + \frac{1}{2}i$

8. $\frac{2}{5} + \frac{1}{5}i$ **9.** $-\frac{7}{625} - \frac{24}{625}i$ **10.** $-\frac{11}{25} + \frac{2}{25}i$ **11.** $\frac{1 - \sqrt{3}}{4} + \frac{1 + \sqrt{3}}{4}i$ **12.** $-\frac{1}{26} - \frac{5}{26}i$

13. $-\frac{1}{10} + \frac{1}{10}i$ **14.** $-\frac{2}{5}$ **15.** (a) $-1 - 2i$ (b) $-\frac{3}{25} - \frac{4}{25}i$

17. (a)   (b)   (c)   (d)

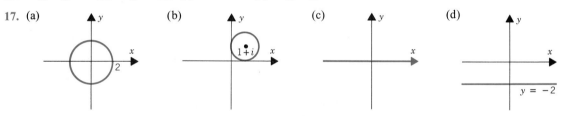

18. (a)   (b)   (c)   (d)

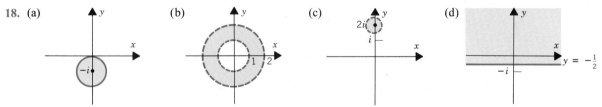

19. (a) $-y$ (b) $-x$ (c) y (d) x **20.** (b) $-i$ **23.** (a) $\dfrac{x_1 x_2 + y_1 y_2}{x_2^2 + y_2^2}$ (b) $\dfrac{x_2 y_1 - x_1 y_2}{x_2^2 + y_2^2}$

27. (c) Yes, if $z \neq 0$.   **28.** $x_1 = i, x_2 = -i$ **29.** $x_1 = 1 + i, x_2 = 1 - i$

30. $x_1 = \frac{1}{2} + i, x_2 = 2, x_3 = \frac{1}{2} - i$

31. $x_1 = i, x_2 = 0, x_3 = -i$ **32.** $x_1 = -(1 + i)t, x_2 = t$ **33.** $x_1 = (1 + i)t, x_2 = 2t$

34. $x_1 = -(1 - i)t, x_2 = -it, x_3 = t$ **35.** (a) $\begin{bmatrix} i & 2 \\ -1 & i \end{bmatrix}$ (b) $\begin{bmatrix} 0 & 1 \\ -i & 2i \end{bmatrix}$

38. (a) $\begin{bmatrix} -i & -2 - 2i & -1 + i \\ 1 & 2 & -i \\ i & i & 1 \end{bmatrix}$ (b) $\begin{bmatrix} 1 + i & -i & 1 \\ -7 + 6i & 5 - i & 1 + 4i \\ 1 + 2i & -i & 1 \end{bmatrix}$

EXERCISE SET 10.3 (page 543)

1. (a) 0 (b) $\pi/2$ (c) $-\pi/2$ (d) $\pi/4$ (e) $2\pi/3$ (f) $-\pi/4$ 2. (a) $5\pi/3$ (b) $-\pi/3$ (c) $5\pi/3$

3. (a) $2\left[\cos\left(\dfrac{\pi}{2}\right) + i\sin\left(\dfrac{\pi}{2}\right)\right]$ (b) $4[\cos\pi + i\sin\pi]$

   (c) $5\sqrt{2}\left[\cos\left(\dfrac{\pi}{4}\right) + i\sin\left(\dfrac{\pi}{4}\right)\right]$ (d) $12\left[\cos\left(\dfrac{2\pi}{3}\right) + i\sin\left(\dfrac{2\pi}{3}\right)\right]$

   (e) $3\sqrt{2}\left[\cos\left(-\dfrac{3\pi}{4}\right) + i\sin\left(-\dfrac{3\pi}{4}\right)\right]$ (f) $4\left[\cos\left(-\dfrac{\pi}{6}\right) + i\sin\left(-\dfrac{\pi}{6}\right)\right]$

4. (a) $6\left[\cos\left(\dfrac{5\pi}{12}\right) + i\sin\left(\dfrac{5\pi}{12}\right)\right]$ (b) $\dfrac{2}{3}\left[\cos\left(\dfrac{\pi}{12}\right) + i\sin\left(\dfrac{\pi}{12}\right)\right]$

   (c) $\dfrac{3}{2}\left[\cos\left(-\dfrac{\pi}{12}\right) + i\sin\left(-\dfrac{\pi}{12}\right)\right]$ (d) $\dfrac{32}{9}\left[\cos\left(\dfrac{11\pi}{12}\right) + i\sin\left(\dfrac{11\pi}{12}\right)\right]$

5. 1   6. (a) -64 (b) $-i$ (c) $-64\sqrt{3} - 64i$ (d) $-\dfrac{1 + \sqrt{3}i}{2048}$

7. (a)

(b)

(c)

(d)

(e)

(f)

8.

9.

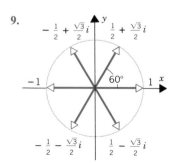

10. $\sqrt[4]{2}\left[\cos\left(\dfrac{\pi}{8}\right) + i\sin\left(\dfrac{\pi}{8}\right)\right], \sqrt[4]{2}\left[\cos\left(\dfrac{9\pi}{8}\right) + i\sin\left(\dfrac{9\pi}{8}\right)\right]$

11. (a) $\pm 2, \pm 2i$ (b) $\pm(2 + 2i), \pm(2 - 2i)$

12. The roots are $\pm(2^{1/4} + 2^{1/4}i)$, $\pm(2^{1/4} - 2^{1/4}i)$ and the factorization is $z^4 + 8 = (z^2 - 2^{5/4}z + 2^{3/2})\cdot(z^2 + 2^{5/4}z + 2^{3/2})$.

13. Rotates z clockwise by $90°$ 14. (a) 16 (b) $\dfrac{i}{4^9}$

15. (a) $\operatorname{Re}(z) = -3, \operatorname{Im}(z) = 0$ (b) $\operatorname{Re}(z) = -3, \operatorname{Im}(z) = 0$
    (c) $\operatorname{Re}(z) = 0, \operatorname{Im}(z) = -\sqrt{2}$ (d) $\operatorname{Re}(z) = -3, \operatorname{Im}(z) = 0$

EXERCISE SET 10.4 (page 549)

1. (a) $(3i, -i, -2 - i, 4)$ (b) $(3 + 2i, -1 - 2i, -3 + 5i, -i)$ 2. $(2 + i, 0, -3 + i, -4i)$
   (c) $(-1 - 2i, 2i, 2 - i, -1)$ (d) $(-3 + 9i, 3 - 3i, -3 - 6i, 12 + 3i)$
   (e) $(-3 + 2i, 3, -3 - 3i, i)$ (f) $(-1 - 5i, 3i, 4, -5)$

3. $c_1 = -2 - i, c_2 = 0, c_3 = 2 - i$ 5. (a) $\sqrt{2}$ (b) $2\sqrt{3}$ (c) $\sqrt{10}$ (d) $\sqrt{37}$

6. (a) $\sqrt{43}$ (b) $\sqrt{10} + \sqrt{29}$ (c) $\sqrt{10} + \sqrt{10}i$ (d) $\sqrt{699}$ (e) $\left(\dfrac{1 + i}{\sqrt{6}}, \dfrac{2i}{\sqrt{6}}, 0\right)$ (f) 1

8. All k such that $|k| = \frac{1}{5}$ 9. (a) 3 (b) $2 - 27i$ (c) $-5 - 10i$

10. The set is a vector space under the given operations.

11. Not a vector space. Axiom 6 fails; that is, the set is not closed under scalar multiplication. (Multiply by i, for example.)

12. No, R^n is not closed under scalar multiplication. (Multiply a nonzero vector in R^n by i.)

13. (a)   14. (b)   15. (a), (d)   16. (a), (b), (d)

17. (a) $(3 - 2i)\mathbf{u} + (3 - i)\mathbf{v} + (1 + 2i)\mathbf{w}$ (b) $(2 + i)\mathbf{u} + (-1 + i)\mathbf{v} + (-1 - i)\mathbf{w}$
    (c) $0\mathbf{u} + 0\mathbf{v} + 0\mathbf{w}$ (d) $(-5 - 4i)\mathbf{u} + (5 + 2i)\mathbf{v} + (2 + 4i)\mathbf{w}$

18. (a) Yes (b) No (c) Yes (d) No   19. (a), (b), (c)

20. (a) $\mathbf{u}_2 = i\mathbf{u}_1$ (b) Three vectors in a two-dimensional space   (c) A is a scalar multiple of B.

21. (b), (c)   22. $\mathbf{f} - 3\mathbf{g} - 3\mathbf{h} = \mathbf{0}$ 23. (a) Three vectors in a two-dimensional space   24. (a), (b)
                                                  (b) Two vectors in a three-dimensional space

25. (a), (b), (c), (d)   26. $(-1 - i, 1)$; dimension $= 1$ 27. $(1, 1 - i)$; dimension $= 1$

28. $(3 + 6i, -3i, 1)$; dimension $= 1$ 29. $(\frac{5}{2}i, -\frac{1}{2}, 1, 0), (-\frac{1}{4}, \frac{3}{4}i, 0, 1)$; dimension $= 2$

EXERCISE SET 10.5 (page 555)

2. (a) -12 (b) 0 (c) $2i$ (d) 37 4. (a) $-4 + 5i$ (b) 0 (c) $4 - 4i$ (d) 42

5. (a) Axiom 4 fails. (b) Axiom 4 fails.   (c) Axioms 2 and 3 fail.   (d) Axioms 1 and 4 fail.
   (e) This is an inner product.   6. $-9 - 5i$ 7. No. Axioms 1 and 4 fail.

9. (a) $\sqrt{21}$ (b) $\sqrt{10}$ (c) $\sqrt{10}$ (d) 0 10. (a) $\sqrt{10}$ (b) 2 (c) $\sqrt{5}$ (d) 0

11. (a) $\sqrt{2}$ (b) $2\sqrt{3}$ (c) 5 (d) 0 12. (a) $3\sqrt{10}$ (b) $\sqrt{14}$ 13. (a) $\sqrt{10}$ (b) $2\sqrt{5}$

14. (a) 2 (b) $2\sqrt{2}$ **15.** (a) $2\sqrt{3}$ (b) $2\sqrt{2}$ **16.** (a) $7\sqrt{2}$ (b) $2\sqrt{3}$ **17.** (a) $-\frac{8}{3}i$ (b) None

18. (a), (b), (c) **20.** (b) **21.** (b), (c)

23. $\left(\dfrac{i}{\sqrt{2}}, 0, 0, \dfrac{i}{\sqrt{2}}\right), \left(-\dfrac{i}{\sqrt{6}}, 0, \dfrac{2i}{\sqrt{6}}, \dfrac{i}{\sqrt{6}}\right), \left(\dfrac{2i}{\sqrt{21}}, \dfrac{3i}{\sqrt{21}}, \dfrac{2i}{\sqrt{21}}, \dfrac{-2i}{\sqrt{21}}\right), \left(-\dfrac{i}{\sqrt{7}}, \dfrac{2i}{\sqrt{7}}, -\dfrac{i}{\sqrt{7}}, \dfrac{i}{\sqrt{7}}\right)$

24. (a) $\mathbf{v}_1 = \left(\dfrac{i}{\sqrt{10}}, -\dfrac{3i}{\sqrt{10}}\right), \mathbf{v}_2 = \left(\dfrac{3i}{\sqrt{10}}, \dfrac{i}{\sqrt{10}}\right)$ (b) $\mathbf{v}_1 = (i, 0), \mathbf{v}_2 = (0, -i)$

25. (a) $\mathbf{v}_1 = \left(\dfrac{i}{\sqrt{3}}, \dfrac{i}{\sqrt{3}}, \dfrac{i}{\sqrt{3}}\right), \mathbf{v}_2 = \left(-\dfrac{i}{\sqrt{2}}, \dfrac{i}{\sqrt{2}}, 0\right), \mathbf{v}_3 = \left(\dfrac{i}{\sqrt{6}}, \dfrac{i}{\sqrt{6}}, -\dfrac{2i}{\sqrt{6}}\right)$

(b) $\mathbf{v}_1 = (i, 0, 0), \mathbf{v}_2 = \left(0, \dfrac{7i}{\sqrt{53}}, \dfrac{-2i}{\sqrt{53}}\right), \mathbf{v}_3 = \left(0, \dfrac{2i}{\sqrt{53}}, \dfrac{7i}{\sqrt{53}}\right)$

26. $\left(0, \dfrac{2i}{\sqrt{5}}, \dfrac{i}{\sqrt{5}}, 0\right), \left(\dfrac{5i}{\sqrt{30}}, -\dfrac{i}{\sqrt{30}}, \dfrac{2i}{\sqrt{30}}, 0\right), \left(\dfrac{i}{\sqrt{10}}, \dfrac{i}{\sqrt{10}}, -\dfrac{2i}{\sqrt{10}}, -\dfrac{2i}{\sqrt{10}}\right), \left(\dfrac{i}{\sqrt{15}}, \dfrac{i}{\sqrt{15}}, -\dfrac{2i}{\sqrt{15}}, \dfrac{3i}{\sqrt{15}}\right)$

27. $\mathbf{v}_1 = \left(0, \dfrac{i}{\sqrt{3}}, \dfrac{1-i}{\sqrt{3}}\right), \mathbf{v}_2 = \left(-\dfrac{3i}{\sqrt{15}}, \dfrac{2}{\sqrt{15}}, \dfrac{1+i}{\sqrt{15}}\right)$

28. $\mathbf{w}_1 = \left(-\dfrac{5i}{4}, -\dfrac{i}{4}, \dfrac{5i}{4}, \dfrac{9i}{4}\right), \mathbf{w}_2 = \left(\dfrac{i}{4}, \dfrac{9i}{4}, \dfrac{19i}{4}, -\dfrac{9i}{4}\right)$ **36.** $\mathbf{u} = -\sqrt{3}i\mathbf{v}_1 + \dfrac{3}{\sqrt{6}}\mathbf{v}_2 - \dfrac{1}{\sqrt{2}}\mathbf{v}_3$

EXERCISE SET 10.6 (page 566)

1. (a) $\begin{bmatrix} -2i & 4 & 5-i \\ 1+i & 3-i & 0 \end{bmatrix}$ (b) $\begin{bmatrix} -2i & 4 & -i \\ 1+i & 5+7i & 3 \\ -1-i & i & 1 \end{bmatrix}$ (c) $\begin{bmatrix} -7i \\ 0 \\ 3i \end{bmatrix}$ (d) $\begin{bmatrix} \bar{a}_{11} & \bar{a}_{21} \\ \bar{a}_{12} & \bar{a}_{22} \\ \bar{a}_{13} & \bar{a}_{23} \end{bmatrix}$

2. (b), (d), (e) **3.** $k = 3 + 5i, l = i, m = 2 - 4i$ **4.** (a), (b)

5. (a) $A^{-1} = \begin{bmatrix} \dfrac{3}{5} & -\dfrac{4}{5} \\ -\dfrac{4}{5}i & -\dfrac{3}{5}i \end{bmatrix}$ (b) $A^{-1} = \begin{bmatrix} \dfrac{1}{\sqrt{2}} & \dfrac{-1+i}{2} \\ \dfrac{1}{\sqrt{2}} & \dfrac{1-i}{2} \end{bmatrix}$ (c) $A^{-1} = \begin{bmatrix} \frac{1}{4}(\sqrt{3}-i) & \frac{1}{4}(1-i\sqrt{3}) \\ \frac{1}{4}(1+\sqrt{3}i) & \frac{1}{4}(1+\sqrt{3}i) \end{bmatrix}$

(d) $A^{-1} = \begin{bmatrix} \dfrac{1-i}{2} & -\dfrac{i}{\sqrt{3}} & \dfrac{3-i}{2\sqrt{15}} \\ -\dfrac{1}{2} & \dfrac{1}{\sqrt{3}} & \dfrac{4-3i}{2\sqrt{15}} \\ \dfrac{1}{2} & \dfrac{i}{\sqrt{3}} & -\dfrac{5i}{2\sqrt{15}} \end{bmatrix}$ **7.** $P = \begin{bmatrix} \dfrac{-1+i}{\sqrt{3}} & \dfrac{1-i}{\sqrt{6}} \\ \dfrac{1}{\sqrt{3}} & \dfrac{2}{\sqrt{6}} \end{bmatrix}$; $P^{-1}AP = \begin{bmatrix} 3 & 0 \\ 0 & 6 \end{bmatrix}$

8. $P = \begin{bmatrix} -\dfrac{i}{\sqrt{2}} & \dfrac{i}{\sqrt{2}} \\ \dfrac{1}{\sqrt{2}} & \dfrac{1}{\sqrt{2}} \end{bmatrix}$; $P^{-1}AP = \begin{bmatrix} 4 & 0 \\ 0 & 2 \end{bmatrix}$

9. $P = \begin{bmatrix} -\dfrac{1+i}{\sqrt{6}} & \dfrac{1+i}{\sqrt{3}} \\[2mm] \dfrac{2}{\sqrt{6}} & \dfrac{1}{\sqrt{3}} \end{bmatrix}$; $P^{-1}AP = \begin{bmatrix} 2 & 0 \\ 0 & 8 \end{bmatrix}$

10. $P = \begin{bmatrix} -\dfrac{2}{\sqrt{14}} & \dfrac{5}{\sqrt{35}} \\[2mm] \dfrac{3-i}{\sqrt{14}} & \dfrac{3-i}{\sqrt{35}} \end{bmatrix}$; $P^{-1}AP = \begin{bmatrix} -5 & 0 \\ 0 & 2 \end{bmatrix}$

11. $P = \begin{bmatrix} 0 & 1 & 0 \\[1mm] -\dfrac{1-i}{\sqrt{6}} & 0 & \dfrac{1-i}{\sqrt{3}} \\[2mm] \dfrac{2}{\sqrt{6}} & 0 & \dfrac{1}{\sqrt{3}} \end{bmatrix}$; $P^{-1}AP = \begin{bmatrix} 1 & 0 & 0 \\ 0 & 5 & 0 \\ 0 & 0 & -2 \end{bmatrix}$

12. $P = \begin{bmatrix} \dfrac{i}{\sqrt{2}} & 0 & -\dfrac{i}{\sqrt{2}} \\[2mm] -\dfrac{1}{2} & \dfrac{1}{\sqrt{2}} & -\dfrac{1}{2} \\[2mm] \dfrac{1}{2} & \dfrac{1}{\sqrt{2}} & \dfrac{1}{2} \end{bmatrix}$; $P^{-1}AP = \begin{bmatrix} 1 & 0 & 0 \\ 0 & 2 & 0 \\ 0 & 0 & 3 \end{bmatrix}$

13. $\lambda = 2 \pm i\sqrt{15}$; no, since A has complex entries.

14. $\begin{bmatrix} 0 & i \\ -i & 0 \end{bmatrix}$ is one possibility.

CHAPTER 10 SUPPLEMENTARY EXERCISES (page 569)

3. $\begin{bmatrix} -i \\ 1 \\ 0 \end{bmatrix}, \begin{bmatrix} 1 \\ 0 \\ 1 \end{bmatrix}$ is one possibility.   5. $\lambda = 1, \omega, \omega^2 \, (= \overline{\omega})$ 10. (b) Dimension $= 2$

INDEX